**Springer Textbooks in Earth Sciences,
Geography and Environment**

The Springer Textbooks series publishes a broad portfolio of textbooks on Earth Sciences, Geography and Environmental Science. Springer textbooks provide comprehensive introductions as well as in-depth knowledge for advanced studies. A clear, reader-friendly layout and features such as end-of-chapter summaries, work examples, exercises, and glossaries help the reader to access the subject. Springer textbooks are essential for students, researchers and applied scientists.

William Bajjali

ArcGIS Pro and ArcGIS Online

Applications in Water and Environmental Sciences

William Bajjali
Natural Sciences Department
University of Wisconsin
Superior, WI, USA

ISSN 2510-1307   ISSN 2510-1315 (electronic)
Springer Textbooks in Earth Sciences, Geography and Environment
ISBN 978-3-031-42229-4   ISBN 978-3-031-42227-0 (eBook)
https://doi.org/10.1007/978-3-031-42227-0

© The Editor(s) (if applicable) and The Author(s), under exclusive license to Springer Nature Switzerland AG 2023
This work is subject to copyright. All rights are solely and exclusively licensed by the Publisher, whether the whole or part of the material is concerned, specifically the rights of translation, reprinting, reuse of illustrations, recitation, broadcasting, reproduction on microfilms or in any other physical way, and transmission or information storage and retrieval, electronic adaptation, computer software, or by similar or dissimilar methodology now known or hereafter developed.
The use of general descriptive names, registered names, trademarks, service marks, etc. in this publication does not imply, even in the absence of a specific statement, that such names are exempt from the relevant protective laws and regulations and therefore free for general use.
The publisher, the authors, and the editors are safe to assume that the advice and information in this book are believed to be true and accurate at the date of publication. Neither the publisher nor the authors or the editors give a warranty, expressed or implied, with respect to the material contained herein or for any errors or omissions that may have been made. The publisher remains neutral with regard to jurisdictional claims in published maps and institutional affiliations.

This Springer imprint is published by the registered company Springer Nature Switzerland AG
The registered company address is: Gewerbestrasse 11, 6330 Cham, Switzerland

Paper in this product is recyclable.

I would like to dedicate this textbook to my beloved family, whose unwavering support, inspiration, and encouragement have been a constant throughout my journey.

To all the students, both past and present, who have entrusted me with their learning, I am grateful for your trust, and it has motivated me to strive for clarity, effectiveness, and innovative teaching methods.

Lastly, I want to express my deepest gratitude to the numerous researchers, authors, and educators in the field of spatial analysis. It is because of their groundbreaking work that this textbook has been made possible. Your contributions to the field continue to inspire future generations.

May this textbook be a source of knowledge and inspiration, empowering both learners and educators, and may it make a valuable contribution to the advancement of education in the realm of GIS applications.
With sincere appreciation,

Preface

The study of geographic information system (GIS) applications is enlightening, challenging, and very interesting. This workbook was created as a guide to students and professionals on the applications of GIS in the geoscience field. GIS applications are now considered to be an important course in the curriculum of undergraduate geoscience, the environment, and some fields of engineering programs. It is the result of more than 30 years of experience in applying GIS technology to water resources and environmental problems. The databases and the applications used in the text reflect real-world problems from different environmental settings that have been gathered from the author's works in the USA and Middle East. Each chapter presents a different set of scenarios and case studies that include an environmental problem that needs a solution. A step-by-step approach was adopted to provide answers and solutions to the presented problems for the scenarios presented.

The textbook is intended to be an introductory course to those who want to get to know and work with ArcGIS Pro. ArcGIS Pro is a new generation of ESRI software that works with different platforms than the previous ArcGIS Desktop. ArcGIS Pro is a multiple-threaded 64-bit application and ESRI's geospatial cloud that supports advanced data analysis. ArcGIS Online is a platform for **web GIS** accessed through a website in a browser. All analysis performed in ArcGIS Online is processed and stored on web servers, and the data created by GIS users will be stored on those servers. ArcGIS Online has a different platform than ArcGIS Pro, but they do communicate with each other, and this depends on GIS user licenses of both products. The licensing issue of ArcGIS Online is explained in Chap. 16. Data can be used in one platform while working in the other, such as adding data directly from ArcGIS Online into ArcGIS Pro. In this case, the ArcGIS Online data are still hosted on web servers, but they are only displayed within ArcGIS Pro.

The primary focus of the textbook is on ArcGIS Pro and ArcGIS Online. The textbook can be used at an undergraduate level or as a dual-level undergraduate/graduate course, or it can be used as a self-study workbook for professionals in the field of geoscience. The textbook aims to teach students and professionals mapping and spatial analysis skills using ArcGIS Pro and ArcGIS Online, as well as state-of-the-art methodologies to acquire, visualize, and analyze data. The textbook has 17 chapters that cover many topics related to GIS applications. Chapter 1 introduces the concept of the vector and raster in GIS and their advantages and disadvantages. Chapter 2 introduces the components of the ArcGIS Pro platform, the new concept of the ArcGIS Pro project, and data integration and symbolizations. Chapter 3 teaches how to work with a variety of ArcGIS Pro tools, classify data, perform different quantitative and qualitative classifications, and generate color-coded maps and state-of-the-art layouts. Chapter 4 addresses coordinate systems, map projections, and reprojections. The chapter discusses the types of coordinate systems: geographic coordinate systems (GCS) and various types of projected coordinate systems. The concept of the datum, projection on the fly, projection of the GCS into Universal Transverse Mercator (UTM), State Plane coordinate (SPC), and various types of customize coordinate systems from the Middle East is discussed. Projection of the raster and geo-referencing are elucidated using real-world examples. Chapter 5 deals with building a geodatabase, importing different data sources in the geodatabase, creating feature classes and feature datasets, and creating relationships between objects and a relationship class between feature class attribute tables and standalone tables. Chapter 6 discusses data editing and topology by thoroughly explaining the simple, advanced, and topological editing process using diverse real-world examples. The use of the simple editing tools to edit existing features and how to fix some common digitizing errors such as overshoot and undershoots, generalize line feature, smooth polygon feature, merging, splitting, and reshaping feature classes. It demonstrates data editing work using topology rules. These rules address understanding coincident and shared geometry, feature creation, building a geodatabase topology, and identifying and fixing topology errors. Chapter 7 discusses geoprocessing, which refers to the tools and processes used to generate derived datasets from other data using a set of tools. Geoprocessing is a very important tool in ArcGIS Pro and plays a fundamental role in spatial analysis. Geoprocessing is discussed with examples based on extracting (clip, erase, and split), combining features (merge, append, dissolve, and buffer), and combining geometries and attributes (union, intersect, and spatial join). Chapter 8 discusses site suitability and data modeling through two real-world scenarios. Model 1 shows how to use different aspects of functionality in GIS to find the most suitable area for building a greenhouse. Model 2 demonstrates how to find the most suitable location to build a nuclear power plant using ModelBuilder in ArcGIS Pro. Chapter 9 offers two instances of geocoding applications that only use 5-digit codes. In the first illustration, a specific Wisconsin state destination post office's zip code was used. Based on the 5-digit number, a locator tool was used to geocode the wells (point layers). The second illustration made use of the zip code's associated street address. The software can accept a different kind of zip code if the user requests to use one. Chapter 10 explains the use of raster analysis in GIS; this chapter is divided into three sections: section 1 is about raster data download and raster dataset conversion, section 2 is about raster projection and processing raster dataset, and section 3 describes the terrain analysis. The user performed different exercises of various GIS functions dealing with raster analysis, such as creating hillshades, contours, vertical profiles, deriving viewshed, slopes, aspects, mosaic images, and clip images. Chapter 11 addresses spatial interpolation techniques using groundwater salinity affected by salt intrusion. Trend analysis was used to determine the direction of salinity distribution along the coast of the Sultanate of Oman. Global polynomial interpolation (GPI), inverse distance weighting (IDW), and kriging interpolation techniques were used to study the effect of the dam on improving water quality along the coast of Oman. Chapter 12 explains the use of hydrology tools in spatial analysis to delineate watersheds using a raster digital elevation model. All tools that are important to watershed creation, such as the flow direction, sink, fill, and flow accumulation tools, are implemented and discussed. Chapter 13 discusses the use of geostatistical analysis to obtain meaningful information related to groundwater data in terms of distribution and patterns. The intention in the textbook is to focus on GIS applications rather than emphasizing complex mathematics and statistics. Nevertheless, some of the tools, such as measuring geographic distribution, analysis patterns, and mapping clusters, are explained and implemented using real-world groundwater data. Chapter 14 discusses the proximity and network analysis techniques, which are important functions in Network Analyst Extension of the ArcGIS Pro. A wide range of topics related to the distance and movement along a linear route are covered. The Near, Point Distance, and Desire line (Spider Diagram) tools were used to verify if the dam in the Dhuleil area recharges the aquifer and improves its water quality. Network Analyst was used to overcome natural barriers such as hills, lakes, or areas where there is absolutely no network of street systems. The network analyst used the actual distance that is associated with the street feature, which is an important feature in the application. The network analyst was used to find the amount of time it requires a water truck to supply the towns with portable water supplies and find the actual path and time that the water truck will take from each well to each town. Chapter 15 explains the 3-D Analyst that is designed to perform different types of analysis and make the map look real and easy to comprehend. ArcGIS Pro worked with data in 2-D and 3-D environments from within the same application. The chapter discusses how to create a triangular irregular network from contours and use it as an elevation source to display features in a 3-D environment. Different layers were integrated from 2-D into 3-D and extruded using a local scene and displayed over the WorldElevation3D/Terrain3D. An animation was created from bookmarks, and a video was created from the 3-D map animation by flying over the scene. Time tracking was used to visualize the movement of subsurface contamination. Chapter 16 introduces ArcGIS Online and ArcGIS StoryMaps. The chapter covers managing the data in ArcGIS Pro and publishing it as web layers to ArcGIS Online. Build a web map by utilizing services in ArcGIS Online and configuring the layers using the Map Viewer. Chapter 17 covers how to create the Instant app in ArcGIS Online. A map was created in ArcGIS Pro and was shared as a web map and web layers in ArcGIS Online. The web map was opened, analyzed, and configured in Map Viewer. The web map was used to create Instant App using different layers that pinpoint the locations of earthquakes, tsunamis, wells, schools, and hospitals that can be used as emergency facilities in terms of shelter, medical treatment, and potable drinking water or identify what exists within a distance from a specific address in the state of California. The Instant App was published and shared with the whole world community.

Note on Data

The textbook includes three types of data that have been used in the exercises of the book's chapters. The first type of data presented is from actual field data gathered by the author and taken from his published work. The second type of data is manipulated based on real information gathered from different projects in the Middle East. The data were modified with the aim of protecting privacy and rules that govern these projects. The third type of data is public domain information and freely available from the Internet to any GIS user. Chapters that included public domain data in the exercises document the data sources.

Superior, WI, USA   William Bajjali

Acknowledgments

I am immensely thankful to the students and colleagues I have had the honor of collaborating with during my academic and professional path. Their exceptional contributions in educational settings, research endeavors, and practical fieldwork have been indispensable in the development of this textbook. I am particularly appreciative of their participation in diverse projects carried out in the United States, Canada, Oman, and Jordan.

Contents

1 Introduction to ArcGIS Pro 1
   What Is GIS? 2
   GIS Infrastructure 3
   Spatial Data Representation 3
   Introduction to ArcGIS Pro 8
   Authorize ArcGIS Pro to Work Offline 8
   Start ArcGIS Pro 9
   Lesson 1: Explore the Vector Data 10
   Lesson 2: Explore the Raster Data 11

2 Working With ArcGIS Pro 15
   Create a New ArcGIS Pro Project 15
   Connect to a Folder 18
   Data Integration in ArcGIS PRO 19
   Find Places on a Map 23
   Labeling 26
   Display Layers Using Visibility Range 28
   Create Thumbnail 31
   Set Map Extent 32

3 Map Classification and Layout 35
   Creating Map and Data Classification 36
   Data Integration 36
   Import Styles from ArcGIS Desktop 39
   Import Styles from Web 40
   Data Symbolization 40
   Classify the Salinity of Groundwater Using the Natural Break Method 43
   Classify the Salinity of Groundwater Using the Quantile Method 44
   Classify the Salinity of Groundwater Using the Equal Intervals Method 46
   Classify the Salinity of Groundwater Using a Manual Method 47
   Sharing the Projects and the Data Using Package Project 48
   Create Page Layout for Groundwater Classification 51
   Add Guides 53
   Insert Layout Elements 55
   Export the Layout as PDF File 61

4 Coordinate Systems and Projections 63
   Geographic Coordinate System (GCS) 63
   Map Projections 63
   Working with GCS 68
   Coordinate Systems and Projections 68
   Assign Datum for the Salinity Feature Class 71
   Georeferencing 73

	Copy Raster.	75
	Georeference the Raster	75
	Project the Raster	79
	Challenge Task	80
5	**Introduction to Geodatabase**	**81**
	Three Types of Geodatabases	81
	Datasets in the Geodatabase	82
	Creating a Geodatabase.	83
	Import Image as a Raster Dataset into File Geodatabase	83
	Create a Feature Dataset in File Geodatabase.	84
	Digitizing Point, Line and Polygon on Screen	87
	Import Data into Geodatabase	90
	Relationship Class.	91
6	**Data Editing and Topology.**	**97**
	Simple Editing	98
	Simple Editing	98
	Digitize the Missing Formation	105
	Update the Area and Perimeter fields in the Geology Attribute Table	107
	Merge Function.	109
	Advance Editing	110
	Fixing Overshoots and Undershoots.	110
	Topological Editing Using Geodatabase.	113
	Build the Topology and Set the Rule	115
	Second Approach: Map Topology.	119
	Fix Watershed Using Topology (Geodatabase).	123
7	**Geoprocessing**	**129**
	Working with Geoprocessing Tools	130
	Extract a Feature Class	136
	Merge Tool	139
	Buffer Tool and Select By Location	139
	Select by Location.	140
	Create an Artificial Water Reservoir for Irrigation Purposes	141
	Erase Tool	143
8	**Site Suitability and Modeling.**	**145**
	ModelBuilder	145
	Site Suitability and Modeling.	146
	Model 1.	146
	The Criteria to Build the Greenhouse.	146
	Model 2: ModelBuilder.	150
	Work with a Model: Find Best Suitable Location.	151
	The Criteria to Find Suitable Location to Build Nuclear Power Plant	151
	Draw a Connection Interactively Between the Variable and Tool	160
	Validate and Run the Modelbuilder in Model Window	161
	Model Output	162
	Change a Model Name and Label.	163
	Intermediate Data	164
	Create Model Tool	164
	Model Parameters.	166
	Run the Model Tool	168

9	**Geocoding**	169
	Part 1: Geocoding Based on Zip Code	170
	Launch ArcGIS Pro and Connect to Data	170
	Create Locator	171
	Locate Places and Addresses	171
	Geocode the Addresses	172
	Wells in Each Zip Code and Average Nitrate Concentration	173
	Symbolizing	174
	Part 2: Geocoding Based on Street Address	175
	Integrate Excel Table	176
	Build the Street Locator Using Create Locator	177
	Test Your Address Locator	178
	Geocode Well Owner by Address	179
	Examination of the Geocoding Results	180
	Changing the Basemap	180
	Match the Unmatched Addresses	181
10	**Raster Format**	183
	Section 1: Explore and Download Digital Elevation Model (DEM)	183
	Section 2: Projection and Processing Raster Dataset	183
	Section 3: Terrain Analysis	184
	Download Dem Image from Usgs Webpage	184
	Download the Dem of the City of Burnsville in Minnesota	184
	Launch ArcGIS Pro	185
	Folder Connection	185
	Exploring Digital Elevation Models	186
	Convert Floating Raster into Integer Raster	187
	Section 2: Projection And Processing Raster Dataset	188
	Project the Dem of Amman-Zarqa Basin	188
	Clip the Raster	189
	Merge Raster Datasets (MOSAIC)	190
	Resample an Image	192
	Classify an Image	193
	Convert Vector Feature into Raster	194
	Section 3: Terrain Analysis	196
	Create Hillshade and Contour for the Dhuleil DEM	196
	Create Contours	198
	Create Vertical Profile	199
	Select a Stream	199
	Ready to Use Toolbox	200
	Profile Tool	200
	Create Vertical Profile	201
	Customize the Chart Properties	201
	Create Visibility Map	202
	Create a New Feature Class	203
	New Field: Offseta	203
	Run the Visbility Tool with the New Field "OFFSETA"	205
	Line of Sight Analysis	205
	Line of Sight	206
	Line of Sight for VisibFreq Map Without OFFSETA	206
	Line of Sight for VisibFreq25 Map with OFFSETA	207
	Create Linear Lines of Sight	208
	Slope and Aspect	209

Derive the Slope 210
Classify the Slope into Six Classes. 210
Reclassify Slope 211
Add New Field for Slope_Reclass 213
Derive Aspect Layer 214
Combine Two Images: Slope and Geology. 215
Find Best Area to Build the Lysimeter 217
Extract by Attribute tool 217
Raster Calculator Tool 219
Calculate the Area of the Lysimeter 220

11 Spatial Interpolation. 223
Method of Interpolation 224
Trend Surface Analysis. 224
Inverse Distance Weighting (IDW). 224
Global Polynomial (GP) 225
Kriging 225
Connect to Data and Data Integration. 226
Density of Groundwater Well 227
Trend Analysis 229
Calculate the Coordinates of the Borehole Layer. 230
Salinity Trend Using Scatter Plot 231
Save as Layer Files 233
Interpolation 234
Global Polynomial Interpolation. 234
Convert the Gpi Layer Into Raster Format 235
Classify the GPI Map 236
Inverse Distance Weighting. 237
Conversion of the Idw into Raster Format 238
Classify the IDW Map 239
Interpolation Using Kriging 240
Convert the Kriging Layer into A Raster Format 241
Classify the Ordinarykriging Raster 242

12 Watershed Delineation 243
Flow Direction 243
Flow Accumulation. 244
Stream Link. 244
Delineate Watershed Based on a Pour Point. 245
Data Connection and Integration 245
Integrate Layers into the Watershed Map 246
Delineating the Watershed 246
Step 1: Run the Flow Direction Tool 246
Step 2: Identify the Locations of the Sink (Sink Tool) 247
Zonal Statistics 248
Step 3: Run the Fill Tool. 249
Step 4: Run the Flow Direction Tool 249
Step 5: Create a Flow Accumulation Raster 250
Build Raster Attribute Table Tool 251
Run the Flow Accumulation and Save the Output as Float 251
Convert the Float Raster into Integer Raster. 252
Step 6: Create Source Raster to Delineate WatersheD 252
Source Raster 253
Step 7: Delineate Watershed 254

	Point-Based Watershed	255
	Convert Pourshed Raster into a Vector	256
	Generating the Stream Network	257
	Create Stream Order	257
13	**Geostatistical Analysis**	**259**
	Measuring Geographic Distribution Toolset	259
	Measuring Geographic Distribution	260
	Data Connection and Integration	261
	Symbolizing the Three Layers	261
	Mean Center	262
	Mean Center with Weight	263
	Standard Distance and Mean Center	264
	Distance Between Khaldiyah Dam and the Center of the Buffer	265
	Identify an Injection Well	266
	Select by Location and Select by Attributes	266
	Near Tool	267
	Analyzing Pattern Toolset: Identify Pattern Based on Location	268
	Average Nearest Neighbor Tool	268
	Null Hypotheses	269
	Identify Pattern Based on Values (Getis-Ord General G)	272
	Find the Ideal Distance	273
	High/Low Clustering (Getis-Ord General G)	274
	Spatial Autocorrelation (Global Moran's I)	275
	Spatial Join Between Grid_1000 and Well Layers	276
	Spatial Autocorrelation (Global Moran's I)	277
	Cluster and Outlier Analysis (Anselin Local Moran I)	279
	Run Cluster and Outlier Analysis (Anselin Local Moran I)	280
	Hot Spot Analysis (Getis-Ord GI*)	282
	Summary Statistics	284
14	**Proximity and Network Analysis**	**287**
	Part I – Proximity Analysis	287
	Proximity Analysis in Vector Format	287
	GIS Approach to Solve Scenario 1	288
	Buffer the WWTP in the SAMRA Region	289
	Select Wells Outside the WWTP_ Buffer in the Samra Region	289
	Select by Location	289
	Select by Attributes	290
	Add New Field to the Well Layer	291
	Calculate Field	292
	Buffer the Stream in the Region	293
	Select Wells in the Dhuleil Region with Low TDS and NO3	294
	Hide Wells Using Definition Query	296
	Convert Well Shapefile into Feature Classes in Geodatabase	297
	Use the Near Tool	298
	Run Near Tool	298
	Save Region, City, and Borehole as a Layer	299
	Create a New Field in the Borehole	300
	Generate Spider Diagram (Desire Lines)	300
	Multi-Ring Buffer Around Hay Arnous City	302
	Select Cities Inside Each Buffer Zone	304

Proximity Analysis in Raster and Vector Formats	305
Euclidean Distance	305
Euclidian Distance	306
Classify the Dam Distance Raster	307
Generate Near Table Tool	308
Join Two Tables Based on Common Field	309
Definition Query	310
Well Classification	311
Salinity (TDS) Classification	311
Nitrate (NO_3) Classification	311
Part II – Network Analyst	312
Create Network Dataset	315
Customize and Build the Network	316
Explore the Network	319
First Approach	320
Second Approach	320
Run New Service Area	321
Add Facilities	322
Set the Cutoff Time	323
True Path and Total Time Between the Wells and each Town	324
Add Facilities	325
Add an Incident	326
Classify the Straight Lines Routes	328
Create the Route Layer	329
Create Stops	330
Run Analysis	331

15 3-D Visualization 333

Surfaces and Z Values	333
Create Triangular Irregular Network From Contour Line	334
Data Connection and Integration	334
Change the Symbols of the TIN	336
Drape Layers onto Dhuleil_TIN	337
Add Data to the Scene	337
Extrude Layers	338
On-Screen Navigator	339
Heading Mode and Full Control Mode	339
Apply a Realistic Layer	341
Extrude the Well and the WWTP Layers	341
Display the Street, Valley and Farm Layers in the Scene	342
Change Elevation Units	342
Navigate Underground and Change Vertical Exaggeration	343
Use the Dhuleil TIN as an Elevation Source	344
Create Animation From Bookmarks	344
Create a Bookmark	344
Use Bookmarks to Create an Animation	345
Create a Video	346
Fly Through a Map or Scene	347
Time Tracking	348
Enable Time and Visualize the Data	349
Configure the Time Slider to Play Back the Data	350
Create an Animation and Export a Video	351
Add Surface Information and Elevation Profile	352

	Elevation Profile	354
	Elevation Profile - Interactive Placement	354
	Elevation Profile - From Layer	355
	Update the Elevation Profile Graph	356

16 Working with ArcGIS Online and StoryMap App 357
Section 1: Publish Data to ARCGIS Online 357
Step 1: Sign in to ArcGIS Online 357
Step 2: Publish a Feature Service 358
What Is My Content? 358
Step 3: Edit the Description and Term of Use 361
Step 4: Open the Feature Service 363
Content and Setting Toolbar 364
Rename the Layers 365
Change Layer Symbology 366
Save the Web Map 369
Share an App Using ArcGIS Online 370
Share the Web Map 370
Create Story Map 372
Work with Media 375
Work with Immersive – Sidecar Docked 380
Work with Immersive – Sidecar Floating 385
Electrical Conductivity 386
Dissolve Oxygen (DO) 386
    Add the Third and Fourth Graphs 387
Hydrogen ION Concentration (pH) 387
Temperature 387
Conclusion 388
Review the Story 389
Publish and Share 391

17 Instant App – Emergency Situations in California 393
ArcGIS Pro Approach 393
Change the Name and the Basemap 393
Connect to Data and Add Layers to the Map 394
Set the Extent 395
Update Label Properties and Configure Pop-up Windows 396
Configure Pop-ups 397
    Challenge Task 399
Share a Web Map and a Web Feature Layer 399
Configure the Web Feature Layer 400
Share a Web Map 401
Publish a Geo APP Using an Instant App Template 402
First Approach to Build Geo App 403
Examination of the Attribute Table of the Wells 404
Configure the Pop-up Window of the Earthquake 404
Configure the Pop-up of the Earthquake 405
Save the Web Map 407
Create a Web App 408
Configure Your Instant App 410
Type a Specific Address 414

References - Data Source Credits 417

Index 421

Introduction to ArcGIS Pro

GIS has great value in our time, as it is a comprehensive information system that evolved and is still developing in parallel with advancing technology. This era of human kind is characterized as an information age, where the whole world is experiencing and interacting with a new revolution, changing our traditional way of looking at things and conducting business with a completely different approach. The emphases are on technology and its use in every activity ranging from agriculture, industry, business, social, research and education. The advancement in technology changed our world and our approaches to meet our new needs that rely completely on technology and data. The value of information in our time has become vital and important for development. GIS itself is an important module of the information system. The economies of all industrial countries and many other nations all over the globe have become more dependent on services. This means that the current economy relies increasingly on computers, networking, accurate information and data. This shift required a mass of skilled laborers that could deal with the technology and data processing.

GIS technology is not an exception in regard to its use in water resources, geology, and environment-related problems. It is a powerful tool for developing solutions for many applications ranging from creating a color-coded geological map and interpolating the water quality of groundwater aquifers to managing water resources on a local or regional scale.

Water is the most precious and valuable resource and is vital for socioeconomic growth and sustainability of the environment. In some arid countries, water resources are limited, scarce, and mainly sourced from groundwater. In some Middle Eastern countries, surface water is limited to a few river systems and intermittent streams that are associated with rain during the winter. Precipitation is vital and the primary source of recharge for various groundwater aquifers in these regions.

Groundwater in the region has been utilized through wells tapping various water-bearing formations to provide more fresh water to supply the increased demand for water supply and irrigation. This practice negatively affected the whole hydrogeological setting of the basins. For example, total water withdrawal in the region (Israel, Jordan, and Palestinian territories) in 1994 was approximately 3050 million cubic meters. The estimated total renewable water supply that is practically available in the region is approximately 2400 million cubic meters per year. The water deficit is pumped from the aquifers without being replenished. This practice caused the groundwater level to decline dramatically in some well fields, up to 20 m, which caused some major springs in 1990 to cease completely in the Azraq basin, Jordan.

Therefore, the management of water resources has become a major effort for governments in the region. Various ministries, water institutions, and private companies worldwide are using GIS as a tool to manage water resources in their countries. GIS can be used to capture data and develop hydrologic datasets for all components of water resources. These include understanding the region's hydrology, mapping sources of contamination, preparing water quality and water–rock interaction maps, delineating watershed areas, and more.

Supplementary Information The online version contains supplementary material available at https://doi.org/10.1007/978-3-031-42227-0_1.

What Is GIS?

GIS is a computer system that creates, manages, analyzes, and maps data that is attached to unique locations. It enables users to capture, store, manipulate, analyze, and present spatial or geographic data. The location data, along with all types of information associated with it, provide a foundation for mapping and analysis that is used in virtually every field (ESRI). ArcGIS Pro and ArcGIS Online provide users with the flexibility to work with data either locally on their computers or online. ArcGIS Pro uses a new type of geodatabase for storing data that can be saved on your PC. Additionally, ArcGIS Pro allows users to share their data in different formats, such as geodatabase, shapefile, and Excel, on ArcGIS Online. This online data can be used to create maps, a web, and apps, as explained in Chaps. 16 and 17, and can also be synchronized back to ArcGIS Pro. ArcGIS Online also provides a rich collection of data for users to start their projects. The data includes high-resolution imagery for most of the world, basemaps for reference, boundaries and places, demographics and lifestyle data, transportation data, Earth observation data, and much more.

With the advancement in technology, a GIS map is dynamic, means that the map can be modified in a very little time, and can be stored, displayed, and printed out quickly and efficiently. GIS is a new methodology in science and applications; it is a new profession and a new business.

GIS refers to three integrated parts.

1. Geographic: The geographical location of the real world (coordinate system)
2. Information: The database
3. Systems: The hardware and software

GIS Description

A GIS (Geographic Information System) is a computer-based tool that helps us visualize information with patterns and relationships that are not otherwise apparent. The ability to ask complex questions about data and analyze many features at once and then instantly see the results on a map is what makes GIS a powerful tool for creating information. GIS can be used in many disciplines, such as resource management, criminology, urban planning, marketing, and transportation. GIS is a useful tool for researchers and scientists, and it plays a vital role in scientific research such as in environmental science, earth sciences, and other fields (Oxford Bibliographies, 2017).

What Can a GIS Do?

A GIS performs six fundamental operations that make it a useful tool for finding solutions to real-world problems. Throughout this course, you will gain experience with the ArcGIS tools used for these operations.

1. **Capture data:** You can add data from many sources to a GIS, and you can also create your own data from scratch. You will learn about getting data into a GIS in Modules 4 and 5.
2. **Store data:** You can store and manage information about the real world in ways that make sense for your application. You will learn about organizing data in Module 4.
3. **Query data:** You can ask complex questions about features based on their attributes or their location and get quick results. You will gain experience with querying in Module 6.
4. **Analyze data:** You can integrate multiple datasets to find features that meet specific criteria and create information useful for problem solving. You will perform analysis in Modules 6 and 7.
5. **Display data:** You can display features based on their attributes, a powerful feature you will come to appreciate. You will learn how to symbolize features in different ways in the next module (Module 2).
6. **Present data:** You can create and distribute high-quality maps, graphs, and reports to present your analysis results in a compelling way to your audience. You will learn how to create a report in Module 6 and how to design an effective map in Module 8.

GIS is a computerized system that deals with spatial data in terms of the following:

1. *Storage*: Digital and database storage (CAD format and dbs format).
2. *Management of Data*: Integration of the database into the GIS system (AutoCAD, Dbf format).
3. *Retrieval*: The capacity to view the various database data formats (AutoCAD, Dbf, Shapefile, and Coverage).

4. ***Conversion***: Convert different sets of data from one form to another (shapefile to feature geodatabase).
5. ***Analysis***: Manipulating data to produce new information (geology map).
6. ***Modeling***: Simplifying the data and its process (fuzzy logic model)
7. ***Display***: Presenting the output works (maps and reports).

Organization

GIS is a complete system that consists of sophisticated hardware and software and can provide organizations with a powerful tool for managing and analyzing data in a geospatial context, leading to better decision-making, increased efficiency, and improved communication and resource management.

ArcGIS Pro is capable of accommodating different data types, allowing users to work with a broad range of items including databases, file-based datasets, and web services. Additionally, it enables the inclusion of shared maps, layers, tools, and other resources into a project. The ArcGIS Online portal also supports various items such as feature layers, Layer Packages, Map Packages, Project Packages, geoprocessing packages, Map files, imagery layers, Elevation layers, KML layers, layer files, OGC web feature service feature layer, OGC web tile service tile layer, CAD drawings, and more.

GIS Infrastructure

1. Hardware: The machine where the GIS can be run (computer, digitizer, plotter, printer).
2. Software: The program needed to run the GIS (ArcGIS and its extensions)
3. Data: The digital and database (information)
4. Organization & People: This is the most important part of the GIS structure. GIS is too important and so costly that it cannot be considered just equipment. It requires organization and staff to utilize this technology. Unfortunately, many organizations treat GIS as equipment rather than an important tool.

GIS Principles

1. The computer is an unavoidable technology in our time. We are living in the digital age, which has become an important element in nearly all professions.
2. Computer training in most scientific disciplines is essential. Without this technology, all professionals will be handicapped.
3. GIS is an inevitable technology that will be used in all scientific fields. GIS has become the accepted and standard means of using spatial data.
4. GIS is more accurate, flexible, object efficient, and rapid fun than the traditional method of spatial data inventory.
5. GIS is replacing traditional cartography. Much of the traditional "pen & ink" cartography performed by skilled draftspersons and artists is being replaced by GIS.
6. GIS is opening new horizons. New modes of analysis and applications are constantly being discovered.

Spatial Data Representation

Spatial data are a fundamental component in any GIS environment. The data are based on the perception of the world as being occupied by features. Each feature is an entity that can be described by its attribute or property, and its location on earth can be mapped using a spatial reference. The most common representation of spatial data that measures the landscape is using discrete data (vector model) and continuous data (raster models). The data models are a set of rules used to describe and represent real world features in GIS software.

Vector Data Model

The vector data model is a representation of the world of distinct features that have definite boundaries and identities and a specific shape using points, lines, and polygons. Vector data are structured with two specific elements (node, vertex) and coordinates. This model is useful for storing data that have discrete boundaries, such as groundwater wells, streams, and lakes. Each entity has a dimension, boundary and location. For example, a well has a specific measurement, and its location can be described using a coordinate system such as latitude-longitude. The following represents the three fundamental vector types that exist in GIS.

Point A point entity is simply a location that can be described using the coordinate system (longitude, latitude or X, Y). The point has no actual spatial dimension and has no actual length and width but has a specific location in space (single coordinate pair). Points can be represented by different symbols. Points generally specify features that are too small to show properly at a given scale. For example, buildings, schools, or a small farm at a scale of 1:25,000 can be represented as a point.

Figure 1.1 shows the locations of five groundwater wells, with each well representing a point feature. Table 1.1 shows the coordinate locations of the wells in (X, Y). The coordinate system allows users to integrate the wells into GIS and make them subject to mapping. The well feature is associated with an attribute table. The attribute of each well has information related to the depth and the yield of each well.

Line A spatial feature that is given a precise location that can be described by a series of coordinate pairs. Each line is stored by the sequence of the first and last point together with the associated table attribute of this line. Line is a one-dimensional feature and has length but no width. Line is a linear feature such as rivers, pipelines, and fences. The more points used to create the line, the greater the detail. The recent requirement that the line features include topology means that the system stores one end of the line as the starting point and the other as the end point, giving the line "direction".

Figure 1.2 shows four pipelines (A, B, C, and D). Each pipeline is represented by a line that has its first and last node to distinguish its location. Each line has attributes of length and discharge. Note that each node has coordinates (X, Y) stored in another table (Table 1.2).

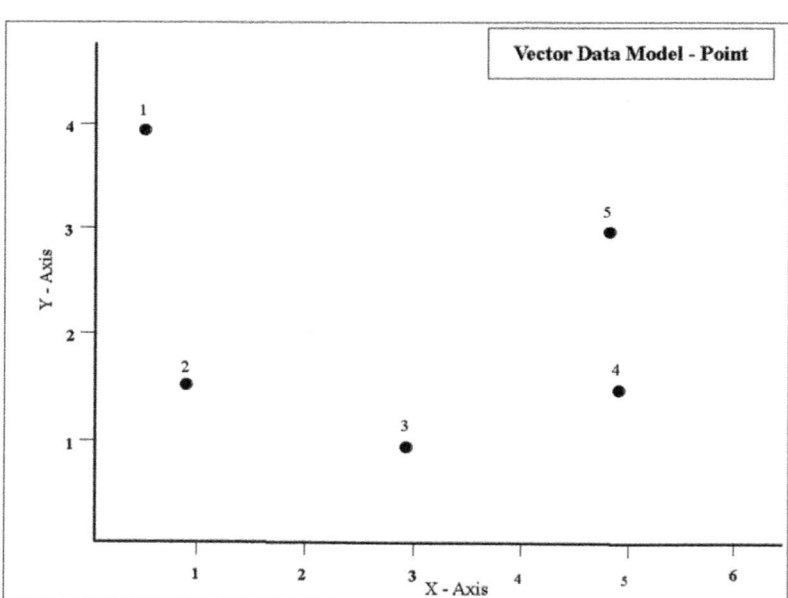

Fig. 1.1 Point feature representation

Table 1.1 Attribute table of point features

No	Well	X	Y	Well	Depth (m)	Yield (m^3/h)
1	WAJ-1	0.5	4.0	WAJ-1	78	90
2	WAJ-2	1.0	1.5	WAJ-2	48	68
3	WAJ-3	3.0	1.0	WAJ-3	35	54
4	WAJ-4	5.0	1.5	WAJ-4	58	75
5	WAJ-5	5.0	2.7	WAJ-5	55	75

Fig. 1.2 Line feature representation

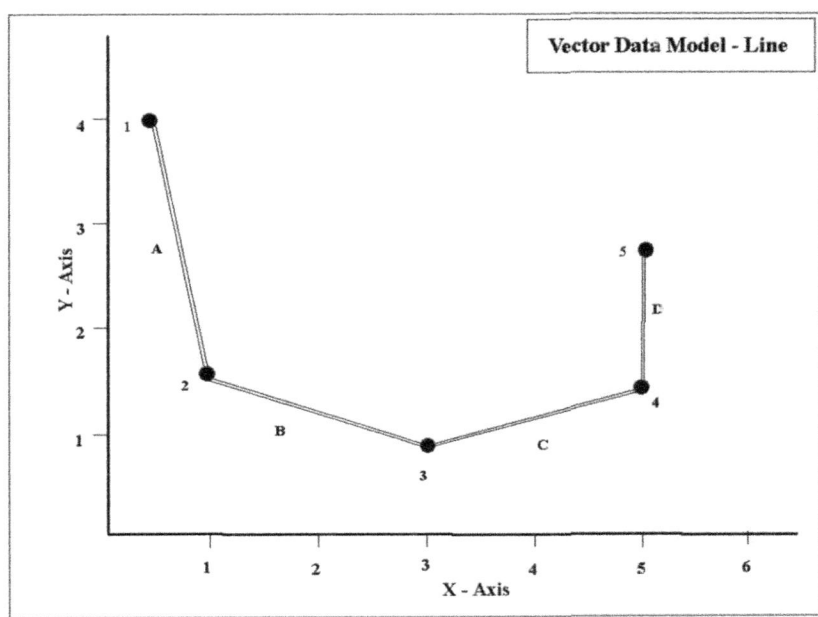

Table 1.2 Attribute table of line feature

Node no	X	Y	Line	1st node	Last node	Line	Length	Discharge
1	0.5	4.0	A	1	2	A	25.5	5
2	1.0	1.5	B	2	3	B	20.6	4
3	3.0	1.0	C	3	4	C	20.6	4
4	5.0	1.5	D	4	5	D	15.0	3
5	5.0	2.7						

Polygon The polygon is an area fully encompassed by a series of connected lines. The first point in the polygon is equal to the last point. A polygon is a 2-D feature with at least three sides, and because lines have direction, the area that falls within the lines compromises the polygon, and the perimeter can be calculated. All of the data points that form the perimeter of the polygon must connect to form an unbroken line. Polygons are often irregularly shaped, such as parcels, lakes, and political boundaries.

Figure 1.3 shows polygon A, which represents an agricultural field. The polygon has its first and last node in node number 1 to settle its location. Each node has coordinates (X, Y) that are stored in another table (Table 1.3). Aside from location attributes, the polygon has associated attributes of area and crop.

Features on maps have spatial relationships that show how those features are related to each other in space. The most important spatial relationships are as follows:

1. **Distance**: This measures the distance from one feature to another in the GIS map. The distance concept is an important relationship, as the distance between features can be measured in any unit regardless of the map's coordinate system.
2. **Distribution**: This is the collective location of features where relationships can show the feature among themselves or their spatial relationships with other features in the map.
3. **Density**: This is the number of features per unit area or simply how close features are to each other.
4. **Pattern**: This is the consistent arrangement of a feature.

Raster Data Model
A representation of an area or region as a surface divided into a grid of cells (Fig. 1.4). It is useful for storing data that vary continuously, such as in an aerial photograph, a satellite image, a surface of humidity, or a digital elevation model (DEM) (Fig. 1.5).

Fig. 1.3 Polygon feature representation

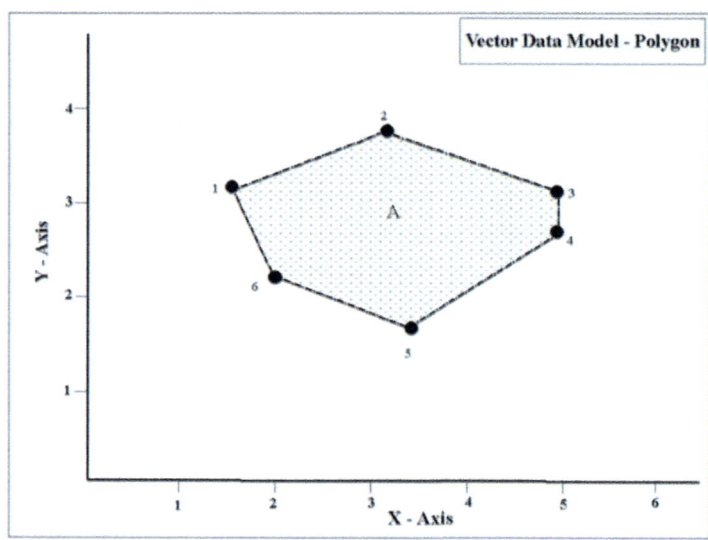

Table 1.3 Attribute table of the polygon feature

Node no	X	Y	Polygon	Node	Polygon	Area	Crop
1	1.6	3.1	A	1, 2, 3, 4, 5, 6	A	520	Tomato
2	3.2	3.8					
3	5.0	3.1					
4	5.0	2.6					
5	3.4	1.8					
6	2.0	2.2					

Fig. 1.4 Depicts the basics of the raster data model. The cell is the minimum mapping unit and the smallest size at which any landscape feature can be represented

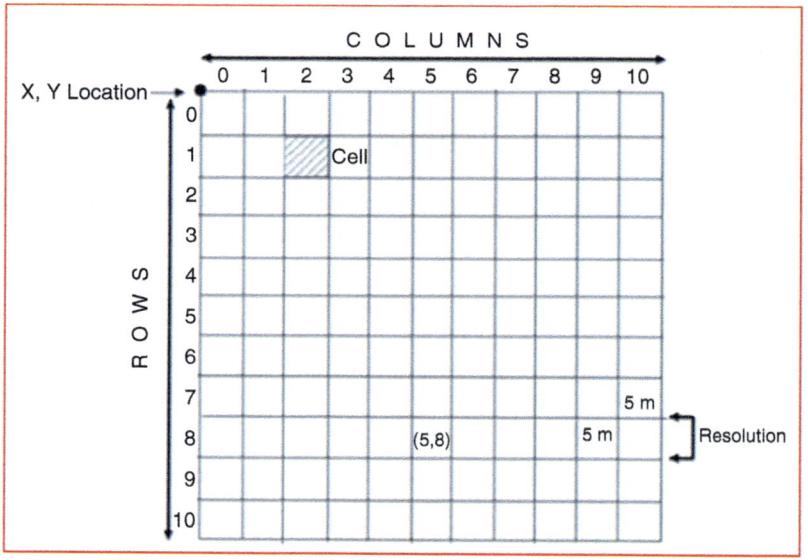

These cells in the raster dataset are used as building blocks for creating points, lines, and polygons. In the raster data model, points, lines, and polygons are represented by grid cells. The location of each cell in the grid is determined by two things: 1) the origin of the grid (the upper left-hand, which is (0, 0) and the resolution (size of the cell). The resolution is determined by measuring one side of the square cell. For example, a raster model with a cell representing 5 m by 5 m (25 m^2) in the real world would be said to have a spatial resolution of 5 m. Each cell in the raster carries a single value, which represents the characteristic of the spatial phenomenon at a location denoted by its row and column. The precision of raster data is determined by the resolution of the grid dataset. The data type for that cell value can be either integer or floating-point.

Fig. 1.5 Aerial Photograph of Faxon Creek, Superior, WI and DEM Jafr, Jordan

The raster model averages all values within a given cell to yield a single value. Various techniques are used to assign cell codes, such as presence-absence, cell center, dominant area, and percent coverage. The more area covered per cell, the less accurate the associated data values. The area covered by each cell determines the spatial resolution of the raster model from which it is derived. Raster coding produces spatial inaccuracies as the shape of features is forced into an artificial grid cell format. Therefore, there is no way to know where any small feature occurs within the cell, as the location according to the raster format is simply the entire cell. If the raster cell represents 100 m by 100 m, the cell represents a well that has a 0.5-m diameter. The whole cell in this case represents the well, which makes the raster format imprecise.

Advantages and Disadvantages of the Raster and Vector Model
There are several advantages and disadvantages for using either the raster or vector data model for storing and displaying spatial data.

Raster Model: Advantages

1. Simple data structure
2. Efficient for remotely sensed or scanned data
3. Simple spatial analysis procedures

Raster Model: Disadvantages

1. Requires greater storage space on a computer
2. Depending on pixel size, graphical output may be less pleasing
3. Projection transformations are more difficult
4. Difficult to represent topological relationships

Vector Model: Advantages

1. Data can be represented in their original resolution without generalization.
2. Requires less disk storage space
3. Topological relationships are readily maintained
4. Graphical outp.ut closely resembles hand-drawn maps

Vector Model: Disadvantages

1. More complex data structure
2. Inefficient for remotely sensed data
3. Some spatial analysis procedures are complex and process intensive
4. Overlaying multiple vector maps is often time consuming

GIS Project

To carry a GIS project, users need to integrate spatial data into GIS software, where the data can have a vector or raster dataset. GIS data come from many resources

1. Hard copy maps
2. Digital files
3. Imagery
4. GPS
5. Excel, text delimited, and dbf files
6. Reports

GIS analysis is based on a database, which is powerful and important in GIS.

Introduction to ArcGIS Pro

ArcGIS Pro is an advanced GIS application from Esri. The program performs spatial analysis, and users can share their work with ArcGIS Online or the ArcGIS Enterprise portal. ArcGIS Pro consists of different components: maps, scenes, layouts, data, tables, and different tools that are structured in a project(Links to an external site).

ArcGIS Pro is tightly linked with ArcGIS Online, which is a cloud-based system built to encourage the sharing of GIS data and other resources between organizations and users. Both ArcGIS Pro and ArcGIS online required an account from ESRI that provided access to more data and tools. ArcGIS Pro can also run offline. ArcGIS Pro has the capability of ArcGIS desktop (ArcMap, ArcCatalog, and ArcToolbox). It can Import ArcMap document, integrate ArcScene, and ArcGlobe. It also maintains Python and ModelBuilder. ArcGIS Pro continues to include various extensions in much the same way as ArcGIS desktop.

ArcGIS Pro has a more modern GUI, and its performance improved in terms of analysis, 2-D and 3-D integration, labeling, symbology, creating layouts, editing, and sharing. ArcGIS Pro does not accept certain files such as Personal Database or ArcInfo Coverage but can import ArcMap documents (*.mxd). By default, a project is stored in its own system folder with the extension (*.**aprx**). A project also has its own geodatabase (a file with the extension (*.**gdb**) and its own toolbox (a file with the extension (*.**tbx**).

When you start ArcGIS Pro, you can create a project(Links to an external site.) from one of the four system templates. Each template creates a project file that starts the application in a different state. For example, a project created from the Map template starts with a map view containing a basemap layer. You can also start without a template (Links to an external site.). This allows you to work in ArcGIS Pro without saving a project file.

New projects can also be started from project templates (Links to an external site.) made by you or shared with you by colleagues. A project template is a customized starting state for a project. Recently, used templates appear on the start page. You can also browse to templates. There are various ways to open your saved projects(Links to an external site.). Any project you have used recently is accessible from the start page. You can browse to other saved projects to open them. You can also pin a project to the start page to make sure it is always readily available.

Authorize ArcGIS Pro to Work Offline

Users can use ArcGIS Pro to work offline, which allows users to use the application without being signed in to a licensing portal. In this case, you can only use the application on one machine. This is the machine you use to take your license offline.

If you work in an environment with periodic online access, you can sign in to an active portal to exchange content with ArcGIS Online when access is available. Even when ArcGIS Pro is authorized to work offline, you must sign in to an active portal to get

1. Content from ArcGIS Online
2. Share content to ArcGIS Online.

Start ArcGIS Pro

To start ArcGIS Pro, you must be licensed as a name user and have an account with ESRI. If you are a student, your instructor will create a user account for you; if you are a professional, your administrator at your organization will create the account. Each user must have a login name and password.

First time you start ArcGIS Pro, you must enter your ArcGIS online by typing in any browser www.arcgis.com(Links to an external site.) "Sign In" use your login name and password. Signing in to your organization also allows you to access your content, group content, and your organization's content in your project. You can sign in to additional portals to use their content as well. You must also sign in to publish and share with your organization.

Open a Project
Launch ArcGIS Pro and log in using your account.
   Under **New Project** in the right click **Open another project.**

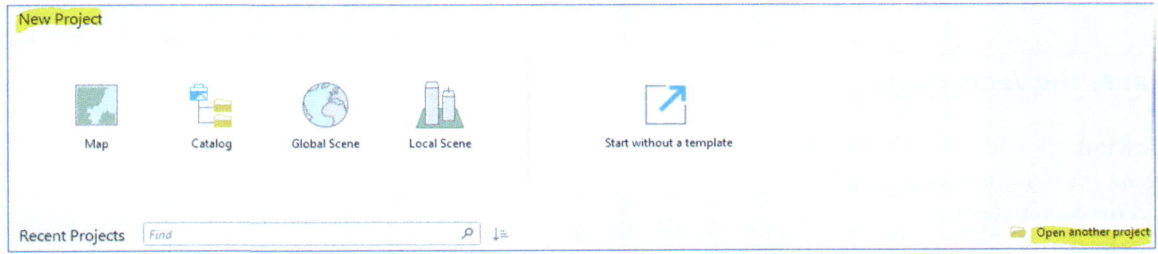

   Browse to \\ENV_Water\Ch01 click **Ch01.aprx** and click OK.
   Ch01 – ArcGIS Pro open with an interface and consists of the following:

1. Ribbon
2. Views
3. Panes

Ribbon The ribbon at the top of the application window displays and organizes functionality into a series of tabs. Some of these tabs (core tabs) are always present. Others (contextual tabs) appear when the application is in a particular state.

Catalog View Catalog View is in the middle of the program and consists of map view. The map view is a widow that displays a map. Every view has a tab that can be used to close the view or drag it to a different position. A project may have many views, which can be opened and closed as needed. The tab of the active view is blue, and currently, there are no open maps.

Panes There are two panes, and both are dockable windows. The first pane is called the **Contents pane** (left), which displays the contents of a view (empty now). In the textbook we will abreviated it as CP. The second pane is called the Catalog pane (right), which displays the contents of a project or portal or commands and settings related to an area of functionality, such as the Symbology and Geoprocessing panes (you will work with them in the coming chapters).

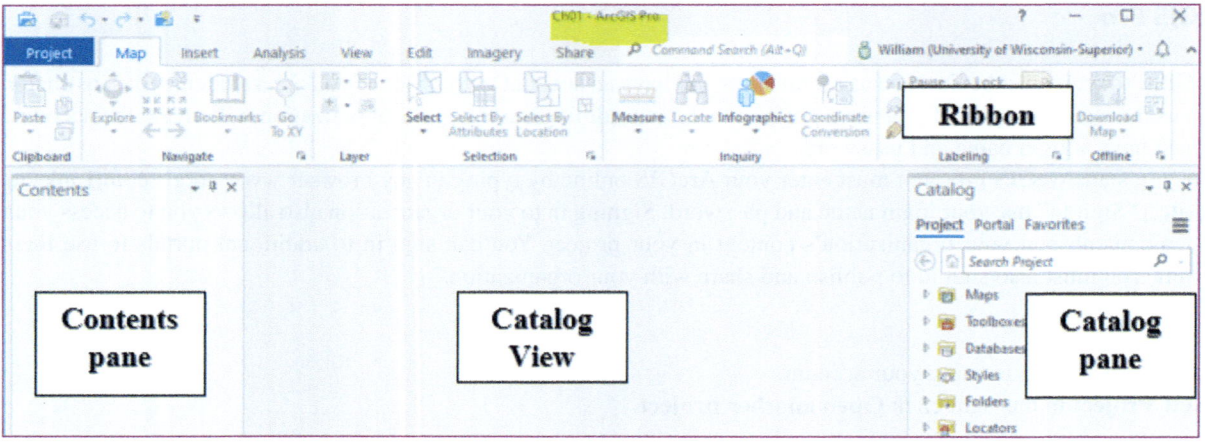

Lesson 1: Explore the Vector Data

Open the Vector Map
In the **Catalog** pane expands the **Maps**, which consist of two maps: **Raster** and **Vector**.
   D-click the **Vector** Map to open it

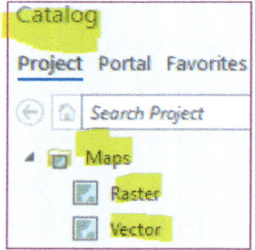

Result The following happens

1. The vector map is displayed in the **Contents** pane, and the **vector** map consists of four layers: **SamplingSite**, **NewtonCreek**, **MurphyOil**, and **HogIsland**. The Content pane works like ArcMap in ArcGIS desktop. In the Content Map, two layers also display the **World Topographic Map** and **World Hillshade**. The **World Topographic Map** is a basemap that serves as a reference map used as a background for your work. In the coming chapters, you will choose a different type of basemap. The **World Hillshade** is an elevation map from the living atlas of ESRI.
2. The **Map view**, the project opens with a map centered on Newton Creek in Superior, Wisconsin. The window containing the map is a map view, and the blue highlighted tab "**Vector**" at the top of the View window indicates that the Vector view map is active.

Lesson 2: Explore the Raster Data

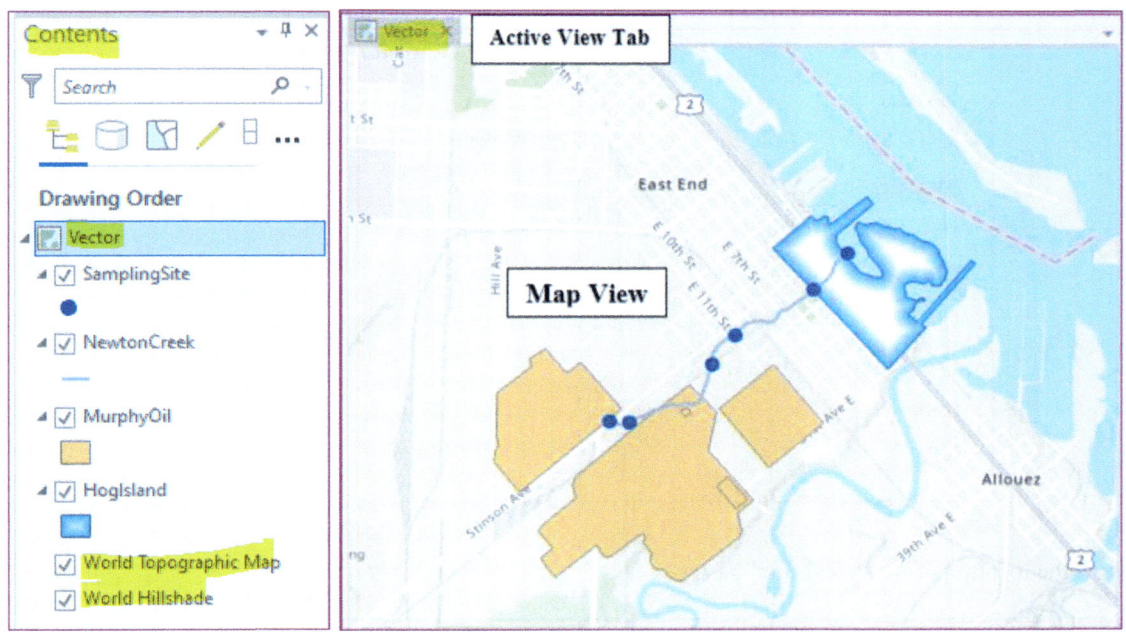

Layers in the **vector** map consist of points (**SamplingSite**), lines (**NewtonCreek**), and polygons (**MurphyOil** and **HogIslan**d). These layers can be classified and symbolized by different symbols. In the map, the 4 layers displayed above the World Topographic map are symbolized in different symbols and colors. The map shows that Newton Creek starts from Murphy Oil Inc. and discharges into Hog Island. Five sampling sites were used as sampling locations to monitor the quality of the water creek.

Lesson 2: Explore the Raster Data

Open the Raster Map
In the **Catalog** pane expands the **Maps**, and and double-D-click the **Raster** Map to open it.
Result The **vector** map is closed from the Contents pane, and the raster map is open in the Contents pane and consists of the elevation layer "**elevlidar**" of Superior, WI. Both **the vector** map and **raster** map are displayed as two tabs above the **map view**. The **raster** tab is blue, which indicates that the raster view map is active.

In the Contents pane, r-click "**elevlidar**", click Properties, and click **Source** tab.
The Source tab includes six sections that provide detailed information about the "**elevlidar**" raster.

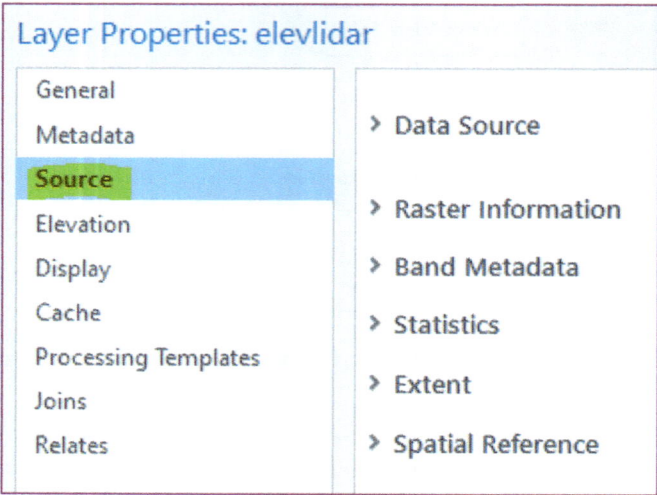

- **Data Source**: Show the location of the **"elevlidar"** raster

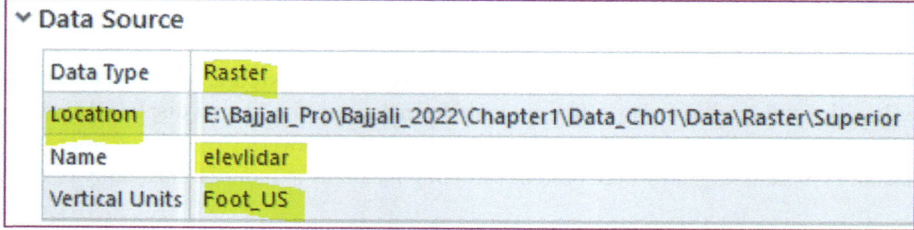

- **Raster Information**: This shows that the **"elevlidar"** raster has 9989 columns and 5105 rows. The raster consists of 1 band, and the cell size (resolution) is 6.289 × 6.289 feet. It also indicates that the pixel type is a floating point, meaning that the attribute table cannot be opened. The pixel depth of the raster is 32 Bit.

Raster Information	
Columns	9989
Rows	5105
Number of Bands	1
Cell Size X	6.289612
Cell Size Y	6.289612
Uncompressed Size	194.53 MB
Format	GRID
Source Type	Generic
Pixel Type	floating point
Pixel Depth	32 Bit
NoData Value	-3.4028231e+38
Colormap	absent
Pyramids	level: 5, resampling: Nearest Neighbor
Compression	None
Mensuration Capabilities	Basic

Lesson 2: Explore the Raster Data

- **Band Metadata**: it shows the information about the raster
- **Statistics**: This shows the minimum, maximum, mean, and std. deviation of the raster elevation.

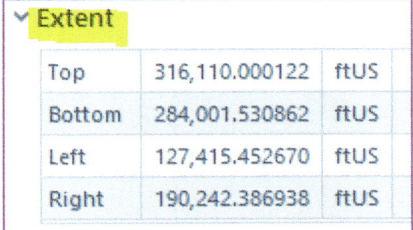

- **Extent**: this shows the coordinate extent in feet.

Extent		
Top	316,110.000122	ftUS
Bottom	284,001.530862	ftUS
Left	127,415.452670	ftUS
Right	190,242.386938	ftUS

- **Spatial Reference**: shows that the coordinate of the **"elevlidar"** raster is registered in Customized Transverse Mercator and the datum is **D custom**. It also provides information about the parameters of the projection, such as false easting and false northing.

Spatial Reference	
Projected Coordinate System	dc_4
Projection	Transverse Mercator
Authority	Custom
Linear Unit	US Survey Feet (0.3048006096012192)
False Easting	194000.0
False Northing	0.0
Central Meridian	-91.91666667
Scale Factor	0.999994968
Latitude Of Origin	45.88333333

Click OK after exploring the layer properties of the "**elevlidar**" raster.

Save the Project
It is a good practice to save your changes periodically while you work. You can save the changes to your project or the project as follows:

- Click the Save button on the **Quick Access toolbar** at the top of the app.
- Click the Project tab on the ribbon and click Save.
- Press Ctrl + S.

1. Save your project
2. Exit ArcGIS Pro (Project tab/Exit)

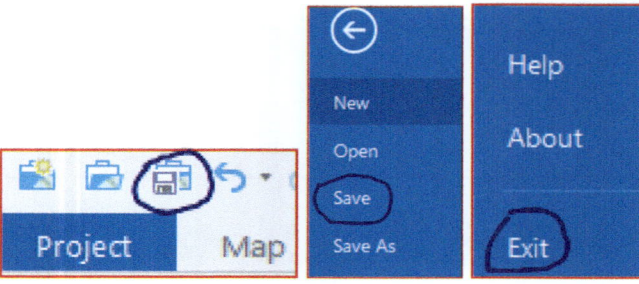

Working With ArcGIS Pro

2

The benefit from the use of geographic information system (GIS) software is tremendous and ranges from managing transportation in a dense city to finding and modeling sophisticated environmental problems. GIS is used all over the world to achieve various tasks from managing the environment and offering better service. It allows the user to carry out research and study practically everything, such as land, climate, environment, natural resources, and population.

Chapter 2 introduces the fundamental concepts of GIS and the major functionality contained in ArcGIS Pro. You will work with a variety of ArcGIS Pro tools, and you will learn how to create color coded maps, query, and solve a variety of spatial problems.

ArcGIS Pro has the capability of ArcGIS desktop and beyond but has a different platform. All the functions are performed in ArcGIS Pro, and there is no need to open separate programs such as ArcGIS Desktop. ArcGIS Pro maintained the Python, Model Builder, and various extensions in the same way as Arc GIS desktop.

ArcGIS Pro is a ribbon-based application. Many commands are available from the ribbon at the top of the ArcGIS Pro window; more advanced or specialized functionality is found on panes that can be opened as needed. The ribbon is at the top of the application window to display and organize functionality into a series of tabs. Ribbons have two types of tabs: core tabs are always present, and contextual tabs appear when the application is in a specific state. For example, a set of contextual feature layer tabs appear when a feature layer is selected in the content pane. Users can also customize the ribbon by creating tabs and choosing which commands appear on them.

ArcGIS Pro allows you to store multiple items, such as maps, layouts, tables, and charts, in a single project and work with them as needed. The application also responds contextually to your work. Tabs on the ribbon change depending on the type of item you are working with.

ArcGIS Pro offers several types of texts that users can use in maps. The main types are labels, annotation, text map notes, and graphic text in a layout and graphic text on the map. A label is a piece of text that is automatically positioned and whose text is based on feature attributes. Labels are the easiest way to add text to a map for each feature. ArcGIS Pro has the **Standard Label Engine** and the **Maplex Label Engine**, which provides further capabilities for placing labels.

GIS is made up of layers that make maps in GIS. You can add as many layers as you want, and the layers may contain features or images.

This section will focus on how to start the ArcGISPro project, connect to folder, data integration, data symbolizing, and labeling.

Create a New ArcGIS Pro Project

Before you can start working and making a spatial map, you must first create a **project** using a template. The template helps to support specific layers, tasks, and apps and can contain maps, layouts, scenes, and other items. It can also contain connections to data stored in folders, databases, and servers. Maps, layers, and other GIS content can also be added from portals such as your ArcGIS organization or ArcGIS Living Atlas of the World. Content you create in ArcGIS Pro can also be shared to your portal. The created project contains maps, databases, toolboxes, and other folders that are helpful when making the

---

Supplementary Information The online version contains supplementary material available at https://doi.org/10.1007/978-3-031-42227-0_2.

map. ArcGIS Pro automatically creates a default file geodatabase and default toolbox in the project's home folder. The file geodatabase is the default spatial data container.

Working with ArcGIS Pro

1. Open ArcGIS Pro

Note: if this is your first-time using ArcGIS Pro, you must sign in using your licensed ArcGIS account. ArcGIS Pro allows you to create Map, Catalog, Global Scene, Local Scene templates or you can start without a template. In this chapter, you will create a **Map template.**

2. Under "**New Project**"
3. Click **Map**
4. Fill the Create a New Project dialog box as follows:
5. Name: Ch02
6. Location: \\ENV_Water (a folder in your computer or server)
7. Click OK

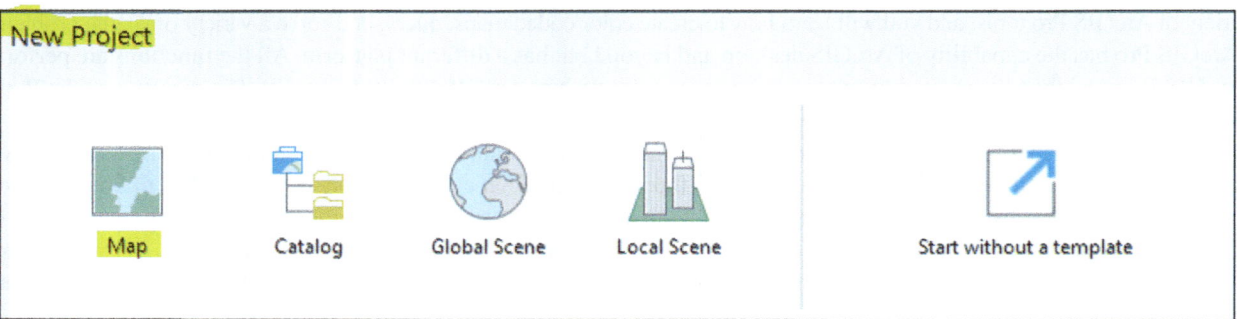

Note If you open the folder "Ch02" under **ENV_Water**, you will see 4 items inside: **Ch02.gdb**, **Index**, **Ch02.aprx**, and **Ch02.atbx**. If you do not see the extensions of the 3-files (i.e., Ch02.gdb), go to Window Explorer, click View tab check "File Name Extensions"

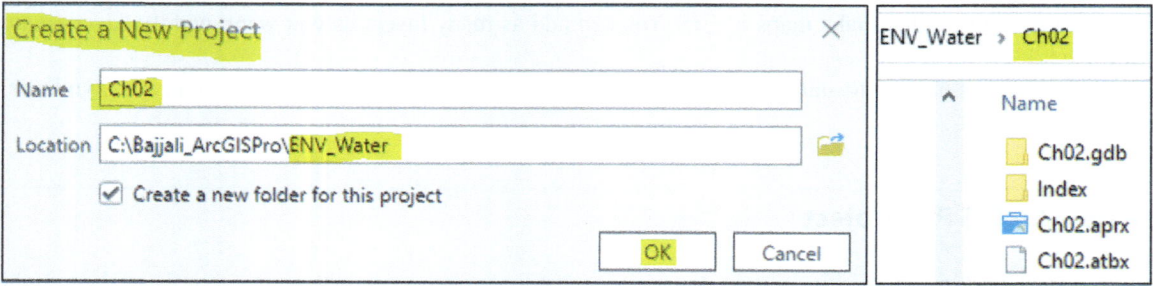

ArcGIS Pro will open with **Contents pane (CP)** to the left (analogous to table of content in ArcMap) and **Catalog pane** (analogous to Catalog window in ArcMap) to the right, and the **View window** in the middle, which displays the maps and is considered the primary work area of ArcGIS Pro.

The **View window** displays the **World Topographic Map** and the **World Hillshade**. The **World Topographic Map** is a basemap. A basemap cannot be symbolized or saved on your local drive but can be replaced by another basemap within the basemap gallery in ArcGIS Pro (Map tab/Layer group).

Create a New ArcGIS Pro Project

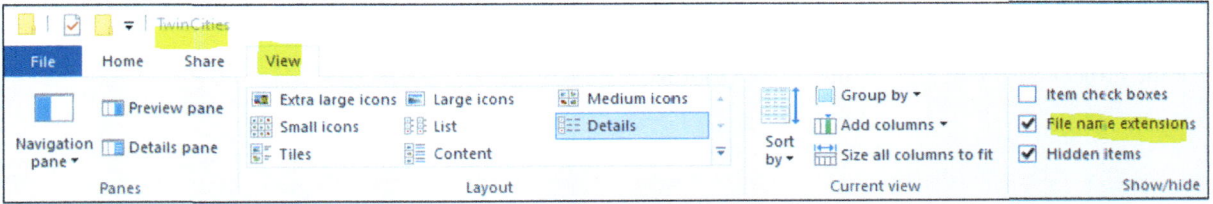

The **CP** includes the **Map**, which is empty now (you will populate it later) and the basemap **"World Topographic Map"** showing part of the world's continents. ESRI provides several types of basemaps that can be used in a project. The basemap has no attributes, but it offers a graphical background for the study area, as it covers the whole earth.

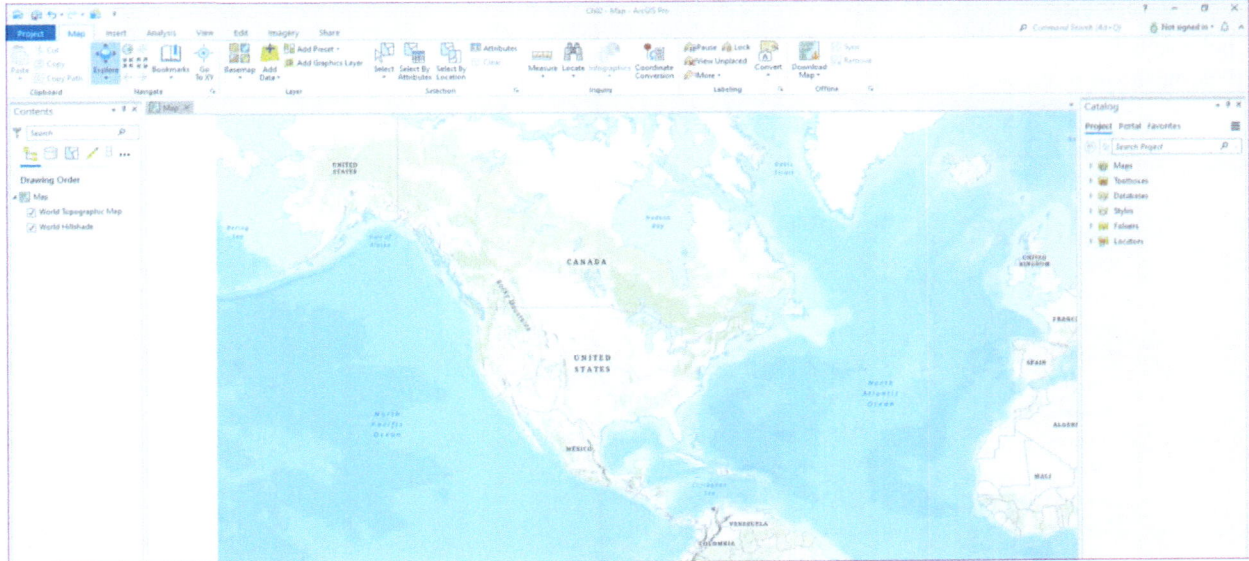

The **Catalog pane** will open with the following files:

- Maps
- Toolboxes
- Databases
- Styles
- Folders
- Locators

Note If you do not see the **Content** pane, **Catalog** pane, or **Catalog View** click **View** tab and, in the **Windows** group, click the button of the one that you want to open

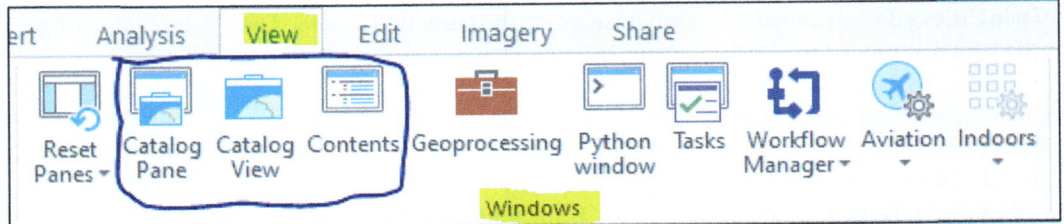

Identify the Location of the Project

6. Click the Project tab on the ribbon, then click **Options**

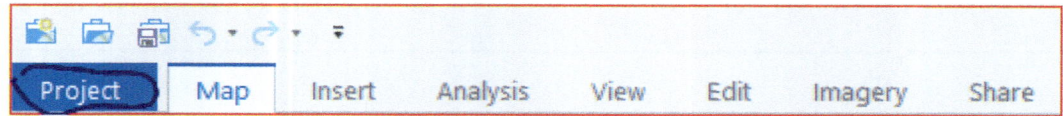

A dialog box displays that the name of the project is "**Ch02**", and the project resides in **Ch02** in the **ENV_Water** folder. The default database "**Ch02.gdb**" and the default toolbox "**Ch02.atbx**" and both also reside in the **Ch02** folder.

7. Click Cancel
8. Click the arrow to go back to ArcGIS Pro

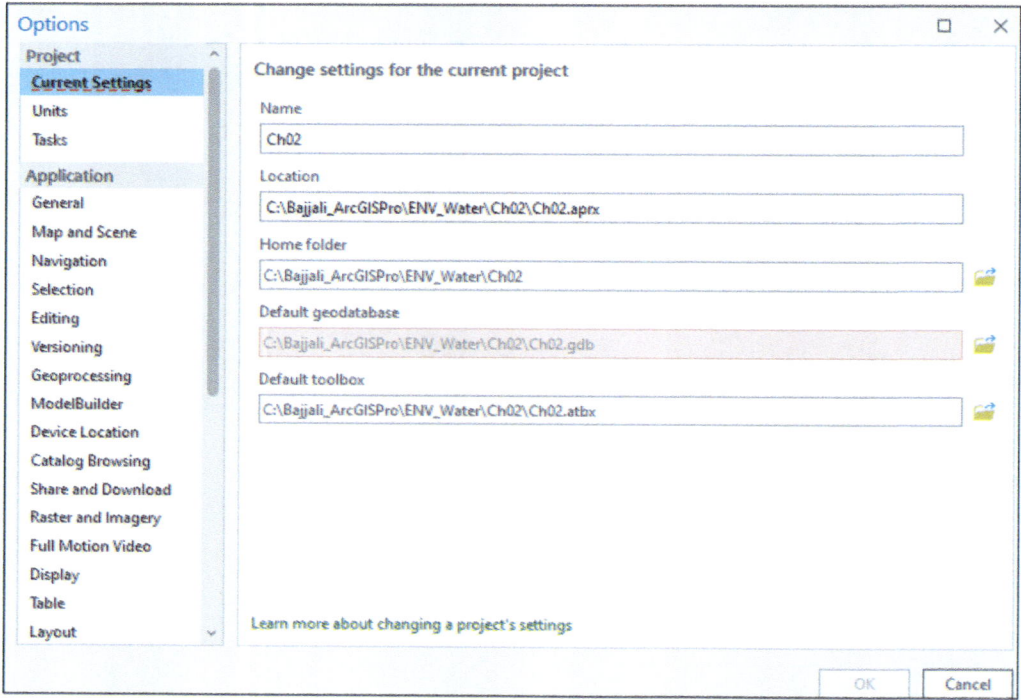

Connect to a Folder

This will allow users to connect to folders that contain data. The data can be in a folder on a local drive or network computer. The data in the folder provide the items needed to complete a project. The data for this project are feature classes in a file geodatabase "**TwinCities.gdb**" residing in the **Database** inside the **Data** folder in the **Data_Ch02** folder. The "**TwinCities.gdb**" contains the files that needed to be accessed more often in the project.

9. Make sure that the **Catalog** pane is open
10. Click the **Insert** tab on the ribbon, in the **Project** group, click **Add Folder**. OR in the **Catalog** pane, right-click **Folders** and click **Add Folder Connection**
11. Browse to \\Database\ Data_Ch08 folder, highlight **Data_Ch02** and click **OK**.

Data Integration in ArcGIS PRO

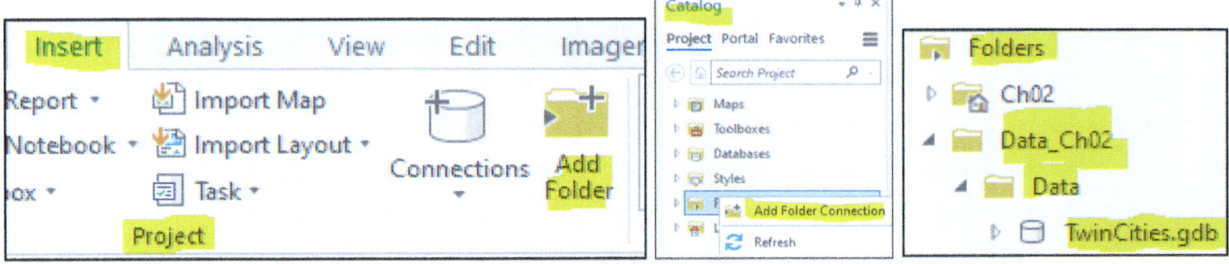

Result The folder connection "**Data_Ch02**" appears in the **Catalog** pane in the **Folders** category.

12. In Catalog pane, under the **Folders** expand "**Data**" folder, it contains the geodatabase file "**TwinCities.gdb**".

Note The file geodatabase "**TwinCities.gdb**" has 7 feature classes, and some of them will be used in the project.

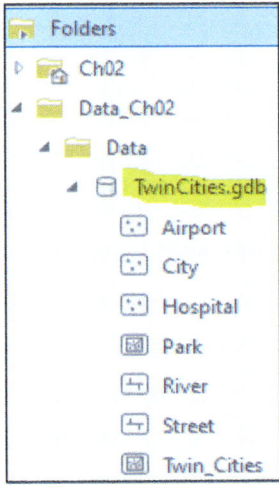

Data Integration in ArcGIS PRO

Now you are going to integrate feature classes from the "**TwinCities.gdb**".

13. In **Catalog** pane expand the "**Data**" under the "**Folders**" and then expand the "**TwinCities.gdb**"
14. Click Ctrl on the keyboard and click the **Airport**, **City**, **Hospital**, **Street**, and **Twin_Cities** to select them and then drag them into the map display area.

Result Five layers are displayed above the topographic map in the map display area and are added to CP below the **Map**

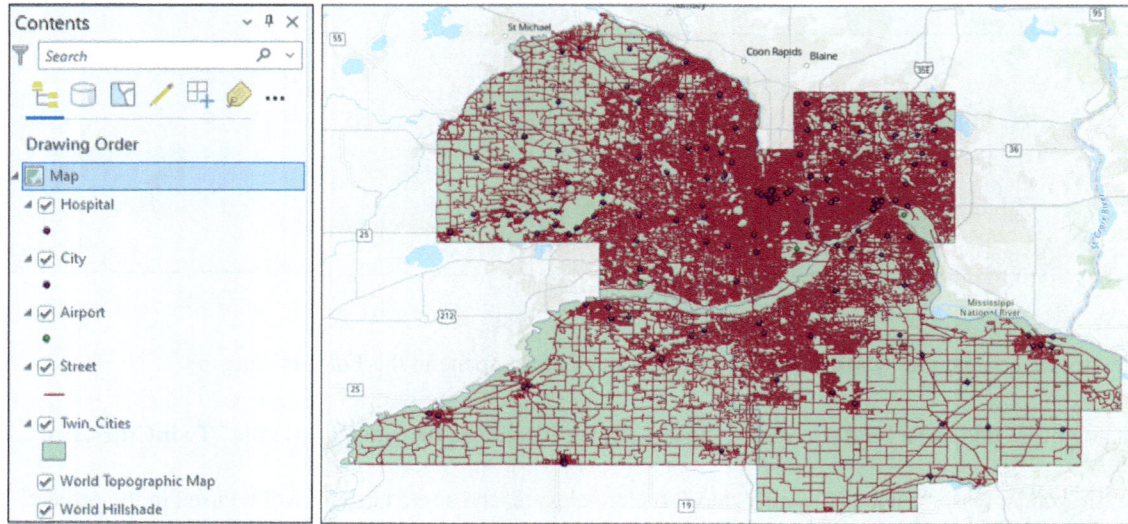

Save the Project

Note: It is a good practice to save your changes periodically while you work. You can save the changes to your project as follows:

15. Save your project, you can save the project in different approaches.

- Click the Save button on the **Quick Access toolbar** at the top of the app.
- Click the Project tab on the ribbon and click Save.
- Press Ctrl + S.

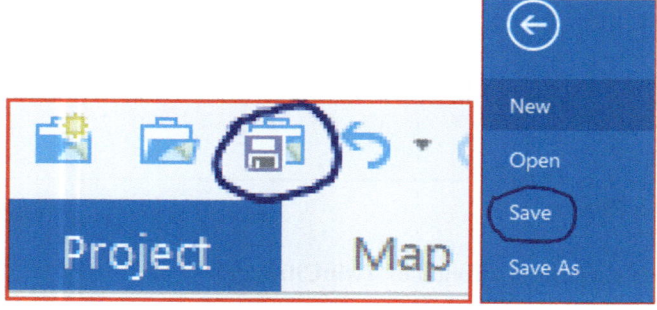

Explore the Navigate Group in the Map Tab

When you work with maps and scenes, you can zoom in and out and move around in the case of scenes, move up, down, and look around.

In the **Map** tab on the ribbon, in the **Navigate** group, there are various icons

The **Explore, Full Extent, Fixed Zoom In, Fixed Zoom Out, Previous Extent, Bookmarks**, and **Go To XY** tool.

Exploration Map

The default tool for maps and scenes is the **Explore** tool, which is used to move, pan, identify feature attributes via a pop-up, and zoom in and out of maps and scenes.

Data Integration in ArcGIS PRO

When you pan through a map or scene with the mouse, the pointer becomes a hand. Right-click and hold the mouse button and move the mouse up or down to zoom. The pointer becomes a magnifying glass when you click the right mouse button.

Some Explore tool capabilities are described in the image below.

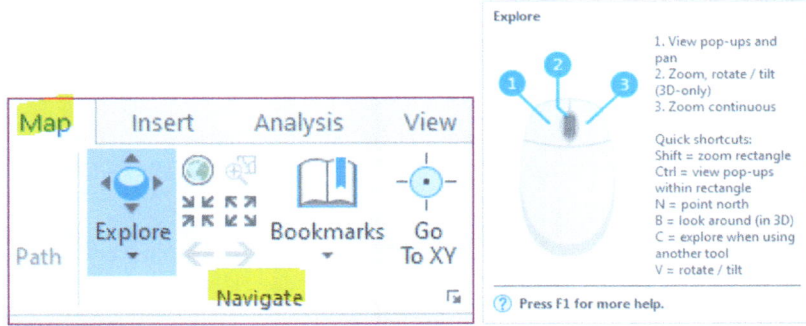

Zoom In/Out: scroll the mouse wheel to zoom in to or out from the cursor position
Zoom In: click **Shift** on the keyboard and draw a rectangle on the map where you want to **Zoom In**. To go to the previous extent right-click the mouse and point to "**Previous Extent**"
Pan: click at any location on the map and then drag the map to pan

Search Information About a Feature Class
To find a feature(s) with a specific attribute, type the attribute you are looking for in the search box in the **Locate** pane. You are going to find information about the Minneapolis-St. Paul Airport (MSP).

16. In the **Map** tab on the ribbon, in the **Inquiry** group, click the drop-down of Locate and select **Layer Search**. The **Locate** dialog box display
17. Select Layer Search Tab and click the drop-down arrow of **Options** (to the left of the window search)
18. Click the arrow to the right of **Search in layer** point to **Airport** and click the arrow and select **IATA**.
19. Type in the search window "**MSP**" and Enter

Result The MSP display

20. Right-click the "**IATA**" below the Airport and point to **Show Details**
21. A Pop-up display showing the attribute information of the MSP

Note You can also Zoom to, Pan to, Flash, and Add to Selection

22. Click **X** in the search window to remove the **MSP**

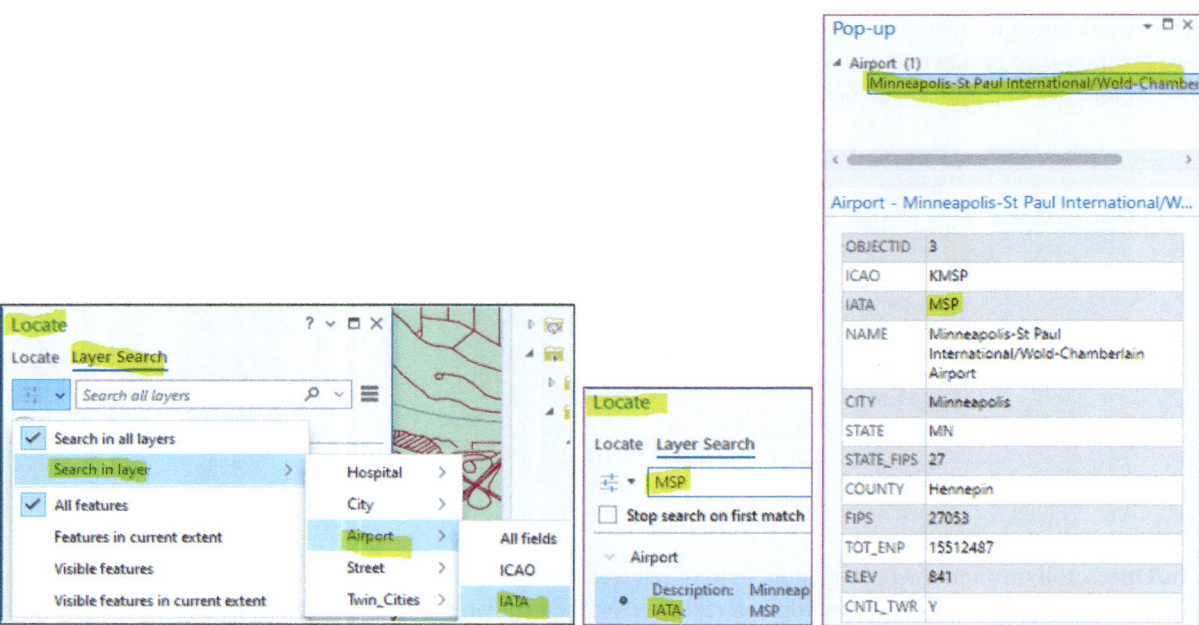

Locate and Address

23. In the Locate dialog box, click the **Locate** tab, and click the drop-down arrow of the **Options** and select **ArcGIS World Geocoding Service**.

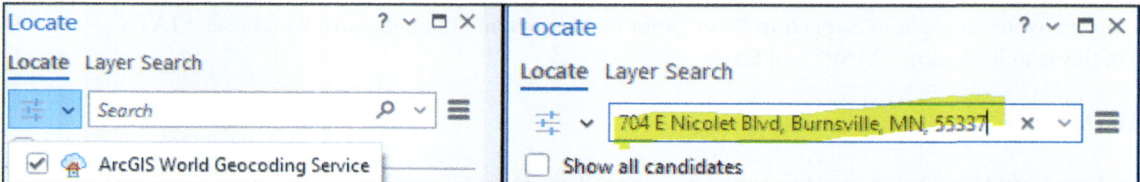

24. In the **search** window type "**704 E Nicolet Blvd, Burnsville, MN, 55337**" and Enter
25. The address will be displayed on the map and a pop-up information display about the address
26. Close Locate dialog box

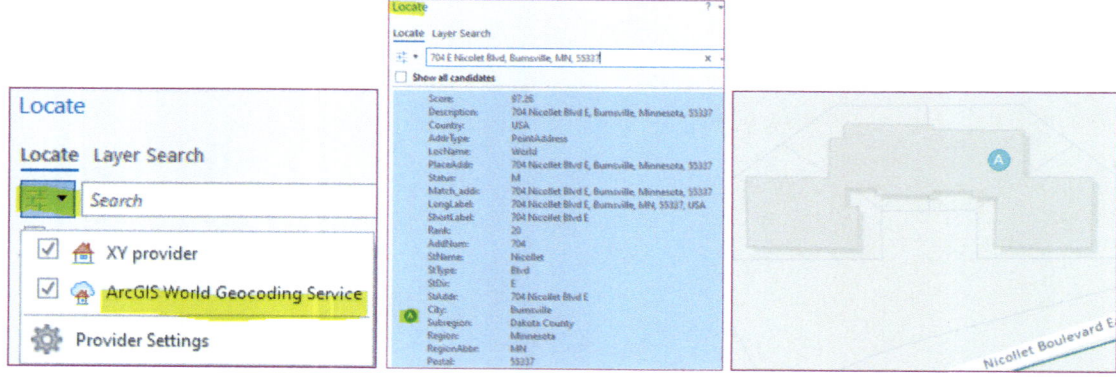

Find Places on a Map

Finding places or locations on a map is a fundamental function of a GIS. You can enter the coordinates in longitude-latitude or latitude-longitude order. In GIS, X is longitude, and Y is latitude. You are going to use decimal degrees (DD). The coordinate will be entered using W and N characters after the numeric values (you can use the character also before the numeric values)

27. In the **Map** tab on the ribbon, in the **Inquiry** group, click the Locate icon. The **Locate** dialog box display
28. Make sure the **Locate** tab in the **Locate** dialog box is selected
29. Click the drop-down arrow of the **Options** and select XY provider
30. Type in the search window "93.26 W, 44.75 N" and Enter
31. R-click the **A** text in the **Locate** dialog box and point to **Add Graphic**

Result A point added to the map and the **Locate Graphic Layer** is added to the CP.

If you want to delete the added point, right click the **Locate Graphic Layer** in the CP and **Remove**.

32. Close the Locate dialog box and click the **Save** button on the **Quick Access toolbar**.

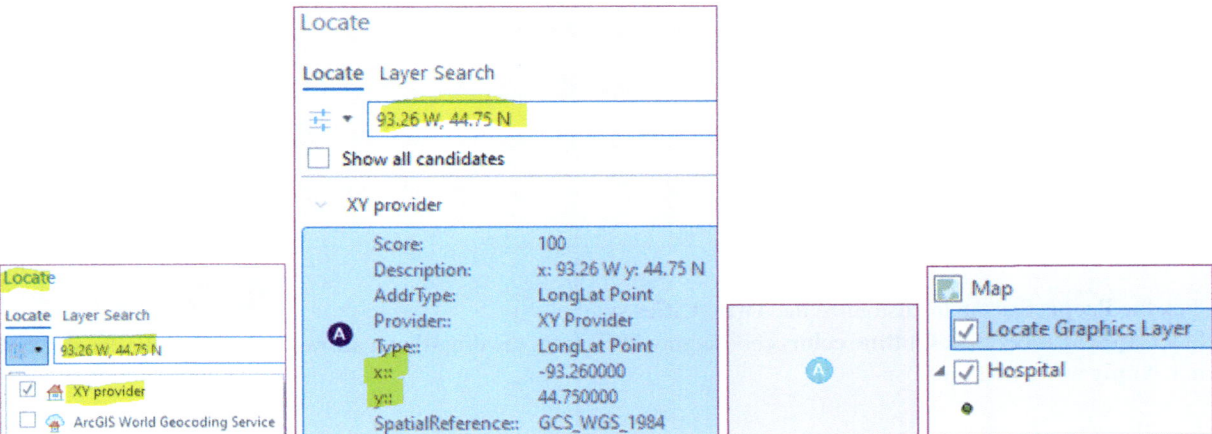

Change the Symbol of a Feature Class

33. Click Ctrl on the keyboard and in the **CP**, check the box of the **Hospital**, all the feature class will be unchecked
34. In the CP, check on only the **Twin_Cities, then** right click it and point to the **Zoom To Layer**

Change the Color of the Twin_Cities

35. In the CP highlight the **Twin_Cities** feature
36. The **Feature Layer** tab appears on the ribbon
37. Under **Feature Layer** in the **Drawing** group, click the arrow below **Symbology** and choose **Single Symbol**.

The **Symbology – Twin_Cities** pane appears, allowing you to change the symbol of the layer
    The Symbology pane consists of 5 icons

- Primary symbology
- Vary Symbology by attribute
- Symbol layer drawing
- Display filters
- Advanced Symbology options

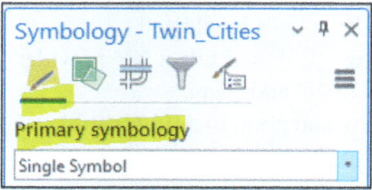

38. Ensure that the **Primary symbology** is selected
39. Click the symbol to open the "**Format Polygon Symbol**"
40. From the Gallery tab, select Black Outline (1pt) under ArcGIS 2D

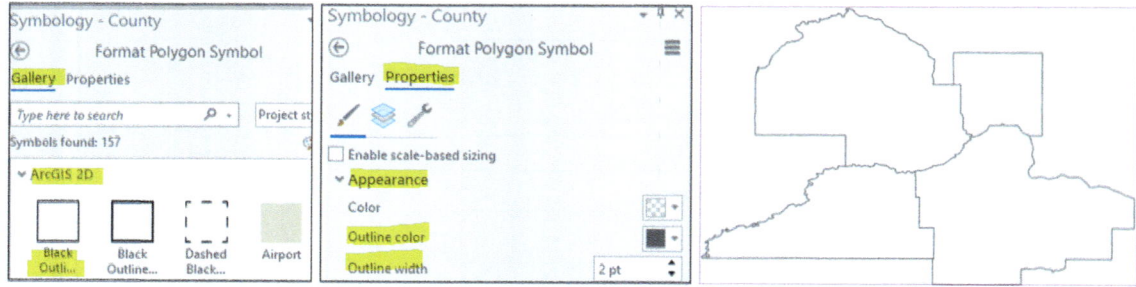

41. Click the **Properties** tab to customize the **Twin_Cities** symbol
42. Under **Appearance**, click **Outline color**, choose gray **70%** from the drop-down arrow. Make the outline width **2 pt** and click **Apply** at the bottom

Change the Symbol of the Hospital

43. In the CP, check the **Hospital** and click on its symbol
44. In the **Format Point Symbol** pane, ensure that the **Gallery** tab is selected.
45. In the search window type **Hospital**, enter
46. Choose 2nd symbol under ArcGIS 2D (the symbol will be changed in the map)
47. Click **Properties** tab in the Format Point Symbol, under **Appearance**, change the **Size** to 10 and click **Apply**

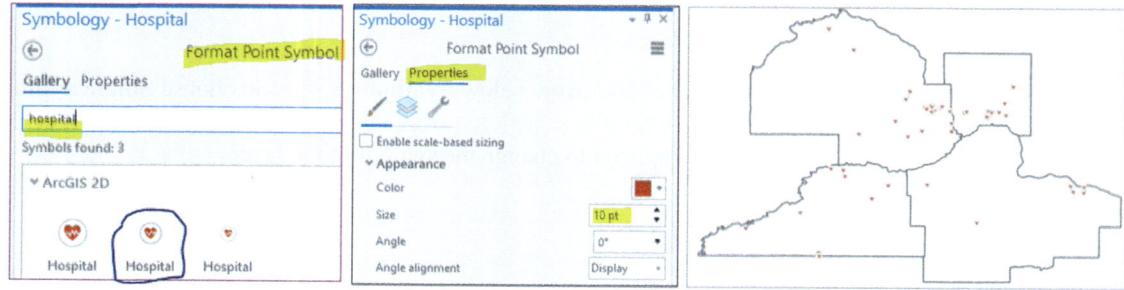

Find Places on a Map   25

Change the Symbol of the Airport

48. In the CP, check and highlight the **Airport** and click on its symbol.
49. In the **Format Point Symbol** pane, ensure that the **Gallery** tab is selected.
50. In the search window type **Airport**, click "Enter" (3 symbols display), to the right select All Styles from the drop-down arrow (more symbols will display)
51. Select the **Airport** Symbol under **Icon Points**
52. Click **Properties** tab, change the Size to **12** and click Apply

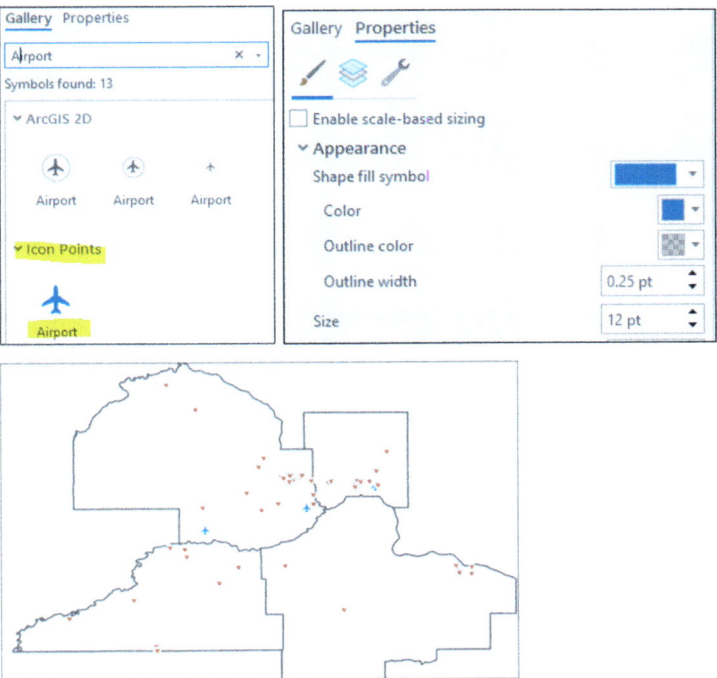

Challenge Yourself

Change the symbol of the **City** and **Street** using the table below

Feature class	Symbol	Color	Width
Street	Minor road	Gray 70%	1
City	Square 3	Pink	1

Change the Name of the Map

53. In the CP, right-click the "**Map**", select **Properties**, change the name to MN, click OK
54. In the CP select the **Twin_Cities** layer and press **F2** key, the **Twin Cities** became editable type **Counties**

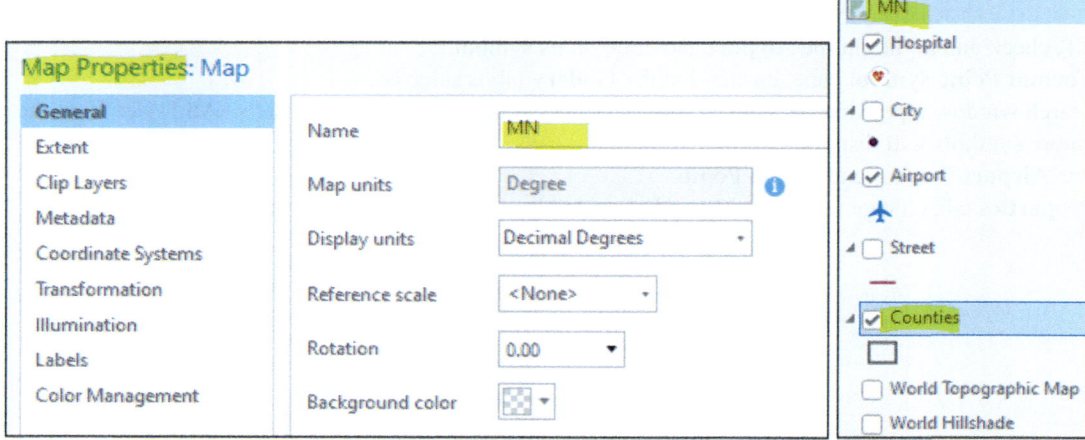

Labeling

In ArcGIS Pro, labeling means inserting text for layer features in maps or scenes. A label is derived from one field or more in the layer attribute table. In ArcGIS Pro, the label positions are generated automatically, the labels are not selectable, and the user cannot edit the display properties of individual labels. Maplex Label Engine is the default in ArcGIS Pro

Labeling and Font Changing

1. In the CP, right-click the **Counties** feature class and choose "**Label**", then right-click the **Counties** feature class again and choose "**Labeling Properties**"
2. In the Label Class – **Counties**, click the **Symbol** tab and expand the "**Appearance**", change the **Font name** to **Times New Roman**, **Size**: 12, **Color**: red, and then click **Apply**
3. Close the Label Class – **Counties.**

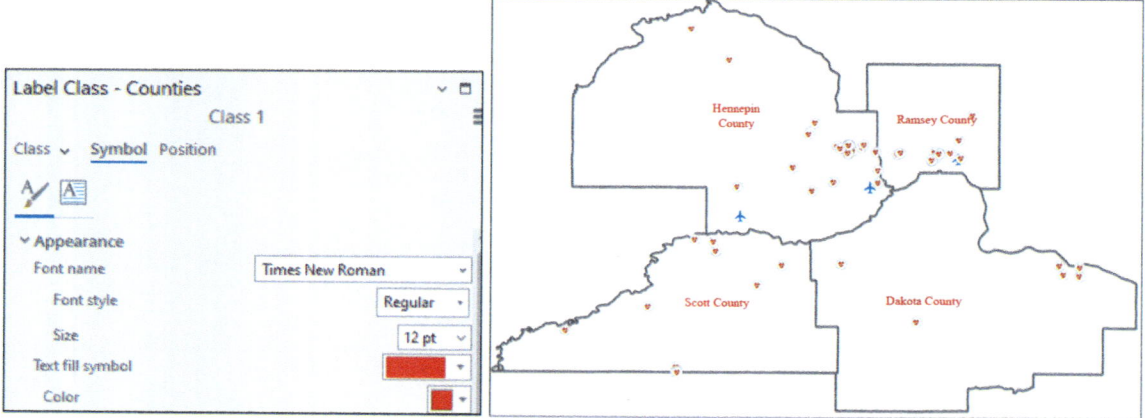

Label the Airport Using the Labeling Tab

4. In the CP, highlight the **Airport**, and click the **Labeling** tab in the feature layer contextual tab to access six groups
1. Layer group
2. Label Class group
3. Visibility Range group

Labeling

4. Text Symbol group
5. Label Placement group
6. Map group

Each group is associated with different function tools and provides access to the most common labeling functions (font, style, SQL, visibility range, and others).

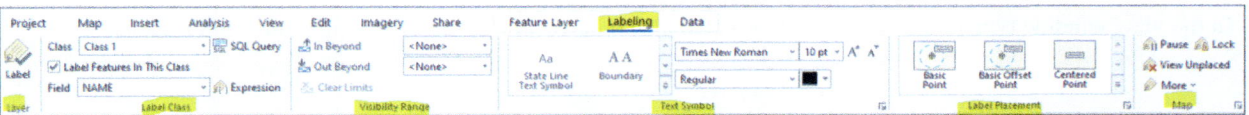

The **Text Symbol** and **Label Placement** groups have an arrow drop-down that opens the Label Class pane

5. In the CP, r-click the **Airport** and click the **Attribute Table**.
6. The Attribute Table open has **3 records** and **13 fields**, and the **IATA** field will be used for labeling.
7. In the Airport attribute table, click X in the upper right corner to close the table

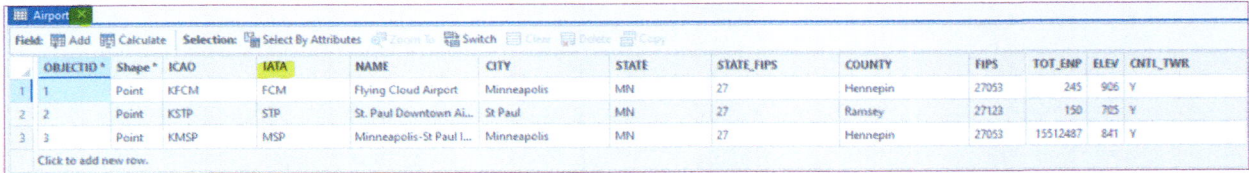

8. In the CP, ensure that the **Airport** layer is highlighted to activate the **Feature Layer Contextual** tab
9. Click the **Labeling** tab in the **Feature Layer Contextual** tab
10. In the **Label Class** group make sure **Field** is set to "IATA"
11. In the **Layer** group, click the **Label** icon

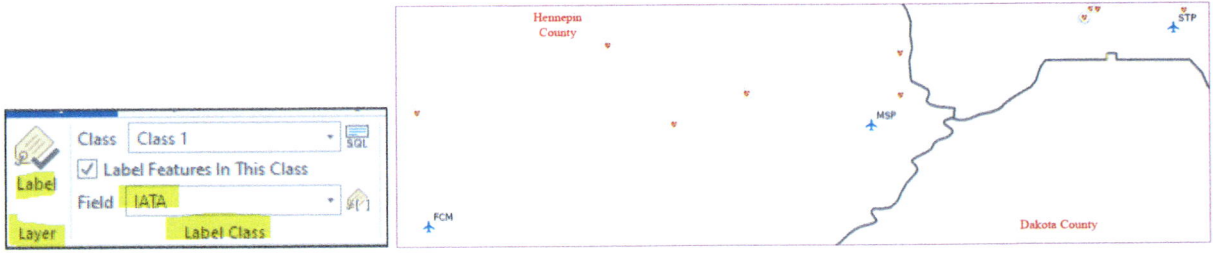

Result The Airports will be labeled

12. In the **Text Symbol** group, change the font to Time New Roman, size 12, style Bold, color: Blue

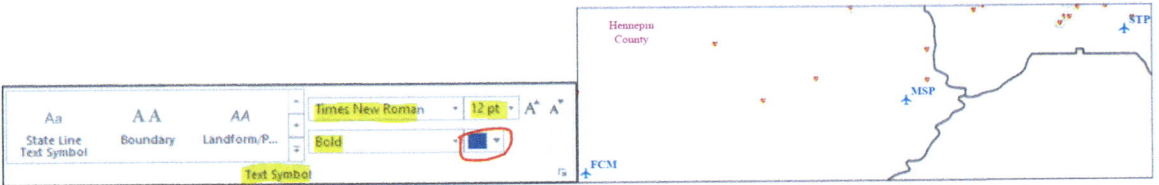

Label the River

13. Open the **TwinCities.gdb** under the **Folder** in the **Catalog** pane and drag the **River** into the map
14. In CP, r-click the **River**, and point to the **Zoom To Layer**
15. In CP, click the symbol of the **River**; in the **Symbology** pane, click the Gallery tab.
16. In the search window type river and then enter
17. Choose **Water (line)** under ArcGIS 2D
18. In the view window type in the **scale window 100,000** and enter (scale is located in the lower left corner)

19. Click **Labeling** tab in **Feature Layer Contextual** tab and in the **Label Placement** group
20. Click the 3rd drop-down arrow to see all the options and choose **Water (Line)** under **Line**.
21. In the **Layer** group, click the **Label** icon
22. In the **Text Symbol** group, change the font into Time New Roman, Size 10, and Italic

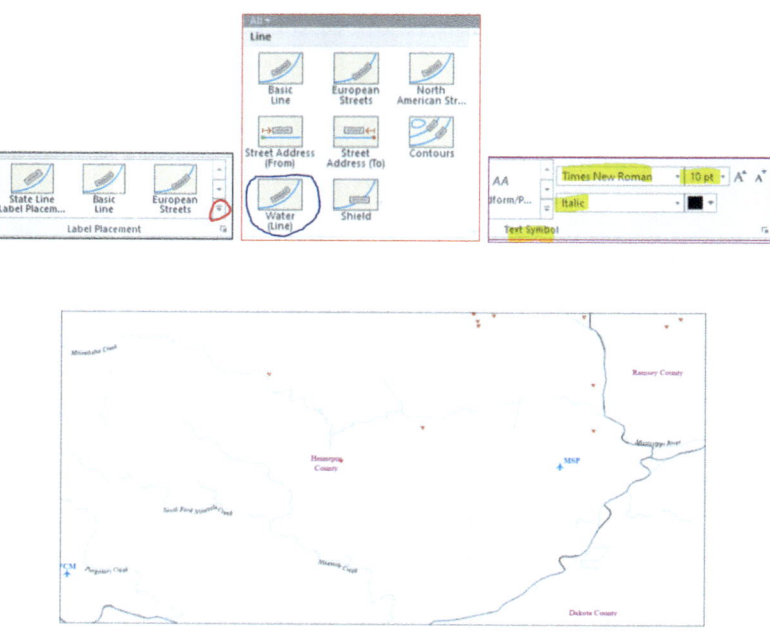

Display Layers Using Visibility Range

A layer that is checked in the Contents pane is drawn in the map or scene. When you zoom out or zoom in, it may become difficult to see more detailed information, or the information may become too coarse, especially if the map or scene contains several layers. Setting a visible scale range helps organize the layers in the map at different scales.

The **Labeling** tab in the **Feature Layer Contextual** tab has the "**Visibility Range**" group. The "**Visibility Range**" group includes the "**Out Beyond**" and the "**In Beyond**" scale ranges. Setting the "**Out Beyond**" scale range applies the smallest desired map scale at which the layer is visible in the display. On the other hand, "**In Beyond**" applies the largest desired visible map scale. Your "**Out Beyond**" map scale value must be larger than your "**In Beyond**" map scale value, or they can be the same if you want the layer to be visible at only one scale.

For example, setting a **scale** in the scale window to a bigger scale than the 200,000 (i.e., 150,000), the labels remain visible in the display. If you set the scale to a smaller scale than the 200,000 (i.e., 250,000) the labels disappear

Display Layers Using Visibility Range   29

23. Make sure that the **River** is highlighted in the CP and the Labeling tab in the **Feature Layer Contextual** tab is selected, also make sure that the scale in the display window is **100,000** (scale located in the lower left corner).
24. In the visibility range group, the "**Out** Beyond" was set to 100,000 from the drop-down arrow (this scale is similar to the map scale).
25. Change the "**Out** Beyond" to 200,000 (this scale is smaller than the map scale) and then enter.

Result The labels of the **Rivers** remain because the 200,000 scale is smaller than the original scale of the map (100,000).

26. In CP, right-click the **Counties** layer and **Zoom To Layer**

Result The label of the river disappears, because the scale of the map become 434,218 smaller (may be in your computer, it's different depending on computer configuration) and this scale is now beyond the 200,000 scale

27. In the CP, select the River layer
28. In the display window type in the **scale window type 190,000** and enter (this scale is bigger than the 200,000-scale set in the "**Out Beyond** ")
29. In CP, right-click the **Twin Cities** layer and **Zoom To Layer**
30. **Save** the project (Ctrl S)

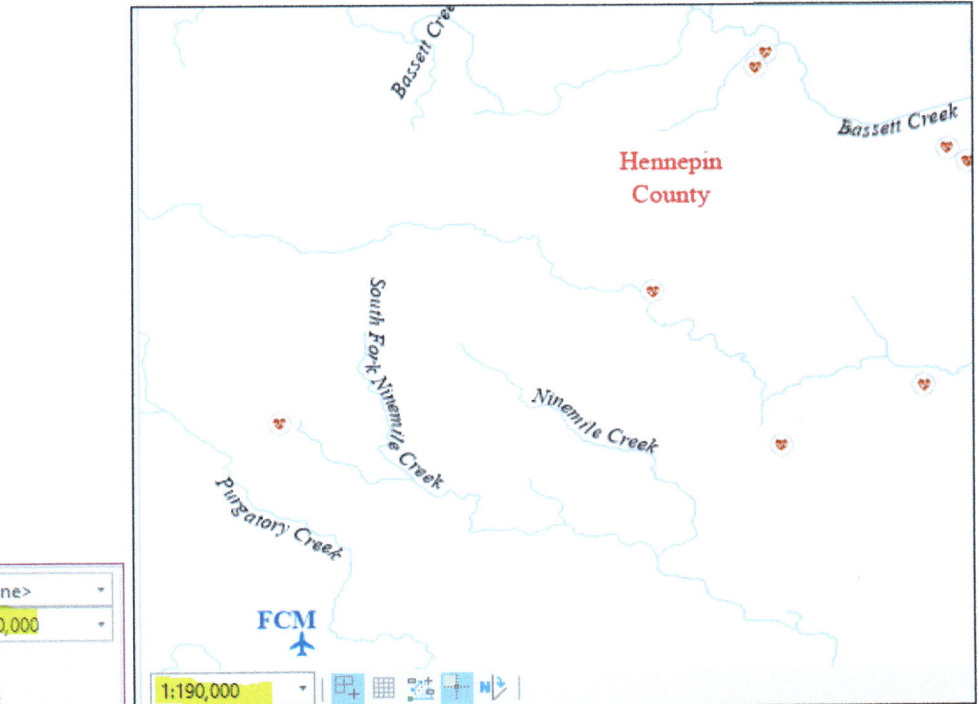

Park Scale-Based Symbol Classes

When the Park is symbolized using "**unique value**" symbology, GIS users can specify the **visible scale range** for each symbol class. This is an effective strategy to limit the amount of detailed data at smaller scales without having to make multiple versions of the layer. This approach is one strategy to control which features are drawn at which scales.

31. Open the **TwinCities.gdb** under the **Folder** in the **Catalog** pane and drag the **Park** into the map
32. The Scale in the lower left corner window is 1: 474,087 (may be in your computer different depend on the computer configuration)
33.
34. In CP, right-click **Park** layer and click **Attribute Table**.
35. Park has **193 records** and **12 fields**; the **CATEGORY** field has 3 variables and will be used to classify the parks.
36. Close the attribute table of the Park by clicking on the X on the top left part of the table
37. In the CP, highlight the **Park** layer and r-click it and select **Symbology**
38. In the Symbology – Park under Primary symbology from drop-down arrow, select Unique Values and Field 1: **CATEGORY**
39. Click the drop-down arrow of **Color scheme** and check **Show names** and select **Basic Random**
40. Under **Classes** tab, change the names under **Label** as below
41. To change the label double click the text under the label and change it

Value	Label	Color
Park Reserve	PR	Yucca Yellow
Regional Park	RP	Lepidolite Lilac
Special Recreational Feature	SR	Ginger Pink

42. R-click the color of **PR** and change the color to "**Yucca Yellow**" (R1C5)
43. R-click the color of RP and change the color to "**Lepidolite Lilac**" (R1C11)
44. R-click the color of SR and change the color to "**Ginger Pink**" (R3C12)
45. Click the drop-down arrow of **More** and turn off "**Show all other values**"

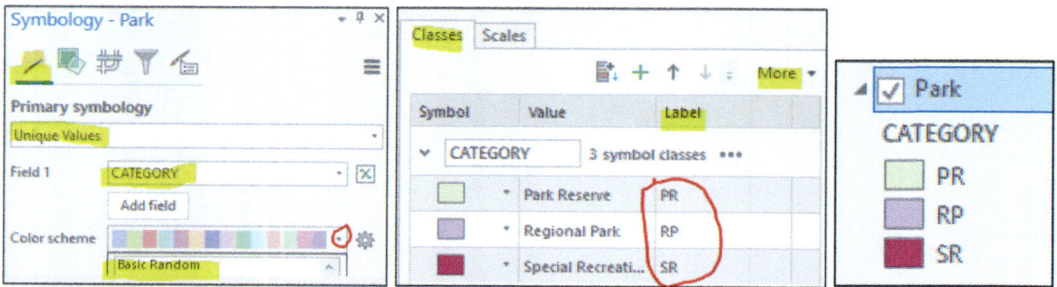

Different Symbols for Different Map Scales

46. Click the **Scales** tab in the **Symbology - Park** pane
47. Keep the scale of **Park Reserve** untouched
48. Move the right stop on the **slider scale** of the **Regional Park** to the left (scale 1:100,000) and move the left stop to the right (scale 1:500,000)
49. Keep the scale of **Special Recreation Feature** untouched

Create Thumbnail

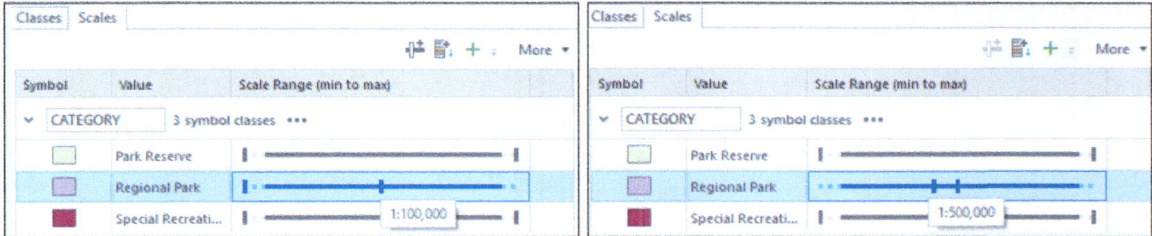

50. In the display window type in the **scale window 90,000** (scale located in the lower left corner)
51. The **Regional Park** feature disappears from the map because the 90,000 scale is larger than the 100,000 and 500,000 scales.

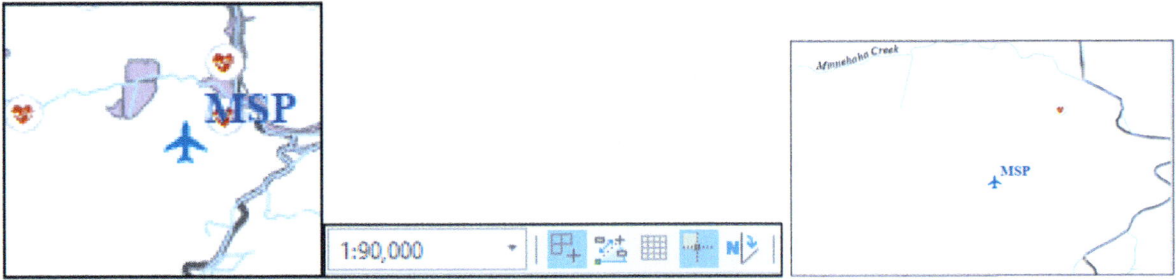

52. In CP, right-click the **Counties** layer and **Zoom To Layer**
53. **Save** the project (Ctrl S)

Create Thumbnail

A bookmark is a navigation shortcut to a position on a map or perspective in a scene to return to later or share with others. Bookmarks can be used to create keyframes in an animation (which will be used in the coming chapters).

54. The **MN** map zooms in on the **MSP** airport location (scale approximately 1:125,000).

55. On the **Map** tab, in the **Navigate** group, click drop-down arrow of the **Bookmarks** and click **New Bookmark**.
56. Name: **MSP**
57. Description: Minneapolis-Saint Paul is the main airport in MN
58. Click OK
59. In CP, r-click **Counties** and Zoom To Layer

60. On the **Map** tab, in the **Navigate** group, click drop-down arrow of the **Bookmarks** and select MSP

Result The map zooms to the **MSP** location

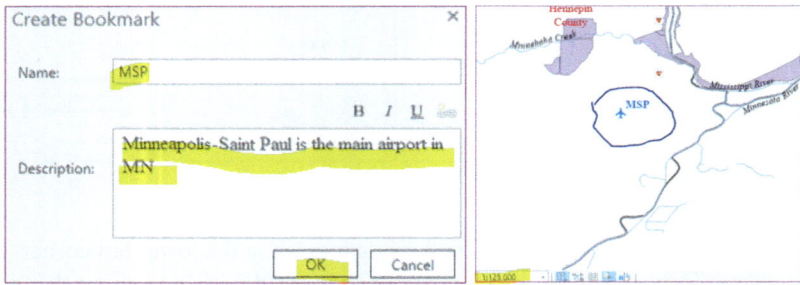

Delete the Thumbnail
To delete the thumbnail, do the following:

61. On the **Map** tab, in the **Navigate** group, click the drop-down arrow of the **Bookmarks** and select Manage Bookmarks.
62. In the Bookmarks pane, hover over the bookmark and click the red x to remove the MSP bookmark.

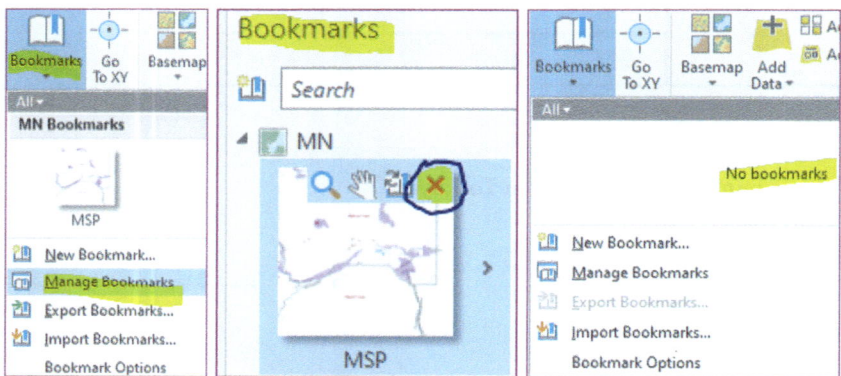

Result The MSP bookmark was permanently removed from the project.

Set Map Extent

The map extent is defined by a set of coordinates that outline the area of the map or scene in the project. By default, the extent is the spatial extent covered by all features in all layers in the map or scene. This is a dynamic extent. It updates accordingly when data are added or removed from the map or scene. GIS users can set a custom map extent

63. On the **Map** tab in the **Navigate** group, click **Full Extent** button
64. The MN map will zoom to the whole world.

Set Map Extent   33

Set the Extent to Twin Cities layer

Now you want to set the **Extent** to the extent of the **Counties** layer

65. In CP, right-click the **MN** map and click **Properties** to open the Map Properties dialog box.
66. On the **Extent** tab, click **Use a custom extent**.
67. Choose **Counties** below the **Extent of a layer** to generate extent coordinates.
68. Click OK to close the Map Properties dialog box.
69. Click the **Full Extent** button again
70. The MN map will zoom to the **Counties** layer

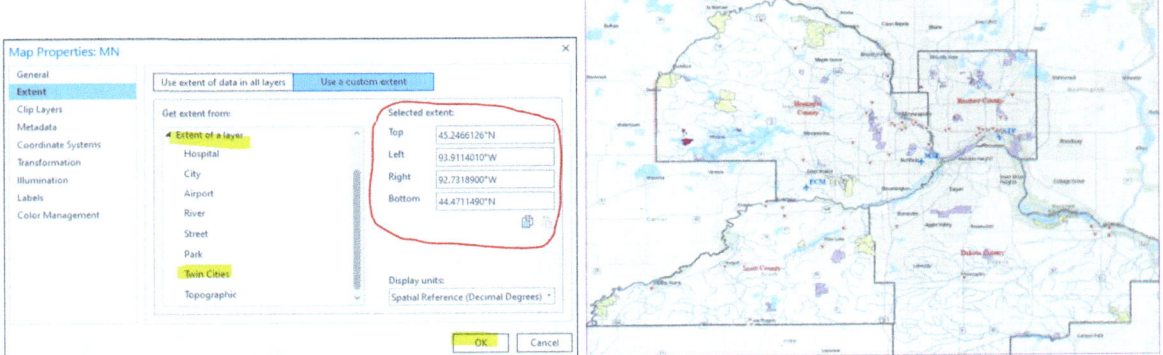

Challenge Tasks

Bookmark the Crow River
Label Streets

Question
What style did you pick? Why?

Map Classification and Layout 3

The map layer that represents features in GIS has more than its location and shape. The GIS layer can be associated with different information. For a river, this might include its name, length, depth, and water quality. For a county, this might include its population, ethnic group, household, income, age, and others. The information associated with a feature in a GIS is called an attribute. For example, population can be an attribute of a city, country, and other features. Feature attributes are stored in an attribute table. In an attribute table, each feature is a record, and each attribute is a field. The attributes for all the features in a layer are stored in the same attribute table.

FID	Shape *	NAME	POP2004	MALES	FEMALES	AGE_5_17	AGE_18_21	AGE_22_29
0	Polygon	Bayfield	15866	7590	7423	2906	520	877
1	Polygon	Taylor	19905	9966	9714	4184	917	1622
2	Polygon	Marinette	44538	21415	21969	7979	2388	3069
3	Polygon	Langlade	21190	10291	10449	3936	833	1526
4	Polygon	Washburn	17108	8071	7965	2991	617	1029
5	Polygon	Burnett	16433	7897	7777	2700	607	988

This attribute table for a layer of county stores each feature's ID number (FID), Shape, NAME, POP2004 (Population in 2004), MALES, FEMALES, age of populations and others. A feature on a GIS map is linked to its record in the attribute table by a unique numerical identifier (ID). Every feature in a layer has an identifier. Because features on the map are linked to their records in the table, you can click a feature on the map and see the attributes stored for it in the table. When you select a record in the table, the linked feature on the map is automatically selected as well and vice versa.

When you add data to ArcGIS Pro, the program assigns random colors for the layer symbols. You can change the colors and assign a color of your choice to make the map easy to view. When applying proper symbols and classification to the map in GIS, the map becomes easy to understand. ArcGIS Pro offers diverse symbol and label styles that users can use on maps and can modify them, so maps look just as desired. Features can also be symbolized based on an attribute. Maps on which features have been symbolized based on an attribute often convey more detail and clarification. For instance, road lines could be symbolized by a type attribute to indicate different roads, such as highways, interstates, or major roads. Individual well locations could be symbolized by a yield attribute to show the capacity of wells by discharge in m^3/h.

The type of symbology depends on whether an attribute's values are text or numbers. The numbers represent counts, amounts, rates, or measures. When a layer is symbolized based on an attribute with a text value, features are represented with a different symbol. Exactly how the symbols differ from one another depends on what you are mapping. For instance, if you were symbolizing geology according to the outcrop formations, you might use polygon symbols with different shades or color to represent the different formations. However, if you were mapping streams according to base flow, you might show streams with permanent base flow as a solid line and the intermittent stream as a dashed line.

Supplementary Information The online version contains supplementary material available at https://doi.org/10.1007/978-3-031-42227-0_3.

© The Author(s), under exclusive license to Springer Nature Switzerland AG 2023
W. Bajjali, *ArcGIS Pro and ArcGIS Online*, Springer Textbooks in Earth Sciences, Geography and Environment,
https://doi.org/10.1007/978-3-031-42227-0_3

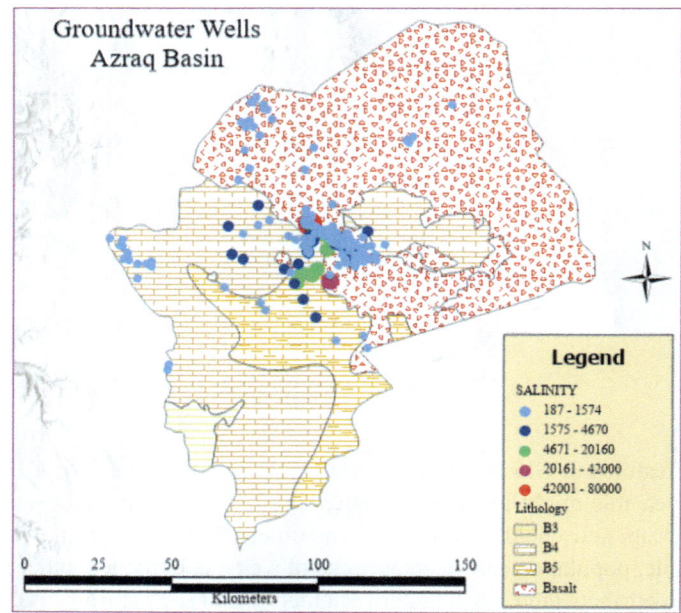

The map shows the groundwater wells in the Azraq Basin – Jordan

Feature quantities are typically represented on a map by creating groups of features with classes and assigning a special symbol to each class. The most common ways to symbolize quantities are **graduated symbols** and **graduated colors**.

Displaying features in a graduated sequence allows the map to visualize the distribution patterns in quantity data. For example, the groundwater map above is symbolized based on graduated symbols. This classification shows a quantitative difference mapped well features by varying the size of the symbol. The groundwater wells are classified into ranges that are each then assigned a symbol size to represent the range. The classification of the groundwater has five classes, and five different symbol sizes and colors are assigned. The symbols are drawn with colors ranging from cyan, blue, green, pink, and red. The red wells can be interpreted to represent greater salinity values than the wells with green and blue colors. Likewise, cyan has smaller symbols and represents a lower salinity than the wells with larger symbols.

Creating Map and Data Classification

Scenario 3-1 You are a geologist working for the Ministry of Water and Irrigation in the Azraq Basin. You have given a task to prepare a map showing the geology and the range of the groundwater salinity in the basin.

Creating a Geology Map Using Text Attributes

Data classifications are useful for representing continuous data in logically defined categories for use with mapping. In this chapter, the geology layer is a polygon and has a qualitative attribute called "lithology". The lithology field consists of different local geological codes that will be used to create a color-coded map. The map depicts the outcropping geological formations in the Azraq basin.

Data Integration

1. Launch ArcGIS Pro
2. Click **Open another project** (upper right) browse to **Ch03** under **Env_Water** and select **Ch03.aprx** and click OK

Result The **Ch03.aprx** open and you see the **Data_Ch3** folder is already added to the **Folder** in the Catalog pane

Data Integration 37

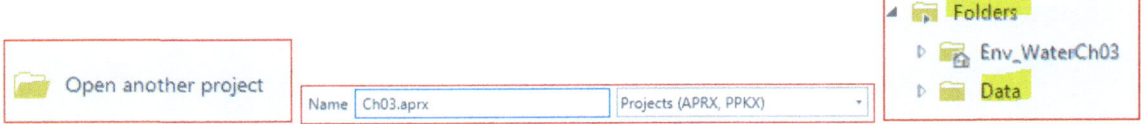

Rename the Map and Call It the "*Azraq Basin*"

3. In the CP, r-click on the Map and click on **Properties**.
4. Click on the General tab.
5. In the **Name** box, select the **Map** and click delete and type in **Azraq Basin**.
6. Then, click OK
7. In Catalog pane, open the **Data** under **Folder\Data_Ch03**, select the **Fault.shp**, **Geology.shp**, and **Well.shp** by holding the control key "**Ctrl**" and drag them to the **Map View**.

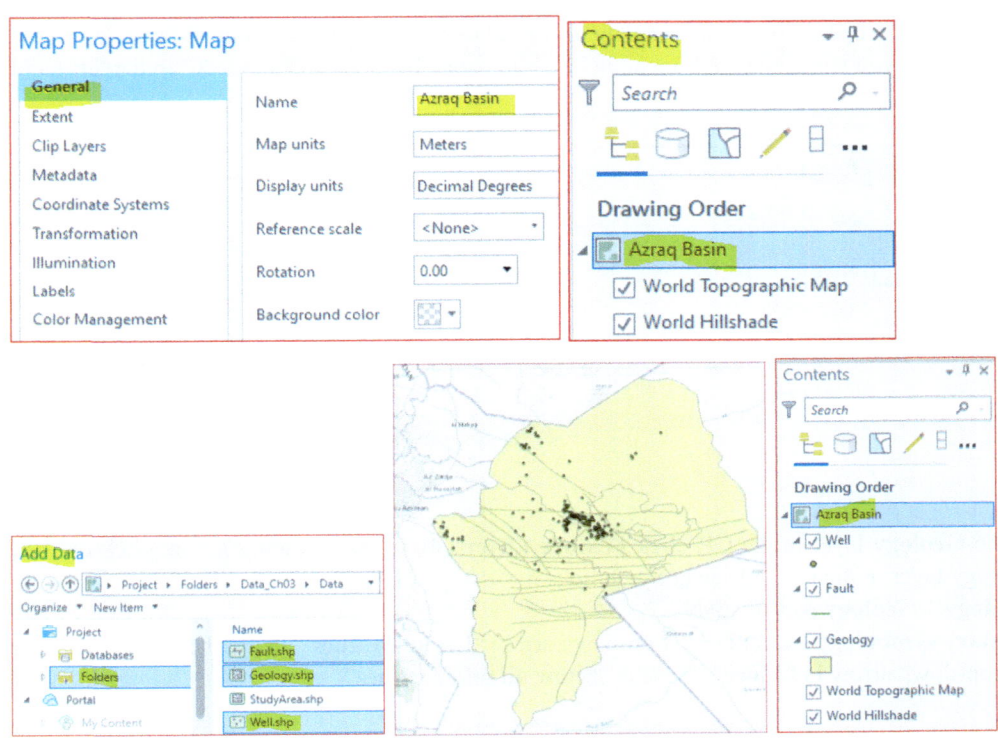

8. Save the Project by clicking the Save button on the **Quick Access** toolbar

Set the Extent to the Geology Layer

9. In CP, r-click **Azraq Basin** map, click **Properties** and select the **Extent** tab, and click "**Use a custom extent**"
10. Under Get extent from: select **Geology**, then press OK
11. Save the Project by clicking the Save button on the **Quick Access** toolbar

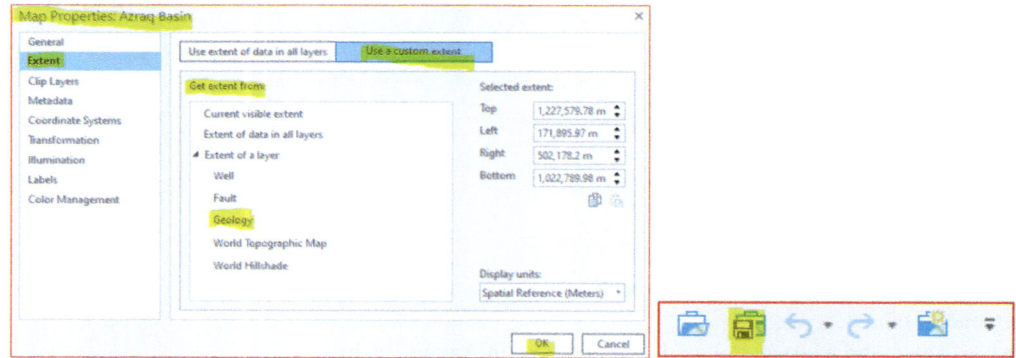

Classify the Geology Layer Using the Lithology Field

12. In the CP, r-click on the **Geology** layer, click on the "**Attribute Table**"
13. The attribute table will open. The "**Lithology**" field is used to symbolize the **geological** layer.
14. Close the table after you have done looking at the attribute table, by clicking on the "X" in the left-hand corner.

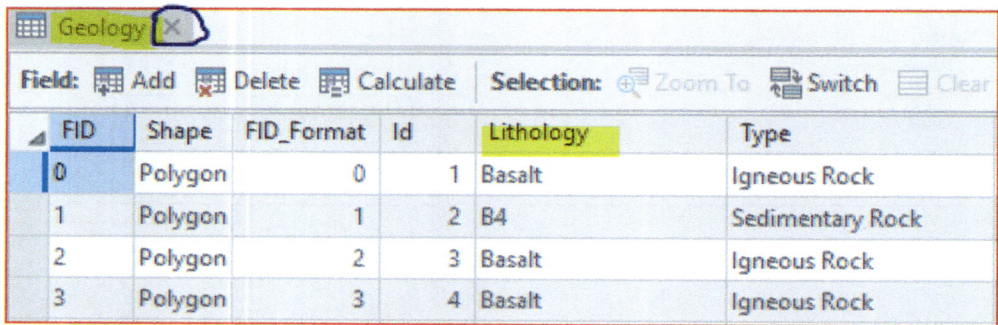

15. Highlight the **Geology** layer in the CP, click **Feature Layer** tab, in the **Drawing** group, click on the **Symbology** (or r-click Geology layer in the CP and select Symbology)
16. The **Symbology – Geology** pane display
17. On the **Primary symbology**, select from drop-down arrow "**Unique Values**" and in the Field 1, choose "**Lithology**"
18. Click the drop-down arrow in "**More**" uncheck "**Show all other values**"

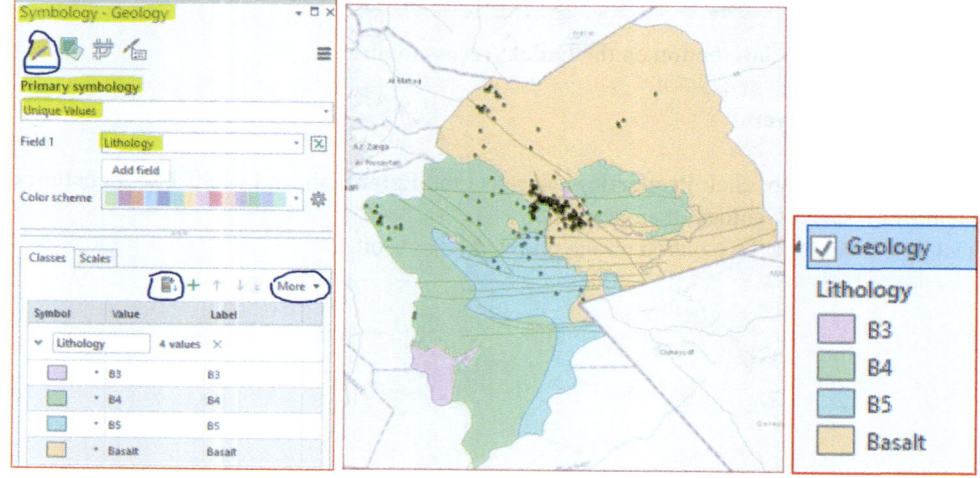

Importing geology symbols (Styles) from ArcMap

Styles are collections of symbols and other map components (colors, color schemes, label placements, and layout items) that can promote consistency and standardization across related map scenes and layouts. Styles are stored locally or in a portal as single files but can be added to a project. The ArcGIS Pro styles have a **.stylx** file extension.

The styles that are authored by Esri and included in the ArcGIS Pro installation. Only a subset of system styles is added to each new project by default. These default systems styles are listed in the Catalog pane in the **Project** tab under the **Styles** folder and consist of the following:

1. ArcGIS 2D
2. ArcGIS 3D
3. ArcGIS Colors
4. ColorBrewer Schemes (RGB)

You can add any of the other system styles to a project to have their contents also appear in the galleries. A unique aspect of system styles is that you can search for symbols from within all system styles, even when they are not added to a project.

Note Styles can be imported from **ArcGIS Desktop** or from the **web** if it is shared to a portal.

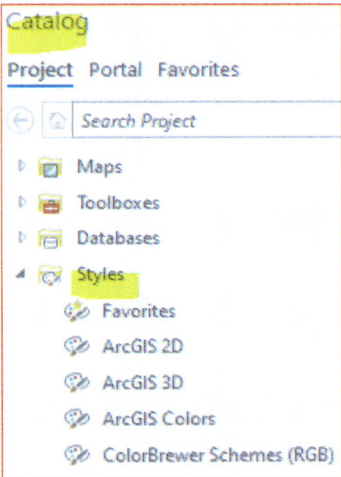

Import Styles from ArcGIS Desktop

If ArcGIS Desktop is installed on your computer, you can import any of the styles to the ArcGIS Pro project.

20. On the Insert tab, in the Styles group, click Import
21. Browse to C:\ProgramFiles(x86)\ArcGIS\Desktop10.8\Styles
22. Choose "**Geology 24K**" style and click OK.

Result The "Geology 24K" is added to Catalog pane under the Styles folder

23. R-click the "**Geology 24K**" style under the **Styles** in the **Catalog** pane and **Remove**.

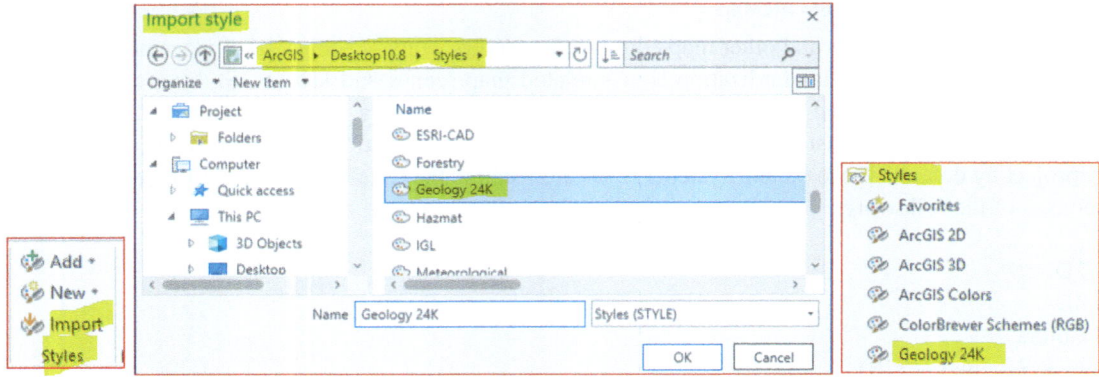

Import Styles from Web

Styles can also be imported into an ArcGIS Pro project from the Web (ESRI server)

19. Click the **View** tab, click **Catalog View**, in the **Contents** pane, click **ArcGIS Online** under **Portal**
20. In the **Search** window box, enter "**geology stylx**" and click **Enter**

Note All portal styles that match the search criteria will be returned, including "**Geology 24K**".

21. Right-click on "**Geology 24K**", click Add Style.

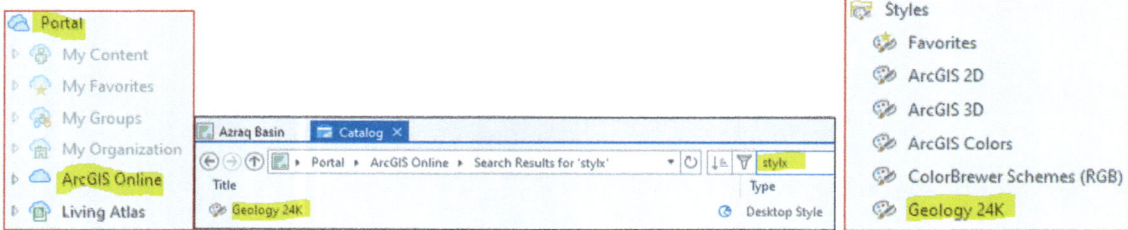

Result The "**Geology 24K**" is added to Azraq Project under "**Styles**" in Catalog pane.

Data Symbolization

Vector spatial data exist as points, lines, and areas. Representing these features, combined with their attributes, often means encoding something more complex than just geographic location. In this section, you will symbolize the **Geology** layer based on the **Lithology** field in the attribute table.

22. Above the **Map View**, click the **Azraq Basin** tab to make it active to display the **Azraq** map.

Data Symbolization   41

23. In the CP, click on the **B3** symbol of the **Geology** layer and it will take you to the **Format Polygon Symbol – B3** in the **Symbology – Geology** pane.
24. Make sure the **Gallery** tab selected and scroll down and select "*624 Carbonaceous Shale*",
25. Click the **Properties** tab, under **Appearance** click the **Color**, under "Geology 24K", pick "*Quaternary 2*" (2, 1) and make the outline color "**Black**", outline width 0.4, and then click **Apply**.
26. Repeat the previous step and change the lithology of B4, B5, and Basalt using the table below.

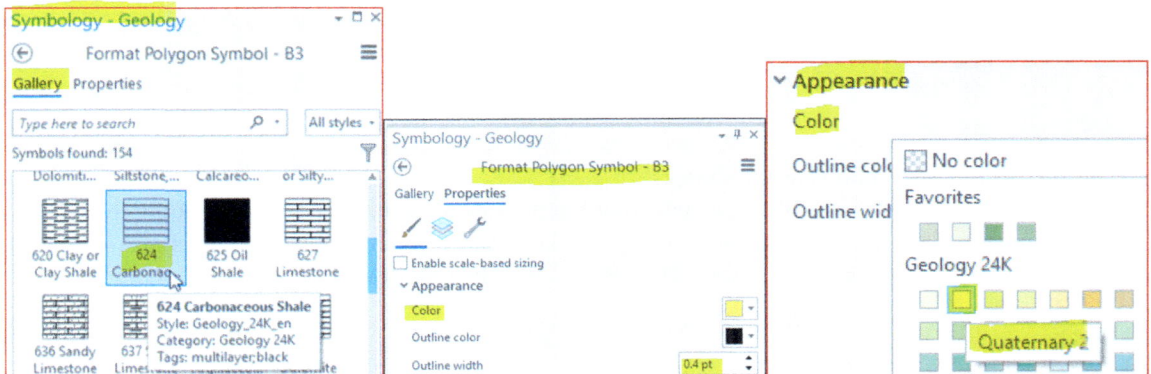

Lithology	Symbol	Color	Outline Color	Outline Width
B3	624 Carbonaceos Shale	Quaternary 2 (2,1)	Black	0.4
B4	627 Limestone	Tertiary 4 (8,1)	Black	0.4
B5	638 Argillaceous Limestone	Tertiary 8 (12,1)	Black	0.4
Basalt	407 Igneous	Volcanic 6 (5,8)	Black	0.4

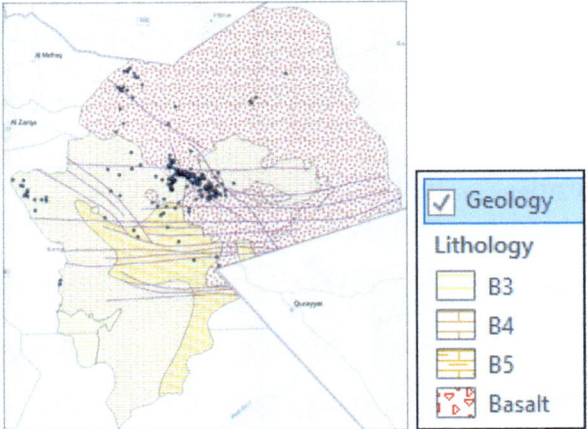

27. Save your project (Ctrl + S)

Save Geology Feature Class as Layer
A layer can exist outside of your map or project as a layer file (.lyrx). This allows others to access the layers you have built. You can share layers over the network or by email. When you add a layer file to a map, it draws exactly as it was saved, provided the data referenced by the layer is accessible

28. In the CP, r-click **Geology** layer, point to **Sharing** and choose **Save As Layer File**.
29. Save the "**Geology.lyrx**" under Folder in **Ch03** in a new folder call it **Layers** (you must create it)

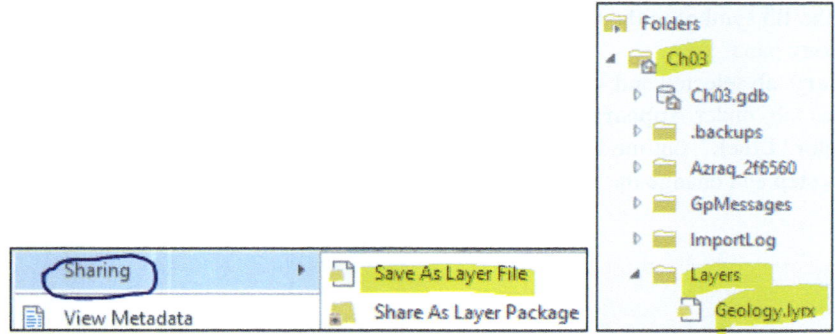

Creating a Salinity Map Using Numeric Attributes

ArcGIS Pro provides many types of classification and color ramps, which can be used to highlight different aspects of the data. When classifying the data, you can use one of many standard classification methods provided in ArcGIS Pro, or you can manually define your own custom class ranges. This section uses the following classification:

1. Natural Break
2. Quantile
3. Equal Interval
4. Manual

Create a New Map and Call It Salinity Using the Natural Break

A **Map** in ArcGIS Pro represents a collection of tabular and symbolized geographic layers and persists information such as the coordinate system and various other metadata.

1. Click **Insert** tab on the ribbon, in the **Project** group, click **New Map**
2. In the **Contents** pane rename the **Map** and call it **Salinity Natural Break**
3. In **Catalog** pane, expand the **Ch03** folder under the Folder
4. Drag the **Geology.lyrx** in the Map View
5. Under **Folder**, open **Data_Ch03\Data** and drag **Well.shp** into the Map View.

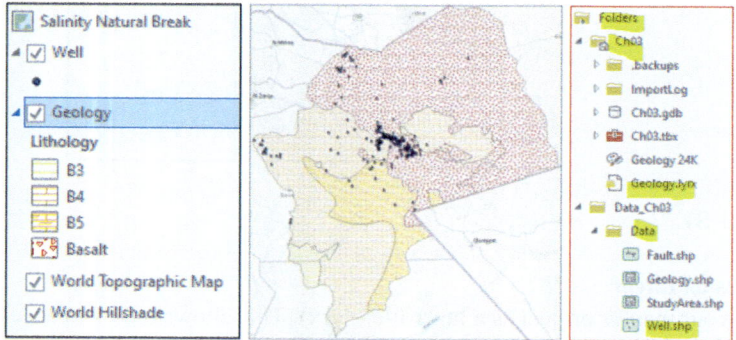

Result Geology and **Well** are added into the CP.

Classify the Salinity of Groundwater Using the Natural Break Method

The **natural breaks** (Jenks) classification is designed to place variable values into naturally occurring datasets. The features are divided into classes whose boundaries are set where there are relatively large differences in the data values. The set of data is classified by finding points that minimize the within-class sum of squared differences and maximize the between-group sums of squared differences. The advantage of this classification is that it identifies actual classes within the dataset, which is useful to create true representations of the actual salinity of the groundwater wells.

1. In the CP, r-click on the **Well** layer, open "**Attribute Table**".
2. The "**SALINITY**" field in the attribute table will be used to classify the **Well** layer.

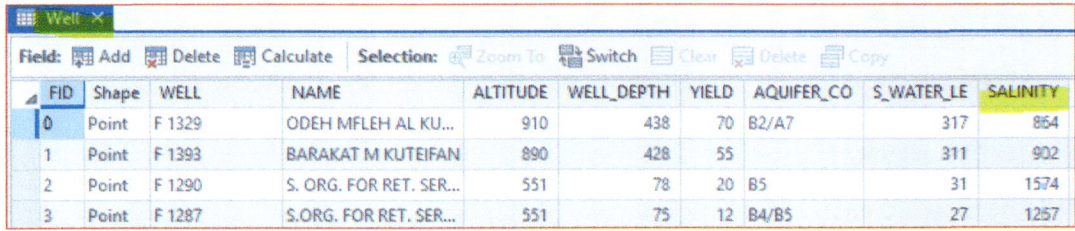

3. R-click the "**SALINITY**" field and select "Sort Ascending"

Question   What is the highest and lowest salinity?

4. Close the attribute table of the **Well**.
5. In the CP, r-click on the **Well** layer, click the **Symbology**.
6. The **Symbology – Well** pane of the **Well** open
7. Under **Primary symbology,** select "**Graduate Symbols**" and fill it as follows:
8. Field: SALINITY
9. Method: Natural Breaks (Jenks)
10. Classes   5
11. Minimum size: 6 pt
12. Maximum size: 16 pt
13. To the right of the **Template**, click the symbol, and choose **"Circle 1"** (under ArcGIS 2D)
14. Click the back arrow in the **Symbology-Well** pane
15. In **Classes** tab, under the Symbol, r-click the biggest symbol (bottom), change the color to **red**, r-click the 2nd symbol from bottom and change the color to **pink**, change the 3rd to **green**, the 4th into **blue** and the top into **cyan**

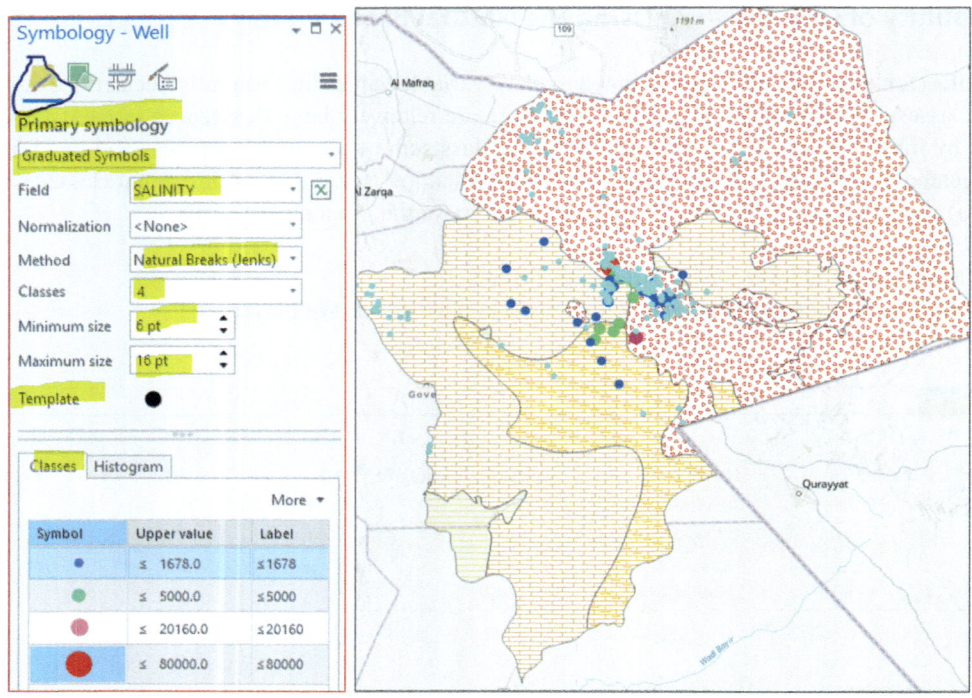

16. In the Symbology – Well pane, click the Advanced symbology options (5th icon)
17. Open the Format labels, under **Alignment**, check "**Show thousands separators**"

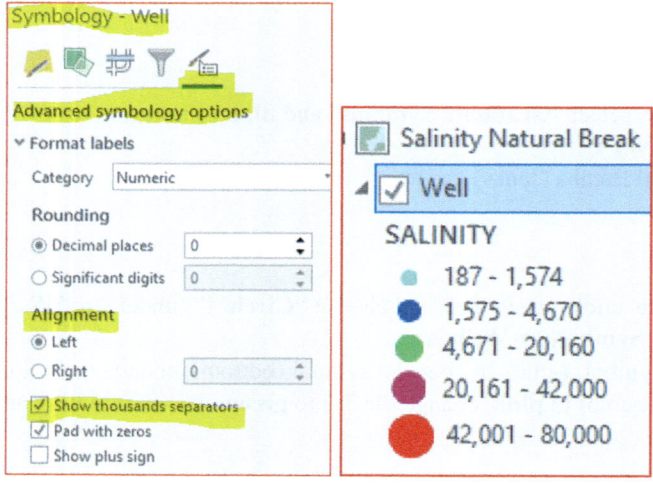

Result After you have changed all the colors and made the appropriate symbol size changes, your dialog box should look like this or very similar.

18. Save your project (Ctrl + S)

Classify the Salinity of Groundwater Using the Quantile Method

In the **quantile** method, each class contains an equal number of features, and this method is mainly used in homogenous data. This method is useful when you want to emphasize the relative location of highly saline wells among other low salinity wells. This method sometimes provides misleading results because the groundwater salinity in the wells is grouped in equal numbers in each class. To avoid this phenomenon, the number of classes can be increased.

Classify the Salinity of Groundwater Using the Quantile Method

19. Click **Insert** tab on the ribbon, in the **Project** group, click **New Map**
20. In the Contents pane rename it to **Salinity Quantile**
21. In **Catalog** pane, expand the **Ch03** folder under the **Folder**
22. Drag the Geology.lyrx into the Map View
23. Under **Folder**, open **Data_Ch03\Data** and drag **Well.shp** into the Map View.

Result The **Geology. lyrx** and **Well.shp** added to the Map View and in the CP.

Note If you did not close the Symbology pane from the previous step, it will be available for you to continue

24. If the Symbology – Well pane is close r-click the well and choose Symbology
25. Fill the Symbology – Well pane as below
    (a) Under Primary Symbology: Graduate Symbols
    (b) Field: SALINITY
    (c) Method: Quantile
    (d) Classes   5
    (e) Minimum size: 6 pt
    (f) Maximum size: 16 pt
26. To the right of the **Template**, click the symbol, and choose **"Circle 1"** (under ArcGIS 2D)
27. Click the back arrow in the **Symbology-Well** pane
28. In **Classes** tab, under the Symbol, r-click the biggest symbol (bottom), change the color to **red**, r-click the 2nd symbol from bottom and change the color to **pink**, change the 3rd to **green**, the 4th into **blue** and the top into **cyan**
29. In the Symbology - Well pane, click the Advanced symbology options (5th icon)
30. Open the Format labels, under **Alignment**, check **"Show thousands separators"**

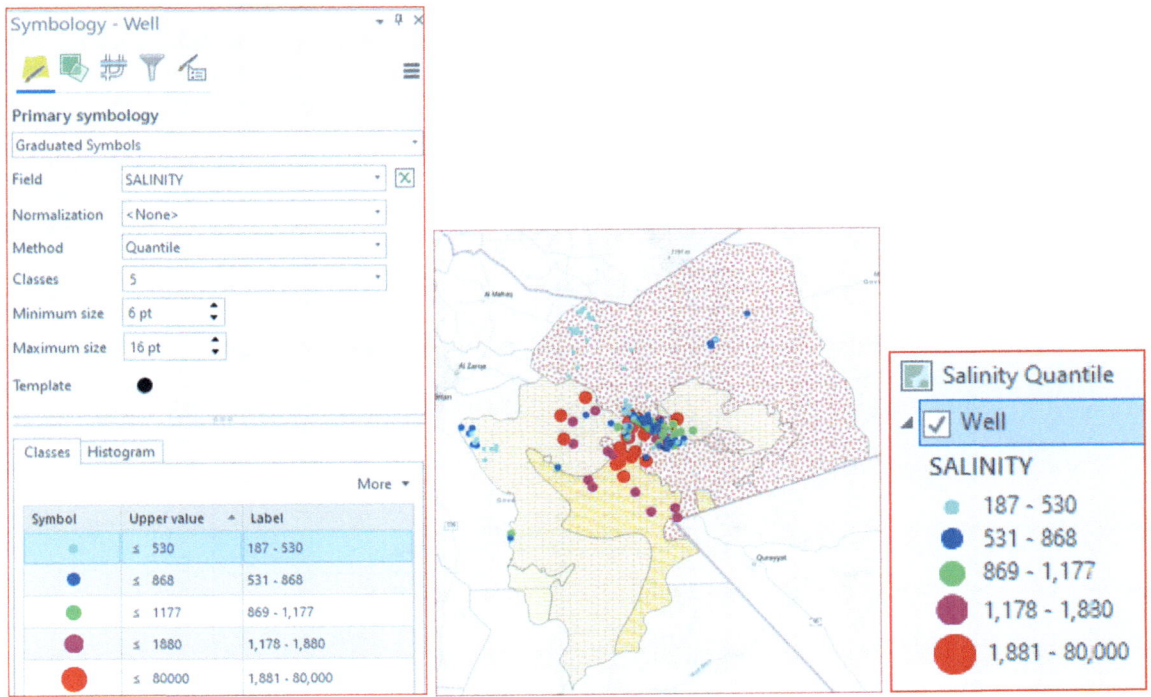

31. Save your project.

Classify the Salinity of Groundwater Using the Equal Intervals Method

The **equal interval** method generates classes that have equal ranges. This is an advantage, as each class will be equally represented on the map. The equal interval method is best used in recognizable data ranges but not for heterogeneous data such as in the case of groundwater salinity in the Azraq basin.

32. Click **Insert** tab on the ribbon, in the **Project** group, click **New Map**
33. In the Contents pane rename it to **Salinity Equal Interval**
34. In **Catalog** pane, expand the **Ch03** folder under the **Folder**
35. Drag the **Geology.lyrx** into the Map View
36. Under **Folder**, open **Data_Ch03\Data** and drag **Well.shp** into the Map View.

Result The **Geology. lyrx** and **Well.shp** added to the Map View and in the CP.

Note If you did not close the Symbology pane from the previous step, it will be available for you to continue

37. If the **Symbology – Well** pane is close r-click the **Well** and choose Symbology
38. Fill the **Symbology – Well** pane as below
    (a) Under Primary Symbology: Graduate Symbols
    (b) Field: SALINITY
    (c) Method: Equal Interval
    (d) Classes   5
    (e) Minimum size: 6 pt
    (f) Maximum size: 16 pt

39. To the right of the **Template**, click the symbol, and choose **"Circle 1"** (under ArcGIS 2D)

40. Click the back arrow in the **Symbology-Well** pane
41. In **Classes** tab, under the Symbol, r-click the biggest symbol (bottom), change the color to **red**, r-click the 2nd symbol from bottom and change the color to **pink**, change the 3rd to **green**, the 4th into **blue** and the top into **cyan**

Practice Apply the thousand separators to the classes

42. Save your Project

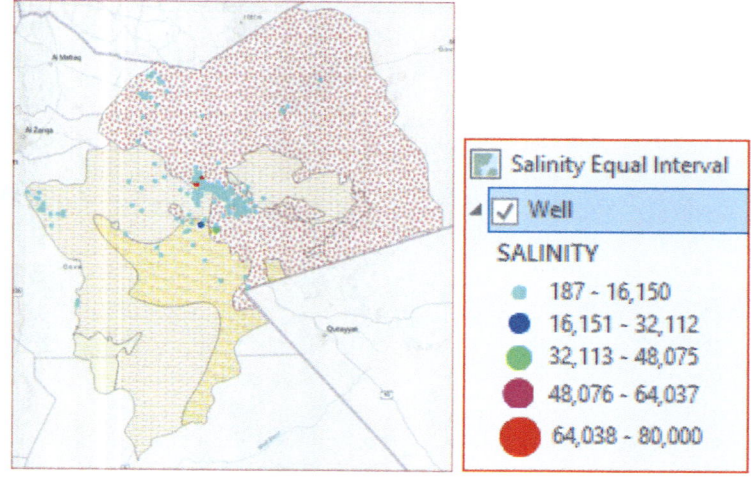

Classify the Salinity of Groundwater Using a Manual Method

This method allows users to use their own classes manually by setting the class ranges that are appropriate for the dataset. This method allows us to classify the water salinity based on the known drinking water quality standard or the water–rock interaction standard. The method is appropriate for classifying the salinity of groundwater because it allows us to emphasize features with specific values; for example, wells that are highly saline at certain locations can be excluded as sources for drinking or irrigation.

43. Click **Insert** tab on the ribbon, in the **Project** group, click **New Map**
44. In the Contents pane rename it to **Salinity Manual**
45. In **Catalog** pane, expand the **Ch03** folder under the **Folder**
46. Drag the Geology.lyrx into the Map View
47. Under **Folder**, open **Data_Ch03\Data** and drag **Well.shp** into the Map View.

Result Geology.shp and **Well.shp were** added to the Map View and in the CP.

Note If you didn't close the Symbology pane from previous step, it will be available for you to continue (if it is close r-click the **Well** and point to Symbology)

48. Fill the Symbology pane as follows:
    - (a) Under Primary symbology: Graduate Symbols
    - (b) Field: SALINITY
    - (c) Method: Manual Interval
    - (d) Classes   5
    - (e) Minimum size: 6 pt
    - (f) Maximum size: 16 pt

49. To the right of the **Template**, click the symbol, and choose **"Circle 1"** (under ArcGIS 2D)

50. Click the back arrow in the **Symbology-Well** pane
51. In the Symbology – Well pane, in the **Classes** tab, under the **Upper value**, delete the number 1574 and type 1000 and change the color of its symbol to **cyan**.
52. Continuing by changing the number as in the table below

Symbol color	Upper value
Cyan	≤ 1000
Blue	≤ 2000
Green	≤ 3000
Pink	≤ 4000
Red	≤ 80,000

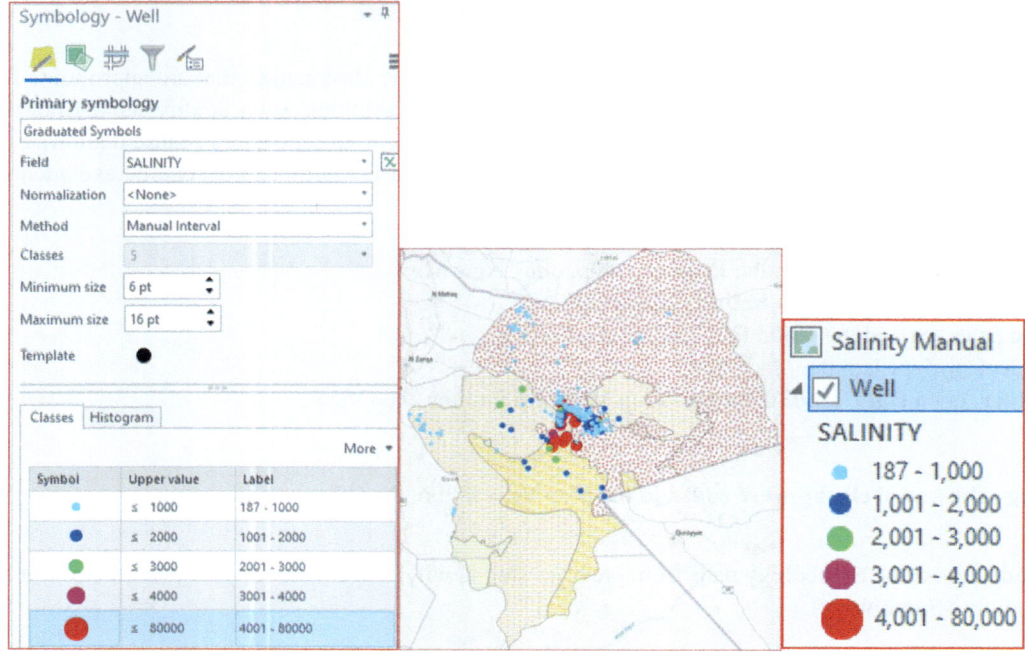

53. Save you project

Sharing the Projects and the Data Using Package Project

Project packages (.ppkx) make it easy to share complete projects. A project package is a file that contains all maps and the data referenced by its layers, as well as folder connections, toolboxes, geoprocessing history, and attachments. Project packages can be used for sharing projects between colleagues in a work group, across departments in an organization, or with any other ArcGIS users through ArcGIS Online. The project packages will be saved as a file so that they can be accessed from the local computer or server.

54. Above the Map View, switch to the Azraq Basin tab

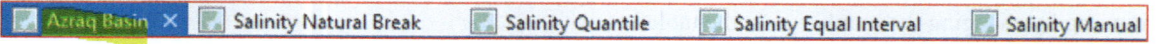

55. Click the **Share** tab on the ribbon, in the **Package** group, click **New Project Package**, the **Package Project** pane appears
56. In the **Package Project** dialog box, in the Package tab, under Start Packaging, check "**Save package to file**"
57. Name: browse to \\Env_Water\Ch03**Azraq.ppkx** and click **Save**
58. In the Summary window, type Groundwater salinity classification in the Azraq basin, Jordan
59. In the Tag window, type Groundwater, Azraq, Jordan, Geology, Salinity, Classification, and Enter after typing each word

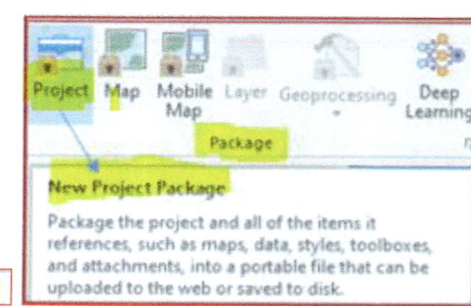

Sharing the Projects and the Data Using Package Project

60. Check below the **Tag** window:
    (a) Share outside of organization
    (b) Include Toolboxes
    (c) Include History Items
61. Under **Finish Packaging**, click **Analyze** tab (if there are no problem a message display "No errors or warning found"
62. Click the **Package** tab next to the **Analyze** tab (a message display stating a successfully created package)

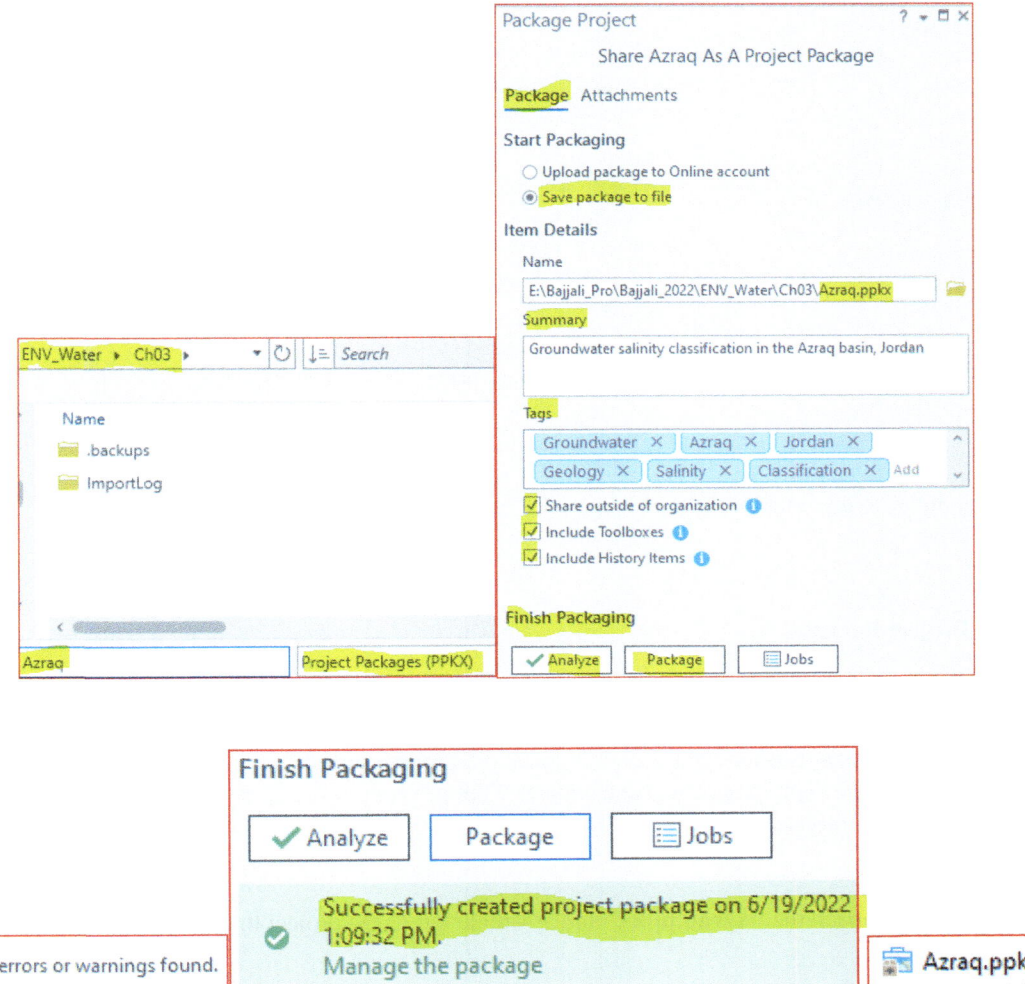

Result The project package "**Azraq.ppkx**" created in the **ENV_Water\Ch03** folder

63. Close the Package Project pane
64. Save the Project

Unpack Project Package in ArcGIS Pro

It is recommended before unpacking the project package to know in advance the location where you want to unpack it.

65. Click **Project** tab on the ribbon and select **Options** tab
66. From the **Options** dialog box, click **Share and Download** tab
67. Under Unpacking, check "Ask where to save before unpacking"
68. Click OK

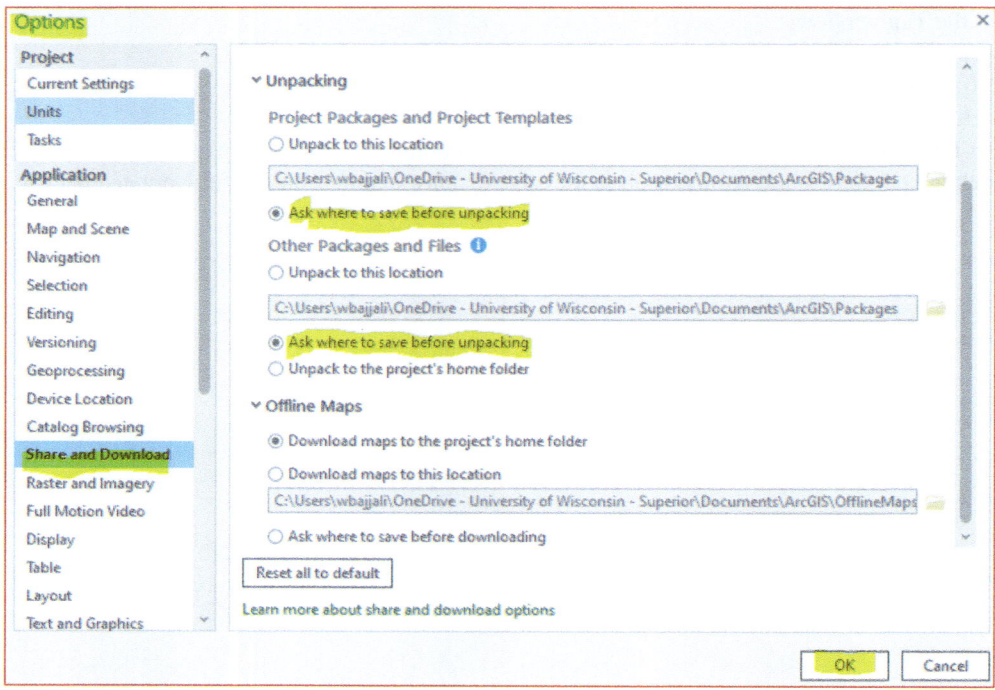

69. Click the back arrow to go to the ArcGIS Pro project
70. Exit ArcGIS Pro

Open the Azraq Project Package

71. In your computer, open **Window Explorer** and browse to **Azraq.ppkx** (\\Ch03)
72. D-click on "**Azraq.ppkx**"
73. ArcGIS Pro open and ask you where you want to unpack the "**Azraq.ppkx**"
74. Pick a location where you want to unpack, for example: \\Temp\Azraq
75. Click OK and save your project

Result ArcGIS Pro will unpack the "**Azraq.ppkx**" and assemble the Azraq project that contains 5 maps:

- Azraq Basin
- Salinity Natural Break
- Salinity Quantile
- Salinity Equal Interval
- Salinity Manual

Create Page Layout for Groundwater Classification   51

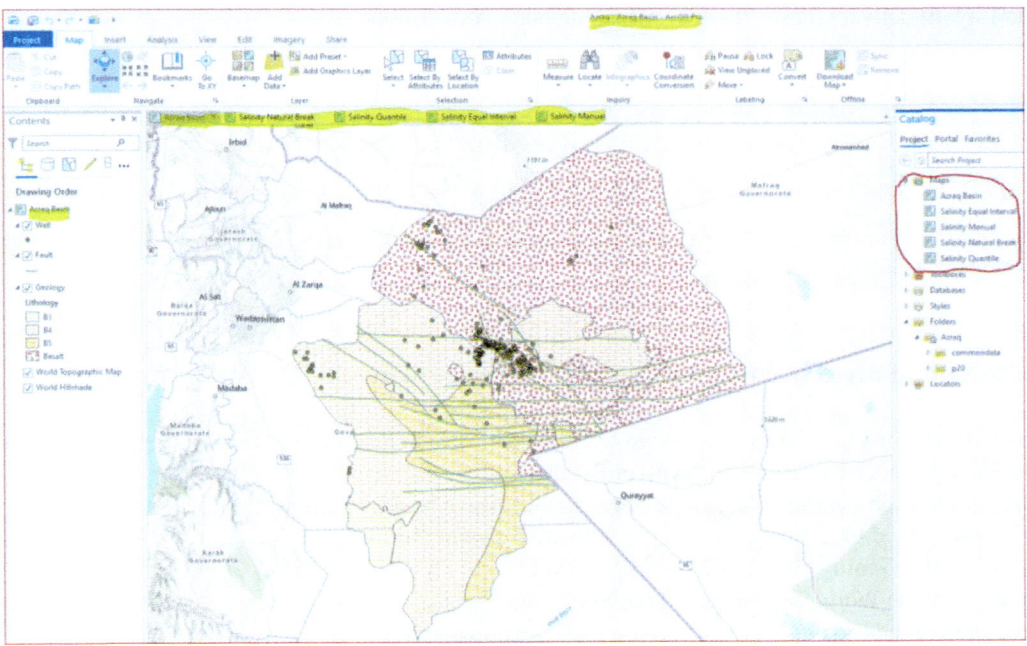

Create Page Layout for Groundwater Classification

A page layout (or layout) is a collection of map elements organized on a virtual page designed for map printing. Common map elements include one or more map frames, a scale bar, a north arrow, a map title, descriptive text, and a legend. For geographic reference, you can add grids or graticules. ArcGIS Pro allows users to create multiple layouts in a project

Scenario 3-2 Your boss asked you to create an 8.5 by 11 inches layout page map that includes the 4 types of classifications: Jenks, quantile, equal interval and manual classification. The layout page should be a landscape orientation in pdf format. The map document will be distributed to the shareholder of Azraq Basing during the annual briefing about the groundwater quality.

Create a Layout

1. **Insert** tab on the ribbon, in the **Project** group, click **New Layout** drop-down arrow, under **ANSI-Landscape**, select **Letter 8.5" x 11"** template
2. The **Layout** is added to the CP and above the Map View.
3. Rename the "Layout" Groundwater Classification Layout"

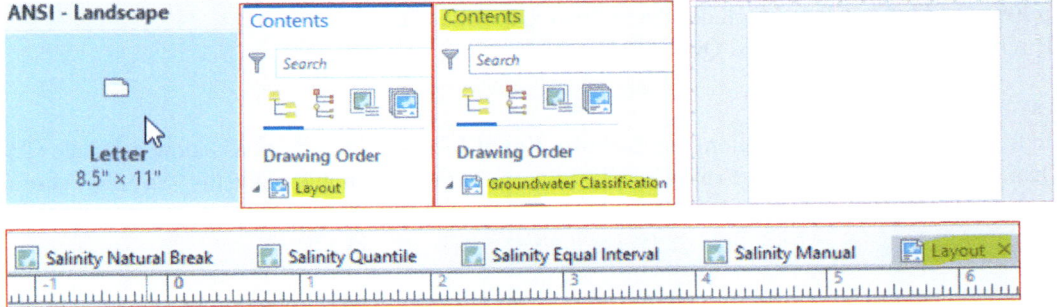

Note Once the **Layout** active, you see the **Layout** tab display on the ribbon

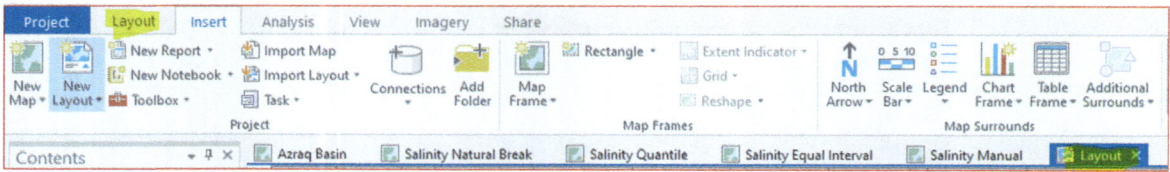

Add the Salinity Natural Break to the Layout

4. **Insert** tab on the ribbon in the **Map Frames** group, open the **Map Frame** drop-down arrow and choose the map that has a scale (1:933,664) under the "**Salinity Natural Break**"
5. Draw a box for the frame in the left-top side of the layout (leave healthy margin) to place other items
6. The **Map Frame** added to CP and includes the **Salinity Natural Break Map**
7. In the CP, r-click the **Geology** layer and Zoom To Layer
8. In the CP, uncheck the World Topographic Map
9. Click on the map in the Layout, and you can use the selection handles to resize the map.
10. In the CP, rename the Map Frame "**Natural Break Frame**"
11. Save the project

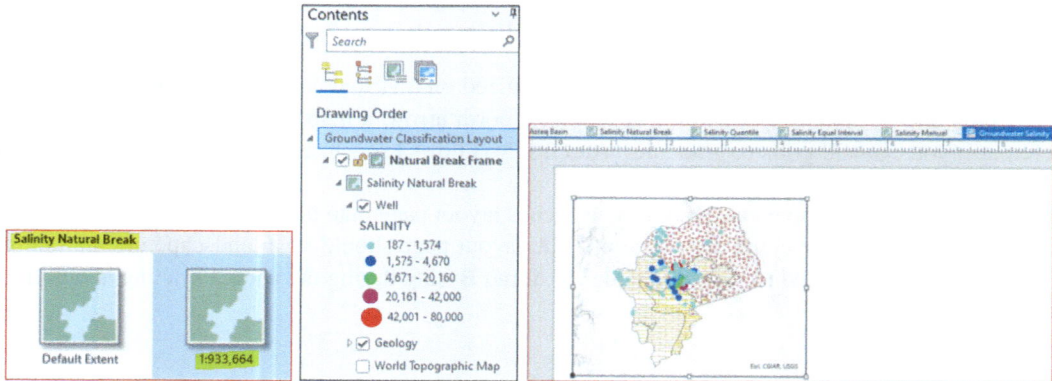

Add the Salinity Quantile to the Layout

12. **Insert** tab on the ribbon, in the **Map Frames** group, open the **Map Frame** drop-down arrow and choose the map that has a scale (1:933,664) under the "**Salinity Quantile**"
13. Draw a box for the frame in the right-top side of the layout (leave healthy margin) to place other items
14. In the CP, r-click the **Geology** layer/Zoom To Layer
15. The **Map Frame** added to CP and includes the **Salinity Quantile Map**
16. In the CP, rename the **Map Frame** to "**Quantile Frame**"

Practice Add to the layout in the left-bottom the "**Salinity Equal Interval**" map that has below it a scale (1:933,664) and rename the Data Frame "**Equal Interval Frame**", and add again in the right-bottom side the "**Salinity Manual**" map that has below it a scale (1:933,664) and rename the Data Frame "**Manual Frame**"

17. In the CP, r-click the **Geology** layer in both **Salinity Equal Interval**" map and **Salinity Manual** and Zoom To Layer

Add Guides 53

Result The final layout will be similar to the image below.

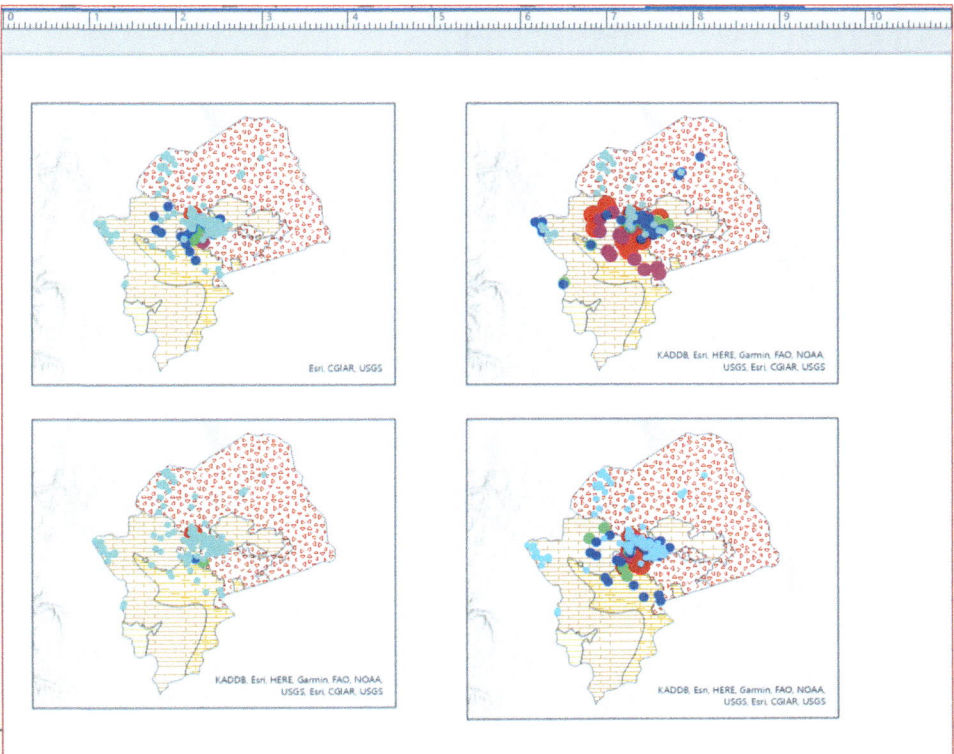

Add Guides

18. Right-click the top **ruler** and click **Add Guides.**

Guides are nonprinting lines that help you align elements on the layout.

19. On the **Add Guides** dialog box, under **Orientation**, click **Both**.
20. Click the **Placement** drop-down menu and click Offset from edge.
21. Replace the value in the **Margin** box with 0.20 in.

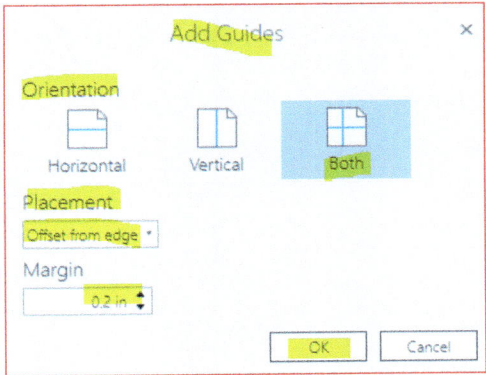

22. Click OK.
23. Place your cursor on a 4-inch vertical ruler, r-click and **Add Guide**
24. In the CP select "**Natural Break Frame**", click the **Map Format** tab on the ribbon, in the **Current Selection** group, select **Map Frame** from the drop-down arrow

Note You can select the **Natural Break Frame** also by selecting it on the layout

25. In the **Size & Position** group, set the following
    (a) X = 0.2 in/enter
    (b) Y = 4 in/enter
    (c) Width = 5.2 in/enter
    (d) Height = 3.7 in/enter

26. Continue setting the size and position for the rest as in the table below.

New name	X	Y	Width	Height
Natural Break Frame	0.2	4	5.2	3.7
Quantile Frame	5.6	4	5.2	3.7
Equal Interval Frame	0.2	0.2	5.2	3.7
Manual Frame	5.6	0.2	5.2	3.7

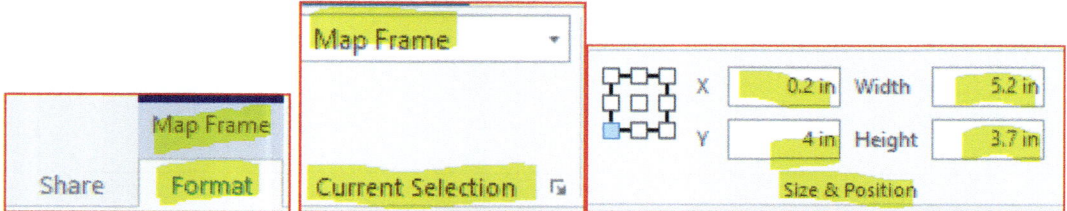

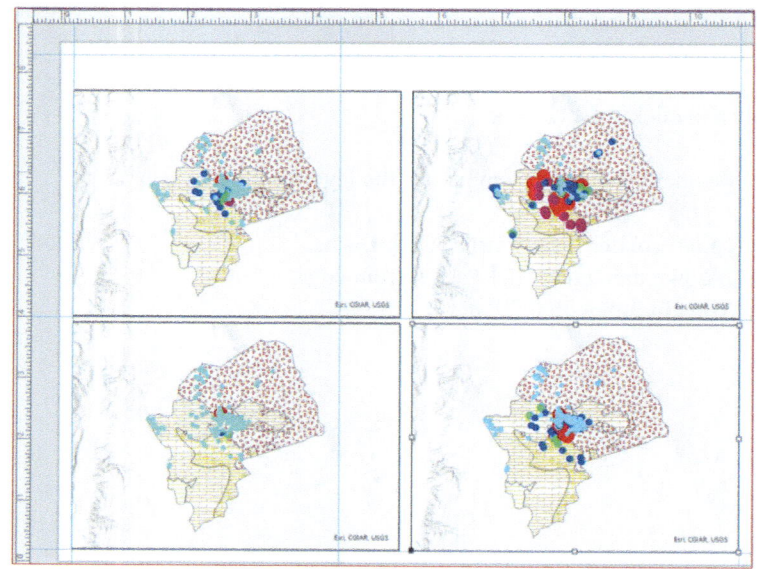

Insert Layout Elements

The layout can include many elements. Some elements such as legend, scale bar, and north arrow associated with map frame. Text and graphics are elements not associated with a map frame. In this section, you are going to insert many elements. First you are going to insert a title for the whole "**Groundwater Classification**" layout in order to reflect the purpose of the layout, and it will be placed at the top of the layout. A text title will be created for each map frame inside the layout. In this situation, each map frame will have its own title. Place your title at any location inside the Map Frame, as you will arrange all your layout elements (legend, scale, north arrow), after you integrate all of them later.

Insert Title for the Layout

27. In the CP highlight Groundwater Classification Layout
28. **Insert** tab on the ribbon, in the **Graphics and Text** group, select **Rectangle Text**
29. Draw a box for the frame at the top of the layout to place the text
30. Type Groundwater Classification
31. Click outside the box to deselect it
32. In the CP, a Text title display, rename the Text to **Layout Title**
33. In the CP, highlight the **Layout Title**, click the **Format** tab on the ribbon
34. In the **Text Symbol** group make the font **Times New Roman** and the size **24** and center it

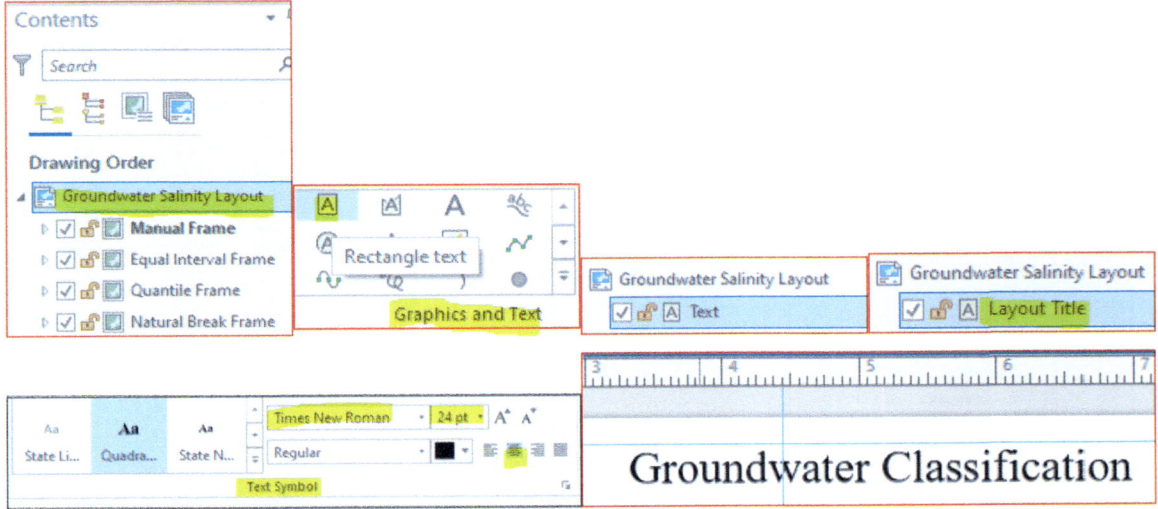

Insert Title for the Natural Break Frame

35. In the CP highlight the **Natural Break Frame**
36. **Insert** tab on the ribbon, in the **Graphics and Text** group, select **Straight Tex**t and click on the top of the **Natural Break Frame** in the layout
37. Type Natural Break Classification
38. Click **Format** tab on the ribbon, click **Text Symbol** group, change the font to Times New Roman and the size 14
39. Place the **Natural Break Frame** in the Layout in the upper left corner
40. In the CP rename the **Text** to Natural Break Classification Title

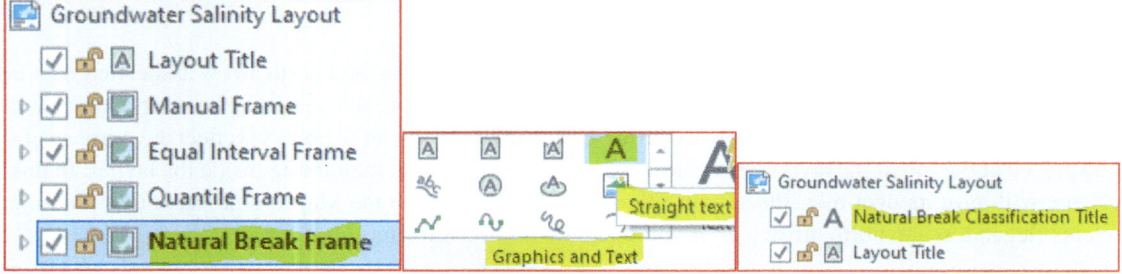

41. Repeat the previous steps and place a title for each data frame as in the table below, and rename the Text tittle

Data frame	Title in the data frame	Change the text in the CP
Natural break frame	Natural break classification	Natural break classification title
Quantile frame	Quantile classification	Quantile classification Title
Equal interval frame	Equal interval classification	Equal interval classification Title
Manual frame	Manual classification	Manual classification Title

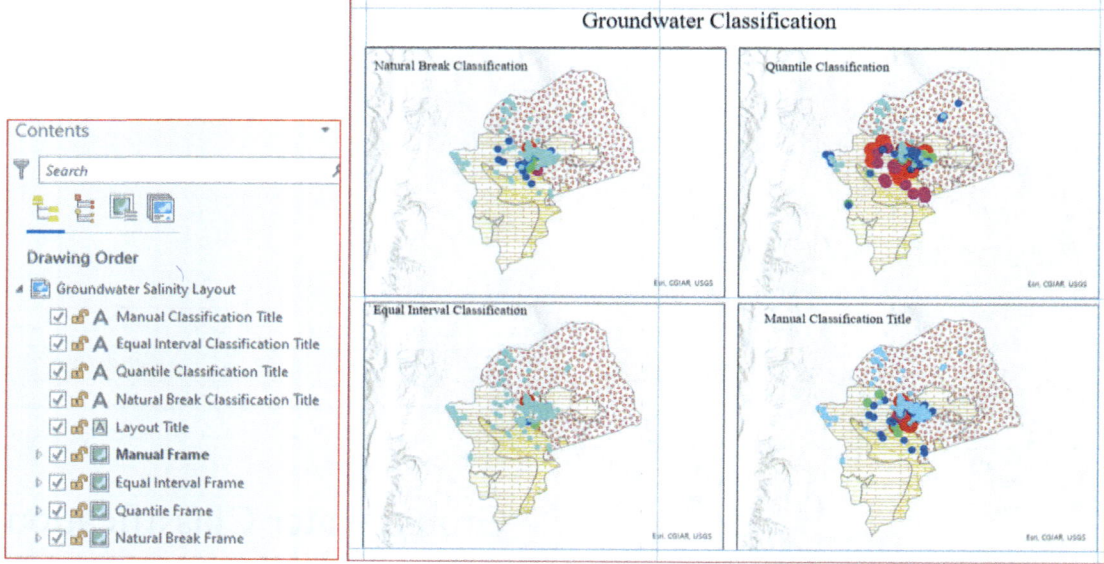

Add a Legend

The legend tells the map reader the meaning of the symbols used to represent features on the map. Legends always display the legend patch set for each feature layer. You are going to add legend for the four data frames. To add the legend, do the following:

42. In the CP, highlight the **Natural Break Frame** or click inside the **Natural Break Frame** in the layout
43. **Insert** tab on the ribbon, in the **Map Surrounds** group, click **Legend** button
44. Draw a box inside **Natural Break Frame** at any location to place the legend

Insert Layout Elements

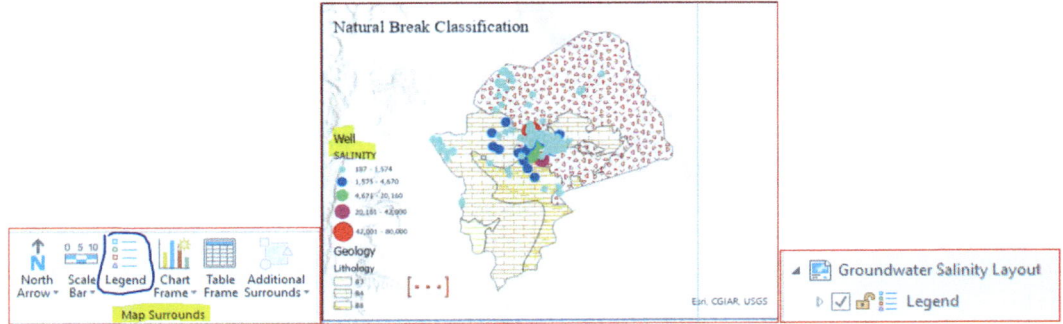

Result A legend is added in the **Natural Break Frame** in the layout and added in the CP.

Modify the Legend

45. In the CP, rename the Legend to **Natural Break Legend**
46. In the CP, r-click **Natural Break Legend** and point to **Properties**
47. The **Format Legend** display, make sure the **Legend** tab selected, below the **Options** under **Legend items** click **Show Properties** button
48. Under **Show**, uncheck **Layer name**
49. Under **Arrangement**, check Keep in single column
50. Under **Sizing**, change the
    (a) Patch width = 12 pt
    (b) Patch height = 10 pt
    (a) check Scale to fit patch size
51. Click the **Text Symbol** tab, click the General tab, and expand the **Appearance**
52. Font Name: Time New Roman
53. Size: 12 pt
54. Click Apply

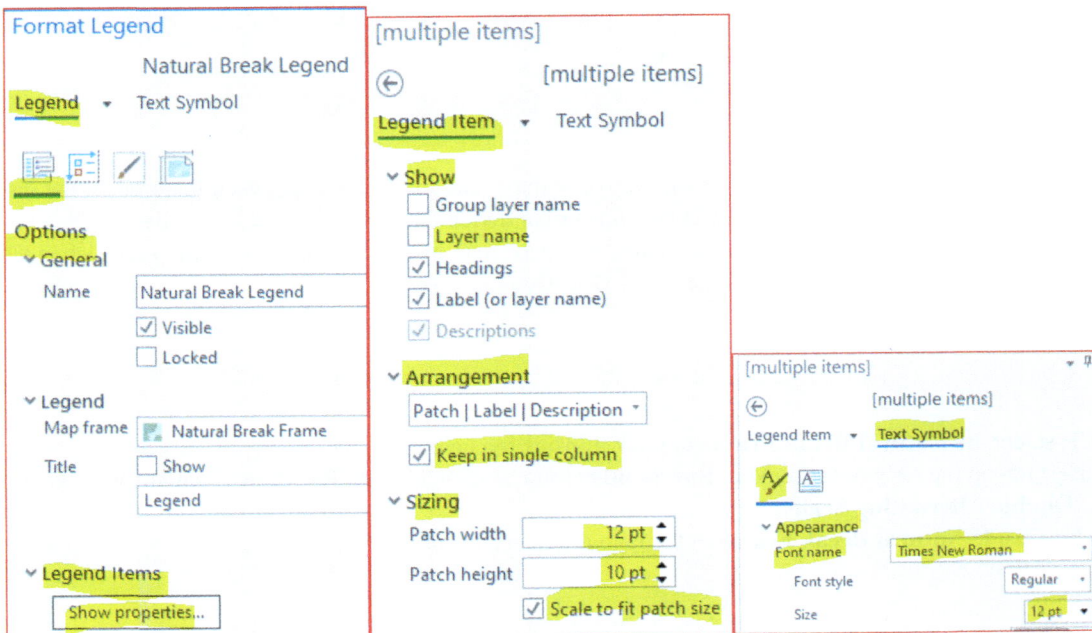

55. Click the arrow to go back
56. Make sure the **Legend** is selected in the **Natural Break Frame** in the layout
57. In the **Format Legend** dialog box, under the Legend tab, click the **Display** button (3rd icon)
58. Under **Border**
59. Symbol: click the color and change it to gray 70% and 1 pt
60. X gap = 0.1
61. Y gap = 0.1
62. Background
63. Symbol: click the color and change it to Arctic White (1,1)
64. X gap = 0.1
65. Y gap = 0.1
66. In the CP expand the Natural Break Frame, the Salinity Natural break, and the Well layer
67. Change the name of "**SALINITY**" to **Salinity (mg/l)** and Enter

Result The changes reflected in the Legend in the layout.

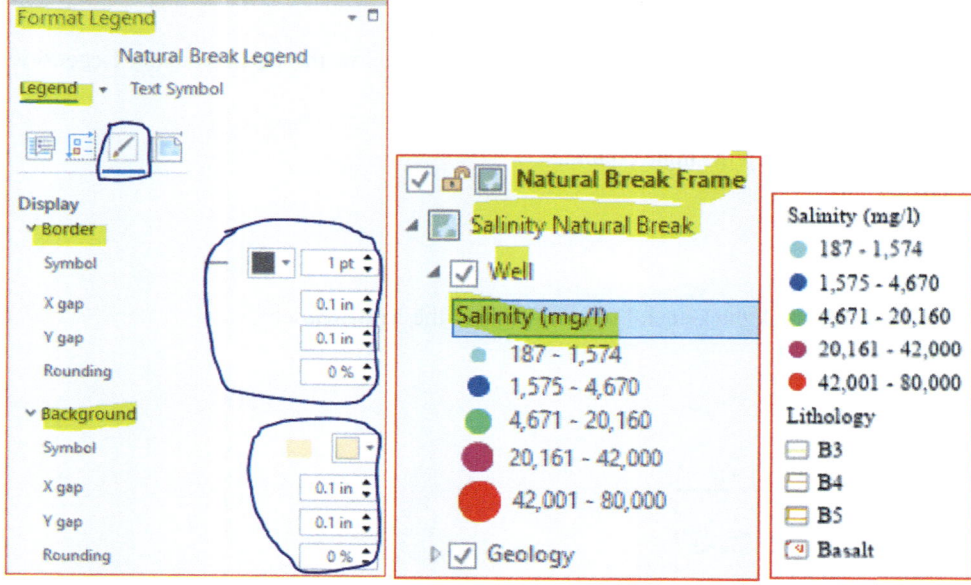

Scale Bar

The scale bar provides a visual indication of the size of features and the distance between features on the map. ArcGIS Pro has different types of scales, and the scale can be a line or bar divided into parts. When a scale bar is added to the layout, it is associated with a map frame and maintains a connection to the map inside the frame. If the map scale changes, the scale bar updates to remain correct.

Insert Scale

68. In the CP, select the **Natural Break Frame** or click it inside the layout.
69. Click Insert tab on the ribbon, in the **Map Surrounds** group, and click **Scale Bar** drop-down arrow from under **Metric** choose "**Double Alternating Scale Bar 2**"
70. Draw a box inside **Natural Break Frame** in the layout to place the scale

Insert Layout Elements 59

Result The **Scale Bar** added to the **Natural Break Frame** in the layout, and it was also added in the **CP**.

71. In the CP, r-click **Scale Bar**, select Properties, the **Format Scale Bar** display,
72. Select Scale Bar tab, select the **Option** button, and fill it as follows:
73. Name: Natural Break Scale Bar
74. Map Units: Kilometers
75. Label text: Kilometers
76. Offset 2pt
77. Label Position: Below center

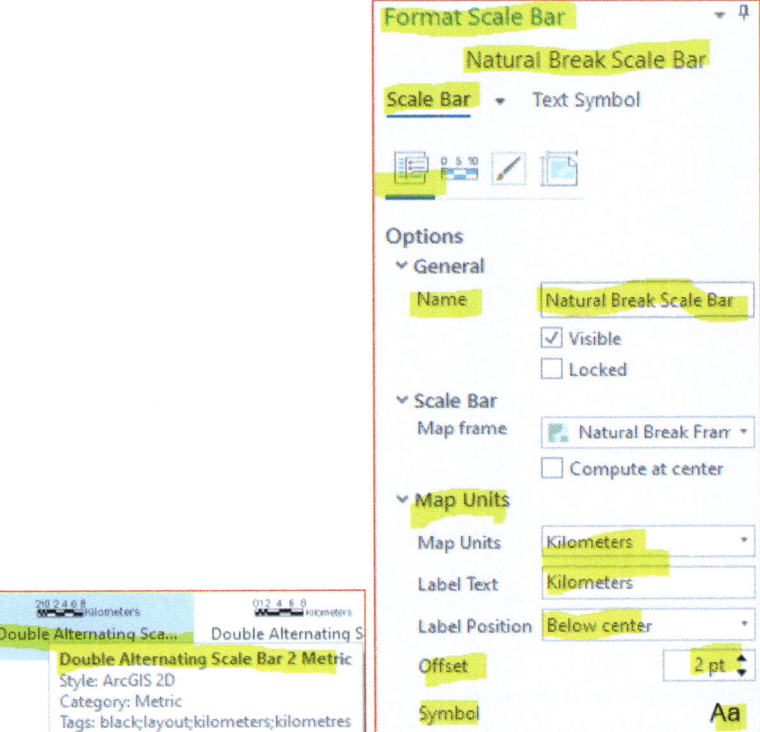

78. In the **Format Scale Bar**, under **Scale Bar** tab, click **Properties** (2nd icon)
79. Under "Fitting Strategy" select Adjust width
80. Under "**Divisions**" set the Division Value to **50**
81. Divisions: 3
82. Subdivisions: 1
83. Uncheck "Show one division before zero"
84. In the **Format Scale Bar**, click the **Text Symbol** tab, in the **General** tab open the Appearance
85. Font name: Time New Roman
86. Size: 10 pt
87. Apply

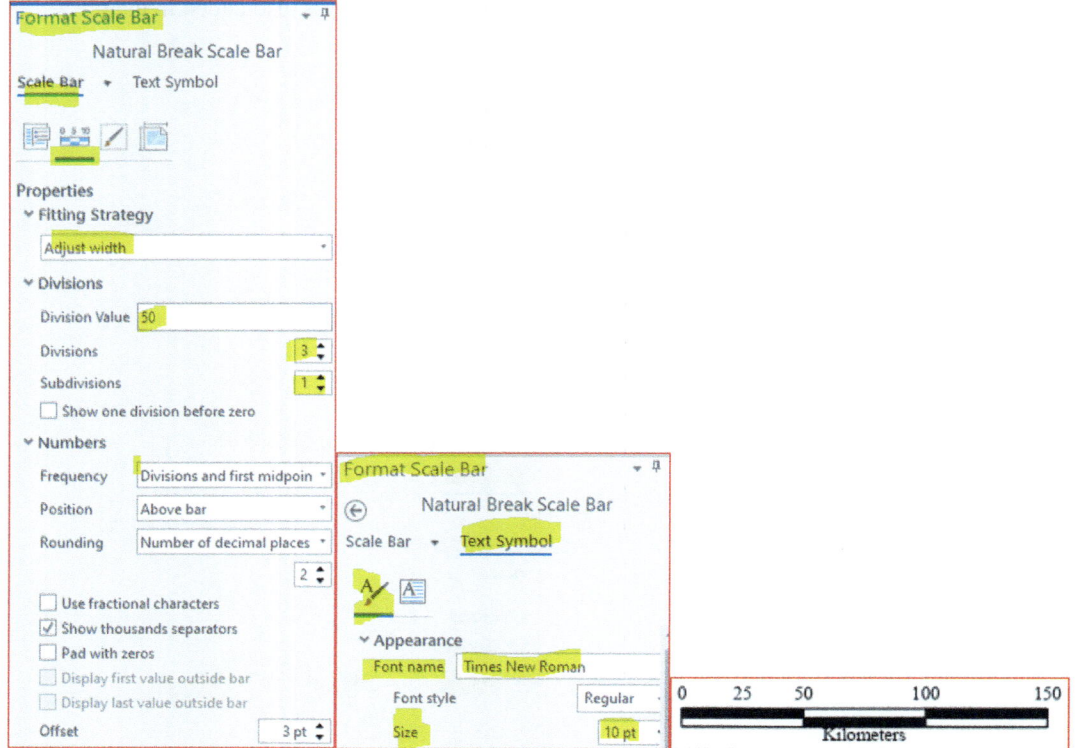

Insert North Arrow

88. Click **Insert tab** on the ribbon, in the **Map Surround**, click the **North Arrow** drop-down arrow and select any north arrow (i.e., **ArcGIS North 2**)
89. Draw a box inside the **Natural Break Frame** in layout to place the **North Arrow**.

Modify the Location of Elements in the Natural Break Frame in the Layout

To interact with the map in the layout, such as selection, editing, and navigation, you must activate the map frame containing that map. There are two ways to activate a map frame:

1. Verify that the map frame you want to activate is the default map frame. Then, on the Layout tab, in the Map group, click Activate
2. In the Contents pane, right-click the map frame, and click Activate

In the activated map frame mode, you can work with the map within the context of the page. The rest of the layout will become unavailable until you click **Close Activation**. The activation can be closed from the **Layout** tab. When you are in activated map frame mode, you have two sets of navigation tools available:

1. To pan and zoom within the map frame, use the map navigation tools on the Map tab.
2. To pan and zoom the page, use the layout navigation tools on the Layout contextual tab.

Insert Layout Elements

Activate the Natural Break Frame

90. In CP, highlight the Natural Break Frame, in the **Layout** tab on the ribbon, in the **Map** group, click **Activate** button.

Result This step will make only the Natural Break Frame active and will freeze the rest

91. In the **Map** tab on the ribbon, in the **Navigate** group, click the **Explore** button to pan the Azraq map to the left.
92. To deactivate, click the Layout tab, in the Map group, and click **Close Activation**.

Arrange the Legend, Scale Bar, and North Arrow in the Natural Break Frame
To arrange the Legend, Scale Bar, and North Arrow in the Natural Break Frame in the layout so they will not overlap with each other, do the following:

93. In CP, highlight the Natural Break Frame to activate it in the layout, click on the **Legend**, it will be selected, move it to the right
94. Click on the **Scale Bar**, it will be selected, move it to the lower right corner of the Natural Break Frame in the layout
95. Click on the **North Bar**, it will be selected, move it to the right of the Scale Bar in the Natural Break Frame in the layout

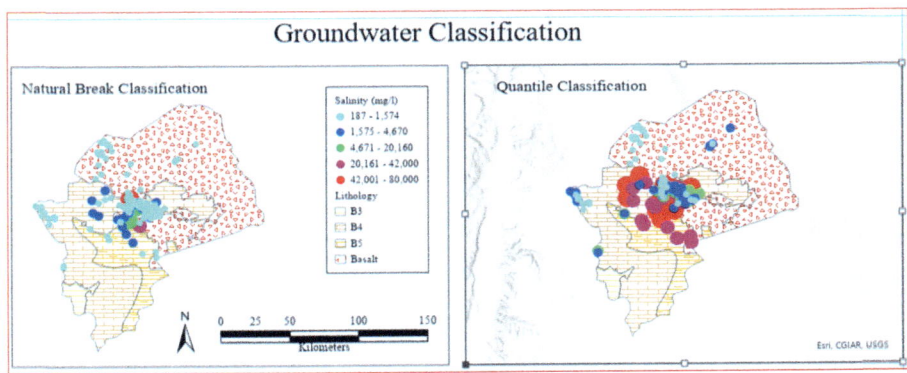

96. Save your Project

Practice Insert legend, scale bar, and north arrow in the Quantile frame, Equal Interval frame, and Manual frame.

Export the Layout as PDF File

97. Click the **Share** tab on the ribbon, in the **Output** group, click Export Layout and fill it as follows:
98. File Type: PDF
99. Name: \\Ch03\Groundwater Salinity Layout.pdf
100. Under Quality, check "Compress vector graphics"
101. Resolution: make it 300 DPI
102. Click Export

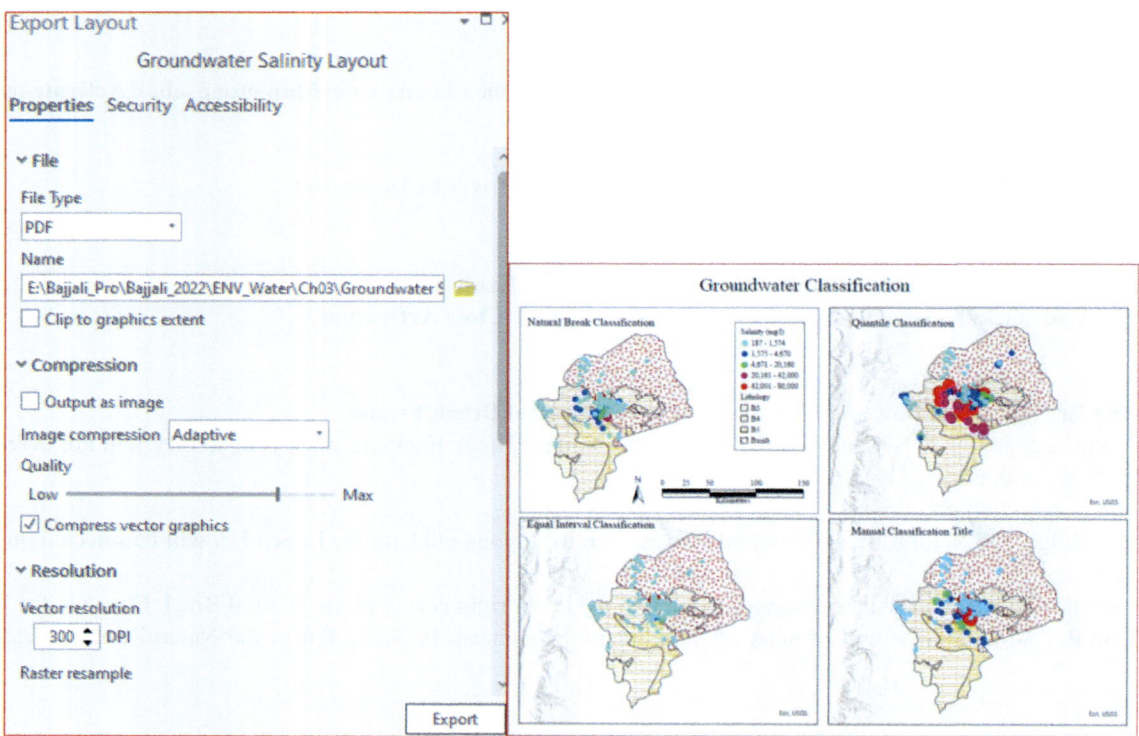

4 Coordinate Systems and Projections

We live on a spherical earth, and the location of any feature can be determined by using a coordinate system.

The latitude and longitude are a 3-D coordinate system or spherical coordinate system that describes the location of features on the Earth's surface. Another type of coordinate system that can be used is called the "projected" coordinate system (Fig. 4.1).

There are various types of plane coordinate systems, which are expressed in x and y coordinates. The plane coordinates are the projection of the sphere from a 3-D view into a 2-D plane view. The latitude and longitude coordinates can be converted directly into different plane coordinates. The conversion from one coordinate to another is extremely important in any GIS work. The conversion from latitude longitude to any plane coordinate is called projection. Conversion from one plane coordinate to another plane coordinate is called reprojection. ArcGIS is a very advanced technology that can accommodate and handle any projections and coordinate conversions.

Geographic Coordinate System (GCS)

The GCS is used to locate and measure the location of any feature on the Earth's surface in terms of latitude and longitude and is based on a 3-D sphere. Earth is divided into two types of lines, **meridians** and **parallels**. Longitude is the line of meridians that run from north to south and measure the East – West locations. The prime meridian runs straight from the North Pole to the South Pole and passes through Greenwich in England. The rest of the meridians are moving away from the prime meridian and are spaced farthest apart on the equator and converge to a single point at the North and South Poles.

Latitudes are lines of parallels and run from east–west and measure locations in the north–south direction (Fig. 4.2). Parallels are equally spaced between the equator and the poles and always parallel to one another, so any two parallels are always the same distance apart all the way around the globe. Parallels & Meridians cross one another at right angles 90°.

The longitude ranges from 0° to 180° east and 0° to −180° west, while the latitude ranges from 0° to 90° north and 0° to -90° south. The east–north orientation is positive, and the west–south orientation is negative.

The latitude longitude is used the same way the X, Y coordinate is used in any plane coordinate. It is used as a reference grid to find the location of features. The origin of the GCS is the point where the prime meridian intersects the equator.

Latitude and longitude can be measured either in degrees, minutes and seconds or by decimal degree. One degree equals 60′ min, and 1-min equals 60″ s.

Map Projections

Map projection is a mathematical formula where the 3-D Earth's view is transformed into a 2-D or plane surface. A map projection is simply a systematic representation of a graticule of latitude and longitude lines on a flat sheet of paper. The progress in computer technology and mapping mathematics makes projection challenging research.

Supplementary Information The online version contains supplementary material available at https://doi.org/10.1007/978-3-031-42227-0_4.

© The Author(s), under exclusive license to Springer Nature Switzerland AG 2023
W. Bajjali, *ArcGIS Pro and ArcGIS Online*, Springer Textbooks in Earth Sciences, Geography and Environment,
https://doi.org/10.1007/978-3-031-42227-0_4

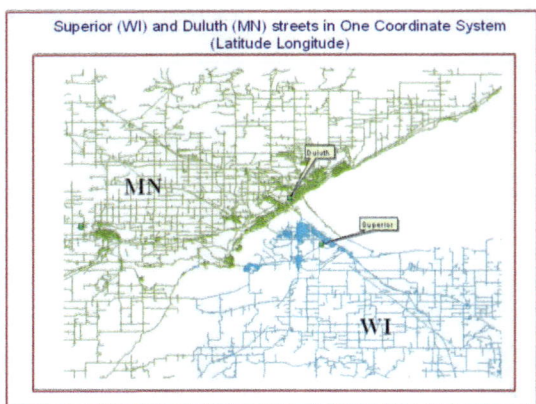 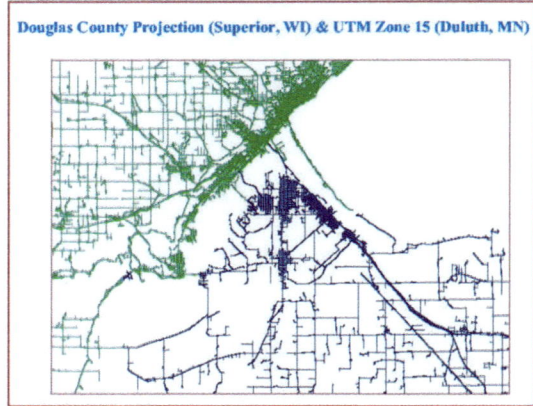

Fig. 4.1 The left map shows the streets of Superior (Wisconsin) and Duluth (Minnesota) registered in latitude longitude. The right map shows that the same file is registered in the UTM zone

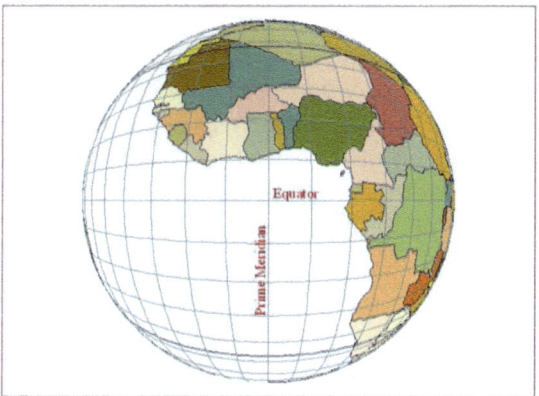

Fig. 4.2 Line of Latitude and Longitude

Projection is used widely in cartography, land information systems, remote sensing and GIS. The map projections generate different types of plane coordinates that are easier to use and work with than the spherical coordinate.

Projection and Distortion
The shape of the earth is quite complicated, so for map projection, the earth is considered an ellipsoid or sphere. The earth ellipsoid is rotated about its minor axis and has semimajor (a) and semiminor (b) axes (Fig. 4.3).

The shape, area, distance and direction of the features on the Earth's surface are correctly shown on a globe. The transformation from the earth surface onto a flat plane surface involves distortion. Parallels, meridians, and the perpendicular intersection of parallels and meridian cannot be duplicated. The major alteration has to do with the angles that will affect the area, shape, distance and directions. There is no ideal projection that retains the major global properties. This leads scientists to generate hundreds of map projections to minimize the distortion and retain at least one of the earth's properties. Therefore, cartographers classified the map projection into major and minor properties. The major is the conformal and equivalent, and the minor is the equidistant and azimuthal. Here are some examples about the major world projection

1. **Conformal or orthomorphic projection** retains the angle and shape of a small area. When this condition occurs, the parallels and meridians intersect at 90°.
2. **Equivalent or equal-area projection** will retain the correct relative size. Thus, in such a projection, the parallels and meridians will not intersect at 90°
3. **Equidistant projection** retains the distance between two points in a map. The scale must be the same as the principal scale on the reference globe from which the transformation was made.
4. **The azimuthal or true-direction projection** represents part of the Earth's directions correctly with a straight line. However, no projection can show direction so that the latitude and longitude are both straight lines.

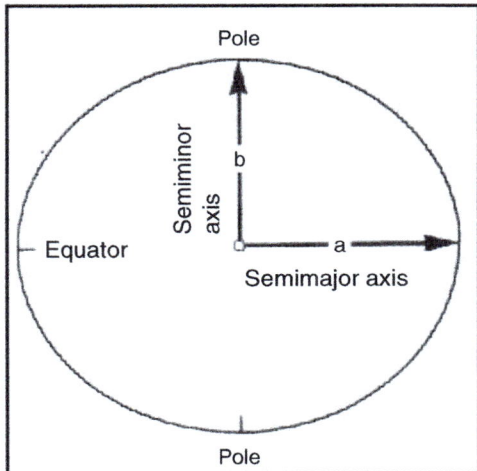

Fig. 4.3 The shape of an ellipsoid

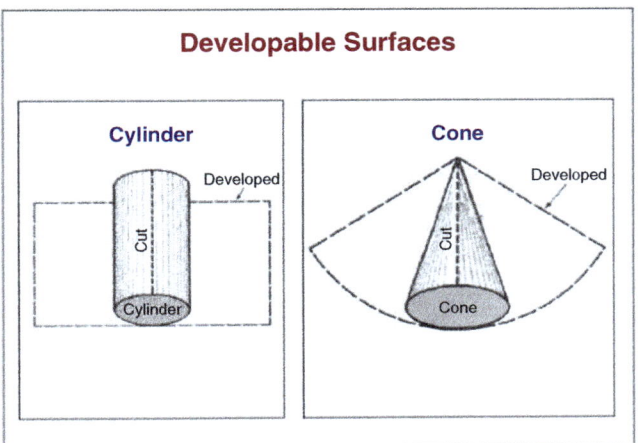

Fig. 4.4 Projection surfaces the cylinder and the cone

Many map projections include more than one property, which is extremely important in a small-scale map.

Geographers use three physical surfaces for the construction of map projections. Developable surfaces (Fig. 4.4) include the cylinder and cone, and nondevelopable surfaces include the plane. The cylinder and cone are not flat at the time the projection is created but can be flattened later by making an appropriate cut in the surface and unrolling it. Without stretching or tearing, when unrolled, distortion of the surface or of the pattern drawn on it is called a developable surface.

Map projection may be produced from three viewpoints: at the center, from infinity, and on the surface of the globe. For example, a light can be used to project the globe on a cylinder. The light can also be placed at any desired location, which gives rise to variations in the map projection.

Changing the location of the light source modifies the characteristic of the resulting projection to the tangent or secant intersect on the cylinder and the cone. The first is the simple (tangent) case, and the second is the secant case.

The simple case results in one line, and the secant case results in two lines. In the simple case, the projection surface (azimuthal plane, cylindrical or conic surface) touches the globe at one point or along one line. In the secant case, the projection surface cuts through the globe to touch the surface at two lines (Fig. 4.5). The line of tangency is called a standard line in map projection. For cylindrical and conic projections, the simple case has one standard line, whereas the secant case has two standard lines. If the standard line along the parallel is called the standard line, and if it follows a meridian, it is called the standard meridian. Along the standard line, there are no distortions because there is a one-to-one relationship between the projection surface and the reference globe. The standard line is identical to the reference globe, and away from the standard line, a distortion occurs.

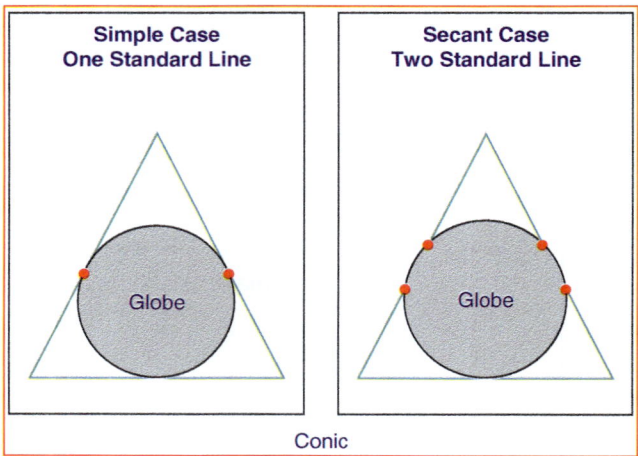

Fig. 4.5 Use the cone to construct a map projection

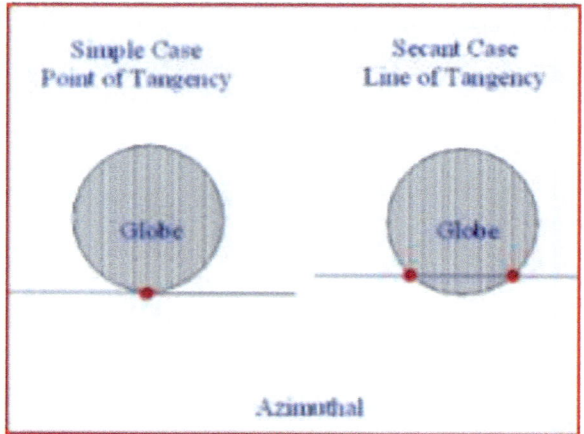

Fig. 4.6 Point and line of tangency

An azimuthal projection has only one point of tangency in the simple case and a line of tangency in the secant case. Therefore, a plane may be tangent at any point on the globe (Fig. 4.6).

The orientation of the cylinder and the cone may be changed as desired and can be normal, transverse, or oblique. The normal orientation is when the cylinder can be placed so its tangent is along the equator, and in the case of the cone, its tangent is along the parallel. Transverse projection occurs when the cylinder or cone is turned 90° from the normal orientation. Oblique projection occurs when the cylinder and cone lie between the normal and transverse positions.

Concept of the Datum

Datum is sets of parameters and ground control points defining local coordinate systems. Because the earth is not a perfect sphere but is somewhat "egg-shaped," geodesists use spheroids and ellipsoids to model the 3-dimensional shape of the earth. Although the Earth can be modeled by an egg-shaped solid, local variations still exist due to the differential thickness of the Earth's crust or differential gravitation due to the density of the crustal materials.

A datum is created to account for these local variations in establishing a coordinate system. Figure 4.7 shows that the earth is an irregular surface (thick black line). A generalized earth-centered coordinate system is called World Geodetic System 84 (WGS84). WGS84 provides a good overall mean solution for all places on Earth. However, for specific local measurements, WGS84 cannot account for local variations. Instead, a local Datum has been developed. For example, the local North American Datum of 1927 (NAD27, dashed red line) more closely fits the Earth's surface in the upper-left quadrant of the Earth's cross-section. NAD27 only fits this quadrant, so to use it in another part of the earth will result in serious errors in

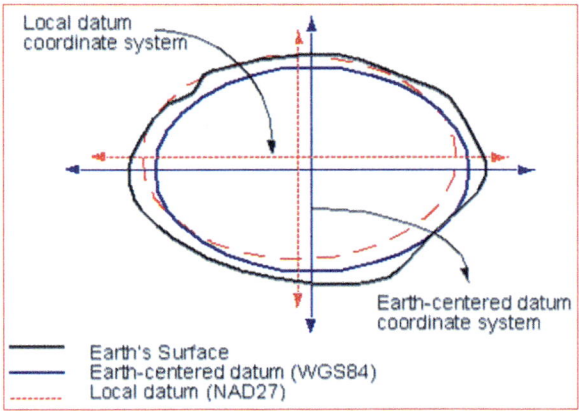

Fig. 4.7 Geodetic Datums: NAD 27, NAD 83 and WGS84. (Image from ESRI)

Base Projection:	Transverse Mercator
Central Meridian:	-91.917
Central Parallel:	45.8833
False Easting:	194000
False Northing:	0.013 U.S. Survey feet
Scale Factor:	1.00004
Unit:	Foot US
Datum:	NAD83

Fig. 4.8 Douglas County of Wisconsin has the following parameters

measurement. For mapping North America, to obtain the most accurate locations and measurements, NAD27 was updated to NAD83, and NAD83 was adjusted in 1991. Any new map created will be based on NAD83.

Projection Parameters
A datum specifies the dimensions of a specific spheroid, a point of origin, an azimuth from the origin to a second point, and the spatial orientation of the spheroid relative to the earth. A GCS assigns unique coordinate values to locations on the surface of a spheroid. The system is usually based on latitude and longitude and is fully specified by a unit of measure (typically degrees), a prime meridian and a datum (e.g., NAD83). A projected coordinate system (PCS) is a combination of a map projection, projection parameters, and an underlying GCS that determines the set of X and Y coordinates assigned to a map.

When a map projection is used as a basis of the coordinate system, an origin should be established first. The point of origin is defined by the central parallel and the central meridian. The central parallel and central meridian in some studies are called the latitude of origin and longitude of center, respectively. For example, Douglas County (Fig. 4.8) has the following projection:

Once map data are projected onto a planar surface, features must be referenced by a planar coordinate system. The latitude-longitude coordinate, which is based on angles measured on a sphere, is not valid for measurements on a plane. Therefore, a Cartesian coordinate system is used, where the origin (0, 0) is toward the lower left of the planar section. The true origin point (0, 0) may or may not be in the proximity of the map data you are using.

The following lessons will be performed in this chapter:

1. Working with GCS
2. Projection on the Fly
3. Projection of the GCS into UTM Zone 15
4. Georeferencing
5. Raster Projection
6. Datum Conflict

Working with GCS

The following two examples show how to calculate the distance in latitude – longitude coordinates.

Application 1 Distance calculation between two points on the map using the Pythagorean Theorem (Fig. 4.9). The theorem is used to find the length of the hypotenuse of a right triangle, a calculation that affords many practical uses in various fields such as land surveying and navigation.

Application 2 Calculate the distance between Superior and Eau Claire in Wisconsin using the latitude longitude. The locations of the two cities are as follows:

1. Superior: −92.06 longitude and 46.70 latitude.
2. Eau Claire: −91.52 longitude and 44.80 latitude.
   Formula of calculation: cos d = sin a * sin b + cos a * cos b * cos c
   a = 46.7 latitude of Superior
   b = 44.8 latitude of Eau Claire
   c = 0.54 this is longitude of Superior – Longitude of Eau Claire
   cos d = sin 46.70 * sin 44.80 + cos 46.70 * cos 44.80 * cos 0.54
   cos d = 0.7277 * 0.7046 + 0.6858 * 0.7095 * 0.9999
   Cos d = 0.999237
   d = 2.237 distance in degree
   2.237 * 69.17 miles = 154.7 mile or (2.237 * 111.32 km = 249 km)

Coordinate Systems and Projections

In this chapter, you are working with coordinate systems and projections. To work with the data, you must open a project package. The package is a compressed file containing GIS data. You learned how to create a project package in Chap. 3. You are going to unpack the package and use its content.

Open the Project Package

1. Open Window Explorer and browse to **Ch04** under **ENV_Water**
2. Double click the project package "**Projection.ppkx**"
3. ArcGIS Pro will open and "**Select unpacking location**" dialog box open

Note See **Chap. 3** on how to set the option to allow users to choose a location to unpack their project package.

4. Browse to **Ch04** under **ENV_Water** and Click OK

Fig. 4.9 Distance calculation between two points

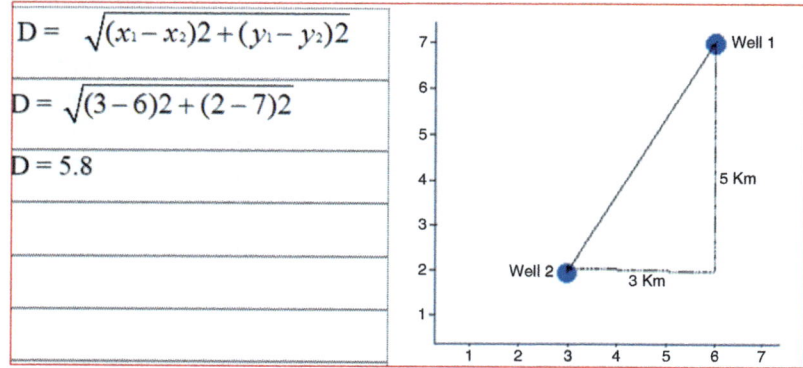

Coordinate Systems and Projections   69

Result The **Define Projection** map open in ArcGIS Pro. The content pane contains the vector point layer (**GPS.shp**) and the image (**lake.tif**). The **Catalog** pane under **Folder**, the **Projection** folder, includes two new folders, **Commondata** and **P20**. The **Commondata** contains the data q1, and inside it are the layer (**GPS.shp**) and the raster (**lake.tif**) **P20** contains **Ch04.tbx** and **Ch04.dbf**. In addition, inside the Ch04 folder, a new folder is created "**Projection_dff02f**" that stores the two folders: **Commondata** and **P20**.

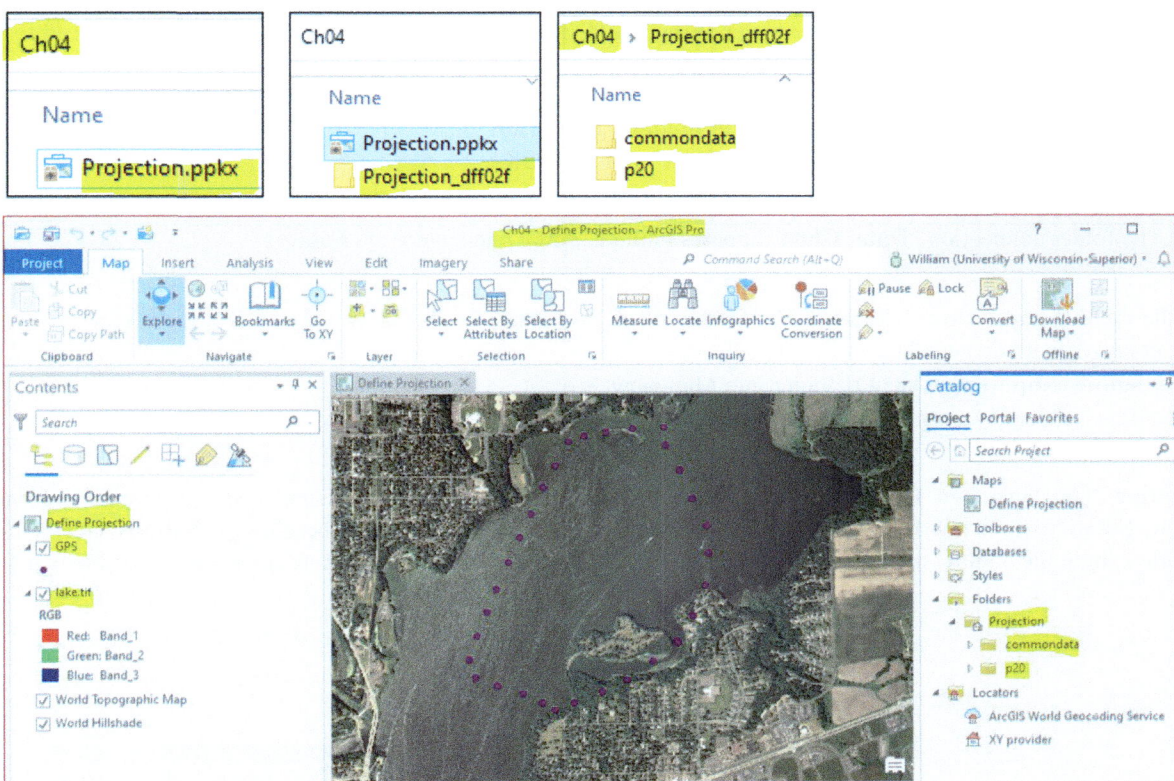

Projection on the Fly
The **Define Projection Map** includes two layers: a "GPS" point vector layer and a "Lake" raster layer. Both layers are displayed above the World Topographic Map in the display area.

5. In the CP, r-click the **GPS**, click **Properties**, and click **Source** tab, in the right panel, open **Spatial Reference**. The coordinate of the **GPS** is **the geographic coordinate system (GCS), and the datum is the North American Datum 1983** (NAD 1983).
6. In the CP, r-click the **Lake.tif**, click **Properties**, and click **Source** tab, in the right panel, open **Spatial Reference**. You will see that the coordinate of the **Lake.tif** is **UTM Zone 15** and the datum is **NAD 1983.**

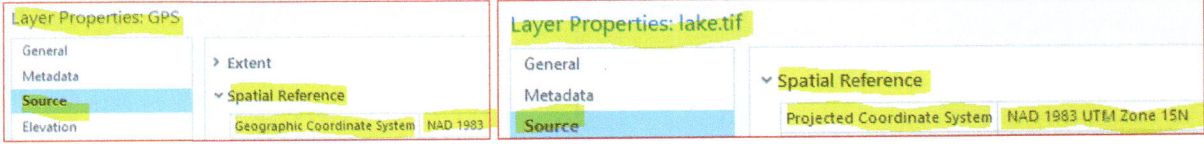

Comment The **GPS** layer is not projected and registered in latitude-longitude and NAD 1983, while the **Lake** raster is projected in **UTM** coordinate system and NAD 1983.

Despite the two layers having different coordinate systems, both layers are aligned with each other. This phenomenon is called projection on the fly. ArcGIS Pro displays both layers in the correct place on the map. Projection on the fly is valid if the layers are registered in any coordinate system and datum.

7. Click the **Save** button on the **Quick Access toolbar** at the top of the app, or **Ctrl+S**.

Connection to the Data

This will allow users to connect to the folder that provides the spatial data needed to complete the project. The data for this project are available in the database folder under **Data_Ch04**. The data folder contains the layers that you need to access more often.

8. In Catalog pane, r-click **Folders**, and click **Add Folder Connection**
9. Browse to **Database** folder highlight "**Data_Ch04**" and click OK

Result The folder connection "**Data_Ch04**" appears in the **Catalog** pane under the **Folders**

10. In the Catalog pane, expand "**Data_Ch04\Data**" folder, it contains 5-subfolders: **Q1** and **Q2**, each folder includes different layers that you need later
11. Drag **Salinity.shp** from \\Data\Q1 folder into Map view
12. A notification display stating "Unknown Coordinate System

Explanation The Unknown Coordinate System means that **Salinity.shp** has missing spatial reference information. Therefore, **Salinity.shp** is not displayed in the extent of **lake.tif** and **GPS.shp**. This is because the **Salinity** layer is missing information about the datum (spatial information). This means that the **Salinity** layer is missing the file "**Salinity.prj**"

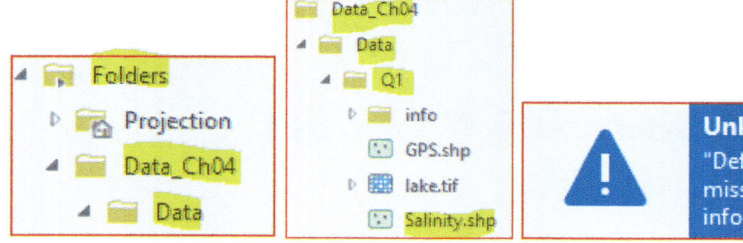

13. In the CP, r-click **Salinity**, point to **Data**, and then **Export Features** and fill it in the Export features pane
14. Input Features: **Salinity**
15. Output Location: **ch04.gdb**
16. Output Name: **Salinity**
17. Click OK
18. Remove the original **Salinity** layer from the CP
19. In the CP, click the symbol of the Salinity, in the Symbology – Salinity, select Gallery tab, and choose Circle 1, then click Properties tab, change the color to cyan, and the size 10, then click Apply

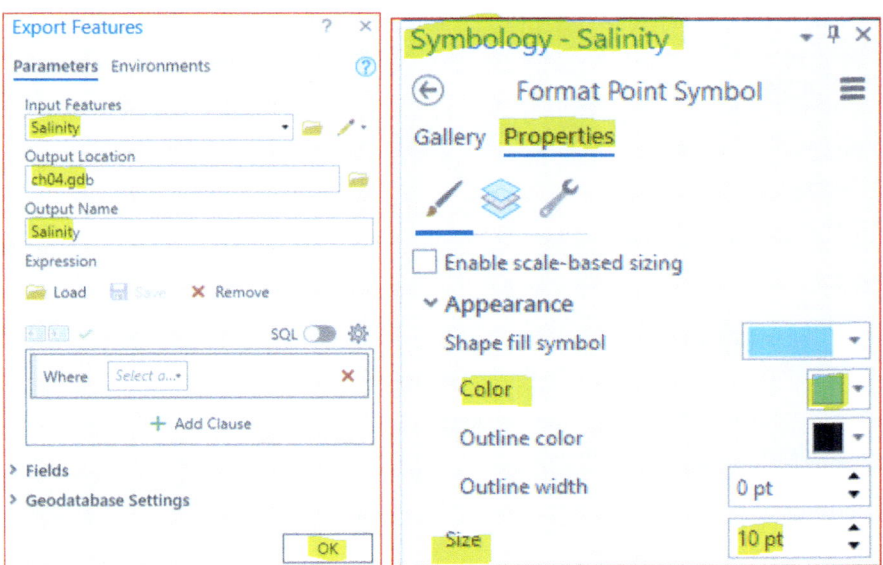

Assign Datum for the Salinity Feature Class

The Salinity Layer includes total dissolved solid (TDS) information about the water in the lake. Therefore, the Salinity layer is supposed to be located inside the lake. The reason it is not located there is because the Salinity layer is missing the datum. In this step, you will assign datum to the Salinity layer, so the Salinity layer will move automatically into its right location inside the lake.

20. Click the **Analysis** tab on the ribbon, in the **Geoprocessing** group, click the **Tool** button
21. Click the **Toolboxes** tab, open **Projections and Transformations**, and click **Define Projection**.
22. In the Geoprocessing pane, under Define Projection, fill it as follows:
23. Input Dataset or Feature Class: drag the **Salinity** from the CP
24. To the **Coordinate System** click the drop-down arrow of the "**Select coordinate system**" (globe)
25. The **Coordinate System** pane, click the drop-down arrow of the "**Add Coordinate System**", click "**Import Coordinate System**", browse to **Folder\Data_Ch04\Data\Q1** and select **GPS.shp**
26. Click Ok
27. The **Salinity** layer moves and displays inside the lake in the widow view.

Result Projection on the fly took place, and the **Salinity** layer is aligned with **lake.tif** and **GPS.shp** because all the layers now have coordinate systems and datums.

28. Click **Project** tab on the ribbon, click **Save As** tab, browse to **p20** under **Project \ Folders \ Projection** and save it as **Define Projection.aprx**
29. Click **Save**

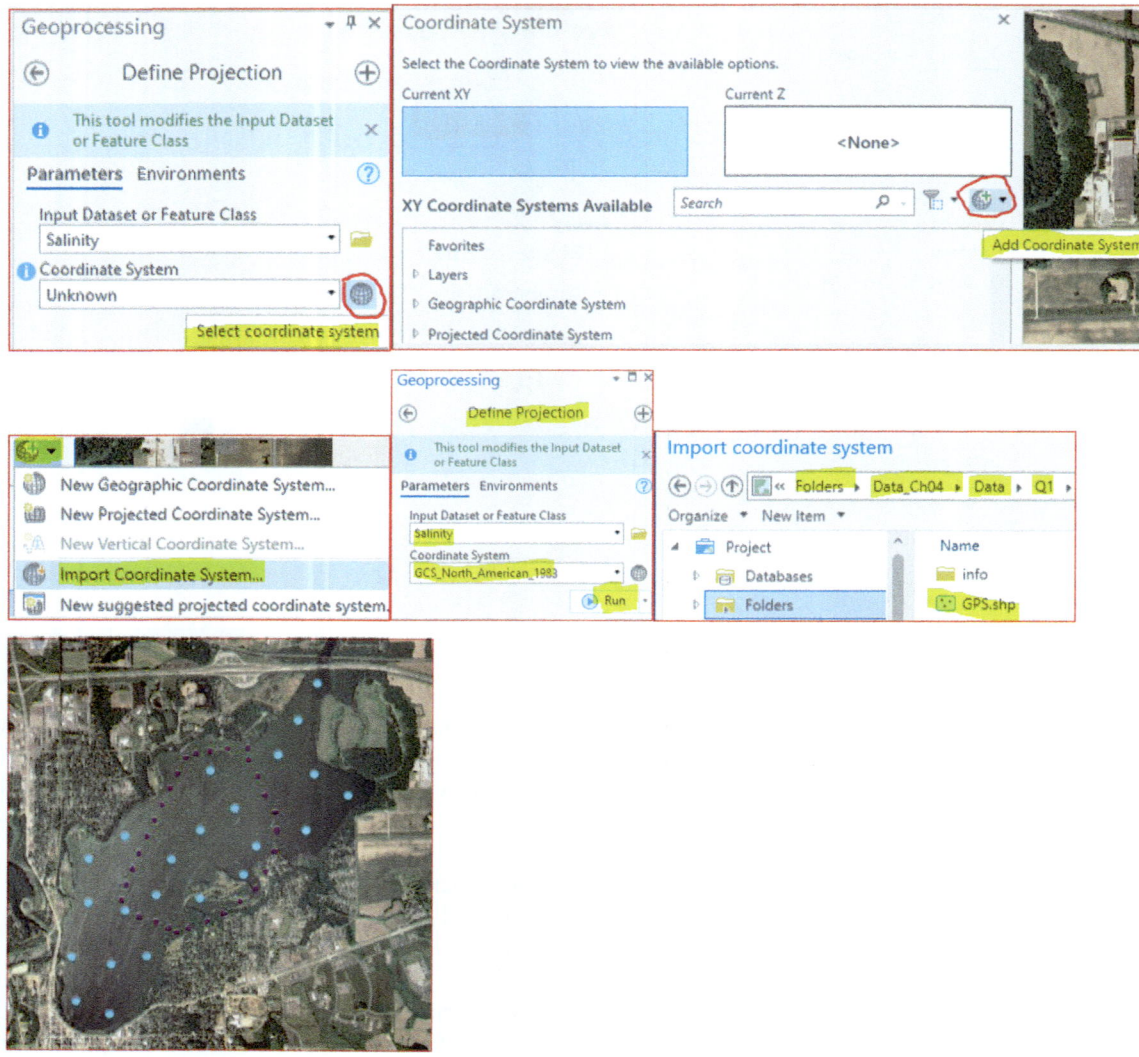

Projection of the GCS into UTM Zone 15

The Project tool converts geographic data from one coordinate system to another. Dissimilar projections cause different types of distortions (see the introduction). You are now going to project the **GPS** from latitude – longitude onto Universal Transverse Mercator (UTM) zone 15 N.

The raster layer (**lake.tif**) and vector layer (**GPS**) in the **Content** pane have the same datum (NAD1983) but different coordinate systems. The **lake** is registered in UTM projection zone 15 N, and the **GPS** is in latitude and longitude.

30. If the **Geoprocessing** pane, click the arrow back in the **Define Projection** pane

Note if the **Define Projection** pane is closed, repeat the previous steps by clicking the **Analysis** tab on the ribbon. In the **Geoprocessing** group, click the **Tool** button and open **Projections and Transformations**.

31. Click **Project** under the Projections and Transformations
32. Fill the Project pane as follows:
33. Input Dataset or Feature Class: drag the **GPS** from the CP
34. Output Dataset or Feature Class: Type **GPS_UTM**
35. Output Coordinate System: click the drop-down arrow and select **lake.tif** to import the coordinate system from **lake.tif** (NAD_1983_UTM_Zone_15 N).
36. Click Run

Georeferencing

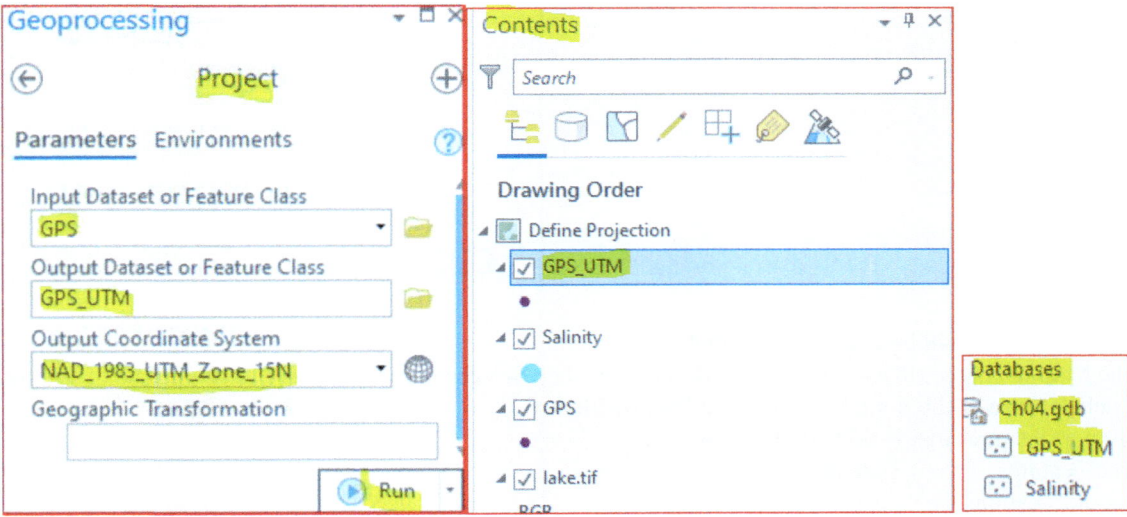

Result GPS_UTM is now registered in **NAD_1983_UTM_Zone**_15 N, saved in the database **Ch04.gdb** and added into the **Content** pane.

37. Save your project (Ctrl + S)

Georeferencing

Raster data are commonly obtained from many sources, such as aerial photographs, satellite images, and scanned maps. Scanned maps and some downloadable images from the internet usually do not contain spatial reference information. Therefore, these images cannot be represented on the map, and their locations will not fit correctly on the surface of the earth. Thus, to use these types of raster data in GIS analysis or as a background image, GIS users will need to use accurate location data to align or georeference the raster data to a map coordinate system.

GIS users georeference the raster define how the data are situated in map coordinates. This process includes assigning a coordinate system that associates the data with a specific location on the earth. Georeferencing raster data allows it to be viewed, queried, and analyzed with other geographic data.

Scenario 4-1 You are an ecologist and found on the internet an image representing Silurian aquifer in east Wisconsin. The image has a false coordinate, and the rest of your digital data are registered to a geographic coordinate (Latitude - Longitude).

GIS Approach
To georeference the scanned image, you will use a vector layer of Wisconsin.
1. Click the **Insert** tab in the ribbon, on the **Project** group, click **New Map** and call it "**Georeference**"
2. In the **Catalog** pane, expand **Folder\Data_Ch04\Data\Q2** and drag **State48.shp** onto the Map View

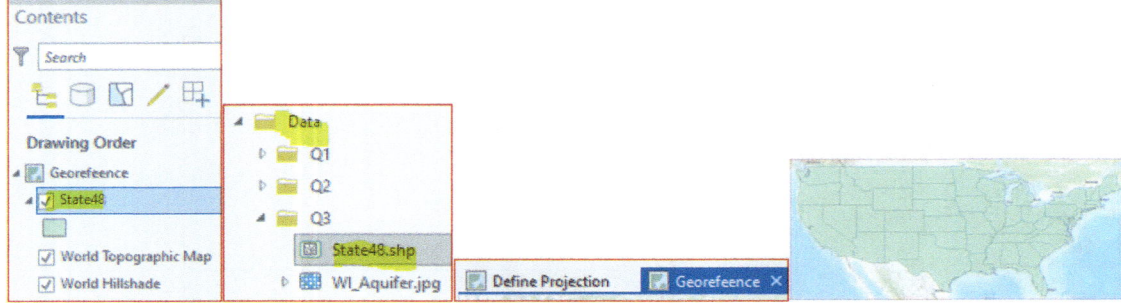

3. In the CP, r-click **State48** layer and click **Label**
4. Click the **Map** tab in the ribbon, on the **Selection** group, click **Select Features** tool and click on state of **Wisconsin**
5. In CP, r-click **State48.shp** point to **Data** select **Export Features**
6. In the Export Features dialog box, fill it as follows:
   (a) Input Features: State48
   (b) Output Location: ch04
   (c) Output Name: WI
7. Click OK

8. In the CP, r-click **State48.shp**, and point to **Remove**
9. In the CP, r-click **WI** layer and point to Zoom To Layer
10. In the CP, click the symbol of WI, in the **Symbology-WI** in the **Gallery** tab, click on Black Outline (1pt). Click **Properties** tab, and click on **Outline color** and make it red
11. Click Apply

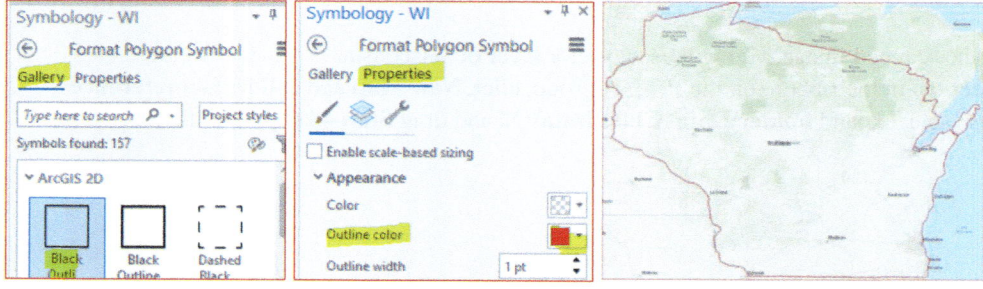

12. Click the **Map** tab in the ribbon, on the **Layer** group, click **Add Data** and browse to Q2 folder, highlight the raster "WI_Aquifer.jpg" and click OK

Result   "WI_Aquifer.jpg" is far from the **WI** layer because the image has a false coordinate system.

Comment If you r-click the "**WI_Aquifer.jpg**" in the CP, click Source, then open the Spatial Reference, you will notice that the image has unknown coordinate system.

Copy Raster

To work with a copy of the original image "**WI_Aquifer.jpg**", you can copy it and save it in the geodatabase **ch04.gdb**. When storing a raster dataset in a geodatabase, the user should not add a file extension to the name of the raster dataset. If you want to store the image as ESRI Grid, this grid has no extension. You are now going to copy "**WI_Aquifer.jpg**" and call it "**Silurian_Aquifer**".

12. Click the **Analysis** tab on the ribbon, in the **Geoprocessing** group, click the **Tool** button
13. Click the **Toolboxes** tab, open **Data Management Tools, Raster, Raster Dataset**, and click **Copy Raster** and fill it as below
    (a) Input Raster: WI_Aquifer.jpg
    (b) Output Raster Dataset: Silurian_Aquifer
    (c) NoData Value   256
    (d) Format: ESRI Grid Format
14. Click Run

Result The "Silurian_Aquifer" image is saved in **Ch04.gdb** and added into the **CP**.

15. In the CP, r-click the "**WI_Aquifer.jpg**" and remove

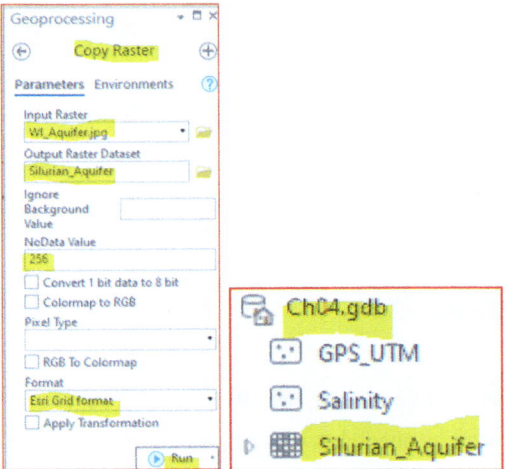

Georeference the Raster

The copied image has no coordinate system, and you are going to assign for it a coordinate system using the Georeference tool and the WI feature class.

16. In the CP, highlight the "**Silurian_Aquifer**" image
17. Click the **Imagery** tab in the ribbon, in the **Alignment** group, and click **Georeference** button to open the **Georeference** tab

The tools on the **Georeference** tab are divided into several groups to help you use the correct tools in the different phases of your georeferencing session. Once you click on the **Georeference** tab, the top right corner of the Map View, will show the "**Silurian_Aquifer**" that will be georeferenced and the RMS, which is empty (Errors), because the georeferencing process didn't start yet.

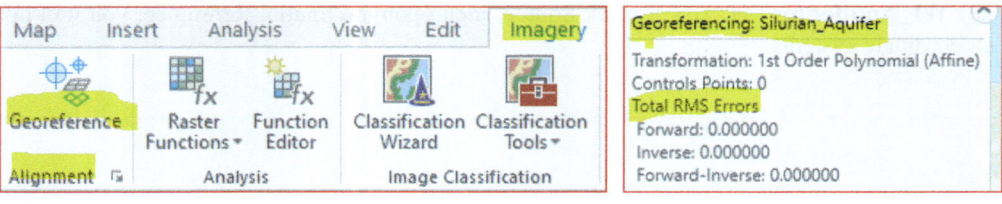

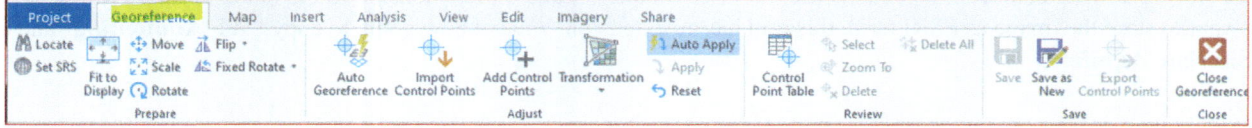

18. In the **Prepare** group, click **Set SRS**.

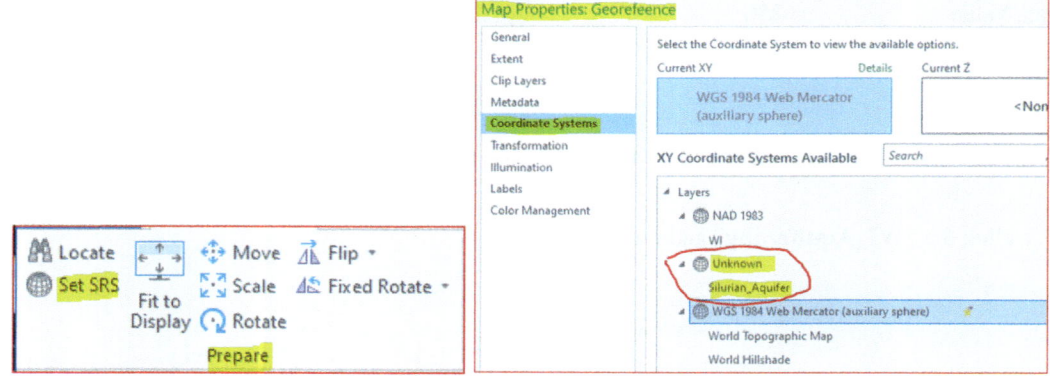

Comment The "**Silurian_Aquifer**" raster dataset does not have a spatial reference, this is seen in the **Map Properties: Georeferencing**. The **Map Properties: Georeferencing** dialog box allows you to choose the coordinate system for the georeferencing session. It also displays the coordinates of the layers in the Contents pane.

19. In the Prepare group, click **Fit to Display**

Result The raster layer you are georeferencing is placed with the current map display. You can also use the **Move**, **Scale**, and **Rotate** tools to place the raster as needed (try all of them to practice).

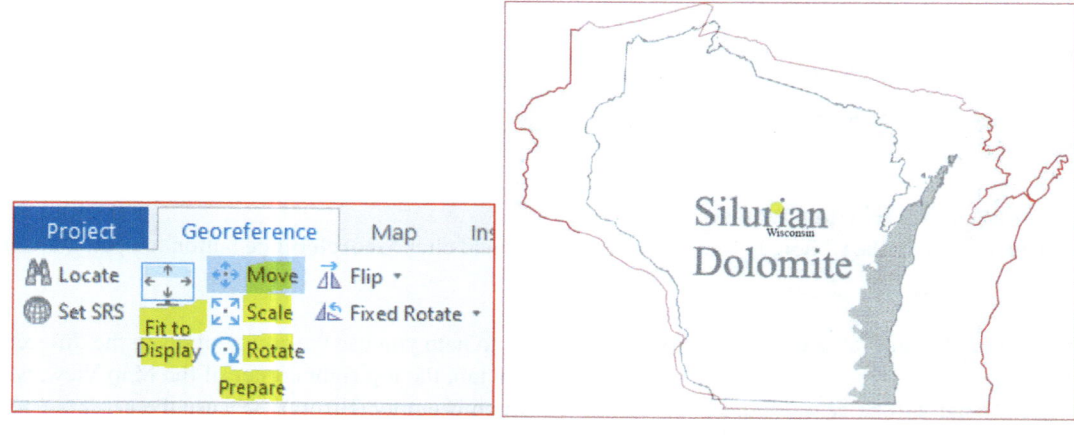

Georeference the Raster

20. In the **CP**, turn off the **World Topographic Map** and **World Hillshade** because they are not going to be used in the process of georeferencing. In this exercise, you keep only **WI**.
21. In the **Adjust** group, click the **Add Control Points** tool to create control points.
22. To add a control point, first click a location on the "**Silurian_Aquifer**" (source layer); then click the same location on the **WI** (target layer) on the map.
23. If you want to zoom or pan, click on **C** letter on the keyboard
24. In the **Review** group, click the **Control Point Table** button to evaluate the residual error for each control point. Currently you have 20 control points, and this is sufficient to georeference your image.

If you find the error is high, highlight it and click delete on the keyboard, the current error is 0.026 and this is acceptable. You can also press the **L** key to switch the transparency of your source raster on and off.

25. On the **Adjust** group, click the drop-down arrow of **Transforms**, choose the transformation you want to use. The transformation depends on the number of control points. In this exercise, use the 2nd order polynomial if you have more than 6 control points.

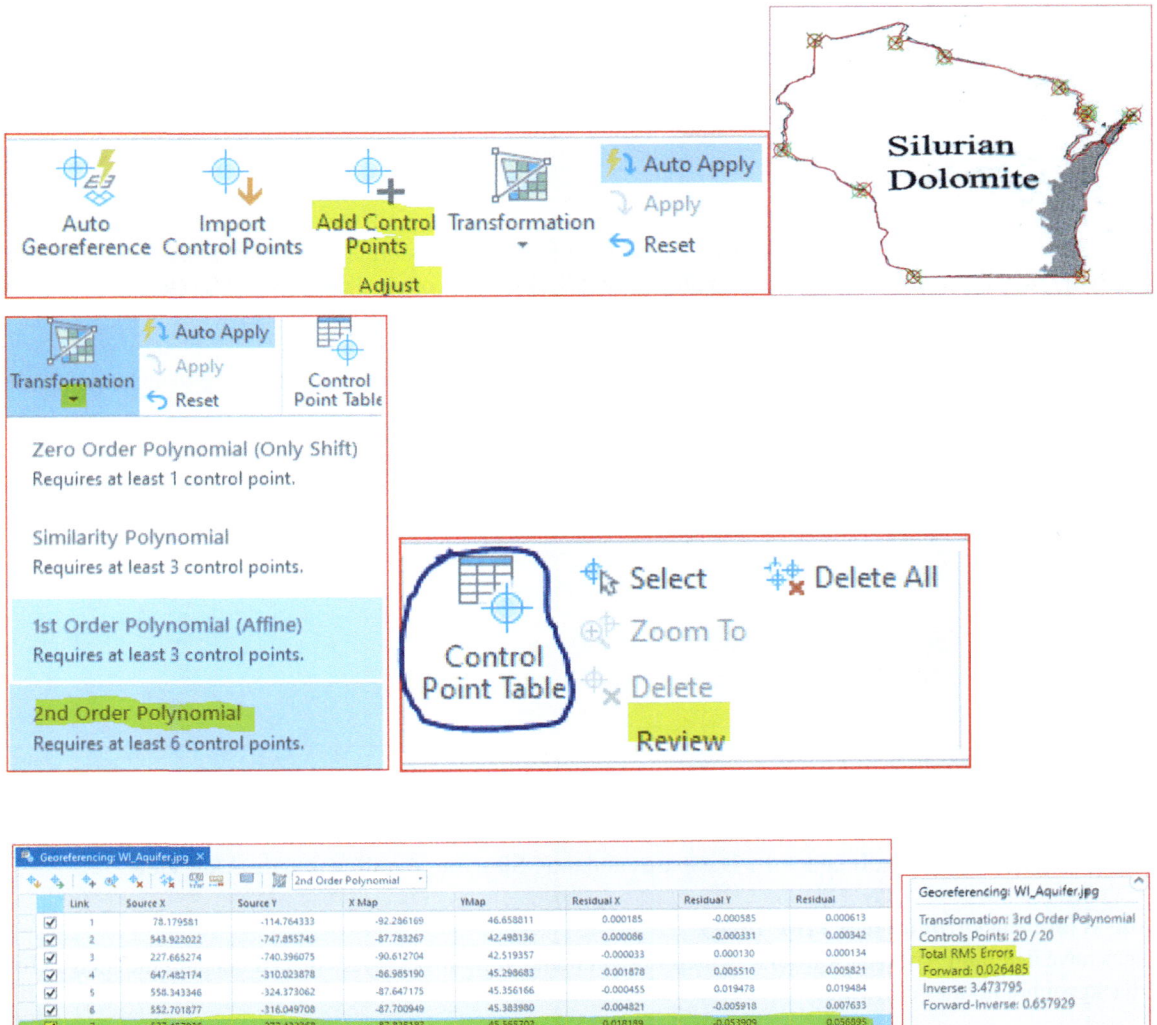

26. When you're satisfied with the current alignment, stop entering control points.
27. In the **Save** group, click **Save** (it will be saved in the geodatabase)

28. In the CP, r-click the "**Silurian_Aquifer**" and remove
29. In the Catalog pane, open the **Databases/Ch04.gdb** and r-click the "**Silurian_Aquifer**"
30. In the Raster Dataset Properties: **Silurian_Aquifer**, scroll down and open the Spatial Reference. You will see that the **Silurian_Aquifer** is registered in **GCS_NAD83**.

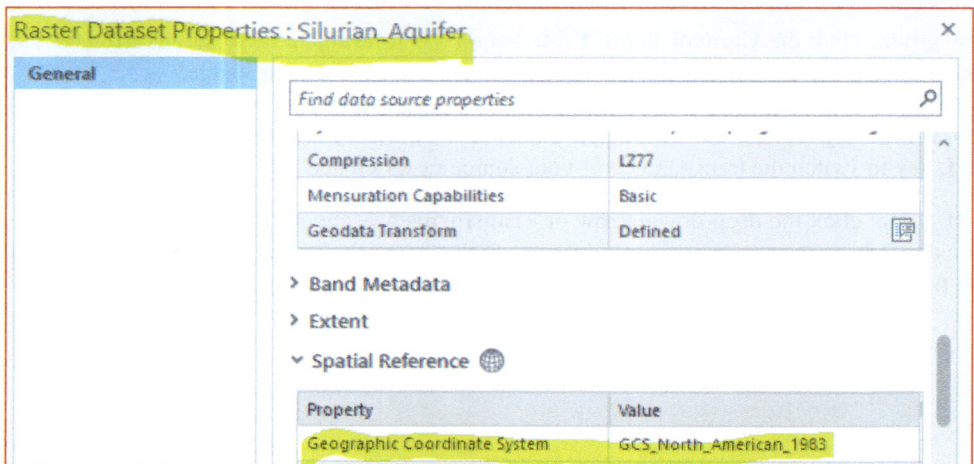

Prepare Raster for Projection

In this section, you will project the raster "**Silurian_Aquifer**" that you have georeferenced in the previous section. You will project the raster from **GCS_North_American_1983** into **NAD 1983 Wisconsin Transverse Mercator**.

31. Click **Insert** tab on the ribbon, in the **Project** group, click **New Map** button
32. Click **Project** tab on the ribbon, click **Save As** tab, browse to **Project \ Folders \ Projection \ p20** and save it as **Project_Raster.aprx**
33. Click **Save**
34. In the **CP**, change the name of the **Map** into **Raster Projection**

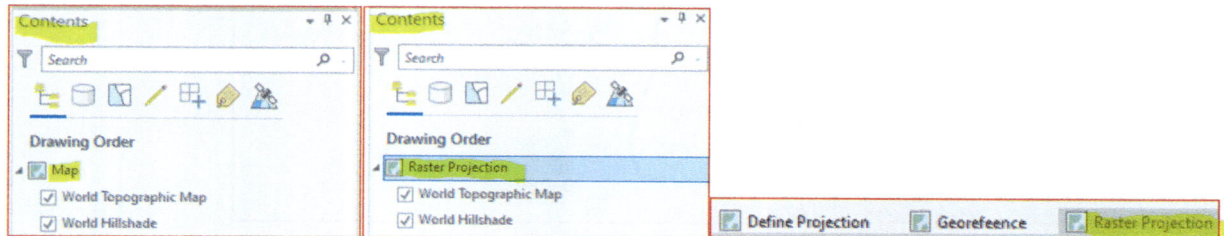

35. In Catalog pane, open **ch04.gdb** under the Databases and drag **Silurian_Aquifer** into the Map View

Remove the White Background of a TIFF Image

Some images have a default white background, which is a hindrance if it blocks layers placed beneath the image. To remove the white background, do the following:

36. In the CP, r-click the **Silurian_Aquifer** image layer, and click Symbology
37. In the Symbology pane, fill it as below
38. Under **Primary symbology**, choose RGB
39. Click the Mask tab.
40. Check the "**Display background value**", and under it, set the value for each band to 255.
41. Set the background color to No Color.

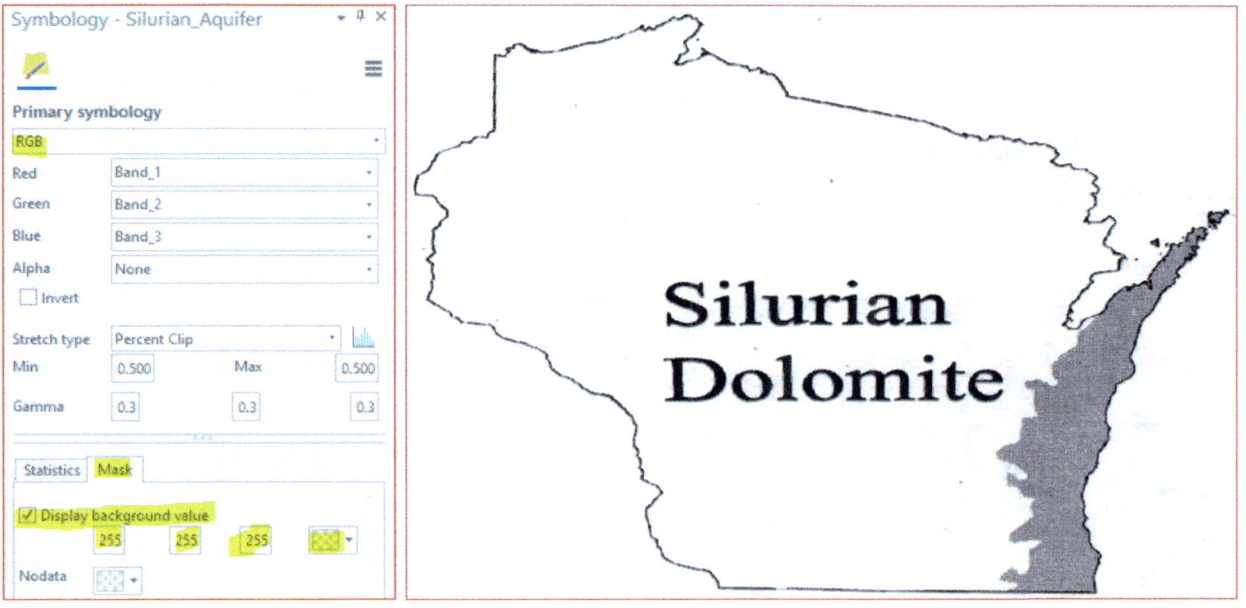

Project the Raster

42. In the **Analysis** tab on the ribbon, in the **Geoprocessing** group, click the **Tools** button
43. In the **Geoprocessing** pane, click the **Toolboxes**, open the **Data Management Tools**, open the **Projections and Transformations**
44. Under **Projection and Transformation** open the **Raster** and click **Project Raster**
45. The **Project Raster** pane display, fill it as below
46. Input Raster: **Silurian_Aquifer**
47. Output Raster Dataset: **Aquifer_WTM**
48. Click Select Coordinate System (Globe)
49. In the Coordinate System dialog box, under XY Coordinate Systems Available
50. Open the Projected Coordinate Systems, open the State Systems, and scroll down to the end and select "**NAD 1983 Wisconsin TM (Meters)**"

51. Click OK
52. Run

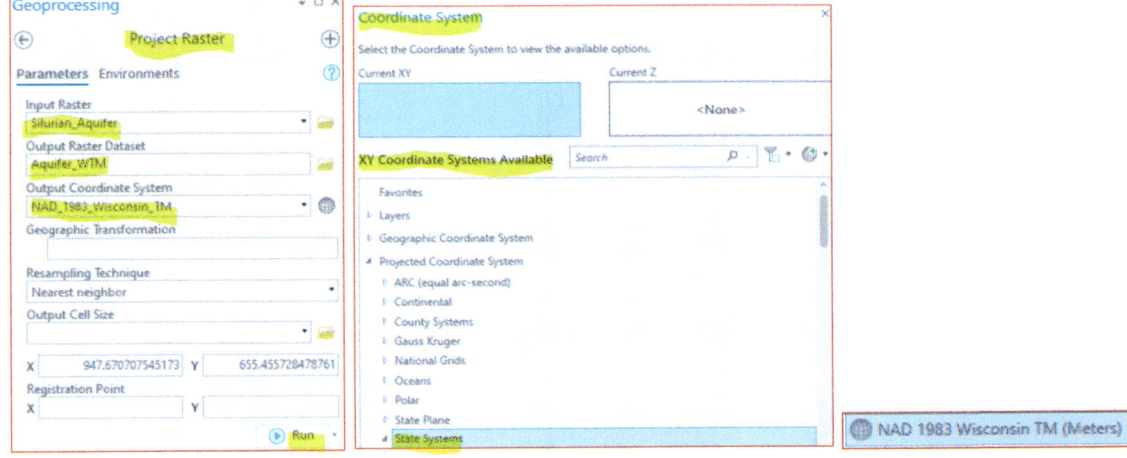

Result The image projected from latitude-longitude into the projected coordinate system and the shape of Wisconsin changed.

53. Save your Project (Ctrl + S)

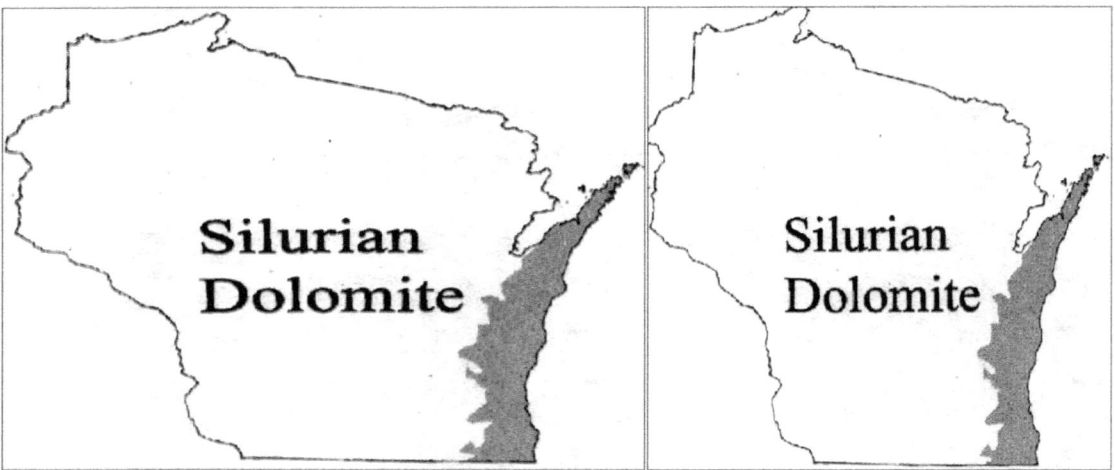

Challenge Task

One of the last steps you completed was selecting a transformation after selecting control points. The instructions for the chapter instructed you to use "2nd Order Polynomial". Return to that point and click through the other transformation options to see the differences. Does it seem like 2nd Order Polynomial was in fact the most accurate based on the RMS?

Introduction to Geodatabase

ArcGIS Pro works with different GIS and non-GIS file formats. The Geodatabase software logic provides the common application logic used throughout ArcGIS for accessing and working with all geographic data in a variety of files and formats. This supports working with the geodatabase, and it includes working with shapefiles, computer-aided drafting (CAD) files, triangulated irregular networks (TINs), grids, imagery, Geography Markup Language (GML) files, and numerous other GIS data sources supported by the ArcGIS Data Interoperability extension.

Geodatabase is a collection of geographic datasets of various types held in a common file system folder or a multiuser relational database management system such as Oracle, PostgreSQL, Microsoft SQL Server, IBM Db2, or SAP HANA. Geodatabases come in many sizes; have varying numbers of users; and can scale from small, single-user databases built on files up to larger workgroup, department, and enterprise geodatabases accessed by many users.

Geodatabase has a lot of meaning in ArcGIS Pro in addition to a collection of datasets.

The geodatabase is the native data structure for ArcGIS desktop and ArcGIS Pro. It is the primary data format used for data editing and management. It is designed to work with and leverage the capabilities of the geodatabase.

The geodatabase physically stores geographic information and uses a database management system or file system. The data can be accessed either through ArcGIS or through a database management system using SQL.

Geodatabases have a comprehensive information model for representing and managing spatial data. The information model is employed as a series of tables holding feature classes and attributes. In addition, advanced GIS data objects add real world behavior, rules for managing spatial integrity, and tools for working with spatial relationships of the core features and attributes.

Three Types of Geodatabases

File geodatabases: A file geodatabase is stored as multiple files in a folder with a **.gdb** extension. Each dataset is contained in a single file. By default, files can grow to 1 TB, but this can be changed to 4 or 256 TB using a configuration keyword. The geodatabase file is the built-in data structure for ArcGIS Pro and is the primary data format used for editing and data management. Geodatabase combines "geo" (spatial data) with "database" to create a central data repository for spatial data storage, management, and analysis.

Mobile geodatabases: A mobile geodatabase is stored in an SQLite database that is entirely contained in a single file and has a geodatabase extension. The size limit for a mobile geodatabase is 2 TB.

Enterprise geodatabases are stored in relational databases. They can be virtually unlimited in size and number of users; the limits differ depending on the database management system (DBMS) vendor.

Supplementary Information The online version contains supplementary material available at https://doi.org/10.1007/978-3-031-42227-0_5.

Datasets in the Geodatabase

A key geodatabase concept is the dataset, and the geodatabase in ArcGIS Pro contains three primary dataset types:
1. **Tables**: A collection of rows, each containing the same fields. Feature classes are tables with shape fields.
2. **Feature classes**: A table with a shape field containing point, line, or polygon geometries for geographic features. Each row is a feature.
3. **Raster datasets**: Contain rasters that represent continuous geographic phenomena.

A geodatabase can be generated from scratch by creating or collecting dataset types. After building a number of these fundamental dataset types, one can add to or extend their geodatabase with more advanced capabilities (such as adding topologies, networks, or subtypes) to model GIS behavior, maintain data integrity, and work with an important set of spatial relationships.

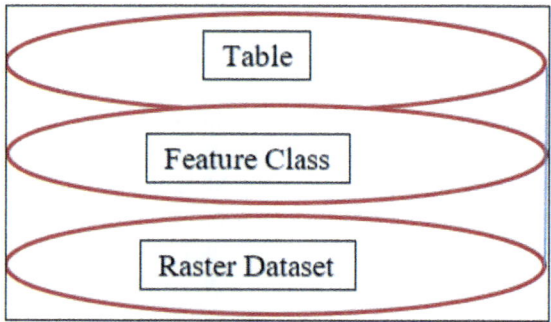

Geodatabase Storage in Tables and Files

Geodatabase storage includes both the schema and rule base for each geographic dataset plus simple tabular storage of the spatial and attribute data. The three primary datasets in the geodatabase (feature classes, attribute tables, and raster datasets), as well as other geodatabase elements, are stored using tables. The spatial representations in geographic datasets are stored as either vector features or rasters. These geometries are stored and managed in fields along with traditional attributes.

The **Geology** feature class is stored as a table. Each row represents one feature, and the **Shape** field holds the polygon geometry for each feature. The polygon value is used to specify that the field contains the coordinates and geometry that define one polygon in each row.

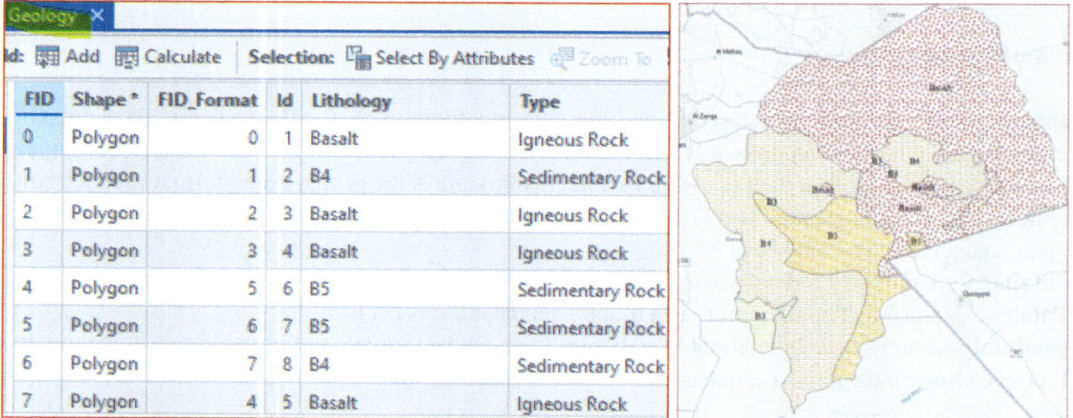

A key geodatabase strategy is to use the database management system (DBMS) to scale GIS datasets to extremely large sizes and numbers of users, for example, to support simple small databases for one or a few users up to instances with hundreds of millions of features and thousands of simultaneous users. Tables provide the primary storage mechanism for geographic datasets. Structured query language (SQL) is strong at querying and processing the rows in tables, and the geodatabase strategy is designed to use these capabilities.

Creating a Geodatabase

This section will allow you to capture data using digital data to create feature classes and store them in a geodatabase.

Scenario 5-1 You are working as a hydrogeologist for the Water Authority, and you are asked to create a file geodatabase and fill it with point, line, and polygon feature classes. You are also going to use the "**Image_Rectify.tif**" image as a source to capture groundwater wells as a point feature class, the "major fault" as a line feature class, and the "KSWTP" as a polygon feature class.

Create File Geodatabase (Dhuleil.gdb)
1. In Window Explorer, browse to **Ch05** under **ENV_Water** and d-click **Ch05.apx** to launch ArcGIS Pro
2. ArcGIS Pro open and in the CP, you see the **World Topographic Map** and the **World Hillshade**
3. In **Catalog** pane, open the Databases, you see there are two databases: **Ch05.gdb** (default) and the **Dhuleil.gdb**
4. In **Catalog** pane, r-click **Folders** and click **Add Folder Connection**
5. Browse to \\Database\, select **Data_Ch05** and click OK
6. In the **CP**, rename the **MAP** to **Dhuleil**, and save the project

Result In the **Catalog** pane, the **Data_Ch05** from **Ch05** added to the **Folder** and the Map will be called **Dhuleil**.

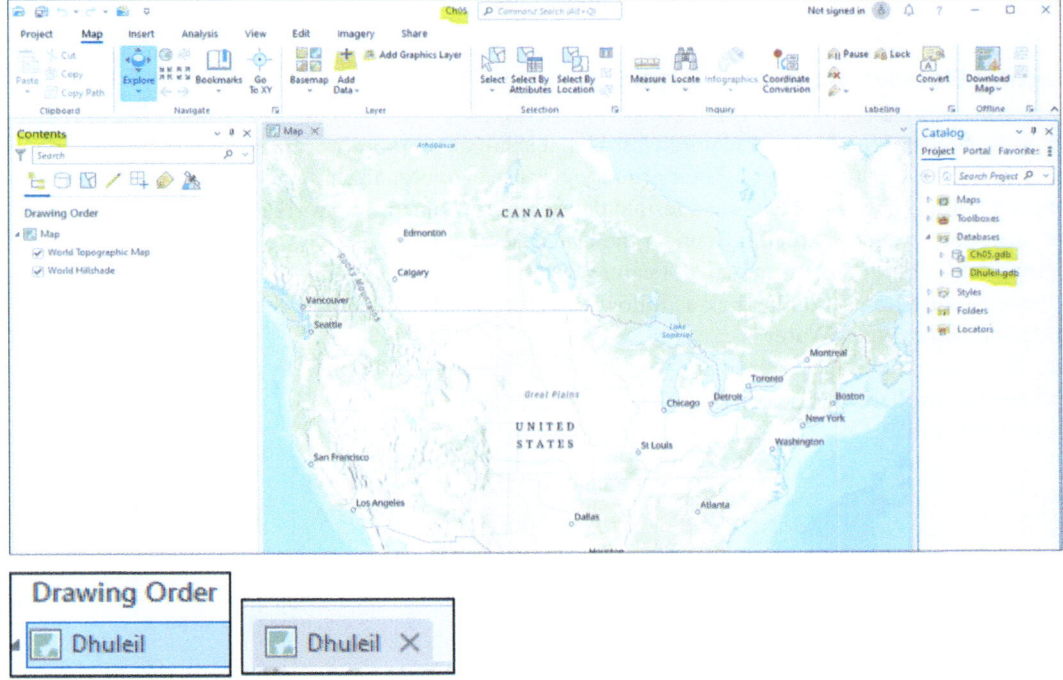

Import Image as a Raster Dataset into File Geodatabase

The term raster dataset refers to any raster data model that is stored in a geodatabase.
7. Click the **Analysis** tab on the ribbon, in the Geoprocessing tool, click the Tools button
8. In the **Geoprocessing** pane, click the **Toolboxes** tab, click the **Conversion Tools**, and open **To Geodatabase**.
9. Click **Raster To Geodatabase** and fill it as below
10. **Input Rasters**: In Catalog pane, browse to the **Image** folder under **Data_Ch05\Data** and choose **Image_Rectify** and click OK
11. **Output Geodatabase**: In Catalog pane, browse to the **Databases** and select **Dhuleil.gdb**
12. Run

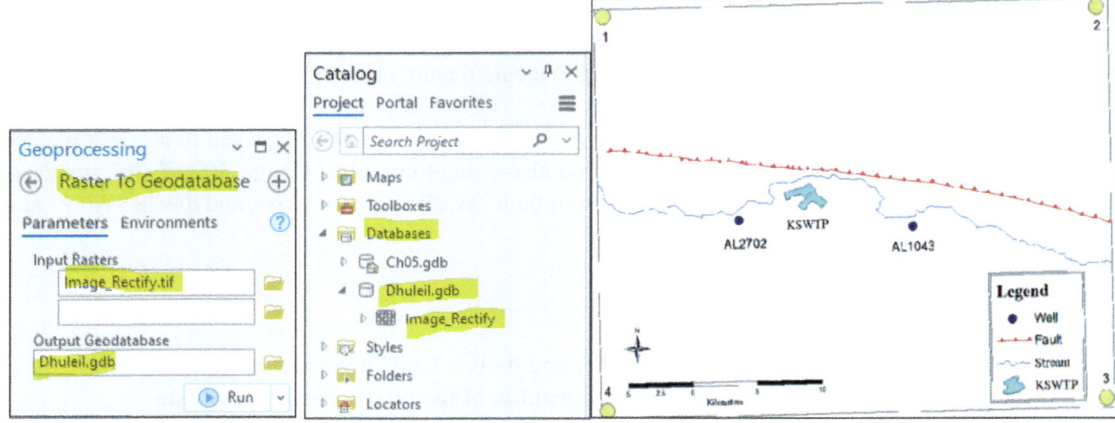

Result Image_Rectify was converted and stored in Dhuleil.gdb.

13. In Catalog pan, open the **Dhuleil.gdb** and drag **Image_Rectify** into the **CP** or **Map View**

Result **Image_Rectify** is displayed in the **Map View** and consists of two wells, stream, fault and KSWTP. In this section, you will use the image as a background to capture feature dataset.

Create a Feature Dataset in File Geodatabase

A feature dataset is a collection of related feature classes that share a common coordinate system. Feature datasets are used to spatially integrate related feature classes. Their primary purpose is to organize related feature classes into a common dataset for building a topology, a network dataset, a terrain dataset, or a geometric network.

14. Open **Catalog** pane, expand **Dhuleil.gdb** under the Databases
15. R-click **Dhuleil.gdb**, point to **New** and click Feature Dataset.
16. Fill the Create Feature Dataset dialog box as follows:
17. Feature Dataset Name: type "**Water**"
18. Coordinate System: click Select coordinate system, open Projected Coordinate System, open UTM, open WGS 1984, open Northern Hemisphere, and select WGS 1984 UTM Zone 36 N
19. Click OK and then run

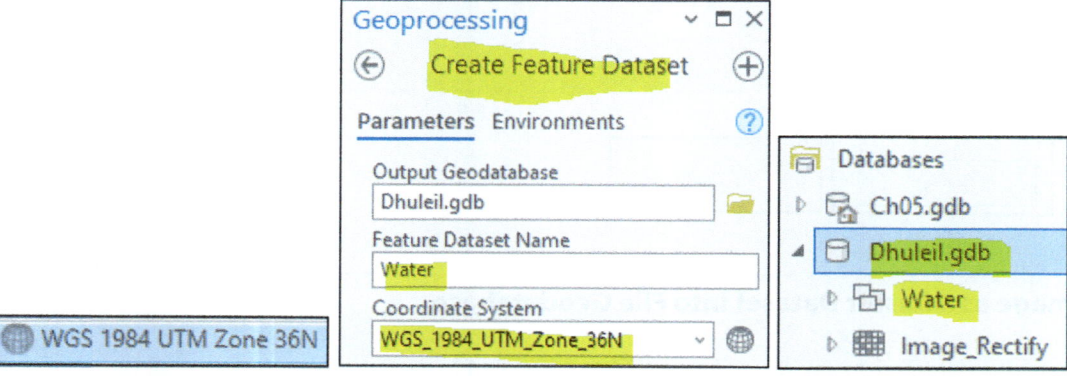

Result The **Water** Feature Dataset is created in **Dhuleil.gdb,** and it is registered in UTM Zone 36 N and associated with the WGS 1984 datum.

Create a Feature Dataset in File Geodatabase 85

Import shapefiles into the Water Feature Dataset
This step will import two shapfiles (**Dam.shp** and **Stream.shp**) into the **Water** feature dataset. The two shapefiles are registered in the coordinate system of **GCS_WGS84**. The **Water** feature dataset is registered in the projected coordinate system (**WGS_1984_UTM Zone 36 N**). Once **Dam.shp** and **Stream.shp** are imported into the **Water** feature dataset, they are automatically converted into the coordinates of the **Water** feature dataset.
20. In the **Catalog** pane, expand Databases, r-click **Water** feature dataset.
21. Point to Import, select Feature Class(es)…
22. Fill the Feature Class to Geodatabase as follows:
23. Input Features: In Catalog pane, browse to the Shapefile folder under the Data folders, highlight **Dam.shp** and **Stream.shp**
24. Click OK
25. Click Run

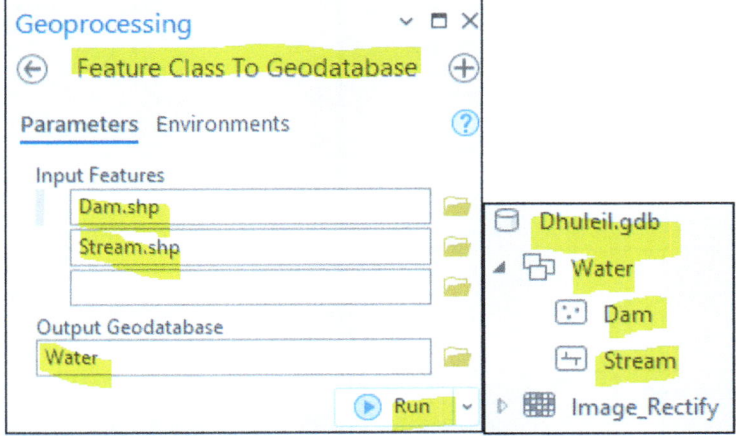

Result   The **Stream** and **Dam** is imported in the **Water feature dataset**.

Create Feature Classes and Assign Spatial Reference
To create a feature class, the file geodatabase must already exist. **Dhuleil.gdb** is created in the Databases. You are going to create **3**-new feature classes and call them **Well (point)**, **Fault (line),** and **Plant (polygon)**.
26. Open **Catalog** pane, expand **Databases**, and r-click **Dhuleil.gdb**
27. Select New and then Feature Class
28. Fill the Create Feature Class dialog box
29. Name: Observation
30. Type of features stored in the feature class: Point
31. Under Geometric Properties   Check Z Value – coordinate include Z values used to store 3D data
32. Click Next
33. Click below Field Name: "Click here to add a new field"
34. Remove the Field and type: Name
35. Under Data Type: Text
36. Length: 10
37. Click Next
38. In the Spatial Reference Dialog box
39. Open Geographic coordinate system, open the World, and select WGS 1984
40. Click Finish

Result   The **Observation** feature class is created and stored in **Dhuleil.gdb**.

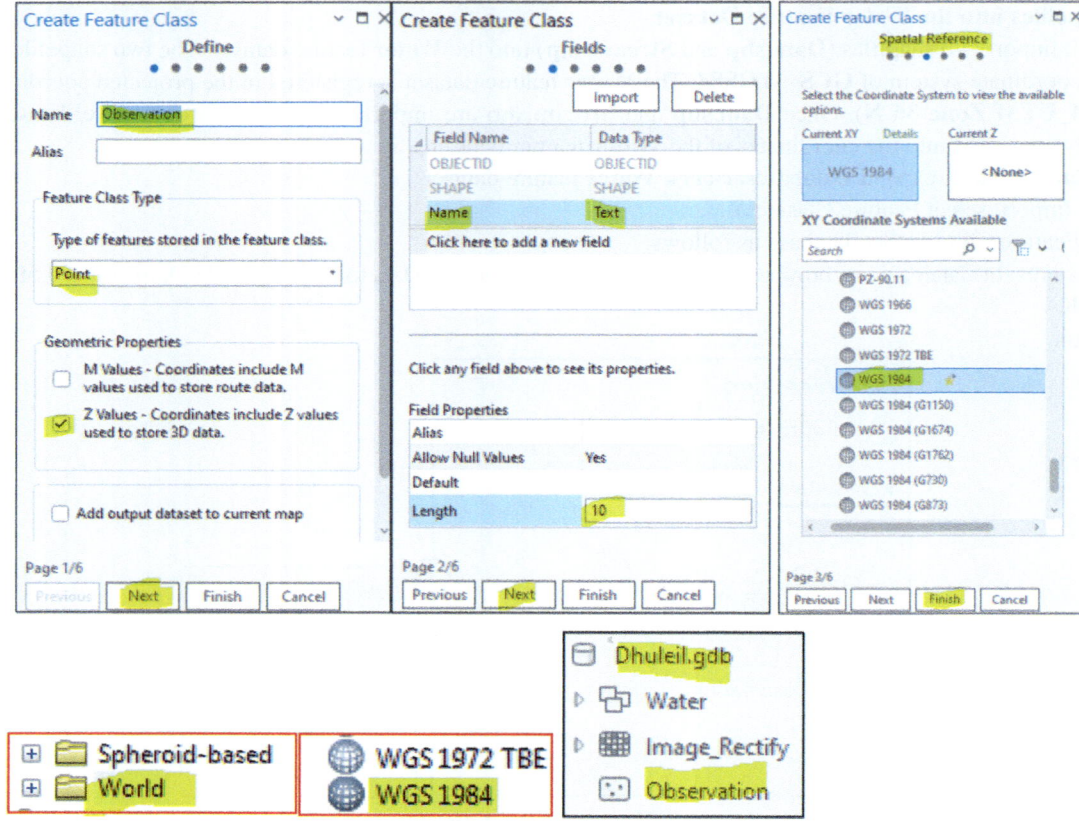

Continue by repeating the previous steps and create a **line** feature class and call it "**Fault**" and **polygon** feature class and call it "**Plant**" and register them into **GCS_WGS_1984**.

Result The **Fault**, **Plant**, and the **Observation** feature classes are integrated into the CP and stored in **Dhuleil.gdb**, but all of them are empty.

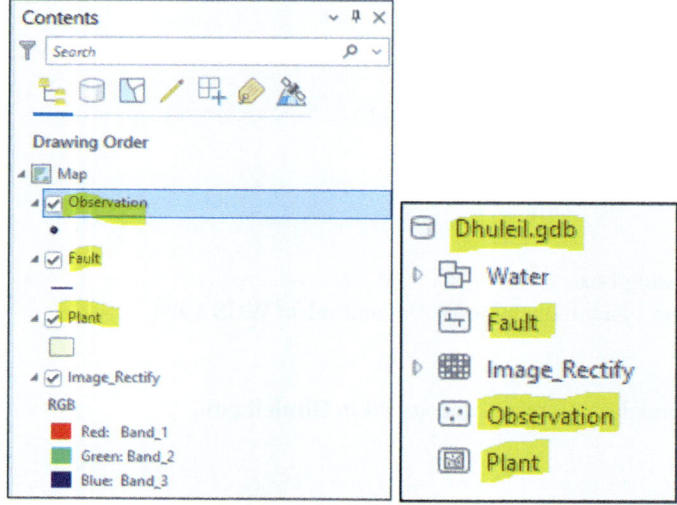

Digitizing Point, Line and Polygon on Screen

On-screen digitizing is a process in which a map is created using another map. This map could be an image, a scanned picture, or a previously digitized map. This technique is used to trace features to create new layers. This practice is similar to traditional tablet digitizing, but rather than using a classical digitizer and a puck, the user creates layers on the computer screen with the mouse and referenced information as a background. On-screen digitizing may also be used in an editing session where the user can update or add new features. The accuracy of the digitized features cannot in any way be higher than the original base image. For accurate tracing, during digitization, the user should zoom in for better viewing. Nevertheless, this does not mean that the newly captured feature will more closely match the real-world coordinate.

In this section, you are going to use the raster dataset (**Image_Rectify**) that are stored in **Dhuleil.gdb** as backdrop to digitize on screen the **Observation** (point), **Fault** (line), and **Plant** (polygon).

Capture the Feature Classes using the Raster Dataset
You are going to digitize the two **Observations**, the **Fault**, and the **Plant** using the **Image_Rectify** raster dataset. The image is registered in latitude-longitude and associated with the Jordanian datum (D_Jordan).

Create the Observation Feature Class
Now you will create two observation wells by digitizing them using the **Raster Dataset** as a background.
39. In the CP, click the symbol of the **Observation**, in the Symbology – Observation dialog box, make sure the Gallery tab is selected and click the "Circle 4"
40. The symbol of the Observation in the CP changed
41. In the CP, r-click the **Observation** and click Attribute Table
42. In the upper-right corner of the table, click the **Option** button, and click the **Fields View**.
43. **Fields: Observation** shows that the Field Name: Name and its type and length are text and 10, respectively
44. Close the **Fields: Observation** and the **Observation** table

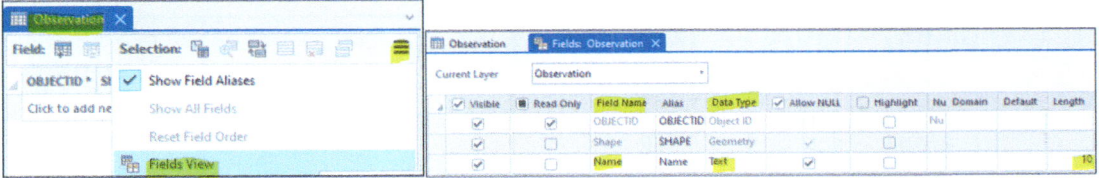

45. On the ribbon, click the **Edit** tab, in the **Features** group, click **Create** button
46. In the **Create Features** pane, click the **Observation** feature template and click the **Active template** (blue arrow) and click the **Point** tool in the **Active Template**
47. Click the point on the map that has the "Al2702" label on the image to create the point feature.
48. Then, click the observation labeled "AL1043" to create the second point.
49. At the bottom of the map and on the **Configure** tool, click **Finish**, or press the F2 key on the keyboard to finish the digitizing.
50. Close the Create Features pane

Result Two points are created on the map, and the second point is selected.

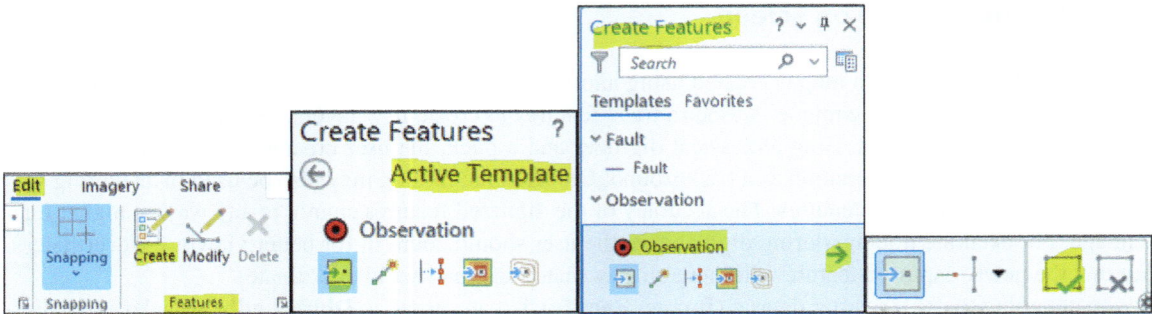

51. In the CP, r-click **Observation** and click **Attribute Table**
52. The last digitized point is still selected, type under Name field "**AL1043**"
51. Type "**Al2702**" in the first line under **Name** and enter
52. On the ribbon, click the **Edit** tab, in the **Manage Edits** group, click **Save** button and then click Yes to save all edits.
53. In the Observation table click "**Clear**" to deselect the 2nd point OR in the ribbon click Map tab, in the Selection group, click Select button to deselect the Point button, and then close the attribute table of **Observation**

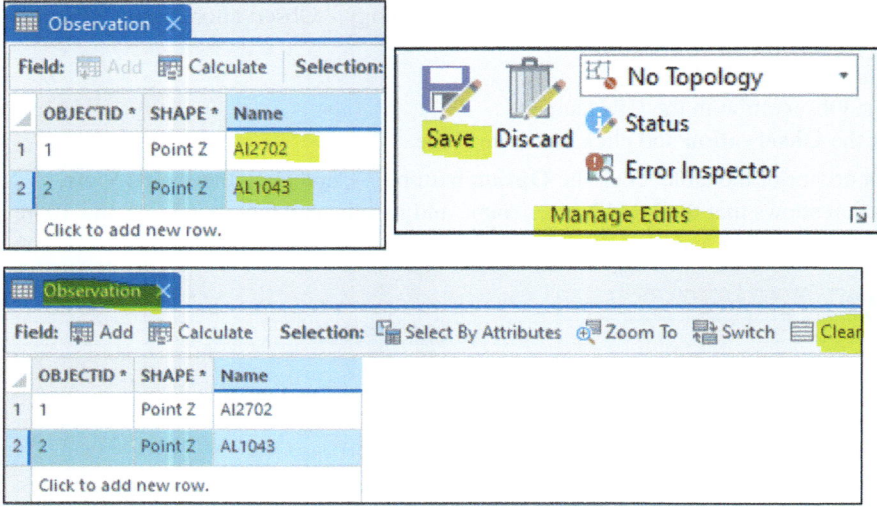

Create the Fault Feature Class
54. In the CP, click the **Fault** symbol, in the **Symbology** pane, select **Gallery** tab and choose "**Dashed 4:1**" symbol, then click **Properties** tab, under the **Appearance** select the red **color**.
55. Click Apply, and close the **Symbology** pane

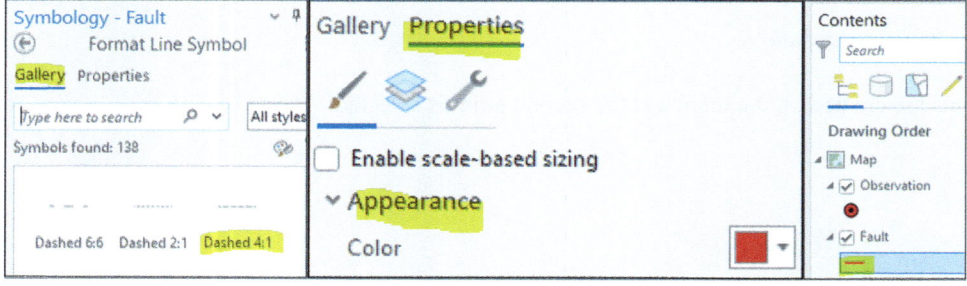

56. On the ribbon, click the **Edit** tab, in the **Features** group, click **Create** button
57. In the **Create Features** pane, click the **Fault** feature template and click the **Active template** (blue arrow) and click the **Line** tool in the **Active Template**

Digitizing Point, Line and Polygon on Screen

58. Using the image as a guide, digitize the **Fault** by clicking the beginning of the **Fault** on the east side and continue clicking on the **Fault** as you go along it to completely digitize it.
59. Once you are done, double click to finish digitizing (or r-click and click finish)
60. On the **Configure** tool, click **Finish**, or press the F2 to finish the digitizing.
61. Close the Create Features pane
62. On the ribbon, click the **Edit** tab, in the **Manage Edits** group, click **Save** button and then click **Yes** to save all edits.
63. On the ribbon click **Map** tab, in the **Selection** group, click Select button to deselect the Point button

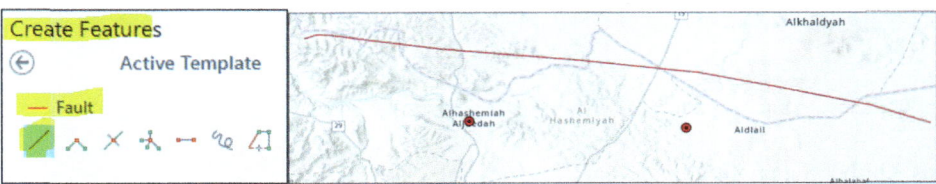

Result   The line is created on the map, and it is selected.

Create the Plant Feature Class
65. In the CP, click the **Plant** symbol, in the **Symbology** pane, select **Gallery** tab and choose "**Water (area)**" symbol
66. In the Symbology pane, select Gallery tab and choose Water (area) symbol
67. Close the Symbology pane
68. Zoom In around the **Plant** in the image
69. On the ribbon, click the **Edit** tab, in the **Features** group, click **Create** button
70. In the **Create Features** pane, click the **Plant** feature template, click the **Active template** (blue arrow) and click the **Polygon** tool in the **Active Template.**
64. Using the image as a guide, place your pointer over the right-east corner of the **Plant** and click once, then click along the outside edge of the image to create the polygon.

71. On the **Configure** tool, click **Finish**, or press the F2 to finish the digitizing.
72. Close the Create Features pane
73. On the ribbon, click the **Edit** tab, in the **Manage Edits** group, click **Save** button and then click **Yes** to save all edits.
74. On the ribbon click **Map** tab, in the **Selection** group, click Select button to deselect the Point button

75. In the CP, r-click **Plant**, click the Attribute Table

76. Click the Option button (in upper-right corner), click **Fields View**
77. In the **Fields: Plant** table, click the last row that says, "**Click here to add a new field**".
78. A new row is created, and the table is filled as follows:
79. Under the Field Name, type Name
80. Double-click the **Alias** column and enter Name.
81. Select the **Data Type** column and choose **Text** from the drop-down menu
82. Under the Length column type 10
83. On the ribbon, in the **Fields** tab, in the **Changes** group, click **Save** button

84. Close **Fields: Plant** and the attribute table of **Plant**
85. In the CP, select the **Plant**, on the ribbon, select **Edit** tab, in the **Selection** group, click the **Attributes** button
86. The Attribute pan dialog box open
87. Under Attributes tab, to the right of the Name, type KSTP
88. Click Apply
89. On the ribbon, click the **Edit** tab, on the Manage Edits group, click **Save** button

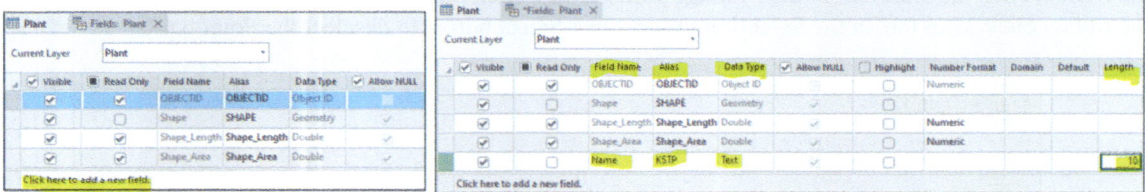

Note KSTP means Khirbet Samra Treatment Plant

90. Open the attribute table of Plant, you will see that **KSTP** added under Name, then close the table
91. In the CP, right click the raster dataset "**Image_Rectify**", remove
92. Save the project

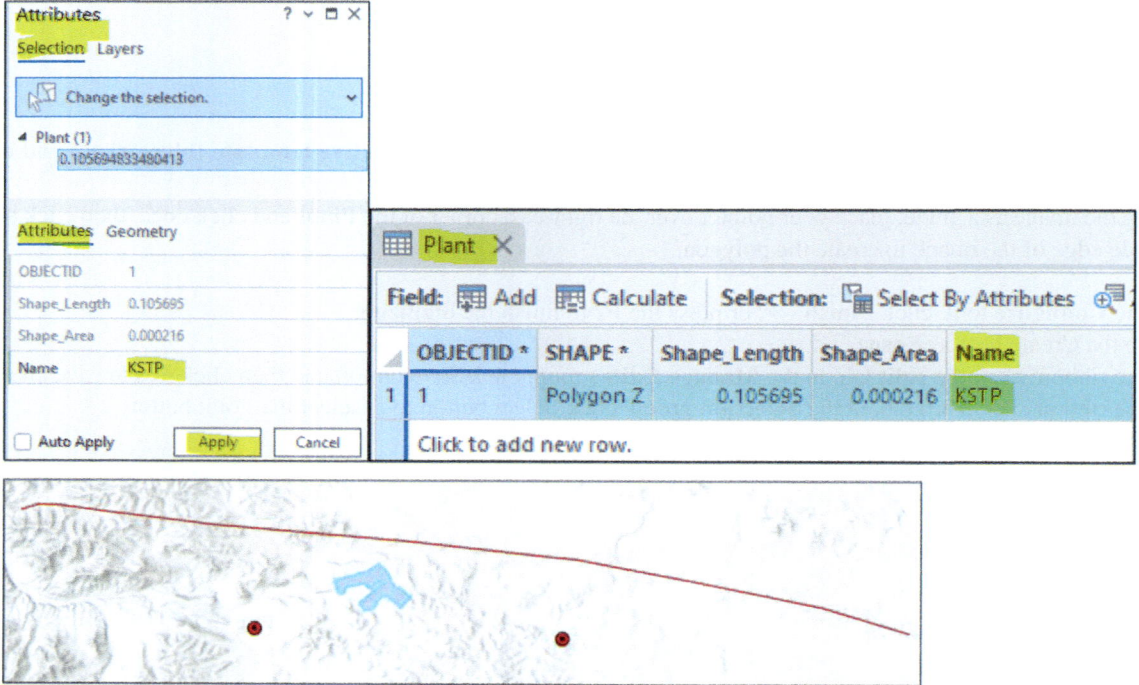

Import Data into Geodatabase

1. **Insert** tab on the ribbon, in the **Project** group, click **New Map**, and call it **Jizzi**
2. In **Catalog** pane, under Database, r-click **Dhuleil.gdb**, point to **Import** and then click **Feature Class(es)**
3. The **Feature Class To Geodatabase** dialog box opens and fills it as follows:
4. Input Features: In Catalog pane, browse to **Jizzi** folder under **Data** folder, highlight the **Catchment** and **Well** shapefiles
5. Click OK
6. Output Geodatabase: **Dhuleil.gdb**
7. Click Run

Relationship Class

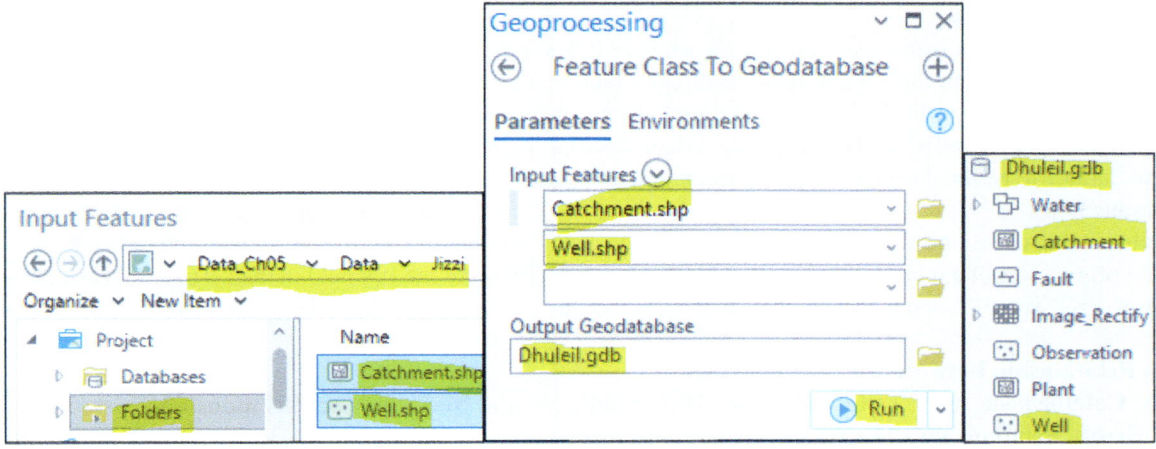

Result The **Catchment** and **Well** feature class will be stored in **Dhuleil.gdb**, if you don't see them, r-click **Dhuleil.gdb** and click **Refresh** or click F5

8. In **Catalog** pane, under Database, r-click **Dhuleil.gdb**, point to **Import** and then click **Table(s)**
9. The **Table To Geodatabase** dialog box open, fill it as below
10. Input Table: In Catalog pane, browse to **Jizzi** folder under **Data** folder, highlight the **Table 1.txt** and **Table 2.txt**
11. Click OK
12. Output Geodatabase: **Dhuleil.gdb**
13. Click Run

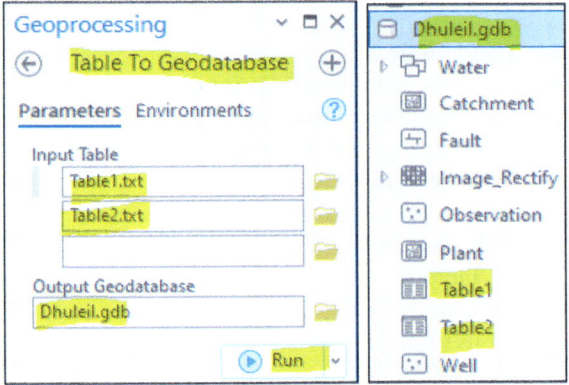

Relationship Class

Relationship classes describe relationships between items in the geodatabase. There are different relationships, such as simple, one-to-one, or one-to-many relationships. These relationships can be used to ensure that the links between items in the database are maintained and up to date. Deleting a feature, such as a groundwater well, can cause the removal of other features, such as a pump or casing records in a related table. The create relationship class tool creates a relationship class to store an association between fields or features in the origin table and the destination table. The Origin Table is the table or feature class that will be associated with the destination table. The destination table is the table or feature class that will be associated with the origin table.

Simple Relationship Class
When you create a relationship class, you specify whether it is simple or composite. In this section, you are going to deal with simple relationship. In this simple relationship, the well feature class, Table 1, and Table 2 in the geodatabase exist independently of each other. In this exercise, the groundwater wells have more data related to water chemical analysis and other data related to the wells' depth and yields. When creating the first relationship class, you choose the Well to be the origin and Table 1 to be the destination. In the second relationship class, you choose Table 1 to be the origin and Table 2 to be the destination.

In a relationship class, objects in the origin match objects in the destination through the values in their key fields. The key field in the origin class of a relationship is called the primary key (PK), and the key field in the destination class is called the foreign key (FK). The FK contains values that match those of the primary key field in the origin class.

Create a Relationship Between the Well and Table 1
14. In the **Catalog** pane, under **Database**, r-click **Dhuleil.gdb**, point to **New** and select **Relationship Class**
15. The **Create Relationship Class** dialog box open, fill it as below
    - (a) Origin Table: Well (stored in Dhuleil.gdb)
    - (b) Destination Table: Table 1 (stored in Dhuleil.gdb)
    - (c) Output Relationship Class: Well_Table 1
    - (d) Relationship Type: Simple
    - (e) Forward Path Label: Well
    - (f) Backward Path Label: Table 1
    - (g) Origin Primary Key: INVEN_No
    - (h) Origin Foreign Key: INVEN_No
16. Accept the other default
17. Click Run

Result The relationship class created and stored in Dhuleil.gdb.

Create a Relationship Between Table 1 and Table 2
18. In the **Catalog** pane, under **Database**, r-click **Dhuleil.gdb**, point to **New** and select **Relationship Class**
19. The **Create Relationship Class** dialog box open, fill it as below
    - (a) Origin Table: Table 1 (stored in Dhuleil.gdb)
    - (b) Destination Table: Table 2 (stored in Dhuleil.gdb)
    - (c) Output Relationship Class: Table 1_Table 2
    - (d) Relationship Type: Simple
    - (e) Forward Path Label: Table 1
    - (f) Backward Path Label: Table 2
    - (g) Origin Primary Key: INVEN_No
    - (h) Origin Foreign Key: INVEN_No
    - (i) Accept the other default
20. Click Run

Result The relationship class created and stored in Dhuleil.gdb.

After establishing the two relationship classes, you will use them to identify the well salinity.

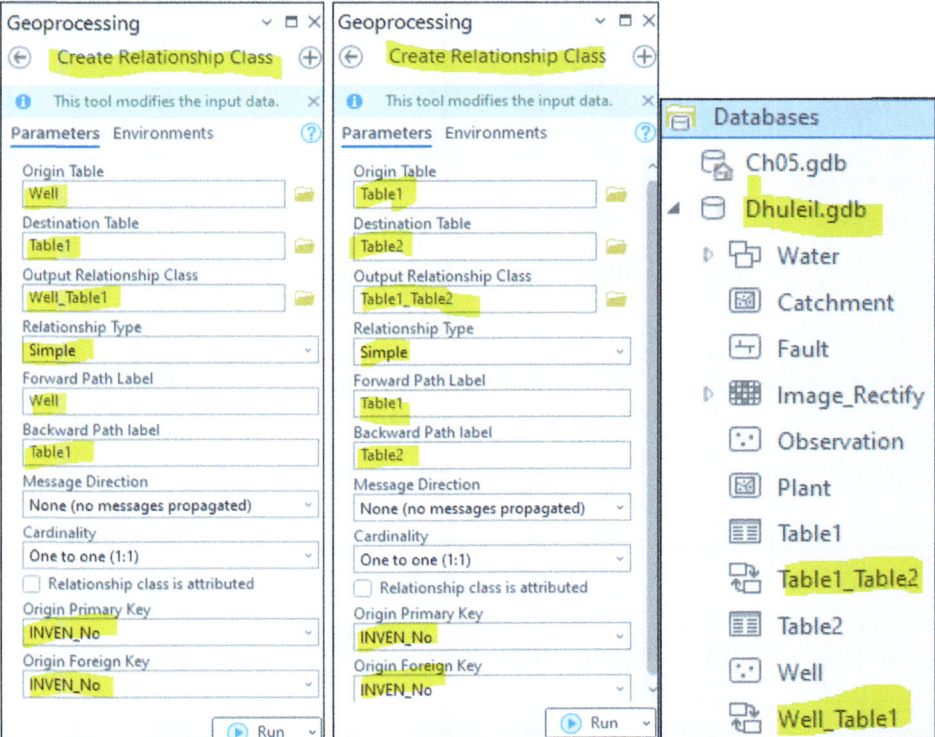

Find the Wells with Salinities Higher Than 500 mg/l

21. In Catalog pane, drag from Dhuleil.gdb the Catchment, Well, Table 1, and Table 2 in the display window
22. In the CP, click the symbol of the Catchment and make it hallow, and click the symbol of the Well and click the Circle 1 and make the color blue

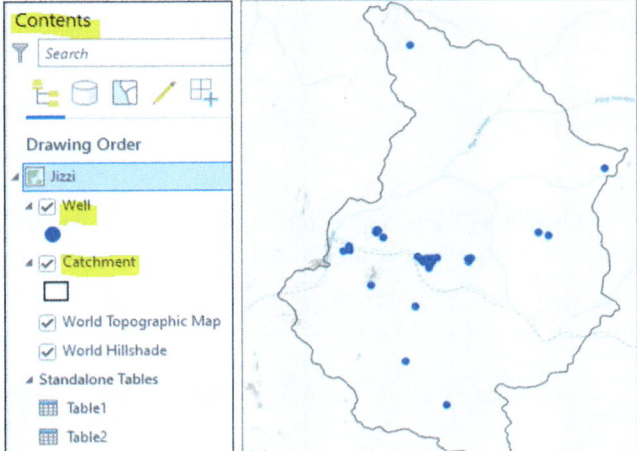

23. In the **CP**, select **Table 1**, click the **Map** tab, in the **Selection** group, and **Select By Attributes**
24. In the Select By Attributes fill, it is as follows:
25. Input Rows: Table 1
26. Selection Type: New selection
27. Write the SQL as follows:
28. where TDS is greater than 500
29. Click Apply and close the **Select By Attributes** dialog box

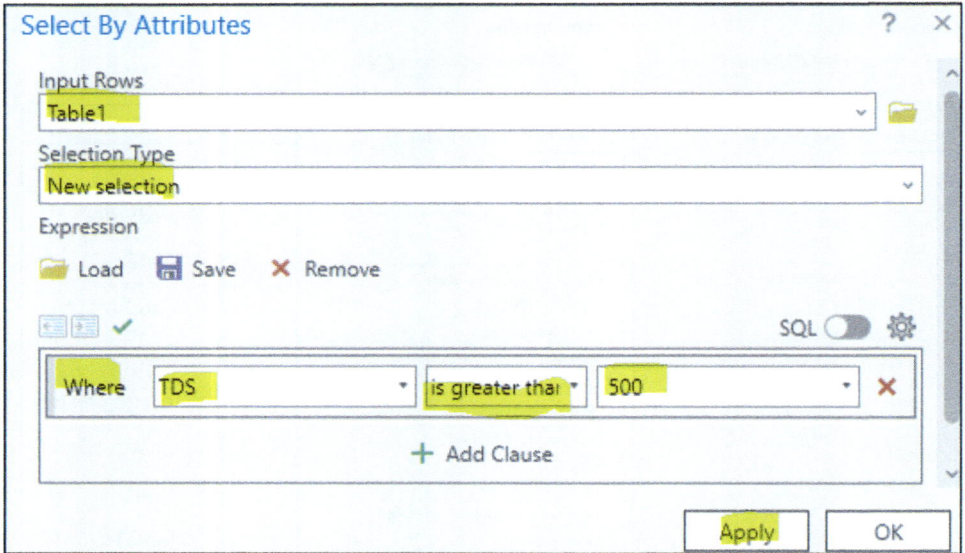

30. In the CP, open the attribute table of Table 1.
31. At the bottom of the table, click "**Show Selected Records**"

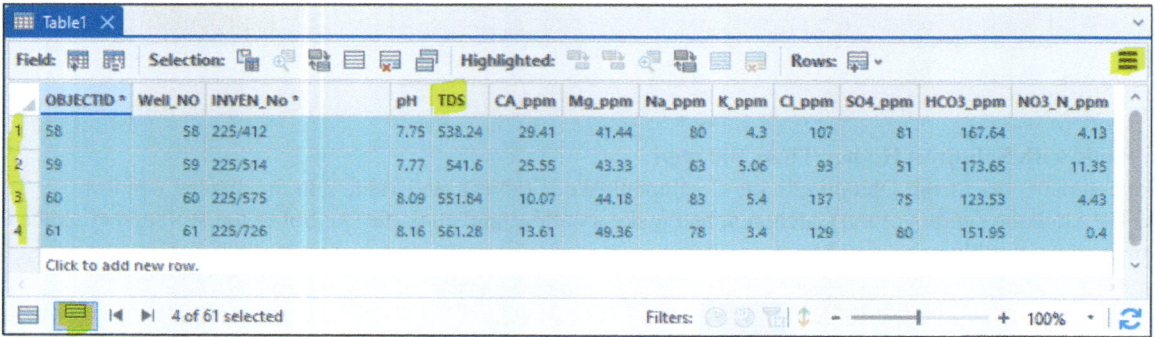

Result Four records selected

32. In the **Table 1** attribute table, click **Option** (3-lines on the top right), point to **Related Data** and select **Well – Table 1**
33. The **Well** attribute table opens and shows that **4 wells** are selected in the table and in the middle of the map in the window display.

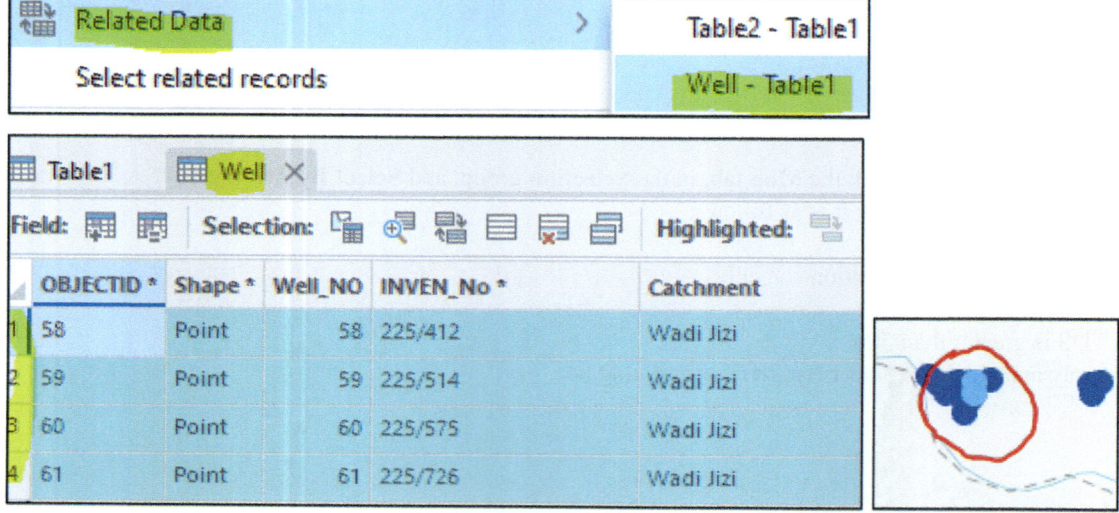

Relationship Class

34. In the table, select the **Table 1** tab
35. In the **Table 1** attribute table and click **Option** (3-lines on the top right), point to **Related Data** and select Table 2 – Table 1
36. Table 2 Attribute table open and 4 records selected

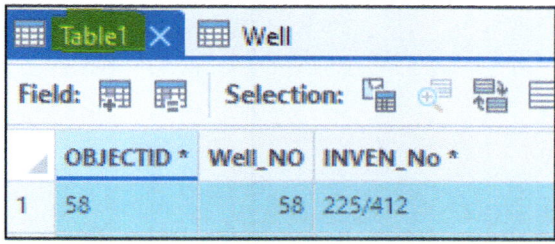

Data Editing and Topology

During data creation, either using a tablet or on-screen, digitizing can generate errors. The error can be due to human error, such as missing a point, line, polygon, or digitizing extra features. Errors can also be generated during scanning, tracing or geo-referencing. An ArcGIS user can edit various types of data, such as feature data stored in shapefiles, geodatabases, and different tabular formats. The editing can include points, lines, polygons, and text.

Editing occurs in an edit session where vector features or tabular attribute information can be created or modified. Start an edit session when you are ready to begin editing, but remember to end the edit session when you're done. If you have more than one data frame in your map, you can only edit the layers in one data frame at a time, even if all data are in the same workspace. The editing of the data can be done if they are either in the same or in different coordinate systems.

Topology is an advanced way to edit data, and it is defined as a data structure that creates connections and describes the spatial relationship between point, line, and polygon features. In other words, the topology is simply the arrangement of how the three different features (point, line, and polygon) share geometry. All spatial elements in a GIS layer are connected in some fashion to each other, which allows the layer to be categorized, queried, manipulated, and stored more efficiently. The topology is also a set of rules, behaviors, and models on how points, lines, and polygons share coincident geometry. For example, two adjacent catchment areas will have a common water divide between them. The set of sub-catchment polygons within each watershed must completely cover the watershed polygon and share edges with the whole catchment boundary. The topology is a useful data structure concept in GIS that allows GIS users to know the location of the feature, what is connected to it, what is surrounded by it, and how to identify spatial relationships with other features. It can also help to get around using the nodes and vertices to accomplish various spatial analysis tasks. In GIS, one can find and trace a route on a map between two cities and measure the distance and time of arrival.

When topology is applied in GIS, a data structure table is built from _nodes_ and _chains_ of the features. The tables are used to determine various relationships, such as what is connected, what is adjacent (left & right), and what is the direction of chains? Topology is applied after digitizing and editing. When data are digitized or created, there is no connection, or relationship, to the feature that has recently been digitized. This means that no informational content associated with point, line, or polygon is available, except location. For example, if you digitize a river and its tributaries, then run the topology, it will build the spatial information. It does this by recognizing the nodes at the end of each digitized stream and creating new nodes at intersections where the river crosses. The end result is that each segment of the river consists of three topological chains separated by a node (Fig. 8.1). One stream segment consists of Arc 1, Arc 2, and Arc 3 using a start node and end node. Arc 1 has node 1 as its starting node and node 4 as its ending node.

Topology offers special information on the data structure, provides powerful functions for spatial analysis and presents a number of advantages to GIS. The topology allows users to calculate the spatial information and property of the features. The spatial property for the point is location (X, Y), the line is the length, and the polygon is the perimeter and area. Topology provides spatial relationships that allow users to query the data and provide spatial analysis when running the network analyst.

Supplementary Information The online version contains supplementary material available at https://doi.org/10.1007/978-3-031-42227-0_6.

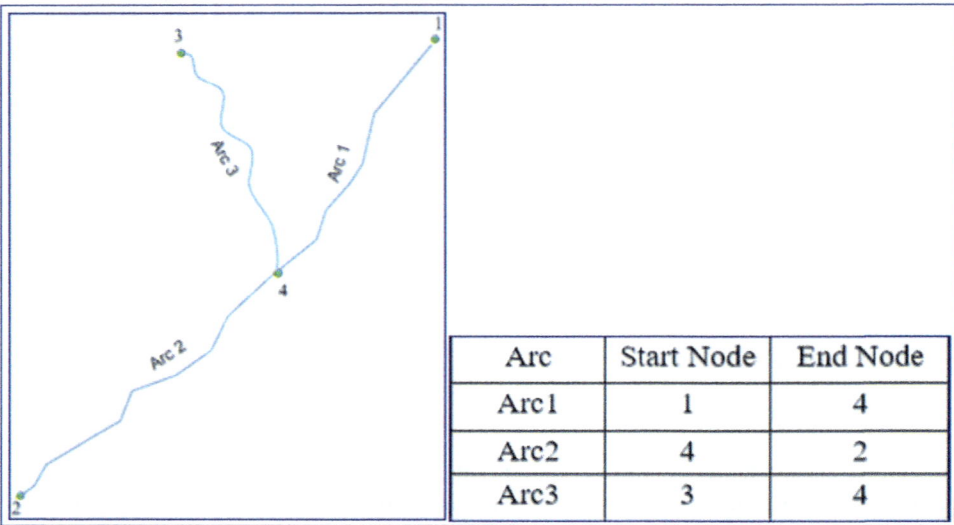

Editing in ArcGISPro

In this chapter, you are going to be introduced to various types of editing, which range from simple, advanced, and all the way to topological editing.

Simple Editing

(a) Delete
(b) Move
(c) Split
(d) Reshape
(e) Modify
(f) Merge

Advanced Editing

(a) Overshoots and undershoots
(b) Generalize feature
(c) Smooth feature

Topological Editing Using Geodatabase

(a) Fix Lines using topology
(b) Fix polygons using topology

Simple Editing

Scenario 6-1 You are giving a shapefile that was digitized from an aerial photograph, and your boss asked you to modify it by deleting and moving some polygons from it.

Open the Project and Connect to Folder
1. Launch ArcGIS Pro and click "**Open another project**"
2. Browse to \\ENV_Water\Ch06 and click Open **Ch06.aprx** and click OK
3. The ArcGIS Pro open and has no data attached to it

Simple Editing 99

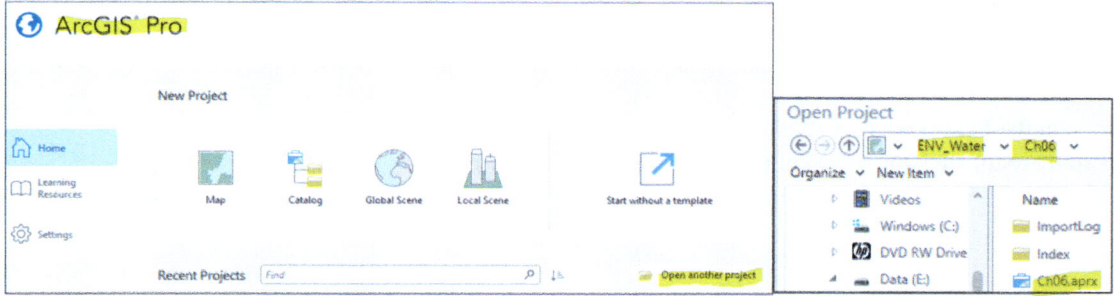

4. In **Catalog** pan, r-click the **Folder**, and point to **Add Folder Connection**
5. Browse to \\ENV_Water**Ch06** and select **Data_Ch06** ((or \\Database\ Data_Ch06) and click **OK**
6. In the **CP**, select the **Map** and press the **F2** key and change the name to **Editing**

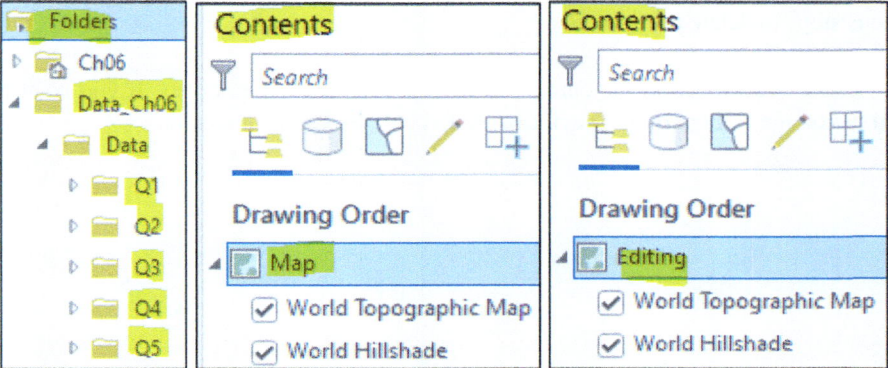

Data Integration and Labeling
7. Open the **Folder** in the **Catalog** pane, expand **Data_Ch06/Data/Q1** and drag **Farm.shp** into the **Map View.**
8. In the CP, r-click **Farm** and point to **Label**
9. The seven polygons of the Farm now are labeled

Delete Function

In this step you are going to delete 2-polygons: **H** and **G** from the **Farm** layer

10. Click the **Edit tab** on the ribbon, in the **Selection** group, click the **Select** button and select polygon **H** located in the northeast of the map.
11. Click **Delete** button in the **Features** group, and click **Yes** to confirm the delete
12. In the **Manage Edits** group, click the **Save** button, click **Yes** to save all edits

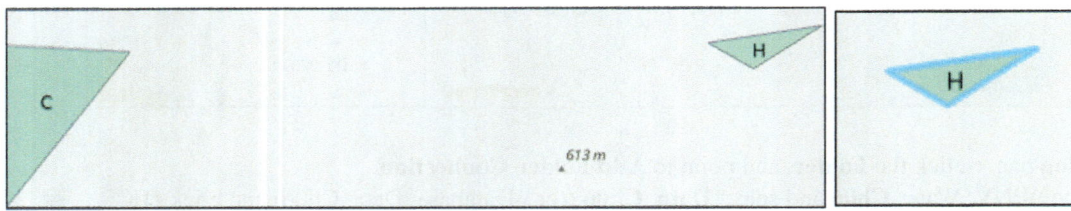

Result   The **H** Polygon has been deleted now.

13. Repeat the previous steps to delete polygon **G**.

Note To remove the selected feature from the current selection click **Ctrl + Select**. If you want to change the **Select** button, click the **Map** tab on the ribbon, in the **Navigate** group click the **Explore** button.

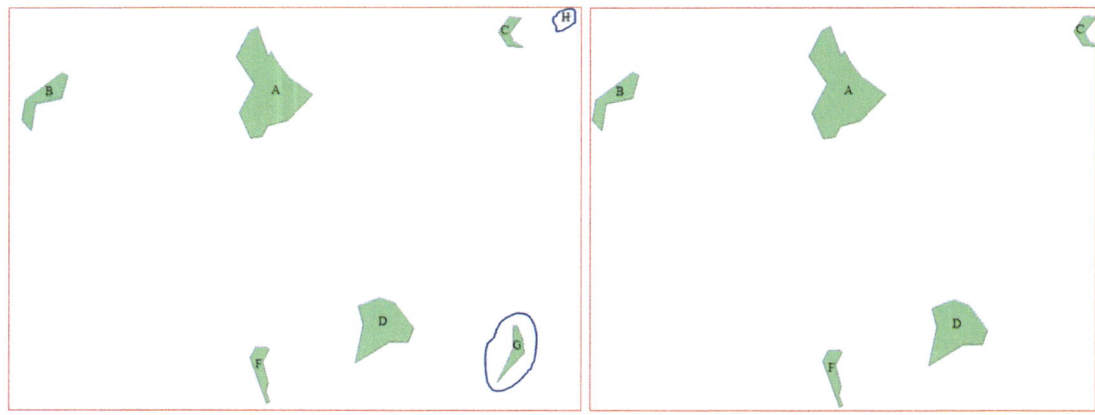

Move Function

In this step, you are going to move two polygons: **D** and **F**.

You have found that the actual location of polygon **D** is between polygon **A** and polygon **B**. The location of polygon **F** is between polygon **A** and polygon **C**. Therefore, you are going to move both polygons **D** and **F** into their appropriate locations.

14. Click the **Edit** tab on the ribbon. In the **Selection** group, click the **Select** button and select polygon **D**.
15. In the **Tools** group, click the **Move** button, polygon **D** will have a dashed line perimeter, and the Modify Features pane also displays
16. Place the cursor inside polygon **D**, drag it and place it between polygons **B** and **A**
17. At the bottom of the map, click **F2** on the **Configure** tool.
18. Repeat the previous step and drag polygon **F** between **A** and **C**.
19. To deselect click the Select icon and click any location

Simple Editing

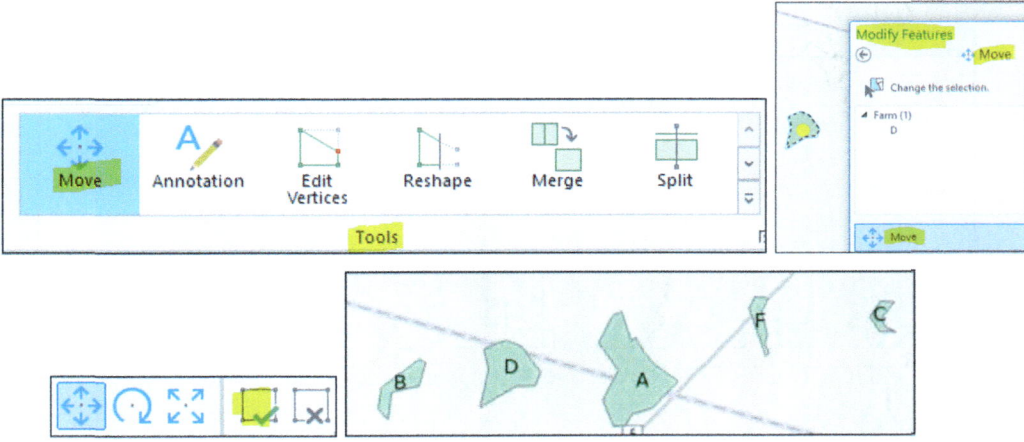

Results   Polygon **D** and **F** moved to their proper locations.

Split Function

Polygon **A** is a large farm land and is used to cultivate one crop (i.e., potato). You have decided to split polygon **An** into two polygons in order to use the land to cultivate two crops: (i.e., potato and tomato).

20. Zoom into polygon **A**
21. You will split polygon A between points 1 and 2, as seen in the figure below.
22. Edit tab/Select polygon A
23. Select **Split** icon in the **Tools** group
24. Click the point corner labeled 2 and then double click the point labeled 1.
25. Click **F2**, the **green check** sign in the **Configure** tool below.

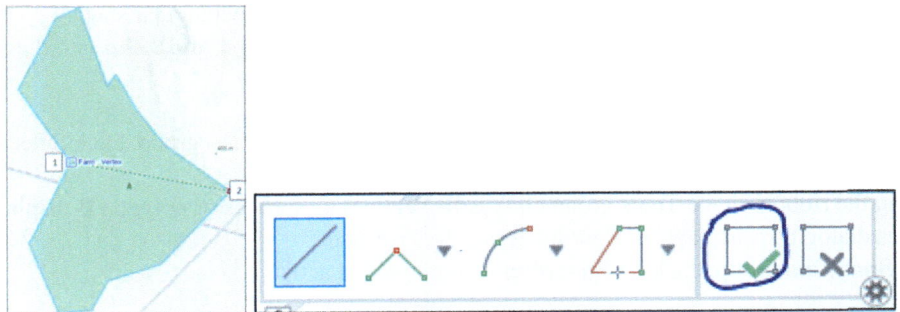

26. In the CP, r-click Farm layer and open the attribute table
27. You will see two records selected and both of them labeled "A"
28. Highlight the last record that is labeled **A**, replace it by typing **M**
29. Then, hit Enter and the **Farm** table
30. Click **Save** icon in the **Manage Edits** group, and click **Yes** to save all edits
31. In **CP**, r-click the **Farm** Zoom To Layer
32. Save your project

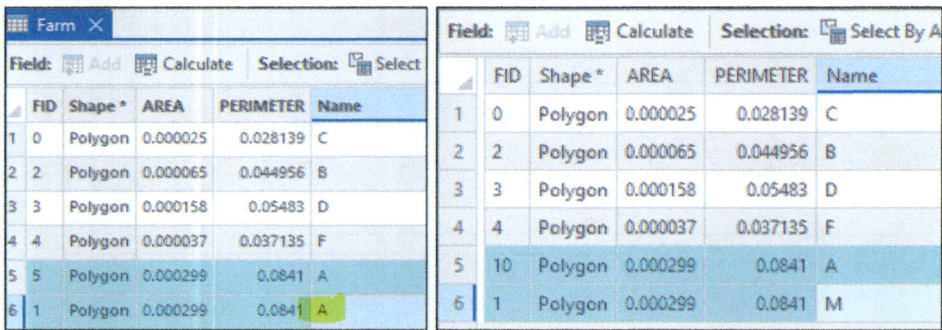

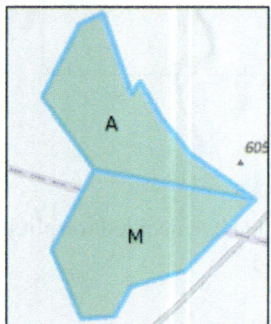

Result   Polygon **A** is now split into two polygons, **A** and **M**.

Reshape Function
The **Farm** layer that has a **B** label is now going to be modified and expanded to make more land available for agriculture. Polygon B is reshaped to fit the size and shape of the **LandB** layer, which has a rectangular shape.
33. Open the **Folder** in the **Catalog** pane, expand **Data_Ch06/Data/Q1** and drag **LandB.shp** into the **Map View.**
34. In CP, right click the **LandB** layer and click Zoom To Layer
35. In CP, click the symbol of the **LandB** layer, then click "Black Outline (1 pt)"
36. In the CP, highlight the **Farm** layer, click **Edit** tab on the ribbon, in the Selection group, click Select button and click on the map the **Farm** land that has **B** label
37. Click **Edit** tab on the ribbon, in the **Tools** group click **Edit Vertices** (7 vertices of polygon **B** displayed)
38. The Edit Vertices dialog box displays the coordinates of the 7 vertices.
39. **Edit** tab, in the **Snapping** group, click on **Snapping** icon to make it active

Note The snapping turn blue means it is active.

40. Click Vertex Snaps to the nearest vertex (3rd icon on Snapping Tool)
41. The polygon will have one node (red) and six vertices (green)

Simple Editing

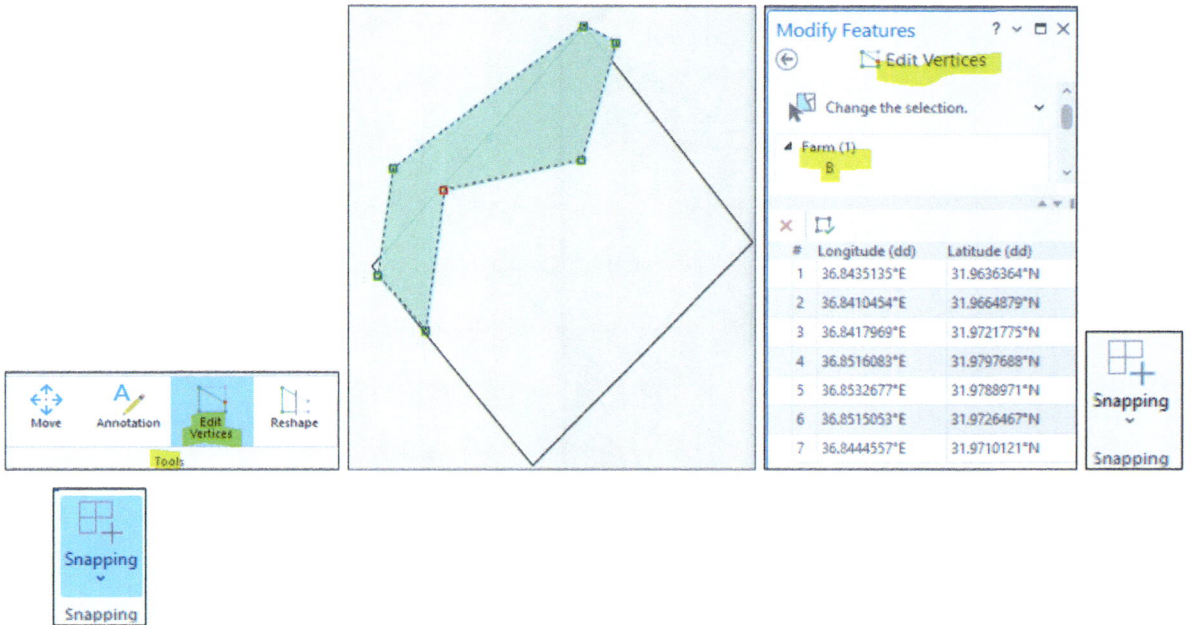

42. Click **Vertex 1** and drag it and place it on the lower left corner of the **LandB**. The vertex will be snapped to the corner (see the sketch below).
43. Click **Vertex 2** and drag it and place it on the upper left corner of the **LandB**.
44. Click **Vertex 5** and drag it and place it on the lower right corner of the **LandB**.
45. Place your Edit Tool above **vertex 3**, right click and delete the vertex
46. Repeat the previous steps and delete vertices 6 and 7.

Note You can also click the **Delete** icon on the configure tool at the bottom of the map.

47. To deselect, click somewhere outside the drawing
48. Click Finish icon on the configure tool at the bottom of the map

Result The farm with label B will be reshaped and will fit the rectangular shape of **LandB.**

49. Click the Save button on the **Manage Edit** tools
50. Close the **Edit Vertices** dialog box
51. In the CP, r-click the Farm and Zoom To Layer and save your project
52. Click the Map tab on the ribbon, and in the Navigate group, click the Explore button to change the cursor from the select shape to hand shape.

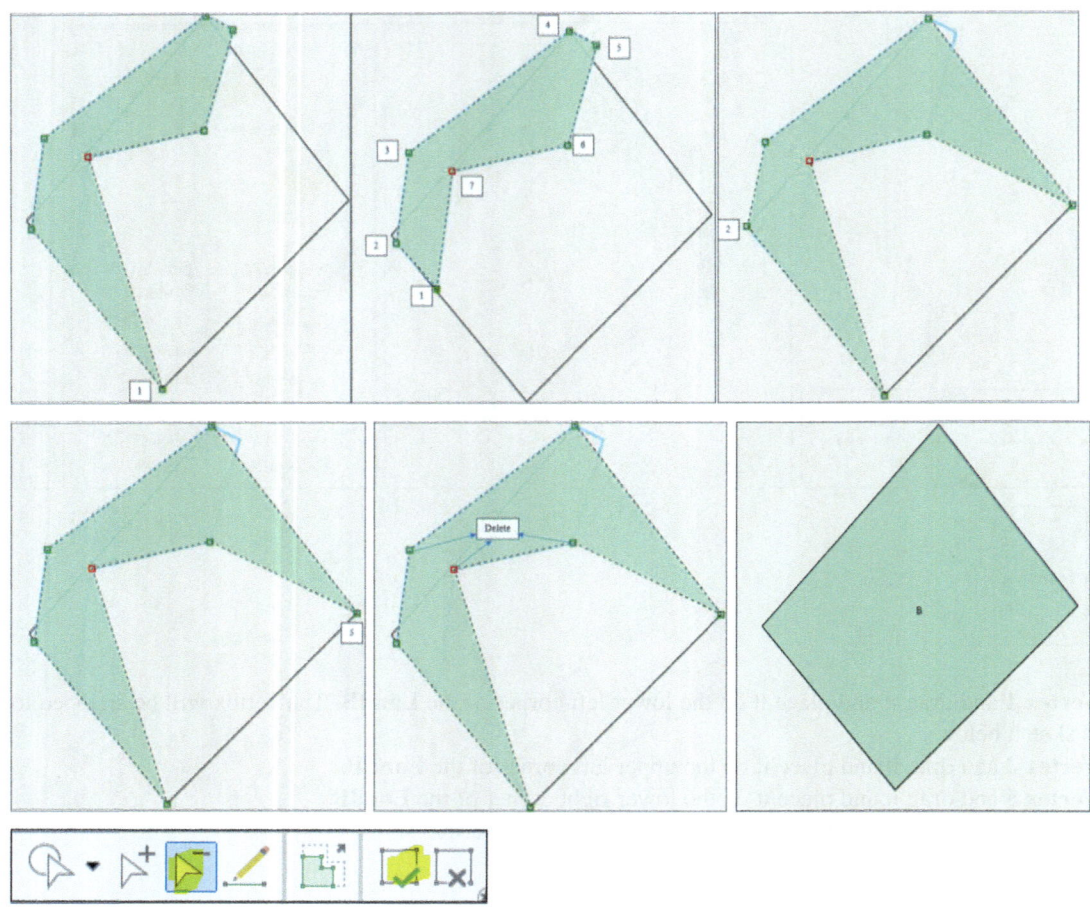

Modify Feature

One of the ways to modify a feature in ArcGISPro is to add features by digitizing and then updating the attribute table.

Scenario 6-2 The geology of Dhuleil was subject to a detailed study to add outcropping formations that were missing in the original map. A group of geologists went to the field, and with the use of GPS, they delineated the outcropping formation that was missing from the original map. As a GIS technician, your duty is to use the new captured data by GPS to update the original geological map "**Geology.shp**".

1. Click **Insert** tab on the ribbon, in the Project group, click **New Map** and call it **Geology**
2. In **Catalog** pane, open Folder\Data_Ch06\Data\Q2, click Ctrl and click **Field_Geology.shp** and **Geology.shp** and drag them into the **Map View**.
3. In the CP, click **Geology.shp** and point to Feature Layer in the ribbon, in the Drawing group, click the drop-down of the **Symbology** button and select **Unique values** under Symbolize your layer by category
4. In the Symbology – Geology dialog box, in **Field 1,** select "GEOLOGY". In Classes Tab, click the drop-down arrow of **More** and uncheck "**Show all other values**"
5. In the CP, click the "**Geology**", click the **Labeling** tab, in **Label Class** group, in the **Field** drop-down window select **Geology**, in the Layer group click **Label** button
6. In the CP, r-click "**Field_Geology**" and point to **Label**

Result   The Geology layer is displayed in two classes (**Basalt** and **Limestone**), and **Field_Geology.shp** is labeled with 1 and 2.

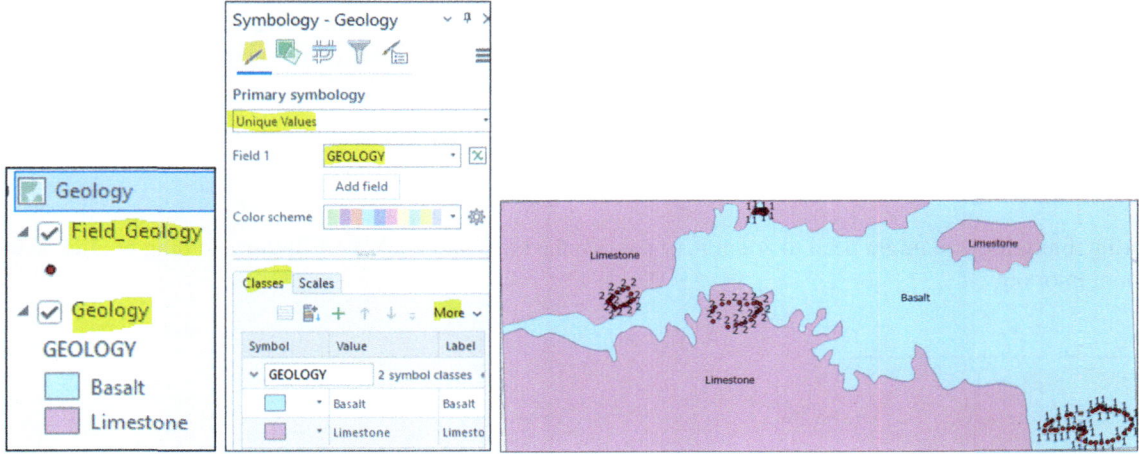

Digitize the Missing Formation

Now, you are going to digitize the **basalt** and **limestone** features using the "**Field_Geology**" layer as a reference. The captured data by GPS are labeled; number **1** represents **limestone**, while number **2** represents **basalt**. The captured limestone crops out above the basalt, and the outcropping basalt is above the limestone.

Digitization of the Limestone Formation
7. In the **CP** highlight **Geology** layer, in the Feature Layer tab on the ribbon, in the **Effect** group, in the **Transparency** window type **60%**
8. In the **Edit** tab on the ribbon, in the **Snapping** group, click the snapping down arrow click on Snapping Toolbar Point snap (1st icon) and Vertex snap (3rd icon) to make them active.
9. Zoom in around number **1** in the lower right corner of the **Geology** layer

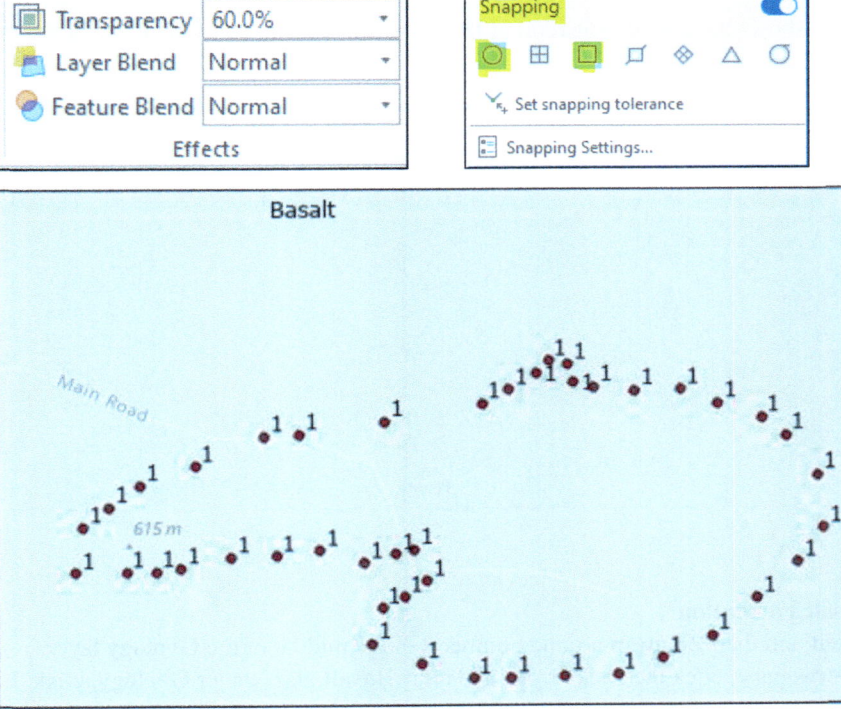

10. Click **Edit** tab, in **Features** group, click **Create** button. The Create Features pane display
11. Click the **Limestone** class under Geology, click the **blue** arrow and then click the **Polygon** symbol
12. Click one point of the **Field_Geology, t**hen click a second point, and continue until you finish all the points, and when you reach the last point, double click to finish digitizing
13. In the configure tool, click the F2 icon
14. In the **Manage Edits** group, click the **Save** button and click **Yes** to **save** all edits
15. Save your project

Note During digitizing you might need to zoom in or pan, on the keyboard click the **C letter** and drag or click **X letter** and drag to zoom in or out

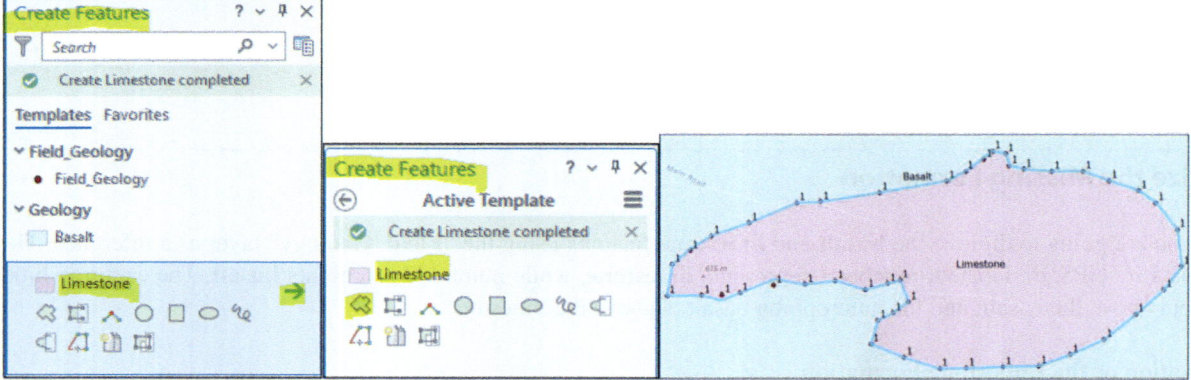

16. Set the **Extent** to **Geology** layer, to set the **Map Extent** refer to **Chap. 2**.
17. In the **Map** tab, in the **Navigate** group, click the **Full Extent** button
18. Zoom into the **limestone** formation that has a **label 1** that is located on the north of the map
19. Repeat the previous steps to finish digitizing the **limestone** formation and save it.

Result The image below shows the final digitization of the **limestone** formation in the north of the map.

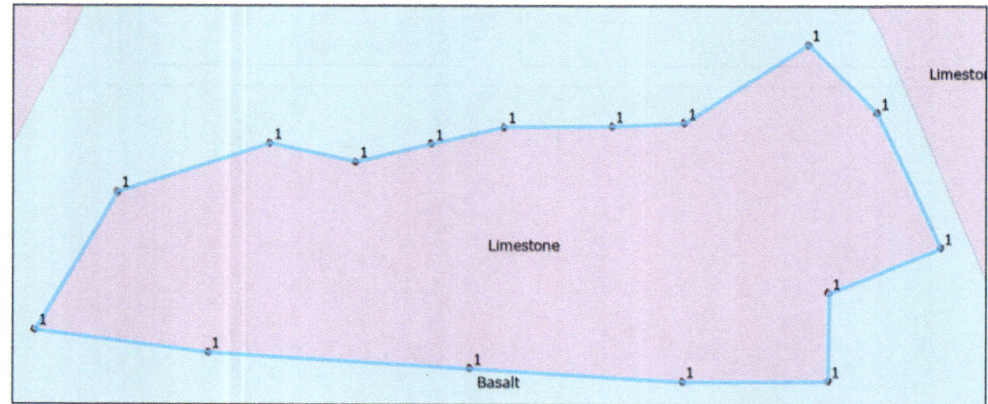

Digitization of the Basalt Formation
20. Click the **Full Extent** and then **Zoom in** around number **2** in the middle of the **Geology** layer.
21. In the **Create Features** pane, click the back arrow, and click **Basalt** class under **Geology**, click the **blue** arrow and then click the **Polygon** symbol

Update the Area and Perimeter fields in the Geology Attribute Table   107

22. Zoom in around the points in the center of the **Geology.shp**
23. Click one point of **Field_Geology.shp**, then click a second point, and continue until you finish all the points. When you reach the last point, double click to finish digitizing.
24. Click the check **green** sign (F2) on the **Configure tool** to apply the digitizing
25. In the **Manage Edits** group, click the **Save** button and click Yes to **save** all edits
26. Save your project
27. Repeat the previous steps to finish digitizing the last **basalt** formation.

Result The image below shows the final digitization of the **basalt** formation in the west of the map.

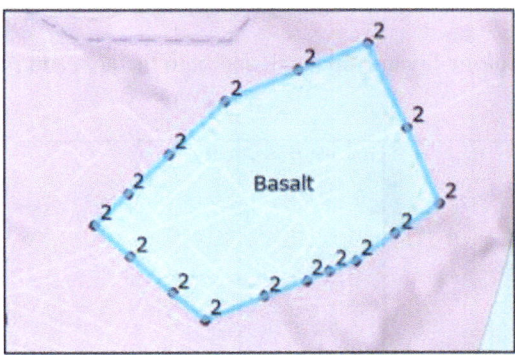

28. In the CP, r-click **Geology.shp** and open the attribute table
29. You will notice that all the limestone and basalt formations are added

Comment When the attribute table of the **Geology** layer, you will notice that the spatial properties (Area and Perimeter) of the Geology layer is not updated in the attribute table. This is because the Geology layer is a shapefile. If the Geologygeological layer has been saved in the geodatabase, the spatial properties will be updated automatically. You will calculate the spatial properties in the next paragraph.

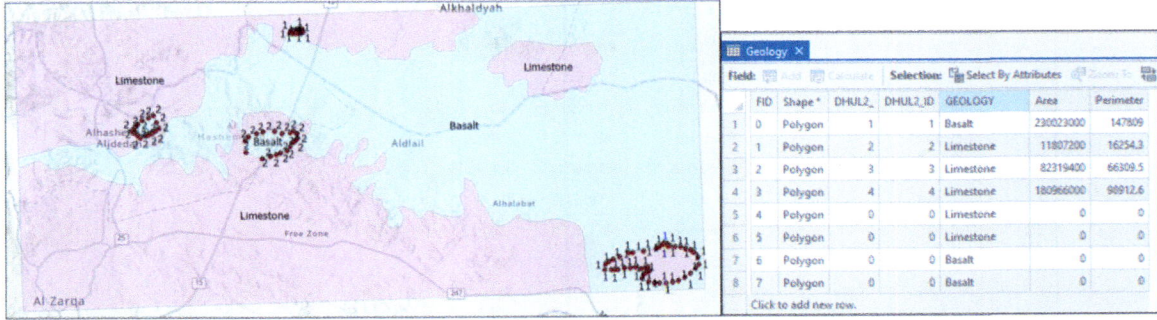

Update the Area and Perimeter fields in the Geology Attribute Table

30. In the CP, r-click **Geology** layer and open the attribute table
31. Right click the **Area** field in the attribute table, and point to **Calculate Geometry**
32. The **Calculate Geometry** dialog box display and fill it as follows:
33. Input Features: **Geology**
34. Field: **Area**
35. Property: **Area** (select from the drop-down)
36. Area Unit: **Square Meters** (select from the drop-down)

37. In the Coordinate System open the drop-down arrow and select "**Geology**"
38. Click OK
39. Right click the **Perimeter** field in the attribute table, and point to **Calculate Geometry**
40. The **Calculate Geometry** dialog box display and fill it as follows:
41. Input Features: **Geology**
42. Field: **Perimeter**
43. Property: **Perimeter length** (select from the drop-down)
44. Length Unit: **Meters** (select from the drop-down)
45. In the Coordinate System: open the drop-down arrow and select "**Geology**"
46. Click OK
47. Close the attribute table of the Geology layer
48. Close the Create Feature pane

Result   The attribute table of the **Geology** layer is updated, and both the area and perimeter of the 4 digitized records are calculated.

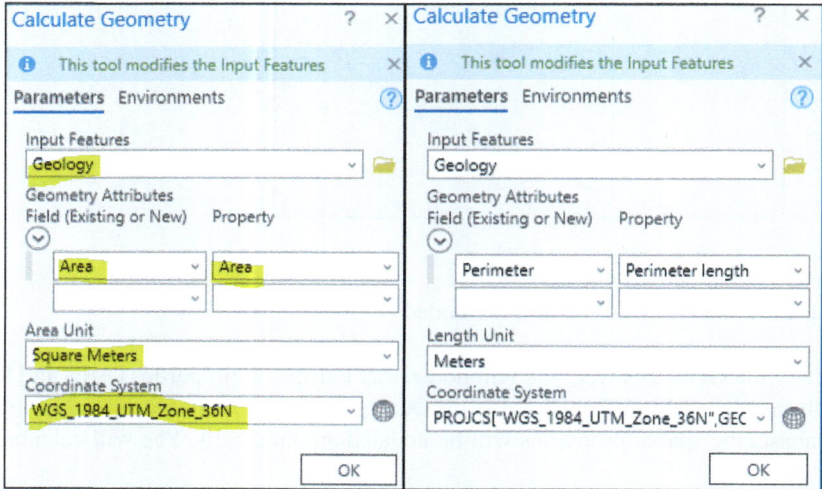

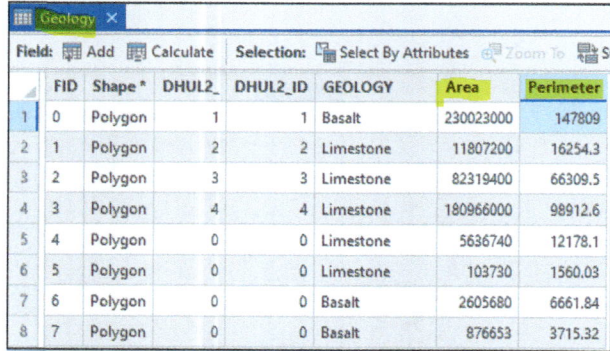

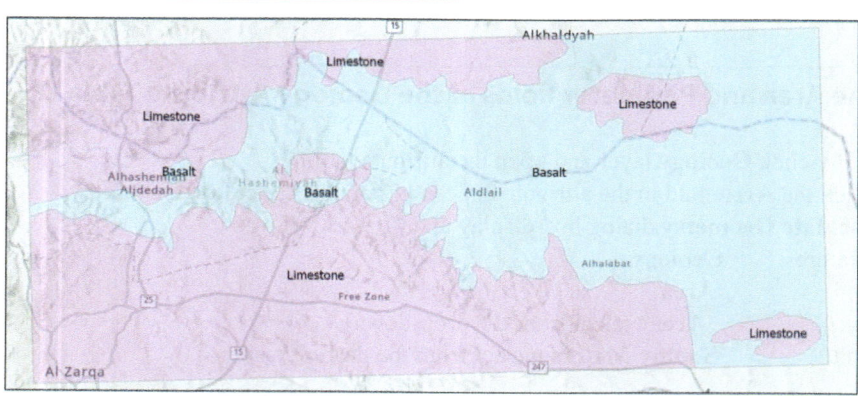

Merge Function

The merge function works with a single layer, and it can group selected records of a line or polygon features into one feature. In this scenario you are going to reduce the numbers of basalt and limestone records into one record each. So instead of having five records for the limestone in the attribute table, you will have only one record; same thing for the basalt features.

49. Click **Insert** tab on the ribbon, in the **Project** group, click **New Map** and call it **Merge**
50. In Catalog pane, under Folder open \\Data\Q2 and drag **Geol_Dhul.shp** into **Map View**
51. Symbolize the **Geol_Dhul** layer based on the GEOLOGY field using the Unique Values as in the above
52. Open the attribute table of the **Geol_Dhul** layer
53. The attribute table of **Geol_Dhul** consists of **8 records**: **5 limestone** and **3 basalt records.**

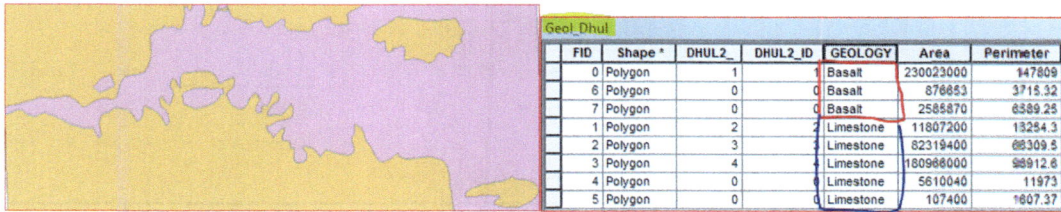

54. Click the **Map** tab on ribbon, in the **Selection** group, click **Select By Attributes**
55. Fill the **Select By Attributes** dialog box as follows:
56. Input Rows: **Geol_Dhul**
57. Selection Type: New Selection
58. Under the SQL make sure that "**Where Geology is equal to Limestone**"
59. Click Apply then OK

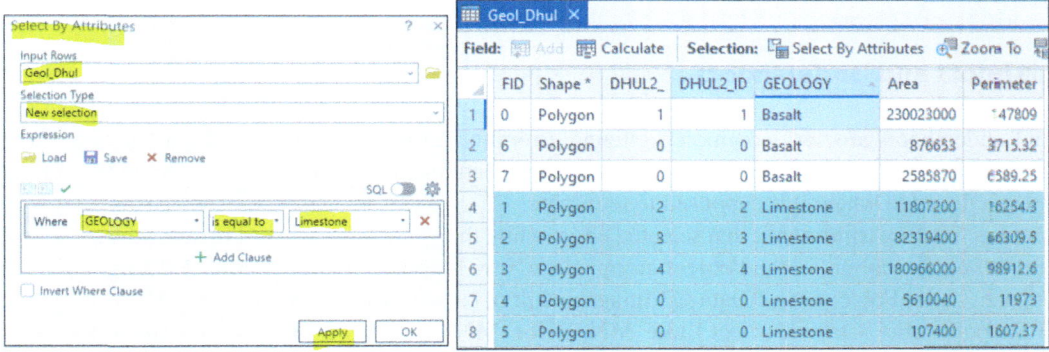

Result   The 5 records of the limestone are selected
60. Click **Edit** tab on the ribbon, in the **Tools** group, and select **Merge** (under construct)
61. The **Merge** pane in the **Modify Features** display
62. Click **Merge** button (at the bottom of the dialog box)
63. Repeat the previous steps and merge the 3 records of Basalt into one layer.
64. Close the attribute table of the **Geol_Dhul.shp** layer

Result   The 5 limestone records and the three basalt records become one record.
65. In the Edit tab, in the Manage Edits group, click the Save button and click Yes to save all edits.
66. Save the project

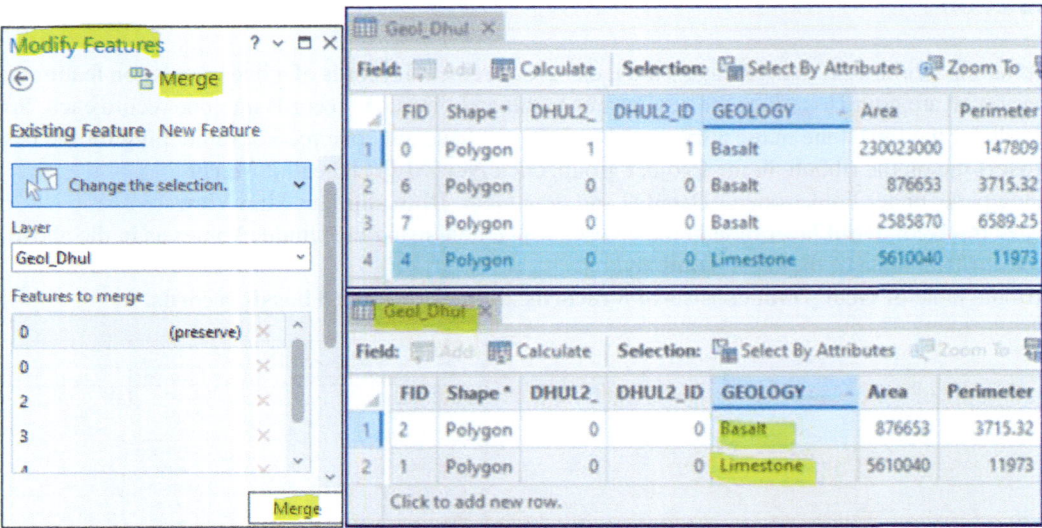

Advance Editing

In this section, guide you will learn how to use the **Advanced Editing** tool in **ArcGIS Pro** to edit existing GIS features, how to fix some common digitizing errors, and how to update the spatial data using some of these advanced tools. The following topics will be covered:

Fixing Overshoots and Undershoots

Overshoots and undershoots are very common digitizing errors that affect the quality of digitized data. Overshoots occur when a line that is supposed to terminate at the edge of another feature extends past the edge. An undershoot occurs when a line does not reach the edge where it is supposed to terminate.

The overshoot is fixed by trimming it to a selected edge, while the undershoot is fixed by extending it to a selected edge.
1. Click **Insert** tab on the ribbon, in the **Project** group, click **New Map** and call it **Advance Editing**
2. In Catalog pane, under Folder open \\Data\Q3\Image and drag **North_Duluth.jpg** into Map View
3. Open \\Q3\Shapefile click Ctrl and select **Lake_MN**, **River_MN**, and **Street_MN** and drag them into the Map View

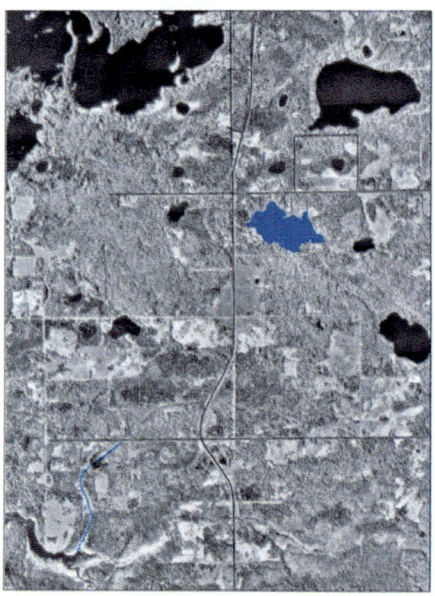

Result   The image of the **North_Duluth.jpg**, **Lake_MN**, and **River_MN** layers display.

Fixing Overshoots and Undershoots

Test Your Knowledge

- Change the symbol of the **Lake_MN** and **River_MN** shapefiles into blue color and the **Street_MN** into dark color.
- Set the **Extent** to **North_Duluth.jpg**

Correct Undershoots of the Street
The undershoot means when the street is not extending far enough to intersect another street. This happens during digitizing, and this could happen if the snap tolerance is not set correctly.

4. In CP, Highlight **Street_MN** and Zoom in around the street located in the northeast of the map
5. Select the undershoot street (Map tab, Selection group)
6. Click **Edit** tab on the ribbon, in **Selection** group click **Select** button and click the undershoot street to select it, in the **Features** group click the **Modify** button
7. The **Modify Features** pan display
8. Under **Reshape** click **Extend or Trim** tool
9. The **Extent or Trim** pan display, select **Extend** tool
10. Click below the undershoot on the **street** that you want the undershoot to connect to it
11. Click again the other undershoot to extend the second undershoot
12. Do not close the **Modify Features** pane

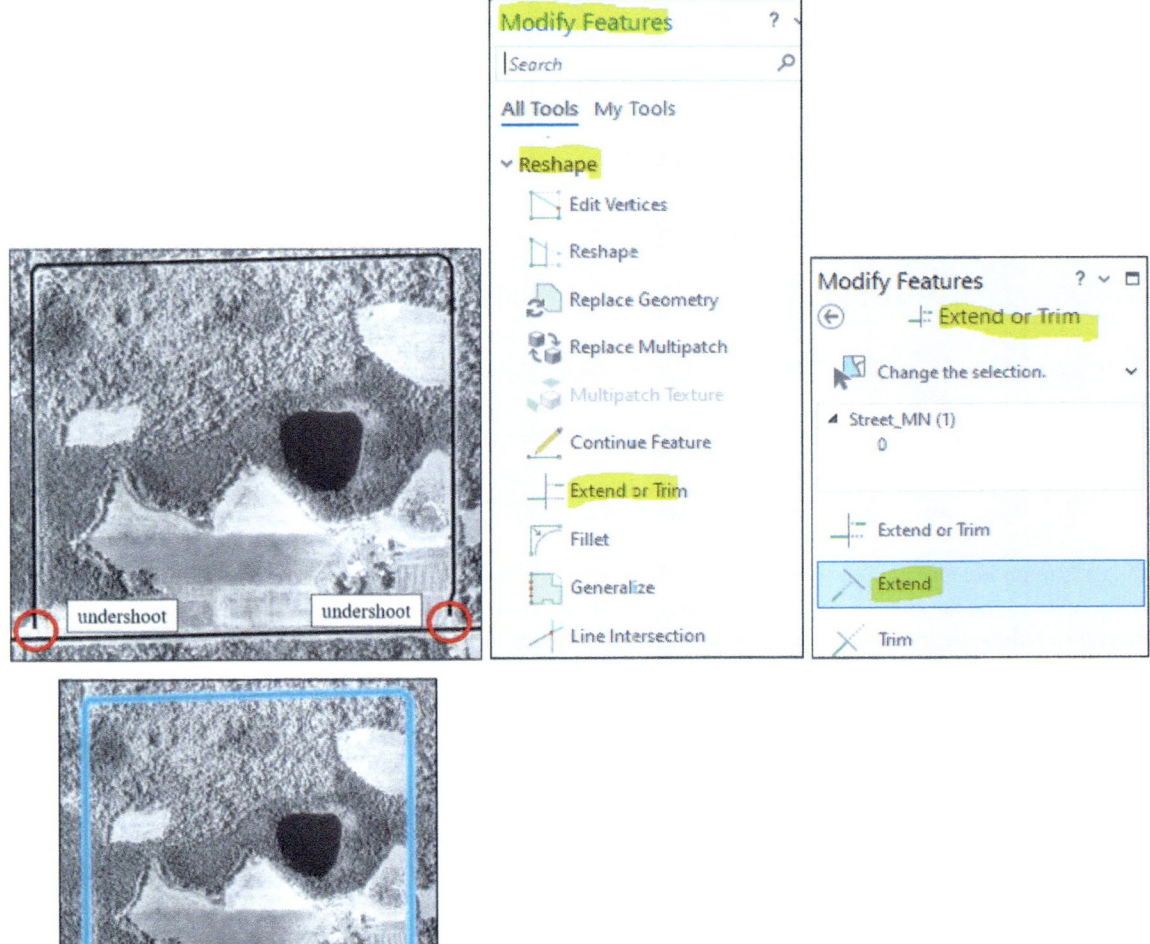

Result   The undershoot street in both locations will be connected to the street below it.
13. Zoom to the full extent and clear the selected features

Correct Overshoots of the Street

Overshoot means that a street past its intersection with another street. This happens as explained in the undershoot.

14. In CP/Highlight **Street_MN** and Zoom in around the street located in the upper middle section of the image, you will notice that the overshoots where two small streets, inside the loop, extend beyond the straight street
15. Click **Edit** tab on the ribbon, in **Selection** group click **Select** button and click the **street** (overshoot) that needs to be trimmed
16. In **Modify Features,** under **Reshape,** under **Extend or Trim** tool, click **Trim** tool
17. Click the street that the overshoot street will connect to.
18. Both overshoots are trimmed off
19. Zoom out to the full extent and clear the selected features
20. Save the Project

Generalize a Stream Feature

Generalizing reduces the number of vertices that describe a feature, so the feature's shape is somewhat less precise. The step will allow you to reduce the number of vertices that describe the river (**River_Mn.shp**).

1. In CP, highlight the **River_MN**, and r-click it and Zoom to Layer
2. Click **Edit** tab on the ribbon, in **Selection** group click **Select** button and click the **river**
3. Click the arrow back in the **Modify Features** pane, under **Reshape**, select the **Generalize** tool
4. Under **Method** select "**Smooth**"
5. Under **Value** type **10 ft** for the **Maximum Allowable Offset** and click **Generalize**

Result The number of vertices has been significantly reduced. If you zoom in closer at any part of the stream, you can see a slight effect on the shape of the stream.

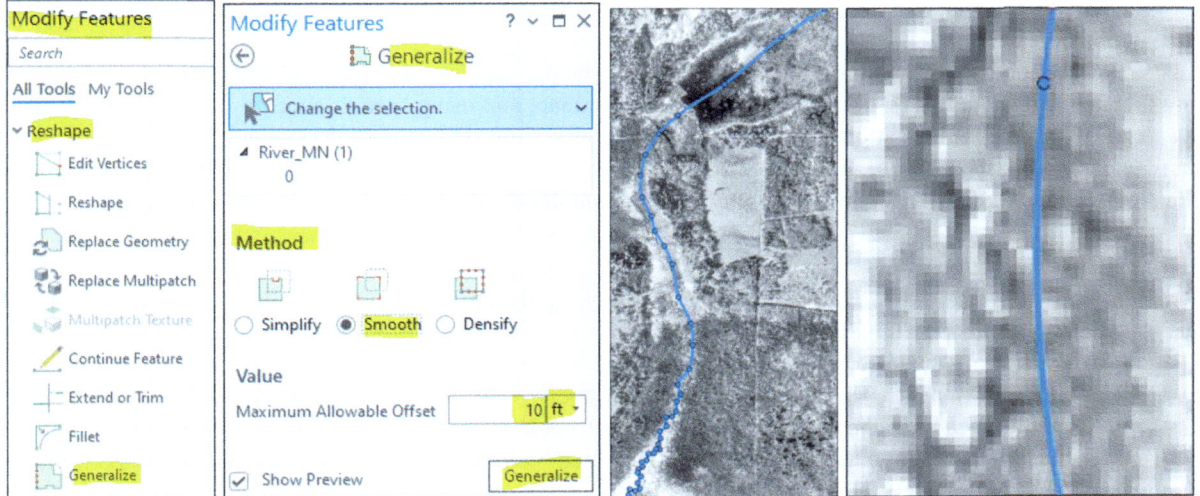

Generalize a Lake Feature

The outline of the lake next to the stream feature appears rough because it was digitized with too few vertices. To improve the appearance of the lake, you will use the Smooth tool.

6. In CP, r-click **Lake_MN** and Zoom to Layer
7. Click the arrow back in the **Modify Features** pane, under **Reshape**, select the **Generalize**
8. Click inside the lake, the Sunshine Lake will be selected
9. Under **Method** select "**Smooth**"
10. Under **Value** type **10** for the **Maximum Allowable Offset** and click **Generalize**
11. Make sure the "Show Preview" is checked
12. The border of the lake looks smoother now.
13. Close Modify Features pane
14. Zoom out to the full extent
15. Save the **Project**

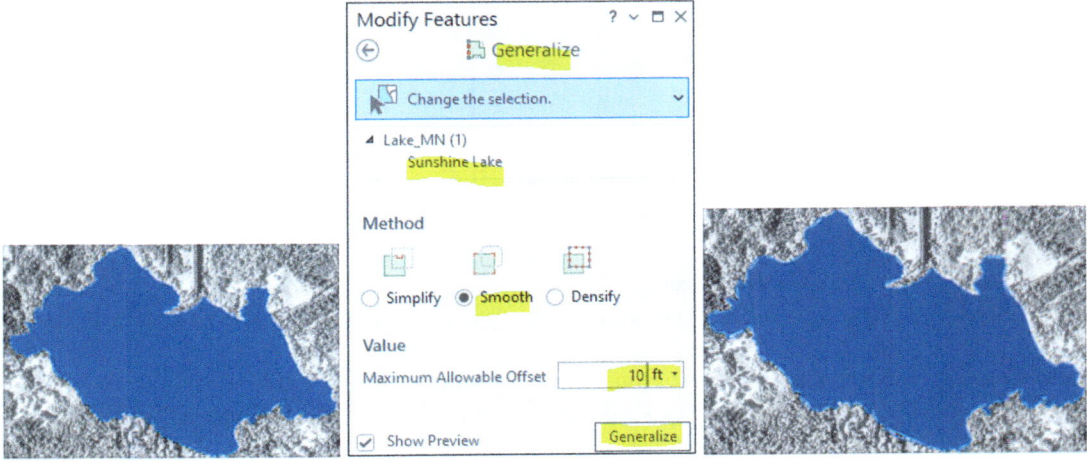

Topological Editing Using Geodatabase

Topology describes the way interrelated features are organized and connected. In ArcGIS Pro, it is implemented as a collection of spatial rules that are used to constrain active edits or audits and maintain the correctness of features based on their position relative to other features. There are two types of topology in ArcGIS Pro: topological editing and geodatabase topology. Both methods help to manage and maintain the accuracy of the feature data with respect to coincidence, adjacency, containment, and connectivity.

In this section, Geodatabase topology will be used. Geodatabase topology includes creating a topology for a dataset, assigning features and spatial rules, validating the features in a map, and using specific tools to fix errors and mark exceptions. Geodatabase topology rules allow you to define relationships between features in the same feature class or subtype or between two feature classes or subtypes. The created rules depend on the spatial relationships that GIS users wish to monitor for the feature classes that participate in the topology and the topology error fixes for points, lines, or polygons.

Fix Fault System Using Topology
The catchment area of Wadi Andam-Halfyan in the Izki region of **Oman** has many wells. The wells are used mainly for domestic water supply and agriculture. The catchment area has revealed a structural style that may have had a profound effect on the geomorphological and hydrological setting of the area. In the catchment, 2 major fault trends have been observed; **Fault A** and **Fault B** are oriented northeast and north, respectively, while **Fault C** is oriented northwest. A detailed field study of the geological structure of the catchment revealed that **Fault A** consists of only one identical fault system and not broken **5-line fault segments**. Thus, **Fault A** should be corrected by joining the five-line segments into **one fault** system. You are going to use the **Geodatabase Topology** to rectify the fault on the map.

Scenario 6-3 **Fault** feature **A** contains gaps between the segments, and the task is to remove the gaps using the **topology** technique in the Geodatabase environment. To perform the task and fix the errors, you will implement two topology techniques.

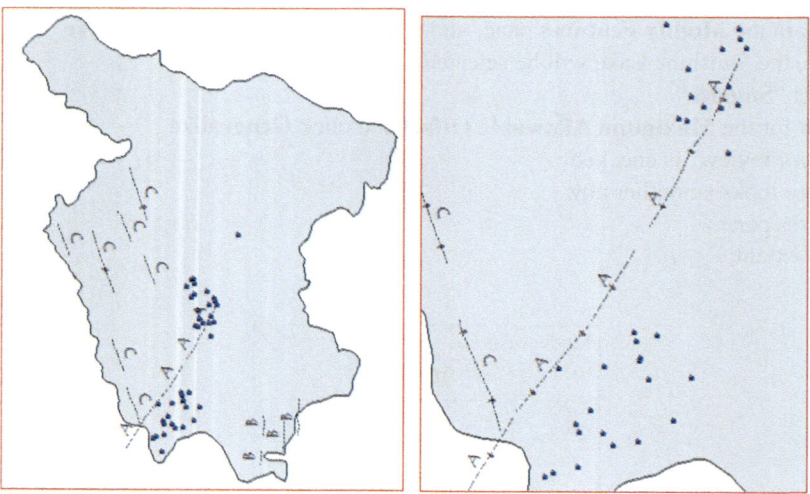

Create Feature Dataset in the File Geodatabase
To use the Geodatabase topology to fix the errors in the database, you must build a feature dataset and set the proper rule to fix the error.
1. Click **Insert** tab on the ribbon, in the **Project** group, click **New Map** and call it **Fault**
2. In Catalog pane, under Geodatabase, right click **Ch06.gdb**, point to New and select **Feature Dataset**
3. In the **Feature Dataset** dialog box fill it as follows:
4. Output Geodatabase: **Ch06.gdb**
5. Feature Dataset Name: **Water**
6. Click the globe (Select coordinate systems), the Coordinate System dialog box open, click the drop-down arrow of Add Coordinate System, Import Coordinate System
7. Browse to **Q4** under **Folder\Data_Ch06\Data** and select **Fault**
8. OK and click Run

Build the Topology and Set the Rule

Result The **Water Feature Dataset** is created and registered in UTM Zone 40 N, and the datum is WGS84.

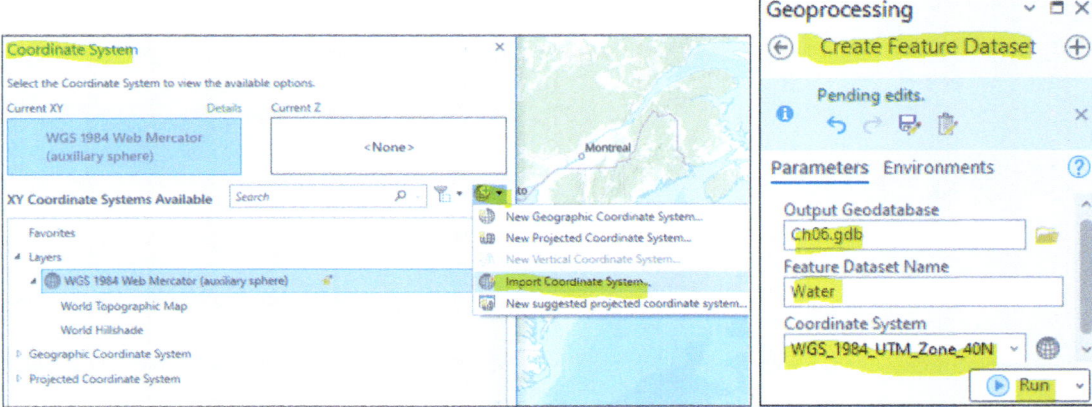

9. In **Catalog** pane, r-click on the **Water** feature dataset, point to **Import** and click **Feature Class(es)**
10. The **Feature Class to Geodatabase** dialog box opens and fills it as follows:
11. Input Features: **Fault** (from **Q4**)
12. Output Geodatabase: **Water**
13. Run
14. Drag the Fault from the Water Feature Dataset into a Map View
15. In the CP, uncheck the World Hillshade
16. Click the symbol of the Fault in the Symbology pane in **Gallery**, under ArcGIS 2D select **2.0 Point** line symbol, in the Properties tab, select red color and Line width 2
17. Click Apply and close the Symbology pane

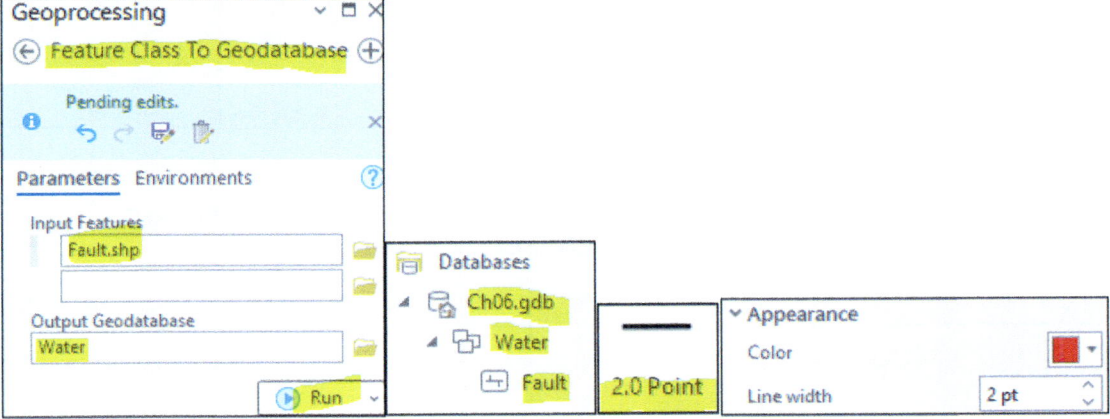

Result The **Fault** feature class is displayed in the CP in red.

Build the Topology and Set the Rule

18. In Catalog pane, under Ch06.gdb, right click the **Water** Feature Dataset, point to **New** and select **Topology**
19. In the create Topology Wizard in the Define left pane fill it as below
20. Topology Name: **Fault_Topology**
21. XY Cluster Tolerance: **0.01 m**
22. Under **Feature Class**, check the **Fault**, accept the default and click **Next**
23. In the **Add Rules**, click on "**Click here to add a new rule**"

24. Select **Fault** from the **Feature Class 1** drop-down arrow
25. From the **Rule** drop-down arrow, select "**Must Not Have Dangle (Line)**"

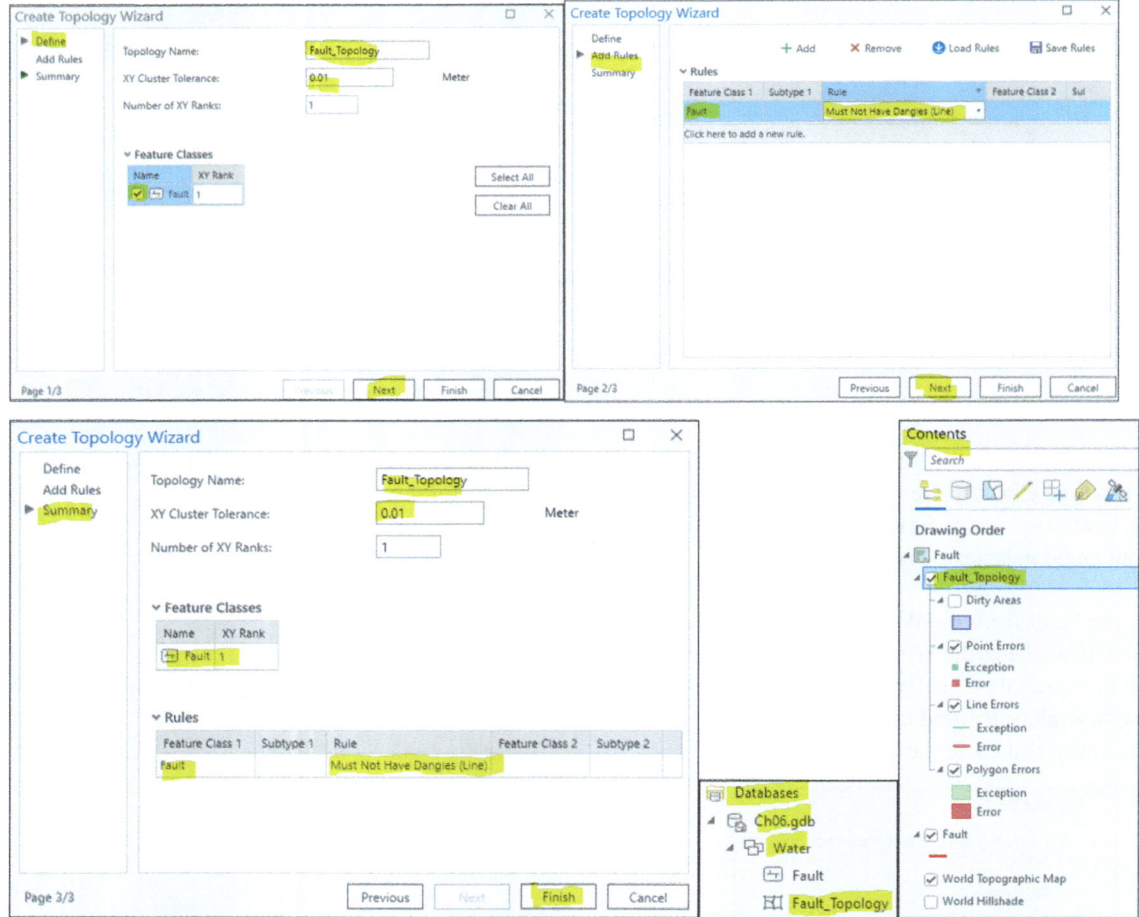

26. Next, then after reading the Summary, click **Finish**
27. In the **Catalog** pane, under the **Water** feature dataset, r-click **Fault_Topology** and click **Properties**
28. In the **Topology Properties**: **Fault_Topology**, in the **Error** tab, click **Generate Summary**
29. The number of **Fault** error display **32**, click OK

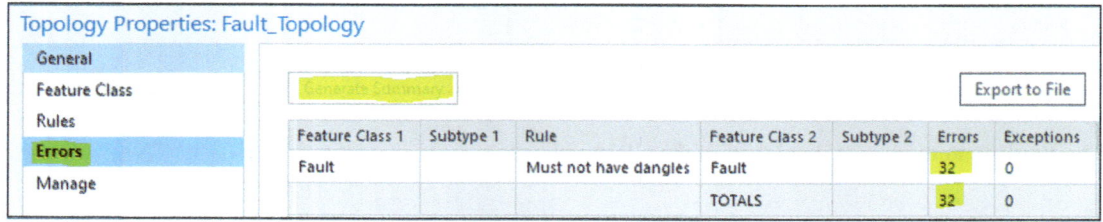

30. In the **Catalog** pane, r-click **Fault_Topology**, and click **Add To Current Map**

Result The **Fault_Topology** is added to the **Water Feature Dataset** and added to the **CP**.

Label the Fault Using Two Fields

It is helpful in this exercise to label the Fault based on the values of two fields. This can be done using label expressions. Displaying the OBJECTID and Name in the labels quickly reveals the name of the Fault and its number that you need later in this exercise.

31. In the CP, r-click the **Fault**, and select **Label** to turn on labeling
32. In the CP, r-click the **Fault**, and select **Labeling Properties**
33. In the **Label Class** pane, click the Language drop-down list, and select **VBScript**.
34. Under Expression, clear the expression box
35. Under **Fields**, double click **Name** to add the Name field to the Expression box, and after the [Name], click space and type **Ampersand** (&), then type **Quote** ("), then click space, then click **Comma** () then click space, then click **Quote** (") and type **Ampersand** (&) and click space.
36. Under Fields, double-click OBJECTID to append the OBJECTID field to the expression.
    [Name] &", "& [OBJECTID]
37. Click the Verify button to validate the expression. A message popped up sating "**Expression is valid**"
38. Click Apply to run the script.

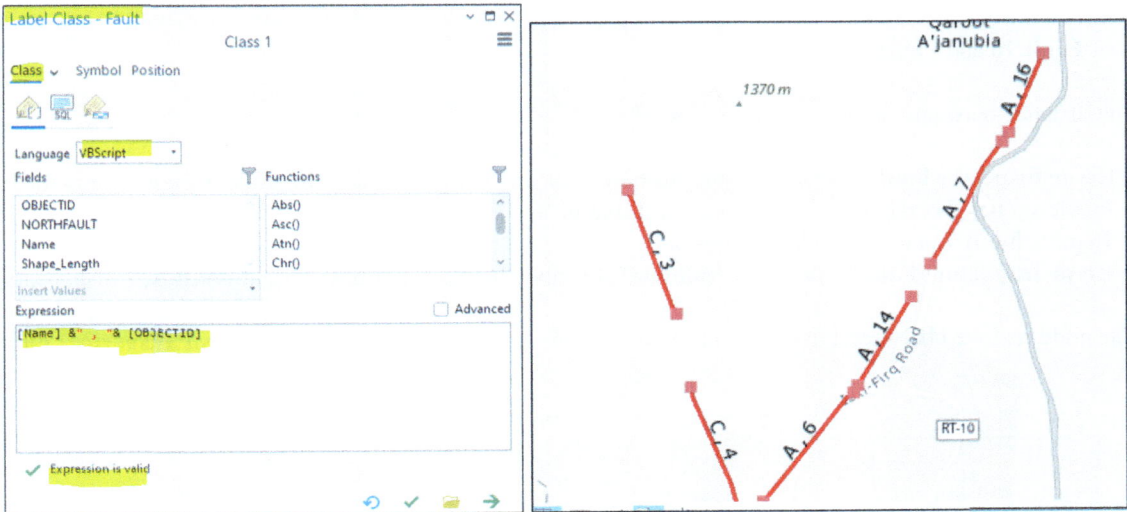

Result The Fault shows each segment labeled based on the values from the **OBJECTID** and **Name** fields.

Validate and Fix Geodatabase Topology

This step validates and fixes geodatabase topology errors. It contains a table viewer for viewing and filtering errors, tabs for inspecting feature geometry, and tools for managing exceptions.

39. Click the **Edit** tab, in the **Manage Edits** group, select from the **No Topology** drop-down/select **Fault_Topology (Geodatabase)** and click **Error Inspector**
40. The **Error Inspector: Fault** table appear
41. In the **Error Inspector: Fault**, click the **Validate** tab

Result The table displays **32 fault** dangling nodes at the perimeter of the three **fault systems** (A, B, and C) in the current map that are validated against all topology rules defined in the **Fault Topology Geodatabase**. The goal is now to fix the dangle in **Fault A** only

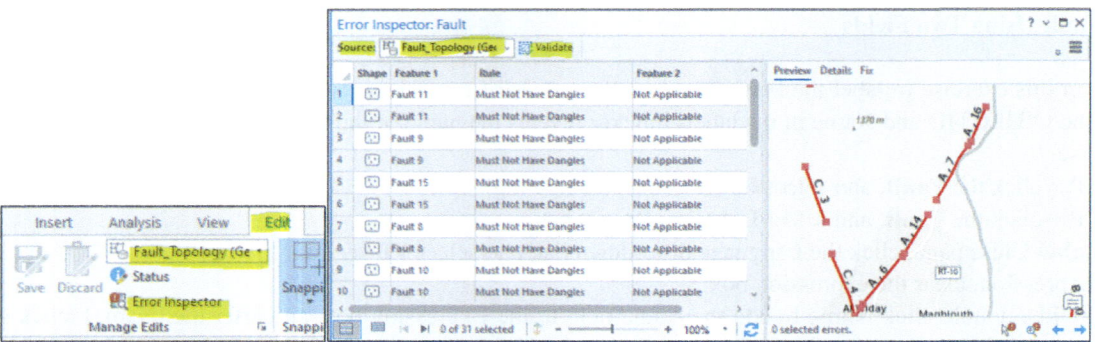

First Approach: Fault Topology (Geodatabase)
Fix Dangle Between Faults 7 and 16

42. In the **Map View** magnified north of **Fault A**, in the upper two segments of the **Fault**, between the two fault segments that are labeled **A,16** and **A,7**, to measure the distance between the two dangles
43. In the **Map View**, click **Map** tab, in the Inquiry group, click the **Measure** button measure the distance between the two nodes of **Fault 16** and **Fault 7**

Result The distance between the two dangles is **150.37 m.**

44. In the **Error Inspector Fault** table, highlight **Fault 16**, the lower dangle of **Fault 16** turns **orange** color in the Preview
45. In the Preview, click **Fix** tab, click **Snap** and type **150.5** m, and Enter
46. **Fault 16** joins **Fault 7,** and one node displays only
47. In the **Error Inspector: Fault** table click "**Validate**", the node between **Fault A, 16** and **Fault A, 7** disappears

Result The node removed between the two Fault segments and Fault 7 disappears from the **Error Inspector: Fault** table

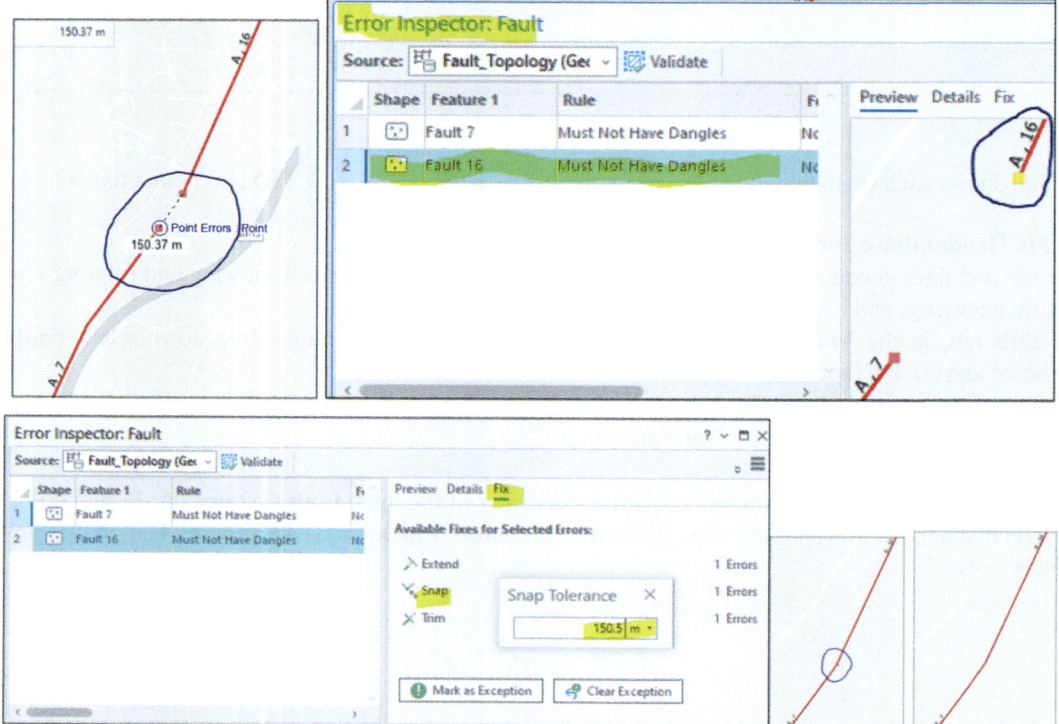

Second Approach: Map Topology

Fix Dangle Between Faults 7 and 14

48. The Map View zooms to the next segments of **Fault A** between the **A, 7** and **A, 14** segments that need to be fixed.
49. In the **Error Inspector: Fault** table, **Fault 7** and **Fault 14** are displayed.
50. As in the previous step, measure the distance between the two nodes of **A,7** and **A,14**.
51. The distance is **509.12 m**.
52. Highlighting **Fault 7** in the **Error Inspector Fault** table, the dangle of **Fault 7** turns **orange** color.
53. In the Preview, click **Fix** tab,509.2click **Snap** and type **509.2** m and Enter
54. **Fault A, 14** joins **Fault A,7** and one node displays only
55. In the **Error Inspector: Fault** table click "**Validate**", the node disappears

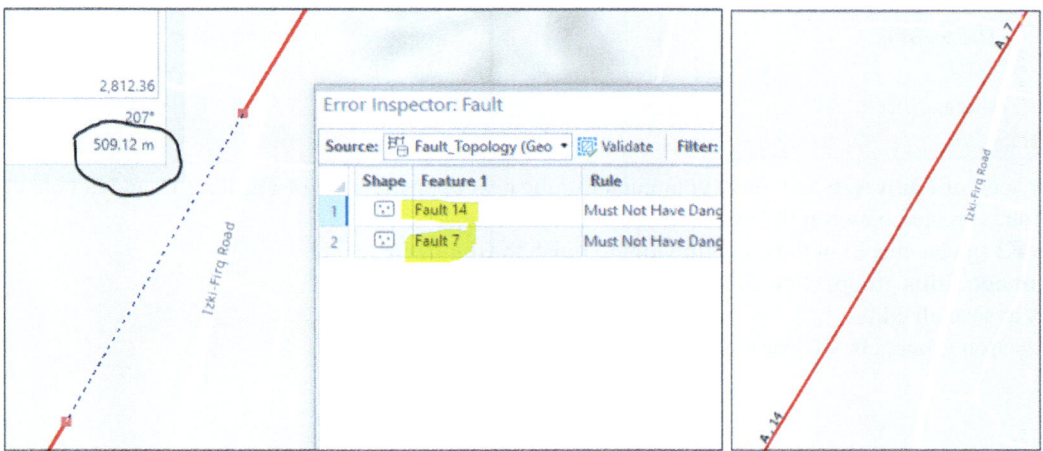

Fix Dangle Between Fault 6 and Fault 14
56. In the Map View zoom in to Fault A between A, 14 and A, 6 segments that need to be fixed
57. In the **Error Inspector: Fault** table **Fault 6** and **Fault 14** display
58. Measure the distance between the two nodes as in the above (**113.97** m)
59. In the **Error Inspector Fault** table, **Fault 14** is highlighted, and the dangle of **Fault 14** turns **orange**.
60. In the Preview, click **Fix** tab, click **Snap** and type **114** m and Enter
61. **Fault 14** joins **Fault 6,** and only one node displays
62. In the **Error Inspector: Fault** table click "**Validate**", the node disappears
63. Click Edit tab on the ribbon, click the Manage Edit group and click Save button
64. Save your project

Second Approach: Map Topology

Fixing Dangles Between Fault 6 and Fault 15
This is a second approach to remove the gaps between the nodes along the fault segments based on the map approach. Topological editing is an editing mode that constrains coincident geometry to an ordered graph of topologically connected edges and nodes. It requires no setup and operates only on visible features that are editable.
   To proceed, you need to make sure that you are in editing mode.
65. Click **Edit** tab on the ribbon, in the **Snapping** group, click the Snapping draw-down arrow and make sure the Point snap (1st icon) is active
66. Click **Edit** tab on the ribbon, in the **Manage Edits** group, select **Map Topology** from the drop-down arrow
67. Zoom in between the 2-nodes of the **Fault A, 6** and **Fault A, 15** that you want to fix the dangle
68. Click **Create** icon in the **Features** group
69. In the Create Features dialog box, select the **Fault**, click the **blue** arrow, and click the Line symbol from the **Active Template**

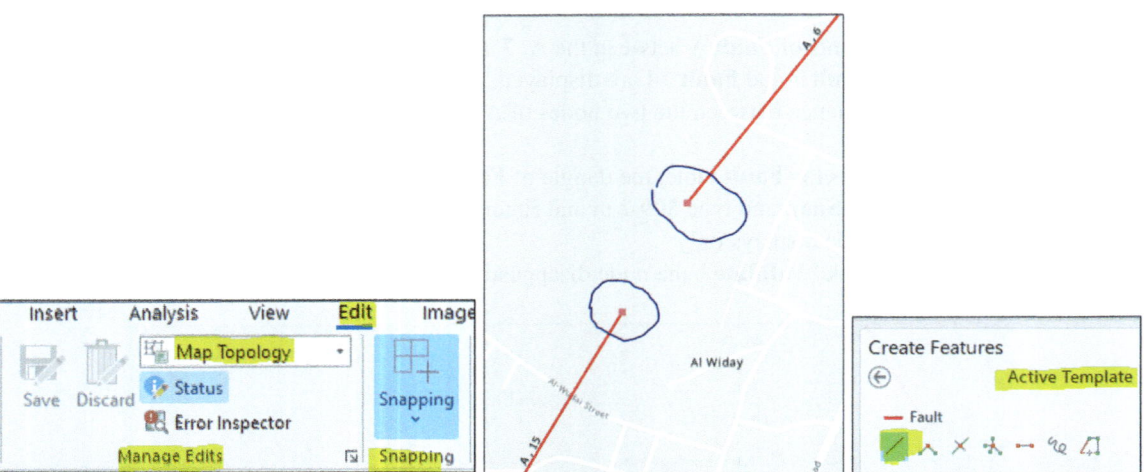

70. Click the node of **Fault A, 6** and move your cursor to the next opposite node of **Fault A, 15** and double click it.
71. A cyan line is created between the two nodes
72. Click the **F2** (green check) in the configure tool to finish the digitizing
73. In the **Manage Edits** group, click the **Save** button
74. Click Yes to save all edits
75. In the Selection group, click Clear to unselect the drawing segments

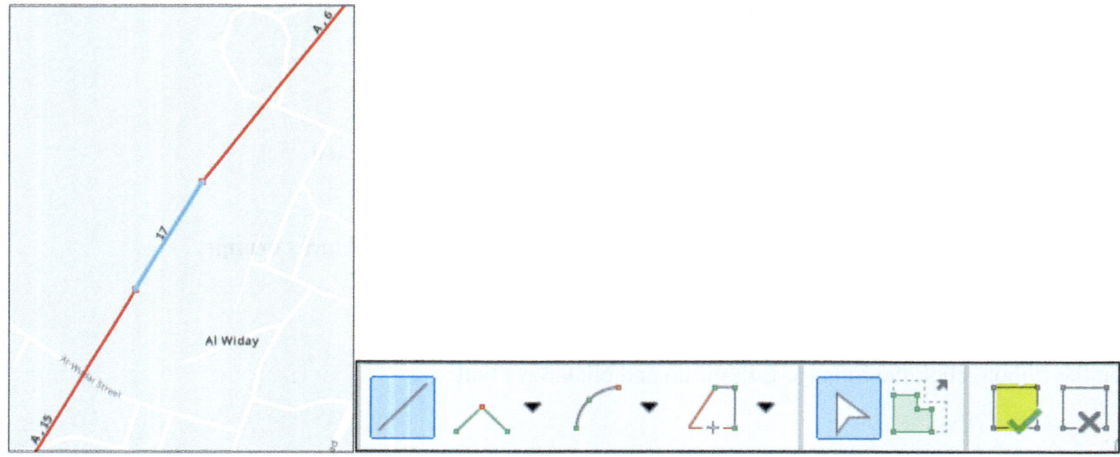

Merge the Segments of Fault A

The Topology tool fixes the dangle errors but does not make the different segments of the fault one record in the attribute table. To make all the segments one record, the **Merge** tool is used.

76. Click **Map** tab on the ribbon, click **Selection** group, and **Select By Attributes**
77. In the **Select By Attributes** dialog box, fill it as follows:
78. Input Rows: Fault
79. Selection Type: New Selection
80. where Name is equal to A.
81. Click Apply

Result Fault A selected in the map view.

82. Click **Edit** tab on the ribbon, in the **Features** group, click **Modify** button
83. In the **Modify Features**, under **Construct**, click Merge tool
84. In the **Modify Features** dialog box, click **Merge** at the bottom of the box

Second Approach: Map Topology 121

Result The five segments of the fault become one line, and you can open the attribute table and see the merge.

85. In the **Manage Edits** group, click the **Save** button
86. Click **Yes** to save all edits
87. In the Selection group, click **Clear** to unselect the drawing segments

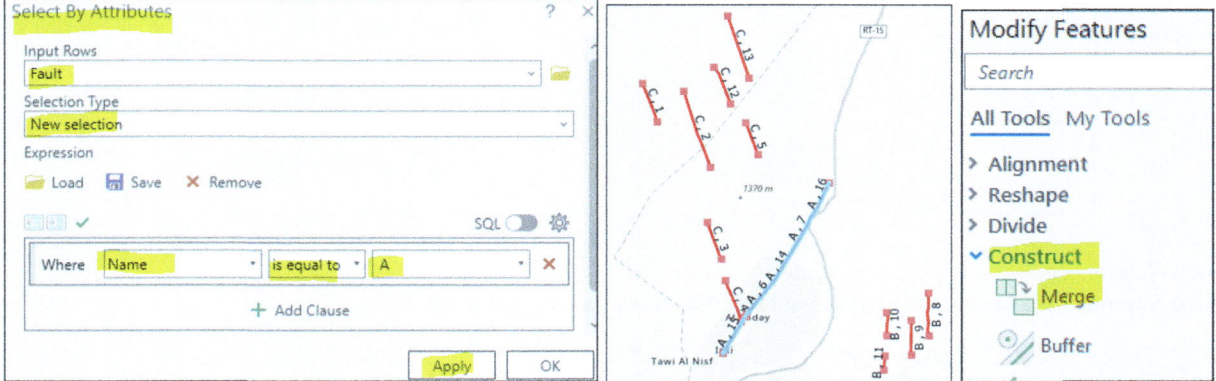

Practice Correcting the gaps along the **Fault C** segment to fix the errors.

Fix Feature Within Feature Class Soil Polygon
These topology rules will define the relationship between features in the same feature class or subtypes. In this exercise, you are going to work with the soil polygon feature class.
1. Click **Insert** tab on the ribbon, in the **Project** group, click **New Map** and call it **Soil**
2. In Catalog pane, under Geodatabase, right click **Ch06.gdb**, point to **New** and select **Feature Dataset**
3. In the **Feature Dataset** dialog box, fill it as follows:
4. Output Geodatabase: **Ch06.gdb**
5. Feature Dataset Name: **Soil**
1. Click the globe (Select coordinate systems), the Coordinate System dialog box open, click the drop-down arrow of **Add Coordinate System**, Import Coordinate System
6. Browse to **Q6** from \\Data\Q6 and select **Soil. shp** from **Q6** folder
7. Click Ok and the Run

Result The **Soil Feature Dataset** is created and registered in **WGS 11984 UTM Zone 40 N.**

8. In **Catalog** pane, r-click on the **Soil** feature dataset, point to **Import** and click **Feature Class(es)**
9. The **Feature Class to Geodatabase** dialog box opens and fills it as follows:
10. Input Features: **Soil** (from **Q6**)
11. Output Geodatabase: **Soil**
12. Run
Building the Topology
13. In **Catalog** pane, right click on **Soil** Feature Dataset, click **New**, and select **Topology**
14. In the Create Topology Wizard, fill it as follows:
15. In the Define pane, Topology Name: **Soil_Topology**
16. XY Cluster Tolerance: **0.01 m**
17. Under Feature Classes, check   **Soil**
18. Click Next
19. In the **Add Rules**, "*Click here to add a new rule*" select **Soil** under **Feature Class 1**

20. Under **Rule** drop-down arrow select "**Must Not Have Gaps (Area)**"
21. Again, "*Click here to add a new rule*" select **Soil** under **Feature Class 1**
22. Under **Rule** drop-down arrow select "**Must Not Overlap (Area)**"

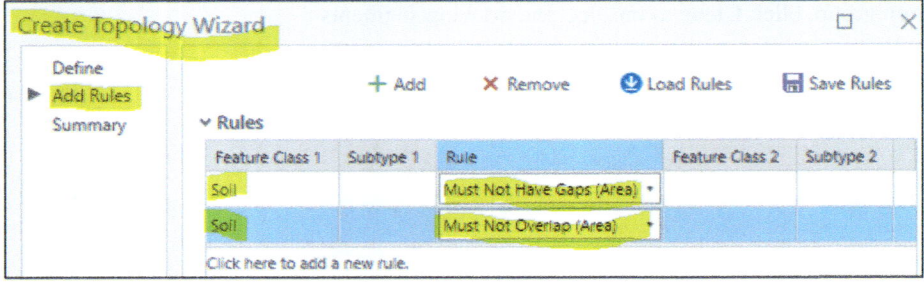

23. Click Next
24. In the Summary make sure every think you set look good, then click **Finish**

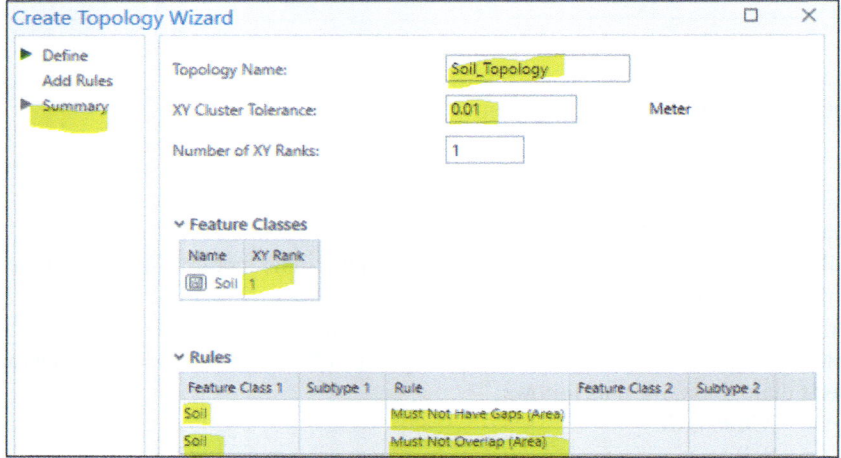

25. In **Catalog** pane, under **Soil** Feature Dataset, r-click **Soil_Topology** and **Add To Current Map**
26. Click **Edit** tab on the ribbon, in **Manage Edits** group, select Soil_Topology (Geodatabase) and click **Error Inspector**
27. The **Error Inspector: Soil** display with the errors and with the **Preview** tab showing the Soil error

Fix Watershed Using Topology (Geodatabase)

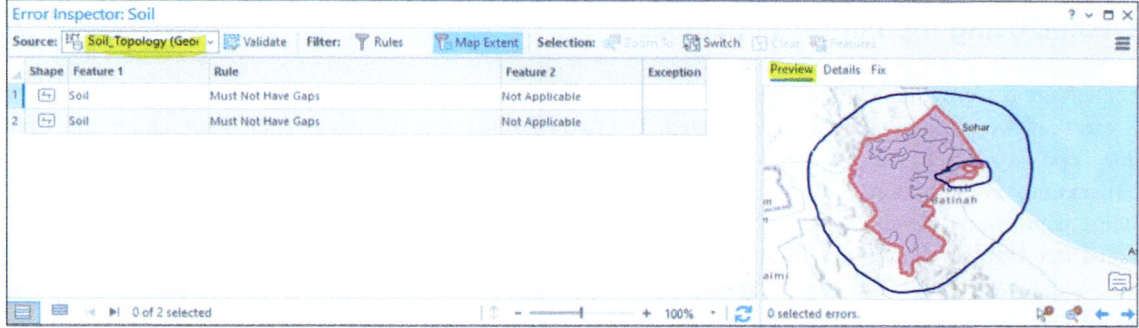

28. Highlight the first row and in the **Preview** window, click **Fix** tab, click **Mark as Exception**

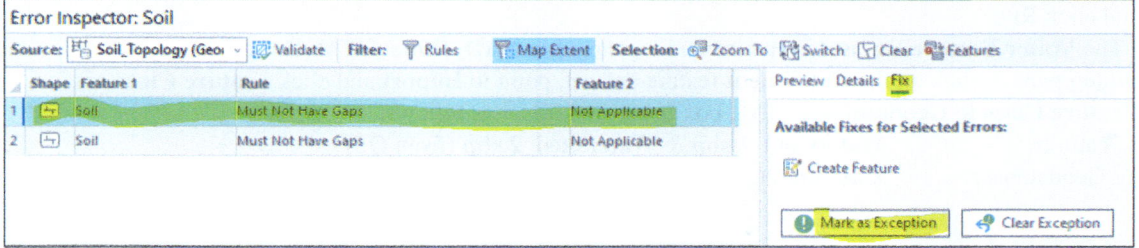

29. Click the next row, and in the **Preview** window, the error will be displayed in the Map View.
30. Click **Fix** tab, and click **Create Feature**

Result The gap disappears, and the error is removed.

31. Close the Error Inspector: Soil
32. Click Edit tab, in the Manage Edits group, click **Save** button
33. Click Yes, to save all edits

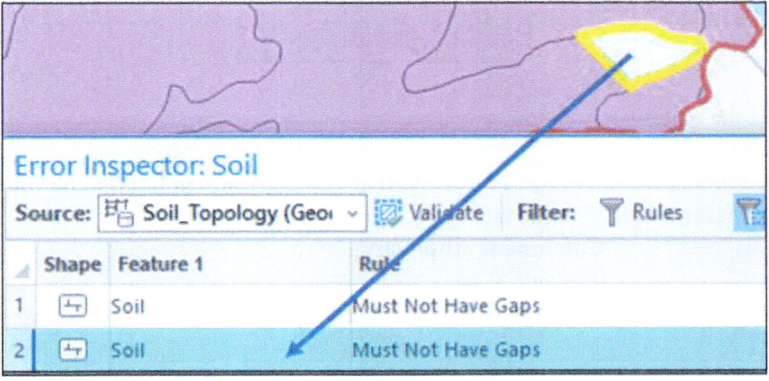

34. Save the Project

Fix Watershed Using Topology (Geodatabase)

Two catchment areas were created using the Hydrology tool in the Spatial Analyst. Two different thresholds were used, and the result was that two catchment areas were created. The two watersheds do not cover each other, and in this exercise, you will use the Topology Tool to make them identical to each other.

2. Click **Insert** tab on the ribbon, in the **Project** group, click **New Map** and call it **Watershed**
3. In Catalog pane, under Geodatabase, right click **Ch06.gdb**, point to **New** and select **Feature Dataset**
4. In the **Feature Dataset** dialog box, fill it as follows:
5. Output Geodatabase: **Ch06.gdb**
6. Feature Dataset Name: **Catchment**
7. Click the globe (Select coordinate systems), the Coordinate System dialog box open, click the drop-down arrow of **Add Coordinate System**, Import Coordinate System
8. Browse to **Q5** under **Folder\Data_Ch06\Data** and select **Watershed_1.shp**
9. OK and click Run

Result   The **Water Feature Dataset** is created and registered in **NAD_/1983_UTM_Zone_15 N.**

10. In **Catalog** pane, r-click on the **Catchment** feature dataset, point to **Import** and click **Feature Class(es)**
11. The **Feature Class to Geodatabase** dialog box opens and fills it as follows:
12. Input Features: **Watershed_1.shp** & **Watershed_2.shp** (from **Q5**)
13. Output Geodatabase: **Catchment**
14. Run

Result   **Watershed_1** & **Watershed_2** will be integrated into the Catchment feature dataset

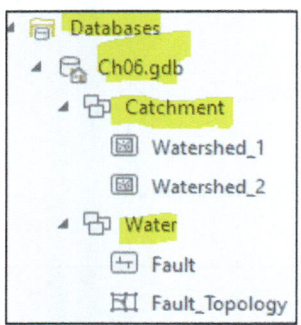

Build the Topology and Set the Rule

15. In Catalog pane, under **Ch06.gdb**, right click on **Catchment** Feature Dataset, point to **New** and select **Topology**
16. In the Create Topology Wizard, in the Define tab, fill it as follows:
17. Topology Name: **Catchment_Topology**
18. XY Cluster Tolerance   0.01 m
19. Under Feature Classes, check   **Watershed_1** & **Watershed_2**
20. Click Next

Fix Watershed Using Topology (Geodatabase)

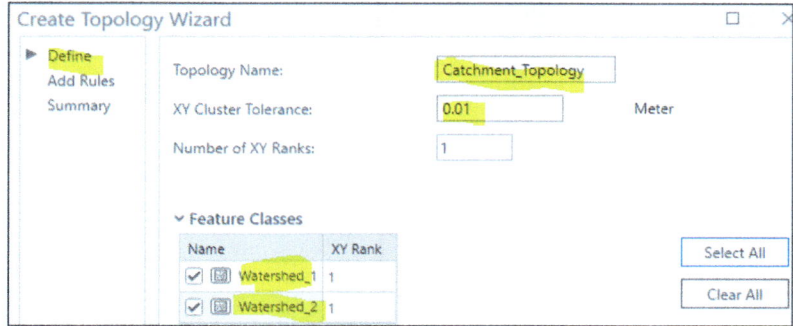

21. In the Add Rules, under **Feature Class 1**, click here to add a new rule, select **Watershed_1**
22. Under **Rule** click and from drop-down arrow select "**Must Cover Each Other (Area-Area)**"
23. Under **Feature Class 2**, check **Watershed_2** and click Next

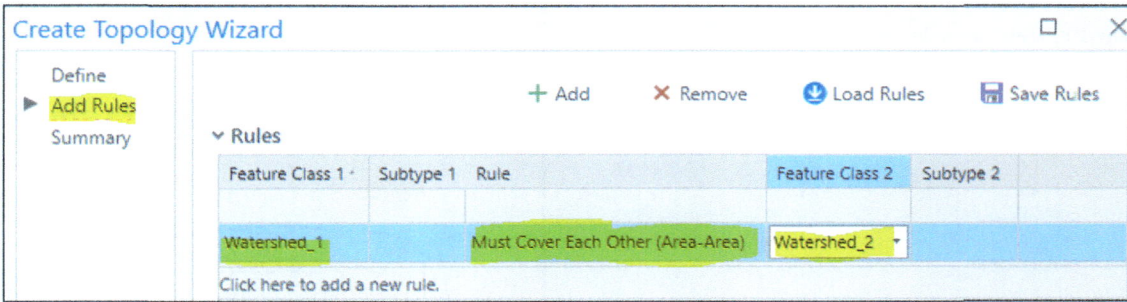

24. Summary of the topology rules displayed
25. Click Finish

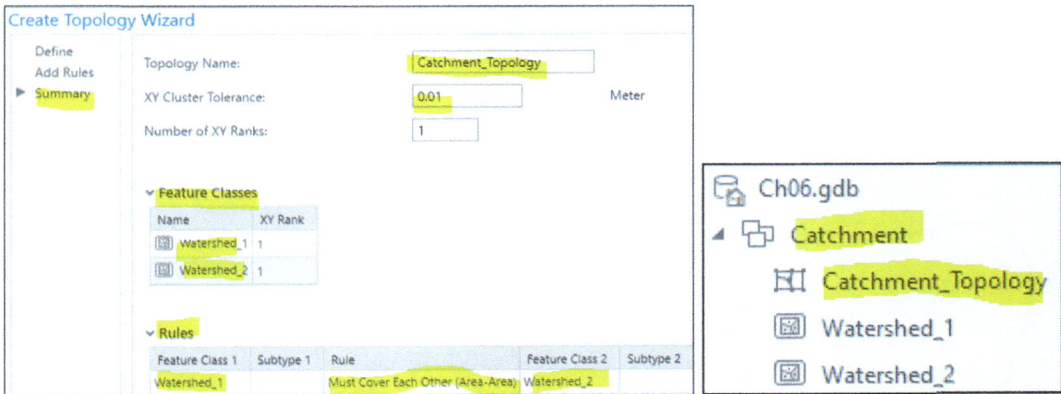

Result The Catchment _Topology added to the Catchment Feature Dataset.

26. In **Catalog** pan, under the **Catchment** Feature Dataset, r-click **Catchment_Topolgy** and click **Add to the Current Map**
27. Click **Edit** tab on the ribbon, in **Manage Edits** group, select **Catchment_Topology** (Geodatabase), and click **Error Inspector**
28. In the **Error Inspector**: **Watershed** table, click the **Validate** tab

Result The **Error Inspector: Watershed** table display with the errors and with the **Preview** tab showing the two catchments

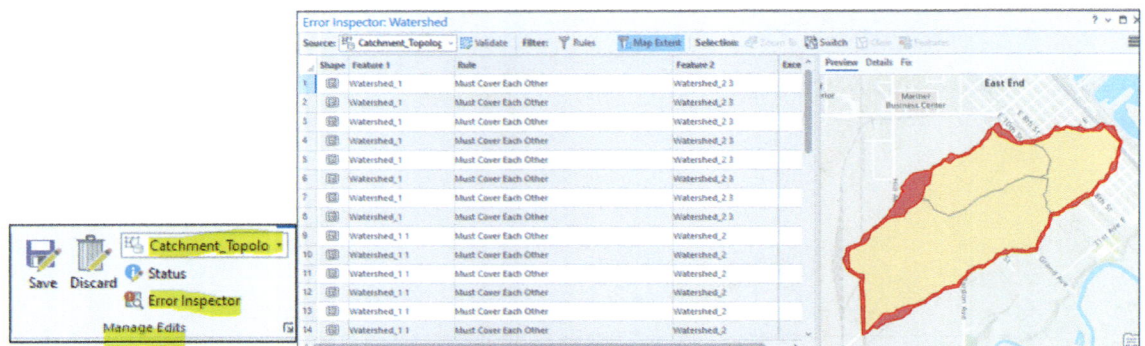

29. Click the **Edit** tab on the ribbon. In the **Selection** group, click the **Select** icon and select the lower left corner in the **Map View**.

Result The catchments in the **Map View**, and some records in the **Error Inspector: Watershed**, and the lower corner of the **window Preview** are selected.

30. In the **Preview** window, click **Fix** tab
31. In the Available Fixes for Selected Errors, click **Remove Overlap**
32. In the **Error Inspector: Watershed**, click Validate tab

Result The overlap was removed.

Another Approach You can select the entire watershed and fix it in one step.

Fix Watershed Using Topology (Geodatabase)

33. Click **Edit** tab on the ribbon, in the **Selection** group, click the **Select** icon and select the two watersheds in the **Map View**
34. In the **Preview** window, click **Fix** tab
35. In the Available Fixes for Selected Errors, click **Remove Overlap**
36. In the **Error Inspector: Watershed**, click Validate tab

Result The two watersheds overlap.

37. In the **Edit** tab, in the **Manage Edits**, click **Save** button
38. Save the project

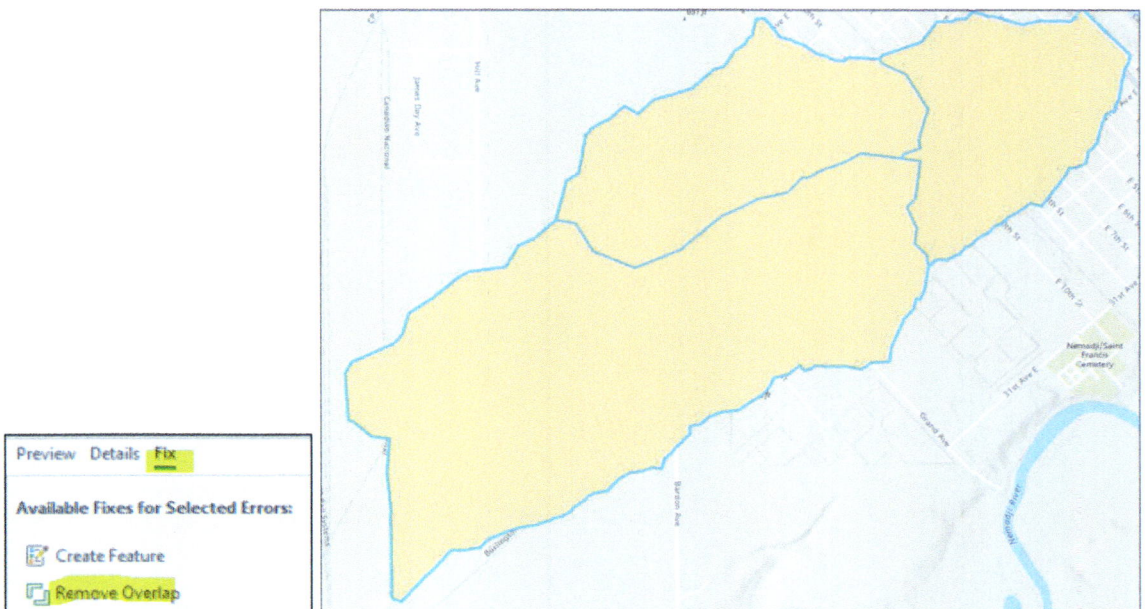

Challenge Task When correcting the topology of a continuous river, would it be best to snap the points or create a line between points?

Geoprocessing

Geoprocessing is a framework and set of tools for processing geographic and related data. The comprehensive suite of geoprocessing tools can be used to perform spatial analysis or manage GIS data in an automated way. Geoprocessing is for everyone that uses ArcGIS Pro. Whether you are a new or advanced user, geoprocessing will likely be an essential part of GIS user day-to-day work.

In GIS analysis, there are different techniques that can be applied to spatial data to solve diverse types of questions. In your GIS work, you might want to find productive groundwater wells in a basin that has high yield and good water quality, or the best location to build a treatment plant, etc. Using the correct geoprocessing tool is essential to obtain the right dataset for the analysis. A typical geoprocessing tool performs an operation on a dataset such as a feature class, raster, or table and creates a resulting output dataset. For example, the buffer tool takes features as input, creates buffer areas around the features to a specified distance, and writes those buffer areas to a new output dataset. As another illustration, if you want to create a new dataset that contains a geographic subset of the features in another dataset, you could use the Clip tool. The tool cut out a piece of one dataset using one or more of the features in another dataset as a cookie cutter.

Geoprocessing refers to the tools and processes used to generate derived datasets from other data using a set of tools. Geoprocessing in GIS is a very important tool in ArcGIS Pro software and plays a fundamental role in spatial analysis. Geoprocessing is a very broad subject in GIS and has many definitions ranging from process study areas for GIS applications to existing data to manipulate GIS data or how GIS data are computed. Geoprocessing can range from simple tasks to very complicated spatial analyses that aid in addressing an important GIS problem. A typical geoprocessing operation takes an input dataset and performs an operation on that dataset, generating new information returns as an output dataset. Geoprocessing tools perform essential operations on a database, such as projections, conversions, data management, and spatial analysis. Some tools modify the attributes or geometry of an input dataset. A few geoprocessing tools have other effects, such as creating selections on layers or generating messages or reports.

The most important geoprocessing tools can be divided based on the following tasks:
1. Extracting features
    (a) Clip
    (b) Erase
    (c) Split
2. Combining features
    (a) Merge
    (b) Append
    (c) Dissolve
    (d) Buffer

Supplementary Information The online version contains supplementary material available at https://doi.org/10.1007/978-3-031-42227-0_7.

3. Combining geometries and attributes
   (a) Union
   (b) Intersect
   (c) Spatial Join

You can use custom geoprocessing tools that are included in a project or build and use your own custom tools. ArcGIS Pro supports custom geoprocessing tools that are built with ModelBuilder or Python. ArcGIS Pro can run be through Toolboxes, which include the Analysis Tools that have a range of operations such as Extract, Overlay, Proximity and Statistics. For example, the overlay and proximity answer basic questions in spatial analysis: "What's on top of what?" and "What's near what?" The first set of tools is discussed in Overlay analysis, and the second set is discussed in Proximity analysis. In ArcGIS Pro, there are two ways to find any tool:
1. Using the search window
2. Browsing the Toolboxes

ArcGIS Pro in the Geoprocessing group includes **Ready To Use** tools, which are ArcGIS Online geoprocessing services that use the hosted data and analysis capabilities of ArcGIS Online. All you need to provide are input features; all the other data necessary for the analysis and the computation is hosted in ArcGIS Online. You can use the **Ready To Use** tools to solve diverse spatial analysis problems, including the following:
1. Profile and viewshed (Elevation)
2. Watershed and stream tracing (Hydrology)

Working with Geoprocessing Tools

In GIS analysis, there are different techniques that can be applied to spatial data to solve diverse types of questions. For example, you might want to find the best groundwater wells in a basin that has high yield and good water quality, or the best location to build a treatment plant, and many more questions. Using the correct geoprocessing tool is essential for the success of the analysis. Geoprocessing in GIS is a very important tool in ArcGIS software and plays a fundamental role in spatial analysis. Geoprocessing is a very broad subject in GIS and has many definitions ranging from process study areas for GIS applications from existing data to manipulation. Geoprocessing can range from simple tasks to very complicated spatial analyses that aid in addressing an important GIS problem. A typical geoprocessing operation takes an input dataset and performs an operation on that dataset, generating new information returns as an output dataset.

The most important geoprocessing tools can be divided based on the following tasks:
4. Extracting features
   (a) Clip
   (b) Erase
   (c) Split
5. Combining features
   (a) Merge
   (b) Append
   (c) Dissolve
   (d) Buffer
6. Combining geometries and attributes
   (a) Union
   (b) Intersect
   (c) Spatial Join

Open the Project and Connect to Folder
1. Launch **ArcGIS Pro** and click "**Open another project**"
2. Browse to \\ENV_Water\Ch07 and click Open **Ch07.aprx** and click OK
3. The ArcGIS Pro open and has no data attached to it

Working with Geoprocessing Tools

4. In **Catalog** pan, r-click the **Folder**, and point to **Add Folder Connection**
5. Browse to \\ENV_Water\CH07, select **Data_Ch07** ((or \\Database\ Data_Ch07) and click **OK**
6. In the **CP**, select the **Map** and press the **F2** key and change the name to **Watershed**

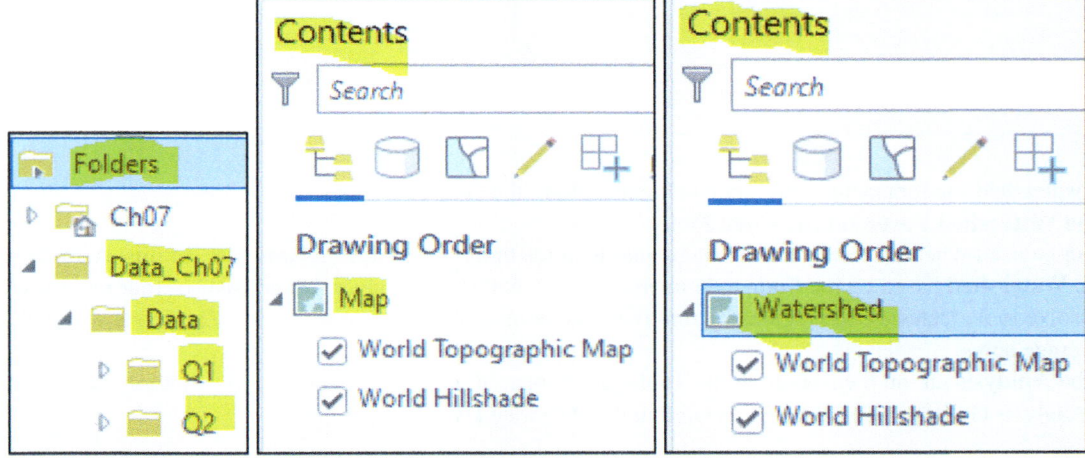

7. Open the **Folder** in the **Catalog** pane and expand **Data_Ch07/Data/Q1** and drag **Watershed.shp** into the **Map View**
8. In CP, r-click the **Watershed** and click **Attribute Table**
9. The **Watershed** attribute table has **8-fields** and **153 records** and one of the fields is called "**Code**". The **Code** field consists of 4 variables A, B, C, and D
10. After exploring the attribute table, close it

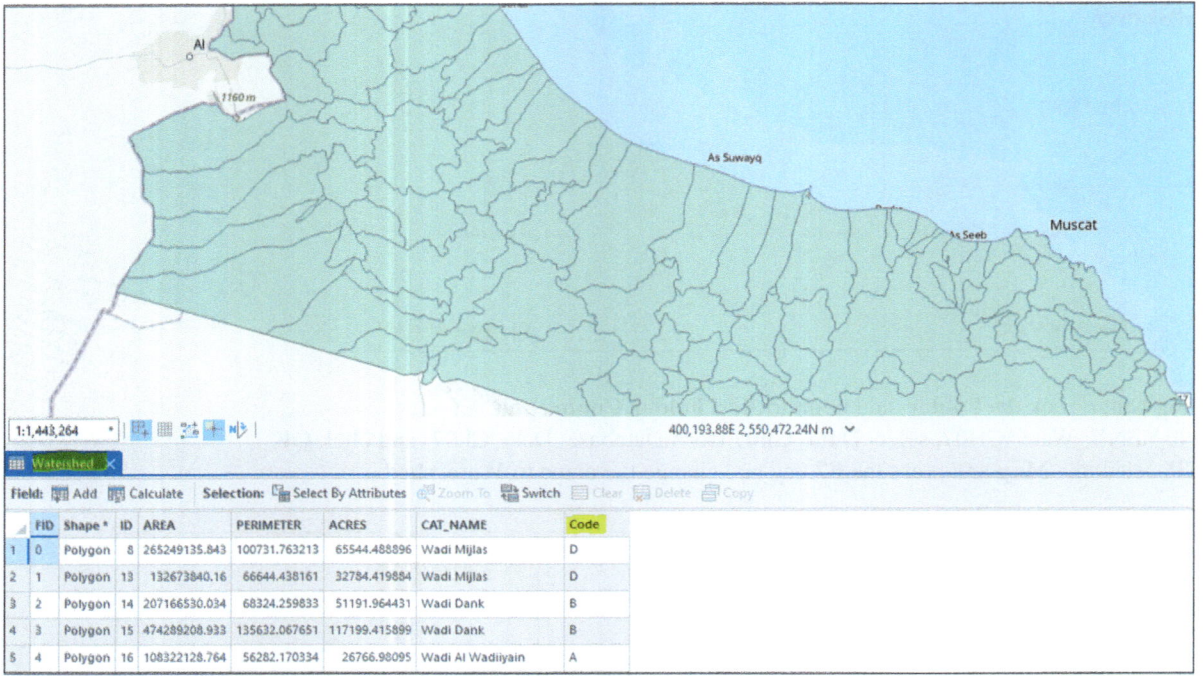

Note The watershed file represents different catchments along the Gulf of Oman

Dissolve the Watershed Based on the Code Field

Dissolve can be used when the user wants to aggregate features based on a specified attribute. In this example, you want to dissolve the **Watershed** based on the *Code* field to create a new feature class layer that consists of four records (A, B, C, and D). The dissolve tool creates 4 code regions by removing the boundaries between the codes. Each code will include the total area and average acres.

11. Click the **Analysis** tab on the ribbon, in the **Tools** group, open the "Analysis Gallery" by clicking the drop-down arrow
12. In the Analysis Gallery under Manage Data/click the **Pairwise Dissolve** button

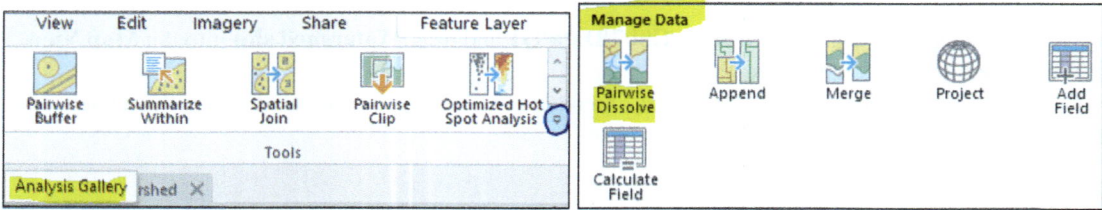

The **Pairwise Dissolve** pan display, under the **Parameters** tab, fills it as follows:

13. Input features: **Watershed**
14. Name your output file   ***Watershed_Code** (will be saved in **Ch07.gdb**)*
15. Dissolve Field(s):   Code
16. Statistical Field(s):
    (a) Area: SUM
    (b) ACRES: MEAN
17. Click Run

Working with Geoprocessing Tools

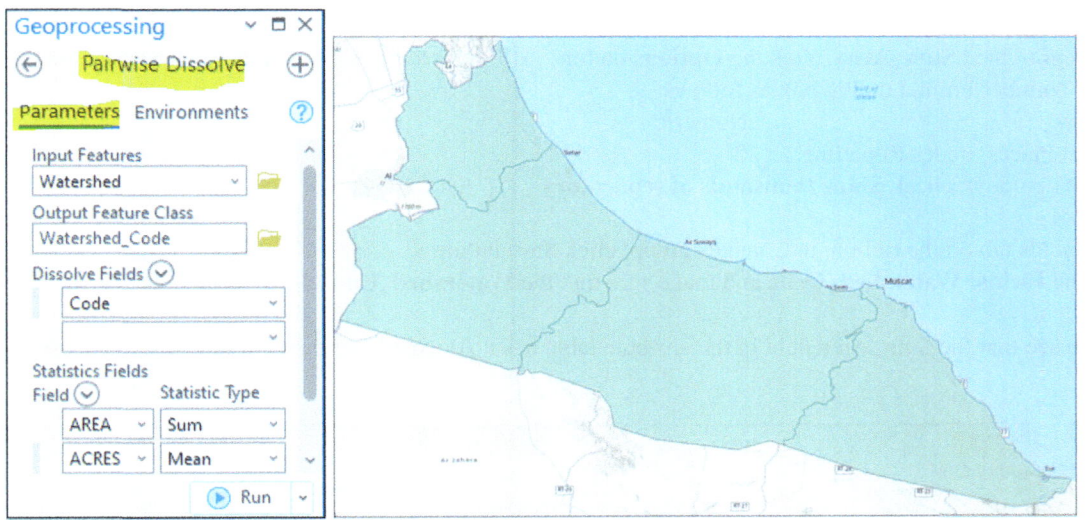

Result   The **Watershed_Code** feature class displays 4 code regions and is listed in the **CP** under the **Watershed** map. It is also stored in the **Catalog** pane under file geodatabase **Ch07.gdb** in the Geodatabase folder.

18. In the CP, r-click Watershed feature class and click Remove

Classify the Watershed_Code Based on the Code Field

19. Open the attribute table of the **Watershed_Code** feature class and examine the new schema of the table. The schema consists of **8 fields** and **4 records**. The dissolve tool reduced the number of records from 153 to 4 records and then closed the attribute table.
20. In the CP, r-click the **Watershed_Code**, click **Symbology**
21. The Symbology – Watershed_Code dialog box display, fill it as below
22. Under Primary symbology, select Unique Values
23. In Field 1:   Code
24. Under Classes tab, click **Add all values** .
25. Click the drop-down of **Color scheme**, check "**Show names**" and select **Basic Random**
26. Click the **More** drop-down and uncheck "**Show all other values**"
27. Close the Symbology pane

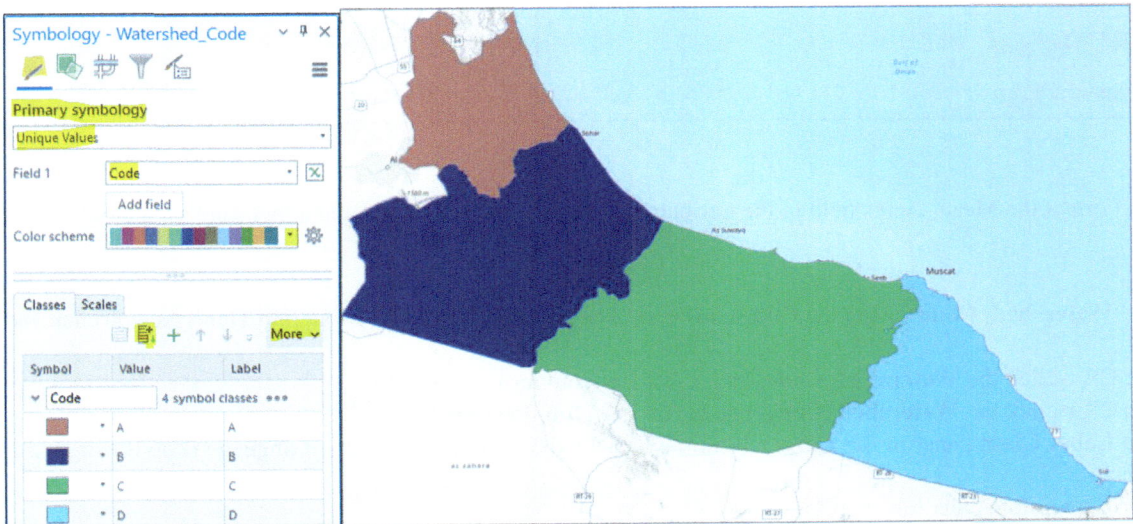

28. Open the attribute table of "**Watershed_Code**", r-click **Sum_Area** and select **Fields**
29. In the **Sum_Area** row, d-click the Numeric under Number Format
30. In the highlighted **Sum_Area**, click the **Options** button … (3-dots) to open the Number Format dialog box
31. Fill the Number Format dialog box as follows:
32. Category   Numeric
33. Decimal Places, under Rounding   0
34. Under Alignment, check **Show thousands of separators**
35. Click OK
36. In the Fields tab on the ribbon, in Changes group, click Save button
37. Close the **Fields: Watershed_Code** and make sure that the **Watershed_Code** attribute table is open

Result You see that the Sum_Area field in the attribute table has no decimal and the thousand separator display

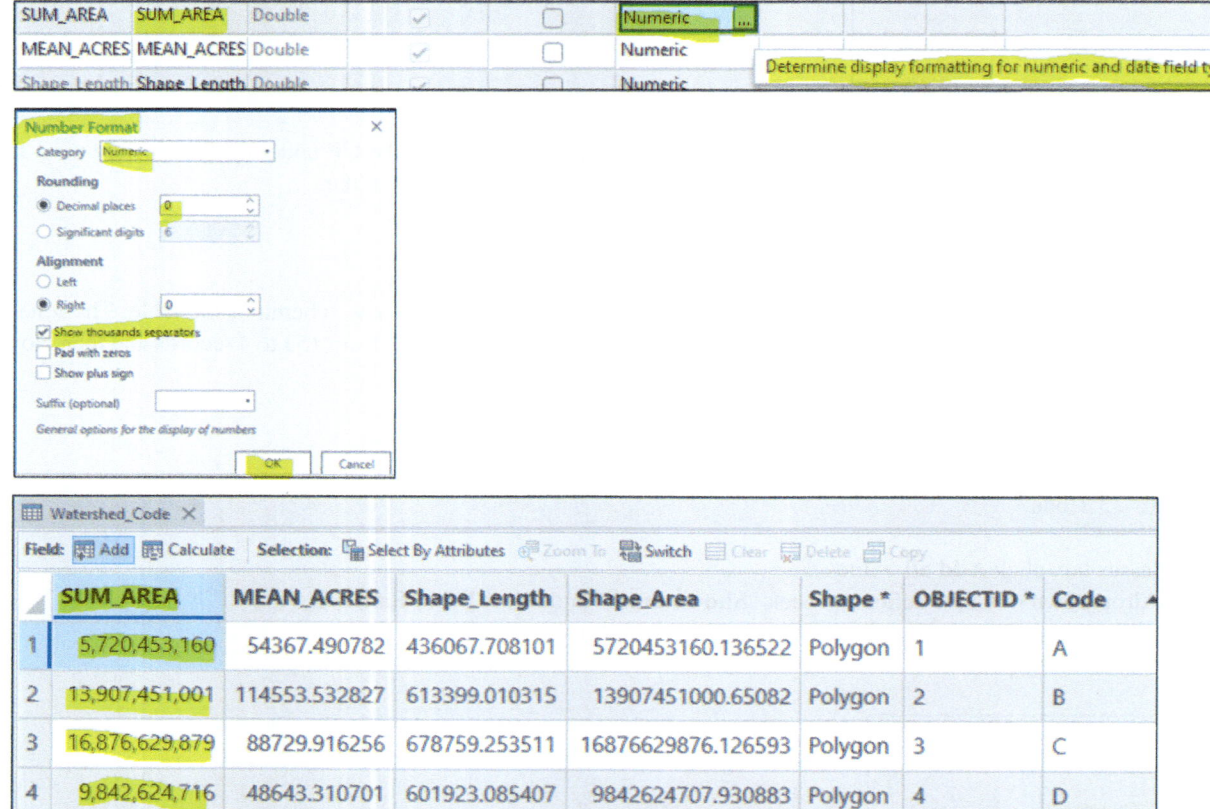

Practice Format the Mean_Acres field in the attribute table to have zero decimal and a thousand separator.

Labeling
Label the "**Watershed_Code**" using the **Code** and **Sum_Aea** fields and change the font into Times New Roman and the font size to 14.
38. In the CP, r-click the **Watershed_Code**, and select **Label** to turn on labeling
39. In the CP, r-click the **Watershed_Code**, and select **Labeling Properties**
40. In the **Label Class** pane, in the **Class** tab, in the Label expression tab, click the Language drop-down list, and select **VBScript**.
41. Under Expression, clear the expression box

Working with Geoprocessing Tools

42. Under **Fields**, double click **Code** to add it to the **Expression** box, and after the [**Code**], click space and type on the keyboard **Ampersand** (&), then type **Quote** ("), then click space, then click **Dash** (-) then click space, then click **Quote** (") and then type **Ampersand** (&) and click space.
43. Under **Fields**, double-click SUM_AREA to append the SUM_AREA field to the Expression box.
    [Code] & "-" & [SUM_AREA]
44. Click the **Verify** button (green check symbol) to see if the expression is correct. A message popped up sating "**Expression is valid**"
45. Click Apply to run the script.
46. Click Position tab, in the Position tab, at the bottom, open the Spread labels, above the Maximum, choose "**Spread words up to a fixed limit**" and to the right make the percentage
47. Close the **Label Class – Watershed_Code**
48. In the CP, click the **Watershed_Code** and click **Labeling** tab, in the Text Symbol, make the font into Times New Roman and the size 14
49. Save your Project

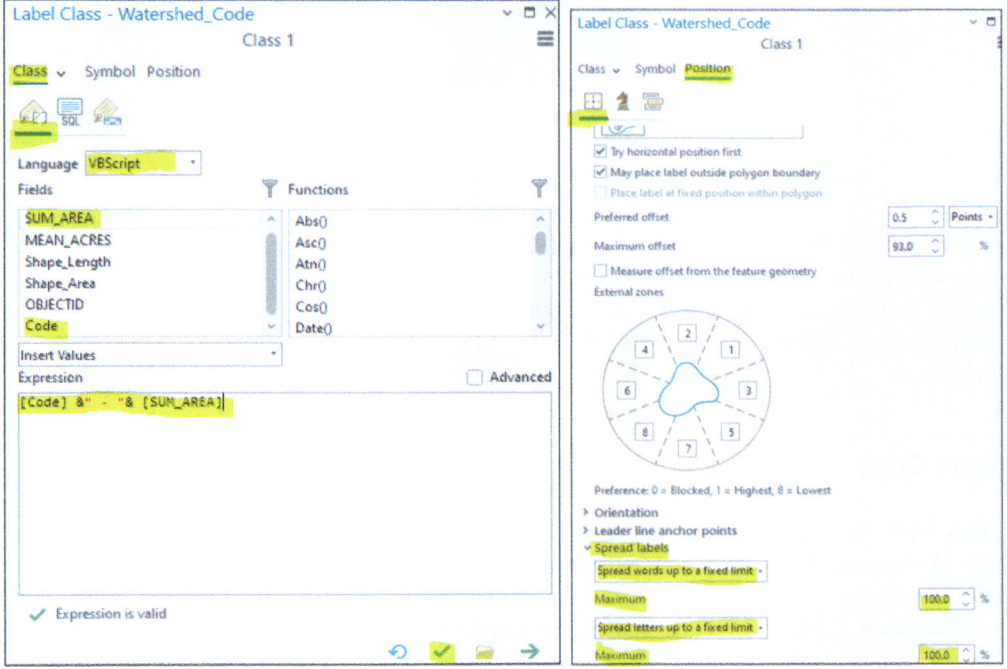

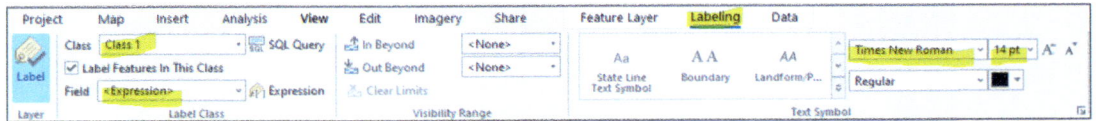

50. Open the attribute table of "**Watershed_Code**", examine the labels and answer the following questions:
    (a) How many records the "**Watershed_Code**" have? Name them
    (b) Which region code has the highest and lowest sum area
    (c) Which region has the highest average ACRES
51. Save the project

Extract a Feature Class

Now, **region C**, which has the largest area, will be saved as an independent watershed.
1. In CP, highlight **Watershed_Code**, and click the **Map** tab on the ribbon, in the **Selection** group, and click **Select** button
2. Click on **RegionC** to select it
3. In CP, r-click **Watershed_Code**, click **Data** and click **Export Features**
4. The **Export Features** pane is shown below
5. Input Features: **Watershed Code**
6. Output Feature Class: **RegionC**
7. OK

Extract a Feature Class

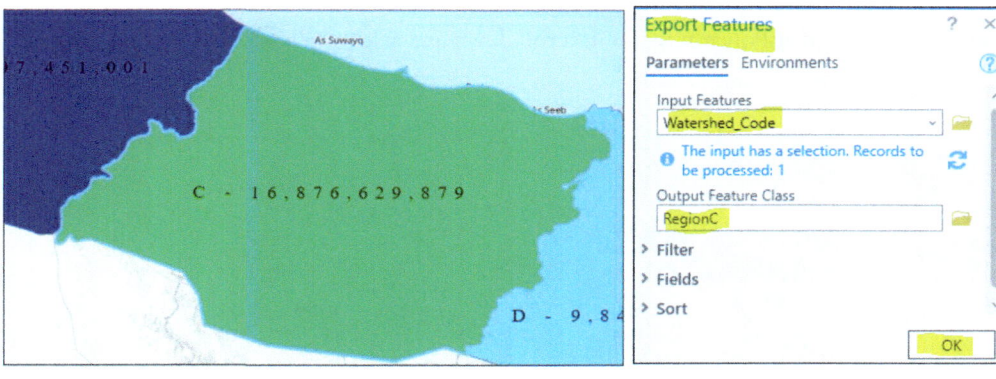

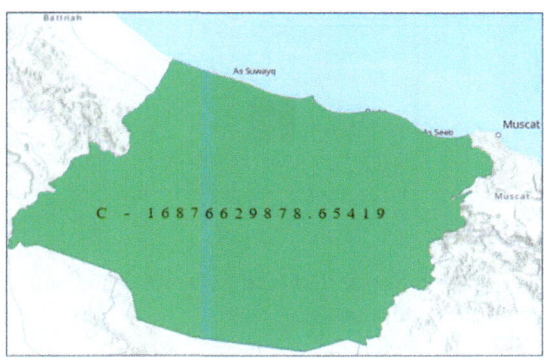

8. Close the Export Features pan and in CP, uncheck the **Watershed_code**
9. Save the project

Clipping a Feature Class
The clip tool cuts out a feature class using another feature class as a cookie cutter. This geoprocessing function is valuable for creating a new dataset. In this section, you are going to clip a network of streams inside Region C.
10. In **Catalog** pane, drag **Stream.shp** from **Q1** under Folder\Ch07_Data\Data into the **Map View**
11. Click the **Analysis** tab on the ribbon, in the **Tools** group, click the **Pairwise Clip** button
12. The **Pairwise Clip** pane display, and fill it as below
13. Input Features: Stream
14. Clip Features: RegionC
15. Output Feature Class: StreamC
16. Click Run
17. Close the **Pairwise Clip** pane
18. In the CP, r-click the **Stream** and remove
19. In the CP, r-click the symbol of the **StreamC** and click **blue** color
20. In the CP, r-click **RegionC** and click **Label**
21. In CP, r-click **RegionC** and click Zoom To Layer

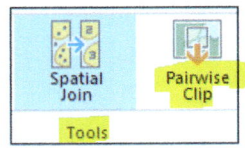
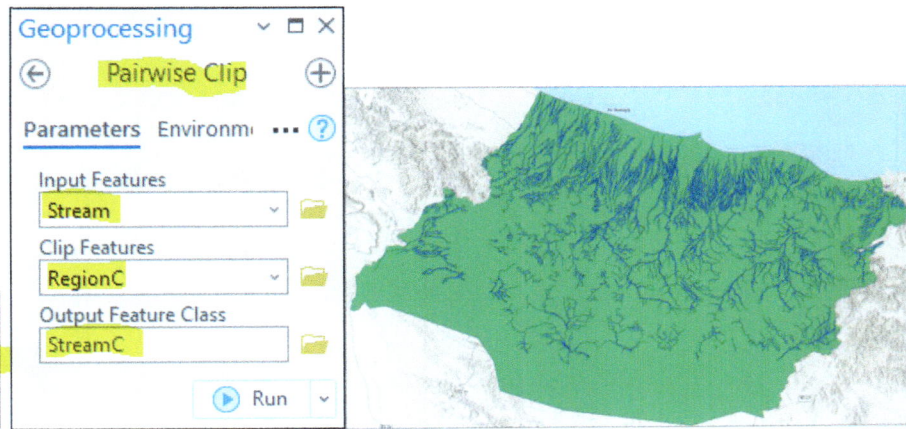

22. In CP, click the symbol of RegionC
23. In the **Format Polygon Symbol**, in the Format Polygon Symbol, scroll down and select "White"
24. Close the Symbology pane
25. Save your Project

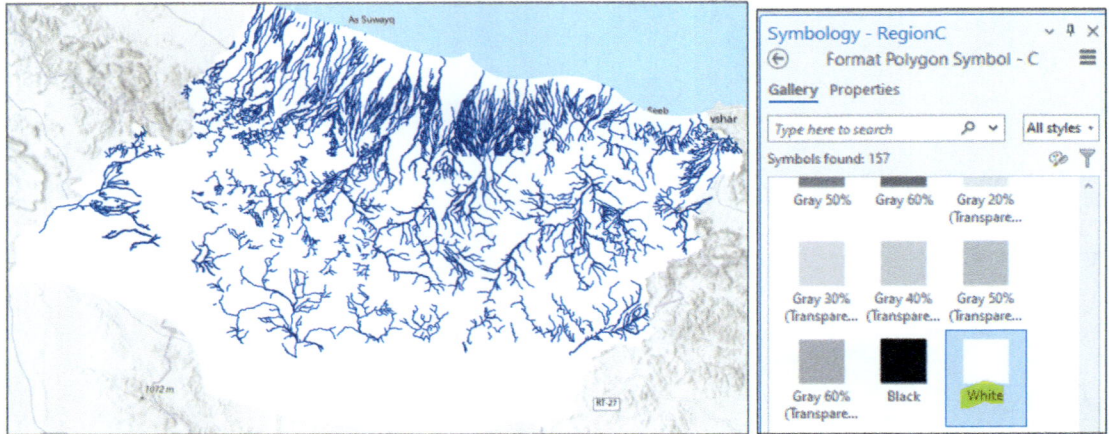

Intersect

The Intersect tool calculates the geometric intersection of any number of feature classes and feature layers. The features, or portion of features, that are common to all inputs will be written to the output feature class.

26. Click **Insert** tab on the ribbon, in **Project** group, click **New Map** and call it **Soil**
27. Open the **Folder** in the **Catalog** pane and expand **Data_Ch07/Data/Q1** and drag **MayhaCatch.shp** and **Soil.shp** into the **Map View**
28. In the **CP**, if **MayhaCatch** is not above the **Soil**, drag it and place it above the **Soil**.
29. Click the **Analysis** tab, in the **Tools** group, click the lower drop-down to display the **Analysis Gallery**, under **Overlay Features**, click **Pairwise Intersect** tool

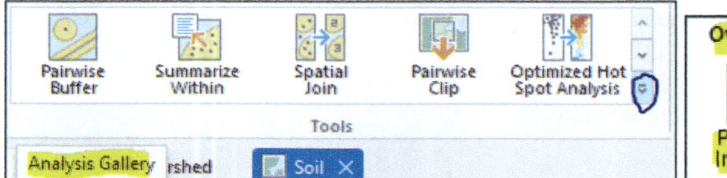 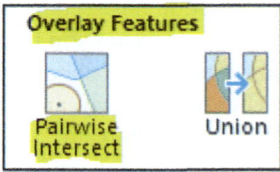

30. The **Pairwise Intersect** pane display, fill it as below
31. Input Features: **MayhaCatch** and **Soil**
32. Output Feature Class: **MayhaCatch_Soil**
33. Accept the other default
34. Run

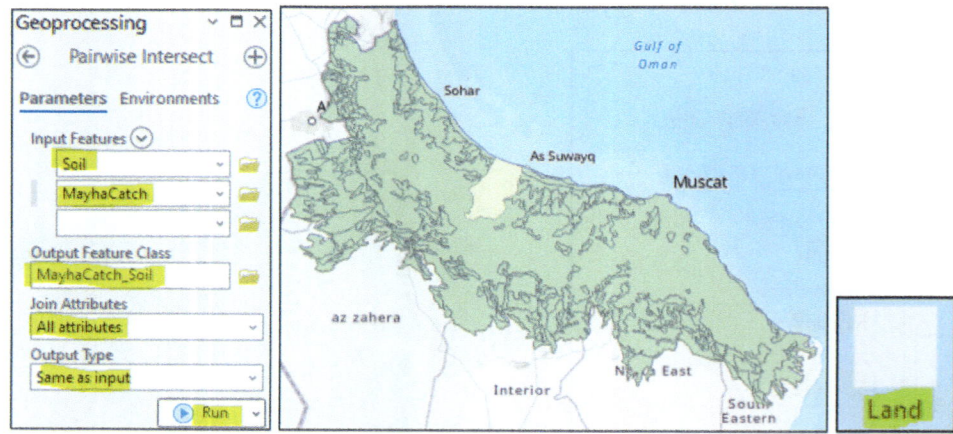

35. In CP, click the symbol of **MayhaCatch_Soil**
36. In the Format Polygon Symbol, scroll down and select **Land**
37. Close the Symbology pane
38. In the CP, remove the **MayhaCatch.shp** and **Soil.shp**
39. Save your Project

Merge Tool

The Merge tool unites numerous input feature classes or datasets into a single, new output feature class or dataset. This merge function can combine point, line, or polygon feature classes or tables. The tool requires that both inputs have the same geometry type.

1. Click **Insert** tab on the ribbon, in **Project** group, click **New Map** and call it **Farm**
2. Open the **Folder** in the **Catalog** pane and expand **Data_Ch07/Data/Q2** and select **AFLAJ.shp, FarmA.shp, FarmB.shp, FarmC.shp,** and **FarmD.shp** and drag them into the **Map View**
3. Click the **Analysis** tab on the ribbon, in the **Tools** group, click the **Analysis Gallery** and under **Manage Data**, select **Merge** tool
4. The **Merge** pane open
5. Fill it as follows:
6. Input Datasets: **FarmA.shp, FarmB.shp, FarmC.shp,** and **FarmD.shp**
7. Output Dataset: **FarmNew**
8. Accept the other default
9. Run
10. Close the Merge pane
11. Save the Project

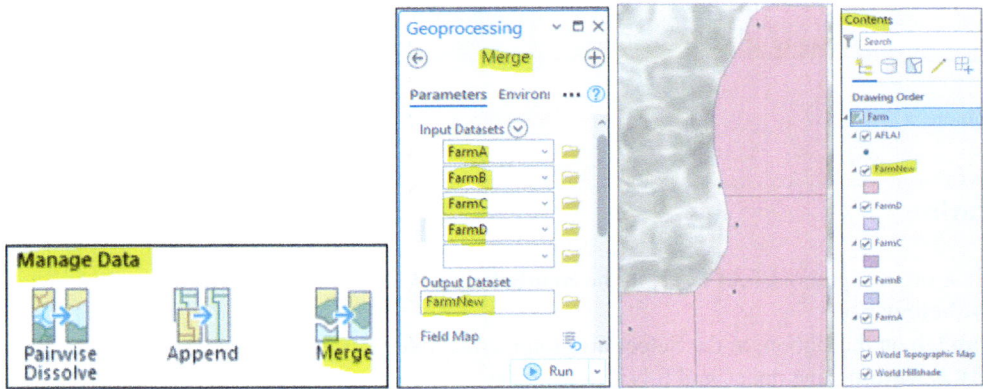

Result The four feature classes (**FarmA, FarmB, FarmC,** and **FarmD**) become one feature class consisting of four records.

Buffer Tool and Select By Location

The "Buffer" is a zone of a specified distance from a selected feature, and it involves the creation of a zone with a specified width around a point, line or area. The result of the buffer is a new polygon, which can be used in queries to determine which entities occur either within or outside the defined buffer zone. The "**Select By Location**" is a special query dialog box that lets you select features based on their location relative to other features.

Scenario 7-1 Your advisor asked you to know how many **Aflaj** are located within 500 meters from the outside boundary of the farm "**FarmNew**". Because these Aflaj will be a source of water for irrigation in the "**FarmNew**".

Buffer Tool
12. Click the **Analysis** tab on the ribbon in the Geoprocessing group, click the **Tools** button
13. In the **Geoprocessing** pane, click the **Toolboxes** tab, open the **Analysis Tools**, then open the **Proximity**, and click the **Buffer** tool
14. Fill the **Buffer** pane as follows:
15. Input Features: FarmNew
16. Output Features Class: FarmNew_Buffer
17. Distance   500
18. Dissolve Type: Dissolve all output features into a single feature
19. Accept the other default
20. Click Run
21. Close the Buffer pane

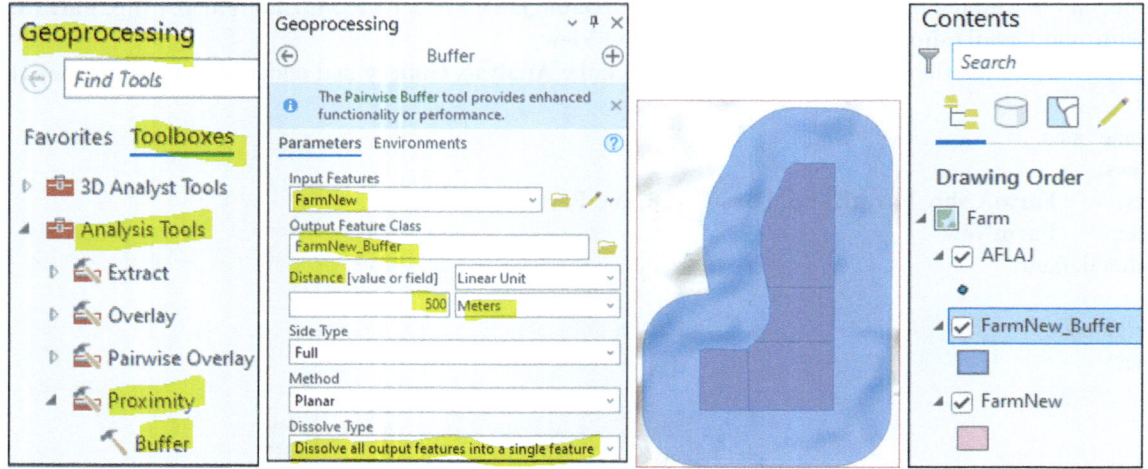

22. Save the project

Select by Location

14. In the CP, click on the symbol of **AFLAJ**, in the **Symbology – AFLAJ** pane, under **Gallery** tab, select **Circle 6**, then close the **Symbology** pane
15. Click the **Map** tab on the ribbon, in the **Selection** group, click **Select By Location** button
16. The **Select By Location** pane opens and fills it as follows:
17. Input Features: *AFLAJ*
18. Relationship: **Completely within**
19. Selecting Features: *FarmNew_buff*er
20. Selection type: **New selection**
21. Click Apply
22. Close the **Select By Location** pane
23. In Cp, r-click *AFLAJ* and open the attribute table after exploring the selected **AFLAJ** close the attribute table

Create an Artificial Water Reservoir for Irrigation Purposes

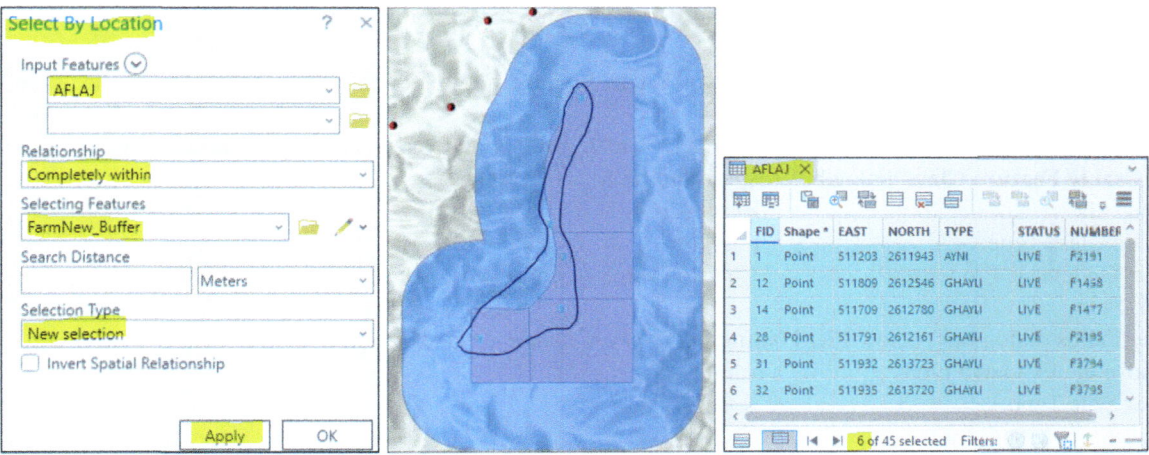

Result Six out of 45 Aflaj inside the buffer zone are selected.

24. In the CP, r-click on **AFLAJ** and point to **Data**, click **Export Features**, and fill the Export Feature dialog box as follows:
25. Input Features: **AFLAJ**
26. Output Feature Class: **AFLAJ_Farm**
27. Click OK
28. In the CP, r-click **AFLAJ** and click **Remove**
29. In the CP, highlight **AFLAJ_Farm**, click **Labeling** tab, in the **Label Class** group, make the Field = NUMBER
30. In the **Layer** group, click the **Label** button
31. In the CP, r-click **AFLAJ** and click **Remove**

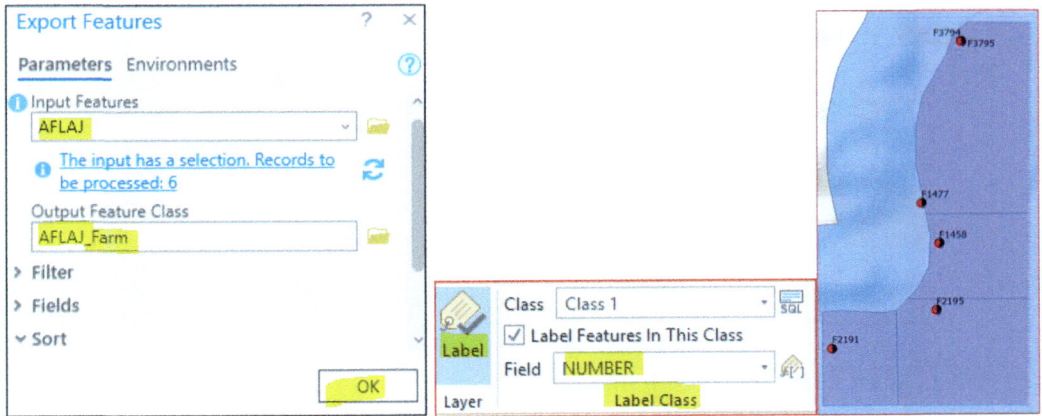

Create an Artificial Water Reservoir for Irrigation Purposes

Scenario 7-2 You are going to create a cylinder shape feature reservoir that could hold at least 40,000 m³ of water. The source of water will come from the **AFLAJ**, which is within half a kilometer of **FarmNew**. The circular reservoir will have a radius and depth of 50 m and 5.1 m, respectively, and it will be placed north of **AFLAJ** no. F1458.

Note Volume of the reservoir cylinder $\pi r^2 h = 3.14 \times 50\text{ m} \times 50\text{ m} \times 5.1\text{ m} = 40{,}035\text{ m}^3$

Create a Circle Reservoir

A circular polygon feature is constructed from two sketched points. The first point defines the center of the circle, and the second point specifies the radius.

31. In the **Catalog** pane, expand "**Databases**" r-click "**Ch07.gdb**", point to **New**, and click **Feature Class**
32. The **Create Feature Class** dialog box display and fill it as follows:
33. Name: **Reservoir**
34. Feature Class Type: **Polygon**
35. Under Geometric Properties   Check Z Value – coordinates include Z values used to store 3D data
36. Click Next
37. In the **Fields** dialog box, below **Field Name**: "*Click here to add a new field*"
38. Remove the Field and type: Name
39. Under Data Type: Text
40. Length: 12
41. Click Next
42. In the Spatial Reference Dialog box, open the drop-down arrow of "Add Coordinate System", click Import Coordinate System, browse to "**Ch07.gdb**" and select "**FarmNew**", click OK
43. Click Finish

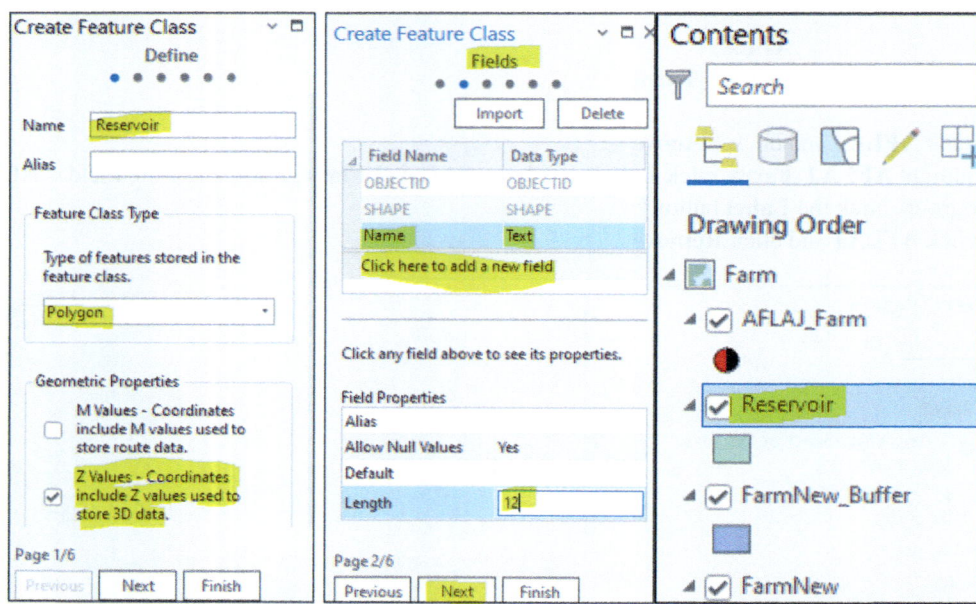

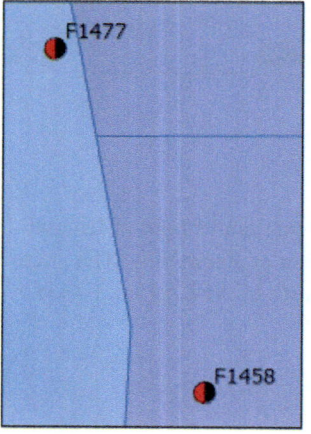

Result The **Reservoir** feature class is created and integrated into the content pane, but it is empty and needs to be digitized.
44. In the **Map view**, zoom in between the two AFLAJs that have the F1458 and F1477 labels.
45. Click the **Edit** tab on the ribbon, in the **Features** group, click **Create** button
46. In the **Create Features** pane, click the **Reservoir** symbol
47. The construction toolbar appears at the bottom of the map. Click the **blue** arrow.
48. The Active Template pane open
49. On the toolbar, click the **Circle** symbol
50. Click the map north of Aflaj F1458 (mid distance between F1458 and F1477) then drag the pointer away and r-click, and select radius, type 50 m and enter
51. Open the Attribute table of the **Reservoir** layer

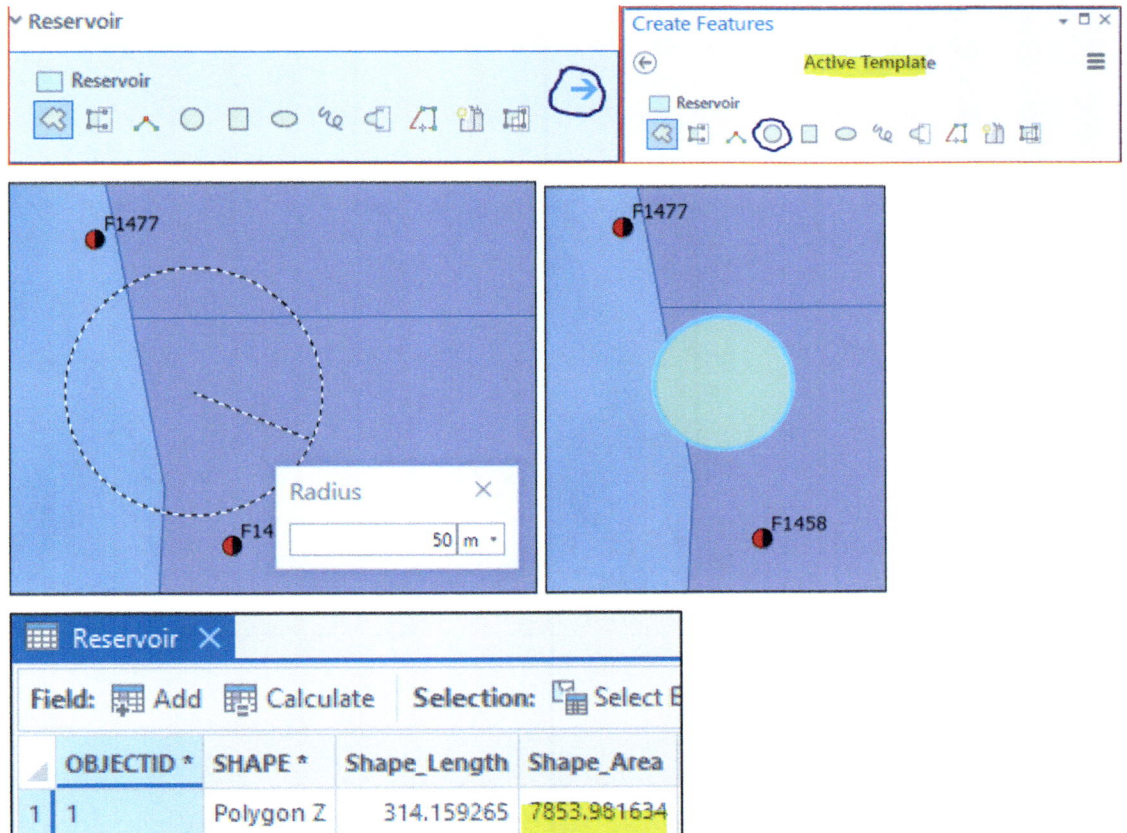

Result The area of the reservoir is 7853.9 m².
52. After examining the area of the **Reservoir**, close the attribute table
53. In the **Manage Edits** group, click **Save** button and click **Yes** to save all edits.
54. Click **Clear** in the **Selection group**, to deselect the **Reservoir** in the Map View
55. Close the **Create Features** pane
56. Save your project

Erase Tool

The Erase tool creates a feature class by overlaying the input features with the polygons of the Erase features. Only those portions of the input features falling outside the erase features, outside of their boundaries, are copied to the output feature class.

Scenario 7-3 The reservoir that has been built inside the new farm will reduce the size of the agricultural area. Therefore, the Erase function is used to remove the area of the reservoir from the total farmland.

57. Click the **Analysis** tab on the ribbon, in the **Geoprocessing** group, click **Tools** button
58. In the **Geoprocessing** pane, select **Toolboxes**, expand **Analysis Tools**
59. Open the **Overlay**, and click the **Erase** tool
60. The **Erase** pane opens and fills as follows:
    (a) Input Features: **FarmNew**
    (b) Erase Features: **Reservoir**
61. Output Feature Class: **FarmNew_Reservoir**
62. Run
63. Close the Erase tool

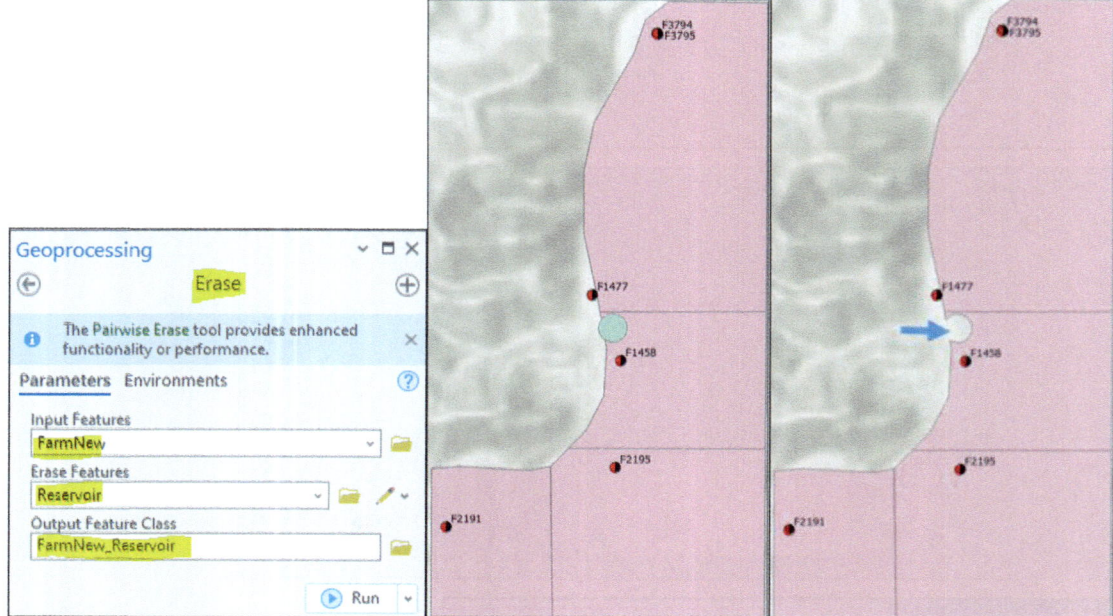

64. Save project and exit ArcGIS Pro

Site Suitability and Modeling

Site selection means finding the location that meets a specific condition or criteria. In ArcGIS Pro, different techniques and tools can be used to find the most suitable site for a particular purpose. One of the straightforward procedures that can be used in GIS is modeling. Modeling helps to generalize or simplify an environmental setting and its processes. For example, to find suitable land for a certain purpose such as irrigation or residential building, you have to apply criteria in your analysis to evaluate where the land is most suitable for that particular use. Understanding the main input functions of suitability modeling guarantees a successful model.

ModelBuilder

In ArcGIS Pro, geoprocessing tools can be used to perform spatial analysis and manage GIS data. ModelBuilder is used to create, edit, and manage geoprocessing models that automate those tools. Models are workflows that string together sequences of geoprocessing tools, feeding the output of one tool into another tool as input. ModelBuilder can also be thought of as a visual programming language for building workflows.

ModelBuilder is a visual programming language for building geoprocessing workflows. Geoprocessing models automate and document your spatial analysis and data management processes. You create and modify geoprocessing models in ModelBuilder, where a model is represented as a diagram that chains together sequences of processes and geoprocessing tools, using the output of one process as the input to another process.

ModelBuilder in ArcGIS Pro allows you to do the following:

- Build a model by adding and connecting data and tools.
- Iteratively process every feature class, raster, file, or table in a workspace.
- Visualize your workflow sequence as an easy-to-understand diagram.
- Run a model step by step, up to a selected step, or run the entire model.
- Make your model into a geoprocessing tool that can be shared or can be used in Python scripting and other models.

The most important elements of the model progression are the following:

1. What is the problem that demands solving?
2. Define a well-articulated criterion for the analysis
3. Gather the necessary data to solve the problem
4. Select the GIS tool needed for the model
5. Create the model to diagram the activity flow

Determine the criteria that allow you to successfully collect the proper data and create the model.

Supplementary Information The online version contains supplementary material available at https://doi.org/10.1007/978-3-031-42227-0_8.

Site Suitability and Modeling

In this chapter, you are going to be introduced to type of approaches to perform modeling. The first approach uses the Toolboxes, and the second approach uses the ModelBuilder.

Model 1

Site suitability is selecting a site that meets one or more necessary criteria, such as location, attributes, area, and more. The model can be performed with a series of individual geoprocessing tools or with a model containing all the necessary tools.

In model 1, you will use different aspects of functionality in GIS to find the most suitable area for building a greenhouse at the Jordan University campus. The data that will be used in model 1 are an image downloaded from Google Earth and then clipped and georeferenced using the Palestine_1923_Palestine_Grid projection. The layers "**Landuse**" and "**Vegetation**" were digitized using the image after it had been rectified.

Scenario 8-1 You are a GIS manager at Jordan University, and you have been asked by the administrator to choose the best location to build a new greenhouse on the north–east region of the campus (Figure below). To build the greenhouse, you must take into consideration different criteria.

Areal image of the study area

The Criteria to Build the Greenhouse

1. The greenhouse should be at least 50 meters away from the sewer pipeline.
2. The greenhouse should be within a code of 400 of the land use layer.
3. The greenhouse should be within a Veg_Code 1 or 2 of the vegetation layer.
4. The potential location should have a minimum area of 8,000 m^2.

Connect to Folder

1. Launch ArcGIS Pro and click "**Open another project**"
2. Browse to \\ENV_Water\Ch08\ and click **Ch08.aprx** and then click OK
3. The ArcGIS Pro open and has no data attached to it

The Criteria to Build the Greenhouse   147

4. In **Catalog** pan, r-click the **Folder**, and point to **Add Folder Connection**
5. Browse to \\ENV_Water\Ch08\ folder and select **Data_Ch08** (or **Database\ Data_Ch08**) and click **OK**
6. In the **CP**, select the **Map** and press the **F2** key and change the name to **Greenhouse**
7. Open the **Folder** in the **Catalog** pane and expand **Data_Ch08/Data/M1** and drag **LandUse, Pipeline, StudyArea**, and **Vegetation** into the **Map View**
8. In the CP, drag **StudyArea** to the bottom
9. In the CP, r-click **Pipeline** and click Symbology.
10. The Symbology – **Pipeline** pane open, click the Symbol and under **Gallery**, choose Water (line), then click the Properties tab change the colors to blue and width 2, click Apply.
11. In the CP, highlight **Vegetation**, click the Symbol in the Symbology pane in the Properties tab, change the color to green and outline width to 0, and click Apply
12. In the CP, highlight **LandUse**, click the Symbol in the Symbology pane, in the Properties tab, change the color to gray and outline width to 0, and click Apply
13. In the CP, highlight **StudyArea**, click the Symbol in the Symbology pane, in the Gallery tab, select Dashed Black Outline (1pt)
14. Close the Symbology pane

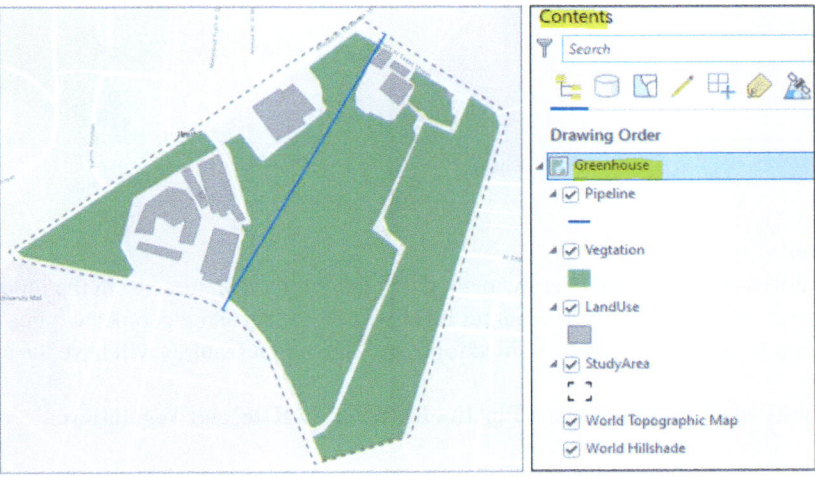

Proximity Analysis: Buffer
The buffer function will create a polygon around the pipeline at our specified distance of 50 m. The buffer shows the area around the pipeline that you cannot use for the Greenhouse, as it is too close to the pipeline.

15. In the CP, highlight the **Pipeline**, and click the **Analysis** tab on the ribbon, in the Geoprocessing group, click the Tools button, the Geoprocessing pane display

16. In the Geoprocessing pane, click the **Toolboxes** and open the Analysis Tools, then open the **Proximity** and click the **Buffer** tool and enter the Buffer dialog box as shown below.
17. Input Features: Pipeline
18. Output Feature Class: PipelineBuffer
19. Distance: 50 m
20. Accept the rest of the default
21. Click Run.

Result The output layer "**PipelineBuffer**" is a polygon feature 50 m around the pipeline, and the attribute table of the layer consists of only one record.

Overlay Analysis: Union

The Union function is a polygon-to-polygon overlay method that takes all the features from the input layers and then calculates the geometric intersection of the layers. The output layer will be of that same geometry type. This means that several polygon feature classes and feature layers can be unified together. The output features will have the attributes of all the input features that overlap.

The Union tool will unify the 3-polygon layers: **PipelineBuffer**, **LandUse**, and **Vegetation.**

22. Click back the black arrow of the Geoprocessing pane, under Analysis Tools, open the Overlay, and click **Union**, enter the Union dialog box as shown below.
23. Input Features: PipelineBuffer, Vegetation, and LandUse
24. Output Feature Class: Union
25. Accept the default of the default
26. Click Run

The Criteria to Build the Greenhouse

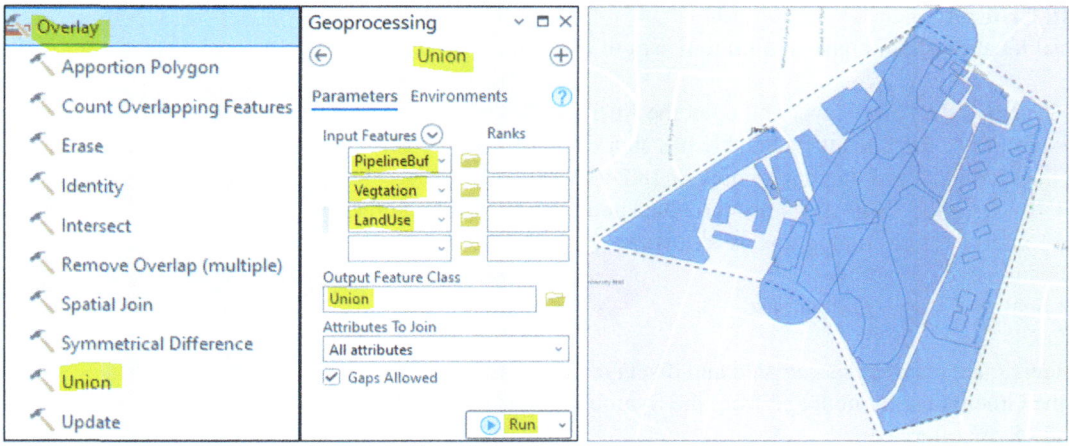

Result The output file "**Union**" includes all fields from the three input layers: **PipelinwBuffer**, **LandUse**, and **Vegetation**. The **Union** feature class will be used to select the criteria to build the Greenhouse at the campus.

The Second and Third Criteria
The Greenhouse should be within a code of 400 of the LandUse layer and within a Veg_Code 1 or 2 of the Vegetation layers.

Select Tool
The select function, typically using a select or **Structured Query Language** (SQL) expression, stores them in an output feature class. The selection ensures that the SQL meets the criteria for the suitability outlined for this assignment.

27. Click back the black arrow of the Geoprocessing pane, under **Analysis Tools**, open the **Extract**, and click **Select**, enter the Select dialog box as shown below.
28. Input Features: Union
29. Output Feature Class: Site
30. After **Where** from the drop-down arrow, choose **Name**, then select **is not equal to**, then select the **Sewage Pipe**
31. Click **+ Add Clause**
32. After **And** from the drop-down arrow, choose **Veg_Code**, then select **is less than or equal to**, then select 2
33. Click **+ Add Clause**
34. After **And** from the drop-down arrow, choose **Code**, then select **is equal to**, then select 400
35. Click Run

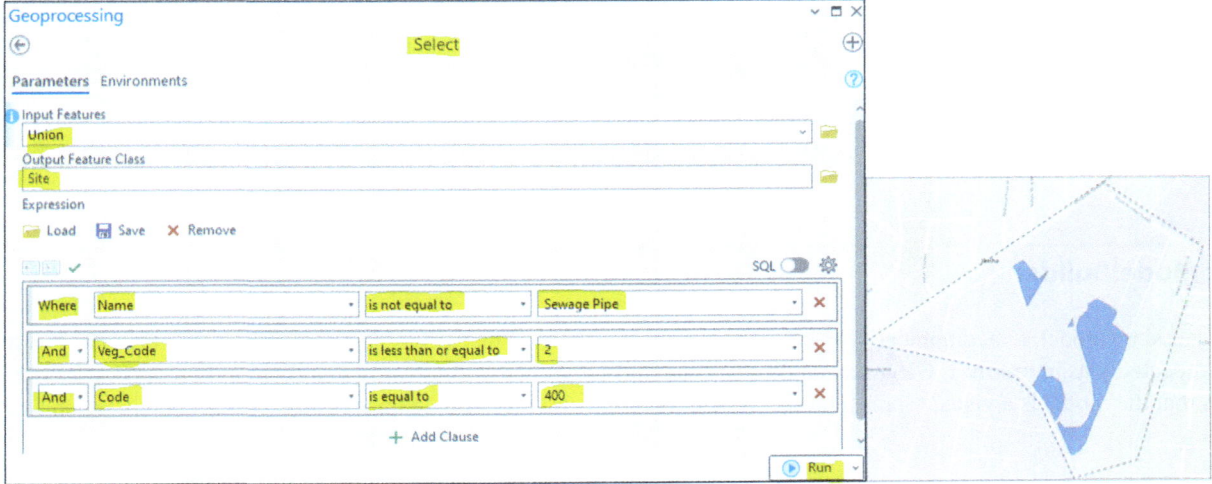

Result The best potential location will be selected.

The Fourth Criteria
The potential location should have a minimum area of 8000 m².

36. In the CP, r-lick the "**Site**" layer and open the Attribute Table.
37. The last record of the Site attribute table has an area of 18,246 m². This land will be used to build the greenhouse on it.
38. Select the second record that has an area of 18,246 m².
39. In the CP, r-click the "**Site**" layer point to Data and click Export Features.
40. Fill the Export Features dialog box as follows:
41. Input Features: Site
42. Output Feature Class: Greenhouse
43. Click OK
44. The Greenhouse feature class created and displayed in the CP
45. Click the Greenhouse symbol and from the Symbology pane, under Gallery tab, select the Cropland
46. Close the Symbology pane
47. Save the project

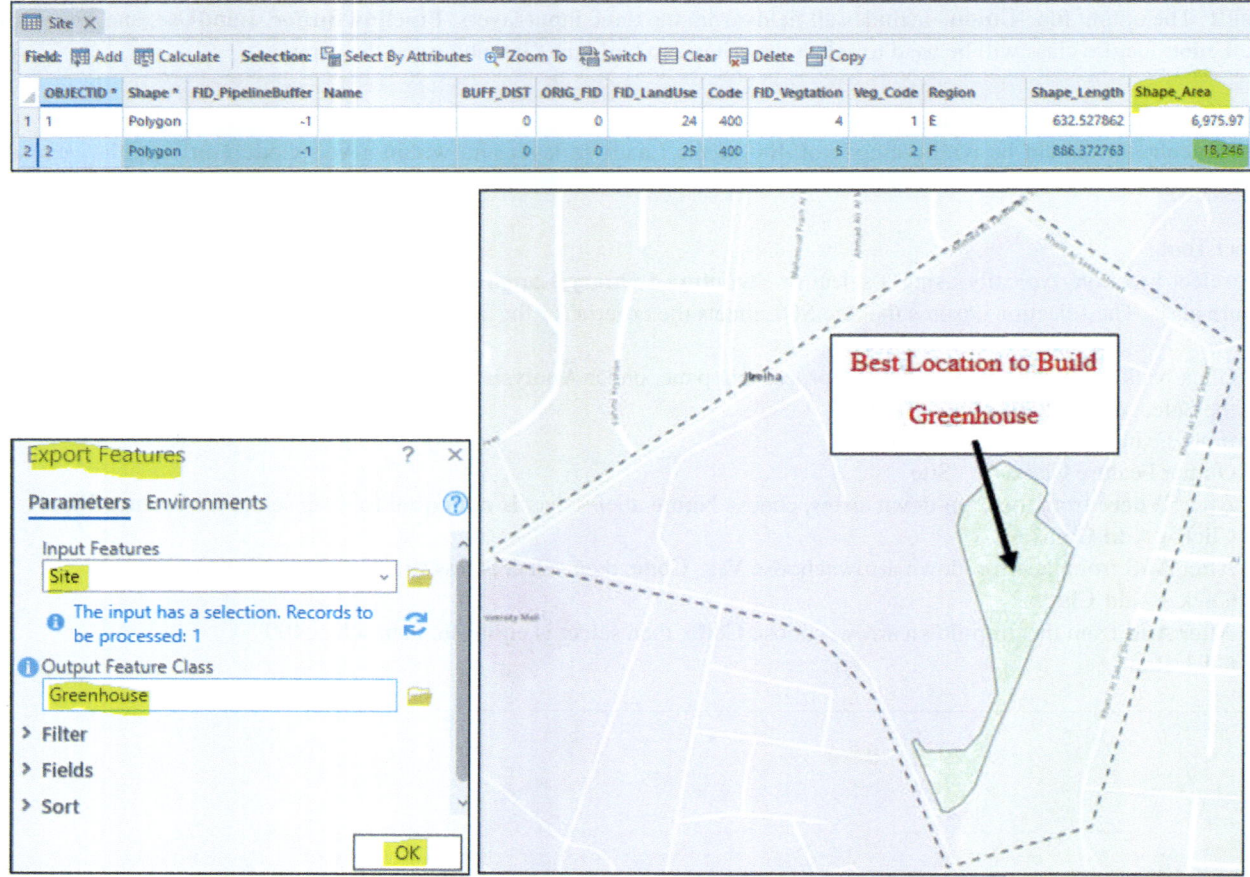

Model 2: ModelBuilder

The geoprocessing model is a graphical way of systematizing analysis. This means instead of running the tools from Toolboxes repeatedly as in **model 1**. GIS users could automate their analysis as workflows through the geoprocessing model. This means that the model is a visual representation of a workflow in which several geoprocessing tools are run in sequence.

In ArcGIS Pro, when you create a model, it opens for editing in ModelBuilder. A new model starts as a blank model view, into which you can add data and tools and connect them to form processes. The new model opens with a temporary welcome message that expires when you add a model element or save the model.

When you create a model in ArcGIS Pro, it opens for editing in ModelBuilder. A new model starts as a blank model view, into which you can add data and tools and connect them to form processes. The new model opens with a temporary welcome message that expires when you add a model element or save the model.

You can create your own tools. Tools you create are called **custom tools** and become an integral part of geoprocessing. You can open and run any tool from the Search or Catalog window, use it in ModelBuilder and the Python window, and call it from another script. Once you have created your own toolbox, you can create a model tool or create a script tool in the toolbox. Any model you create and save in a toolbox becomes a model tool.

In this section, you will create a model and add a process to it. A process is a geoprocessing tool connected to its input and output data. A model with one process is the simplest model that can be run. The output from one tool is often the input to another tool. A geoprocessing model depicts a workflow as a diagram. The model is made up of a process, meaning that each tool is associated with input and output elements. A model process consists of a tool and all variables connected to it. Connector lines indicate the sequence of processing. Many processes can be chained together to create a larger process. A process in a model can be in one of four states: not ready to run, ready to run, running, and has been run.

1. **Not ready to run**: When you initially drag a tool into ModelBuilder, the process is in a not-ready-to-run state because the required parameter values have not been specified. The tool and its inputs and outputs are gray.
2. **Ready to run**: A process is ready to run when the tool has all required parameters filled in. All model elements in the process have color.
3. **Running**: The process is in a running state if the model tools are red.
4. **Has been run**: If you run the model from within ModelBuilder, the tool and derived data elements are displayed with drop shadows, indicating that the process has run, and the derived data have been generated.

Work with a Model: Find Best Suitable Location

Site selection means finding the location that meets a specific condition or criteria. It is a generalized model that can be used in the GIS environment to find the most suitable site for a particular use. The modeling process identifies the main issue that needs to be answered based on specific criteria. For modeling analysis, the main input function to perform the work is to determine the proper data to find the ideal solution. To carry this model in ArcGIS Pro, the exact tools and the procedures to be carried out should be determined and understood in advance for an effective result. Site suitability can be determined by using both raster and vector techniques. This task will be based on applying the vector mode, which relies greatly on proximity and overlay analysis. The main concept will be discerning the sensitive area from the study area and selecting an area that is most suitable.

This exercise allows you to use different aspects of GIS functionality to find the most suitable area for building a **Nuclear Power Plant** (**NPP**) in the Dhuleil Area, Jordan.

The Criteria to Find Suitable Location to Build Nuclear Power Plant

The Dhuleil area in Jordan is proposed to be a location to build a Nuclear Power Plant, as it is an ideal location due to the presence of a plentiful amount of reclaimed water that can be used for cooling purposes. Nevertheless, the main question is: what is the possibility of building a Nuclear Power Plant in the area without affecting the local environment and the water resources?

Dhuleil is an agricultural area that has a major limestone and basalt aquifer. Many wells are tapping into these two permeable formations, and they have been used for irrigation since the 1960s. In the mid-1980s, a major sewage treatment plant, named Khirbet Al-Samra (KSWTP), was built to increase the water resources for irrigation and lessen the use of groundwater, whose water quality and quantity had deteriorated due to extensive use and overexploitation. The area has a network of drainage systems that shifted water from intermittent streams into perennial streams after the KSWTP started to discharge its treated water into the major Zarqa River, which ended up in the King Talal Dam reservoir. The water stored in the reservoir was then released to irrigate the Jordan Rift Valley, which is one of the most important irrigated areas in the country.

The geological structure of the Dhuleil area is highly fractured, and the structure influences the groundwater recharge from precipitation and surface runoff during the wet season. A previous study showed that wells near fault systems have slightly elevated concentrations of contaminants such as nitrate and salinity. Therefore, any NPP should be built at a suitable distance away from the fault system. The surficial geology in the study area consists of basalt, limestone, alluvium, sandstone, siltstone, and marly limestone. The first four formations are considered highly permeable and are generally associated with moderate to high potential rates of local recharge from rain or surface runoff during wet seasons. The marly limestone and siltstone are considered impermeable surface deposits that have very low to low recharge potential with reduced rates of water movement into or out of these formations. Any potential site for building the proposed NNP should be built within these two impermeable layers. Therefore, the surface water and groundwater should be protected from any potential contamination from the proposed NPP.

The Criteria to Find Best Site to Build the NPP

To find the best location, you must use four criteria related to distance from the water body, structure, groundwater well locations, and geological impermeable formations.

1. The NPP should be at least 300 m away from the main stream.
2. The NPP should be at least 200 m away from the fault system.
3. The NPP should be at least 500 m away from the groundwater wells.
4. The NPP should be built within code 2 (Siltstone and Marly Limestone formations)

Build a Geoprocessing Model in Modelbuilder

In this task, you must select a suitable location to build a nuclear power plant based on four criteria using a geoprocessing model in ModelBuilder.

1. Click **Insert** tab on the ribbon, in **Project** group, click **New Map** and call it "NPP Model".
2. Open the **Folder** in the **Catalog** pane and expand **Data_Ch08/Data/M2** and drag the **Fault, GEOL_KS, KSWTP, Stream**, and **Well** into the **Map View**
3. In CP, r-click the **GEOL_KS** and select **Symbology**, In the Symbology pane, under **Primary symbology**, choose **Unique Values**, in the **Field1** choose **Formation**, click **Add all values**, open the More drop-down arrow and uncheck "Show all other values"
4. Use the Table below to change the following layers in the Content pane

Layers	Symbol	Color	Size
Well	Circle	Blue	8
Fault	1.0 Point	Red	1pt
Stream	Water (line)	Apatite Blue	2pt
KSWTP	Waterlines		2pt

The Criteria to Find Suitable Location to Build Nuclear Power Plant 153

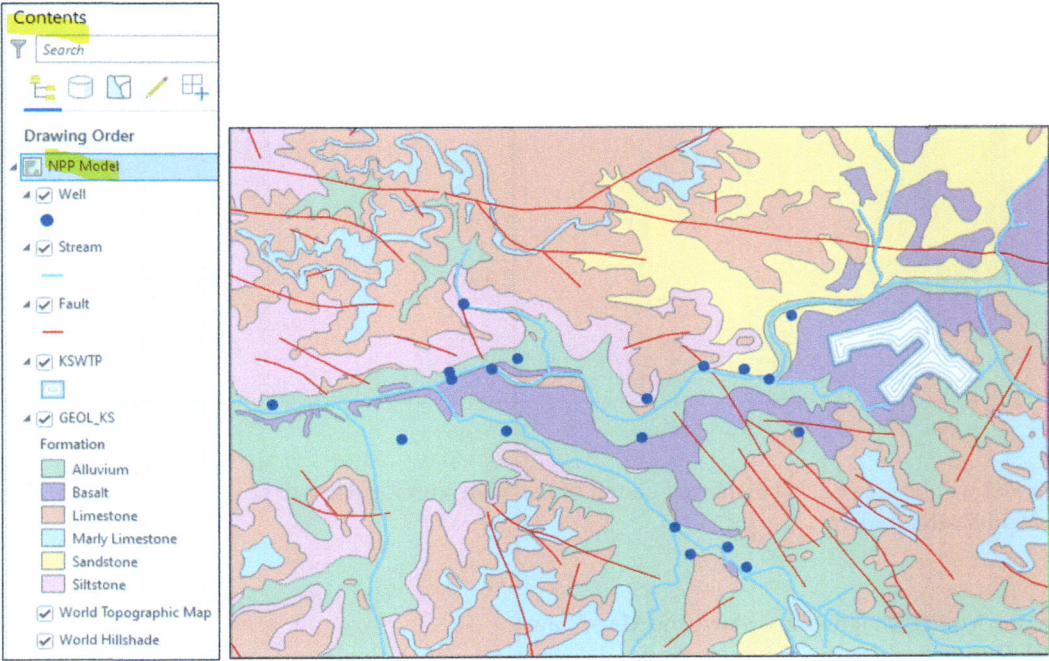

5. Click the Project tab on the ribbon, click Options, click Geoprocessing, ensure settings are as shown below.

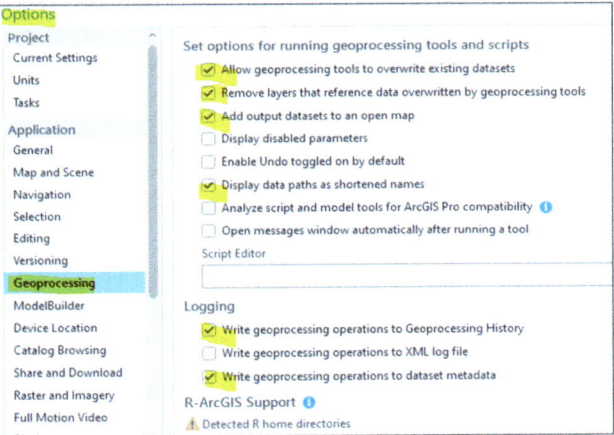

6. Click the back arrow in the upper left corner to return to the project
7. Click the Analysis tab on the ribbon, and in the **Geoprocessing** group, click the **Model Builder** button to launch a **Model** window.

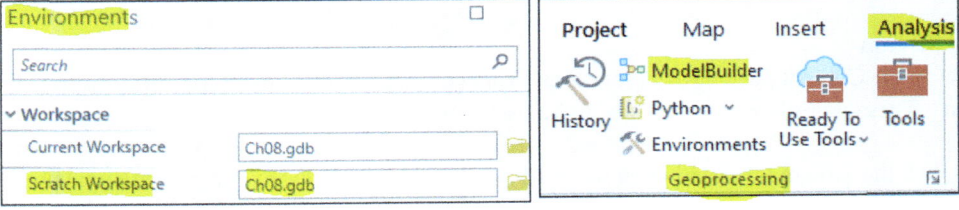

154   8 Site Suitability and Modeling

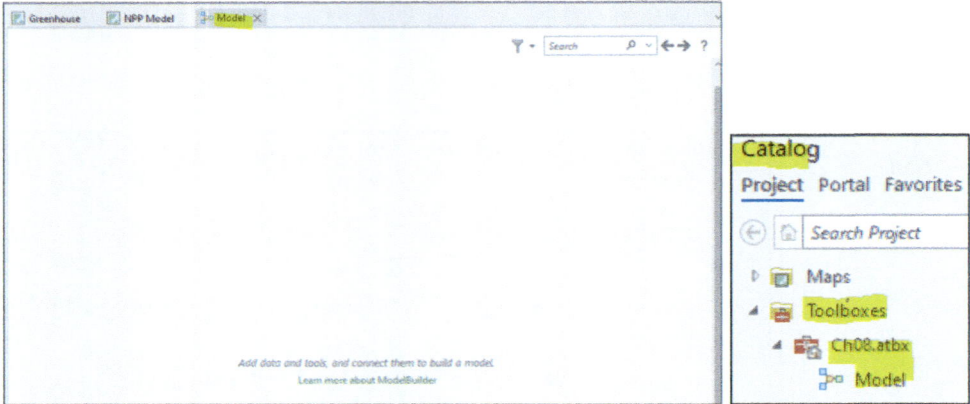

Result The model will be created and displayed on top of the **Map View,** and it will be created in the **Catalog** pane under the **Toolboxes\Ch08.atbx** folders

8. Make sure the **Model** is highlighted above the **Map View**
9. Click **ModelBuilder** tab on the ribbon, in the **Model group**, and click **Environments** button
10. Fill the Environment dialog box as follows:
11. Current Workspace: browse to **M2**
12. Scratch Workspace: **Ch08.gdb**
13. Output Coordinate system: from the window drop-down select **Well**, the ***Palestine_1923_Palestine_Belt*** will display
14. Ok

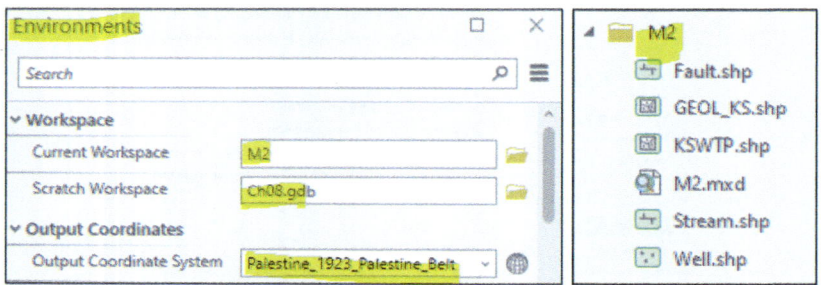

Add Geoprocessing Tools and Add Data

Models work with data, with the output of one geoprocessing tool being used as input to another tool. You can start building your model by first adding the data you want to process. Geoprocessing tools are a fundamental building block of your model. ArcGIS Pro has many geoprocessing tools for accomplishing an extensive number of GIS tasks. Once you know the right tools for the work you are doing, it is straightforward to add those tools to a model.

Select the Main Stream
The **Stream** layer in the Content pane has 25 records and some of these records under the "**Status**" field in the attribute table are called "**Main**". These records will be selected to be part of the model.

15. In the CP, r-click the Stream layer and open the attribute table. After exploring the information, close the attribute table.
16. Make sure the model view is active, click the **ModelBuilder** tab on the ribbon, in the **Insert** group, click **Tools** button
17. In the Geoprocessing pane, click the Toolboxes tab
18. In the search box, type **Select**

The Criteria to Find Suitable Location to Build Nuclear Power Plant   155

19. Drag "**Select Layer By Attribute**" (Data Management) into the Model.
20. Double click the **Select Layer by Attribute** tool in the model window and fill it as follows:
21. Input Rows: Stream
22. Selection Type: New Selection
23. In the SQL expression, fill the statement as "**Where Status is equal to Main**"
24. Click OK.
25. Click **Save** 💾 button on the **Model** group
26. Click **Fit to Window** button in the View group

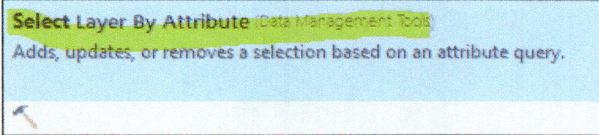

Result The **process** (the tool and the input and the output variables) is in a state **ready-to-run**

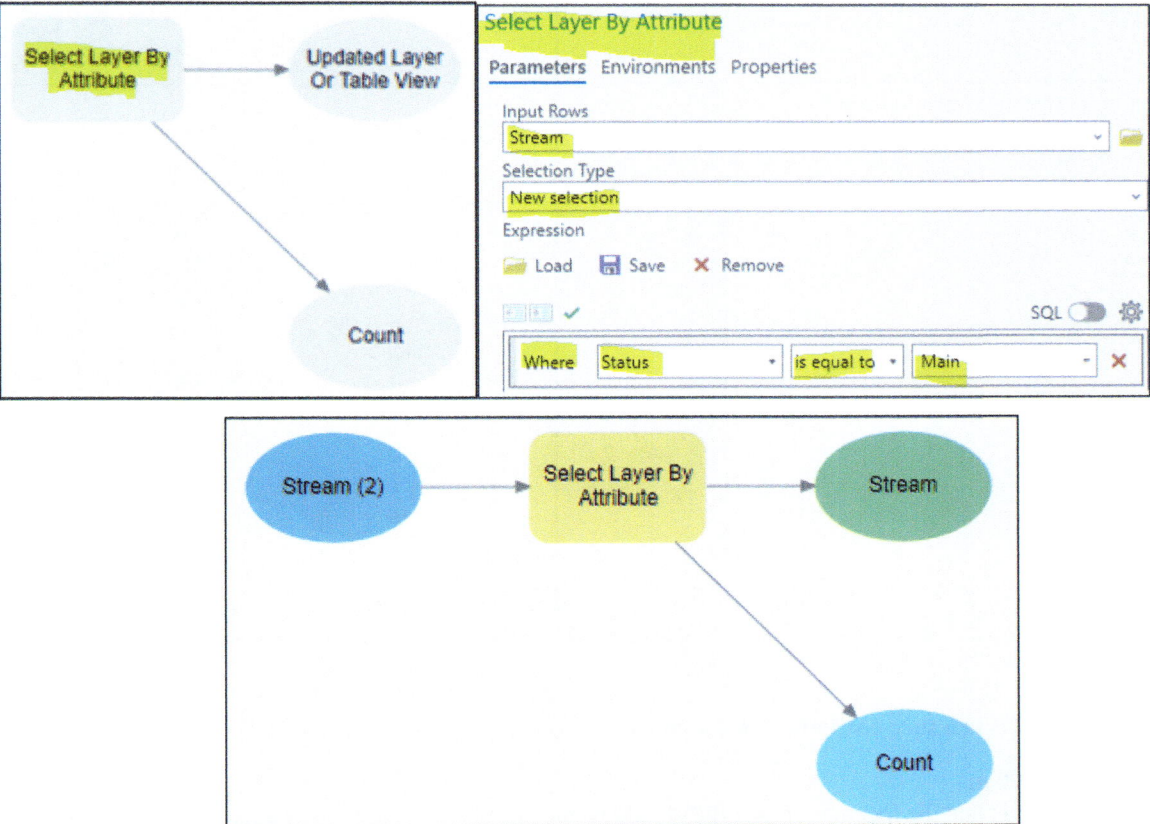

Buffer the Selected Stream: 300 m
The first criterion to find the best site to build the NPP is that the new location should be at least 300 m away from the main stream.

27. In the Geoprocessing pane, in the Search window, type Buffer
28. Drag the Buffer (Analysis Tools) into the Model and place it below **Select Layer By Attribute**.
29. Double-click **Buffer** tool in the "**Model**"
30. Fill the **Buffer** dialog box
    (a) Input Features: Stream (2)
    (b) Output Feature Class: StreamBuffer
    (c) Distance: 300 m
    (d) Accept the rest of the default
31. Click OK.

32. Click **Save** 💾 button on the **Model** group
33. Click **Auto Layout** in the **View** group.

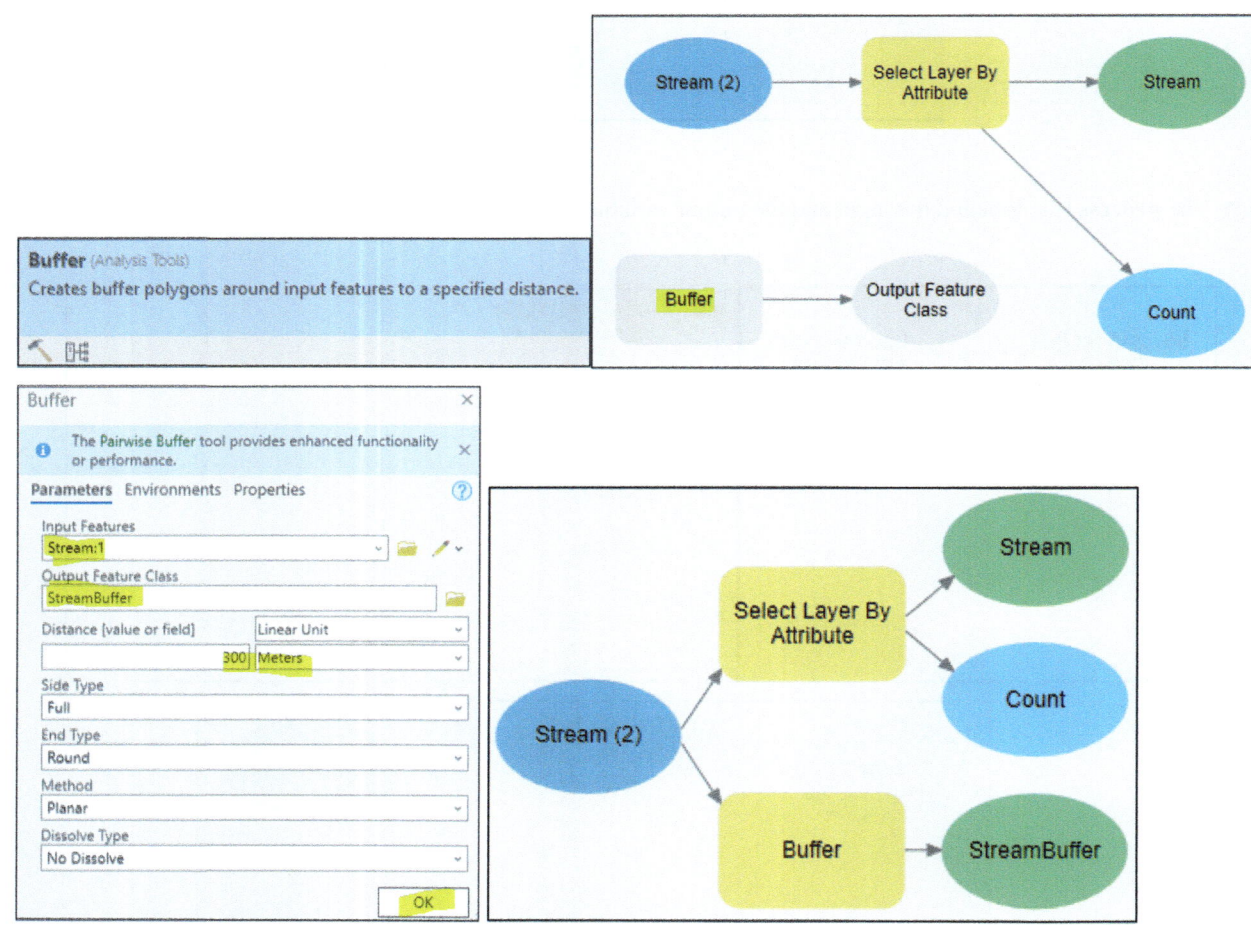

Result The **process** is in a state **ready-to-run** and Stream (2) is connected to the **Buffer** tool.

Buffer the Fault: 200 m
The second criterion is that the location of the NPP should be at least 200 m away from the fault system.

34. In the Geoprocessing pane, drag once again the **Buffer** tool into the **Model**, and place it below the **Buffer** tool
35. Double click the **Buffer (2)** tool and enter inputs as shown below.
36. Fill the Buffer (2) dialog box as follows:

(a) Input Features: Fault
   (b) Output Feature Class: FaultBuffer
   (c) Distance: 200 m
   (d) Accept the rest of the default
37. Click OK

38. Click **Save** button on the **Model** group
39. Click **Auto Layout** in the **View** group.

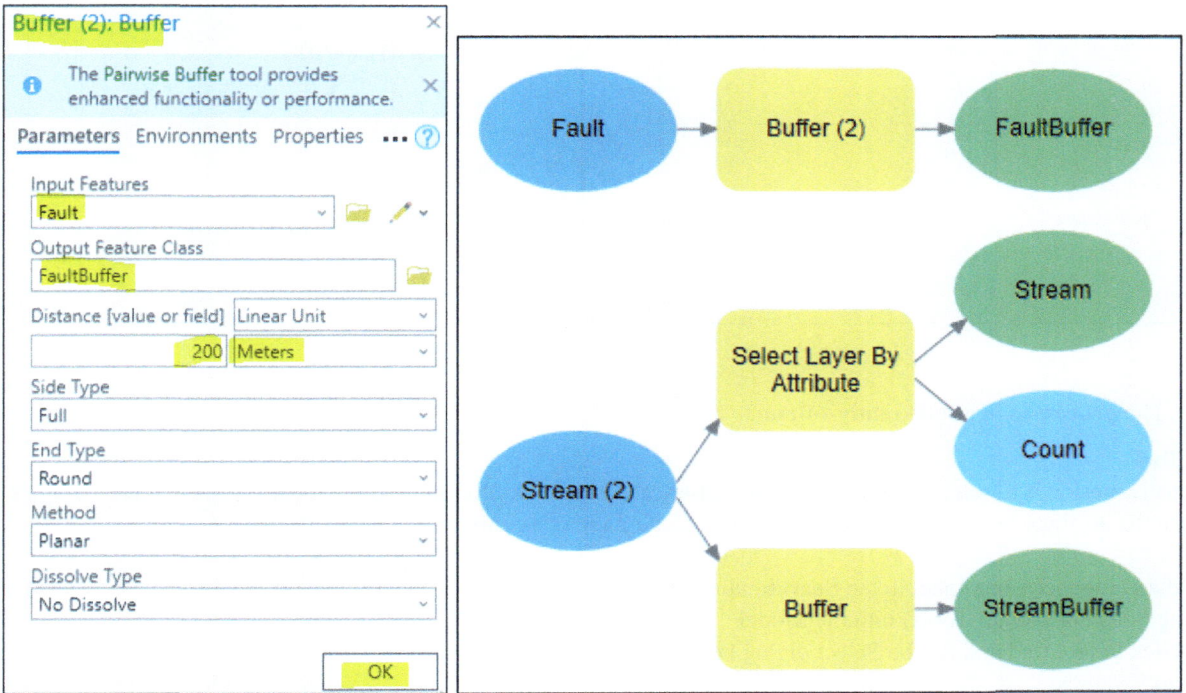

Buffer the Well:500 m
The third criterion is that the location of the NPP should be at least 500 m away from the groundwater wells.

40. In the Geoprocessing pane, drag once again the **Buffer** tool into the **Model**, and place it below the **Buffer** tool
41. Double click the **Buffer (3)** tool and enter inputs as shown below.
42. Fill the Buffer (3) dialog box as follows:
    (a) Input Features: Well
    (b) Output Feature Class: WellBuffer
    (c) Distance: 500 m
    (d) Accept the rest of the default
43. Click OK

44. Click **Save** button on the **Model** group
45. Click **Auto Layout** in the **View** group.

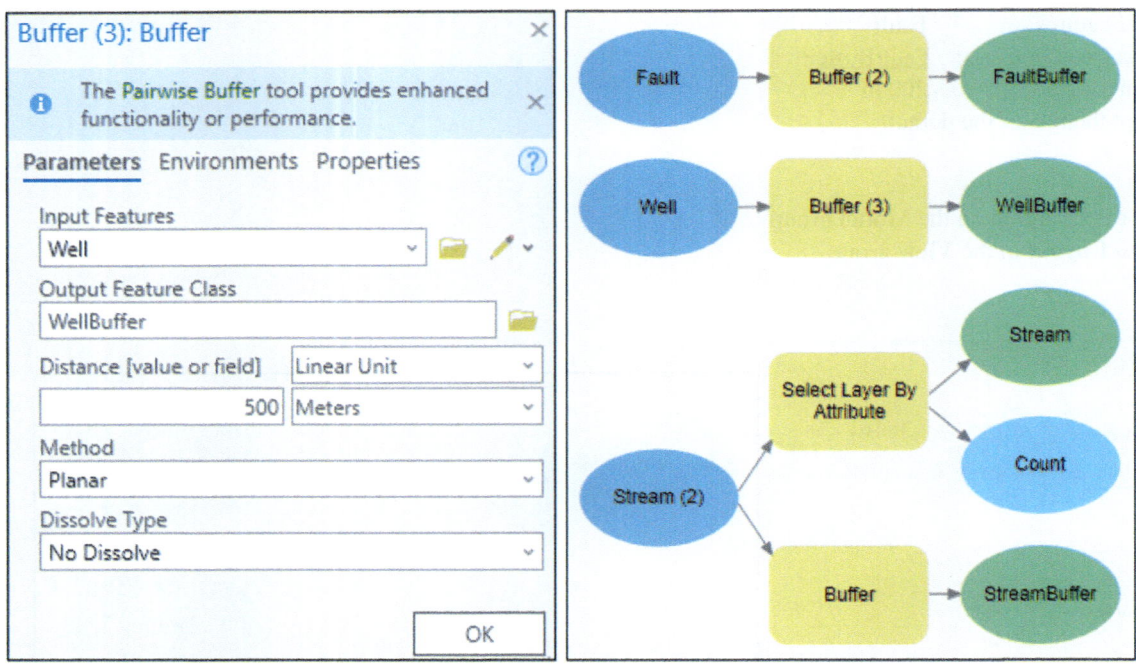

Result The **process** is in a state **ready-to-run**.

Select Tool

The fourth criterion is to select the impermeable geological formations **siltstone** and **marl-limestone** that are associated with code 2 in the attribute table of the **GEOL_KS** layer. Code 2 includes only impermeable formations.

46. In the **Geoprocessing** pane, in the **Search** window, type **Select.**
47. Drag Select (Analysis Tools) into the Model
48. D-click **Select** tool and fill the Select dialog box as below
49. Input Features: GEOL_KS
50. Output Feature Class: GEOL_Code2
51. In the SQL expression, fill the statement as "**Where Code is equal to 2**"
52. Click OK.
53. Click **Save** button on the **Model** group
54. Click **Auto Layout** in the **View** group.

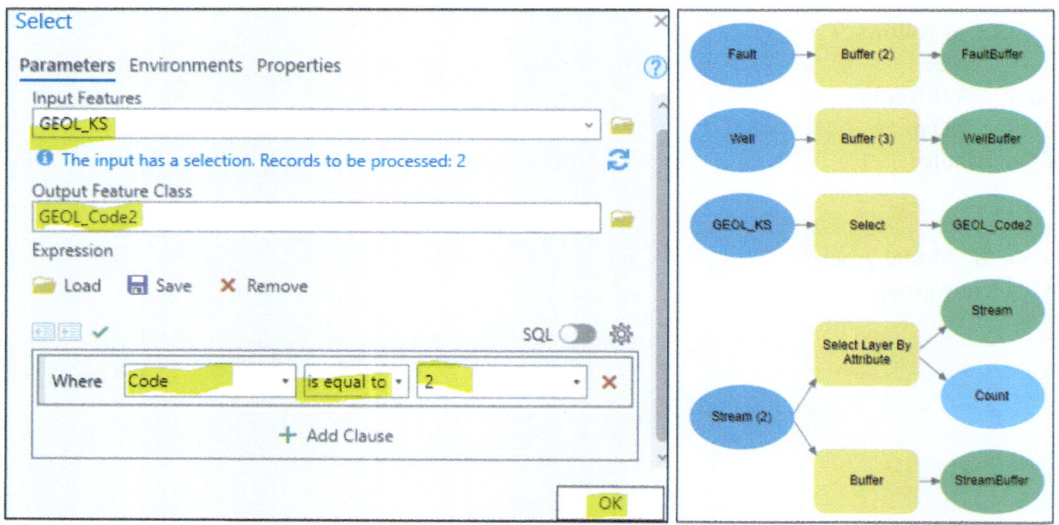

Union the Streambuffer, Faultbuffer, and Wellbuffer

To continue the analysis in the model, you will perform a geometric union of the 3-input feature classes: **StreamBuffer**, **FaultBuffer**, and **WellBuffer**. The Union tool will only work on polygon feature classes. All features and their attributes will be written to the output feature class. This step is required to combine all the buffers that have been created for the wells, faults, and stream and make them one polygon.

55. In the **Geoprocessing** pane, in the **Search** window, type **Union**.
56. Drag **Union** tool into the model
57. D-click **Union** tool and fill the **Union** dialog box as below
58. Input Features: StreamBuffer, FaultBuffer, WellBuffer
59. Output Feature Class: UnionBuffer
60. Attribute to join: All attributes
61. Click OK.
62. Click the **Auto Layout** button in the **View** group.
63. Click **Save** button on the **Model** group
64. Click Full Extent icon

Result The **process** is in a state **ready-to-run.**

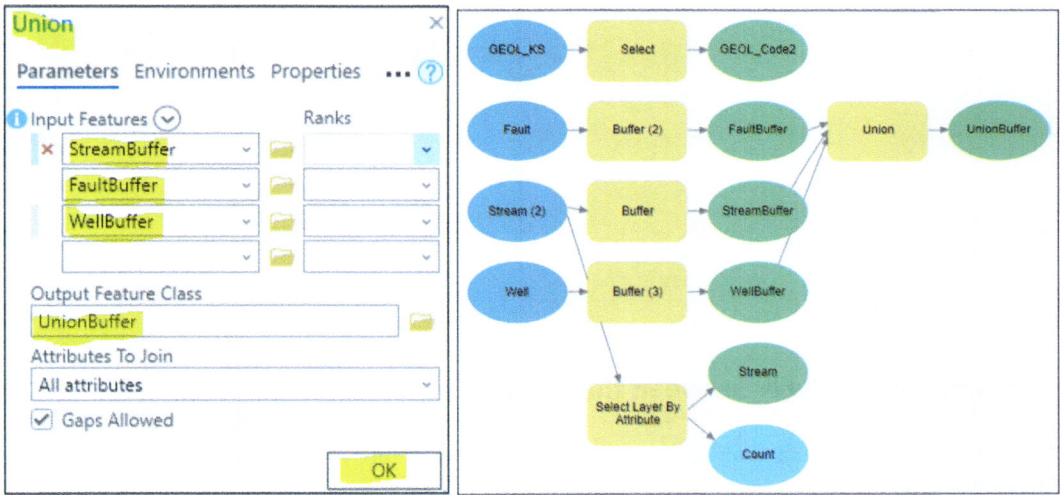

Erase Tool

The Erase tool creates a feature class by overlaying the input feature "**UnionBuffer**" that is created from the 3-input layers with the impermeable geological formation of code 2 in the **GEOL_Code2** layer that has been selected previously using the **Select** tool. Only those portions of the input features that fall outside the boundaries of **UnionBuffer** are copied to the output feature class.

65. In the **Geoprocessing** pane, in the **Search** window, type **Erase**.
66. Drag **Erase** (Analysis Tools) into the Model
67. D-click **Erase** tool and fill it as below
    (a) Input Features: **GEOL_Code2**
    (b) Erase Features: **UnionBuffer**
    (c) Output Feature Class: **Suitable**
68. OK

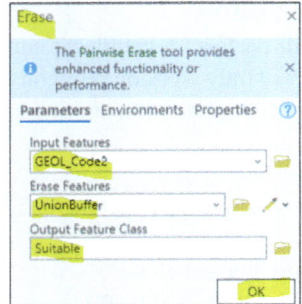

69. Click **Auto Layout** button on **View group** to rearrange all processes
70. Click **Save** button on the **Model** group

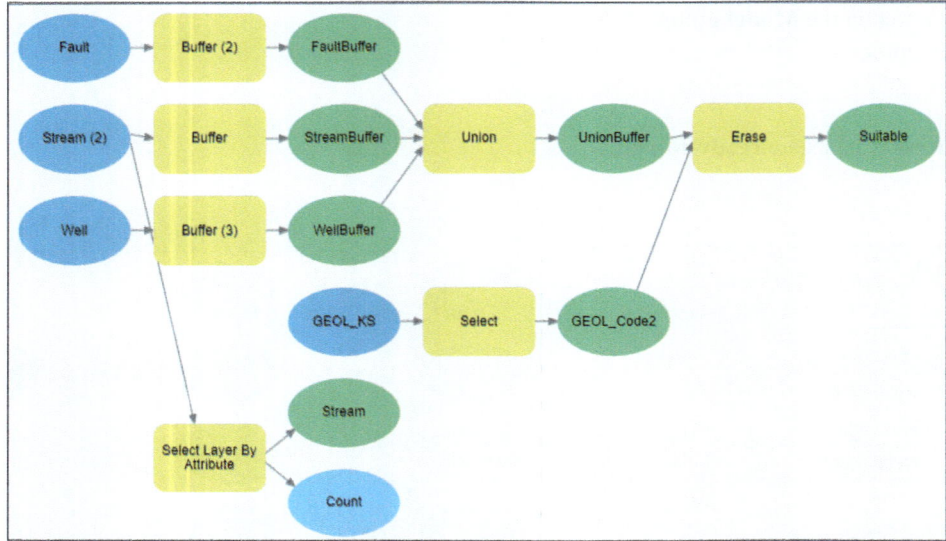

Draw a Connection Interactively Between the Variable and Tool

You can specify the data or other parameter settings for a geoprocessing tool to use by making a connection between the variables and tools in the model. There are two primary ways to make connections:

71. draw a connection interactively or
72. open the tool in the model and specify input parameters using model variables and map layers or by browsing to a dataset.

In this exercise, you will use the draw a connection interactively

73. Place the cursor over the variable (**GEOL_Code2**), the cursor changes to hand pointer
74. Click and hold the left mouse button while you move the cursor to the **Erase** tool
75. From the pop-up menu, select "**Input Features**"
76. Repeat and make a connection line from the **UnionBuffer** to the **Erase** tool
77. From the pop-up menu, select "**Erase Features**"
78. R-click Suitable and check Add to Display
79. Click Save on the Model group

Result The Erase process is connected, and the Model is ready to run

80. Click Auto Layout on the View group.

Validate and Run the Modelbuilder in Model Window

When validating the model, the system ensures that all model processes, which include tool input features and output features, are valid to be run properly. If all the variables and tools are validated, then the model will run. If some tools and their variables are not validated, the model will not run. Therefore, if something is wrong, you should open either the variable or the tool and fix it by providing the correct values. Running a model in ModelBuilder means that you open the model for editing and run it in the ModelBuilder window. You can run a single tool, a sequence of tools, or the entire model.

Alternatively, you can right-click anywhere in the model and select **Run**.

81. On the **ModelBuilder** tab on the ribbon, in the **Run** group, click the Run ▶ button

Results The "**Suitable**" layer is added to the CP, and drop shadows appear around all tools and all output variables. The shadows around the tools and output variables mean that the **Model** is in **the has-been-run** state. The shadow polygons represent suitable locations to build a nuclear power plant.

82. On the **ModelBuilder** tab on the ribbon, in the **Run** group, click the Run ▶ button

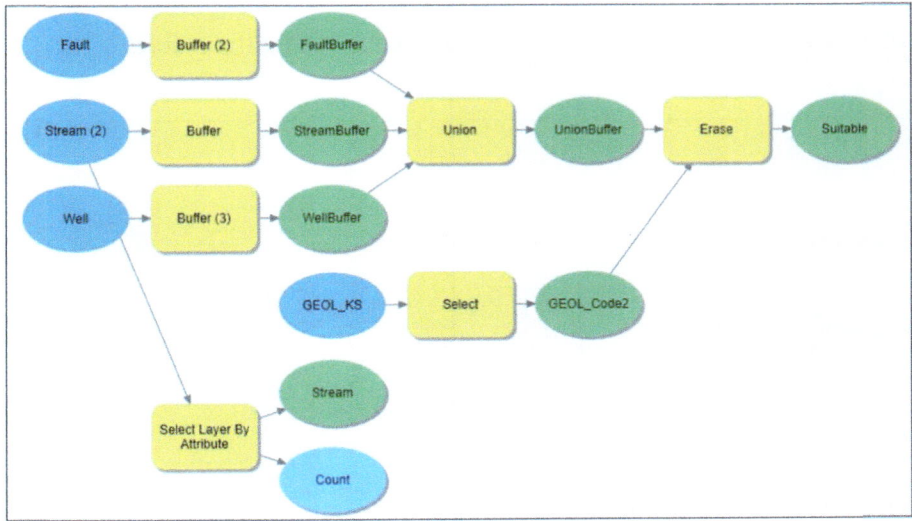

83. In the CP, r-click the "**Suitable**" layer is, you will see it consists of two formations.

Model Output

The **Suitable** layer generated by the model consists of two impermeable formations: the "**Siltstone**" and "**Marly Limestone**". The proposed nuclear power can be built only above these formations. If you examine the location of these two formations, you will notice that the model has selected only the impermeable layers that are located a substantial distance from the wells, faults, and the main stream based on the criteria employed in the scenario.

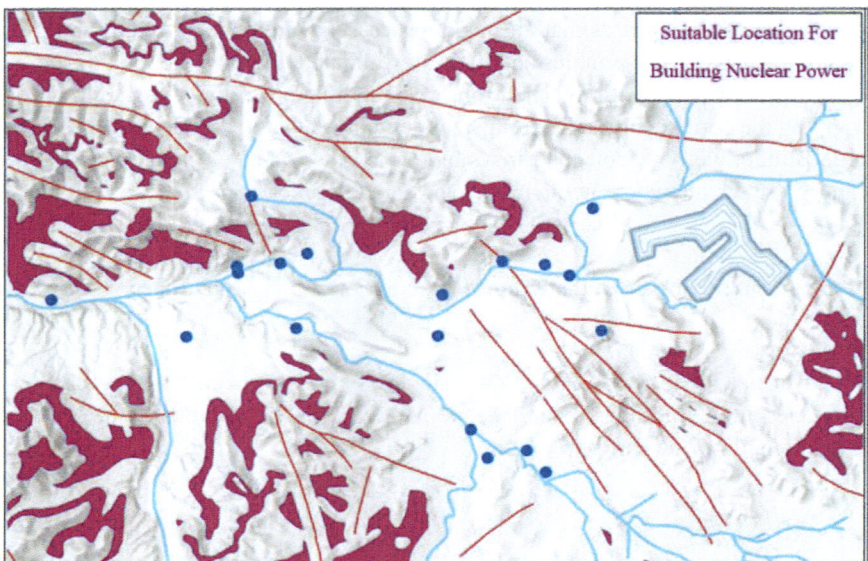

Validate the Model

84. On the **ModelBuilder** tab on the ribbon, in the **Run** group, click the **Validate** button

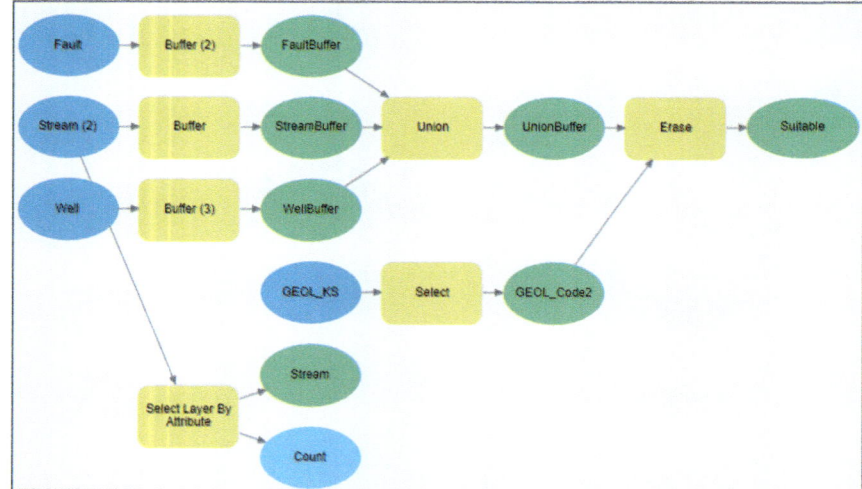

Results The shadow around the tools and the output variables are removed.

85. Once again click the Run button from the **Run** group

Results The shadow around the tools and the output variables will be restored again.

Change a Model Name and Label

Any model created has both a name and a label.

(a) The model label is displayed in the open model view tab, as well as the **Catalog** and **Geoprocessing** panes. The model label can contain spaces and other special characters.
(b) The model name is an internal name used by the system and is used when running the model from Python. The model name cannot contain spaces or other special characters.

You can change both the model name and label in the Model group
The other way is to change the name by right-clicking the model in its toolbox and select **Properties**.

86. In the ModelBuilder on the ribbon, in the Model group, click the Properties button
87. In the **Tool Properties: Model** enter inputs as shown below
88. Name: type "**SiteSelection**"
89. Label: type "**NPP Site Selection**"
90. Description: type " *The Dhuleil area is proposed to be a location to build a Nuclear Power Plant, as it's an ideal location due to the presence of a plentiful amount of reclaimed water that can be used for cooling purposes*"
91. Click OK
92. Click **Save** button, in the **Model** group
93. Close the Model by clicking the X in the Model tab above the Map View

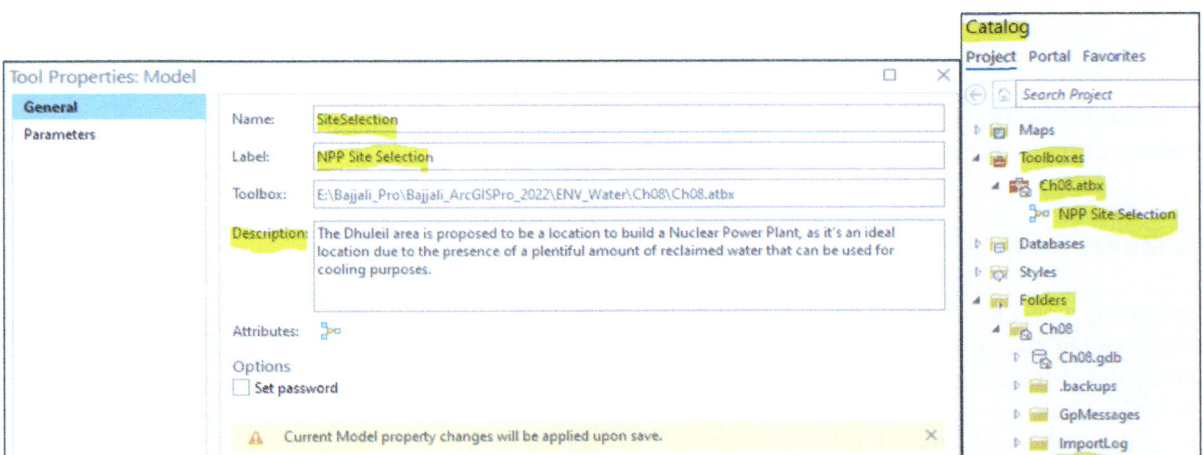

Result The name of the Model will be "**NPP Site Selection**", and this name will be displayed in the **Toolboxes** and the **Folder** in the **Catalog** pane.

Open The NPP Site Selection Model
To view an existing model diagram in ModelBuilder, right-click a model in a toolbox and choose Edit. The model diagram appears in ModelBuilder with the same layout, extent, and appearance with which the model was saved.

94. In the **Catalog** pane, under the Toolboxes, open **Ch08.atbx** and r-click **NPP Site Selection**, and click Edit

Result The **NPP Site Selection** open in the Map View.

Intermediate Data

When a model is run, output data are created for each process in the model (unless the process only modifies or passes the input directly). Some outputs created by the intermediate tools are only needed for generating the final output and have no use after the model has completed. For effective data management, these intermediate outputs can be set as model Intermediate Data and deleted altogether.

For example, the "**WellBuffer**" is only created to select the area that is at least 500 m away from the wells. After that, this variable is no longer needed.

95. In the ModelBuilder tab on the ribbon, in the Run group, click the Intermediate button

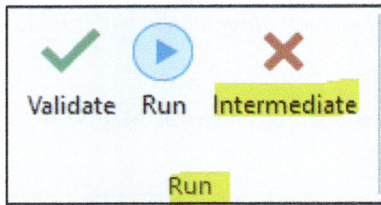

Result The intermediate data deleted from the workspace, the shadow around the tools (Buffer, Select, Union, and Erase) and all the output variables (with the exception of Stream (2) will be removed. This means the following:

- The **Buffer**, **Select**, **Union**, and **Erase** are reset because they generate intermediate data.
- The Select Layer By Attribute does not create intermediate data.

Create Model Tool

A geoprocessing model is saved as a model tool in a toolbox. Model tools can be run like any other geoprocessing tool from the **Geoprocessing** pane and used in other models and Python scripts. You can configure a model tool so you can process different datasets with different settings than those specified inside the model, without actually modifying the model variables in ModelBuilder.

To create a model tool, do the following:

(a) Build and save a model within ModelBuilder.
(b) Set model parameters.
(c) Set model tool properties.
(d) Document the tool.

96. In the **Catalog** pane, under **Toolboxes**, open the **Ch08.atbx** and d-click the **NPP Site Selection**
97. Geoprocessing pane, open with no parameters
98. Close NPP Site Selection dialog box.

Create Model Tool

Result The "**NPP Site Selection**" dialog box indicates that it has no parameters. This is because you built the **NPP Site Selection** without setting any model parameters. Therefore, when you open the model tool in the Geoprocessing pane, the tool dialog box will display no parameters.

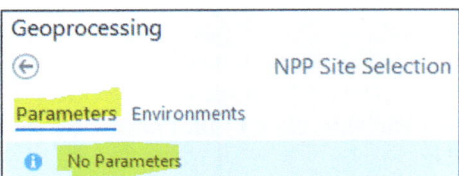

Copy the NPP Site Selection

96. In the **Catalog** pane, under **Toolboxes**, open the **Ch08.atbx** and r-click the **NPP Site Selection** and click Copy
97. R-click **Ch08.atbx** under **Toolboxes** and click Paste
98. Rename "**NPP Site Selection 1**" to "**Model2**"

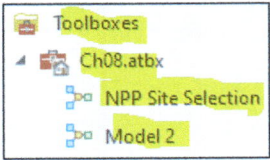

Result Model 2 is copied under **Ch08.atbx**.

99. In **Catalog** pane, r-click the **Databases** and click **New File Geodatabase** and call it **Model2** and click **Save**.
100. The **Model2.gdb** created in the **Databases**.

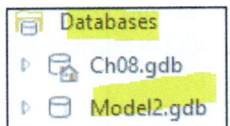

101. In the **Catalog** pane, under **Databases**, r-click **Model2** and click **Edit** to open **Model2**.

Result Model 2 open in the Map View.

102. Click **ModelBuilder** tab on the ribbon, in the **Model** group, click **Environments** button
103. Fill the Environments dialog box as follows:
    (a) Scratch Workspace: **Model2.gdb**
    (b) Accept the other default
104. Click OK to close the dialog box

Model Parameters

To display parameters on the tool dialog box and to add output datasets to a map, you must set model parameters within your model. Once the model parameters have been created, you can run the model as a geoprocessing tool, supplying different values and datasets for its parameters.

105. In the Model, r-click the **Stream** and click **Parameter** to make it model parameter. Then, right click the **Stream** again and click rename and call it **MainStream** and click outside to deselect it.
106. Right click the "**Buffer**", click **Create Variable**, click **From Parameter**, click **Distance**.
107. The **Distance (Value of Field)** display
108. R-click **Distance (Value of Field)** and click **Parameter**
109. In the **View** group, click **Auto Layout**.

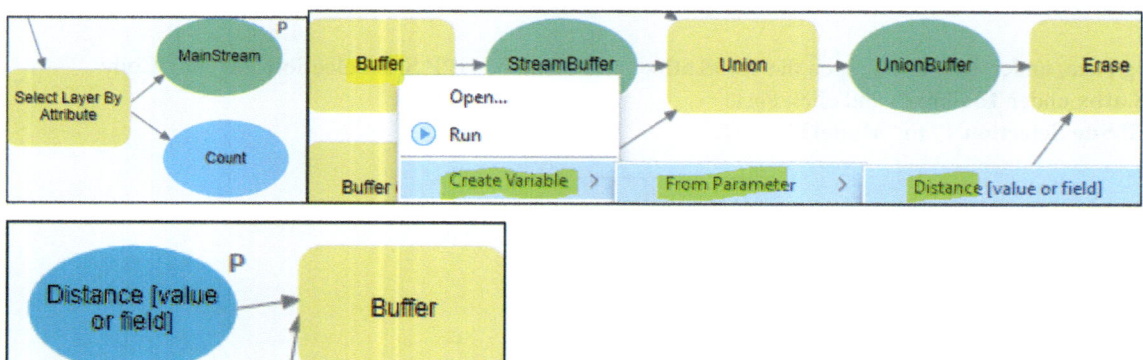

110. R-click the **Fault** and click **Parameter** to make it model parameter.
111. Right click **Buffer2**, click **Create Variable**, click **From Parameter**, and click **Distance**.
112. The **Distance (Value of Field) (2)** display. R-click **Distance (Value of Field)** (2) and click **Parameter**
113. In the **View** group, click **Auto Layout**.
114. R-click the **Well** and click Parameter.
115. R-click **Buffer (3)**, click **Create Variable**, click **From Parameter**, and click **Distance**.
116. The **Distance (Value of Field) (3)** display. R-click **Distance (Value of Field)** (3) and click **Parameter**.
117. In the **View** group, click **Auto Layout**.
118. R-click the variable **StreamBuffer** and click **Parameter**
119. R-click the variable **FaultBuffer** and click **Parameter**
120. R-click the variable **WellBuffer** and click **Parameter**
121. R-click **Union** tool, **Create Variable**, click **From Parameter**, click **Attributes To Join**
122. In the **View** group, click **Auto Layout**.
123. R-click **Attributes To Join** and click **Parameter**
124. R-click the **Union_Buffer** and click **Parameter**
125. R-click **GEOL_Code2** and click **Parameter**
126. R-click **Suitable** and click **Parameter**
127. R-click **Suitable** and make sure the **Add to Display** is checked
128. In the **View** group, click **Auto Layout**.
129. In the **Run** group, click **Validate**
130. Click Save on the Model group, and close the Model
131. the Model and Close

Fault
Buffer (2)
Distance [value or field] (2)
Well
Buffer (3)
Distance [value or field] (3)
Union
UnionBuffer
Erase
Suitable
Select
GEOL_Code2

Distance [value or field]
Buffer
StreamBuffer
Attributes To Join
Union
UnionBuffer
Erase
Suitable
Stream (2)
FaultBuffer
WellBuffer
Select
GEOL_Code2
GEOL_KS
Select Layer By Attribute
MainStream
Count
Fault
Buffer (2)
Distance [value or field] (2)
Well
Buffer (3)
Distance [value or field] (3)

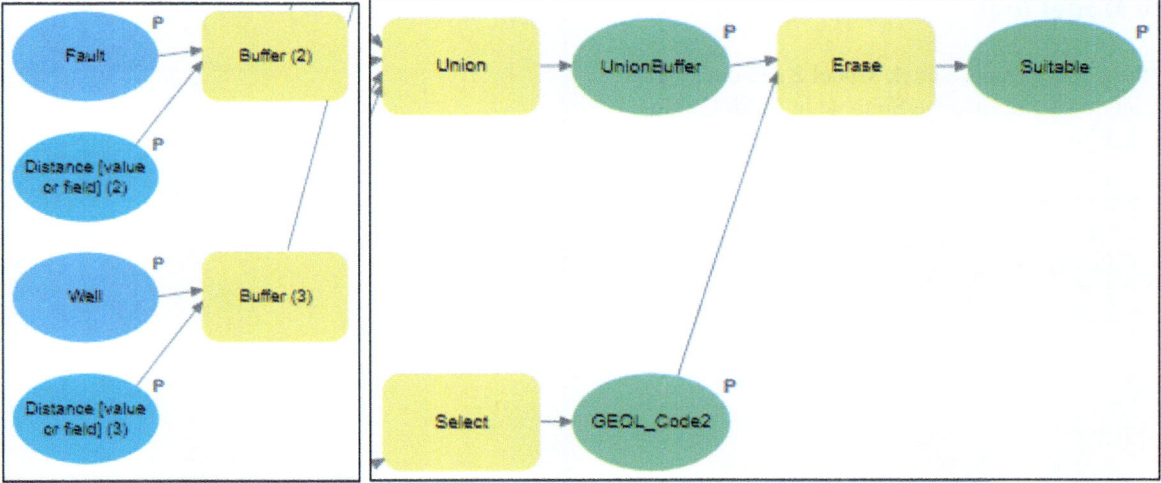

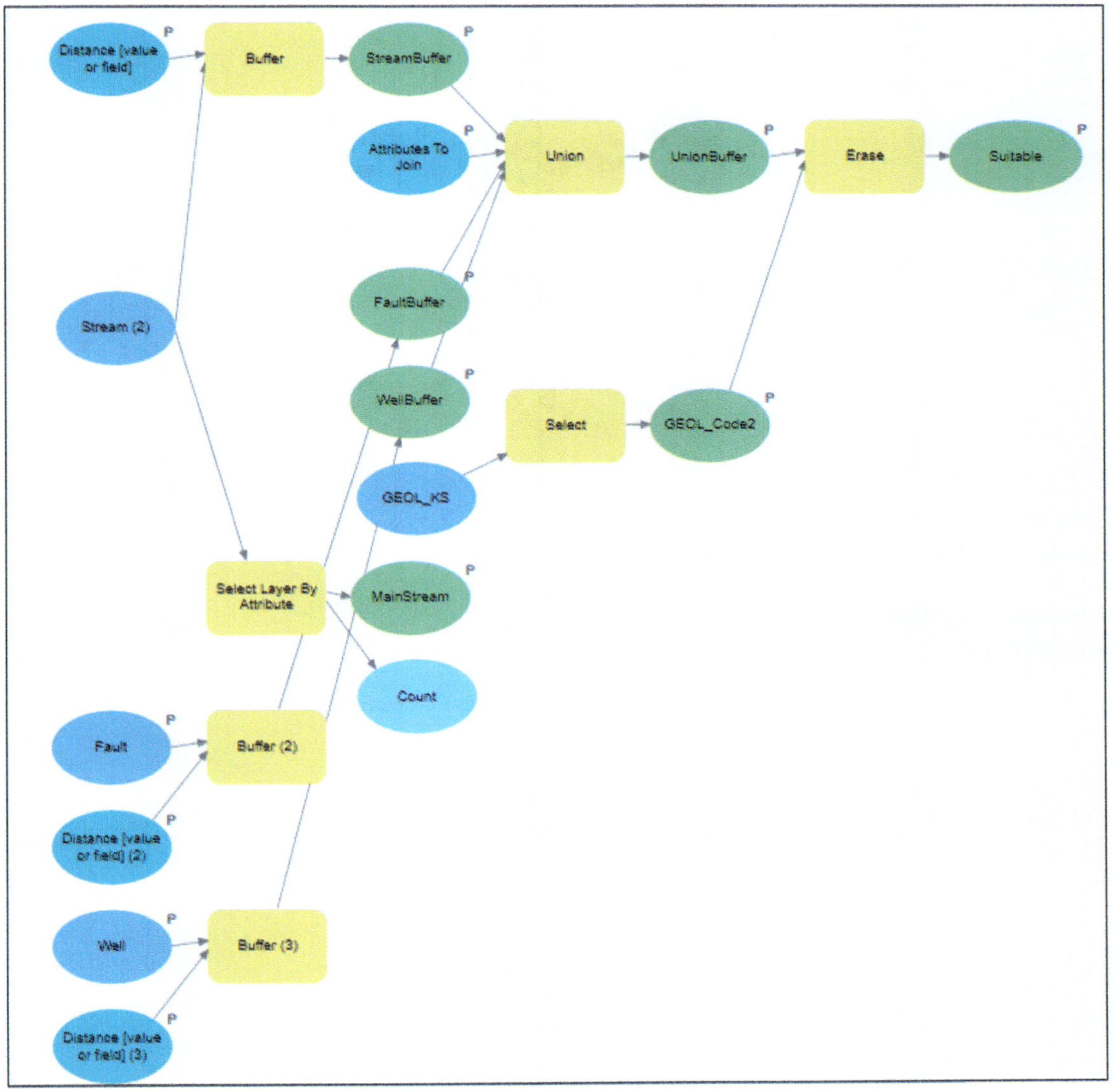

Run the Model Tool

132. In the Catalog Pane, r-click Model 2 and click Open.
133. Scroll down and save the Suitable in **Model2.gdb**
134. Click Run

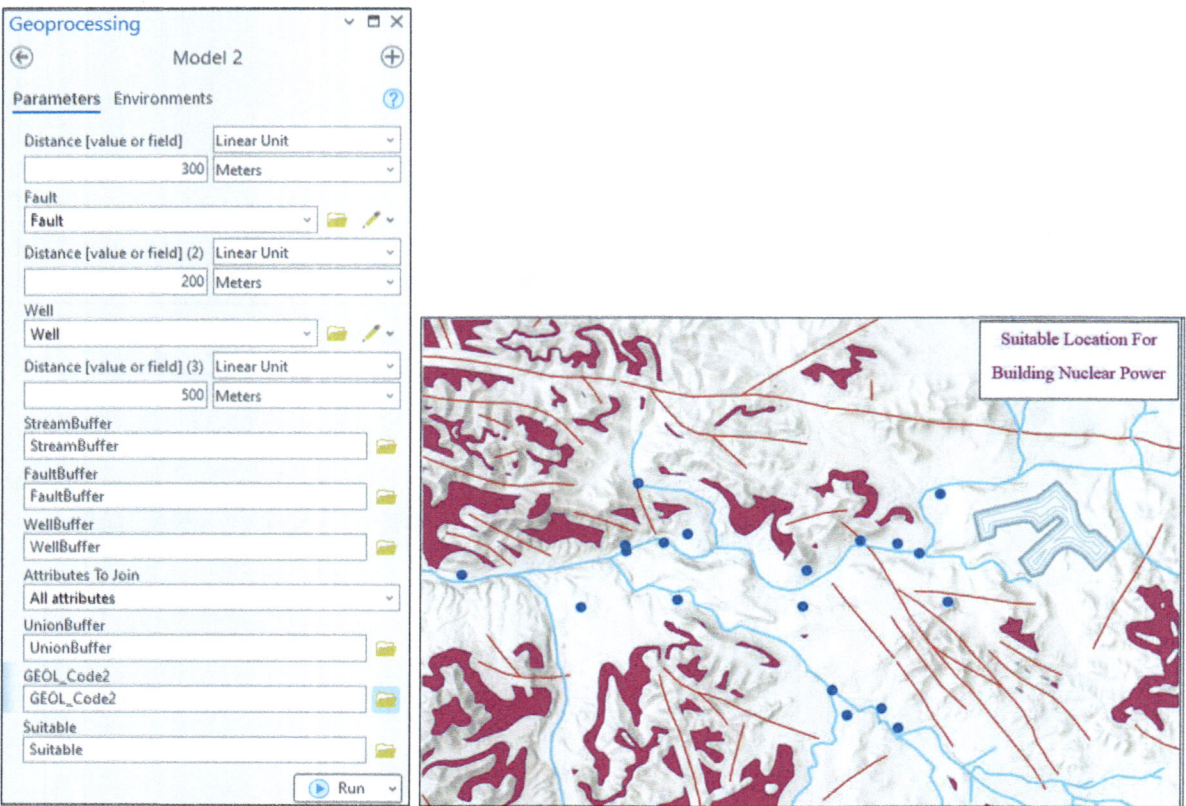

Result All layers, including "**Suitable**", will be added to the **NPP Model** map, and the **Suitable** layer run as a model tool is identical to the model that run within a model.

135. Save the model and the project
136. Close your project.

Geocoding

The Geocoding module allows GIS users to explore and set locator properties and automate geocoding workflows. Geocoding is the process of transforming a description of a location such as a pair of coordinates, an address, or a name of a place to a location on the Earth's surface. You can geocode by entering one location description at a time or by providing many of them at once in a table. The resulting output is a location feature with attributes, which can be used for mapping or spatial analysis. With geocoded addresses, the address locations can be spatially displayed as points in ArcGIS Pro. Geocoding helps users recognize patterns by using some of the analysis tools available with ArcGIS Pro.

Users can geocode addresses that are stored in a table in a single field or multiple fields. A single input field stores the complete address, for example, 712 N 22ND St NE, Superior, WI 54880. Multiple fields are supported if the input addresses are split into multiple fields, such as Address, City, State, and ZIP for a general United States address.

A locator is the tool used to perform geocoding operations. With this module, you can view and modify properties on the locator to tune it to your specific geocoding needs by customizing it for performance or quality. You can also use this module to perform various geocoding operations, including finding the location of a place or address, finding the closest place or address to a given location, or generating a set of autocomplete suggestions for partial input. A locator is a portable file used to perform geocoding, the process of finding addresses and places on a map. Locators contain a snapshot of the reference data used for geocoding, as well as indexes and local addressing knowledge that help return the best match during the geocoding process.

A locator can be accessed as a service on your portal, such as ArcGIS World Geocoding Service, as a service accessed through an AGS server connection, or as a file on disk.

Geocodes a table of addresses. This process requires a table that stores the addresses you want to geocode and an address locator or a composite address locator. This tool matches the addresses against the locator and saves the result for each input record in a new point feature class. When using the ArcGIS World Geocoding Service, this operation may consume credits. Performing geocoding operations using the ArcGIS World Geocoding Service requires an ArcGIS Online organizational subscription, and it consumes credits. The organizational account must have enough credits to complete the entire geocoding request.

In this chapter, you will learn techniques for finding various types of addresses. The user will be introduced to the preparation of geographic data necessary for address matching called reference data. Technically geocoding is a process of using an address locator to enter address text or a table of addresses to find corresponding address locations in a geographic database.

Geocoding requires the following:

1. Address table: This table includes the addresses that need to be converted into a feature class location.
2. Reference data: This is a snapshot of geographic information (point, line, or polygon feature classes) with address information such as streets or feature class
3. Address locator: An address locator lets you convert textual descriptions of locations into geographic features. Address locators are stored and managed in a workspace you choose. The workspace can be a file folder or geodatabase (file or personal geodatabase).

Supplementary Information The online version contains supplementary material available at https://doi.org/10.1007/978-3-031-42227-0_9.

Once you know what you want to find, the next step to prepare for geocoding is to build or locate sources of geographic data for reference data. In this chapter, you will perform geocoding based on two approaches. The first approach uses the zip code, and the second approach uses the street address.

In this exercise, you are going to geocode based on two ways: geocoding based on zip code and geocoding based on street address. The ZIP code is usually associated with the residents and business addresses. Therefore, utilizing the ZIP code address is an easy method to convert the zip codes into a point feature class. Geocoding based on street addresses is the process of converting street addresses into geographic coordinates (latitude and longitude) and then using the points in ArcGIS Pro for further analysis.

Part 1: Geocoding Based on Zip Code

Scenario 1 Approximately three-quarters of the population in Wisconsin and especially those who are living in rural areas rely on groundwater as a source for domestic purposes. Many of these wells are shallow and susceptible to groundwater contamination, especially from septic tanks on owners' properties. As a new employer at the USGS, you have been given a database table of well owners that their wells have a nitrate concentration higher than 20 mg/L. You have been asked to geocode the wells' addresses; based on their zip code and prepare a table showing the average nitrates in the wells in each zip code. This process consists of the following:

(a) Geocode the database file based on the zip code
(b) Summarize the matched addresses based on the zip code
(c) Join the summarized matched addresses with the geocoded layer

Launch ArcGIS Pro and Connect to Data

1. Launch **ArcGIS Pro** and click "**Open another project**"
2. Browse to **\\ENV_Water\ Ch09** folder and click **Ch09.aprx** and then click OK
3. The ArcGIS Pro open and has no data attached to it
4. In **Catalog** pan, r-click the **Folder**, and point to **Add Folder Connection**
5. Browse to **\\ENV_Water\Ch09** folder and select **Data_Ch09** (or **\\Database\ Data_Ch09**) and click **OK**
6. In the **CP**, select the **Map** and press the **F2** key and change the name to **Geocode Zip Code**
7. Open the **Folder** in the **Catalog** pane, expand **Data_Ch09\Data\Q1**, open **GEOCODE.gdb** and drag **Well_Owner** and **ZipCode_WI** into the **Map View.**

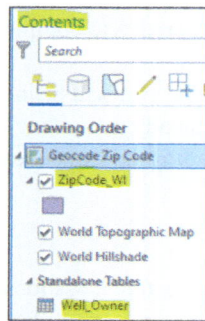 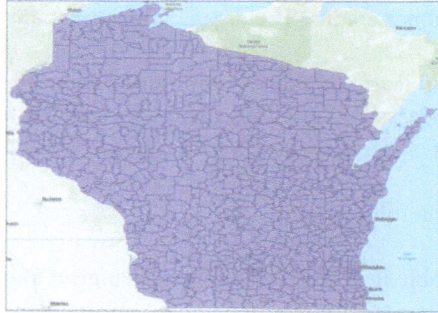

Result The **ZipCode_WI** feature class is the zip code in Wisconsin, and **Well_Owner** represents the name and address of the private well owners.

Locate Places and Addresses

Create Locator

Creates a locator that can find the location of an address or a place, convert a table of addresses or places to a collection of point features, or identify the address of a point location.

8. Click the **Analysis** tab on the ribbon, in the Geoprocessing group, click the **Tools** button
9. The **Geoprocessing** pane appears.
10. Click the **Toolboxes** tab, open **Geocoding** Tools, and click Create Locator
    - (a) Country or Region: United States
    - (b) Primary Table(s):   ZipCode_WI
    - (c) Role: ZIP

    Under Field Mapping
    - (d) Feature ID: ObjectID
    - (e) *ZIP: ZIP
    - (f) Output Locator: ZIPCODE_WI_Create Locator
    - (g) Language Code: English
    - (h) Accept the rest of the default
11. Click Run

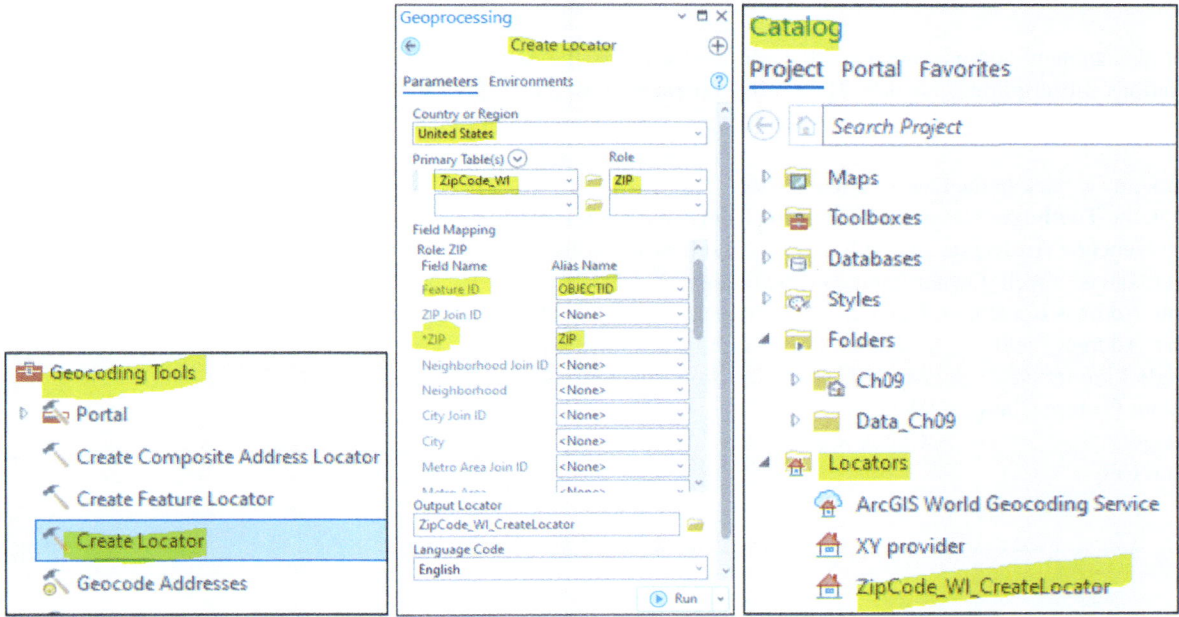

Result The create locator "**ZIPCODE_WI_CreateLocator**" is established and will be stored in **Catalog** pane under **Locators**.

Locate Places and Addresses

You can find addresses and locations on the map by using the **Locate button** located in the **Map** tab.

12. In the **Map** tab on the ribbon, click the **Inquiry** group, click the **Locate** button. The Locate pane display
13. In the **Search** text box in the **Locate** pane, type **54880** and **Enter**

Result The Locate pane displays two locations for "54880". The pink and blue dots generated from the **ArcGIS Online World Geocoding Service** and the "**ZipCode_WI_CreateLocator**" were created in the previous steps. You can use either one of them or both

14. Close to the **Locate** pane, the two locations disappear from the Map View.

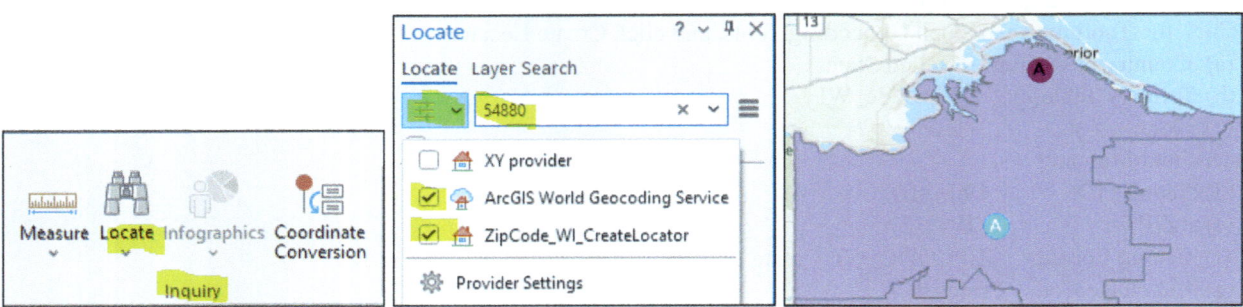

Geocode the Addresses

To geocode a table of addresses using either address locator or the ArcGIS World Geocoding Service, an ArcGIS Online for organizations subscription is needed. The service operates under a credit-based usage model that allows you to pay only for what you use. If you do not already have a subscription, purchase one or request a free trial.

15. Click arrow back in the Geoprocessing pane
16. Click the **Toolboxes** tab, open **Geocoding Tools**, and click **Geocode Addresses**
17. The **Geocode Addresses** dialog box opens and fills it as follows:
18. Input Table: Well_Owner (from Data\Q1\Geocode.gdb)
19. Input Address Locator: ZipCode _WI_CreateLocator (from the drop-down arrow)
20. Input Address Field: Single Field (from the drop-down arrow)
21. Single Line Input: ZipCode (from the drop-down arrow)
22. Output Feature Class: Well
23. Category: Check Postal
24. Click Run

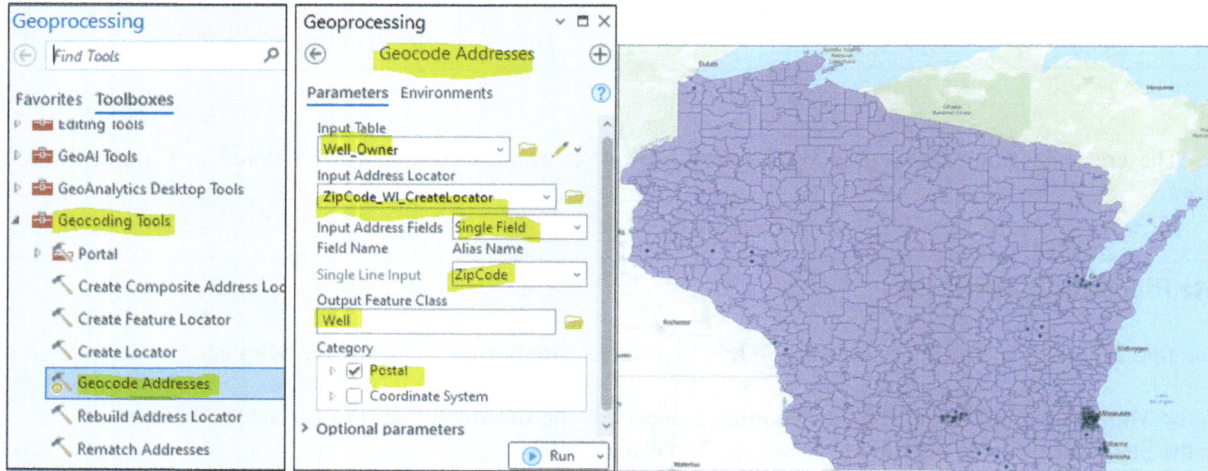

Result The Wells display in the in the Map View and in the CP.

25. In the CP, r-click the **Well** and open the attribute table

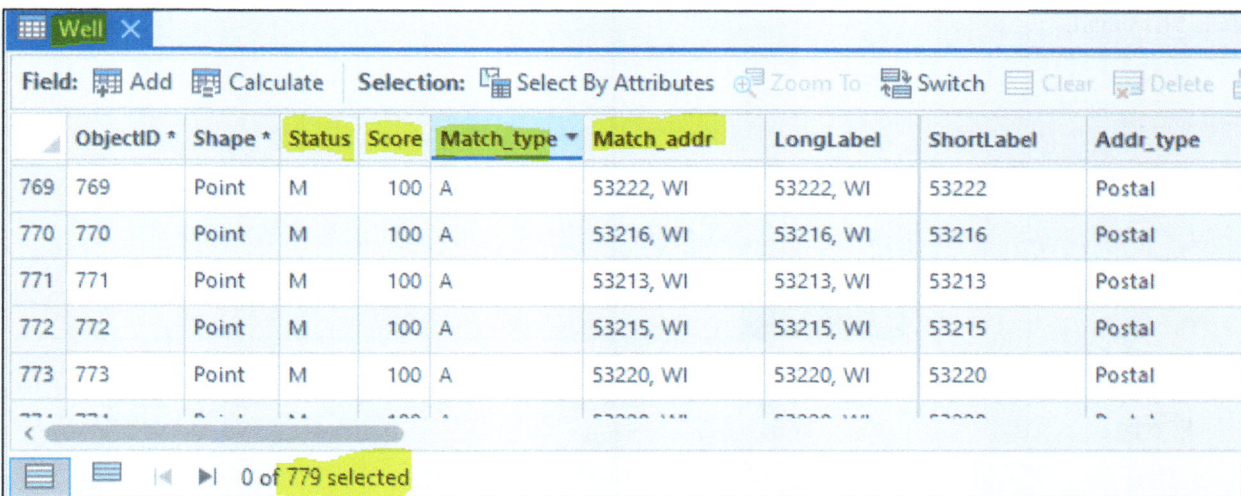

Result The **Well** attribute table is open and has 779 records and many fields, such as **Status, Score, Match_Type, Match_addr**, and all the fields from the **Well_Owner** table that are used for geocoding. The **Status** field indicates that the candidate either matched or did not match, while the **score** field means the percent of matching by the address locator. In this exercise, all the candidates matched 100%.

26. Make sure that the attribute table of the **Well** is open
27. R-click "**Match_adrr**" field and click Sort Ascending
28. Close the attribute table after examining the "**Match_adrr**" field

Comment The attribute table of the **Well** feature class shows that each zip code contains dissimilar number of wells.

Wells in Each Zip Code and Average Nitrate Concentration

This step will show the number of wells and their average nitrate in each zip code in the state of Wisconsin. To obtain information about the number of wells and their average nitrate, the dissolve tool should be used. The dissolve tool aggregates the well feature class based on the "**Match_adrr**" field in the attribute table. The tool will allow you to use statistics of any field present in the attribute table of the well feature class.

29. Click the back arrow in the **Geoprocessing** pane, open the **Toolboxes**, open **Data Management Tools**, then open **Generalization**, and click **Dissolve**
30. The **Dissolve** dialog box is opened and filled in as follows:
    (a) Input Features: Well
    (b) Output Feature Class: Well_Nitrate
    (c) Dissolve Field: Match_addr
    (d) Statistic Field: NO3 and check Mean
    (e) Statistic Field: ZipCode and check Count
31. Accept the other default
32. Click Run

Result The **Well_Nitrate** layer is displayed in the CP, and if you open the attribute table, you see three fields: **Match_addr**, **MEAN_USER_ NO3**, and **COUNT_USER_ZipCode**. The table has been reduced from 779 records to 66 records. The **COUNT_USER_ZipCode** field shows how many wells exist in each zip code. The **MEAN_USER_ NO3** is the average of nitrate in all the wells existing in a zip code. Certain Zip Code includes one well and others include many more. For example, the zip code "**53132**" includes 1 well, and its mean nitrate is 47, while the zip code "**53202**" has 22 wells, and their average nitrate is 51.13 mg/L.

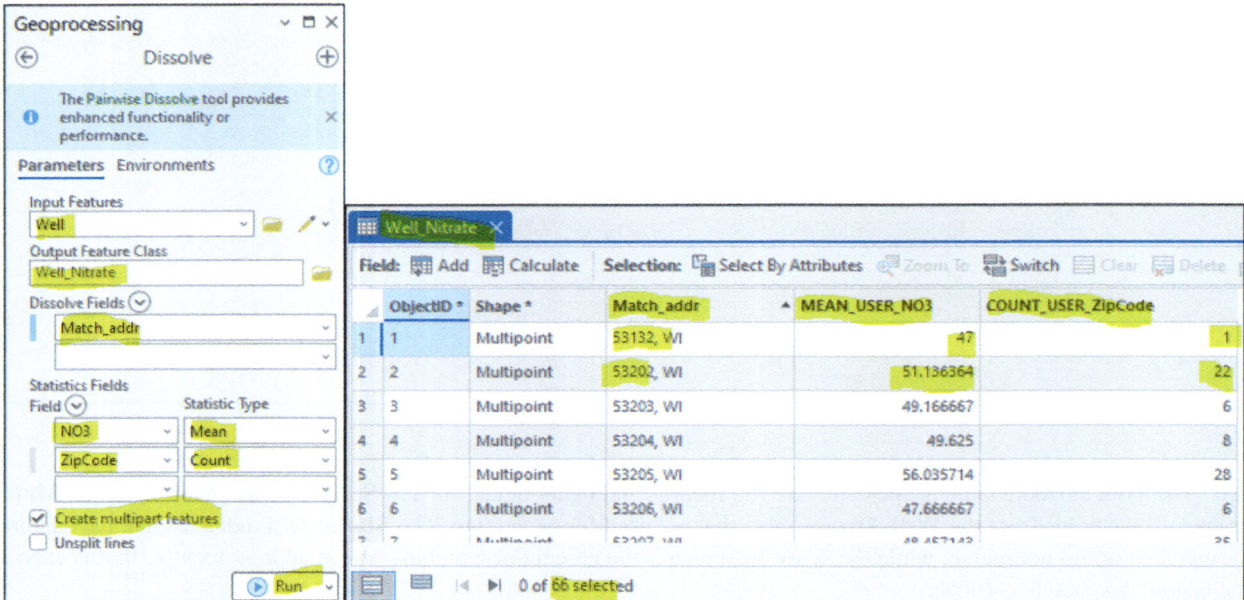

Symbolizing

You will provide symbology for the geocoded wells in the zip codes by using the graduated symbols. This method is used to show a quantitative difference in the nitrate concentration of the well features by varying the size of symbols. Nitrate is classified into ranges that are each then assigned a symbol size to represent the range. The nitrate data will have 4 classes; therefore, four different symbol sizes are assigned different symbol colors. The zip code that has wells with a high average nitrate concentration will have a larger symbol and red color, while the zip code that has wells with a lower average nitrate concentration will have a smaller symbol and blue color.

33. In the CP, r-click **Well_Nitrate**, click **Symbology**
34. In the Symbology – Well Nitrate pane
35. Choose **Graduated Symbols** from the drop-down arrow under the Primary symbology
36. Field: MEAN_USER_NO3
37. Classes: 4
38. Method: Manual Interval
39. Change the symbol size from 8 to 18
40. In Classes tab, under Upper value
41. Type ≤ 40, ≤ 50, ≤ 60, and ≤ 70.5 in the 1st, 2nd, 3rd, and 4th ranges, respectively
42. R-click each symbol and change the color of each symbol from bottom to top into red, pink, green, and blue, as seen below.
43. Click More and uncheck "Show values out of range"
44. Close the Symbology – Well Nitrate pane
45. Save the project

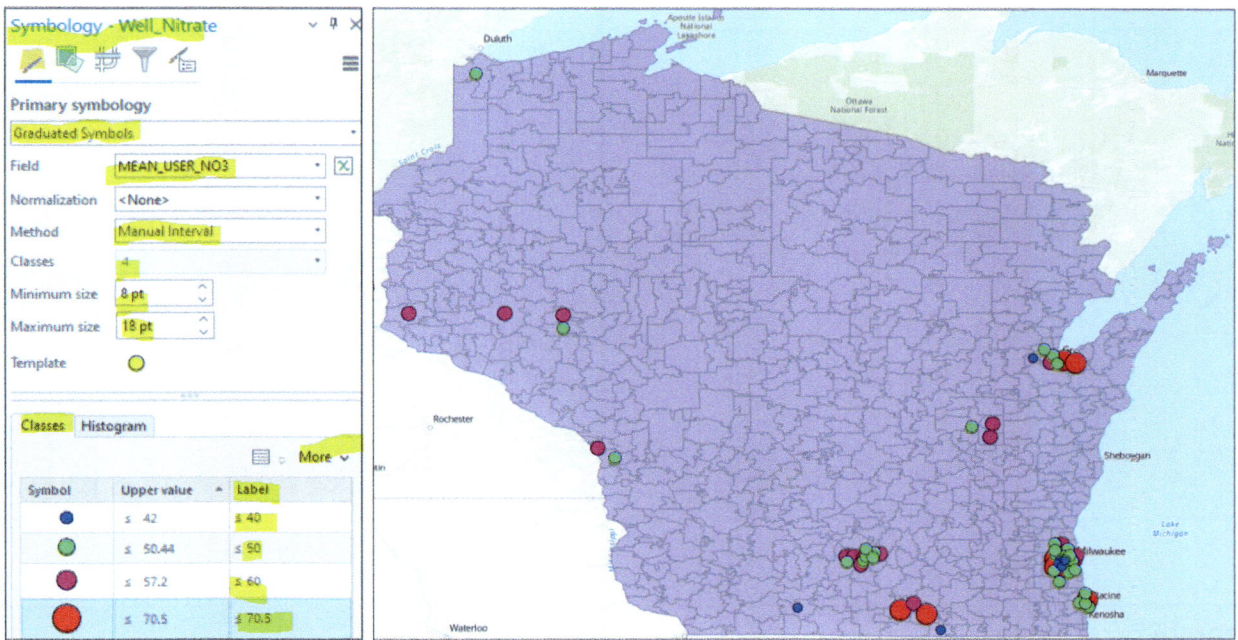

46. Open the attribute table of the **Well_Nitrate** layer and answer the following questions:
    (a) Which Zip code has the highest number of wells and what is the ObjectID numbers?
    (b) Which Zip code has the highest average NO_3 concentration and what is the value?
    (c) Which Zip code has the lowest average NO_3 concentration and what is the value?
    (d) How many Zip codes have average NO_3 concentrations higher than 50 mg/L?

Part 2: Geocoding Based on Street Address

Geocoding is a fundamental part of business data management. Every organization maintains address information for each client. In water resources, Wisconsin private wells have an address. The address is stored in a table that contains the well owner's address, well depth, and other relevant information.

Scenario 2 You are working for Douglas County in the city of Superior. You have been given an Excel table that contains well owners and their addresses. You have been asked to geocode the well address owners by converting them into a well point feature class. The well layer will be used in the analysis to select the deepest wells that are tapping the sandstone aquifer to be used as an alternative source for water supply in case of emergency.

1. Click **Insert** tab on the ribbon, in the **Project** group, click **New Map** and call it **Superior**
2. Open the **Folders** in the **Catalog** pane, expand **Q2** in Data_Ch09\Data select **Street.shp** and drag it into the Map View.
3. In the CP, click the symbol of the **Street** and select "**Arterial Street**" from the Symbology – Street pane from under the Gallery tab

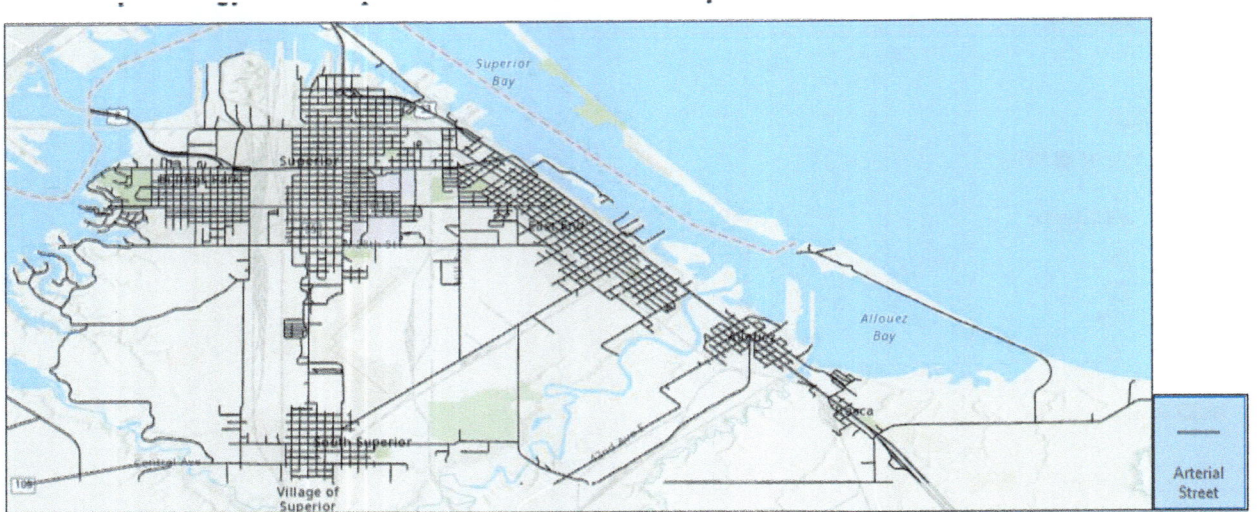

Integrate Excel Table

The Excel table **Well.xls** has information about the wells, and your duty is to convert it into a **table** format in ArcGIS.

4. Click the back arrow in the Geoprocessing pane
5. Click the **Toolboxes** tab, open **Conversion Tools**, and open **Excel**
6. Click **Excel to Table**, the Excel to Table dialog box open, fill it as below
7. Input Excel File: Well.xls (from Data\Q2) and click OK
8. Output Table: Address_Table
9. Sheet: Well
10. Accept the rest of the default
11. Click Run
12. In the CP, r-click Address_Table, and click open, explore it and then close it

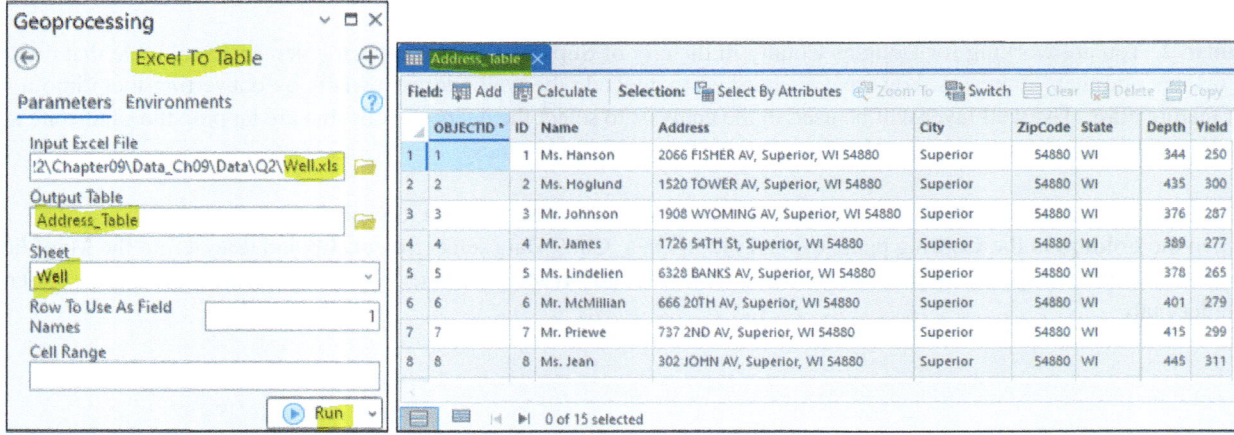

Result The **Well Excel** file is now converted into **Address_Table** and added to the CP, and both files are identical and consist of 15 records.

Build the Street Locator Using Create Locator

Geocoding is performed in ArcGIS Pro with an address locator. An address locator is a dataset stored in either a geodatabase or a file folder that contains information about local conventions for addresses (known as an address locator style) and embedded map data such as street centerlines with address ranges (known as reference data). When geocoding is performed, the address locator interprets an address using the address locator style and finds that address on a map using the reference data. An address locator is created based on a specific locator style. The style determines the type of addresses that can be geocoded, the field mapping for the reference data, and what output information of a match would be returned. It also contains information about how an address is parsed, searching methods for possible matches, and default values for the geocoding options.

13. Click the back arrow of the Geoprocessing pane
14. Click the **Toolboxes** tab, open **Geocoding Tools**, and click **Create Locator**
15. The **Create Locator** dialog box opens and fills it as follows:
16. Country or Region: United States
17. Primary Table(s):   Street
18. Role: Street Address
19. Feature ID: OBJECTID
20. Left House Number From   From_Left
21. Left House Number TO: To_Left
22. Right House Number From: From_Right
23. Right House Number TO: To_Right
24. Prefix Direction: PrfxDir
25. Street Name: STNAME
26. Suffix Type: STTYPE
27. Suffix Direction: SufixDir
28. Left City: City
29. Right City: City
30. Left State: State
31. Right State: State
32. Output Locator: Street_CreateLocator
33. Language Code: English
34. Run

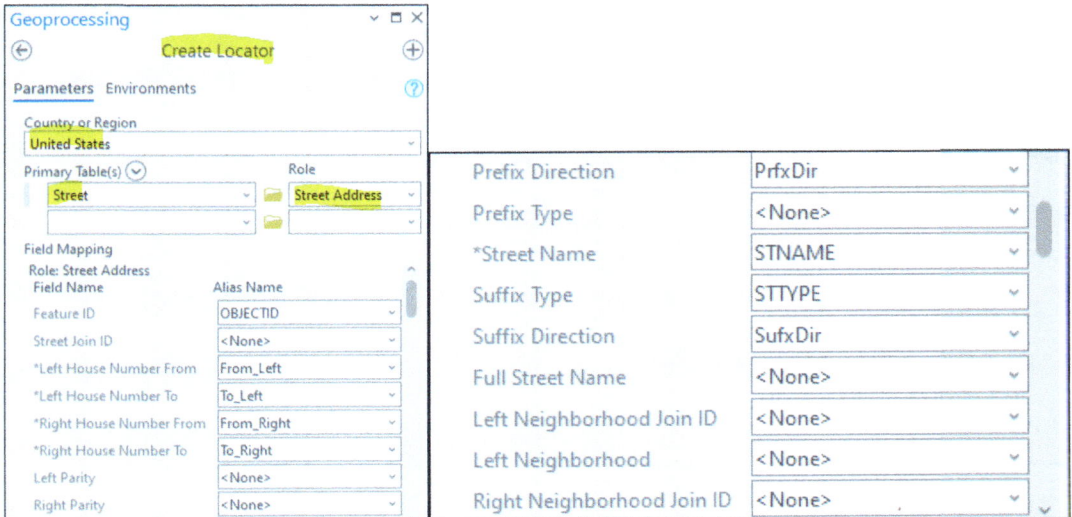

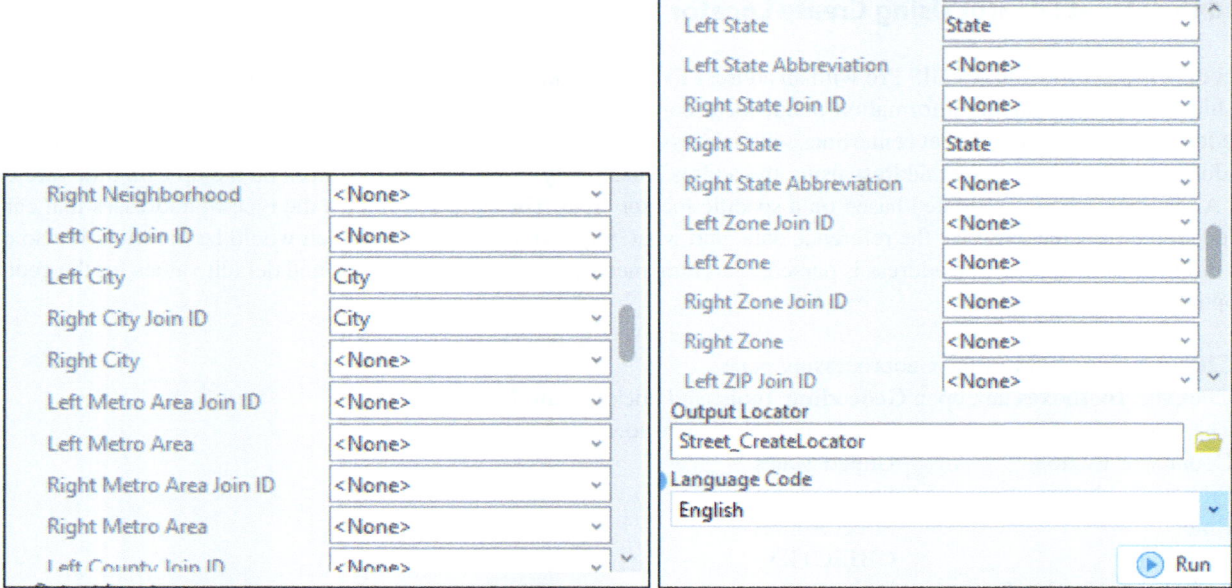

Test Your Address Locator

In this step you will test the "**Street_CreateLocator**" that you just created. You will use the Locate button to quickly search for an address in the city of Superior and display the corresponding location on a map.

35. Click **Map** tab on the ribbon, in **Inquiry** group, click **Locate** button
36. Click the drop-down **Options** and check only the "**Street_CreateLocator**".
37. In the search box, type **712 N 22ND St, Superior** and press Enter.

Result A point display on the map and the **Locate** pop-up display showing information about the address

38. Close the **Locate** pop-up, the graph point of the address **712 N 22ND St, Superior** disappears from the map

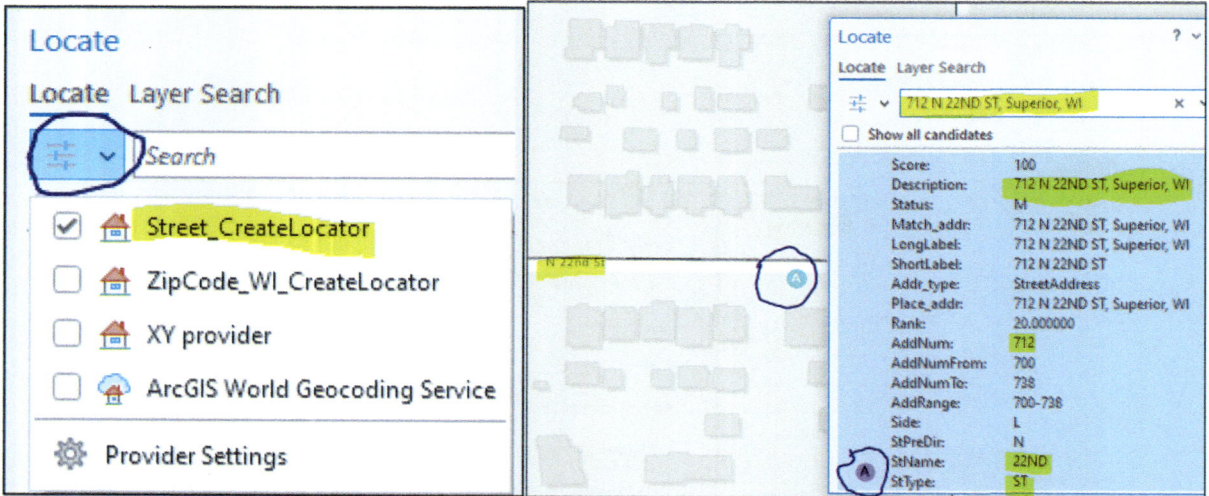

Geocode Well Owner by Address

This step will use the **Address_Table** that stores the name and addresses of the well owners that you want to geocode. The tool matches the stored addresses against the "**Street_CreateLocator**" and saves the result for each input record in a new point feature class.

Note user can use the ArcGIS World Geocoding Service, but this operation consumes credits.

39. In CP, r-click **Address_Table** and click **Geocode Table**
40. Click "**Go to Tool**" tab at the bottom of the **Geocode Table**
41. The **Geocode Table** dialog box display, fill it as below
42. Input Table: **Address_Table**
43. Input Locator: "**Street_CeateLocator**"
44. Input Address Fields: **More than one field**
45. Address or Place: **Address**
46. City: **City**
47. State: **State**
48. Zip: **ZipCode**
49. Output: **Well_Superior** (Ch09.gdb)
50. Category: **Address**
51. Accept the rest of the default
52. Run

Result The Geocoding Completed dialog box display states that **12 addresses matched** and **3 unmatched**. The **Well_Superior** feature class is displayed in the CP.

53. Click **No** for "**Start rematch process?**"
54. Close the **Geocoding Completed** dialog box
55. In CP, click **Well_Superior** symbol of the and symbolize it with **circle1** and **blue** color

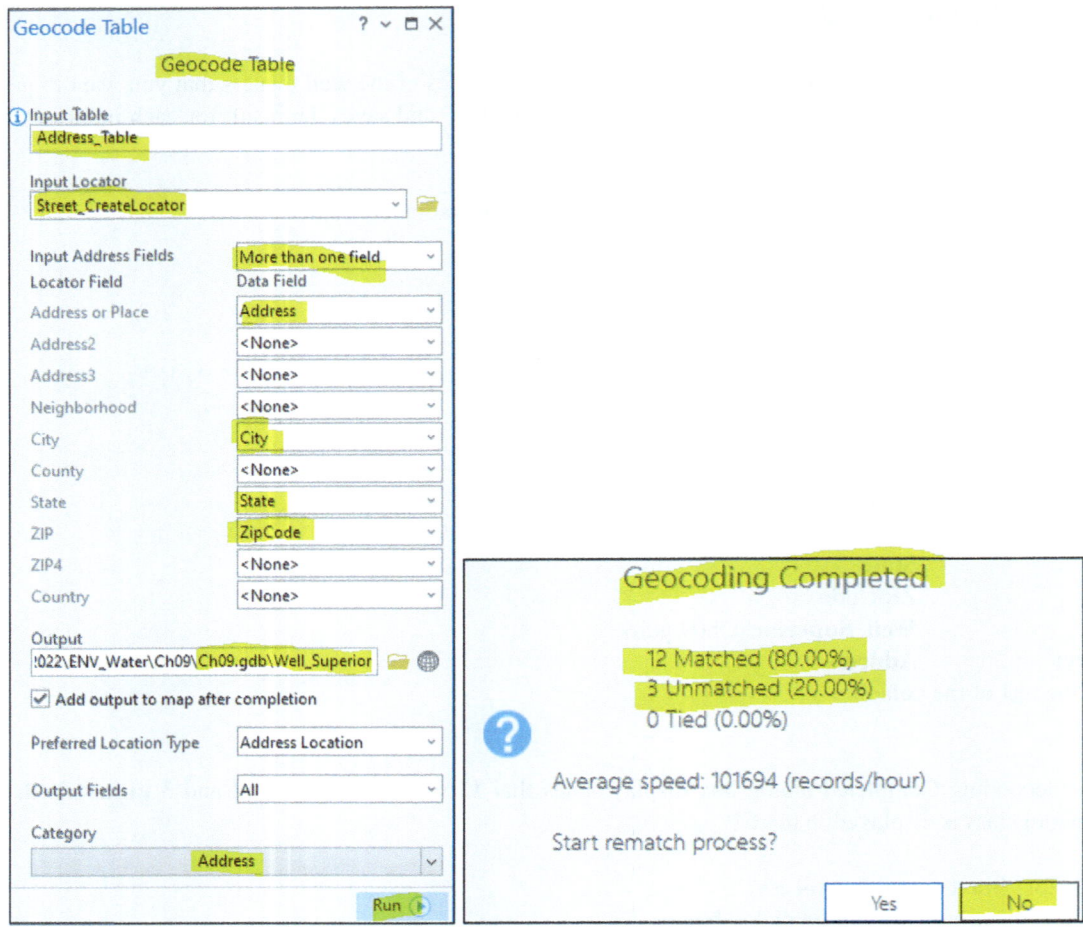

Examination of the Geocoding Results

56. Open the attribute table of the **Well_Superior** feature class

Note Under the **Match_addr** field, the unmatched records are empty. The geocoding algorithm was unable to match the three addresses. You now need to match the unmatched addresses records. You contacted Douglas County regarding the unmatched two addresses of the wells, and they provided you with the correct address (see table below)

Object ID	Wrong address	Correct address
3	666 20TH St, Superior, WI	606 20TH AV E, Superior, WI
4	737 2ND St, Superior, WI	737 2ND AV E, Superior, WI
7	717 N 2ND St, Superior, WI	717 E 2ND St, Superior, WI

Changing the Basemap

You are going to change the basemap from the Basemap gallery and choose the World Street Map
57. In The CP turn off the Street layer
58. Click the **Map** tab on the ribbon, in the **Layer** group, from the drop-down arrow of the **Basemap** select **Streets**

Result The World Street Map display in the CP

Match the Unmatched Addresses

59. In CP, right-click the **Well_Superior** layer and click **Data** and then from the drop-down list select **Rematch Addresses**
60. The **Rematch Addresses** pane and the attribute table for your geocoded feature class appear.

Result The pane opens to **Unmatched** addresses by default, and it shows that there are three unmatched records. The attribute table of the **Well_Superior** feature class also opens automatically. You can use the table to navigate the unmatched data and make selections.

Sometimes an address is not matched because a perfect match cannot be found, but there are close match candidates. The Rematch pane is context sensitive to the attribute table. You can use selection and query shortcuts in the table view and in the Rematch pane by accessing the Menu drop-down menu to more efficiently navigate your data.

61. In the attribute table of **Well_Superior,** select the "**666 20TH St, Superior, WI**"
62. The candidates are displayed in list format on the lower portion of the **Rematch Addresses** pane, and one candidate has been designated with a score of 71.87. The candidate appears as points on the map.
63. In the Rematch Addresses pane, in the "**Address or Place**" change the **666** into **606** and the **St** into **AV** and add **E** after **AV**, it should look like this "**606 20TH AV E, Superior, WI 54880**" then enter
64. The candidate score will be changed from 71.87 to 95
65. In the Rematch Addresses pane, select the second candidate in the list and click **Match** ✓.

Result The candidate is automatically moved to the Matched category in the Rematch pane, and the current view is updated to list the last unmatched address from the **Well_Superior** feature class, which is now the only unmatched address.

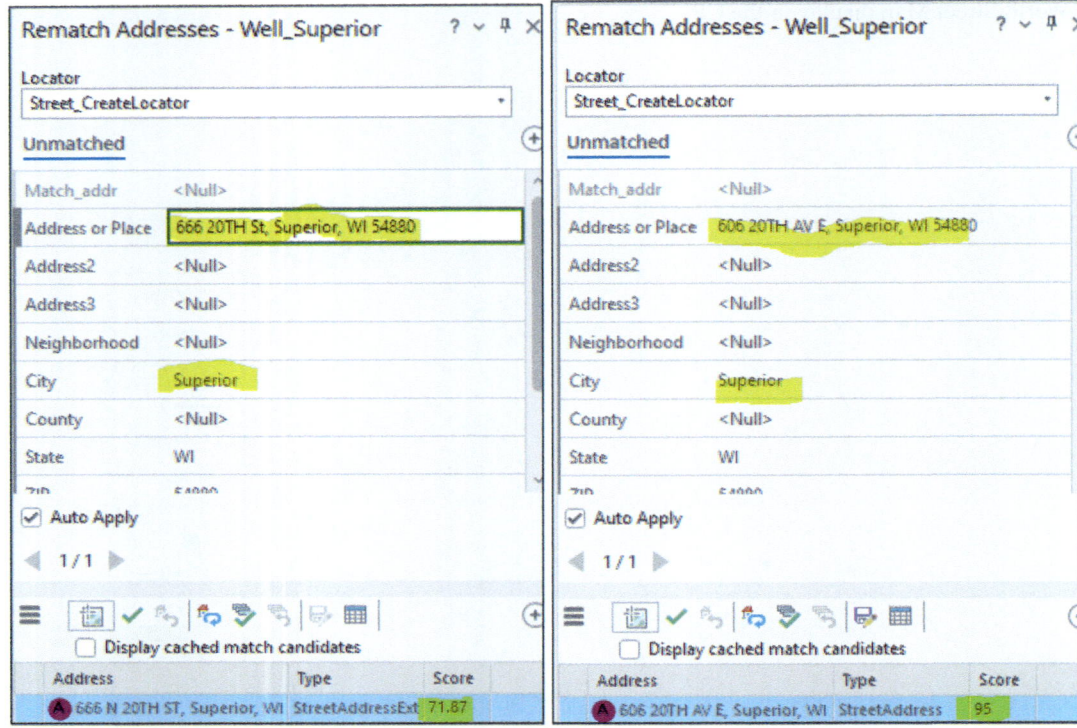

66. In the attribute table of **Well_Superior,** select the "**737 2ND St, Superior, WI 54880**"
67. The candidates are displayed in list format on the lower portion of the **Rematch Addresses** pane, the candidate has been designated with a score of 71.66, and the candidate appears as points on the map.
68. In the Rematch Addresses pane, in the "**Address or Place**" change the **St** into **AV** and type **E** after it, it should look like this "737 2ND AV E, Superior, WI **54880**" then enter
69. The candidate score will change to 94.87
70. n the Rematch Addresses pane, select the candidate in the list and click **Match** ✓.
71. In the attribute table of **Well_Superior,** select the "**717 N 2ND St, Superior, WI 54880**"
72. No candidates display on the lower portion of the **Rematch Addresses** pane
73. In the Rematch Addresses pane, in the "**Address or Place**" change the **N** into **E**, it should look like this "**717 E 2ND St, Superior, WI**" then enter
74. The candidate score will change to 94.87
75. In the Rematch Addresses pane, select the candidate in the list and click **Match** ✓.

Result The candidate is automatically moved to the Matched category in the Rematch pane, and the current view is updated to list the last unmatched address from the **Well_Superior** feature class, which is now the only unmatched address.

76. Close the Rematch pane
77. Save the project
78. Exit ArcGIS Pr0

10 Raster Format

The second type of spatial data that is used in GIS is the raster format, which is one form of organization for spatial data. This type of format is suitable for continuous surfaces, such as temperature. A raster is a regular grid of a mesh of cells (pixels) that is laid over the landscape covering a specific area. The cells of the grid are organized in rows and columns.

The cell is the smallest unit of the raster, and each cell has a value representing information, such as elevation. Raster can be satellite imagery, digital aerial photography, digital pictures, scanned maps, and saved images. The cell is the fundamental unit of analysis in the raster system. A cell represents a location in space. The condition of a given cell is recorded as a numeric value for each cell. The level of detail of features represented by a raster is often dependent on the cell size (spatial resolution). Resolution means detail with which a map shows the shape and location of geometric features such as a lake. Smaller cell sizes result in larger raster datasets to represent an entire surface; therefore, there is a need for greater storage space, which often results in longer processing time.

Feature Representation in Raster Format

A point in a vector representation can be approximately transformed to a single cell in a raster representation. Likewise, a vector line can be approximately transformed to a sequence of raster cells lying along that line, and a vector polygon can be approximately transformed to a zone of raster cells overlaying the polygon area. Like vector format, raster provides procedures for deriving new information by transforming or making associations of information from existing layers. GIS analysis using raster data is commonly used in environmental assessments. Raster processes are more commonly used because they can be significantly faster computationally than vector processes. This chapter consists of four sections, and each section includes different GIS exercises that deal with raster datasets.

Section 1: Explore and Download Digital Elevation Model (DEM)

1. Explore the downloaded image
2. Convert image from float to integer

Section 2: Projection and Processing Raster Dataset

1. Project an image
2. Clip an image
3. Merge raster datasets (Mosaic)
4. Resample an image
5. Classify an image
6. Convert Vector Feature into Raster

Supplementary Information The online version contains supplementary material available at https://doi.org/10.1007/978-3-031-42227-0_10.

Section 3: Terrain Analysis

1. Create Hillshade
2. Create Contour
3. Create Vertical Profile
4. Create Visibility map
5. Create Line of Sight
6. Derive Slope and Aspect
7. Reclassify the Slope
8. Combine the Slope and Geology
9. Find the best location to build the lysimeter

Chapter 10: Working with Raster

The internet has plenty of digital and nondigital data that GIS users can obtain free data that can be used in a project and conduct meaningful research. The data can be found on the web pages of several federal, state, and local governments in addition to private companies.

In this section, you are going to download digital elevation model (DEM) from the National Map Viewer organized by the USGS web page. The website provides applications and web map services for "Topographic Information for the Nation". This information includes topographic maps and geographic information system (GIS) data for elevation, hydrography, watersheds, geographic names, land cover, and more.

Download Dem Image from Usgs Webpage

A DEM is the digital representation of the ground surface elevation with respect to a reference datum excluding surface objects such as buildings, trees, and any other surface features. DEMs are created from a variety of sources. USGS DEMs were derived from topographic maps. DEMs are used to determine terrain attributes such as elevation at any point, slope, aspect, watershed and more.

Download the Dem of the City of Burnsville in Minnesota

To download DEM, click in the link below:
https://apps.nationalmap.gov/downloader/#/

1. In the **Search** window, type "**Burnsville, MN, USA**" and Enter
2. In the left panel, check "**Elevation Product (3DEP)**"
3. Make sure the Datasets tab is highlighted (at the top, below USGS)
4. Click "**Search Products**"

Result One image display

5. Click "**Download Link (TIF)**"
6. The "**USGS_13_n45w094_20160331.tif**" image downloaded in the download folder
7. Move the "**USGS_13_n45w094_20160331.tif**" image in **Data_Ch10**

Note If for a reason or another, you cannot download the image, the image can be accessed in the download folder in the Burnsville folder in **Data_Ch10**.

Download the Dem of the City of Burnsville in Minnesota

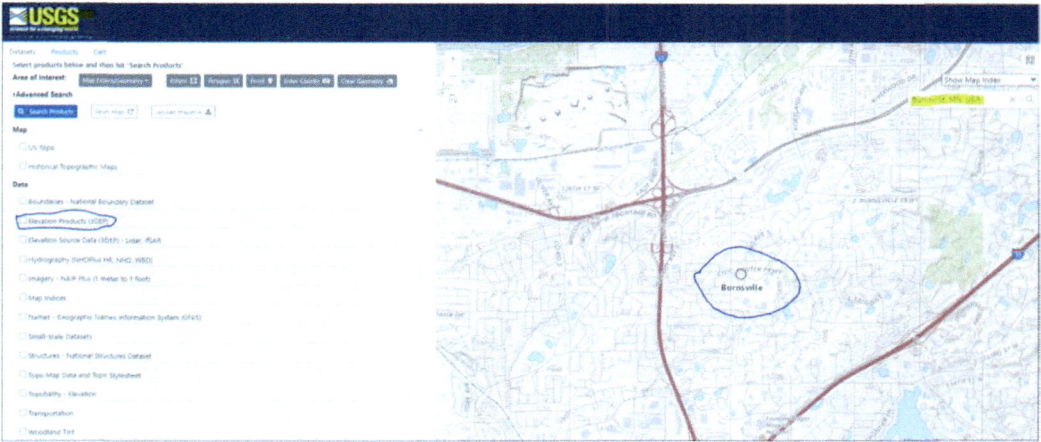

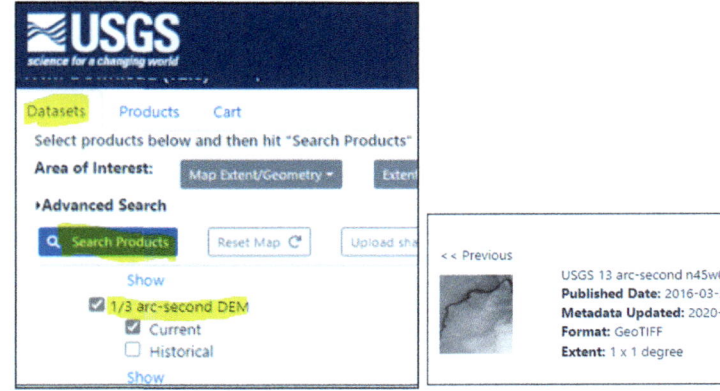

Launch ArcGIS Pro

8. Launch ArcGIS Pro
9. Open another project (upper-right) browse to \\Env_Water\Ch10 select **Ch10.aprx** and click OK

Result Ch10.aprx open and it is empty

10. In the CP, change the name of the **Map** to **DEM**.

Folder Connection

To access the data of the project, connect to the original folder "**Data_Ch10**"

11. Open the **Catalog** pane and expand the **Folders**
12. R-click **Folders** and click **Add Folder Connection**
13. Browse to (or **Database\ Data_Ch10**) open it and highlight **Data_Ch10** and click OK
14. Highlight it and click OK
15. In the Catalog pane, expand "**Data_Ch10**" under the **Folder**.
16. Save your project

Result The "**Data_Ch10**" folder includes 4 folders: **Data01**, **Data02**, **Data03**, and **Download**. All the folders contain data that you need later for the analysis.

Exploring Digital Elevation Models

A DEM is a raster representation of a continuous surface, usually referencing the surface of the earth. The accuracy of these data is determined primarily by the resolution (the distance between sample points). Other factors affecting accuracy are data type (integer or floating point) and the actual sampling of the surface when creating the original DEM.

17. In Catalog pane, expand the "**Data_Ch10**" folder in the Folder
18. Expand the **Download** and then the **Burnsville** folder and drag **USGS_13_n45w094_20160331.tif** into the Map View.
19. In CP rename "**USGS_13_n45w094_20160331.tif**" to "**Burnsville**"

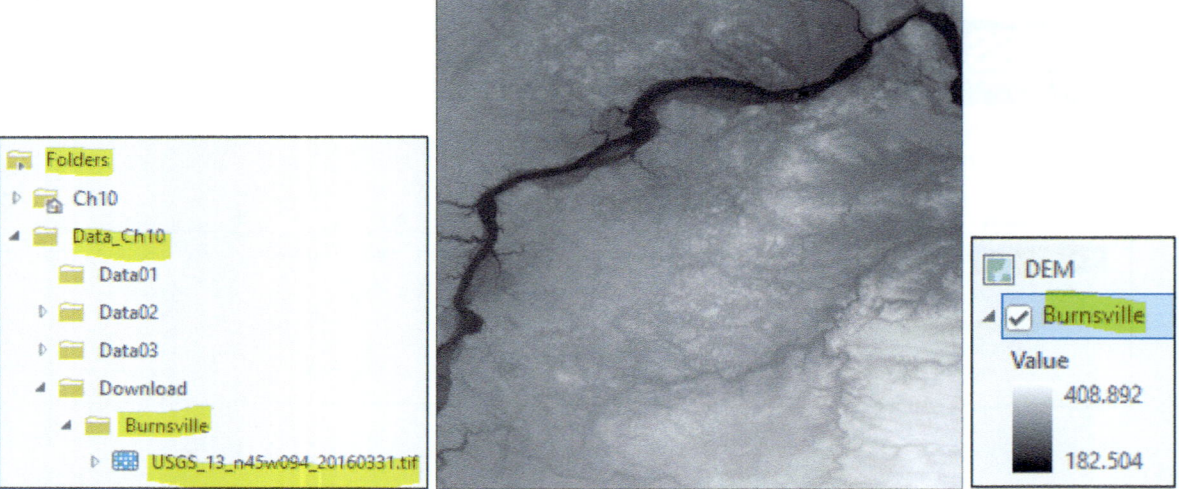

Result The DEM display in the Map View and Contents pane

20. In the CP, r-click "**Burnsville**" choose **Properties** and click the **Source** tab
21. The **Layer Property: Burnsville** consists of six sections, and each section tab provides information about the **Burnsville** DEM
    A. **Data source**: This shows the data type (raster), location on your computer, and its name
    B. **Raster Information**: This shows that the DEM has 10812 columns and rows and consists of 1 band, and the cell size (resolution) is $9.25 \times 10^{-5} \times 9.25 \times 10^{-5}$. This indicates that the pixel type is a floating point, meaning that the attribute table of the raster cannot be opened in ArcGIS Pro. The DEM format is TIFF, and the pixel depth is 32 bit.
    C. **Band Metadata**: Because the Burnsville DEM consists of 1 band, the Metadata is available for this band.
    D. **Statistics**: This shows the min, max, mean, and standard deviation of elevation.
    E. **Extent**: this shows the coordinate extent in deg (decimal degree)
    F. **Spatial Reference**: this shows that the DEM coordinate is registered in GCS and the datum is North American Datum 1983

Convert Floating Raster into Integer Raster

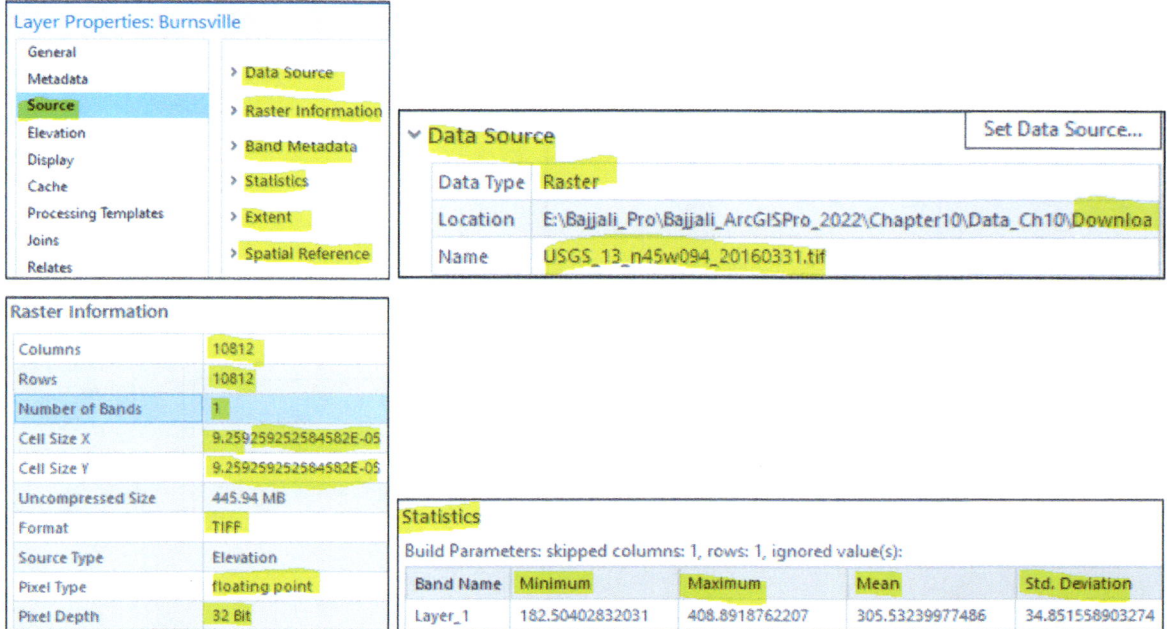

22. Click OK to Examine **Layer Properties: Burnsville**

Convert Floating Raster into Integer Raster

This exercise converts the pixel type of the "**Burnsville**" image from floating point input values into integer point input values using the "**INT**" tool. Burnsville is a floating point and has no attribute table, and its cell values are between 182.504 feet (lower elevation) and 408.892 feet (higher elevation). The "**Int**" tool converts each cell value of a raster to an integer by truncation.

23. Click **Analysis** tab on the ribbon, in the **Geoprocessing** group, click **Tools** button"
24. In the **Geoprocessing** pane click the **Toolboxes** tab
25. Scroll down and open the **Spatial Analyst Tools** and then the **Math** Tools
26. Click the **Int** Tool and fill it as follows:
27. Input raster or constant value: **Burnsville**
28. Output raster: **Burnsville_Int** (save in Ch10.gdb)
29. Run
30. Save your project

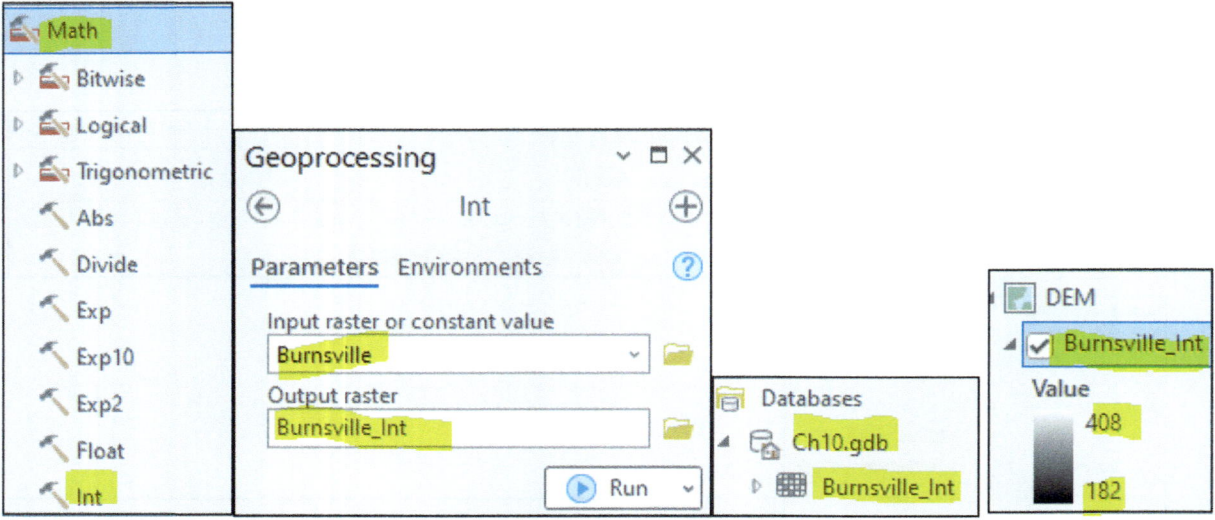

Result The "**Burnsville_Int**" display in the CP under the DEM with integer elevation and it appears also in the Catalog pane in the **Ch10.gdb**. The cell value of a raster is truncated, and the elevation is now between 182 feet (min) and 408 feet (max).

Section 2: Projection And Processing Raster Dataset

Project the Dem of Amman-Zarqa Basin

ArcGIS Pro can perform projection of vectors and rasters. This exercise addresses a raster that is registered in the international coordinate systems of Jordan. Jordan uses specific datums and different projected coordinate systems. The raster will be projected from latitude longitude into the Palestine Projection, which is based on a customized **Transverse Mercator**. The parameters of the projection are listed below:

False_Easting:	170251.55500000
False_Northing:	1126867.90900000
Central_Meridian:	35.21208056
Scale_Factor:	1.00000000
Latitude of origin:	31.73409694
Linear Unit:	Meter

In ArcGIS Pro, the projection for Jordan is stored in the **National Grids/Asia**. In this exercise, you are going to project the image of "**AZ_DEM**" from GCS_WGS_1984 into Palestine Projection.
31. Click **Insert** tab on the ribbon, in the **Project** group, click **New Map**
32. In the CP, change the name from "**Map**" to "**Projection**"
33. In Catalog pane drag the **az_dem** from **Data_Ch10\Data01** folder under **Folders** in Map View
34. The **az_dem** has GCS_WGS_1984 (Check Layer Properties/Source)

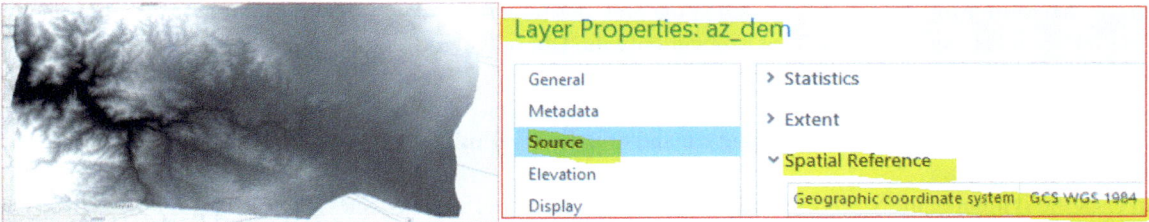

35. In the Geoprocessing pane, click the back arrow
36. Scroll down and open the **Data Management Tools** and the **Projections and Transformations** and then open the **Raster** tools
37. Click the **Project Raster** Tool and fill it as follows:
38. Input Raster: **az_dem**
39. Output Raster Dataset: az_dem_utm
40. Output Coordinate System: Click the **Select coordinate system** (Globe icon)
41. Open **Projected Coordinate System**, open **National Grids**, and then open **Asia**
42. Scroll down and select **Palestine 1923 Palestine Belt** and then click OK
43. Geographic Transformation: automatically the "**Palestine_1923_To_WGS_1984_1**" selected
44. Accept the rest of the default
45. Run

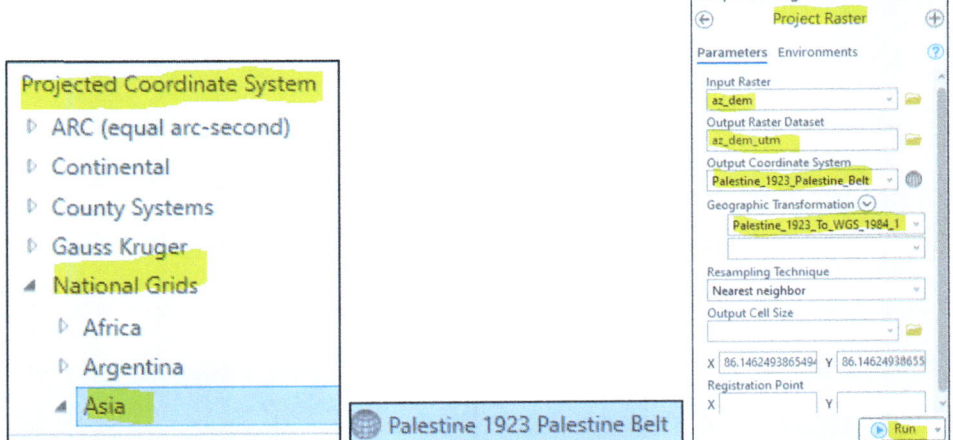

46. Save your project

Result The **az_dem_utm** display in the Map View and the CP

Question 1: What is the resolution of the **az_dem_utm**?
Question 2: What is the Datum of the **az_dem_utm**?
Question 3: Where the **az_dem_utm** image saved?

Clip the Raster

Clipping a raster can be done using a rectangular shape according to the extents defined or using a polygon feature class. The shape defining the clip can clip the extent of the raster or clip out an area within the raster.

The image "**az_dem_utm**" that was projected in the previous section covers a large area of the Amman-Zarqa Basin, and now you are interested in using the **Clip Tool** to cover only the **Dhuleil** area. To clip the image, you are going to use a shapefile "**StudyDhuleil.shp**" that will generate the clipped image.

47. In the Geoprocessing pane, click the Catalog tab and drag **StudyDhuleil.shp** from the **Data_Ch10\Data01** folder under **Folders** in the Map View.
48. In **Cp** click the symbol of the **StudyDhuleil**, in the **Symbology** pane, select **Black Outline** (1pt) and close the **Symbology** pane
49. In the **Geoprocessing** pane click the back arrow and scroll down and open the **Data Management Tools**, open **Raster** Tools and then **Raster Processing**
50. Click the **Clip Raster** Tool and fill it as follows:
51. Input Raster: **az_dem_utm**
52. Output Extent: **StudyDhuleil**
53. Check **Input Features for Clipping Geometry**
54. Output Raster Dataset: **Dhuleil**
55. Accept the rest as the default
56. Run
57. Save the project

Result The image is now clipped to the shape of the study area.

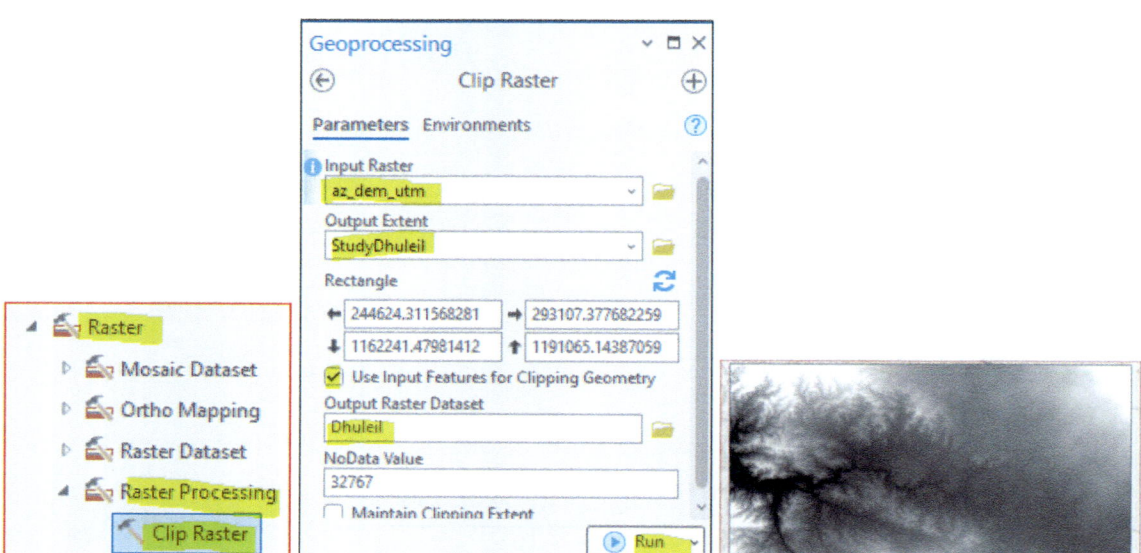

Merge Raster Datasets (MOSAIC)

The Merge Rasters function represents a grouped collection of rasters. It is useful when you have multiple rasters that you want treated as a single item, for example, to calculate the same statistics for all, or to treat as one image. This is useful for some satellite imagery that is stored as separate tiles to reduce the file size of each but should be considered part of the same image.

The "**Mosaic To New Raster**" tool merges multiple raster datasets into a new raster dataset. To merge the raster datasets, they must have

(a) Same number of bands
(b) Same pixel type
(c) Same pixel depth

When merging the raster dataset in a file format, the extension should be specified. There are various extensions that users can choose from, and the most popular formats are *.bil* (ESRI BIL), *.bmp* (BMP), *.gif* (GIF), *.png* (PNG), *.tif* (TIFF), *.jpg* (JPEG), and *.img* (ERDAS IMAGINE). The extensions will not be added to the name of the raster when the raster dataset is stored in a geodatabase. In this example, we merge two rasters, **Dhuleil** and **kt_dam**.

58. Click **Insert** tab on the ribbon, in the **Project** group, click **New Map**
59. In the CP, change the name from "**Map**" to "**Mosaic**"
60. In Catalog pane drag the **kt_dam** raster from Data_Ch10\Data01 folder under **Folders**
61. In the CP, r-click **kt_dam** raster and select Symbology
62. In the Symbology pane change the Unique Values into "**Stretch**"
63. Close the **Symbology** pane

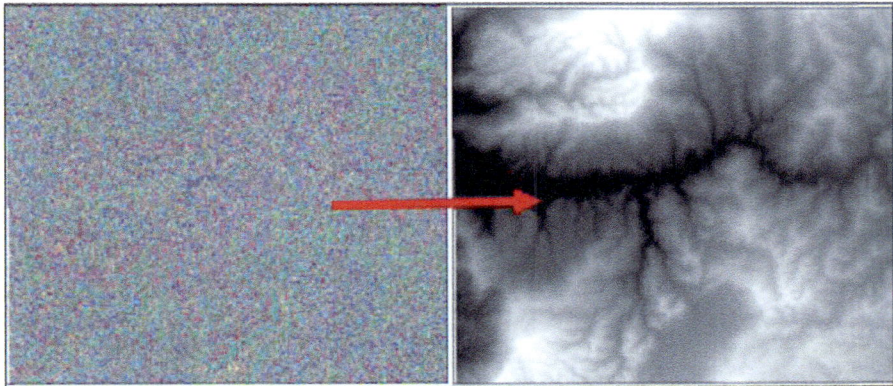

64. In Catalog pane drag the **Dhuleil** raster from **Data_Ch10.gdb** under **Databases**
65. Zoom to see both rasters in the Map View
66. R-click the **Dhuleil** raster, click **Properties**, **Source** tab and then **Raster Information** and write down the following:
    (a) Number of bands:
    (b) Pixel Type:
    (c) Pixel depth:
67. Repeat the same for the **kt_dam** raster
    (d) Number of bands:
    (e) Pixel Type:
    (f) Pixel depth
68. In the Geoprocessing pane, click the back arrow
69. Scroll down and open the **Data Management Tools** and the **Projections and Transformations** and then open the **Raster** tools
70. Scroll down and open the **Data Management Tools** and the **Raster** and then open the **Raster Dataset**
71. Click the **Mosaic To New Raster** Tool and fill it as follows:
72. Input Rasters: kt_dam and Dhuleil
73. Output Location: Ch10.gdb
74. Raster Dataset Name with Extension: Dhuleil_KTDam.tif
75. Spatial Reference for Raster: From the drop-down arrow select kt_dam

Note the coordinate is Palestine_1923_Palestine_Belt

76. Pixel Type: 16_BIT_SIGNED
77. Number of Bands: 1
78. Accept the rest of the default
79. Run
80. Close the **Mosaic To New Raster** pane
81. Save your Project

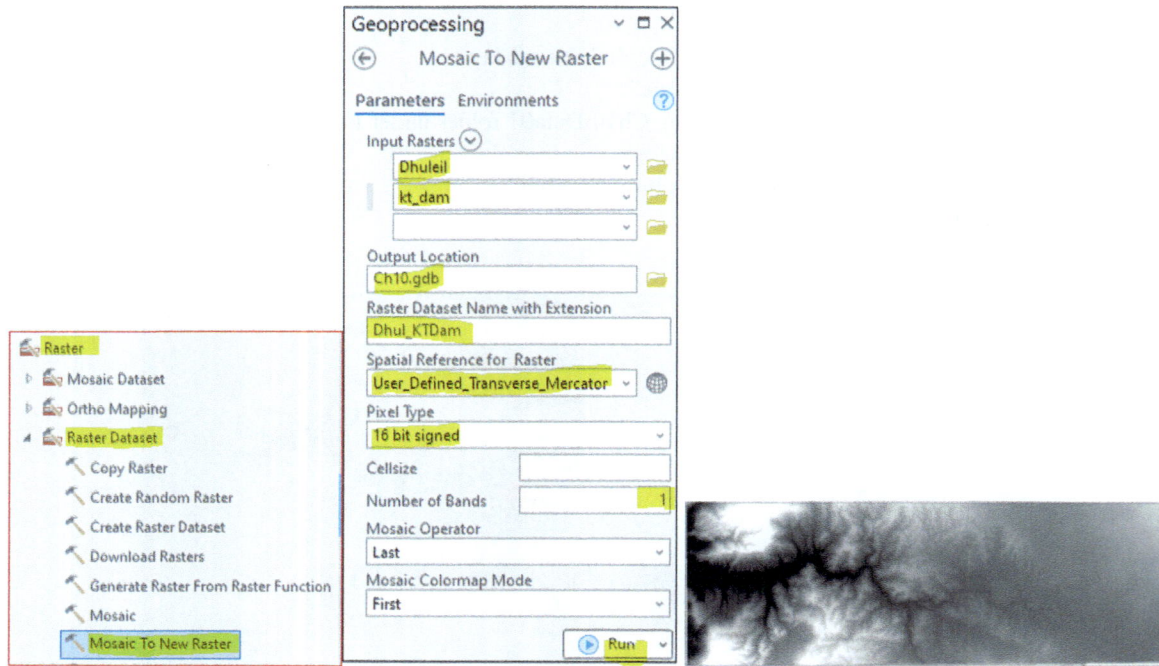

Result The mosaic raster "**Dhul_KTDam**" created from **kt_dam** and **Dhuleil** as both rasters merge together and become one raster.

Resample an Image

The resampling tool allows the user to change the resolution of the raster. The cell size can be changed either to a higher or lower resolution, but the extent of the raster dataset will remain the same. The output raster can be saved in any of the following formats: BIL, BIP, BMP, BSQ, DAT, Esri Grid, GIF, IMG, JPEG, JPEG 2000, PNG, TIFF, or any geodatabase raster dataset.

There are four options for the resampling technique parameter: nearest, majority, bilinear, and cubic. In this exercise, you are going to change the resolution (cell size from lower to higher) from **86.146** meters to **5** meters.

82. Click **Insert** tab on the ribbon, in the **Project** group, click **New Map**
83. In the CP, change the name from "**Map**" to "**Resample**"
84. In Catalog pane drag, the **Dhul_KTDam** raster from **Ch10.gdb** under **Databases**
85. In the CP, r-click **Dhul_KTDam** raster, click **Properties** and click the **Source** tab
86. Open the **Raster Information** you see that the **Cell Size** of the X and Y is **86.146** meters

Classify an Image

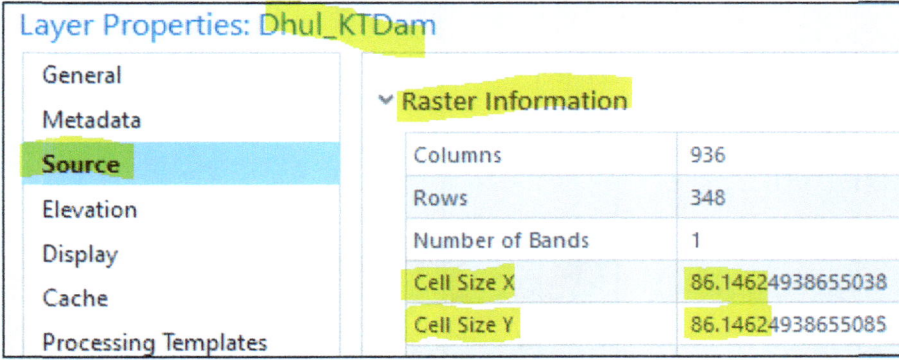

87. Close the Layer Properties dialog box
88. In the Geoprocessing pane, click the back arrow
89. Scroll down and open the **Data Management Tools** and then the **Raster** folder and then open the **Raster Processing**
90. Click the **Resample** tool and fill the Resample dialog box as follows:
91. Input Raster: **Dhul_KTDam**
92. Output Raster Dataset: **Dhul_KTDam5**
93. Output Cell Size: X = 5, Y = 5
94. Resampling Techniques: Bilinear
95. Run
96. In the CP, r-click **Dhul_KTDam5** raster, click **Properties** and click the **Source** tab

Result The cell size of **Dhul_KTDam5** is now 5 meters.

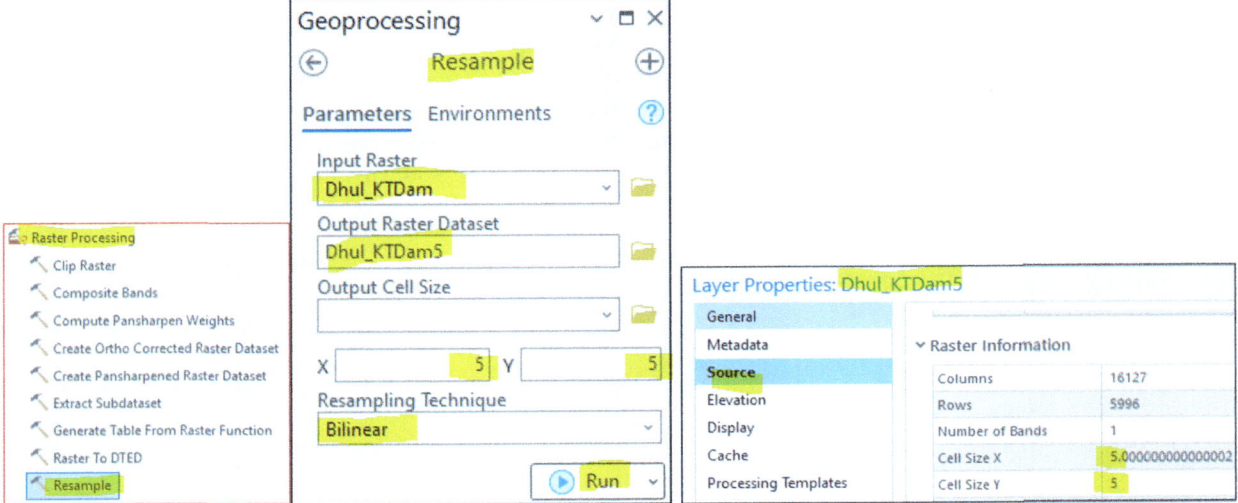

Classify an Image

The classification provides different colors to the DEM image

97. In the CP, r-click **Dhul_KTDam5** image and click Symbology
98. Make sure the **Stretch** is selected under the **Primary symbology**

99. Open the drop-down arrow of Color scheme and check Show names
100. Scroll down and select Elevation 1
101. Close the Symbology pane and save your project

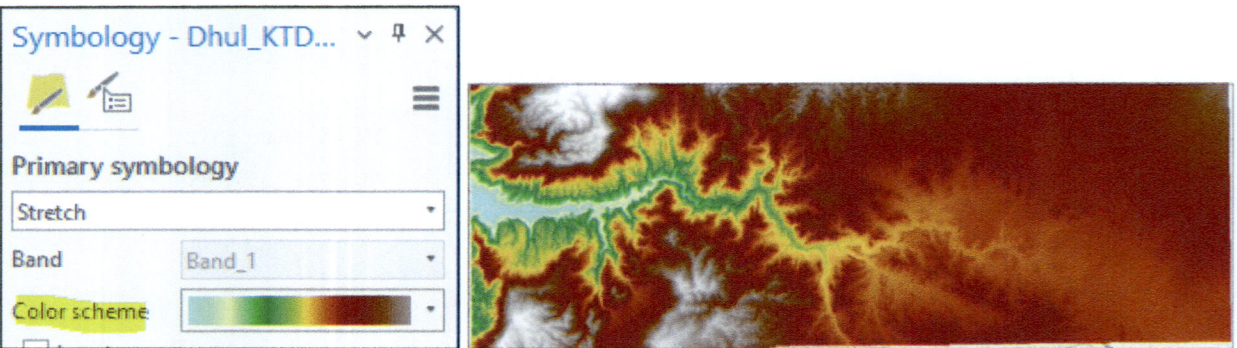

Result the color scheme changed to Elevation 1

Convert Vector Feature into Raster

The Feature to Raster tool in the Conversion Tools converts features to a raster dataset. Any feature class (geodatabase, shapefile) containing point, line, or polygon features can be converted to a raster dataset. This tool is similar to the polygon to raster, and it always uses the cell center to decide the value of a raster pixel. The tool will be used to convert the geology of the **Dhuleil** shapefile into a raster.

1. Click **Insert** tab on the ribbon, in the **Project** group, click **New Map**
2. In the CP, change the name from "**Map**" to "**Conversion**"
3. In Catalog pane drag the **Geology.shp** from **Data01** under **Folders**
4. R-click **Geology** layer click **Symbology** and under **Primary symbology** select **Unique Values**
5. In Field 1 select from the drop-down arrow "**Lithology**" if necessary click Add All Values
6. Open the More and uncheck "Show all other values"
7. Close the Symbology pane

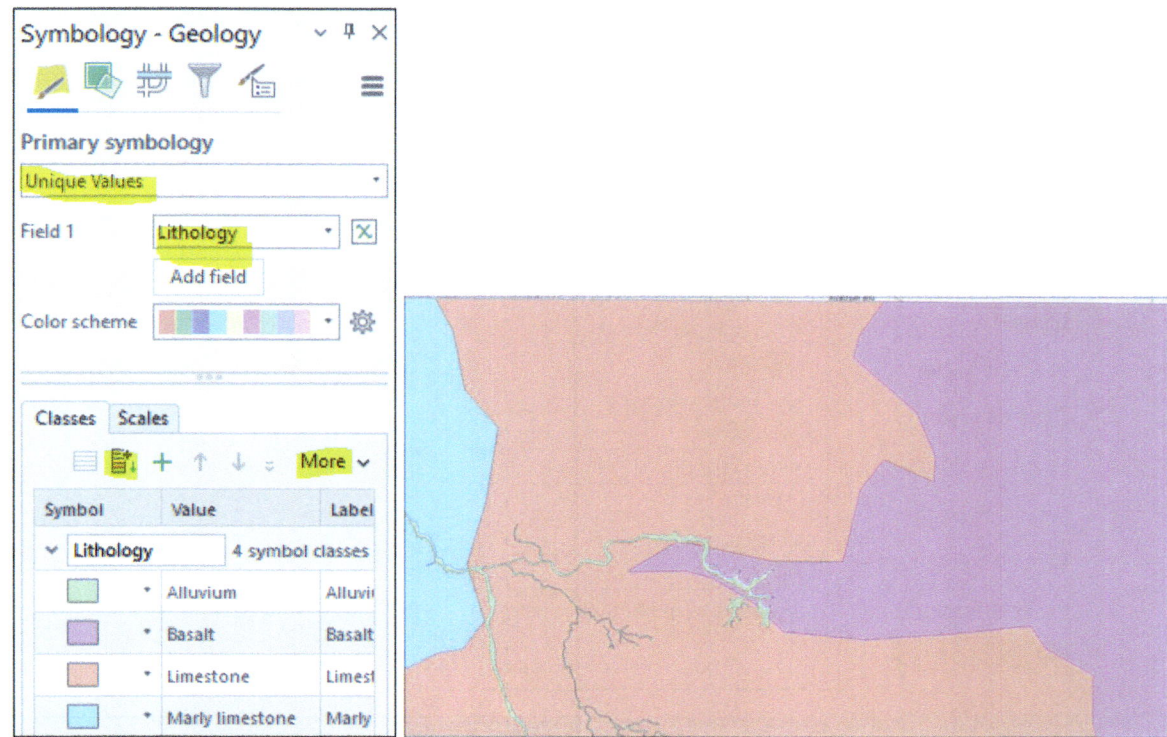

8. In the Geoprocessing pane, click the back arrow
9. Scroll down and open the **Conversion Tools** and then open **To Raster**
10. folder and then open the **Raster Processing**
11. Click Polygon to Raster tool and fill the dialog box as below
12. Input Features: Geology
13. Value field: Lithology
14. Output Raster Dataset: Geology
15. Cell assignment type: Cell Center
16. Cellsize: 100
17. Check Raster attribute table
18. Run

Result The "**Geology**" raster is created with the lithology that consists of 4 geological formations and added to CP and to the Map View. The created raster has an attribute table

19. Close the **Polygon to Raster** pane and save your project

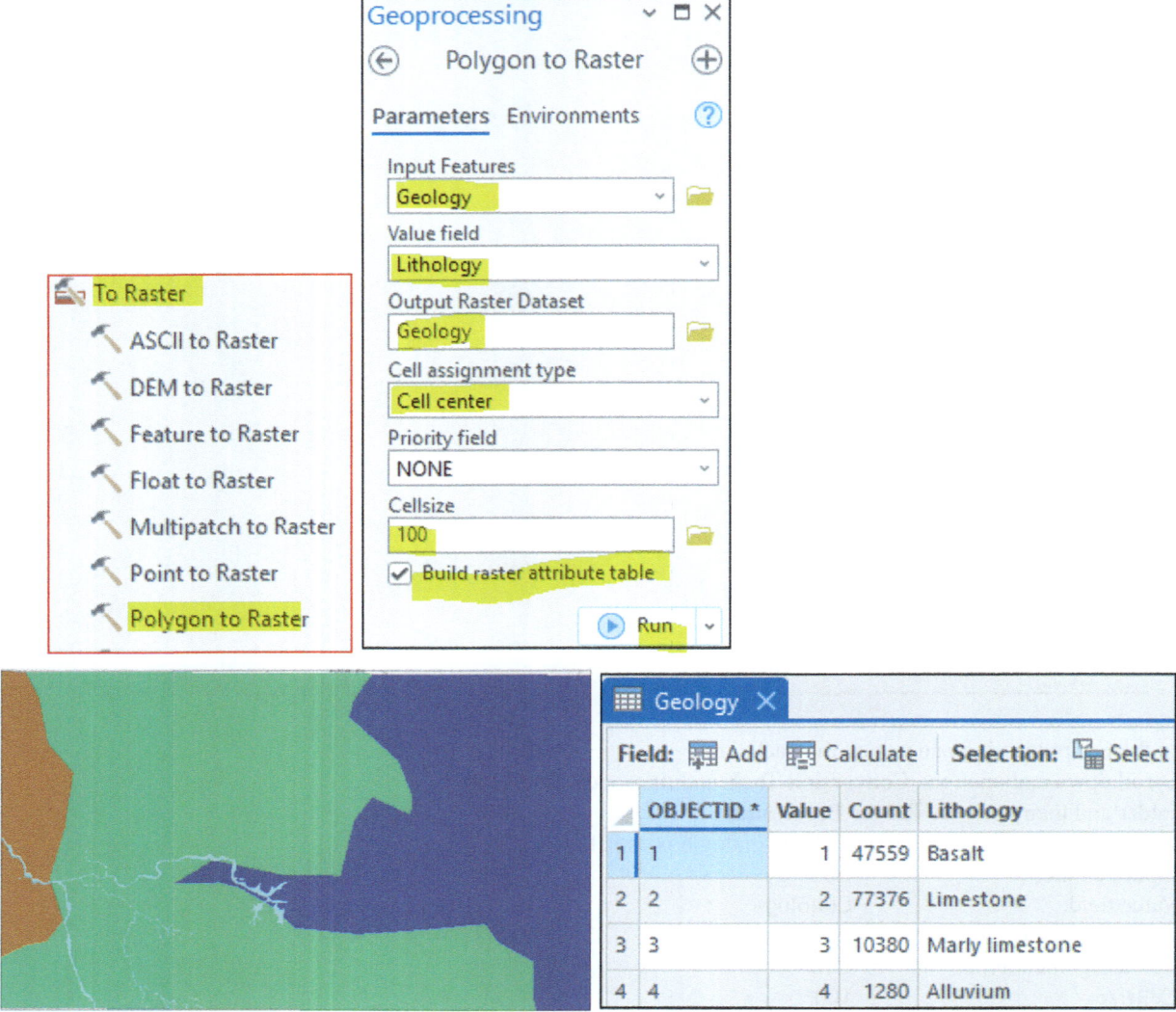

Section 3: Terrain Analysis

Create Hillshade and Contour for the Dhuleil DEM

DEM is an excellent raster format that can be used for different terrain analyses. In this exercise, you are going to work with **Dhuleil DEM**. The Dhuleil area is part of the Amman-Zarqa Basin, which is considered one of the most important groundwater basins in Jordan. The hillshade tool in ArcGIS Pro creates a shaded relief raster from a DEM. The DEM contains all the 3D information about the terrain, but it does not look like a 3D object. To obtain a better expression at the terrain, it is possible to calculate a hillshade, which is a raster format with a 3D-looking image. The **Hillshade** is a hypothetical illumination of a surface based on a given azimuth and altitude for the sun. It creates a 3-D effect that provides a sense of visual relief for the terrain and is considered the most common way to visualize texture. Using a hillshade enhances the topography of the landscape.

Section 3: Terrain Analysis

1. Click **Insert** tab on the ribbon, in the **Project** group, click **New Map**
2. In the CP, change the name from "**Map**" to "**Hillshade**"
3. In Catalog pane drag the **Dhuleil** from **Ch10.gdb** under **Databases**
4. R-click **Dhuleil** image and then Symbology
5. Make sure the **Stretch** is selected under the **Primary symbology**
6. Open the drop-down arrow of **Color scheme**, check Show names and select **Elevation 1**
7. Close the Symbology pane and save your project
8. In the Geoprocessing pane, click the back arrow
9. Scroll down and open the **Spatial Analyst Tools** and then open **Surface** folder and then click the **Hillshade**
10. Fill the dialog box of the **Hillshade** as follows:
    (a) Input raster: **Dhuleil**
    (b) Output raster: **HillShade**
    (c) Accept the Default   315 (azimuth), 45 (altitude), and 1 (Z factor)
11. Run
12. In the CP, select the **Hillshade** and click the **Raster Layer** tab on the ribbon
13. In the **Effects** group, adjust the **Transparency** slider
14. Save your project

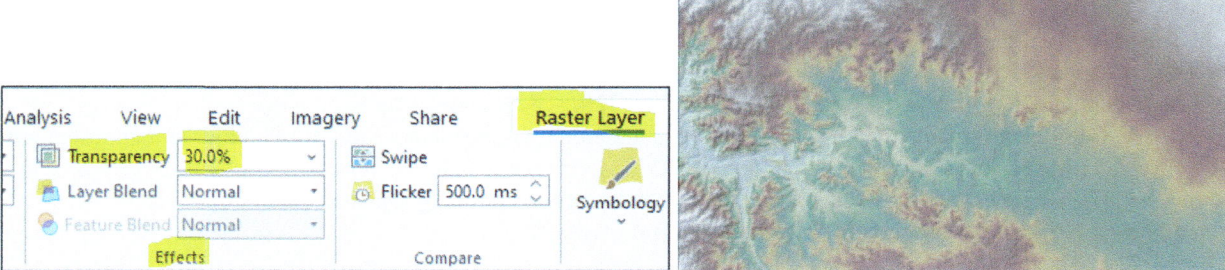

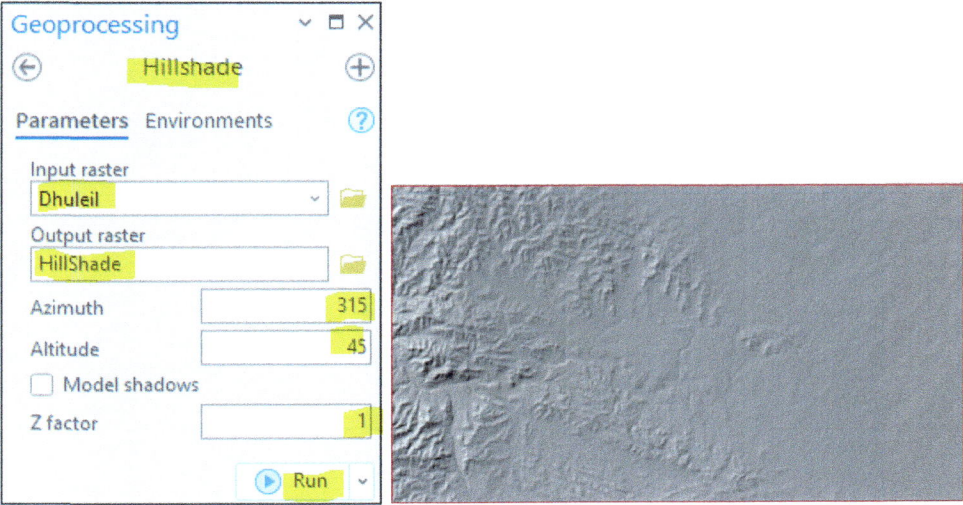

Result Transparency was applied to the Hillshade raster, which allowed us to see the symbology through the hillshade, yielding a three-dimensional effect.

Create Contours

The Contour tool creates isolines (contour lines) from the DEM raster and is commonly used to represent surface elevations on maps. A contour is a line through all contiguous points with equal height values. In this section, 25-meter interval contours are created using the Dhuleil DEM, which is registered in the UTM coordinate system. The elevation of the Dhuleil DEM ranges from 404 to 905 m above sea level.

15. In the Geoprocessing pane, click the back arrow
16. In the **Surface** folder under **Spatial Analyst Tools,** click the **Contour** tool
17. Fill the Contour dialog box as follows:
    - (d) Input Surface: **Dhuleil**
    - (e) Output feature Class: **Contour25**
    - (f) Contour Interval: 25
    - (g) Base Contour: 400
    - (h) Contour type: Contour
18. Accept the rest of the default
19. Run
20. In CP, r-click **Contour25** and click **Symbology**
21. In the **Symbology** pane and under Primary symbology select **Graduate Colors**.
22. **Field = Contour**
23. **Classes = 5**
24. **Method = Manual Interval**
25. In **Classes** tab under **Upper value** type ≤ 525, ≤ 625, ≤ 725, ≤ 825, and ≤ 925
26. Use **More** drop-down arrow, click Format all symbols and in the **Gallery** tab, search for Contour and choose **Contour Topographic, Index**
27. In the Symbology pane click the back arrow and click **Advanced symbology options** icon
28. Open Format labels under **Rounding** make the **Decimal places** "0"

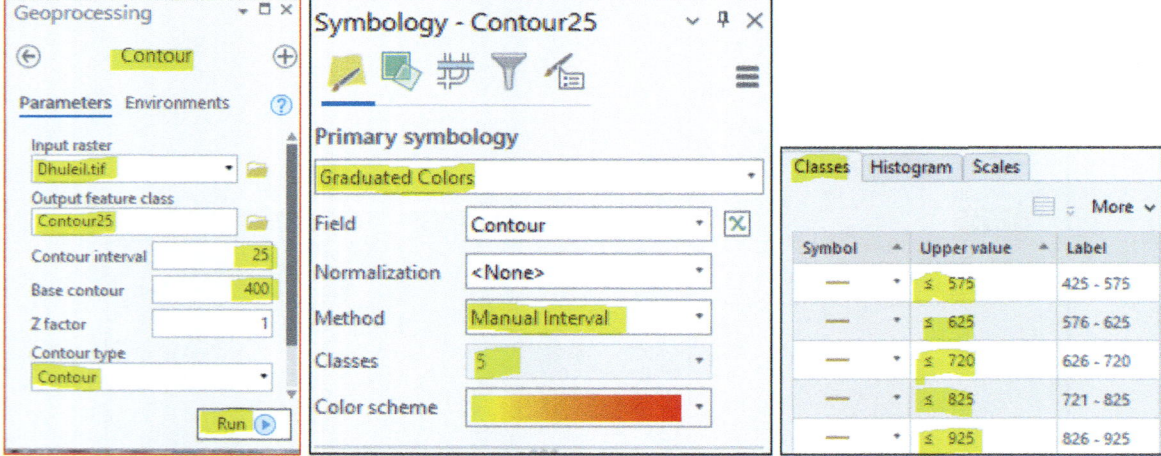

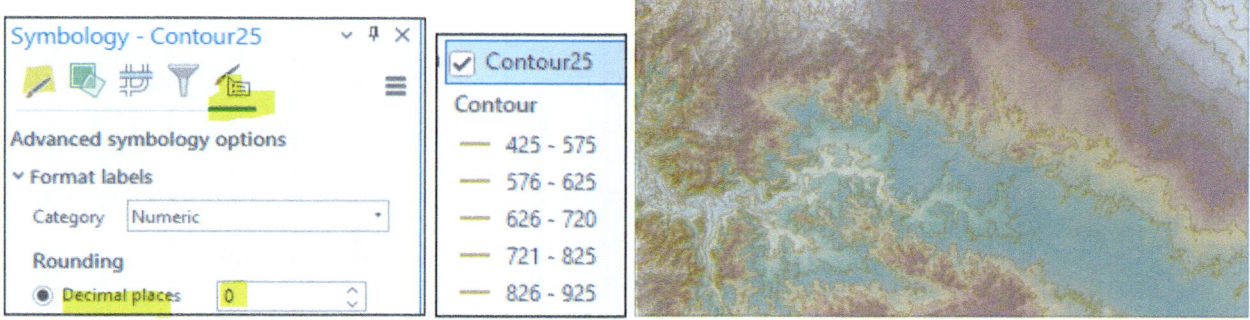

Create Vertical Profile

Profile is a useful GIS operation for terrain analysis and is a very effective tool for viewing the landscape form. It is created by drawing a line across an elevation image, reading elevation along the line, and then plotting the shape of the terrain. In geology, creating a profile is very useful to understand the form of the land and the outcropping formations and the river morphology in terms of shape and form. In this exercise, you want to create a profile for one stream in the Dhuleil area.

To generate the vertical profile, you must use the "**Profile**" tool, which generates a line feature with elevation values extracted along the input line feature. The profile tool can be found in the "**Ready To Use**" toolbox, which is part of ArcGIS Online Geoprecessing services. The "**Ready To Use**" toolbox consists of 3-Toolsets, and the Profile tool is part of the Elevation toolset. The Profile tool runs remotely on a server associated with ArcGIS Online and consumes 0.01 credit from your account.

(a) Elevation
(b) Hydrology
(c) Network Analyst

Select a Stream

29. Drag the **Stream.shp** to the **Hillshade** map from the **Data01** folder
30. In the CP, ensure that the **Stream** is placed above the **Contour25** layer and click the symbol of the **Stream** and choose from the **Symbology** pane the **Water** (line) symbol
31. In the CP, uncheck the **Contour25** layer
32. In the CP, highlight the **Stream**
33. Click the **Map** tab on the ribbon, in the **Selection** group and **Select By Attributes** button
34. In the Select By Attributes dialog box, fill it as follows:
    (a) Input Rows: Stream
    (b) Selection type: New selection
    (c) The expression should be "**Where Stream_ID is equal 48**"
35. Click Apply
36. Close the Select By Attributes dialog box

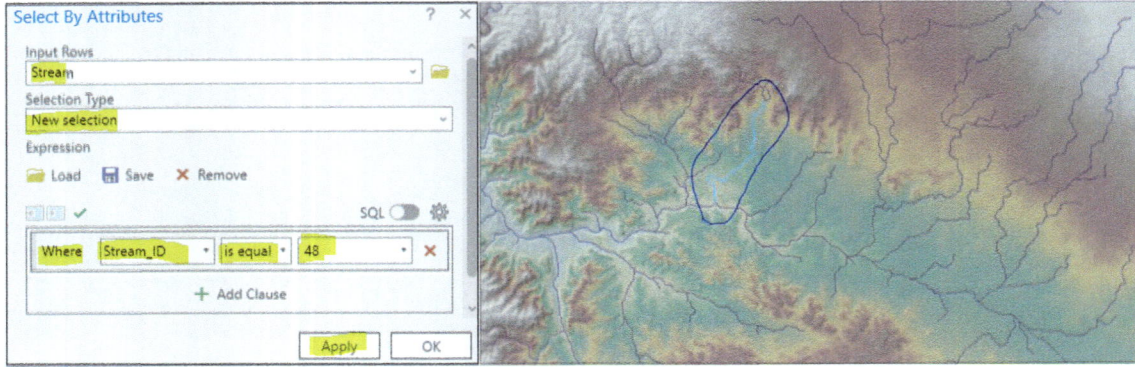

Result The 48 Stream ID is now selected.

Ready to Use Toolbox

Ready To Use Tools are ArcGIS Online geoprocessing services that use the hosted data and analysis capabilities of **ArcGIS Online**. All you need to provide are input features; all the other data necessary for the analysis and the computation is hosted in ArcGIS Online.

You can use the **Ready To Use Tools** to solve diverse spatial analysis problems such as the **Profile** tool. **Ready To Use Tools** are also available in the **Ready To Use Tools** gallery on the **Analysis** tab in the Geoprocessing group. In order to see the **Ready To Use Tools**, you have to sign to your account in **ArcGIS Online**.

Profile Tool

The **Profile** tool returns elevation profiles for the input line features

37. Click the **Analysis** tab, in the **Geoprocessing** group, open the **Ready To Use Tools** button (make sure that you are logged in ArcGIS Online)
38. Click the "**Profile**" tool, the **Profile** dialog box opens, and fill it as follows:
    (a) Input Line Features: **Stream**
    (b) Profile ID Field: **Stream_ID**
    (c) DEM Resolution: **FINEST**
    (d) Accept the rest of the default
39. Run

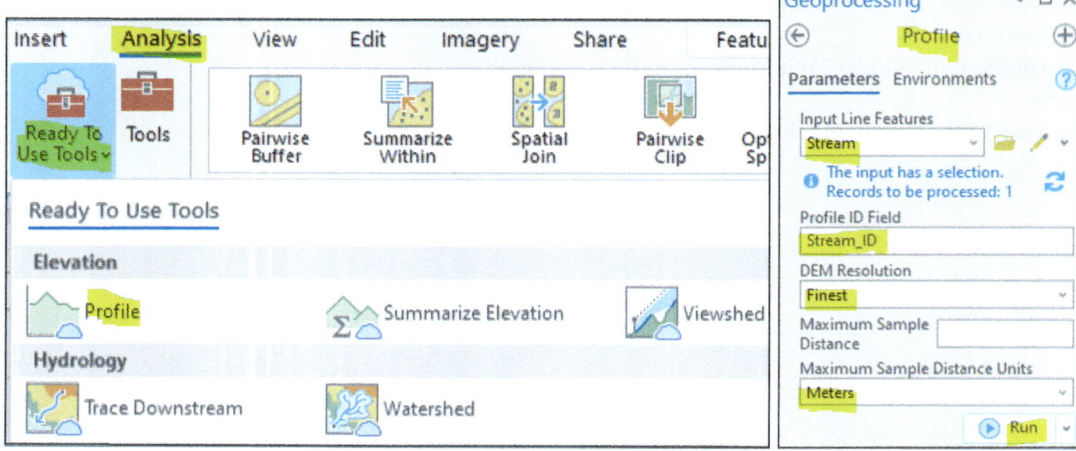

Result A line feature class called "feature_set" was created in Ch10.gdb, and an **output profile** layer was displayed in the **CP** and in the **Map View.**

40. In the CP rename the **Output Profile** to "Tributary"
41. In the CP, r-click the **Stream**, click **Selection** and then **Zoom to Selection**

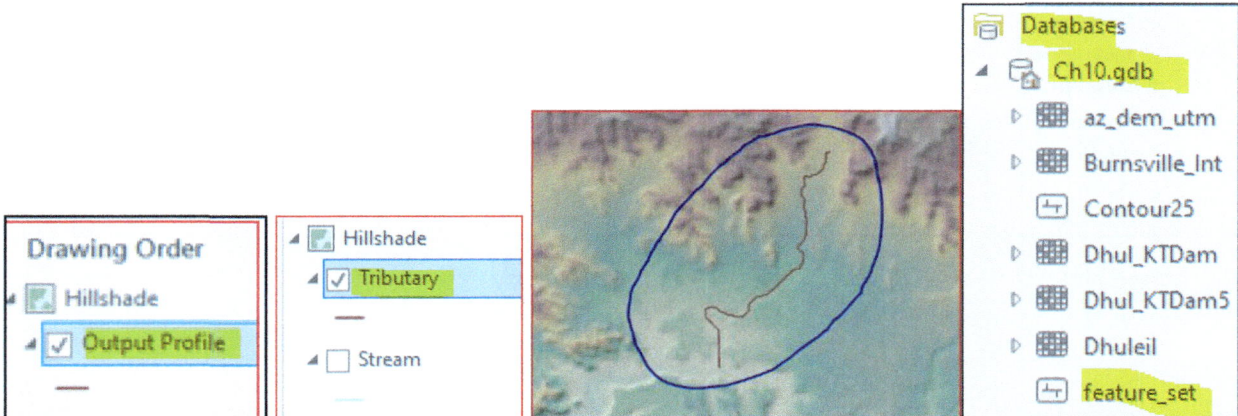

Create Vertical Profile

Profile is a useful GIS operation for terrain analysis and is a very effective tool for viewing the landscape form. It is created by drawing a line across an elevation image, reading elevation along the line, and then plotting the shape of the terrain. In geology, creating a profile is very useful to understand the form of the land and the outcropping formations and the river morphology in terms of shape and form. In this exercise, you want to create a profile for one stream in the Dhuleil area.

42. In the CP, uncheck the **Stream** layer and highlight the **Tributary** layer
43. Click the symbol of the **Tributary** layer and make its color blue
44. Click the **Data** tab on the ribbon, in the **Visualize** group open the drop-down arrow in Create Chart and select "**Profile Graph**"
45. In the CP, r-click **Tributary**, **Create Chart** and select **Profile Graph.**

Result Three things happen: (a) in the CP, a chart called Profile Graph Tributary is added; (b) a profile graph for the tributary display in the Map View; and (c) Chart Properties of the Tributary pane display.

Customize the Chart Properties

46. In the **Chart Properties**, click the **Axes** tab and fill it as follows:
47. Horizontal Axes: Kilometers
48. Vertical Axes: Meters
49. Click the General tab:
50. Chart title: Vertical Profile of Stream Tributary No 48
51. X-axis title: Kilometer
52. Y-axis title: Elevation (m)
53. Close the Chart Properties of the Tributary pane
54. Save the project

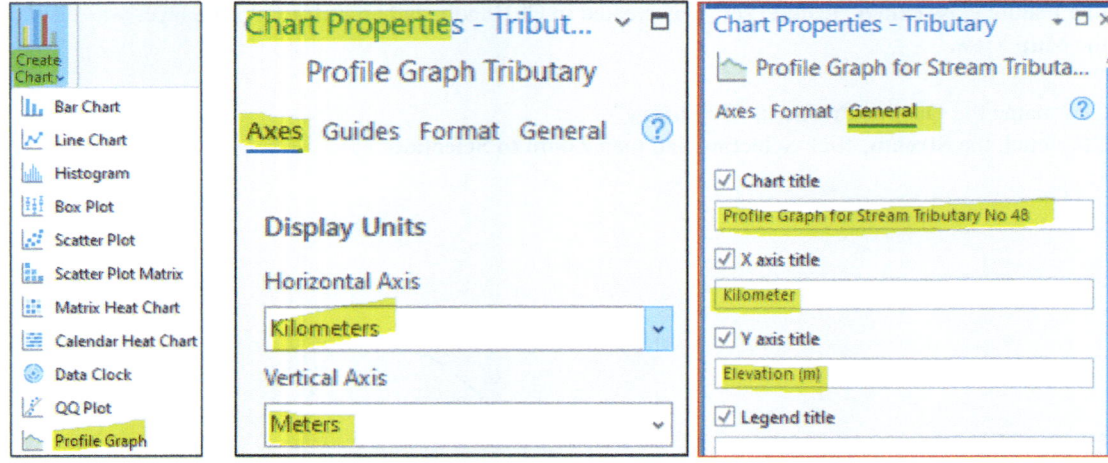

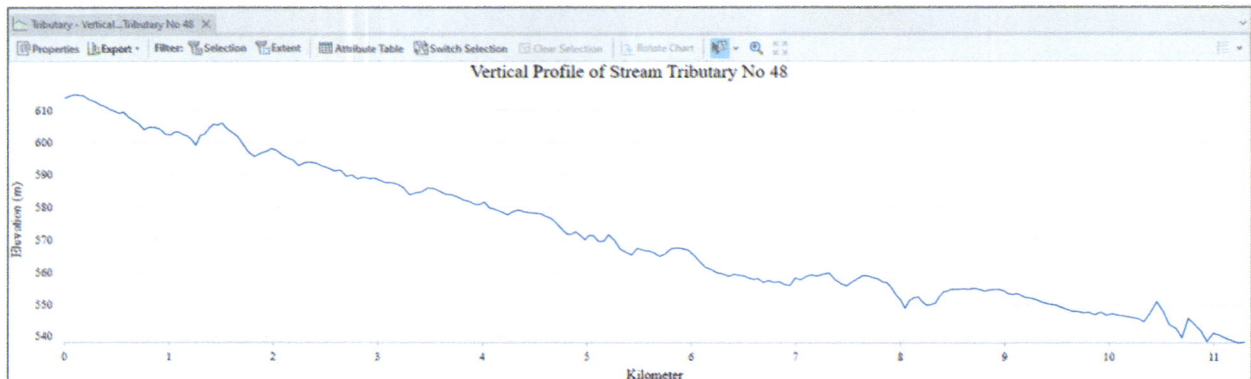

Result the modification adopted in the profile graph of the tributaries

Create Visibility Map

The Visibility tool can show what locations in the raster are visible from a specific location and how many observable locations it is visible from. The visibility map is based on two types of analysis: the first type is frequency visibility analysis, which determines which raster surface locations are visible to a set of observers, and the second is observer visibility analysis, which identifies which observers are visible from each raster surface location.

55. Click **Insert** tab on the ribbon, in the **Project** group, click **New Map**
56. In the CP, change the name from "**Map**" to "**Visibility**"
57. In Catalog pane drag the **Dhuleil** from **Ch10.gdb** under **Databases** and the **KSCentroid.shp** from the **Data01** folder
58. In the CP, click the symbol of the **KSCentroid** and choose **Circle 4** and close the Symbology pane.
59. In the **Geoprocessing** pane click the back arrow and click the Toolboxes
60. Scroll down and open the **Spatial Analyst Tools** and open the **Surface** folder
61. Click **Visibility** tool and fill the Visibility dialog box as below
62. Input raster: **Dhuleil**
63. Input point or polyline observer features: **KSCentroid**
64. Output raster: **VisibFreq**
65. Analysis type: Frequency
66. Accept the result of the default
67. Click Run

68. Open the Attribute table of the **VisibFreq**, you see to records

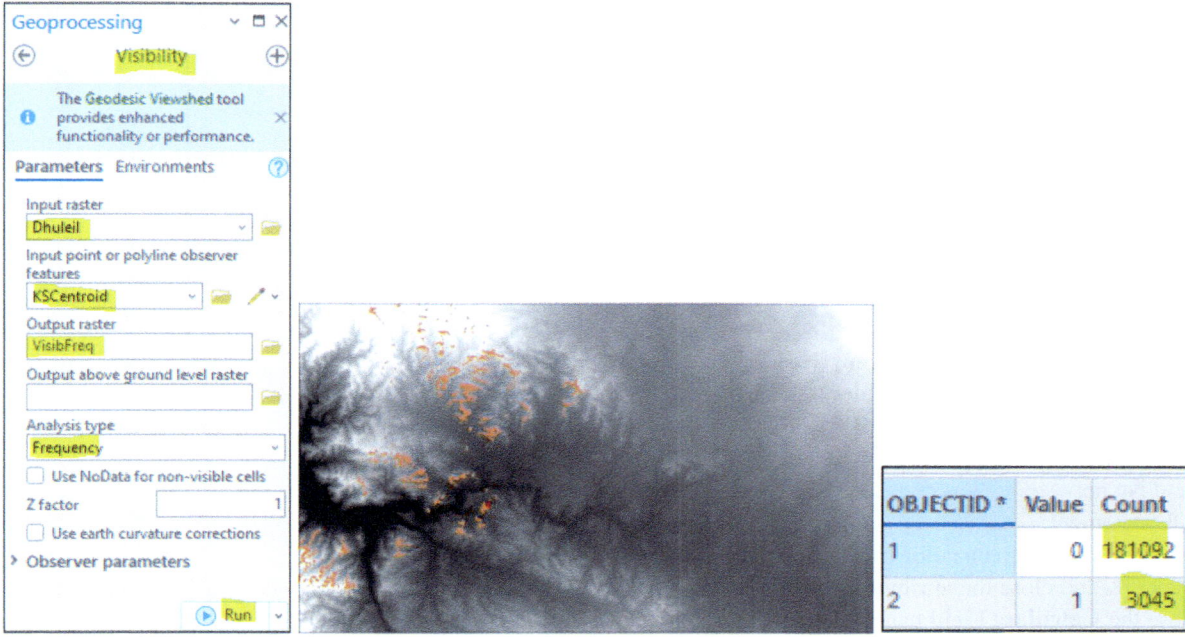

Result The **VisibFreq** map is displayed in the Content pane and the Map View. The attribute table of **VisibFreq** has a **value** field with two records, "0" and "1". **Value 1** indicates the area that can be seen from the **KSCentroid** location, while **Value 0** is the area that cannot be seen from the **KSCentroid** location.

Create a New Feature Class

69. In the CP, r-click **KSCentroid** click **Data** and select **Export Features**
70. The Export Features dialog box open, fill it as below
71. Input Features: **KSCentroid**
72. Output Feature Class: **KSCent_New**
73. Accept the default
74. Click OK

New Field: Offseta

You will add 25-meters height to the **OFFSETA** field in the **KSCent_New** feature class. An OFFSETA field will be added to the observation point (**KSCent_New**). The OFFSETA field indicates a vertical distance in surface units to be added to the z value of the observation point.

75. In the CP, r-click **KSCent_New**, click **Data Design** and select **Fields**. This opens the fields view, displaying the layer's fields in a tabular arrangement
76. Click the last row in the view that says "**Click here to add a new field**"
77. Enter a name "**OFFSETA**" under the **Field Name**
78. Under **Data Type** column choose from the drop-down menu "**Short**"
79. Under **Number Format** click and, in the **Number Format** in the **Category**, select "**Numeric**" and select "**0**" for the "Decimal places" and click OK

80. Click the **Field** tab, in the **Changes** group, click the **Save** button
81. Close the **Fields: KSCent_New** tab above the Map View

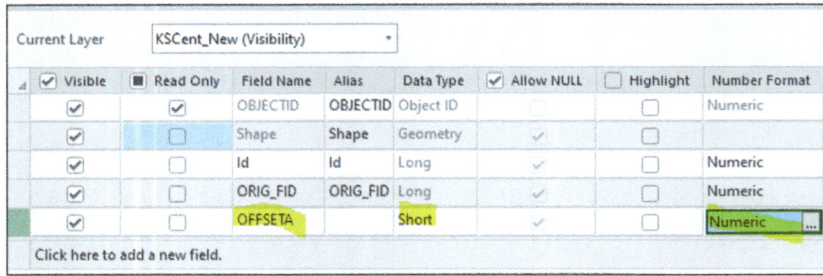

82. Open the attribute table of **KSCent_New**
83. R-click **OFFSETA** and click **Calculate Field**
84. In the **Calculate Field** pane fill it as follows:
85. Input Table: KSCent
86. Field Name: OFFSETA
87. SQL statement: OFFSETA = 25
88. Click the Verify (green color check) you get Expression is valid
89. Click OK and close the attribute table of **KSCent_New**
90. Close the **Calculate Field** pane and the attribute table of **KSCENT_New**

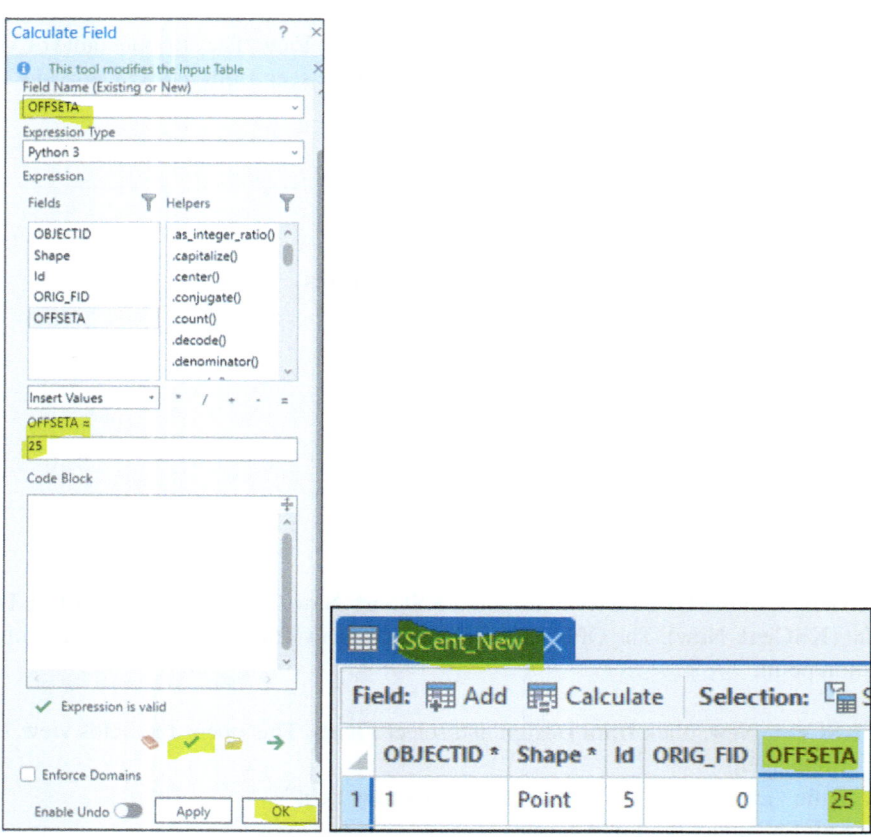

Line of Sight Analysis

Run the Visbility Tool with the New Field "OFFSETA"

91. Click the back arrow in the Geoprocessing pane
92. In Toolboxes click Spatial Analyst Tools and open the Surface folder
93. Click **Visibility** and fill it as below
94. Input raster: **Dhuleil**
95. Input point or polyline observer features: **KSCent_New**
96. Output raster: **VisibFreq25**
97. Analysis type: Frequency
98. Accept the result of the default
99. Analysis type: Frequency
100. Run
101. Save the project

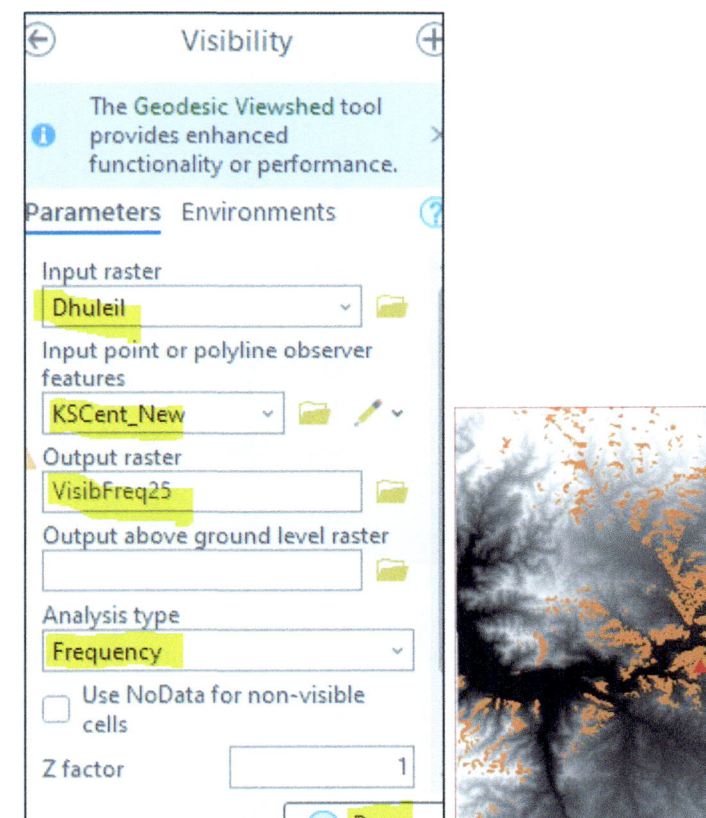
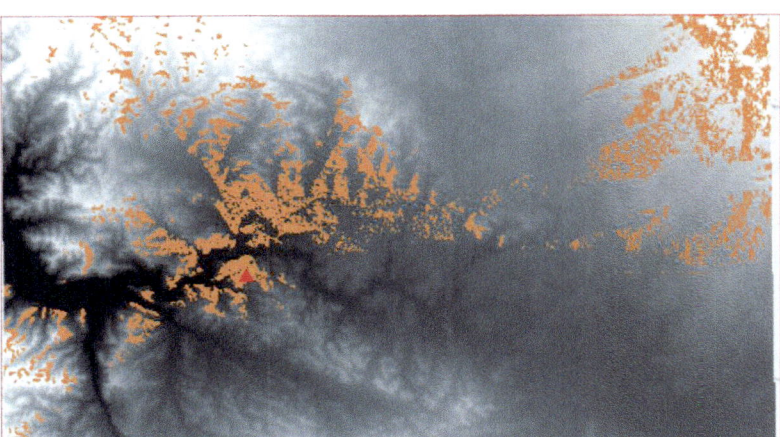

Result the visibility map display and the visible area increase due to the added 25 meter height

Line of Sight Analysis

Line-of-sight analysis determines whether two points in the raster DEM are visible. To use this tool, a line should be drawn between two locations. In this task, you will use first the "**Construct Sight Lines**" and then the "**Line of Sight**". In this exercise, you will use two observation points: the Lufi Dam (**Luhfi_Dam**) and the treatment plant (**KSCent_New**). The **Construct Sight Lines** tool will create a line between the two points.

102. In Catalog pane drag the **Luhfi_Dam.shp** from **Data01** under **Folder**
103. In the CP, click the symbol of **Luhfi_Dam** and search for a dam symbol and choose one that you prefer
104. In the Geoprocessing click the back arrow, click in the **Toolboxes** and expand the **3D Analyst** Tools and open the **Visibility** tools
105. Click the **Construct Sight Lines** and fill the pane as follows:
106. Observer Points: **Luhfi_Dam**
107. Target Features: **KSCent_New**
108. Output: **Sightline** (save it in Ch10.gdb)
109. Sampling Method: 2D distance
110. Accept the rest of the default
111. Run

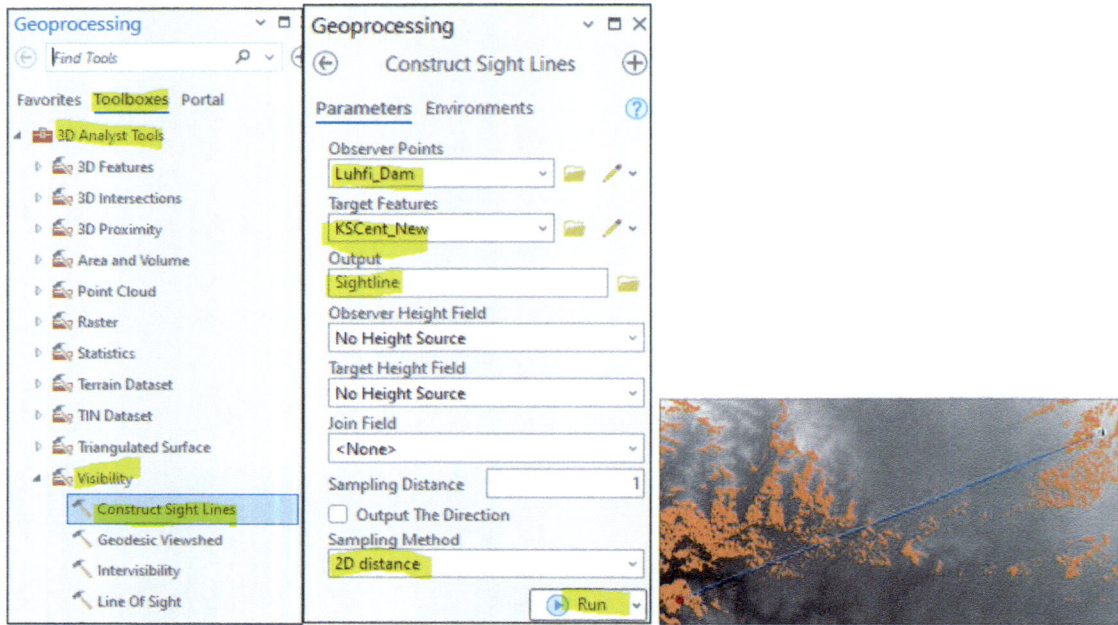

Result The **Sightline** feature class was created as a line between **Luhfi_Dam** and **KSCent_New**. The **Sightline** is stored in Ch10.gdb and displayed in the CP and Map View.

Line of Sight

The **Line of Sight** divides the created line between the two observation points (**Luhfi_Dam** and **KSCent_New**) into segments that are visible (green color) from one point and segments that are invisible (red color). The line of sight can then be compared with the two visibility maps created previously.

Line of Sight for VisibFreq Map Without OFFSETA

112. On the **Construct Sight Lines** pane click the back arrow
113. Click the **Line of Sight** tool under the **Visibility** tools
114. The Line Of Sight pane display, fill it as below
115. Input Surface: **VisibFreq**
116. Input Line Features: **Sightline**

Line of Sight   207

117. Output Feature Class: **LineVisibFreq**
118. Accept the rest of the default
119. Run
120. Do not close the **Line of Sight** pane

Result the VisibleLine feature class created and displayed with two colors: red means not seen and green means visible. This means that if you are standing on the **Luhfi_Dam** and looking toward **KSCentroid,** you can see from your location a small part, and the rest will not be seen.

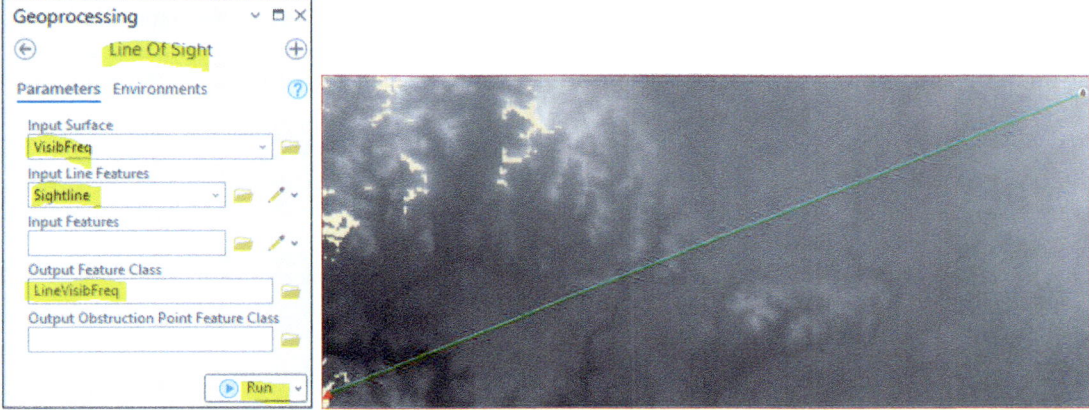

Line of Sight for VisibFreq25 Map with OFFSETA

121. On the **Line of Sight** pane, fill it as below
122. Input Surface: **VisibFreq25**
123. Input Line Features: **Sightline**
124. Output Feature Class: **LineVisibFreq25**
125. Run
126. Save your project

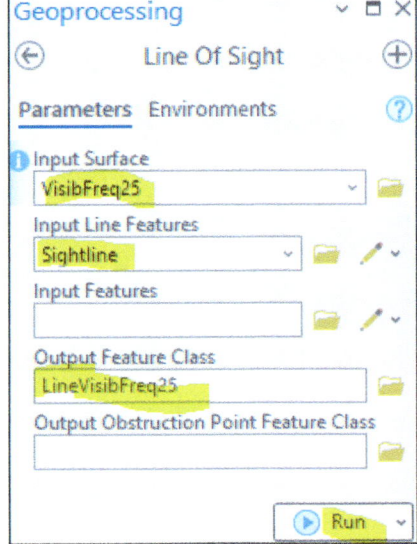
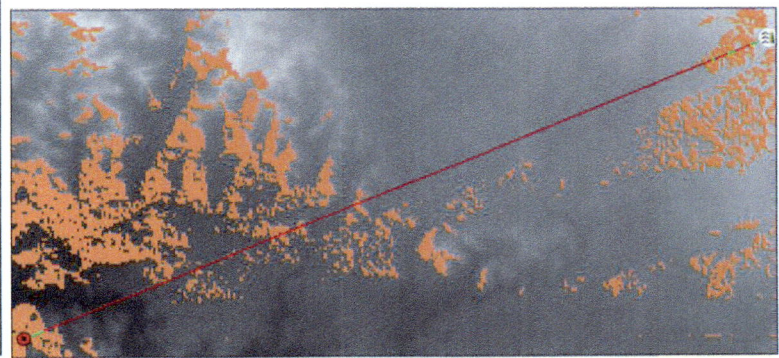

Result The VisibleLine25 feature class is created and displayed with two colors: the red color is dominant, which means that the view between the **Luhfi_Dam** and **KSCent_New** is not seen. The green color means visible. The green color is only seen close to the location of **Luhfi_Dam** and **KSCent_New**. The line of sight is correlated nicely with **VisibFreq25**.

Create Linear Lines of Sight

There is a way to create the line of sight directly by using the Create linear lines of sight. The **Linear Line of Sight** tool shows the surface that is visible between one or more observer and target locations. Use this tool to locate observation posts tasked with monitoring a specific location or for test siting of radio antennas. The results show the number of visible targets, sight lines, and a line of sight. The tool requires surface information to perform the calculations; therefore, you have to have an input surface available or you can add observer and target locations manually by entering known coordinates or by selecting locations from the map. In this exercise, you will add two locations manually.

127. Click the Analysis tab on the ribbon, in the Workflows group, click Visibility Analysis
128. In the Visibility Analysis pane, click the **Linear Line Of Sight** tab and fill it as follows:
129. Input Surface: VisibFreq
130. Observer Points: Enter Manually
131. Click the **Observer Map Point Tool** and click a point in the upper left corner of the map

Result On the map, observer points are marked with blue circles.

132. Target Points: Enter Manually.
133. Click Target Map Point Tool and click a point in the lower right corner of the map.

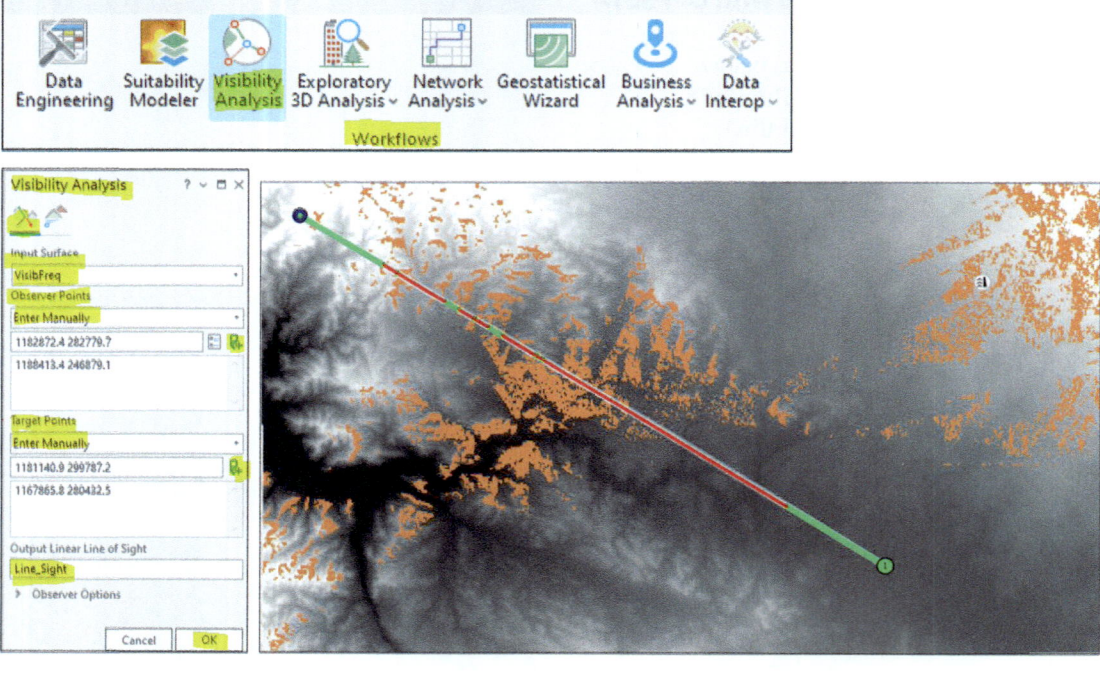

Slope and Aspect

Result On the map, target points are marked with red squares.

134. Output Linear Line of Sight **Line_Sight**
135. OK
136. Click the **Visibility Analysis** pane

Result The Linear Line of Sight tool produces the **Line_Sight** feature dataset in **Ch10.gdb** with four **feature classes**. The value entered in the Output Linear Line of Sight text box is used to name the feature dataset and is prepended to the feature classes within the feature dataset. The **Linear Line of Sight** tool output is added to the **Contents** pane in a group layer. The name used for the feature dataset is also used for the group layer.

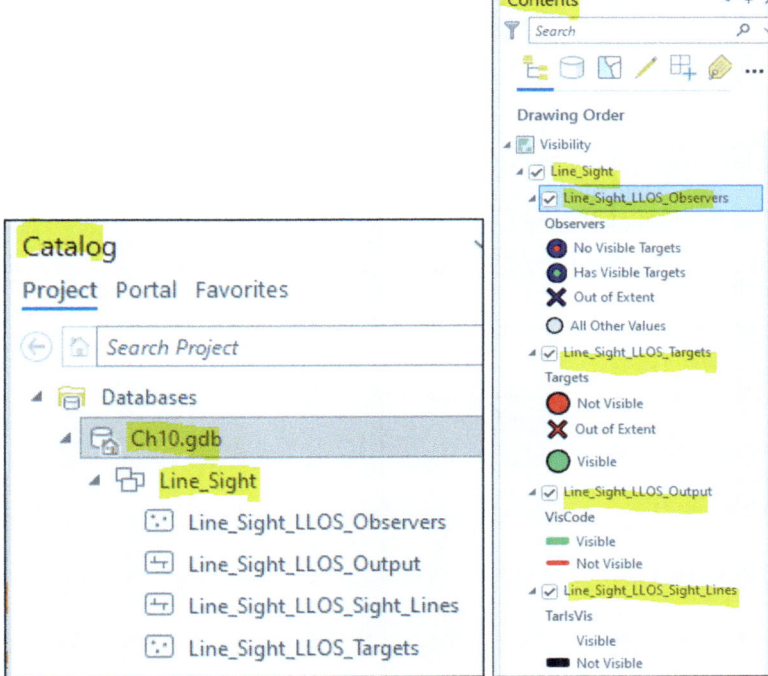

Slope and Aspect

The slope is the incline or steepness of the ground surface of a terrain, and it can be measured in degrees from horizontal (0–90) or percent slope (rise over run multiplied by 100). For example, a slope of 45 degrees equals 100 percent slope. As the slope angle approaches vertical (90 degrees), the percent slope approaches infinity. The slope for a cell in a raster is the steepest slope of a plane defined by the cell and its eight surrounding neighbors.

Slope: degree of angle CBE (α)

$$Slope = \left(\frac{Rise}{Run}\right) * 100 = \left(\frac{CE}{BC}\right) * 100\%$$

Example: CE = 20, BC = 60; $Slope = \left(\frac{20}{60}\right) * 100\% = 33.3\%$

Note if the rise is more than run, the slope will be more than 100%)

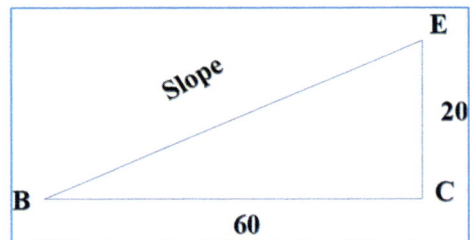

Derive the Slope

1- Click **Insert** tab on the ribbon, in the **Project** group, click **New Map**
2- In the CP, change the name from "**Map**" to "**Slope**"
3- In Catalog pane drag the **Dhuleil** from **Ch10.gdb** under **Databases**
4- In the Geoprocessing click the back arrow, click in the **Toolboxes** and expand the **Spatial Analyst Tools** and open the **Surface** tools
5- Click the **Slope** tool
6- The **Slope** pane dialog box display, fill it as below
7- Input raster: Dhuleil
8- Output raster: Dhuleil_Slope
9- Output measurement: Percent rise
10- Accept the rest of default
11- Run

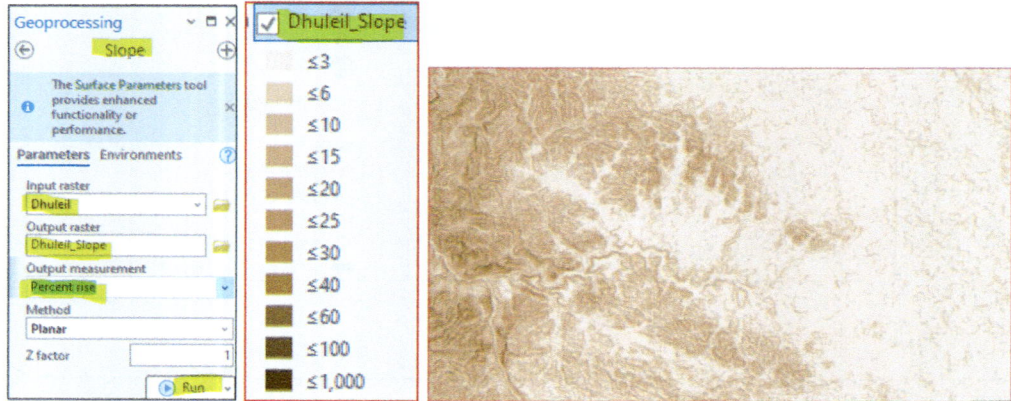

Result The slope displays 11 classes, and the last class is equal to or less than 1,000. This value indicates a very steep slope that corresponds to a degree of angle between 84 and 90 degrees.

Q1: What is the range of percent slope values in **Dhuleil_Slope**?
Quiz: can you run the slope again with "DEGREE" as output measurement?

Classify the Slope into Six Classes

12- In the CP, r-click **Dhuleil_Slope** and click **Symbology**
13- In the Symbology pane, select **Classify** under the Primary symbology
14- Classes   7
15- Method   Manual Interval

Reclassify Slope

16- In Classes tab, under Upper value change the values from top to bottom to ≤5, ≤10, ≤15, ≤20, ≤25, ≤30, and last number ≤51 (see image below)
17- Change the Color scheme to Aspect and close the Symbology pane
18- Save the project

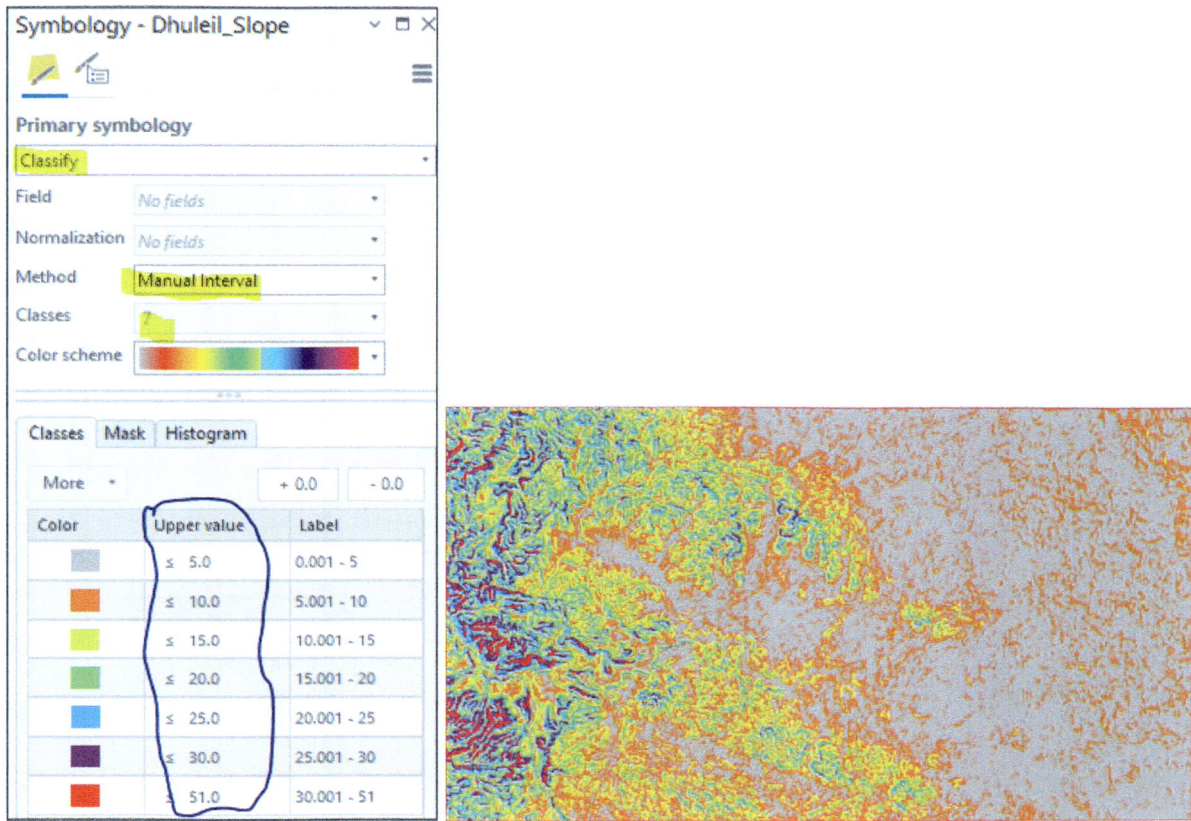

Question: Where is the highest and the lowest slope on the map?

Reclassify Slope

The Reclassify tool allows you to change the range of values in the slope. At the same time, it generates an integer raster, which will allow you to see the attribute table and perform further analysis. You are going to reclassify the **Dhuleil_Slope** into 6 classes (5, 10, 15, 20, 30 and 55)

20. In the Geoprocessing click the back arrow, open the **Spatial Analyst Tools** and then the **Reclass** tools and click **Reclassify**
21. The Reclassify pane open, fill it as below
22. Input raster: **Dhuleil_Slope**
23. Reclass Field: **Value**
24. Under **Reclassification** click the **Classify** tab, make the classes **6** and the method **Manual Interval** and click OK
25. Enter the following values as seen in the table below:

Start	End	New
0	5	1
5	10	2
10	15	3
15	20	4
20	30	5
30	55	6

Note when you finish entering 1 in the first row, click enter in your keyboard to open a new row

26. Output raster **Slope_Reclass**
27. Check "**Change missing values to NoData**"
28. Run
29. Save the project

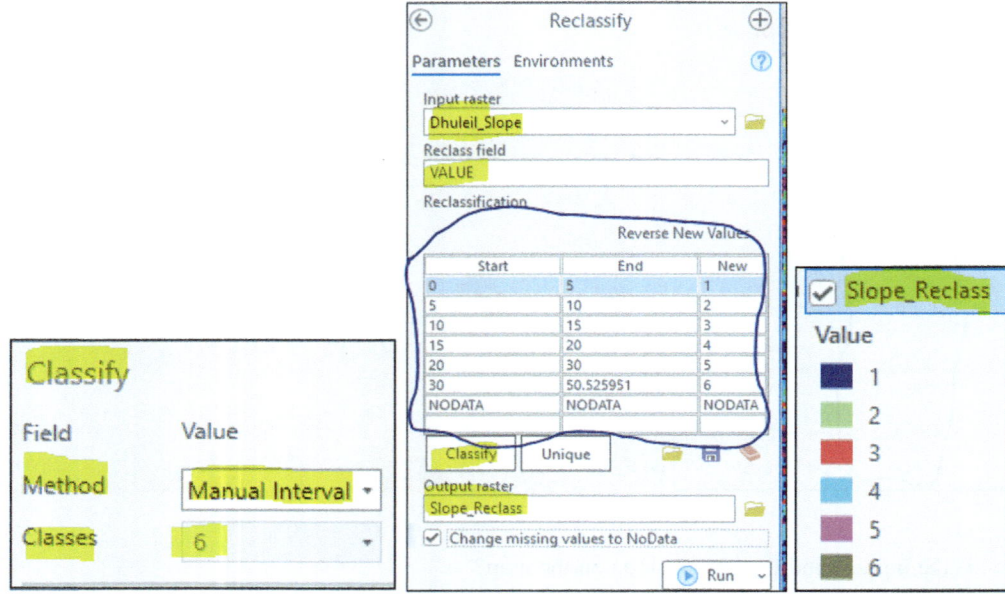

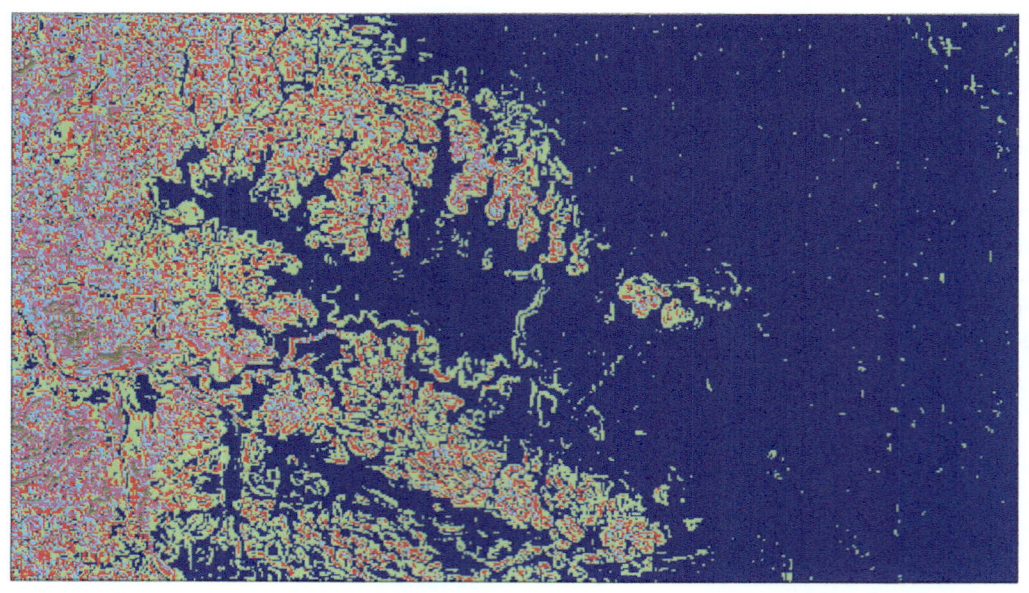

Add New Field for Slope_Reclass

You are going to add a new field to the **Slope_Reclass**. The new field will show the slope range in each class.

30. In the CP, r-click **Slope_Reclass** and click **Data Design** then click **Fields**
31. Click the last row in the view that says "**Click here to add a new field**"
32. Enter a name "**Slope**" under the **Field Name**
33. Under **Data Type** column choose from the drop-down menu "Text"
34. Under **Length**, type 12
35. Click on your keyboard the **Tab** to accept your entering
36. Click the **Field** tab and in **Changes** group click the **Save** button
37. Close "**Fields: Slope Reclass**"

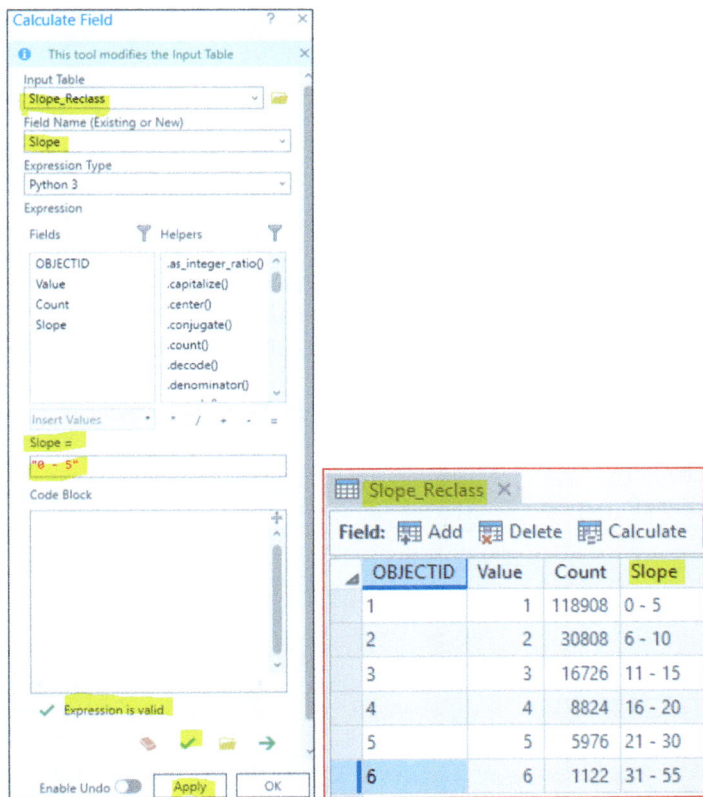

38. Open the attribute table of "**Slope Reclass**"
39. The "**Slope**" field is empty ad you are going to add text into it
40. Populate the "**Slope**" field by doing the following:
41. Highlight the first row then r-click "**Slope**" field and click **Calculate Field**
42. The **Calculate Field** pane open
43. Under Slope = type **"0 – 5"**
44. At the bottom of the Calculate Field pane click the **Verify** icon
45. You will get a message stating that the Expression is valid
46. Click Apply
47. Repeat the previous step for the remaining rows as shown below.
48. Second row "6 – 10" and click Apply
49. Third row "11-15" and click Apply
50. Fourth row "16-20" and click Apply
51. Fifth row "21-30" and click Apply

52. Sixth row "30 – 55" and click Apply
53. Click OK when you finish to close the Calculate Field pane
54. Open Attribute Table of **Slope_Reclass**, you will see that the Slope field is populated
55. Close the **Slope_Reclass** attribute table
56. Save your project

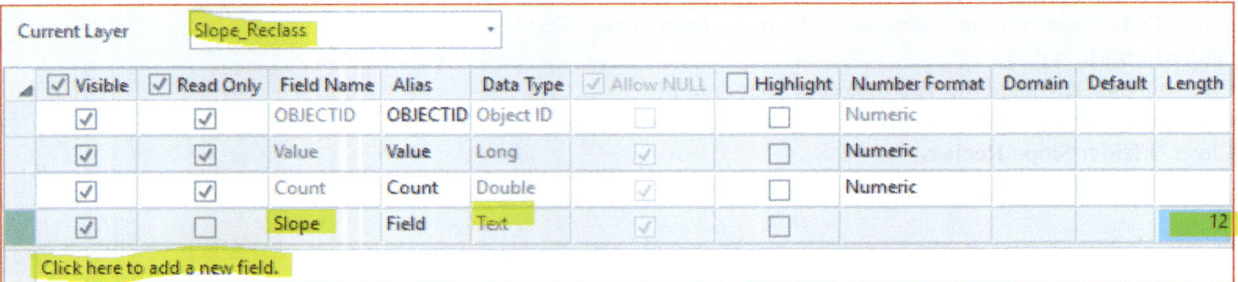

Derive Aspect Layer

The **aspect** raster map indicates the direction that slopes are facing. The compass direction that a topographic slope faces is usually measured in degrees starting from north. Aspect can be generated from continuous elevation surfaces. For example, the aspect recorded for a TIN face is the steepest downslope direction of the face, and the aspect of a cell in a raster is the steepest downslope direction of a plane defined by the cell and its eight surrounding neighbors.

56- In the Geoprocessing click the back arrow, click the **Aspect** in the **Surface** tools
57- The Aspect pane open, fill it as below
58- Input raster: Dhuleil
59- Output raster: Dhuleil_Aspect
60- Accept the default
61- Run

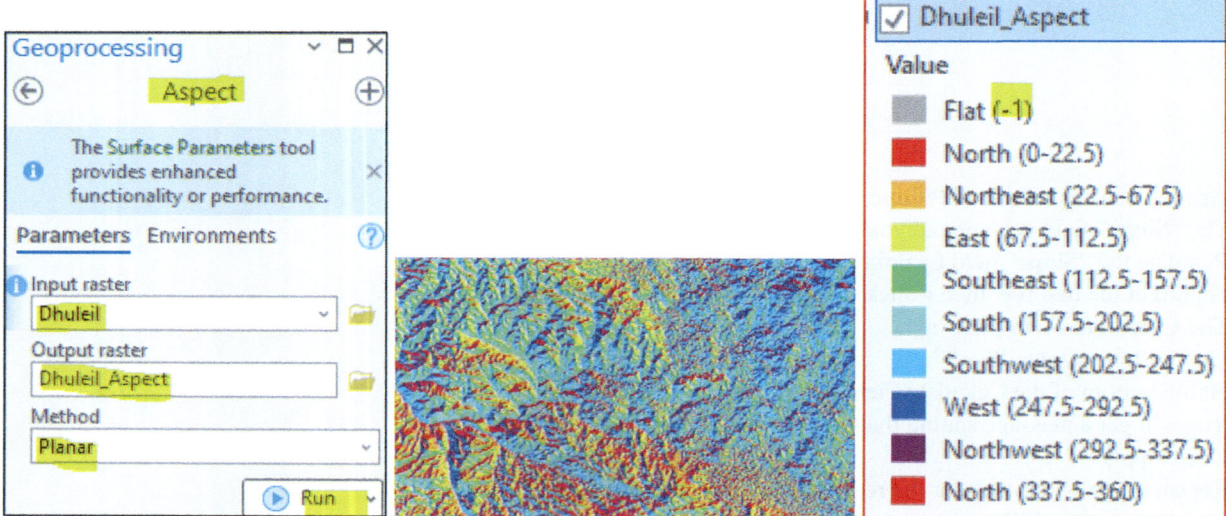

Result Aspect is measured counterclockwise in degrees from 0 (north) to 360 (north), coming full circle). The value of each cell in an aspect grid indicates the direction in which the cell's slope faces. Flat slopes have no direction and are given a value of -1. Nine principal directions are created for the **Dhuleil_Aspect**

Combine Two Images: Slope and Geology

The Combine tool takes multiple input rasters and assigns a new value for each unique combination of input values in the output raster. The original cell values from each of the inputs are recorded in the attribute table of the output raster. Additional items are added to the output raster's attribute table, one for each input raster.

In the image below, two rasters were input into the **Combine** function. Note that each unique combination of values from the two input rasters receives a unique value in the output raster. Two additional fields are added to the output raster attribute table containing the original values from the two input rasters that created the unique combination. Thus, the parentage of the output values can be traced back to the original rasters. Note that if a cell contains **NoData** in any of the input rasters, that location will receive **NoData** for the output. There is no limit to the number of rasters that can be combined; however, there is a practical limit. If there are many rasters all having many different zones, a greater number of unique combinations will be created, resulting in a large attribute table.

Raster01	Raster02	Output
1 1 0 0 ▨ 1 2 2 4 0 0 2 4 0 1 1	0 1 1 0 3 3 1 2 ▨ 0 0 2 3 2 1 0	1 2 3 4 ▨ 5 6 7 4 4 7 8 9 2 1

Raster 01	Raster 02	Output Raster
1	0	1
1	1	2
0	1	3
0	0	4
1	3	5
2	1	6
2	2	7
0	0	4
0	0	4
2	2	7
4	3	8
0	2	9
1	1	2
1	0	1

Scenario 10-1 The Water Authority decided to install a **lysimeter** to estimate the vertical infiltration rate of surficial geology in the **Dhuleil** area. The greatest factor controlling infiltration is the amount and characteristics of the precipitation. In general, rain falling on steeply sloped land runs off more quickly and infiltrates less than water falling on flat land. In addition, some outcropping formations allow rainwater to infiltrate at a higher rate downgradient and recharge the permeable formations. Alluvium and limestone are permeable formations and allow portions of rainwater to infiltrate and contribute to

groundwater recharge. Therefore, the aim is to find an area with a slope less than 10% that consists of limestone and alluvium. Slopes with values of 1 and 2 have slopes equal to or less than 10%.

To perform the analysis, the first step is to combine the two rasters: **Slope** and **Geology** together and make them one rater with the joined attribute table.

62- Click **Insert** tab on the ribbon, in the **Project** group click **New Map**
63- Change the name from "**Map**" to "**Geology**"
64- In Catalog pane drag the **Geology.tif** and **Slope.tif** from the **Data02** folder into the Map View
65- Open the attribute table of the **Geology.tif** and **Slope.tif** and familiarize yourself with the attribute of both rasters

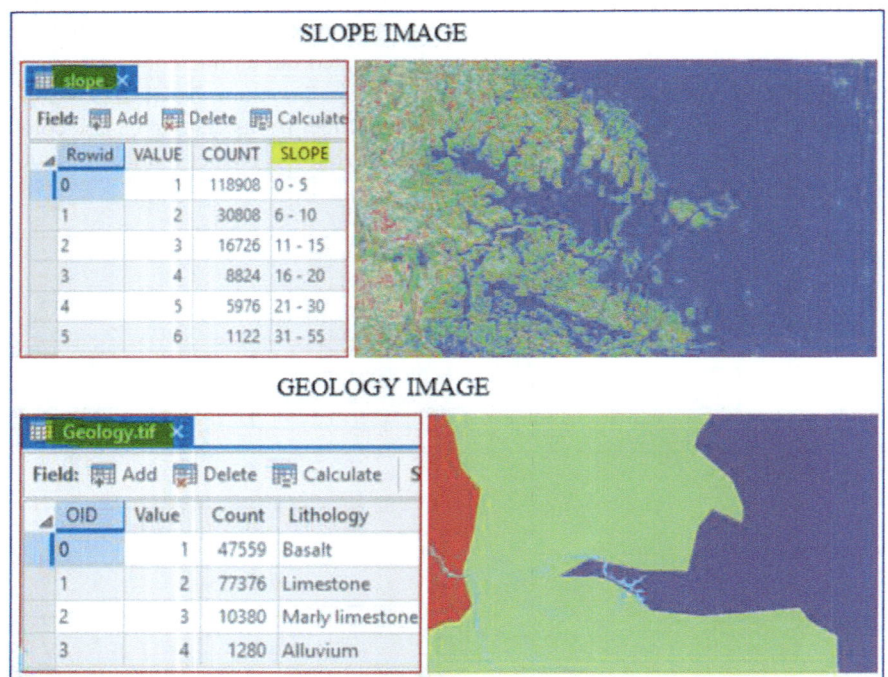

66- When you done, close the attribute table of both rasters
67- In the Geoprocessing click the back arrow, open the **Spatial Analyst Tools** and then the **Local** tools and click **Combine**
68- The **Combine** pane dialog box opens and fills it as follows:
69- Input raster: **Slope.tif** & **Geology.tif**
70- Output raster: **Geol_Slope**
71- Run
72- Close the Combine pane
73- Open the attribute table of "**Geol_Slope**" and familiarize yourself with the output

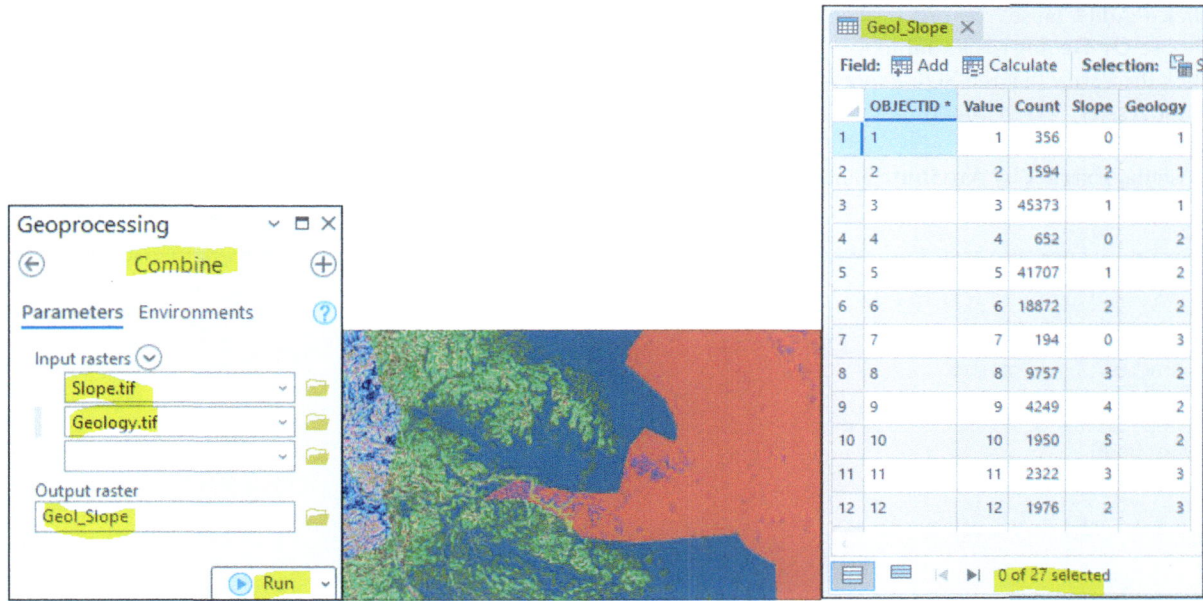

Result The combined output raster "**Geol_Slope**" created with 27 records and the attribute table unites the geology and the slope in one image and one table.

Note To see the image in color make sure the primary symbology of the "**Geol_Slope**" is "**Unique Value**"

Find Best Area to Build the Lysimeter

The **lysimeter** should be built in an area with a slope less than 10% and consisting of limestone and alluvium. Slopes with values of 1 and 2 have slopes equal to or less than 10%. To find the areas of the outcropping limestone and alluvium formations that have slopes less than 10%, certain tools should be used in the spatial analyst tools, and this can be performed using two different approaches:

(a) **Extract by Attribute**
(b) **Raster Calculator**

Both approaches can be used to find the best location to install the lysimeter.

Extract by Attribute tool

This approach allows you to extract the cells of a raster based on a logical query.

74- In the Geoprocessing click the back arrow, open the **Spatial Analyst Tools** and then the **Extraction** and click **Extract by Attributes**
75- In the Extract by Attributes dialog box, fill it as follows:
76- Input raster: **Geol_Slope**
77- Choose: **Where Slope is less than or equal to 2**
78- Click + Add Clause
79- Choose **And Geology is equal to 2**

80- Click + Add Clause
81- Choose **Or Geology is equal to 4**
82- Click the verify that the SQL expression is valid (green check symbol icon)
83- Output raster: **BestLocation**
84- Run
85- Close the **Extract by Attribute** pane

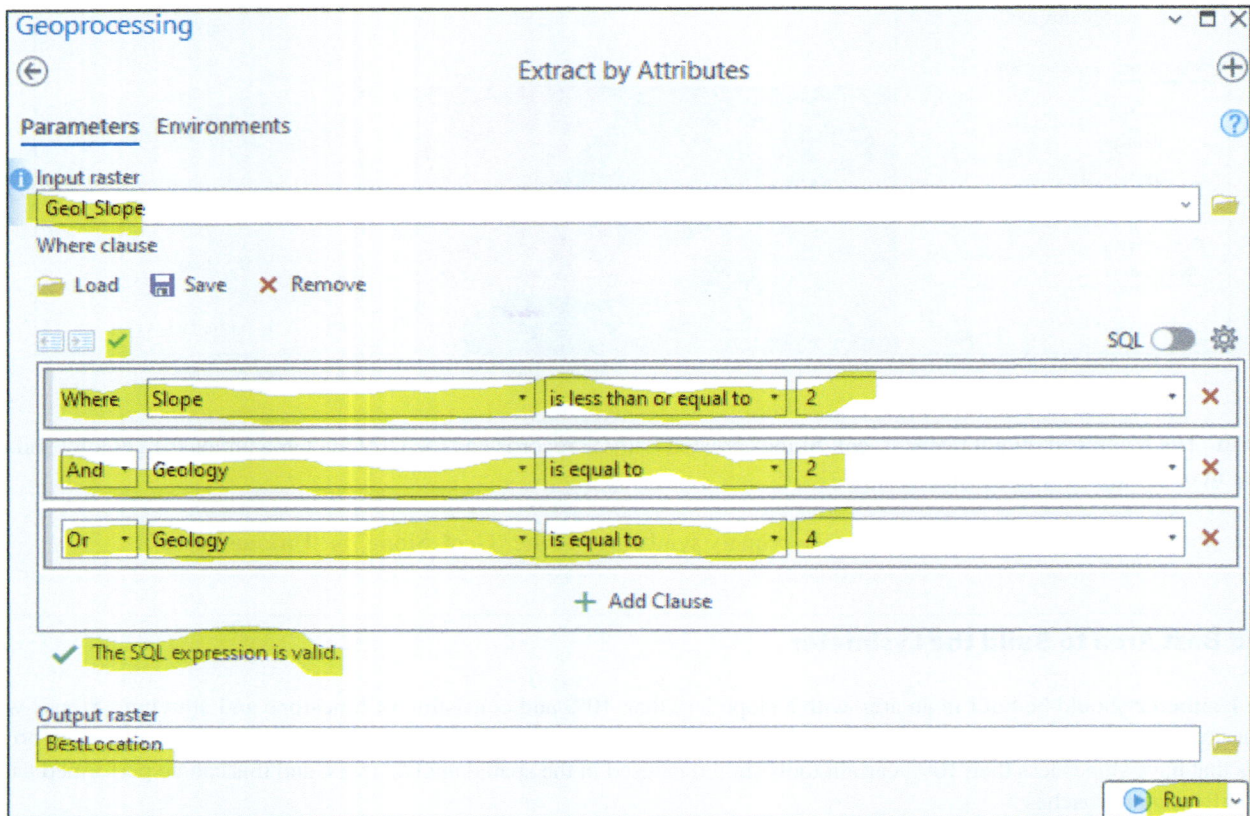

Result The BestLocation raster display consists of 10 records where the lysimeter can be built.

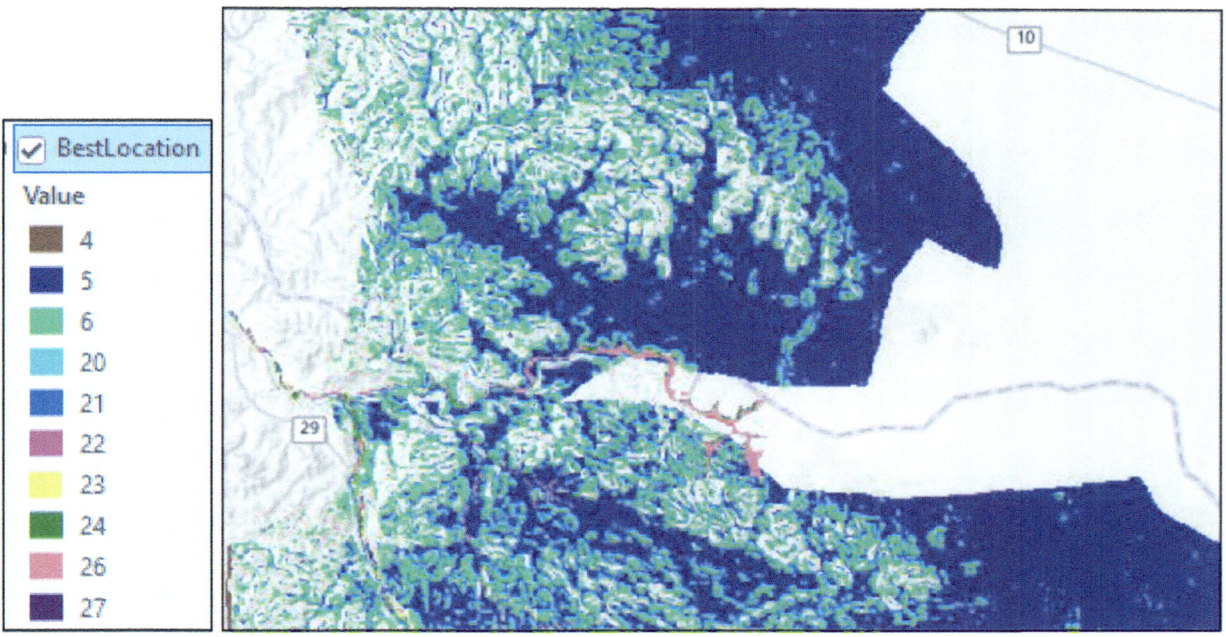

Raster Calculator Tool

In this example, you might use the **Raster Calculator** in the Map Algebra to select the land that has a slope less than 10% and consists of limestone and alluvium formation using the two rasters: **Geology** and **Slope**.

86- In the Geoprocessing click the back arrow, in the **Spatial Analyst Tools** open the **Map Algebra** and click **Raster Calculator**
87- In the **Raster Calculator** dialog box, type the SQL statement as follows:

Combine("Geol_Slope" <=2) & ("Geology.tif" ==2) | ("Geology.tif" ==4)

88- Output raster: **LysLocation**
89- Run
90- Save your project
91- Close the raster calculator pane

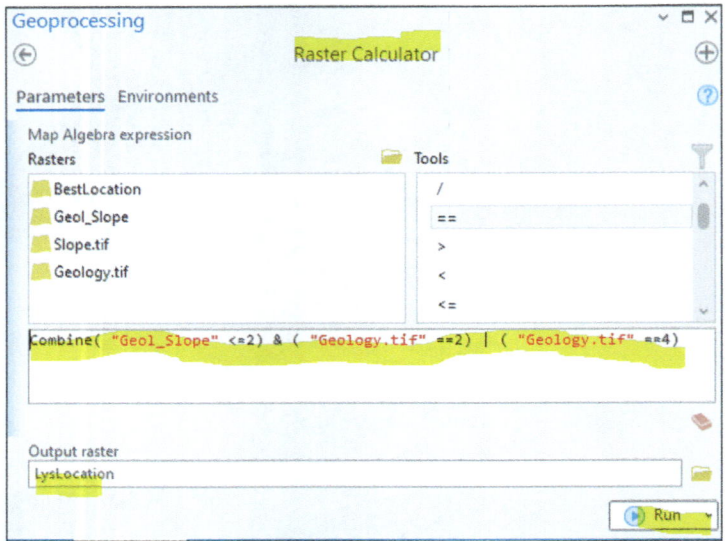

Result **LysLocation** is a raster with two records. Class no. 1 is the location where the lysometer will be installed, and class 0 is the area that is not suitable for lysometer installation.

Calculate the Area of the Lysimeter

93. In the CP, r-click **LysLocation** and click **Data Design** then click **Fields**
94. Click the last row in the view that says "**Click here to add a new field**"
92- Enter a name "**Percent**" under the Field Name
93- Enter a name "**Percent**" under the Alias
94- Under Data Type column choose from the drop-down menu "**Double**"
95- Under **Number Format** choose **Numeric** in the category and make the decimal places equal to **2** and click OK
96- Click on the keyboard tab to accept your entering
97- Click the **Field** tab in the **Changes** group and click the **Save** button
98- Close "Fields: LysLocation"
99- Open the attribute table of **LysLocation** (The layer has 2 records; the count of the value 0 has 57969 cells and the value 1 has 78687 cells)
100- R-click field "**Percent**" and point to **Calculate Field**
101- Fill the **Calculate Field** dialog box as follows:

Calculate the Area of the Lysimeter

102- Input Table: LysLocation
103- Field Name: Percent
104- Under Percent = type (**!Count! / (57969 + 78687) * 100**)
105- Click Apply and then OK to close the Calculate Field dialog box
106- Save the project and Exit ArcGIS Pro

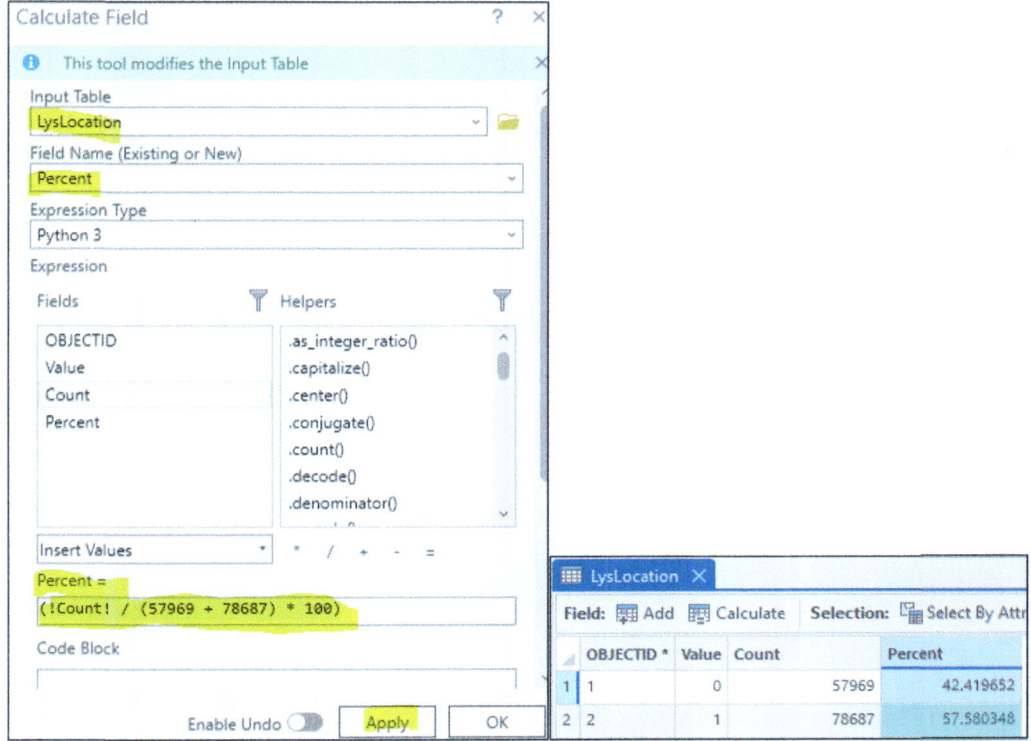

Results The percentage of the area that is suitable for building a Lysimeter is 57.58%.

Spatial Interpolation

11

Spatial interpolation (SI) is a term used to estimate a value of a data variable at an unsampled site from measurements made in proximity to or within a range of available data. This technique is based on Tobler's First Law of Geography, which states that points close together in space are more likely to have similar values than points that are far apart. Use a neighborhood of sample points to estimate a value at an unsampled location (Figure below).

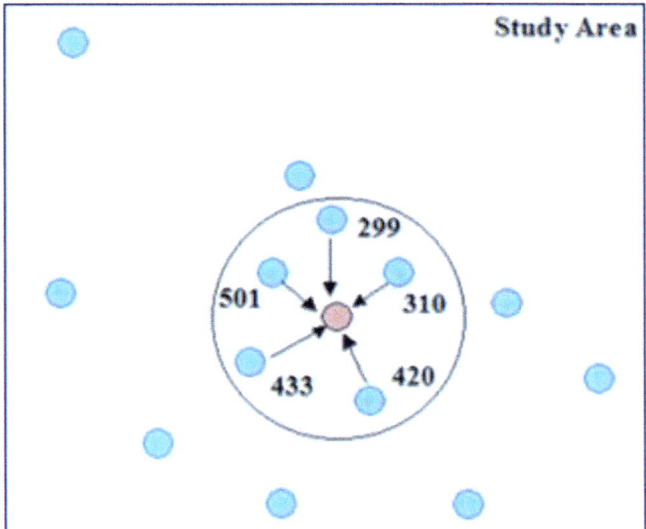

Interpolation uses a neighborhood of sample points of known values (blue color) to estimate a value at an unsampled location (rose color). This method of estimation uses a specific radius from the unsampled point.

Various interpolation techniques are used. These techniques use sample values and X and Y coordinates to estimate the value of an unsampled point. In general, different methods will generate dissimilar results with the same input data, and no method is more accurate than others under all conditions. Users seeking accuracy should take into consideration several point samples and knowledge of the study area.

To produce a continuous representation of the phenomenon in question, interpolation makes use of sampling data. The sampling data are accurate and qualitative. There are various GIS methods that use the interpolation method. Deterministic interpolation techniques create surfaces from measured points. They are based on the extent of the similarity, and an example of such methods is inverse distance weighted (IDW). There is also a degree of smoothing, such as the trend surface analysis method. Geostatistical interpolation techniques such as kriging are based on statistics and are used for more advanced prediction surface modeling. That also includes errors or uncertainty of predictions. The kriging method is based on the theory of regionalized variables and is performed by placing an evenly spaced grid over the area for which we have known values and

Supplementary Information The online version contains supplementary material available at https://doi.org/10.1007/978-3-031-42227-0_11.

© The Author(s), under exclusive license to Springer Nature Switzerland AG 2023
W. Bajjali, *ArcGIS Pro and ArcGIS Online*, Springer Textbooks in Earth Sciences, Geography and Environment,
https://doi.org/10.1007/978-3-031-42227-0_11

can obtain an estimated surface. The basic idea of kriging interpolation is that every unknown point can be estimated by the weighted sum of the known points within a certain radius.

Method of Interpolation

Trend Surface Analysis

Trend surface analysis is a simple way to describe large variations, and its function is to find general tendencies of the sample data rather than to model a surface precisely. The trend analysis calculates the coefficients of a best-fit polynomial surface to fit a set of spatially distributed data points.

In one dimension (1-D): z varies as a linear function of x

$$Z = b_0 + b_1 x + e$$

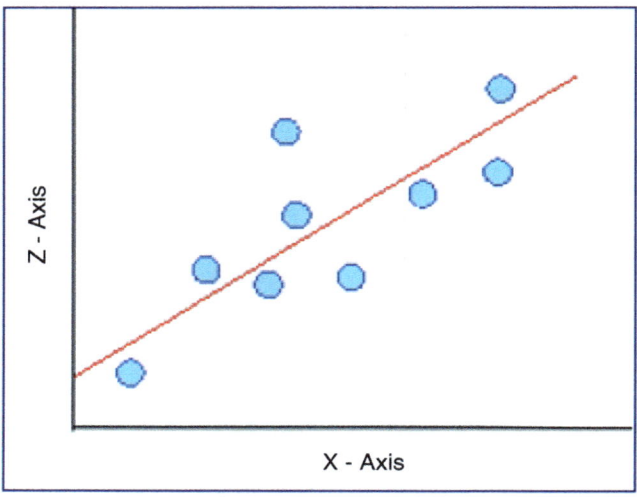

In two dimensions (2-D): z varies as a linear function of x and y

$$Z = b_0 + b_1 x + b_2 y + e$$

where Z is the interpolated parameter.

X and Y are the coordinates of the wells.

The B coefficient is estimated from the control points.

The aim of this method is to develop a general kind of spatial distribution of an observable fact. The surface can be modeled using a linear or trend surface. Linear trends describe only the major direction and rate of change, while the trend surface provides progressively more complex descriptions of spatial patterns.

Inverse Distance Weighting (IDW)

Inverse distance weighting is a very popular technique in GIS and is considered one of the simplest interpolation methods. There are a variety of methods that use weighted moving averages of points within a zone of influence. Interpolation techniques in which interpolation estimates are made based on values at nearby locations weighted only by distance from the interpolation location (figure below).

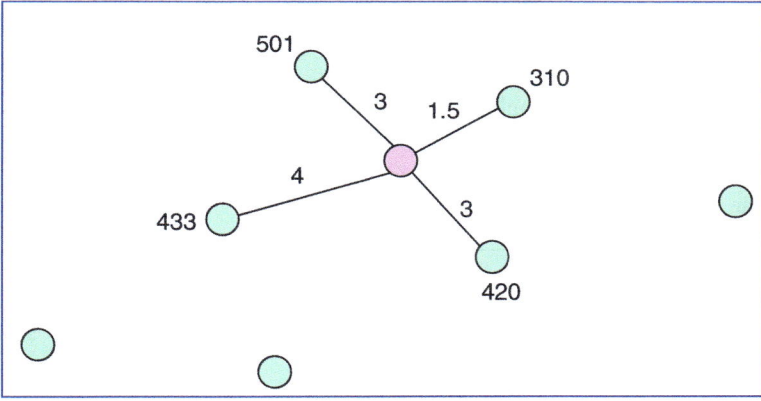

IDW: Closest 4 neighbors
In general, the simplified formula for IDW is:

$$V_0 = \frac{\sum_{i=1}^{n}\left(\frac{Vi}{Di}\right)}{\sum_{i=1}^{n}\left(\frac{1}{Di}\right)}$$

where V_0 is the predictable value at point 0, Vi is the V value at control point i, Di is the distance between control point i and 0, and n is the number of known values used in the evaluation.

The weights are a decreasing function of distance, and the user has control over the mathematical form of the weighting function. The size of the neighborhood can be expressed as a radius or several points.

Global Polynomial (GP)

Global Polynomial or GP fits a smooth surface that is defined by a polynomial to the input sample points such as the TDS field in the attribute table of the well layer. The GP is similar to taking a piece of paper and fitting it in between the raised TDS values. The result from GP interpolation is a smooth surface that represents gradual trends in the surface over the area of interest. It is used by fitting a surface to the sample points when the surface varies slowly from region to region over the area of interest. While examining and/or removing the effects of long-range or global trends. In such circumstances, the technique is often referred to as trend surface analysis.

Kriging

Using geostatistical techniques, surfaces incorporating the statistical properties of the measured data can be created. Geostatistics is based on statistics. These techniques produce not only prediction surfaces but also error or uncertainty surfaces, giving you an indication of how good the predictions are. Many methods are associated with geostatistics, but they are all in the kriging family. Ordinary, simple, universal, probability, indicator, and disjunctive kriging, along with their counterparts in cokriging, are all available in the Geostatistical Analyst. Not only do these kriging methods create predictions and error surfaces, they can also produce probability and quantile output maps depending on user needs. Kriging is the estimation procedure using known values and a semivariogram to determine unknown values. The procedures involved in kriging incorporate measures of error and uncertainty when determining estimations. Based on the semivariogram used, optimal weights are assigned to unknown values to calculate the unknown ones. Since the variogram changes with distance, the weights depend on the known sample distribution. The basic equation used in ordinary kriging is as follows:

$$K(d) = \frac{1}{2n}\sum_{n}^{i=1}\left(Z(xi) - Z(xi+d)\right)^2$$

where d is the distance between known points, n is the number of pairs of samples separated by d, and Z is the attribute value (elevation of known points). The equation indicates that the semivariance is expected to increase as d increases.

One of the most popular approaches is ordinary kriging, which will be applied in this study. Ordinary kriging assumes the following model: $\mathbf{Z(s) = \mu + \varepsilon(s)}$, where μ is an unknown constant.

One of the main issues concerning ordinary kriging is whether the assumption of a constant mean is reasonable. Sometimes there are good scientific reasons to reject this assumption. However, as a simple prediction method, it has remarkable flexibility. The following figure is an example in one spatial dimension:

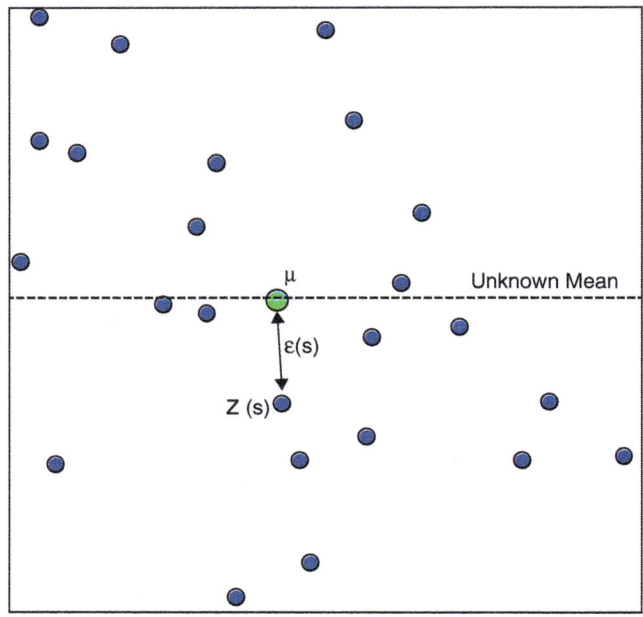

The data are consistent with TDS values collected from the Maawil watershed in Oman. The well locations look like they are distributed randomly. The data are simulated from the ordinary kriging model with a constant mean μ. The true but unknown mean is given by the dashed line. Thus, ordinary kriging can be used for data that seem to have a trend. There is no way to decide, based on the data alone, whether the observed pattern is the result of autocorrelation—among the errors ε(s) with μ constant—or trend, with μ(s) changing with s.

Ordinary kriging can use either semivariograms or covariances (which are the mathematical forms used to express autocorrelation), use transformations and remove trends, and allow for measurement error.

Connect to Data and Data Integration

1. Launch ArcGIS Pro
2. Click Open another project (upper-right) browse to **Env_Water\Ch11** select **Ch11.aprx** and click OK

The **Ch11.aprx** open and the Content pane include the **Map**, which is empty. The Map View displays the World Topographic Map and World Hillshade.

In the **Catalog** pan r-click **Folders** and click **Add Folder Connection** browse to \\Env_Water\Ch11 (or \\Database\Data_Ch11) open it and highlight Data_Ch11 and click OK

The **Data** folder has 4 shapefiles:

- **Dam.shp**
- **Stream.shp**
- **Watershed.shp**
- **Well.shp**.

The files are registered in UTM zone 40, and the datum is WGS 1972. The watershed contains 1,758 groundwater wells drilled mainly in the upper part of the Maawil watershed area and downstream from two dams. The wells include information about the salinity (TDS) and nitrate (NO_3) concentrations.

4. In the **CP**, select the **Map** and press the **F2** key and change the name to **Groundwater Density**
5. In the **Catalog** pane, open the **Folder** and expand the **Data** under **Data_Ch11** and select the **Dam.shp**, **Stream.shp**, **Watershed.shp** and **Well.shp** and drag them into **Map View**
6. Symbolize the **Well** (use circle 1, blue color, and size 6), the **Dam** (use dam symbol), the **Stream** (use Intermittent Water (line), and line width 1 pt), and the **Watershed** (use Black Outline (1pt))
7. Close the **Symbology** pane

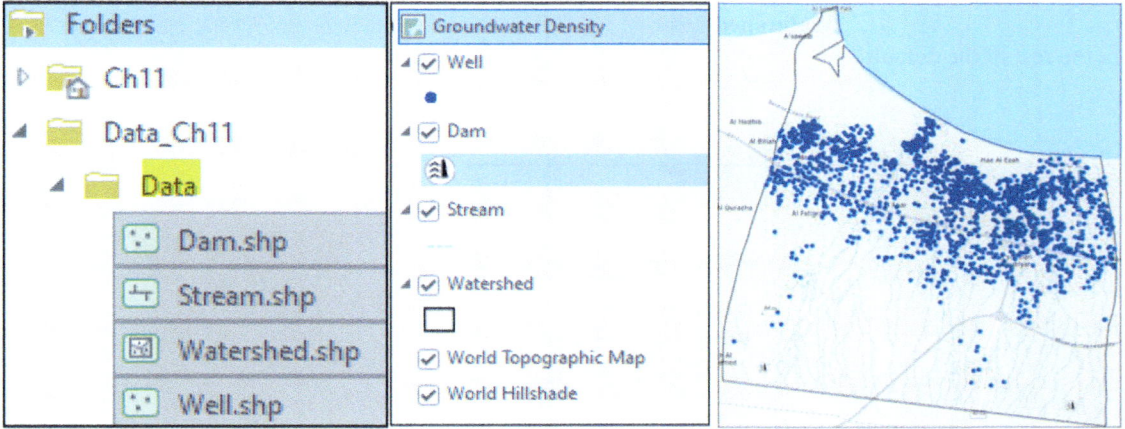

The 4 layers, which are in the northern part of the Sultanate of Oman, are displayed in the Map View with the proper symbology.

8. Save the project

Density of Groundwater Well

The point density calculates a magnitude-per-unit area from point features that fall within a neighborhood around each cell. Adopting larger radii yields a more generalized density raster, and a smaller radius yields a more detailed raster. Only the wells that fall within the neighborhood are considered when calculating the density. If no wells fall within the neighborhood at a cell, that cell is assigned no data (**NoData**).

Scenario 1 You are a hydrogeologist working for the water resources in Oman. You have been given a task to evaluate the groundwater along the coast in the Maawil watershed. You decided first to assess if the densities of the wells influence the salt intrusion and second, if the quality of groundwater downstream of the Maawil and Al-Kabir dams has been improved. The two dams have been built to store the surface runoff produced by rain in the rainy season, and then the rainwater gradually infiltrates downgradient and recharging the aquifer and improves its water quality. To answer these two questions, you decided to use the interpolation technique in GIS environment.

9. Click the **Analysis** tab on the ribbon, in the **Geoprocessing** group, click the **Tools** button, the Geoprocessing pane display
10. In the **Geoprocessing** pane click the **Toolboxes** tab and open the **Spatial Analyst** Tools, then open the **Density** and click the **Point Density** tool
11. The **Point Density** pane dialog box opens, click the **Parameters** tab and fill it as follows:

12. Input point features: **Well**
13. Population field: **TDS**
14. Output raster: **TDS_Density**
15. Output cell size: **50**
16. Neighborhood: **Circle**
17. Radius: **1000**
18. Units type: **Map**
19. Area Units: **Square Kilometers**
20. Click the **Environment** tab
21. Output Coordinate System: click the drop-down arrow and select "**Well**" the following coordinate displays: **WGS_1972_UTM_Zone_40 N**
22. Cell Size Projection Method: **Preserve resolution**
23. Mask: **Watershed**
24. Accept the rest of the default
25. Run

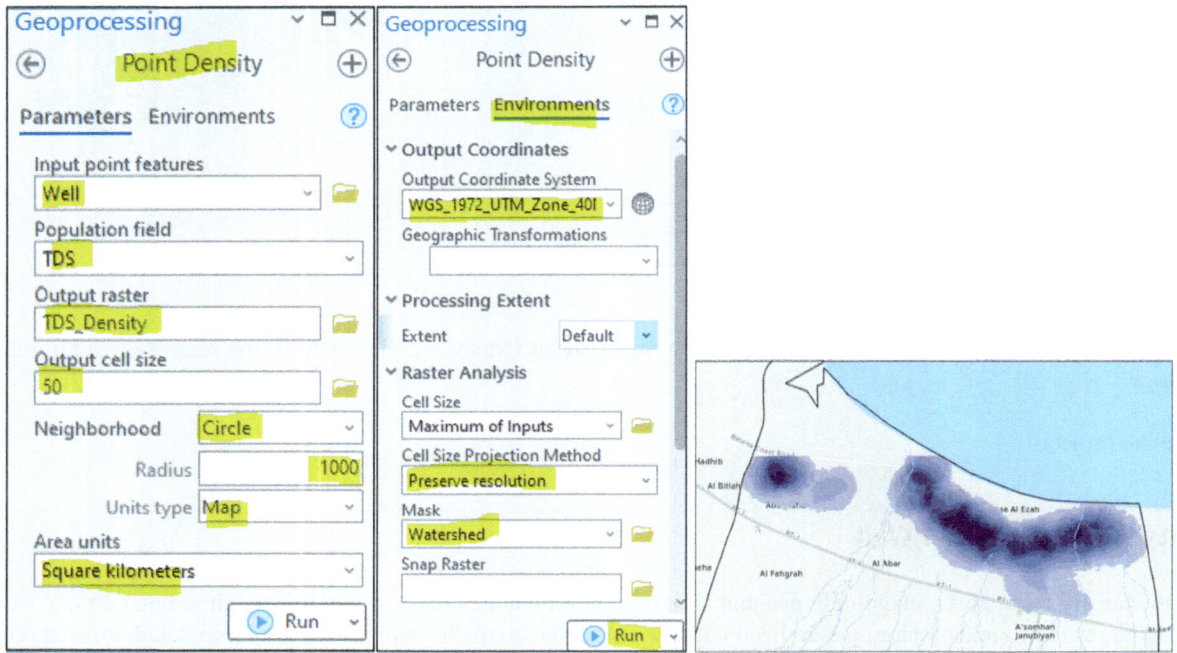

Result The TDS_Density raster map displays and shows the cell densities using the salinity (TDS) field. Wells that have high salinity demonstrate larger densities. It is true that the density of the wells in one location could affect the cone of depression by reducing the water table below the sea level of the Gulf of Oman. This causes the sea saline water to invade the shallow aquifer along the coast and increase its salinity to a higher concentration.

26. In CP, click **TDS_Density** and click Symbology tab below the **Catalog** pane
27. In the Symbology pane under **Primary symbology** select **Classify**
28. In **Classes** choose 5 and the **Metho**d choose **Manual Interval**
29. In the Classes tab under **Upper value** change the concentration and color from top to bottom 1000 (apatite Blue), 2000 (Cretan Blue), 3000 (Medium Apple), 5000 (Ginger Pink), and 30000 (Mars Red).
30. In the Symbology pane, click the **Advanced symbology options** tab

31. Open **Format Labels** and change the **Decimal places** to 0
32. Save the project

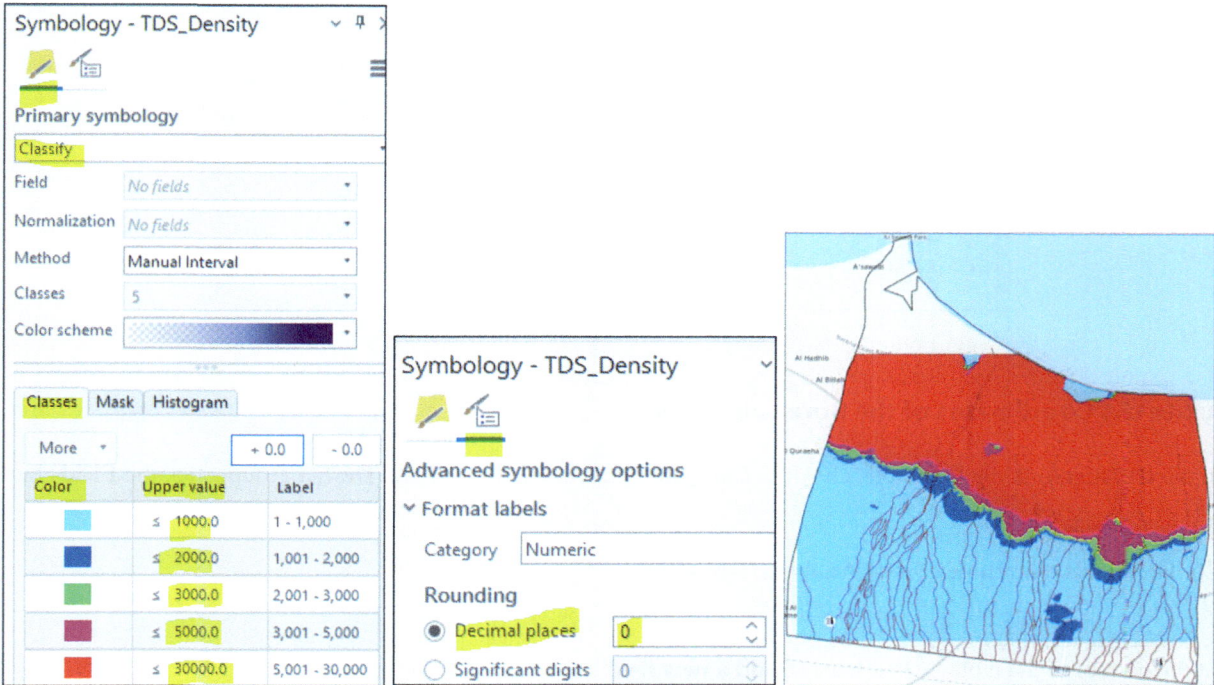

Trend Analysis

The trend analysis tool can help identify trends in the east–west or north–south direction in one of the input variables of the dataset. To perform the trend analysis, you must have the X and Y coordinates for all the wells and then use the scatter plot graph which will show the 1st polynomial order trend by setting the X-axis number to the Y coordinate and the Y-axis number to the TDS value. Trend surface analysis is a useful tool in early data analysis for delineating basic information and trends regarding the distribution of data. This type of analysis will be performed on the TDS to detect any trend in the salinity.

33. Click **Insert** tab on the ribbon, in the **Project** group, click **New Map** button
34. In the CP rename the **Map** to "Trend Analysis"
35. In the **Catalog** pane, integrate the **Well** and **Watershed** from the **Data** folder
36. In the CP, click the symbol of the **Watershed** and use the Black Outline (1pt) symbol from the **Gallery** tab.
37. In the CP, right-click **Well** and point to **Data** and click **Export Features**
38. The Export Features dialog box open, fill it as below
39. Input Features: Well
40. Output Feature Class: Borehole
41. Click OK
42. Remove the Well from the CP
43. In the **CP**, click the symbol of the **Borehole** and make it circle 1, blue and size 3.

Calculate the Coordinates of the Borehole Layer

44. In the **CP**, highlight the **Borehole**, click the **Data** tab in the ribbon, in the **Data Design** group, click the **Fields** button

Result The **Fields: Borehole (Trend Analysis)** table display

45. At the bottom of the table "**Click here to add a new field**"
46. Under **Field Name**, type "**Easting**", then click the tab in the keyboard
47. Under Alias, type "**Easting**", then click the tab twice in the keyboard
48. Under **Data Type**, select "**Double**", then click the tab 3-times in the keyboard
49. Under **Number Format**, click the window and then the 3-dots and select in the **Category Numeric** and make the **decimal places** 4 and click OK
50. Click again "**Click here to add a new field**"
51. Under **Field Name**, type "**Northing**", then click the tab in the keyboard
52. Under **Data Type**, select "**Double**", then click the tab in the keyboard
53. Under **Number Format**, click the 3-dots and select in the Category **Numeric** and make the **decimal places** 4 and click OK
54. In the **Fields** tab on the ribbon, in the **Changes** group, click the "**Save**" button
55. Close the **Field: Borehole** tab above the **Map View** by clicking on the X

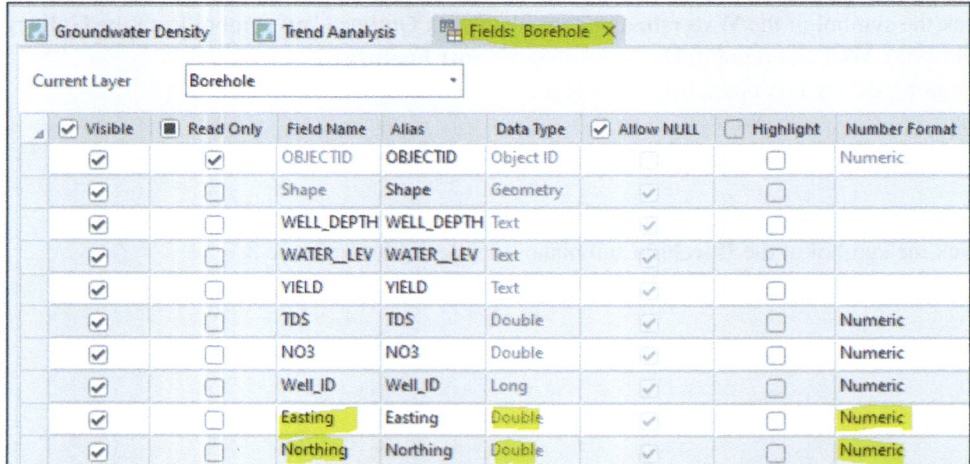

Salinity Trend Using Scatter Plot

56. In the CP, r-click **Borehole** and open the **Attribute Table**
57. R-click the **Easting** header and click **Calculate Geometry**
58. Fill the **Calculate Geometry** pane as follows:
59. Input Features: **Borehole**
60. Target Field: **Easting**
61. Property: **Point x-coordinate**
62. Target Field: **Northing**
63. Property: **Point y-coordinate**
64. Coordinate System: Borehole (display **WGS_1972_UTM_Zone_40 N**)
65. Click OK
66. Close the **Calculate Geometry Attributes** pane then close the **Borehole** attribute table

Result **Easting** and **Northing** are calculated in the projected coordinate system, which is UTM zone 40 and associated with WGS 1972.

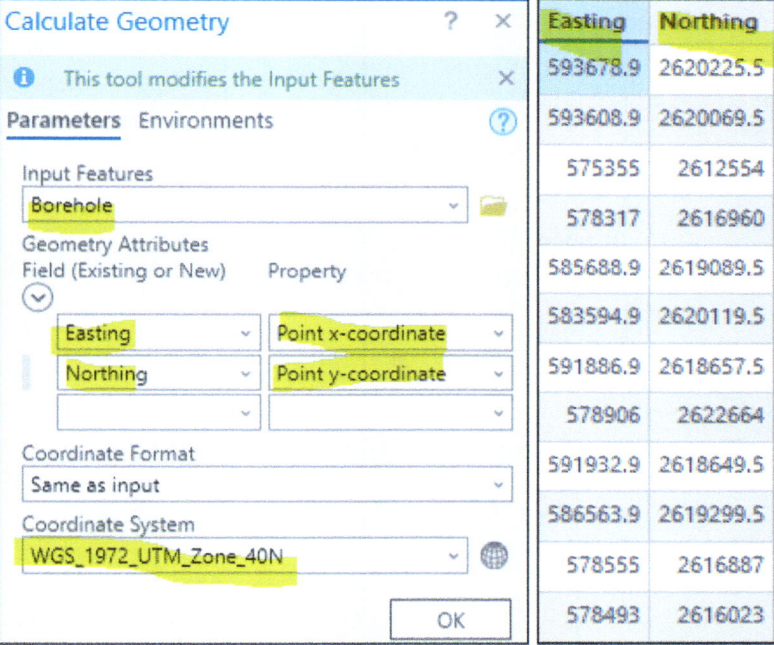

Salinity Trend Using Scatter Plot

Scatter plots visualize the relationship between two numeric variables, where one variable is displayed on the x-axis, and the other variable is displayed on the y-axis. For each record, a point is plotted where the two variables intersect in the chart.

67. In the CP select **Borehole**, click the **Data** tab in the ribbon, in the **Visualize** group, click the drop-down arrow of the **Create Chart** and select **Scatter Plot**
68. In the Chart Properties -Borehole pane fill it as below
69. Click Data tab:
    (a) X-axis Number: Northing
    (b) Y-axis Number: TDS
    (c) Under Statistics check Show linear trend
    (d) Click the linear trend line and change its color to red.
    (e) Symbol size: 3

70. Click Series tab
    (a) change the color of the symbol into **dark amethyst** color (C11, R5)
71. Click Axes tab
    (a) Under the Y-axis, change the maximum to 15000
    (b) Check the Log axis
72. Click the Format tab
    (a) Click the "Text element" icon and highlight **All Text**
    (b) In the **Font** select Times New Roman and the Font Size = 12 pt
73. Click General tab
    (a) Chart Title: Salinity Increase toward the Coast of Oman
    (b) Y axis title: TDS (mg/L)

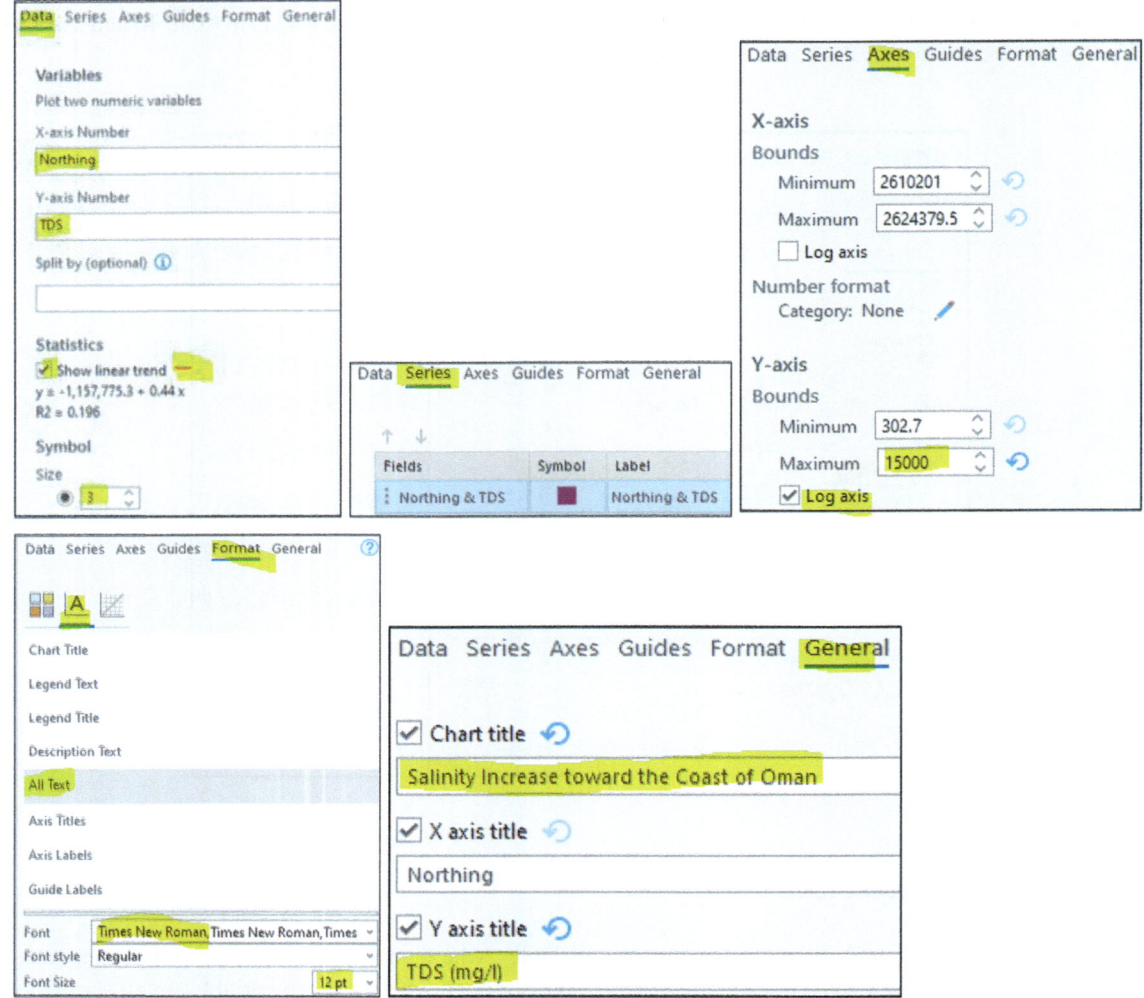

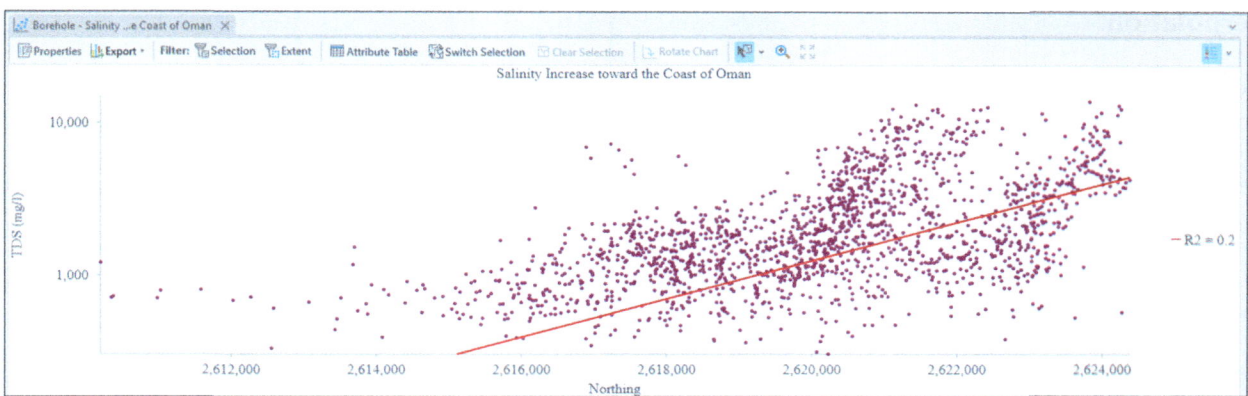

Result The effect takes place directly on the chart and on the Chart in the CP. The scatter plot is displayed in the content pane and the **Map View**. The graph shows the trend projection of the south–north trend (YZ-red). The trend means that the TDS concentration increases toward the Gulf of Oman.

4. Close the **Salinity Increase toward the Coast of Oman Chart**
5. If you want to modify the chart, you can r-click the Chart in the CP and open

Save as Layer Files

Now, you will save the **Borehole** and **Watershed** as layer files. The layer file stores many properties of the input layer, such as symbology. Both layers will be used in the interpolation techniques.

46. In the CP, r-click **Borehole** layer, point to **Sharing** and choose **Save As Layer File**.
47. Save the "**Borehole.lyrx**" in **Catalog** pane in **Data_Ch11** under the **Folder**
48. Repeat the previous step and save the **Watershed.lyrx** in **Data_Ch11** under **Folder.**
49. Save the project

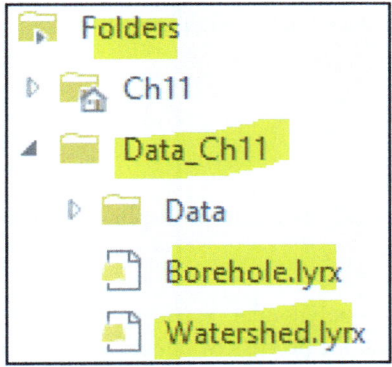

Interpolation

ArcGIS Pro includes different tools to predict values at unmeasured locations. The technique predicts values for cells in a raster from a limited number of sample data points. The data can represent elevation, rainfall, chemical concentrations, population and others. In this paragraph, interpolation tools were used to study the correlation of the TDS (salinity) concentration in groundwater in different watersheds in the Sultanate of Oman. Three types of interpolations will be discussed: global polynomial interpolation (GPI), inverse distance weighting (IDW), and kriging.

Global Polynomial Interpolation

Global polynomial interpolation (GPI) is used to demonstrate how TDS varies in groundwater in the watershed. The output from using this tool is a smooth surface that is defined by a mathematical function (a polynomial) to the borehole points. The GPI is allowed up to 10 order polynomials.

1. Click **Insert** tab on the ribbon, in the **Project** group, click **New Map** button
2. In the CP rename the **Map** to "GPI"
3. In the **Catalog** pane, integrate the **Borehole.lyrx** and **Watershed.lyrx** from the **Data_Ch11** folder
4. Click the **Analysis** in the ribbon, in the **Workflows** group, click the **Geostatistical Wizard**
5. The **Geostatistical Wizard** dialog box displays

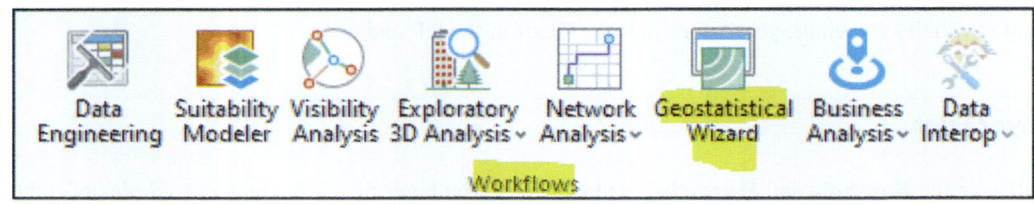

6. Under **Deterministic methods**: select **Global Polynomial Interpolation**
7. Under the input dataset, fill it as follows:
8. Source Dataset: **Borehole**
9. Data Field: **TDS**
10. Click Next
11. Use Mean and click Next

Note 1 This step allows you to choose the order of polynomial from 1 to 9.

12. Select **Power** 1
13. Next
14. The root-mean-square (RMS) is **1936.07**

Comment This step shows three tabs things: the predicted tab displays a scatter plot that show the predicted values versus measured values.

Note 2 you are going to use different order of polynomial. You are going to use the order of polynomial that generates the lower RMS. Therefore, you are going to repeat the process by choosing other powers starting from 2, 3, 4, 5, 6, 7, 8, and 9. The results are shown in the table below.

Convert the Gpi Layer Into Raster Format

Result The best polynomial power is 8, as it generates a lower RMS of 1472.72

15. Go back and select **Power** 8, then click Next
16. Click Finish
17. After reviewing the Method Report, click OK

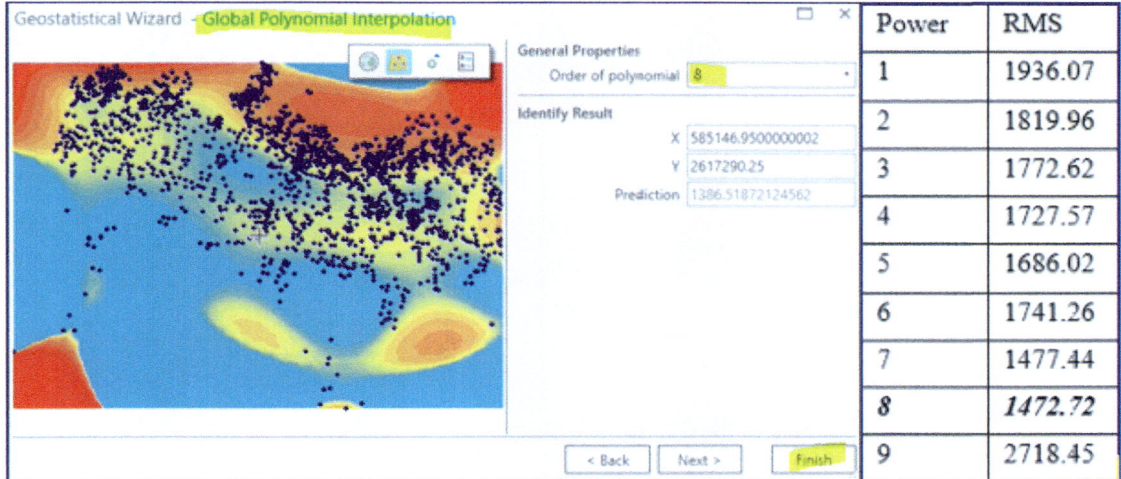

Power	RMS
1	1936.07
2	1819.96
3	1772.62
4	1727.57
5	1686.02
6	1741.26
7	1477.44
8	*1472.72*
9	2718.45

Result The global polynomial interpolation layer is displayed in the CP and has the same area extent as the **borehole** layer.

Convert the Gpi Layer Into Raster Format

Converting the **Global Polynomial Interpolation** layer into raster format is an essential step to clip the interpolated layer to fit the watershed area using the Mask in the Raster Analysis.

18. Right click **Global Polynomial Interpolation** layer, click **Export Layer** and choose **To Raster**
19. The **GA Layer To Rasters** pane open
20. In the **Parameters** tab, fill the dialog box as follows:
    (a) Input geostatistical layer: **Global Polynomial Interpolation**
    (b) Output Raster: **GPI**
    (c) Output cell size: **50**
21. Click the Environments tab and fill it as follows:
    (d) Output Coordinate System: **Borehole**
    (e) Mask Watershed
22. Accept the rest of the default
23. Click Run
24. In the CP, uncheck the **Global Polynomial Interpolation** layer

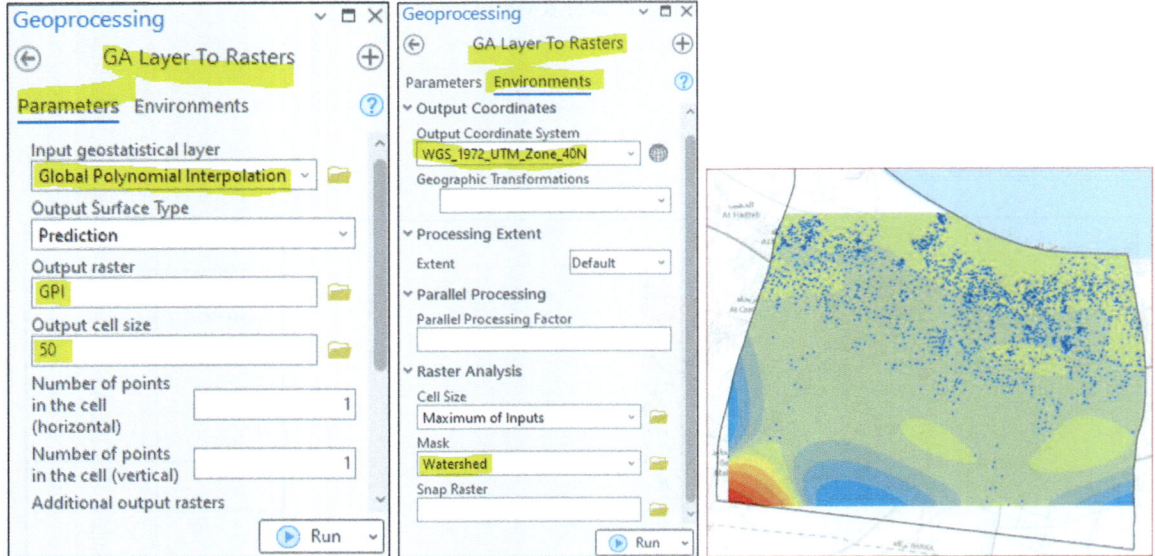

Result The GPI grid is created and added to the CP, and it is clipped to the watershed area.

Classify the GPI Map

You will classify the **GPI** raster into **5 classes** using the **Manual Interval**

25. In the CP, highlight the **GPI** and click the **Symbology** tab in the Catalog pane
26. Under **Primary symbology** select **Classify**
27. Make sure Class 5
28. Method = Manual Interval
29. In the **Classes** tab and under "Upper Value" and Color, replace the values and color from bottom to top
30. 30000 (red), 3000 (pink), 2000 (yellow), 1000 (green), and 500 (blue)
31. In the Symbology pane click the Advanced Symbology Options icon
32. Open Format labels and change the decimal places to 0
33. Under the label, remove the negative value and replace it by 0.
34. Save the project

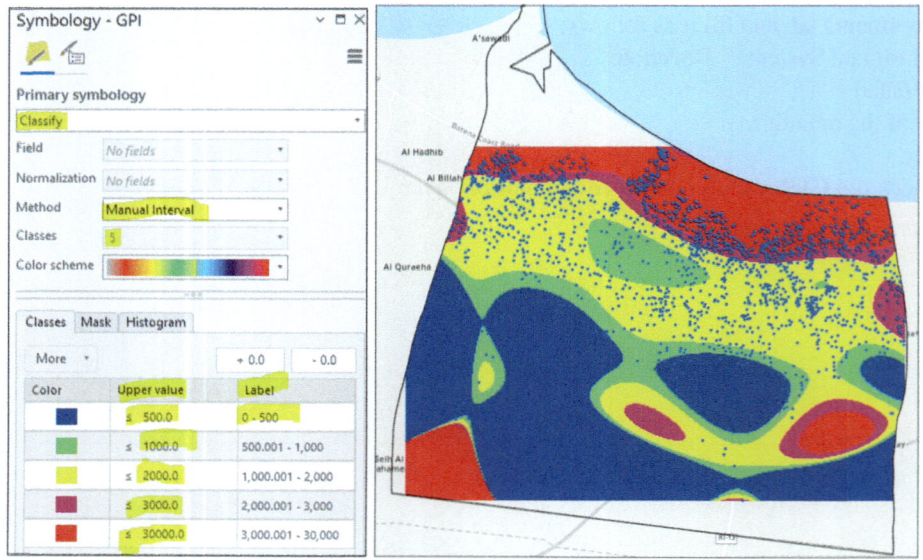

Result The generated GPI salinity maps for the watershed area indicate that the interpolated surface downstream from the two dams was dominated by the low TDS with the exception of the lower left side of the watershed due to the lack of boreholes in this location. The calculated surfaces are highly susceptible to outliers (extremely high and low values), especially at the edges. In general, the water quality demonstrates improvement in the south–north direction.

Challenge integrate the dam layer, symbolize it and comment on the water quality downstream of the dam.

Inverse Distance Weighting

Inverse distance weighted (**IDW**) interpolation determines cell values using a linearly weighted combination of a set of boreholes. The weight is a function of inverse distance. The surface being interpolated should be the location of the dependent variable. In this section, IDW interpolation techniques are employed using the salinity of the boreholes.

1. Click **Insert** tab on the ribbon, in the **Project** group, click **New Map** button
2. In the CP rename the **Map** to "**IDW**"
3. In the **Catalog** pane, drag the **Watershed.lyrx** and then **Borehole.lyrx** from the **Data_Ch11** folder under the Folder into the **Map View**
4. Click the **Analysis** in the ribbon, in the **Workflows** group, click the **Geostatistical Wizard**
5. The **Geostatistical Wizard** dialog box displays
6. Under **Deterministic methods**, check the **Inverse Distance Weighting**
7. Under Input Dataset
8. Source Dataset: **Borehole**
9. Data Field: **TDS**
10. Click Next
11. Check Use Mean
12. Click Next

Note IDW relies mainly on the inverse of the distance raised to a mathematical power. The power parameter lets you control the significance of known points on the interpolated values based on their distance from the output point. It is a positive, real number, and its default value is 2.

13. In the **Power** window "click to optimize" (it changes from 2 to 1)
14. Maximum Neighbors: 15
15. Minimum Neighbors: 5
16. Accept the rest of the default
17. Next
18. The Root-Mean-Square (RMS) is **1273.06**

Result This step shows two things: on the left, a scatter plot (predicted values vs measured values) with three tabs (Predicted, Error, Distribution) and on the right, a summary and table tabs that consist of (Source ID, Included, Measured, Predicted, and Error). You can save the table to the feature class, and it will be saved in Ch11.gdb.

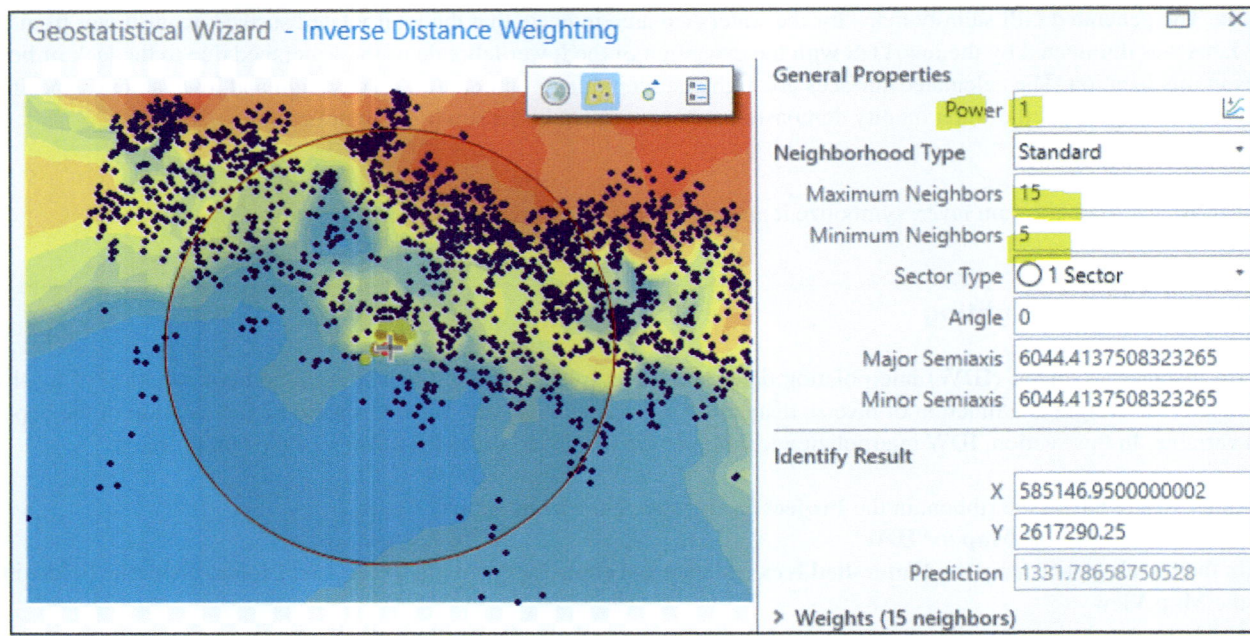

Note Geostatistical Analyst uses power values greater than or equal to 1. When p = 2 (default value), the method is known as inverse distance squared weighted interpolation. However, there is no theoretical justification to prefer this value over others. The effect of changing p should be investigated by previewing the output and examining the cross-validation statistics.

19. Click Finish and then OK

Result The inverse distance weighting layer is displayed in the CP and has the same area extent as the **Borehole** layer.

Conversion of the Idw into Raster Format

Converting the inverse distance weighting layer into raster is an essential step in order to clip the interpolated rater to fit the watershed layer.

20. Right click **Inverse Distance Weighting** layer click **Export Layer** and choose **To Raster**
21. The GA Layer To Rasters pane open
22. Select the **Parameters** tab and fill it as follows:
    (a) Input geostatistical layer: **Inverse Distance Weighting**
    (b) Output Raster: **IDW**
    (c) Output cell size: **50**
    (d) Accept the rest of the default
23. Click the Environments tab and fill it as follows:
    (a) Output Coordinate System: **Borehole**
    (b) Mask: **Watershed**
    (c) Accept the rest of the default
24. Run
25. Close the GA Layer to the Rasters pane and uncheck the Inverse Distance Weighting layer

Classify the IDW Map

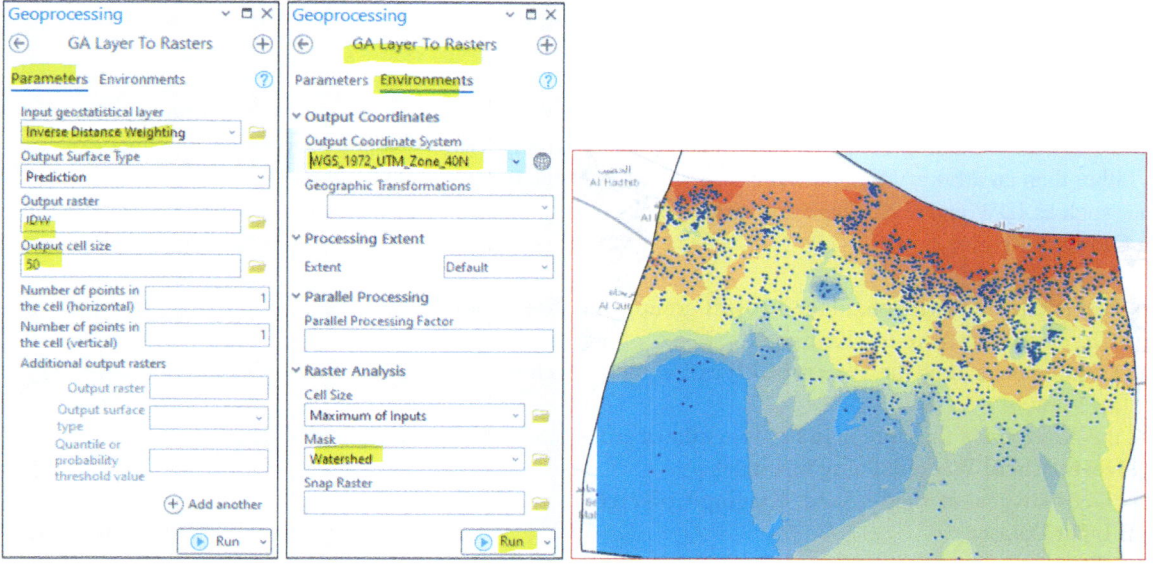

Classify the IDW Map

You will classify the IDW raster into 5 classes using the Manual Interval

1. In the CP, highlight the **IDW** and click the **Symbology** tab in Catalog pane
2. Under **Primary symbology** select **Classify**
3. Make sure Class 5
4. Method = Manual Interval
5. In classes tab and under "Upper Value" and under Color, replace the first number by 500 (blue), 1000 (green), 2000 (yellow), 3000 (pink), and 30000 (red)
6. In the Symbology pane click the Advanced Symbology Options icon
7. Open Format labels and change the decimal places to 0
8. Save the project

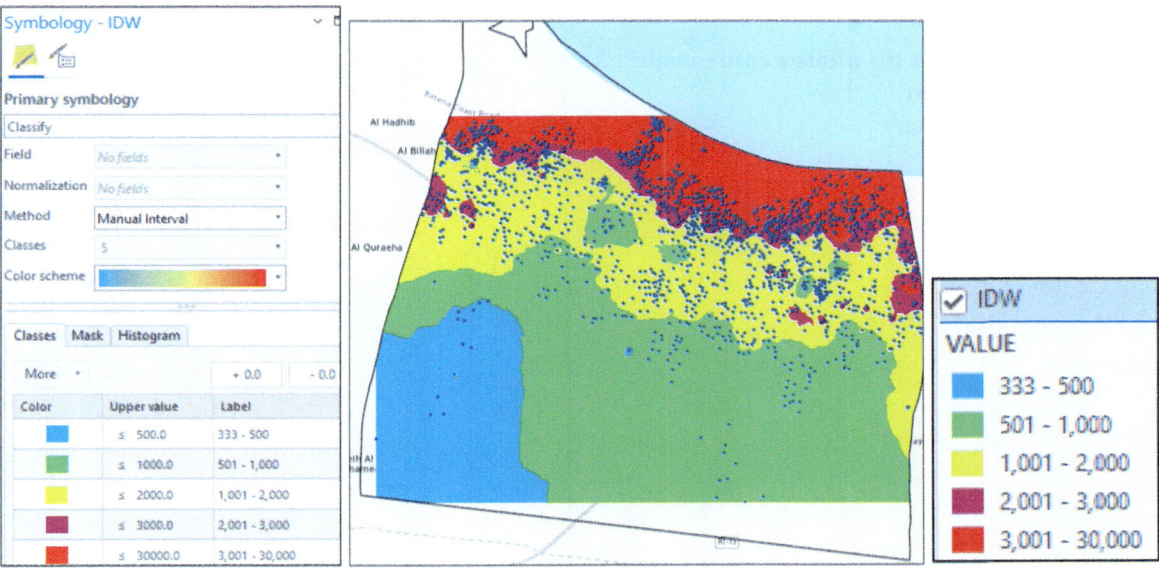

Result The generated IDW salinity maps for the watershed area indicate that the interpolated surface is dominated by high TDS along the coast.

Interpolation Using Kriging

Kriging is the best possible interpolation method based on regression against the observed z values of surrounding data points and weighted according to spatial covariance values. Kriging assigns weights according to a data-driven weighting function rather than an arbitrary function. It is still just an interpolation algorithm and will give very similar results to others techniques such as IDW and GPI. There are various types of kriging, and in this exercise, you will conduct an interpolation based on Ordinary Kriging.

1. Click **Insert** tab on the ribbon, in the **Project** group, click **New Map** button
2. In the CP rename the **Map** to "**Kriging**"
3. In the **Catalog** pane, drag the **Watershed.lyrx** and then **Borehole.lyrx** from the **Data_Ch11** folder under the Folder into the **Map View**
4. Click the **Analysis** in the ribbon, in the **Workflows** group, click the **Geostatistical Wizard**
5. The **Geostatistical Wizard** dialog box displays
6. Under Geostatistical methods: select **Kriging/CoKriging**
7. Under Input Dataset
8. Source Dataset: **Borehole**
9. Data Field: **TDS**
10. Click Next
11. Check Use Mean
12. Click Next
13. Under "**Ordinary Kriging**"
14. Check the "**Prediction**"
15. Click Next

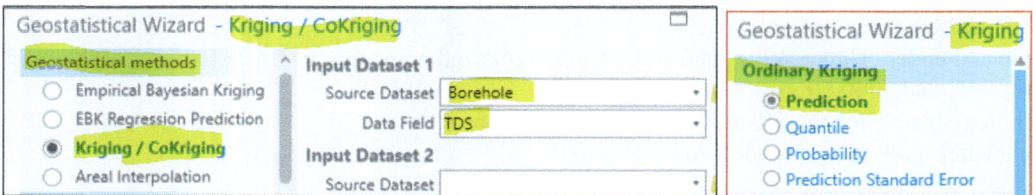

16. In **Optimize model** "click to optimize entire model"
17. **Model #1** select **Gaussian**
18. Accept the rest of the default
19. Next
20. Neighborhood type: **Standard**
21. Maximum Neighbor: 15
22. Minimum Neighbor 5
23. Accept the rest of the default
24. Click Finish and then click OK

Convert the Kriging Layer into A Raster Format

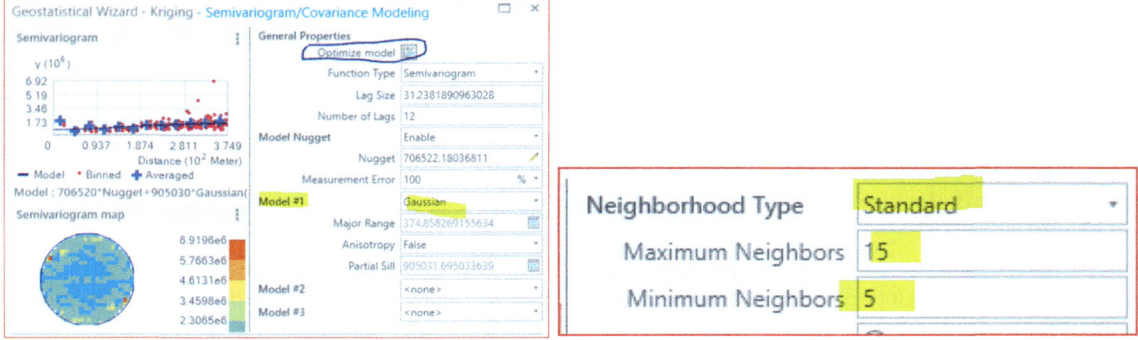

Result The Kriging layer is displayed in the CP and has the same area as the Borehole layer.

Convert the Kriging Layer into A Raster Format

26. Right click **Kriging** in the CP click **Export Layer** and choose **To Raster**
27. The GA Layer To Rasters pane open
28. Select the **Parameters** tab and fill it as follows:
    (a) Input geostatistical layer: Kriging
    (b) Output Raster: OrdinaryKriging
    (c) Output cell size   **50**
    (d) Accept the rest of the default
29. Click the **Environments** tab and fill it as follows:
    (a) Output Coordinate System: **Borehole**
    (b) Mask: **Watershed**
    (c) Accept the rest of the default
25. Run

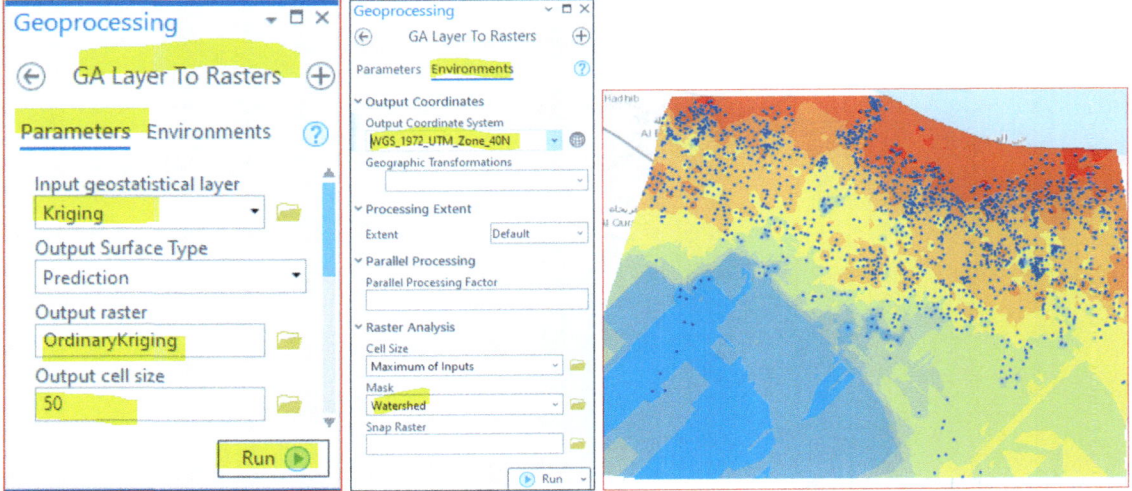

Classify the Ordinarykriging Raster

You will classify the **OrdinaryKriging** raster into 5 classes using the Manual Interval

26. In the CP, highlight the **OrdinayKriging** and click the **Symbology** tab in the Catalog pane
27. Under **Primary symbology** select **Classify**
28. Make sure Class 5
29. Method = Manual Interval
30. In classes tab and under "Upper Value" and under Color, replace the first number by 500 (blue), 1000 (green), 2000 (yellow), 3000 (pink), and 30000 (red)
31. In the Symbology pane click the Advanced Symbology Options icon
32. Open Format labels and change the decimal places to 0
33. Close the Symbology pane
34. Save the project

Question 1 Comment on the salinity interpolation.

Question 2 Run the GPI, IDW, and Kriging using the nitrate (NO3) variable.

Watershed Delineation

12

The watershed is an area of land that serves as a catchment for water. From the watershed, the surface water then enters a common outlet in the form of either a body of water, such as a lake, stream, or wetland, or it infiltrates into the groundwater. It is simply an area that drains surface water from high elevation to low elevation. The watershed is a hydrologic unit that is used to model, as it is considered fundamental to hydrologic designs and is used to aid in the study of the movement, distribution, quality and quantity of water in an area. Watershed analysis is a technique essential in the management, conservation, and planning of Earth's natural resources.

Traditionally, watersheds are created manually from topographic maps by locating the water divide. In ArcGIS Pro, the watershed can be delineated using the Spatial Analyst extension and hydrology tools for watershed delineation.

There are many steps involved in creating the watershed boundary. The delineation of the watershed requires work with the raster DEM of a study area. If the raster DEM is not available and you have a point elevation, you can use one of the interpolation techniques such as IDW, GPI, or kriging in ArcGIS Pro to convert the point elevation into DEM. The watershed can be created using a DEM, as it is considered the main source point to create a watershed model.

After obtaining the DEM, ensure that the raster is depressed. These topographic depressions are also called sinks. Depression is normal in nature and can be generated during the interpolation process of DEM creation. Depression in the DEM occurs when a very low elevation relative to neighboring cells is found, which prevents downslope DEM flow-path routing. These low elevation cells can be removed by increasing their cell value to the lowest overflow point. The table below shows that the cell in row 3 column 2 is a depression.

450	446	441	451	454
447	441	440	446	451
440	339	431	440	445
438	435	422	431	439
431	429	414	422	424

Therefore, to use the raster DEM in watershed delineation, the depressions should be removed using the **Fill** tool.

Flow Direction

The next step is to create a raster grid containing information about flow directions. The Flow Direction tool resides in the hydrology tool. The Flow Direction tool is used to find drainage networks and drainage divides and is determined by the elevation of surrounding cells in the DEM. The water can flow only into one cell, and the GIS model assumes no sinks. If

Supplementary Information The online version contains supplementary material available at https://doi.org/10.1007/978-3-031-42227-0_12.

© The Author(s), under exclusive license to Springer Nature Switzerland AG 2023
W. Bajjali, *ArcGIS Pro and ArcGIS Online*, Springer Textbooks in Earth Sciences, Geography and Environment, https://doi.org/10.1007/978-3-031-42227-0_12

this does not happen in the output grid, the raster DEM sinks. Flow direction is critical in the hydrologic modeling process because it determines the direction of flow for each cell in the land topography. The raster grid created by the Flow Direction tool is based on the D8 flow algorithm. The D8 algorithm is the method for performing flow path analysis for the application of watershed delineation. The method assigns a cell's flow direction to one of its eight surrounding cells that has the steepest gradient. D8 has disadvantages and can be replaced by D∞. The flow direction function is that for each 3 X 3 cell neighborhood, the grid processor stops at the center cell and determines which neighboring cell is lowest. Depending on the direction of flow, the output grid will have a cell value at the center cell. The values for each direction from the center are 1, 2, 4, 8, 16, 32, 64, and 128. For example, if the direction of the steepest drop was to the left of the current processing cell, its flow direction would be coded as 16. The figure below shows the output grid cell values with the center cell, as determined by this matrix:

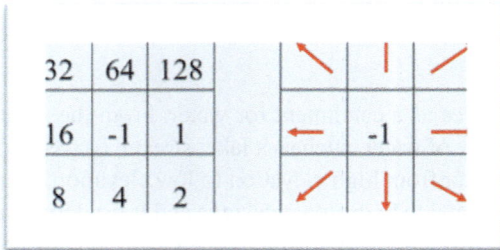

If a cell flows northwest, then in the output grid, the cell in its location will have a value of 32, and if a cell flows southward, then the value will be 4 (figure below).

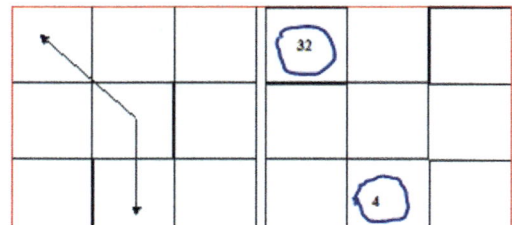

Flow Accumulation

The next step in creating the watershed is to run the flow accumulation function, which is an important step in creating the drainage network and measuring the area of a watershed that contributes runoff to a given cell. Therefore, it is necessary to determine the ultimate flow path based on the direction of flow of every cell on the topography grid. Flow accumulation selects cells with the greatest accumulated flow, which will assist in creating a network of high-flow cells. These high-flow cells should be situated on stream channels and at valley bottoms. Cells that have high accumulation values higher than "**1**" correspond to stream flow, and cells having an accumulation value of "0" correspond to ridgelines. Once flow accumulation is calculated, it is customary to identify those cells with high flow. Higher-flow cells will have a larger value, and the user can select any threshold number (i.e., > 500), which should be close to the network obtained from the traditional method.

Stream Link

After the stream network is established from flow accumulation, each stream section of the stream raster is assigned a unique value, e.g., 1, 2, 3, etc. The intersection of the streams is similar to a node, and the stream section is an arc.

Delineate Watershed Based on a Pour Point

To delineate a watershed, you need to select an outlet cell (Pour Point), which is the lowest point in the watershed, where all flow is directed. The Pour Point can also be any feature, such as a gauge station, dam, bridge, sampling location, confluence of a tributary with a mainstream or any point of interest in the study area. The Pour Point could be a raster or vector. To obtain the watershed, the Pour Point should coincide with the flow path of high flow accumulation values in the flow accumulation raster. Finally, you can use the **Watershed** tool to extract the whole watershed polygons for the Pour Points or a single watershed for a specific stream or tributary. At the end, you can convert your raster watershed to vector so that you can integrate it and align it nicely with the rest of your digital data.

Scenario 1 You are an ecologist working for DNR, and one of your duties is to delineate the catchment area of the CreditRiver region. This catchment will be used as an input function to estimate the recharge amount to the groundwater resources in the area.

Your duty is to do the following:

1. Delineate the watershed based on the whole region.
2. Delineating the watershed based on a single point
3. Convert the watershed raster into vector

Data Connection and Integration

1. Launch ArcGIS Pro
2. Click **Open** another project (upper-right) browse to **Env_Water\Ch12** select **Ch12.aprx** and click OK

Result The **Ch12.aprx** open and the CP includes the Map, which is empty. The Map View displays both the World Topographic Map and World Hillshade.

3. Make sure that the **Catalog** pane is open
4. Click the **Insert** tab on the ribbon, in the **Project** group, click **Add Folder**
5. Browse to \\Env_Water\Ch12 (or **Database\ Data_Ch12**) open it and highlight **Data_Ch12** and click **OK**.

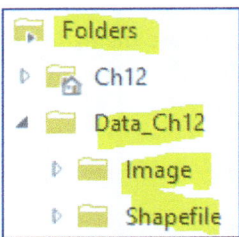

Result Inside the **Data_Ch12** folder, there are two subfolders, **Image** and **Shapefile**, and they contain the required data to delineate the watershed.
6. Save the project

Integrate Layers into the Watershed Map

7. In Catalog pane expand the "**Folders**", the "**Data**" folder under Image highlights **CreditRiver.tif** and drags it into the Map View.

Result The **CreditRiver.tif** DEM is displayed in the Map View and added to CP below the **Map**. The Credit River is in Burnsville, MN, USA.

8. Rename the **Map** to **Watershed**

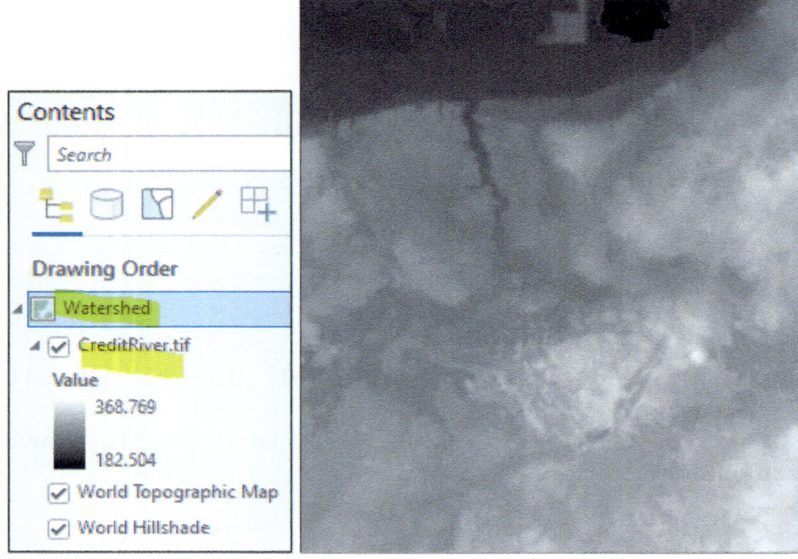

Delineating the Watershed

Delineating the watershed in ArcGIS Pro is required to use the hydrology tool in the Spatial Analyst. Delineation requires different steps using a digital elevation model. In this section, the CreditRiver DEM will be used to delineate the watershed.

Step 1: Run the Flow Direction Tool

This step is very important to run in order to create a raster grid containing the information about flow directions and identify any sinks (depressions) in the DEM raster, as the DEM raster file will be used for performing the task. The **Flow Direction** tool creates a raster of flow direction from each cell to its downslope neighbor, or neighbors, using different methods such as D8. In the D8 method, the direction of flow is determined by the direction of steepest descent, or maximum drop, from each cell.

9. Click Analysis tab on the ribbon in the Geoprocessing group click the Tools button
10. In the Geoprocessing pane, click the Toolboxes tab and scroll down and open Spatial Analyst Tools
11. Open the Hydrology tools and click the Flow Direction
12. In the **Flow Direction** pane, in the **Parameters** tab fill it as follows:

Step 2: Identify the Locations of the Sink (Sink Tool)

    (a) Input surface raster: **CreditRiver.tif**
    (b) Output flow direction raster: **FlowTemp**
    (c) Accept the rest of the default
13. In the Environment tab
    (d) Output Coordinate System: "**CreditRiver.tif**" The coordinate is NAD_1983_UTM_Zone_15 N
    (e) Accept the rest of the default
14. Run

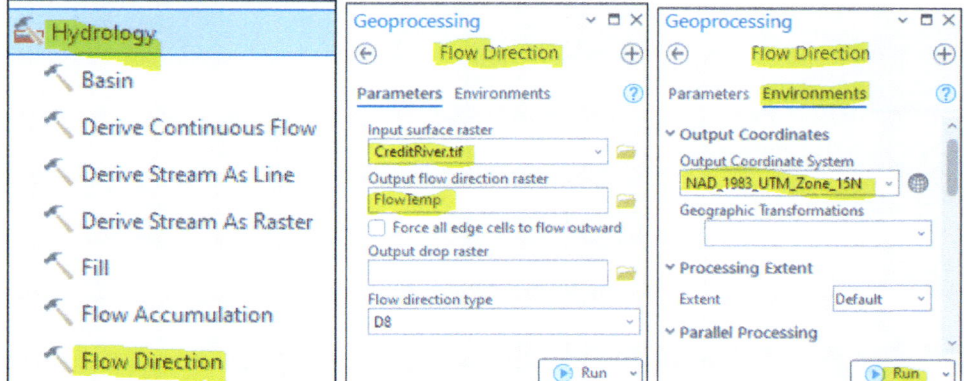

Result The output raster is an integer and contains 163 records, which means that the **CreditRiver** raster has sinks. A sink is a cell or set of spatially connected cells whose flow direction cannot be assigned one of the eight valid values (1, 2, 4, 8, 16, 32, 64, and 128) in a flow direction raster. This can occur when all neighboring cells are higher than the processing cell or when two cells flow into each other, creating a two-cell loop. ...

Step 2: Identify the Locations of the Sink (Sink Tool)

A sink is a cell or set of spatially connected cells whose flow direction cannot be assigned one of the eight valid values in a flow direction raster (1, 2, 4, 8, 16, 32, 64, and 128). The Sink tool will create a raster of the sink coded with depth.

15. Click the back arrow in the Flow Direction pane
16. Click the **Sink** tool and fill the **Sink** dialog box as follows:
17. Input D8 flow direction raster: **FlowTemp**
18. Output raster: **Sink**
19. Run

Comments The output grid shows that the **CreditRiver** has 12402 **sinks**. Some sinks are natural in nature, and some are deficiencies in the DEM during processing. After knowing the number of sinks, we must fill them using the "Fill tool" on the original **CreditRiver** raster.

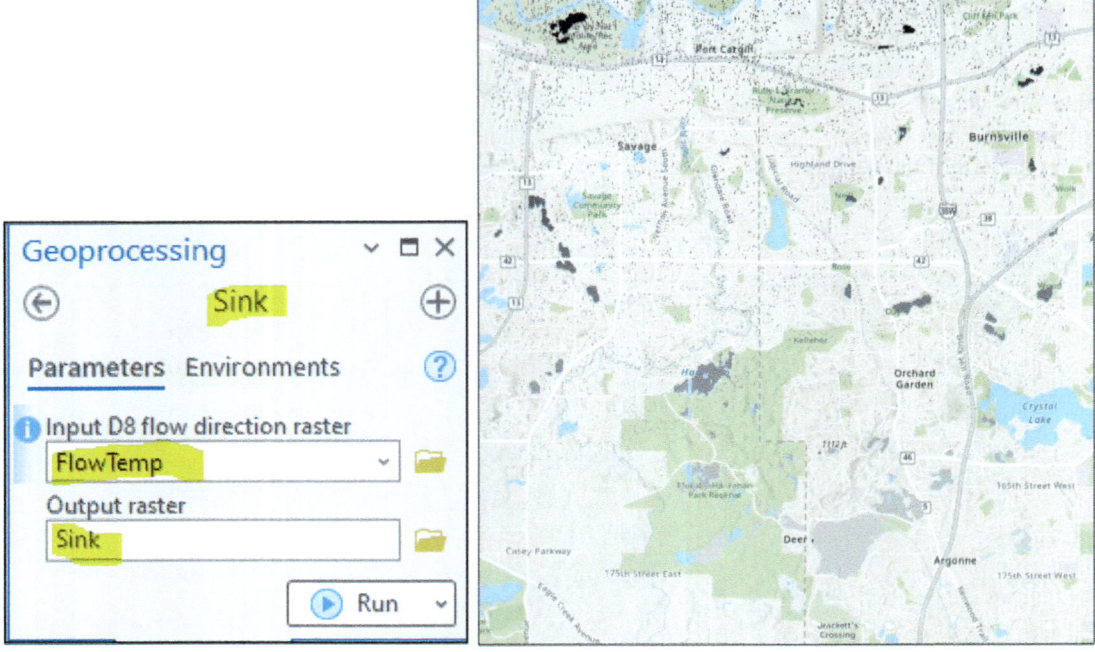

Zonal Statistics

Now you are going to use the Zonal Statistics to create a raster that will identify the minimum elevation of each sink area in the raster

20. Click the back arrow in the **Sink** pane
21. Open Spatial Analyst Tools and open Zonal and click Zonal Statistics
22. The Zonal Statistics open and fill it as follows:
    (a) Input raster or feature zone data: Sink
    (b) Zone field: Value
    (c) Input value raster: CreditRiver
    (d) Output raster: SinkMin
    (e) Statistics type: MINIMUM
23. Accept the rest of the default
24. Click Run

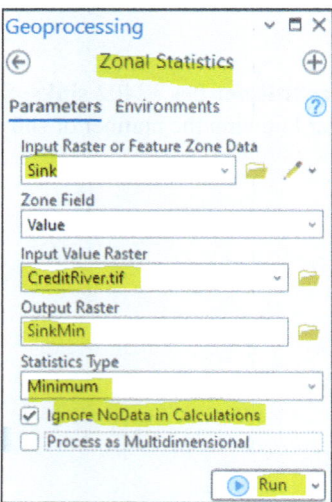

Result The **SinkMin** raster displays an elevation range between 182.50 and 352.46 m above sea level.

Step 3: Run the Fill Tool

The fill tool function fills the sinks in the CreditRiver raster to remove depressions in the data. The tool iterates until all sinks within the specified z limit are filled and can also be used to remove peaks, which are cells with elevations greater than would be expected given the trend of the surrounding surface.

25. Click the back arrow in the **Zonal Statistics** pane
26. In **Spatial Analyst Tools,** open the **Hydrology** tools and click the **Fill** tool.
27. The Fill pane opens. Fill it as follows:
    - (f) Input surface raster: **CreditRiver.tif**
    - (g) Output surface raster: **CreditRiverFill**
28. Run

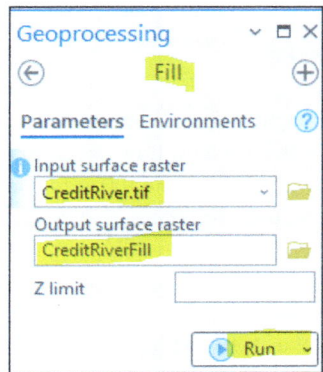

Result Now, **CreditRiverFill** is a raster free of any depressions.

Step 4: Run the Flow Direction Tool

The **Flow Direction** tool will run again after filling all the sinks and this time the output is an integer raster with 8 values (1. 2. 4. 8, 16, 32, 64, and 128).

29. Click the back arrow in the **Fill** pane
30. Click the Flow Direction tool in the Hydrology tools and fill it as follows:
    - (a) Input surface raster: **CreditRiverFill**
    - (b) Output flow direction raster: **Flowdirection**
    - (c) Accept the rest of the default
31. Run

Result Flowdirection is an integer raster with an attribute table of 8 records. Values 16 and 64 have the highest frequency, which means that the direction of the surface flow is west–north.

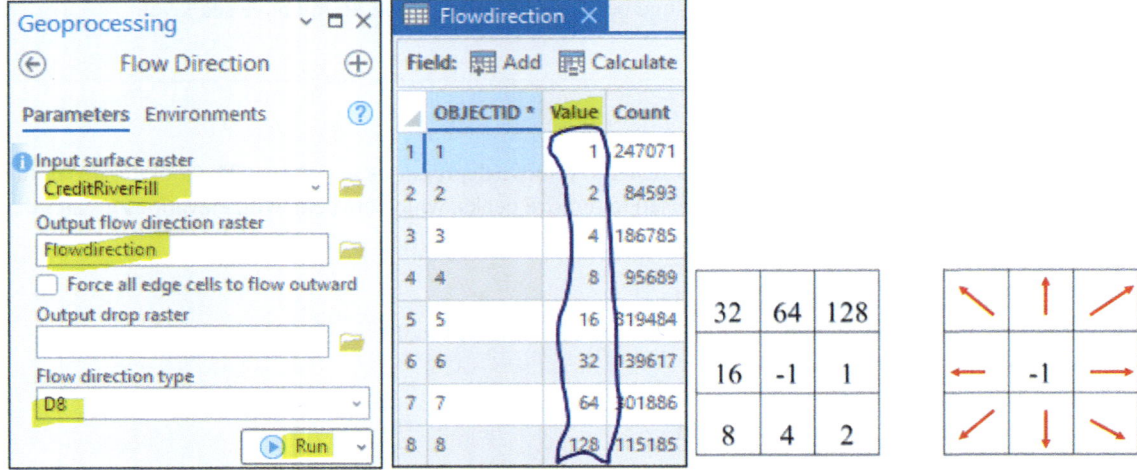

Step 5: Create a Flow Accumulation Raster

This step is important because it will tabulate for each cell the number of upstream cells that will flow into it, and the tabulation will be based on the flow direction raster. Cells with high flow accumulation are areas of concentrated flow and may be used to identify stream channels. Cells with a flow accumulation of 0 are local topographic highs and may be used to identify ridges. The results of **Flow Accumulation** can be used to create a stream network by applying a threshold value to select cells with a high accumulated flow.

32. Click the back arrow in the **Flow Direction** pane
33. Click the **Flow Accumulations** tool and fill it as follows:
34. Input flow direction raster: **Flowdirection**
35. Output accumulation raster: **FlowAccum**
36. Output Data Type: INTEGER
37. Accept the rest of the default
38. Run

Note the output data type of the raster "**FlowAccum**" is supposed to be an integer. This means that the raster should have an attribute table. If the raster has no attribute table, you can build it by using the **Build Raster Attribute Table** tool.

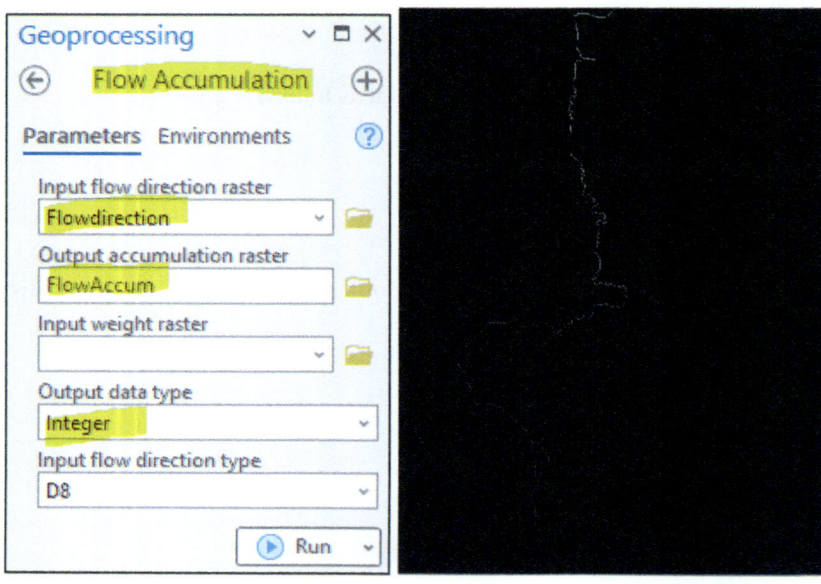

Build Raster Attribute Table Tool

The Build Raster Attribute Table tool creates an attribute table for the "**FlowAccum**" raster in which its pixel type is a signed integer.

39. Click the back arrow in the Flow Accumulation pane
40. Open **Data Management Tools**, open then **Raster** tools and then **Raster Properties**
41. Click **Build Raster Attribute Table** and fill it as follows:
42. Input raster: **FlowAccum**
43. Check the "Overwrite"
44. Run

Result the attribute table created for the **FlowAccum** raster and a stream network displayed

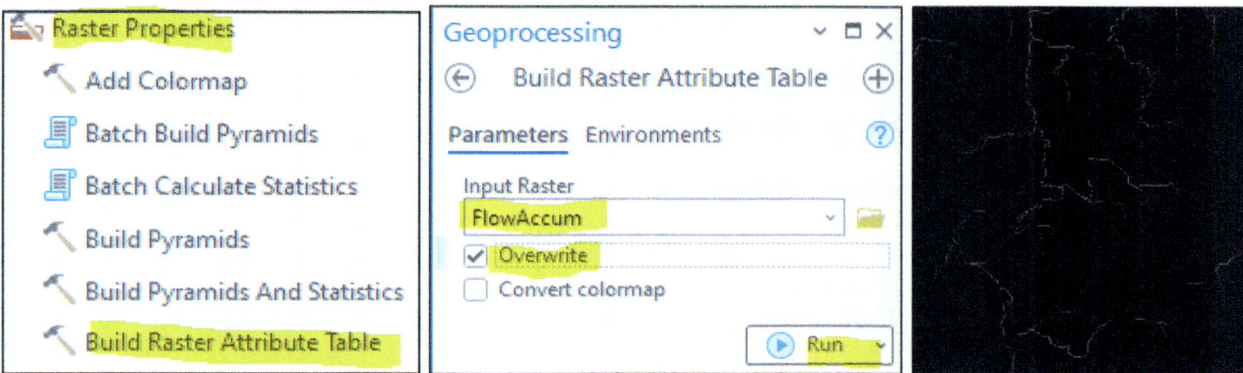

Note
Currently, the building of the attribute table is a bug registered by ESRI. To work around this problem, perform the following: run the Flow accumulation and keep the default (Float), then convert the Float into integer (**Math** tool under **Int** tool) in Spatial Analyst

Run the Flow Accumulation and Save the Output as Float

45. Click the **Analysis** tab on the ribbon in the **Geoprocessing** group, click the **History** button, in the **History** dialog box double click on the Flow Accumulation
46. Fill the **Flow Accumulations** tool as follows:
47. Input flow direction raster: **Flowdirection**
48. Output accumulation raster: **FlowAccumF**
49. Output Data Type: Float
50. Accept the rest of the default
51. Run

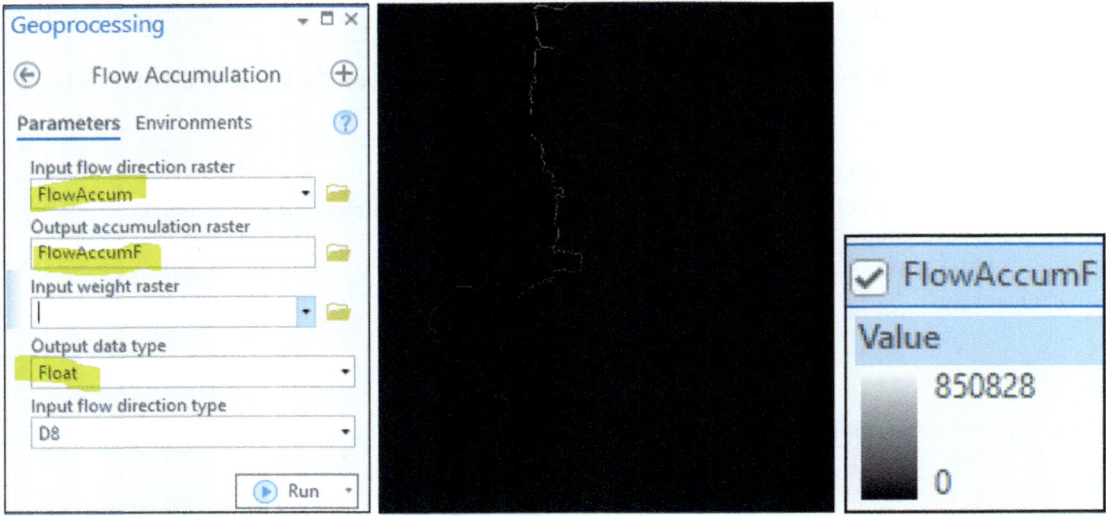

Convert the Float Raster into Integer Raster

52. Click the back arrow in the **Flow Accumulations** pane
53. Open the **Spatial Analyst Tools**, open **Math** tools, and then open the **Trigonometric** tool
54. Click the **Int** tool
55. Input raster or constant value: **FlowAcumF**
56. Output raster: **FlowAcumI**
57. Click Run

Result FlowAccumI created and has an attribute table with 16,996 records.

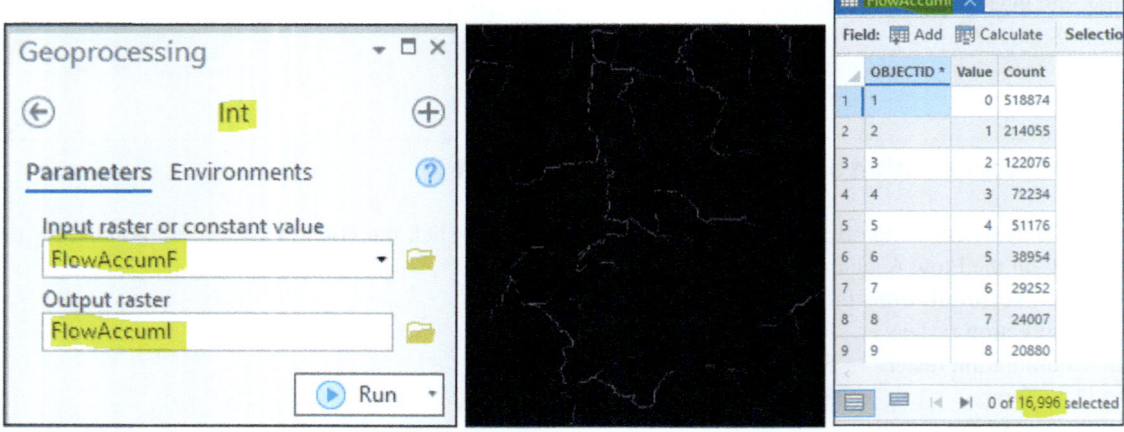

Step 6: Create Source Raster to Delineate WatersheD

The flow accumulation raster will help in deriving the stream network. The stream derivation is based on threshold cell values, which could be 100, 200, 300, or more cells. The 300 cells mean that each cell has a minimum of 300 cells contributing to them. The difference between 300 cells and 100 cells is that 300 cells will generate less dense stream networks, while 100 cells will generate denser streams.

Source Raster

The source raster required two steps:

1st Step: This step is to use the results of flow accumulation to create a stream network by applying a threshold value to select cells with a high accumulated flow. In this example, you are going to use the flow accumulation raster that has more than 500 cells flowing into them. The procedure performs a conditional operation with the Con tool to create a raster where the value 1 represents the stream network.

58. Click the back arrow in the **Int** pane
59. Open the **Spatial Analyst Tools** open the **Conditional tools**
60. Click the **Con** tool
61. Input Conditional raster: **FlowAcumI**
62. Click **+ New Expression** Value is greater than 500
63. After **Where** select Value then select "**is greater than or equal to**" type **500**

"**Where Value is greater than or equal to 500**"

64. Input true raster or constant value: 1
65. Output raster: **Network**
66. Run

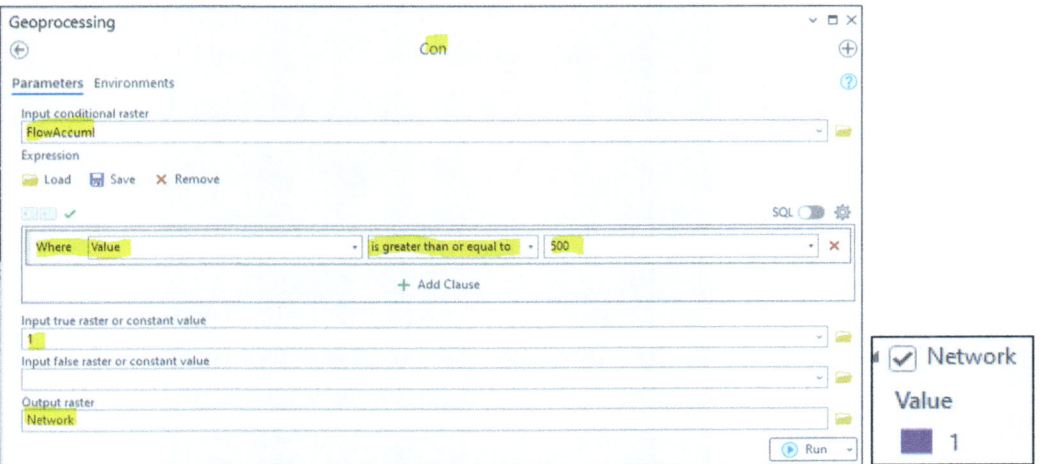

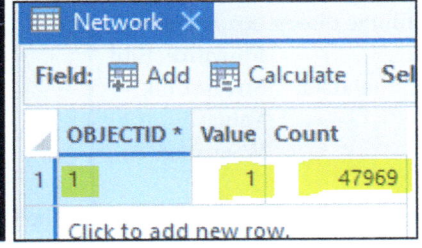

Result The network raster has only 1 record

2nd **step**: Links are the sections of a stream channel connecting two successive junctions, a junction and the outlet, or a junction and the drainage divide. In hydrology, these stream segments are called reaches. A junction is related to a pour point and helps delineate a watershed or drainage sub-basin boundary. In this section, a unique value is assigned to each section of the **network** raster. In other words, each segment intersects with another segment as a unique record.

67. Click the back arrow in the Con pane
68. Open the **Spatial Analyst Tools** and then the **Hydrology** tools
69. Click **Stream Link** tool and fill it as follows:
70. Input stream raster: Network
71. Input flow direction raster: flowdirection
72. Output raster: StreamLink
73. Run

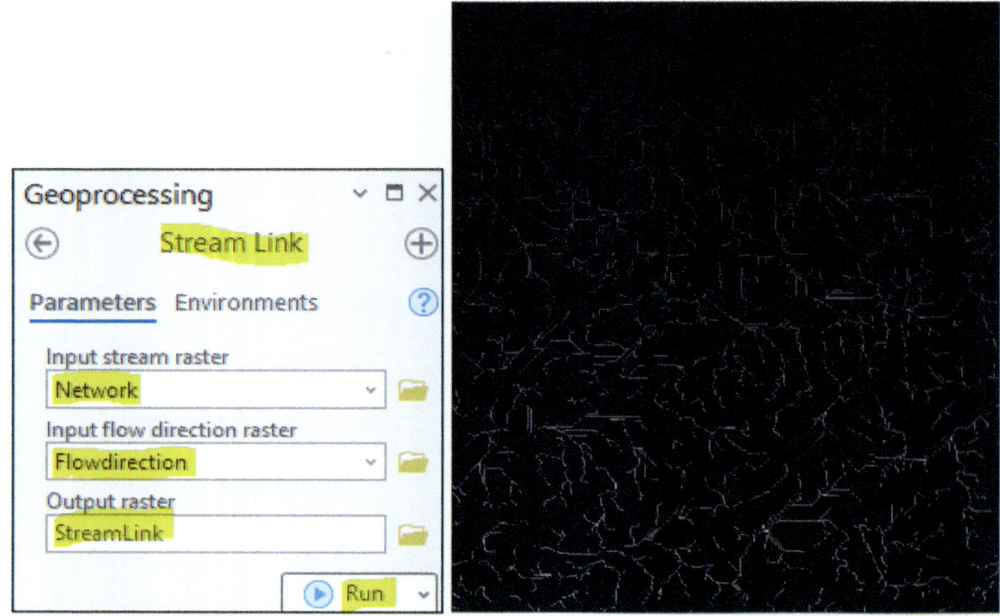

Result The **StreamLink** raster has 1657 segments, and each stream segment is now an independent record and has a "Value" in the attribute table.

Step 7: Delineate Watershed

74. Click the back arrow in the Stream Link pane
75. Open the **Spatial Analyst Tools**, then the **Hydrology** tool
76. Click **Watershed** tool and fill the dialog box as below
77. Input D8 flow direction raster: **Flowdirection**
78. Input raster or feature pour point data: StreamLink
79. Pour point field: Value
80. Output raster;   Watershed
81. Run
82. In the CP, r-click **Watershed** select **Symbology** and change the **Color scheme** into **Basic Random**
83. Close the Symbology pane
84. Save your project

Point-Based Watershed

Results The watershed is created.

Question How many sub-basins are in the watershed?

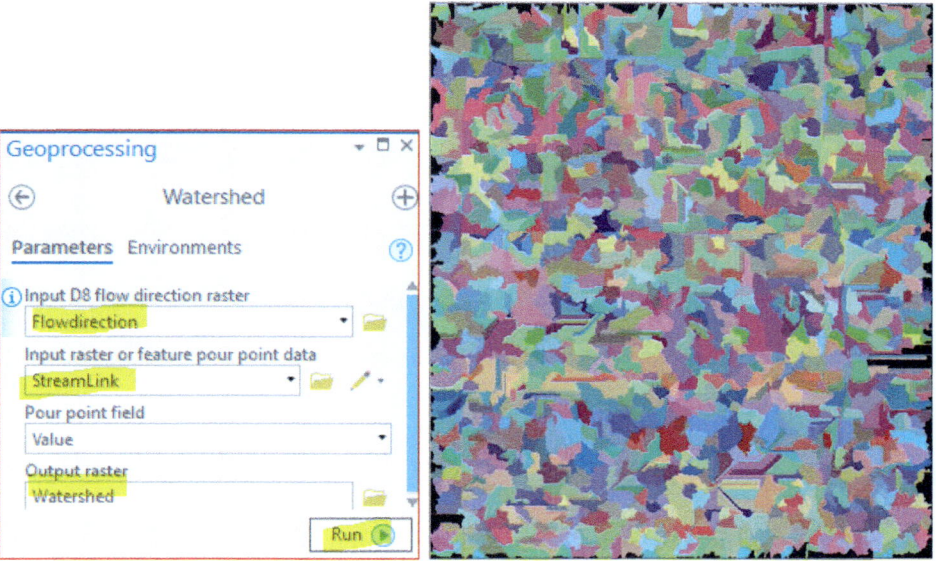

Point-Based Watershed

This approach allows the user to derive a watershed either for each stream or based on a point of interest such as any point of interest along the flow system.

Scenario 2 Your superior asked you to create only one watershed base on the Bridge point shapefile

1. Click **Insert** tab on the ribbon, in the **Project** group, click **New Map** button
2. In the CP rename the **Map** to "**PourShed**"
3. In the **Catalog** pane, integrate the **Flowdirection** from the **Ch12.gdb** under **Databases** and the **Bridge.shp** from the **Shapefile** in the **Data** folder under **Folders**
4. In the CP, click the symbol of the **Bridge** and search for the proper symbol and choose it.
5. Click again the Watershed tool and fill the **Watershed** pane as below.
6. Input flow direction raster: **Flowdirection**
7. Input raster or feature pour point data: Bridge
8. Output raster: PourShed
9. Run
10. Change the color ramp of the **PourShed** into Bathemetry#1 and close the symbology pane

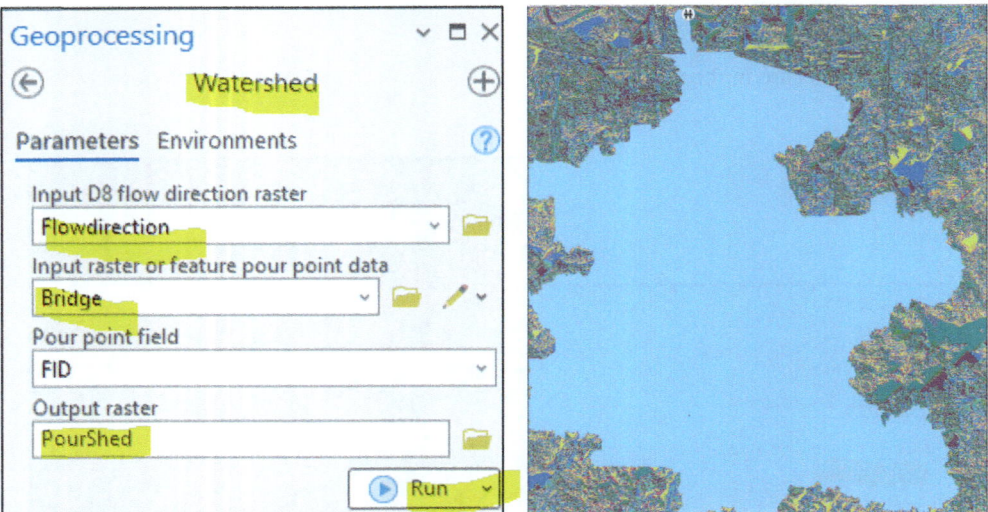

Convert Pourshed Raster into a Vector

Now you are going to vectorize the raster dataset "**PourShed**" by converting it to polygon feature class. If the conversion allows you to choose which attribute field of the input raster dataset will become an attribute in the output polygon feature class.

11. Click the back arrow in the Watershed pane
12. Open the **Conversion Tools** then open **From Raster** and click **Raster to Polygon**
13. Fill the Raster to Polygon as follows:
14. Input raster: **PourShed**
15. Output polygon features: CR_Watershed
16. Click Run

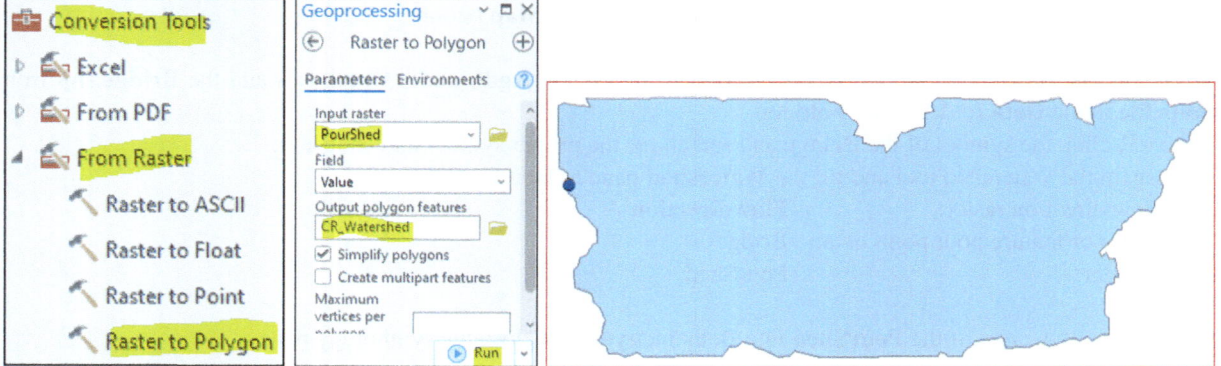

Question what is the total area of the **CR_Watershed**?

Assignment Calculate the amount of recharge to groundwater in cubic meters per year (m³/year). If the average amount of precipitation in the CR_Watershed is 120 mm per year, the vertical infiltration rate to the subsurface water-bearing formation is 5%.

Create Stream Order

Hint

1. Recharge = Area (m^2) x Infiltration Rate (120 x 0.05 mm)
2. Convert the mm into meter (1 mm = 0.001 m)

Choose the right answer from below

(a) 388,653.54 m^3/year
(b) 3,885,535.58 m^3/year
(c) 38,865, 355.81 m^3/year

Generating the Stream Network

This step converts the network raster into a vector file.

1. Click the back arrow in the Raster to Polygon pane
2. Click Raster to Polyline tool
3. Fill the Raster to polyline as follows:
4. Input raster: Network (from Ch12.gdb)
5. Output polyline features: Stream
6. Accept the rest of the default
7. Run

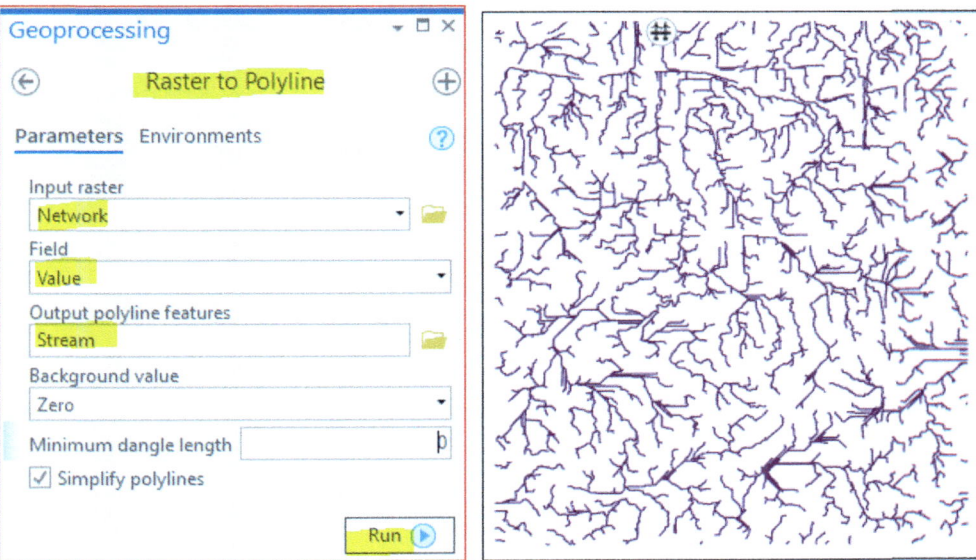

Result A stream network consisting of 3,865 streams and tributaries was created.

Create Stream Order

Stream ordering is a method of assigning a numeric order to links in a stream network. This order is a method for identifying and classifying types of streams based on their numbers of tributaries.

258 12 Watershed Delineation

8. Click the back arrow in the Raster to Polyline pane
9. Open **Spatial Analyst** Tools and then open the **Hydrology** tools
10. Click **Stream Order** tool and fill it as follows:
11. Input stream raster: **Network**
12. Input flow direction raster: **Flowdirection**
13. Output raster: **StreamOrder**
14. Method of stream ordering: **Strahler**
15. Run
16. Close the Geoprocessing pane
17. Save the project and exit the ArcGIS Pro

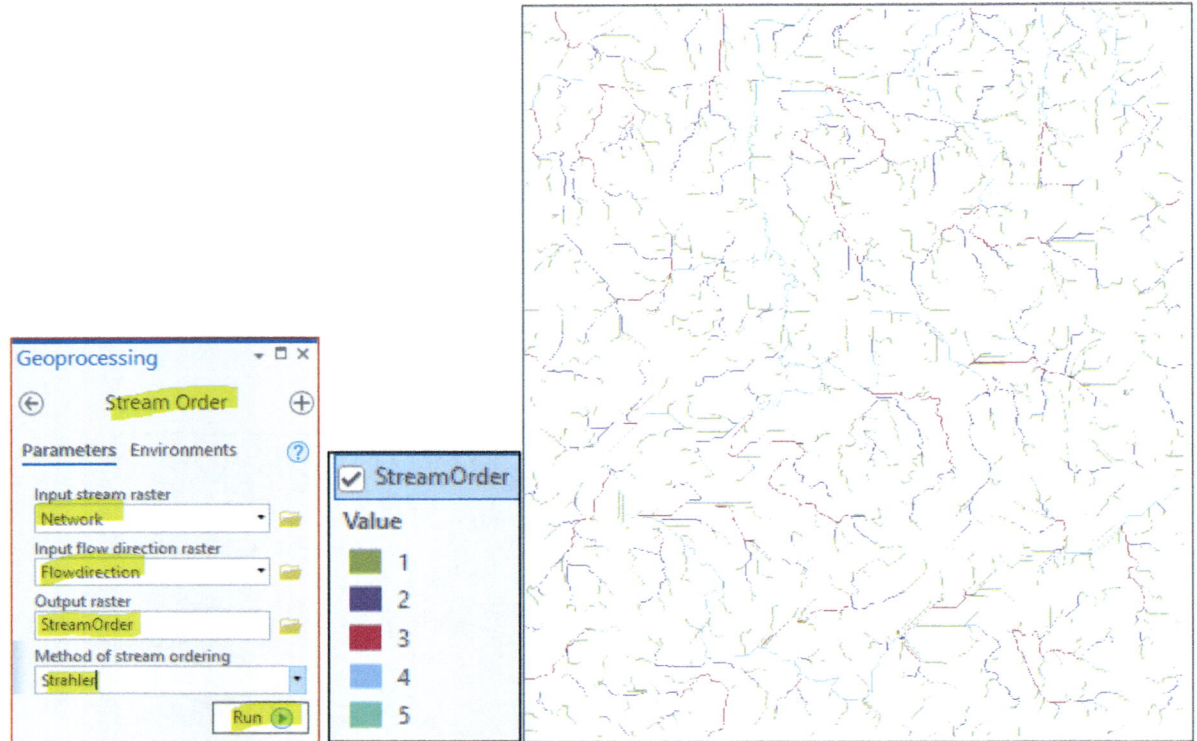

Result Four stream orders are created, where each segment of a stream within the stream network is treated as a node in a tree, with the next segment downstream as its parent. The CreditRiver in Burnsville is a five-order

Geostatistical Analysis

13

Geostatistics is a very useful approach that allows users to obtain meaningful information related to data in terms of its distribution and patterns in GIS. In this chapter, there are some applications of spatial statistics from the GIS environment based on the field of groundwater resources. The intention is to focus on the application of GIS rather than emphasizing complex mathematical and statistical theories. Nevertheless, some of the tools such as Measuring Geographic Distribution, Analysis Patterns, and Mapping Clusters of the Spatial Statistical analysis will be used using groundwater data.

Measuring Geographic Distribution Toolset

Using geographic distribution tools in ArcGIS aims to perform statistical approaches to assist researchers in measuring the distribution of features. The tools allow users, for example, to calculate a value that represents a characteristic of the distribution. For example, the center of groundwater wells tap an aquifer. By doing this, you can see how the wells are dispersed throughout the basin. There are three types of centers that can be calculated: **mean center, median center, and central feature.**

The mean center is the average of the X-coordinate and Y-coordinate values of all features. The resulting X, Y coordinate pair is the mean center. For example, in the Jarash area, there are several wells (Figure below) that spread through the area, and to find the mean center, we calculate the averages of both X and Y coordinates (table below).

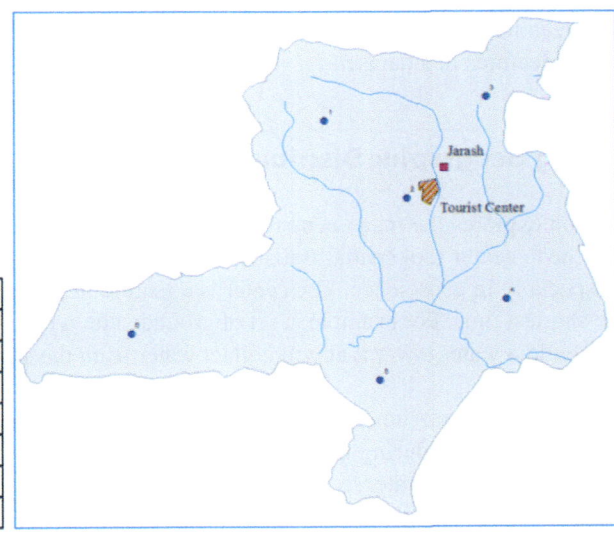

ID	Depth	TDS	X-Coordinate	Y-Coordinate
1	376	1307	228970.98	1189834.73
2	200	4160	232948.09	1185966.58
3	123	3965	236816.24	1191033.32
4	90	1950	237796.90	1180899.85
5	150	2015	231640.55	1176868.26
6	60	1872	219545.77	1179210.94
	Average		231286.42	1183968.95

Supplementary Information The online version contains supplementary material available at https://doi.org/10.1007/978-3-031-42227-0_13

The **Central feature** is the feature associated with the smallest accumulated distance to all other features in a study area. For example, there are 6 wells in the Jarash area (Figure below), and to calculate the central feature, the 6 wells (table below) will be organized into a table. The 6 wells are represented as records and columns, and then the distance between the wells will be recorded. The sum of the total distance of each well from the rest of the wells is then recorded, and the central feature will be the well that has the lowest total distance from all other wells. In the table below, you can see that well No 2 is selected as the central feature.

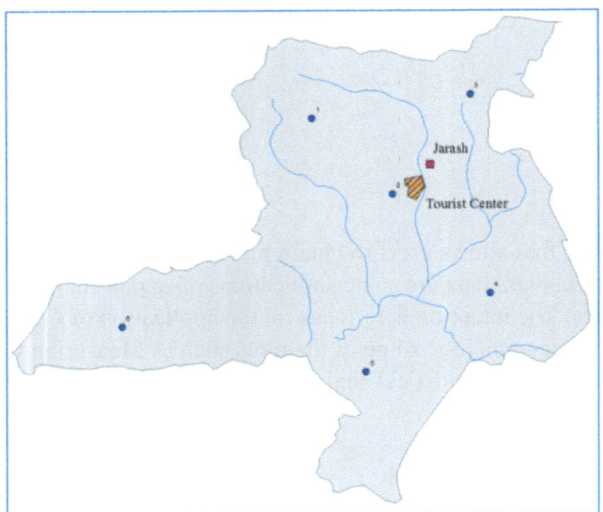

	Well 1	Well 2	Well 3	Well 4	Well 5	Well 6	Sum
Well 1	0.00	5.55	7.94	12.56	13.24	14.20	53.49
Well 2	5.55	0.00	6.37	7.01	9.19	15.00	**43.12**
Well 3	7.94	6.37	0.00	10.18	15.08	20.92	60.49
Well 4	12.56	7.01	10.18	0.00	7.36	18.33	55.44
Well 5	13.24	9.19	15.08	7.36	0.00	12.32	57.19
Well 6	14.20	15.00	20.92	18.33	12.32	0.00	80.77

The **Median Center** is a slightly different way to calculate the middle and is a point in a pattern that minimizes the distance between itself and all other points. The median center identifies the location that minimizes the overall Euclidean distance to the features in a dataset.

Measuring Geographic Distribution

This section explores the mean center with or without weight. To find the center of a randomly distributed feature over an area, the mean center tool spatial statistics tools must be used. Calculating the center has many applications in applied sciences, especially in geoscience. The center is a feature in the middle of a given set of data and can service all other features with the shortest time. For example, a set of groundwater wells is in a particular study area and finding the center of the wells will help build a water tower that will collect water from the surrounding wells faster and with less expense.

Scenario 1 The Water Authority in the Jarash governorate has decided to build an extra water tower in the area to be used as a distribution center during the summer, as during this period, the demand for potable water increases. The water tower should be supplied with water from a high-quality groundwater well. The well should be in the center of the wells that belong to the major cities in the governorate.

You will perform the following:

- Measuring Geographic Distribution Toolset
- Analyzing the Pattern Toolset
- Mapping Clusters

Data Connection and Integration

1. Launch ArcGIS Pro
2. Click **Open another project** (upper-right) browse to \\Env_Water\Ch13 select **Ch13.aprx** and click OK

Result The **Ch13.aprx** open and the CP includes the Map, which is empty. The Map View displays both the World Topographic Map and World Hillshade.

3. Make sure that the **Catalog** pane is open
4. Click the **Insert** tab on the ribbon, in the **Project** group, click **Add Folder**
5. Browse to \\Env_Water\Ch13 ((or \\Database\ Data_Ch13) open it and highlight **Data_Ch13** and click **OK**.
6. In the CP, name your **Map** "Geographic Distribution"

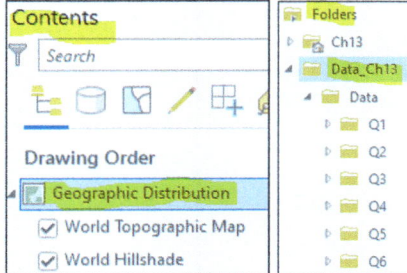

7. In the **Catalog Pane** expand the **Folders**, **Data_Ch13**, **Data**, and then **Q1** folder to integrate **Governorate.shp**, **Town.shp**, and **Well.shp** into the map.

Result The three layers are displayed in the Map View and in the CP. These layers, which show a northern area in Jordan, are unsymbolized.

Symbolizing the Three Layers

8. Symbolize the **Town** layer as Square 3 symbol, pink, and size 6
9. Symbolize the **Governorate** layer as Black Outline (1pt)
10. Now symbolize the **Well** layer based on **Unique Values**, choose **City** in Field 1, click the drop-down of **More** and uncheck "**S**how all other values" and then select "**Format all symbols**" choose Circle 1 symbol, and change the color into red (Burma), green (Jarash), and blue (Jubba) and make the size 7

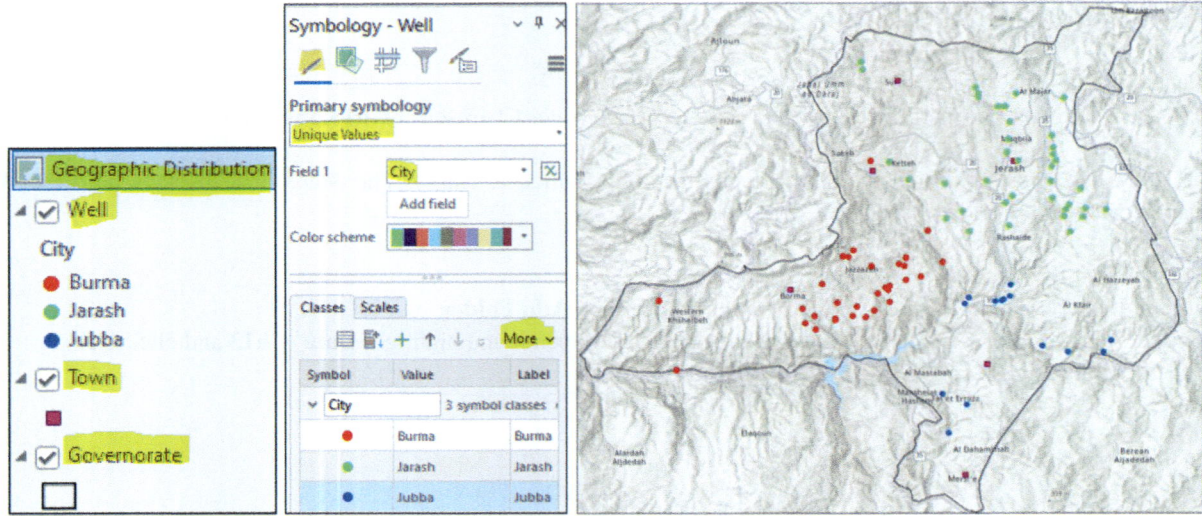

Mean Center

The mean center tool is for tracking changes in the distribution or for comparing the distributions of different types of wells. The mean center will calculate the average X and Y coordinates of all the wells in the 3 cities in the governorate.

11. In the **Analysis** tab on the ribbon, in the **Geoprocessing** group, click the **Tools** icon
12. In the Toolboxes tab, open the **Spatial Statistics Tools** then the **Measuring Geographic Distributions** subcategory
13. Click the **Mean Center** tool and fill it as follows:
14. Input Feature Class:**Well**
15. Output Feature Class:**Well_MeanCenter**
16. Case Field:**City**
17. Accept the rest of the default
18. Click Run

Result The Well_MeanCenter layer is a point feature class displayed in the Map View and consists of three records. Each feature represents a mean center well for each city.

19. In the CP highlight **Well_MeanCenter** and activate the Symbology pane
20. Symbolize the **Well_MeanCenter** based on "Unique Values" from the **City** field, set the symbol of **Well_MeanCenter** to "**Pentagon 1**" and "Size 12"
21. Match the color of the **Well_MeanCenter** to the color of the respective wells

Mean Center with Weight

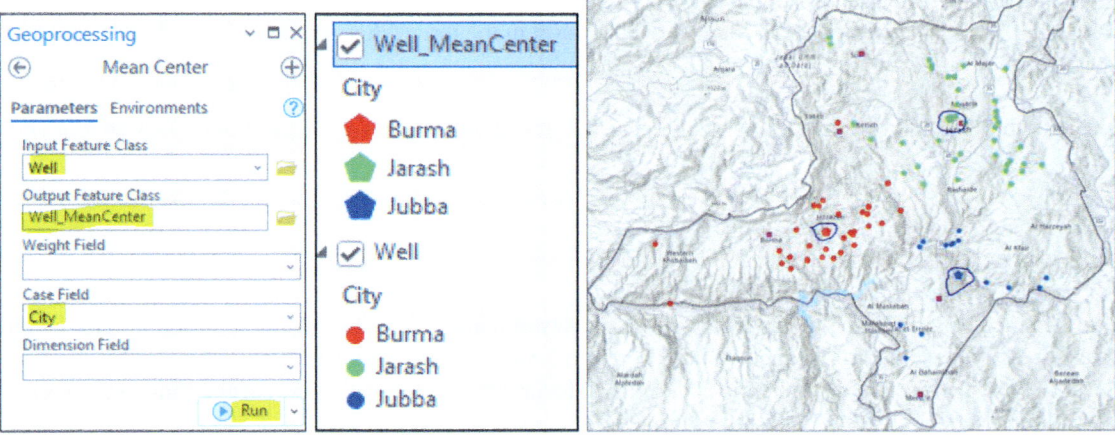

Mean Center with Weight

The next step is to run the **Mean Center** again on the well layer using the **Weight** field. The **Weight** field in the attribute table has 2-values; 1 and 2. Value 2 signifies wells that have TDS and NO_3^- less than 1000 and 45 mg/l, respectively, as well as wells of depths less than 300 m.

22. In the Geoprocessing pane, click the back arrow of the **Mean Center** pane
23. Click the **Mean Center** and fill the dialog box as follows:
24. Input Feature Class:**Well**
25. Output Feature Class:**Well_MeanCenterW**
26. Weight Field:**Weight**
27. Case Field:**City**
28. Accept the rest of the default
29. Click Run
30. Symbolize the **Well_MeanCenterW** layer precisely as **Well_MeanCenter**
31. Save the project

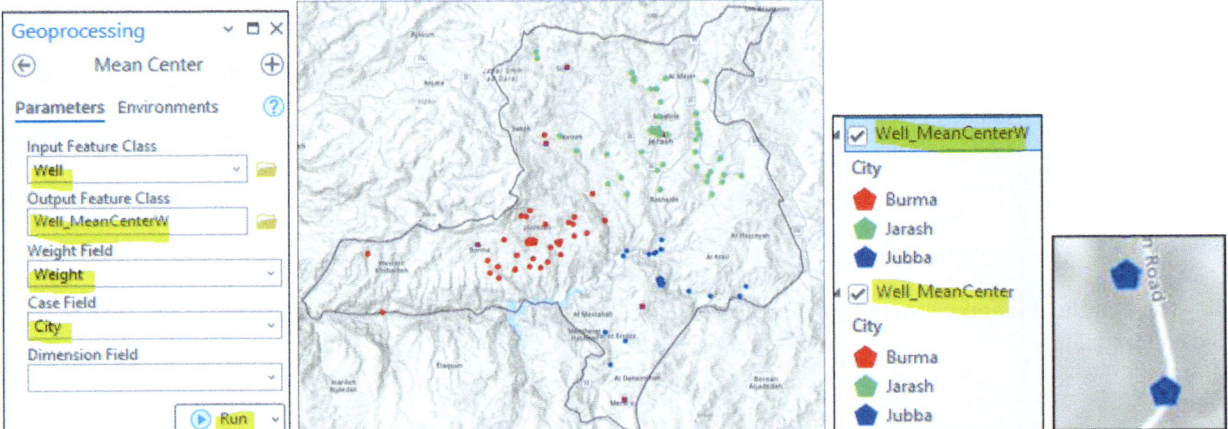

Question
1. Does **Well_Mean_CenterW** align with **Well_MeanCenter**?
2. If they aren't aligned, how far are they from each other?

Hint Use the "Measure" tool (**Map tab, in Inquiry group, Measure button**) to measure the distance between **Well_Mean_Center** and **Well_Mean_CenterW** for each well dataset in the three cities.

Assignment run the **Central Feature** tool and compare that result with the result of the **Mean Center** tool and comment on the result.

Standard Distance and Mean Center

The standard distance calculates the mean center of the displayed features and then draws a buffer around the mean center with a radius equal to the standard distance value. There are 3 values of the standard deviation. The 1st standard deviation covers at least 68% of the sample features, the 2nd standard deviation covers at least 95% of the sample features, and the 3rd standard deviation covers almost 99% of all the samples.

Scenario 2 You are a hydrogeologist in Water Authority and one of your goals is to replenish the groundwater and improve its water quality using an artificial recharge method. To perform this task, you decided to at least select one well that is located within 1 standard deviation from the mean center and near the **Khaldiyah** dam. To execute the assignment, you decided to calculate the standard circle using a "**weight**" criterion. The weight is based on the total depth of the wells, and more emphasis is placed on the wells that are shallower than 100 m. A field called weight is added to the attribute table of the wells.

1. Click **Insert** tab on the ribbon, in the **Project** group, click **New Map** button
2. In the CP rename the **Map** to "**Standard Distance**"
3. In the **Catalog** pane, select **Dam.shp, Geology.shp, Stream.shp,** and **Well.shp** from the **Q2** folder and integrate them into the Map View.
4. Symbolize the **Geology** layer based on "**Unique Values**" using the **Code** field
5. Uncheck "Show all Other Values" from the "More" dropdown menu
6. Change the **Stream** symbol to "Stream" symbol and set the "Line width" to 1
7. Change of the symbol of **Dam** to **Dam** symbol
8. Set the symbol of the **Well** to "Circle 1", size 7, and a blue color
9. In the Geoprocessing pane, click the back arrow of the **Mean Center** pane
10. In the **Spatial Statistics Tools** category and **Measuring Geographic Distributions** subcategory, select **Standard Distance**
11. Fill the **Standard Distance** pane as follows:
12. Input Feature Class:**Well**
13. Output Standard Distance Feature Class:**Well_StandDist**
14. Circle Size:1 standard deviation
15. Weight Field:**Weight**
16. Accept the rest of the default
17. Click Run

Result The **standard distance** tool will calculate the mean center based on the weight and then buffer the mean center to include approximately 68% of the wells in the study area.

Note To continue with the analysis, you must answer the following questions:

(a) How far is the dam from the center of the buffer?
(b) What is the closest well to the dam that can be used in artificial recharge?

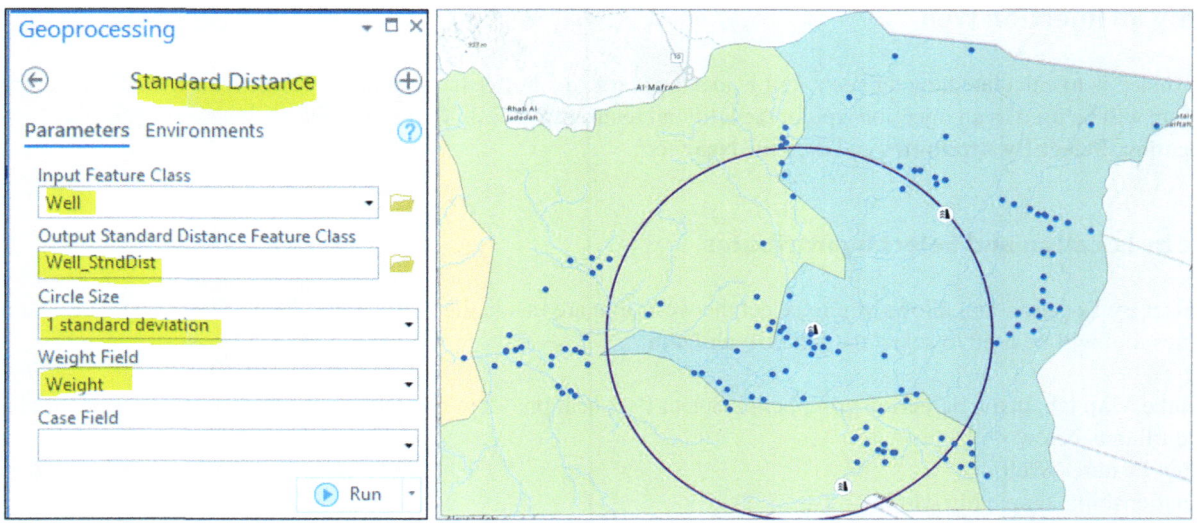

Distance Between Khaldiyah Dam and the Center of the Buffer

To calculate the distance of the **Khaldiyah** Dam from the center of the buffer, you must run the **Mean Center** tool and then use the **Measure** tool to find the distance between the center of the buffer and the **Khaldiyah** Dam.

18. Click the back arrow in the **Standard Distance** pane
19. In the **Measuring Geographic Distributions** subcategory, select **Mean Center**
20. Click the **Mean Center** and fill the dialog box as follows:
21. Input Feature Class: **Well**
22. Output Feature Class: **MeanCenterDam**
23. Weight Field: **Weight**
24. Accept the rest of the default
25. Click Run

Result The **MeanCenterDam** layer is displayed in the Map View at a close distance to the **Khaldiyah Dam**, only 1,135.09 meters from it.

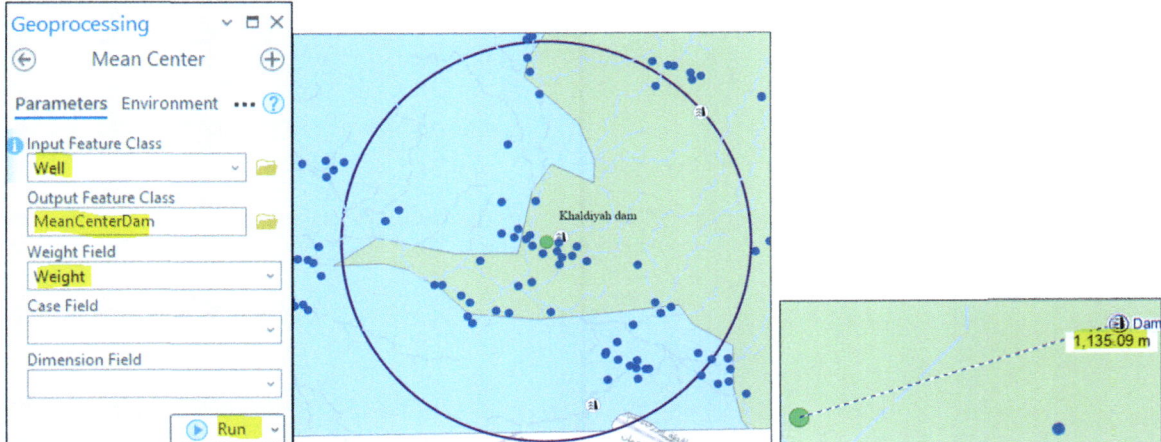

Identify an Injection Well

The next step is to calculate how far each well is located inside the buffer farm from the **Khaldiyah Dam**. The closest well to the dam will be used as an injection well in the artificial recharge process. To perform the analysis, you must use the **Select By Location**, **Select By Attributes** and then the **Near** tool.

Select by Location and Select by Attributes

The Select By Location tool allows you to select the wells that are inside the buffer zone "**Well_StandDist**". The Select By Attributes tool will be used to select the **Khaldiyah Dam**.

26. On the **Map** tab, in the **Selection** group, click Select By Location to open the Select By Location geoprocessing tool and fill it as below.
27. Input Features:Well
28. Relationship:Completely within
29. Selecting Features:Well_StandDist
30. Accept the rest of the default
31. Click Apply and then OK

Result Fifty-three out of 129 wells were selected.

32. In the CP, r-click the Well, point to Data and click Export Features
33. The Export Features geoprocessing tool open, fill it as below
34. Input Features:**Well**
35. Output Feature Class:**Well_Dam**
36. Click OK
37. In the CP, r-click the Well feature class and click Remove

Result The Well_Dam feature class added to Contents pane

38. On the **Map** tab, in the **Selection** group, click Select By Attributes to open the Select By Attributes geoprocessing tool and fill it as below.
39. Input Rows:Dam
40. Selection Type:New selection
41. Select the SQL "**Where RESERVOIR is equal to Khaldiyah**"
42. Click Apply and then OK

Result The **Khaldiyah** Dam is selected.

Near Tool

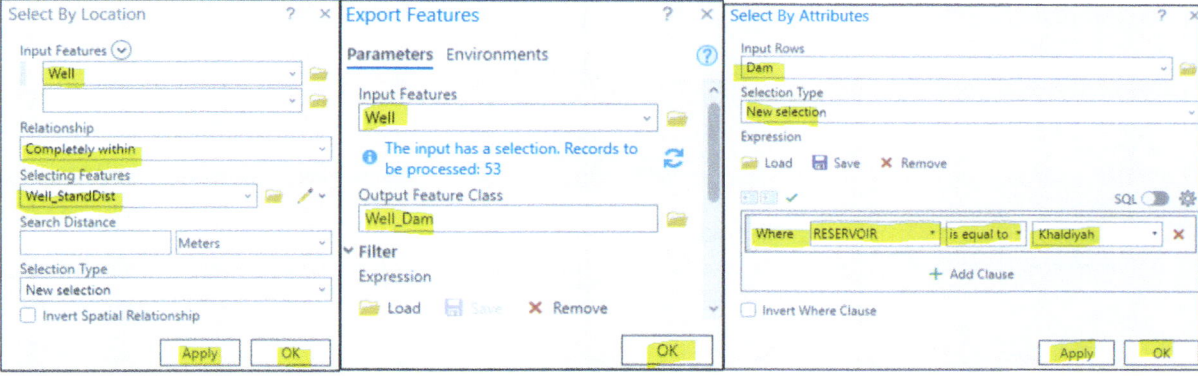

Near Tool

Calculate distance and additional proximity information between the wells in the buffer zone and the dam. Three fields in the attribute table will be added to the **Well_Dam** feature class that will be used to find the closest well to the **Khaldiyah Dam**. The original **Well_Dam** attribute table has 7 records (ObjectID, Shape, Well_ID, TDS, NO3_ppm, Depth, and Weight). The attribute table of the Dam has 4 records (FID, Shape, Reservoir, Year). The Feature ID (FID) of the **Khaldiyah Dam** is selected.

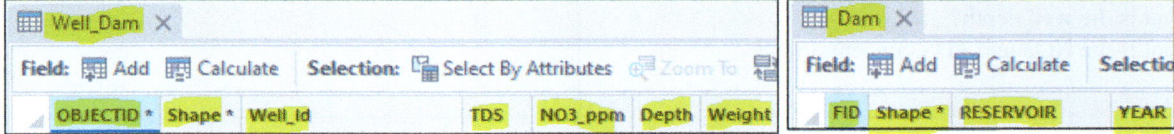

43. In the Geoprocessing pane, click the back arrow
44. Open the **Analysis** Tools and then the **Proximity subcategory** and click the **Near** tool
45. Fill the **Near** geoprocessing pane as follows:
46. Input Features:Well_Dam
47. Near Features:Dam
48. Accept the rest of the default
49. Run
50. Open the attribute table of **Well_Dam**

Result Two fields are created in the attribute table of **Well_Dam**: NEAR_FID and NEAR_DIST. NEAR_FID has one variable: 1 (which signifies the FID of the **Khaldiyah Dam** in the Dam attribute table. NEAR_DIST contains the distance in meters between the wells inside the buffer and the **Khaldiyah Dam**.

51. In the **Well_Dam** attribute table, r-click NEAR_DIST and click **Sort Ascending**

Result The **well** with Well_Id number 109 is the closest to the **Khaldiyah Dam** and is 419.88 meters. This well will be used in artificial recharge

52. Save the project

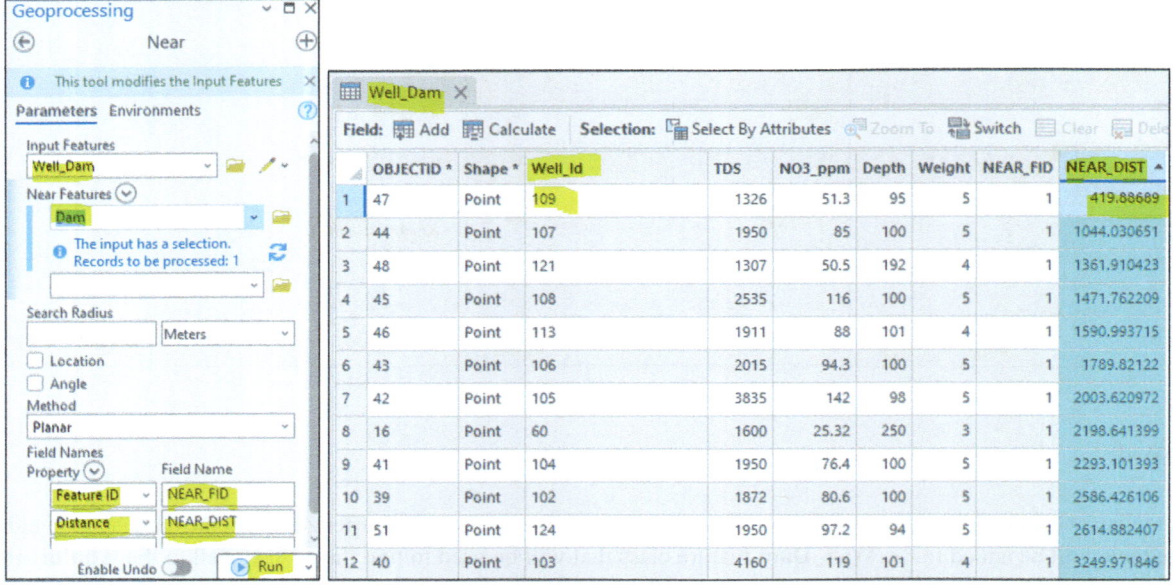

After finding the distance from each well inside the buffer to the **Khaldiyah Dam**, answer the following questions about the closest well to the **Khaldiyah Dam.**
(a) What is the well depth?
(b) What is the NO3 and TDS concentration?

Analyzing Pattern Toolset: Identify Pattern Based on Location

Some statistical analyses aim to identify patterns, trends, and spatial relationships among features in any environment. Whether a certain set of data is more likely to show certain characteristics; Some Spatial Analyst tools can recognize the distribution patterns of geographic layers in a specific study area. In geography, there is a well-documented practice that demonstrates how features located near each other are more similar than features situated farther away from one another (Tobler's First Law of Geography). This idea is common sense; nevertheless, there is always an exception to the rules. For example, the weather in the Jordan Rift Valley, which is approximately 400 meters below sea level, is not like weather in the Ajloun Highlands, which is more than 1,000 meters above sea level. Furthermore, these two locations are only 20 kilometers away from each other. At the same time, the climate of the city of Aqaba is similar to that of the city of Jeddah, even if the two cities are 970 km away from each other.

Average Nearest Neighbor Tool

To identify patterns based on location, users can use the **Average Nearest Neighbor** tool. The tool will detect if features are clustered or dispersed; this tool is used and tested with some degree of confidence level. The statistical approach behind this method is that the tool will measure the distance from each feature in the dataset to its single nearest feature neighbor and then calculate the average distance of all measurements. The tool then creates a hypothetical dataset with the same number of features but placed **randomly** within the **study area**. The tool is then run again, the nearest distance to its nearest neighbor feature is measured, and the average is calculated. The average distance of the random hypothetical data will be assessed with the real data. Two parameters will be generated: I and Z score.

1st **parameter:** the nearest neighbor index (I) is generated as follows:

$$I = \frac{D_r}{D_h}$$

D_r is the calculated average distance of the real data
D_h is the average distance from the hypothetical data

If	$I < 1$	The data show clustering
If	$I > 1$,	The data show dispersion.
If	$I = 1$	The data are randomly distributed

A pattern that falls at a point between dispersed and clustered is said to be random

2nd parameter: the z score will be calculated and is vital to deciding whether to accept or reject the null hypothesis. The z score is associated with the confidence level and is up to the researchers to adopt which confidence level they are willing to test with their hypothesis. Each confidence level is associated with the z score, which is simply a standard deviation. For example, a 90% confidence level has a z score range between -1.65 and + 1.65, and the 95% confidence level has a z score between -1.96 and +1.96 (table below).

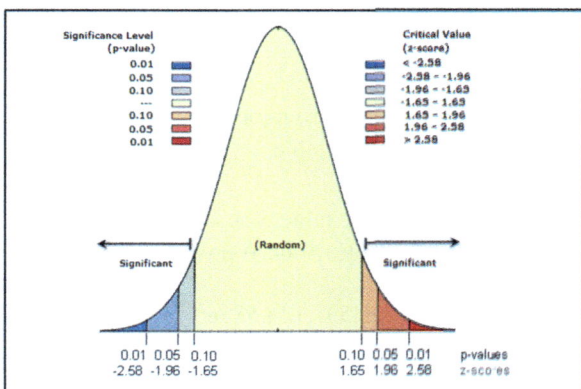

Z-score (Standard Deviations)	p-value (Probability)	Confidence level
-1.65 or +1.65	0.10	0.9
-1.96 or +1.96	0.05	0.95
-2.58 or +2.58	0.01	0.99

Null Hypotheses

In any statistical testing, you have to propose a **null hypothesis** and the null hypothesis states that features in the study area lacking any pattern. This means that the features are not clustered or dispersed but randomly distributed.

Let us assume you are willing to test your hypothesis with 95% confidence level and you are assuming (Null Hypothesis) that the features are **randomly distributed**. After running the test, the Z score value that was generated was between **-1.96 and +1.96,** and the p value was larger than **0.05**. Based on the result, you must **accept** the null hypothesis, which means that your features are randomly distributed. However, if the Z score fell outside that range for example -2.0 or +2.0 standard deviations, you must **reject** the null hypothesis and your observed features are clustering or dispersed.

Scenario 3 You are a hydrogeologist and you have observed a heavy groundwater abstraction from the wells that are used for irrigation in **Wala** catchment area. This practice has dramatically lowered the water table in these wells, which affected the groundwater storage in the whole basin. You have decided to examine if the distance between the wells in the basin are one of the reasons that generate the intense dropdown. The proximity of the wells to each other could affect the zone of influence created by well pumping. Your question is, are the wells in the basin that are used for agriculture randomly distributed or do they have a certain pattern (clustered or dispersed). To determine the answer, you must do the following:

Propose a *null hypothesis* stating that wells in the basin are randomly distributed.
1. Insert a New Map and name it **Nearest Neighbor**
2. From Catalog pane, integrate **WalaWatershed.shp** and **Well.shp** from the **Q3** folder
3. Open Attribute Table of **Well**, you will see it has 333 records, and the **Type** field shows that the wells are used for different purposes. We are interested in selecting the wells that are used for "Irrigation".
4. Close the Well attribute table
5. In the **CP**, right click the **Well** layer and in the "**Properties**" tab click **Definition Query**
6. Click "**+ New definition query**"
7. The query should read as "**Where Type is equal to Irrigation**"
8. Apply the query and exclude all records from the attribute table that do not meet the criteria.
9. Click OK to exit the **Definition Query** dialog box

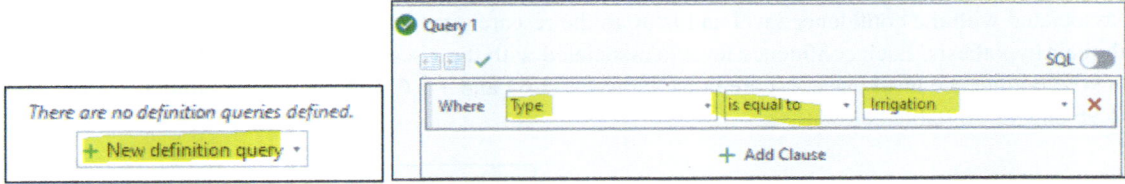

Result The wells used for irrigation are now the only ones displayed, and the rest of the wells are hidden. You can verify that by opening the attribute table.

10. Open the Well attribute table, you see there are only 271 records used for irrigation
11. Open the attribute table of the WalaWatershed, and record the **Shape_Area**

Result The area is 1,803,591,128.58 m^2.

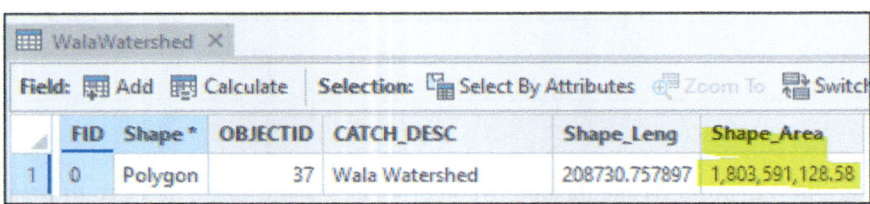

12. In the Geoprocessing pane, click the back arrow of the **Near** pane
13. In the **Spatial Statistics Tools** open the **Analyzing Patterns**
14. Click the **Average Nearest Neighbor** tool, the tool opens, and fill it as follows:
15. Input Feature Class:Well
16. Distance MethodEuclidian
17. Check Generate Report
18. Area:1803591128.58
19. Click Run
20. Once you have run it, click on "View Details" at the bottom of the **Average Nearest Neighbor** pane
21. The **Average Nearest Neighbor** pane open, click on the **Parameters** tab and click on the "**Report File**" link (you can also open it from the folder of the project)

Null Hypotheses

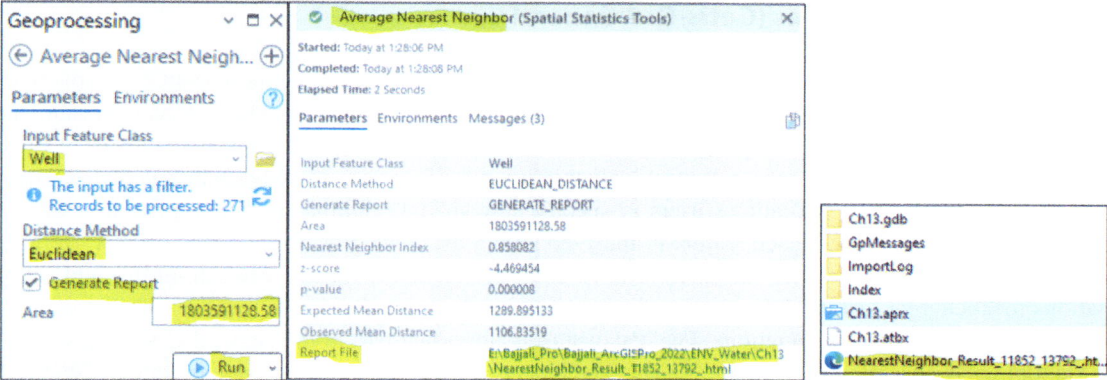

Result A two-tailed normal curve graph will be opened and displayed as an **html** file in your internet browser.

Interpretation The values generated with the normal curve graph will be calculated for you. The generated data are strong evidence that the wells in the basin are clustered and not randomly distributed; therefore, the null hypothesis will be rejected based on three pieces of evidence:

1st **evidence – Z Score**: The calculated Z score in the normal curve is −4.4694 and located on the left tail. This generated value is less than the Z score of the 99% confidence limit presented in the graph (−2.58). The generated Z score (−4.4694) in the curve is in the rejection zone and is smaller than the value of −2.58, therefore we can reject the Null Hypothesis. This confirms that the wells in the basin are clustered and not randomly distributed.

2nd **evidence – p-value**: The calculated **p-value** is 0.000008, which is much less than the **p-value** of the significance level of the left tail (0.005). Because 0.000008 is much less than 0.005, you can reject the Null Hypothesis that the wells are randomly distributed.

Statistical Background: Confidence limits (C = 0.99) + Significant level (α = 0.01) = 1
The significance level in the left tail is equal to 0.005 (0.01/2).

3rd **evidence – I Ratio**: the "I" ratio is also 0.858, which is less than 1. Therefore, you can reject the null hypothesis, and we consider the distribution of the irrigation wells clustered in the basin.

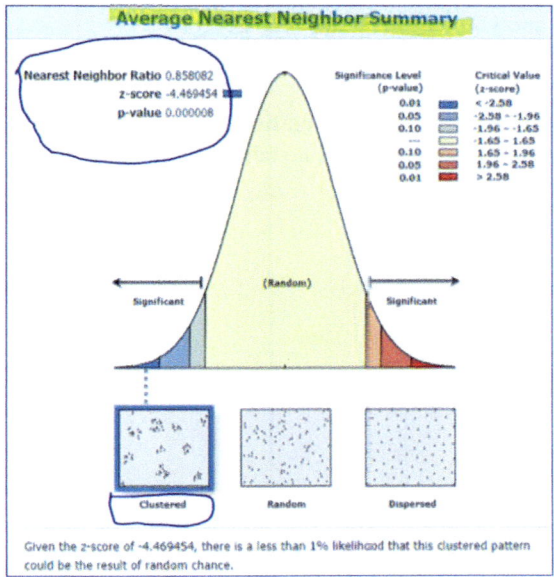

Identify Pattern Based on Values (Getis-Ord General G)

The location of features is not the only aspect determining the clustering but also the values associated with the feature within a crucial distance of each other. Before running the tool for clustering, you must use a tool to find the important distance that will be implemented in testing the clustering. The General G-statistics tool will be used to identify high or low distance values over the entire study area. The distance will reveal whether it is significant or not and will be calculated based on either Euclidean or Manhattan. The tool also allows users to specify how spatial relationships among features are defined. For example, in the "fixed distance band", each feature is analyzed within the context of neighboring features. Neighboring features inside the specified critical distance (distance band or threshold distance) receive a weight of one and exert influence on computations for the target feature. Neighboring features outside the critical distance receive a weight of zero and have no influence on a target feature's computations. The distance is an important part of the General G-statistics, as it will show over which the tool will be ascertained to be significant. The ideal distance will be determined using the "Calculate Distance Band from Neighbor Count" tool.

The "Calculate Distance Band from Neighbor Count" tool returns the minimum, the maximum, and the average distance to the specified Nth nearest neighbor (N is an input parameter) for a set of features, for example, 5 wells.

The General G tool calculates the value of the General G index, Z score and p value for a given input feature class. The Z score and p value are measures of statistical significance that tell you whether to reject the null hypothesis. For this tool, the null hypothesis states that the values associated with the features are randomly distributed. The Z score value means the following:

(a) A Z score near zero indicates no apparent clustering within the study area.
(b) A positive Z score indicates clustering of high values.
(c) A negative Z score indicates clustering of low values.

Scenario 4 In the previous scenario we determined that the wells are clustering, this time you must see if the **Weight** field has an influence on the clustering of the irrigation wells in the watershed and at what distance the clustering taking place. The "**Weight**" field has values from 1 to 5, with 5 representing the most important wells. These wells have the highest yields. In this exercise, you must run the **Calculate Distance Band from Neighbor Count** tool to find the ideal distance to run the General G-statistics. This will show over the tool which will be ascertained to be significant. After identifying the average distance that will return the minimum, the maximum, and the average distance to the 5th nearest neighbor (N = 5 wells), you should use values higher and lower than the average return value and run all of them to decide which distance is ideal to use the General G-statistics.

1. Insert a New Map and rename the Map **General G-Statistics**
2. From Catalog pane, integrate **WalaWatershed.shp** and **Well.shp** from the **Q4** folder
3. Open the Attribute Table of **Well** and you will see it has 333 records
4. In the **CP**, right click the **Well** layer and in the "**Properties**" tab click **Definition Query**
5. Click "**+ New definition query**"
6. The query should read as "**Where Type is equal to Irrigation**"
7. Apply the query and exclude all records from the attribute table that do not meet the criteria.
8. Click OK to exit the **Definition Query** dialog box

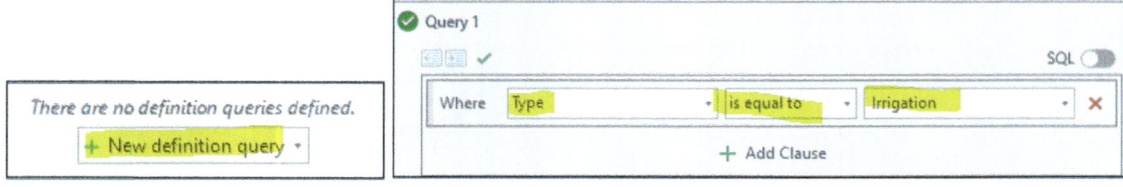

Find the Ideal Distance

Result The wells used for irrigation are now the only ones displayed, and the rest of the wells are hidden. You can verify that by opening the attribute table.

Null Hypothesis Agricultural wells with high ranking values represented by the "**Weight**" field are randomly distributed in the study area.

Find the Ideal Distance

Before performing the General G-statistics, you must run the **Calculate Distance Band from Neighbor Count** tool in order to find the best distance to use with the General G-statistics.

9. In the Geoprocessing pane, click the back arrow of the **Average Nearest Neighbor** pane
10. In the **Spatial Statistics Tools** open the **Utilities** and click **Calculate Distance Band from Neighbor Count** tool
11. The **Calculate Distance Band from Neighbor Count** tool open, fill it as below
12. Input Feature Class:Well
13. Neighbors5
14. Distance MethodEuclidian
15. Click Run
16. Click "View Details" at the bottom of the Geoprocessing pan

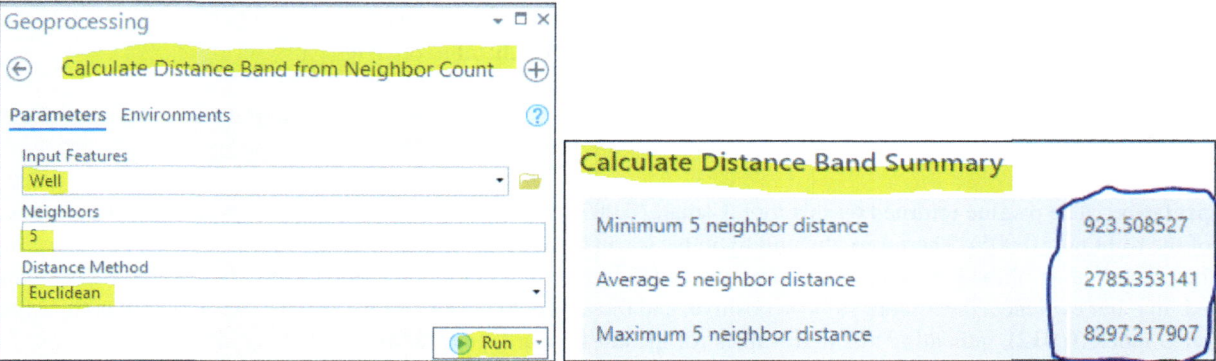

Result The calculated distance band from the neighbor count displays the result of the minimum, average, and maximum of 5 neighbor distances (see below).

(a) Minimum Distance = 923.50
(b) Average Distance = 2785.35
(c) Maximum Distance = 8297.21

The average distance is 2785 m with 5 neighbors; therefore, a lower and higher number should be used to determine which value to use to run the **General G statistical** tool. The distance we want should be from 2000 to 3600 m at 400-m intervals.

17. Close the **Calculate Distance Band from Neighbor Count**

High/Low Clustering (Getis-Ord General G)

Now you will run the **High/Low Clustering (Getis-Ord General G)** tool using different values and the result will be populated in the table below. The High/Low Clustering (Getis-Ord General G) statistic is an inferential statistic, which means that the results of the analysis are interpreted within the context of the null hypothesis. The null hypothesis for the High/Low Clustering (General G) statistic states that there is no **spatial clustering of well weight values**. The output result of the tool produces 5 parameters: Observed General G, Expected General G, Variance, Z score, and p value. The method has no output layer, but a report is created and demonstrates whether the well distribution in the watershed is clustered, dispersed, or random.

18. Click the back arrow in the Geoprocessing pane
19. In the **Geoprocessing** pane under the **Spatial Statistics Tools** category and **Analyzing Patterns** subcategory and click the **High/Low Clustering (Getis-Ord General G)**
20. Fill the **High/Low Clustering (Getis-Ord General G** as shown below
21. Input Feature Class: Well
22. Input Field: Weight
23. Check Generate Report
24. Conceptualizing the Spatial Relationship: Fixed distance band
25. Distance Method: Euclidian
26. Standardization: None
27. Distance Band or Threshold Distance: 2000
28. Click Run
29. Click **View Detail** at the bottom of the Geoprocessing pan icon to view the result
30. Open Ch13 folder and click on **GeneralG_Result_11852_13792_.html** to open it

Result The general G summary displayed 5 values, and GeneralG_Result_11852_13792_.html displayed a normal curve.

Interpretation The p value returned by this tool is small (0.0073), which is much less than the **p-value** of the significance level of the right tail (0.005). Therefore, the null hypothesis can be rejected. Because the z score value is positive and higher than the critical value of the 99% confidence limit presented in the right tail in the graph (2.58), the Null Hypothesis can be rejected. In other evidence, the z score value is positive, and the observed General G index (0.014) is larger than the expected General G index (0.012), indicating that high values for the attribute are clustered in the study area.

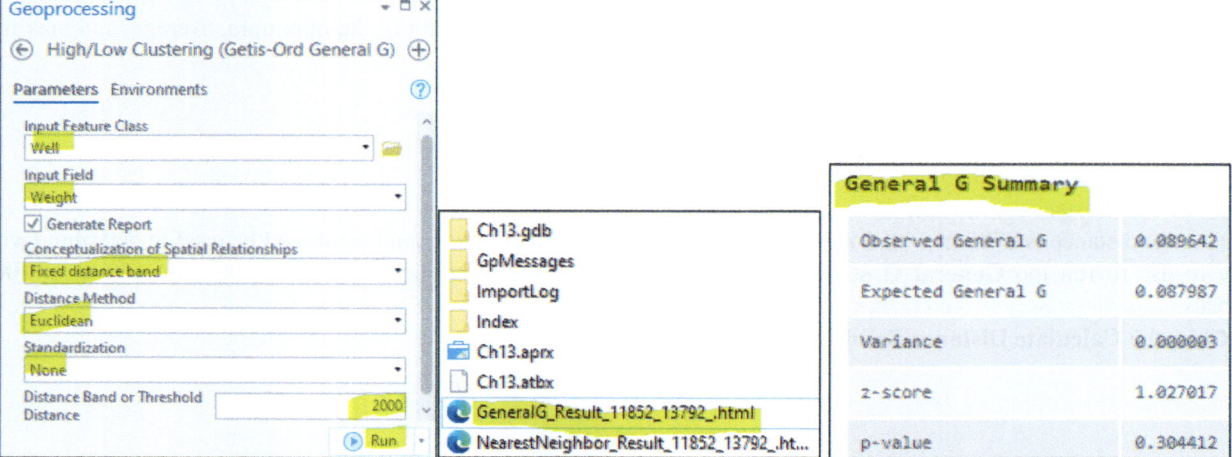

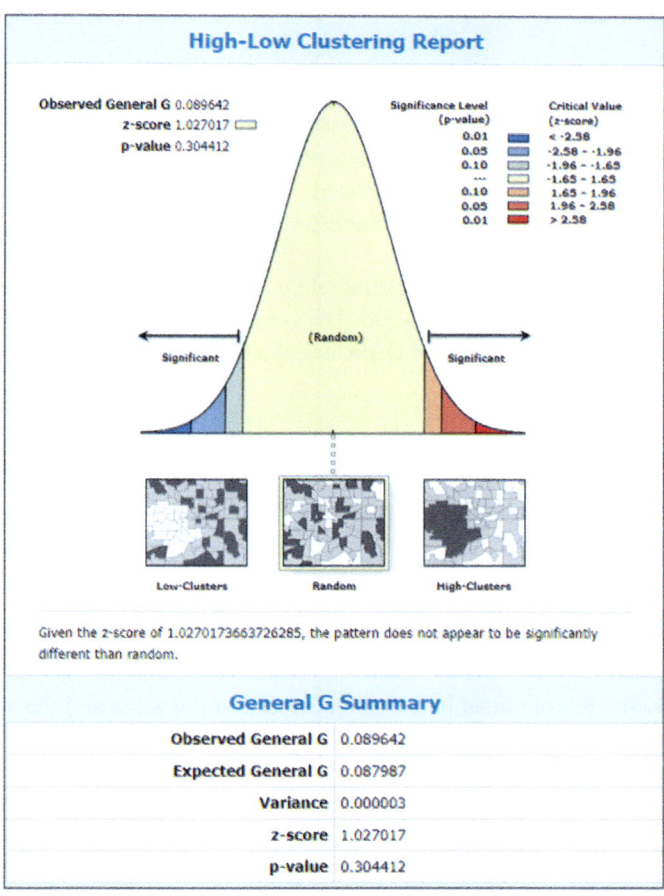

31. Repeat the previous steps while replacing only the distance with 2400, 2800, 3200, 3600, 4000 and keep the rest of the distance as the default.
32. Save your project

Interpretation The best distance to choose is the one that is in the rejection zone and has the highest z score. In this case, 2000 m is the best distance to choose.

High low clustering report

Distance	p value	Z Score	Observed General G	Expected General G	Cluster
2000	0.007	2.67	0.014	0.012	Yes
2400	0.211	1.25	0.018	0.017	No
2800	0.186	1.32	0.024	0.023	No
3200	0.219	1.23	0.029	0.028	No
3600	0.231	1.19	0.037	0.035	No
4000	0.31	1.01	0.044	0.043	No

Spatial Autocorrelation (Global Moran's I)

Global Moran's I index measures the spatial correlation using the feature location and an attribute value together to determine statistically if the data are clustered, dispersed or random. Using the spatial correlation helps define how the variables are arranged in a study area. The tool calculates three important parameters

(a) Moran's I Index value
(b) Z score value
(c) p value.

The results of the analysis are always interpreted within the context of its null hypothesis. For the Global Moran's I statistic, the null hypothesis states that the attribute being analyzed is randomly distributed among the features in the study area.

If Moran's index value is near **+1.0**, clustering is indicated, while an index value near **-1.0** indicates dispersion. The method has no output layer, but a report is created and demonstrates whether the well distribution in the watershed is clustered, dispersed, or random.

The tool will run using the conceptualization of the spatial relationship of **Zone of indifference**, **Euclidian distance**, and **distance bands** of 500, 1000, 1500, 2000, 2500, and 4000. The concept of Zone of indifference in which wells within the specified critical distance (Distance Band or Threshold Distance) of a target well receive a weight of one and influence computations for that well.

Scenario 5 Your supervisor now asked you to look at the density of groundwater wells per block and would like to hear your professional judgment on what distance these well densities cluster at. This information is critical in management because it helps to adjust the rate of pumping of the wells that are located close to each other in the clustering pattern. Your duty is to do the following:

Spatial Join Between Grid_1000 and Well Layers

To prepare the data for analysis, spatial join must be performed between the wells and the study area represented as a grid with a cell dimension of 1 kilometer by 1 kilometer.

1. Insert a New Map and rename the Map **Moran's I Index**
2. From Catalog pane, integrate **Grid_1000.shp**, **WalaWatershed.shp**, and **Well.shp** from the **Q5** folder
3. Click the back arrow in the Geoprocessing pane
4. In the **Geoprocessing** pane under the **Analysis Tools** open the Overlay tools and click on **Spatial Join**.
5. The **Spatial Join** dialog box will be displayed and filled in as follows:
6. Target Features:Grid_1000
7. Join Features:Well
8. Output Feature Class:Well_Grid
9. Join Operation:Join one to one
10. Match Option:Completely contains
11. Click Run

Result The **Well_Grid** is created, and it shows the density of the groundwater wells in the **Wala watershed**. If you open its attribute table, you will find a field called **Join_Count**. Some of the values are zero, which indicates the cells that have no wells. The cells that have one or more wells indicate the number of wells located within each cell. The maximum number of wells found in one cell is 7, and they are in the northern part of the watershed.

Spatial Autocorrelation (Global Moran's I)

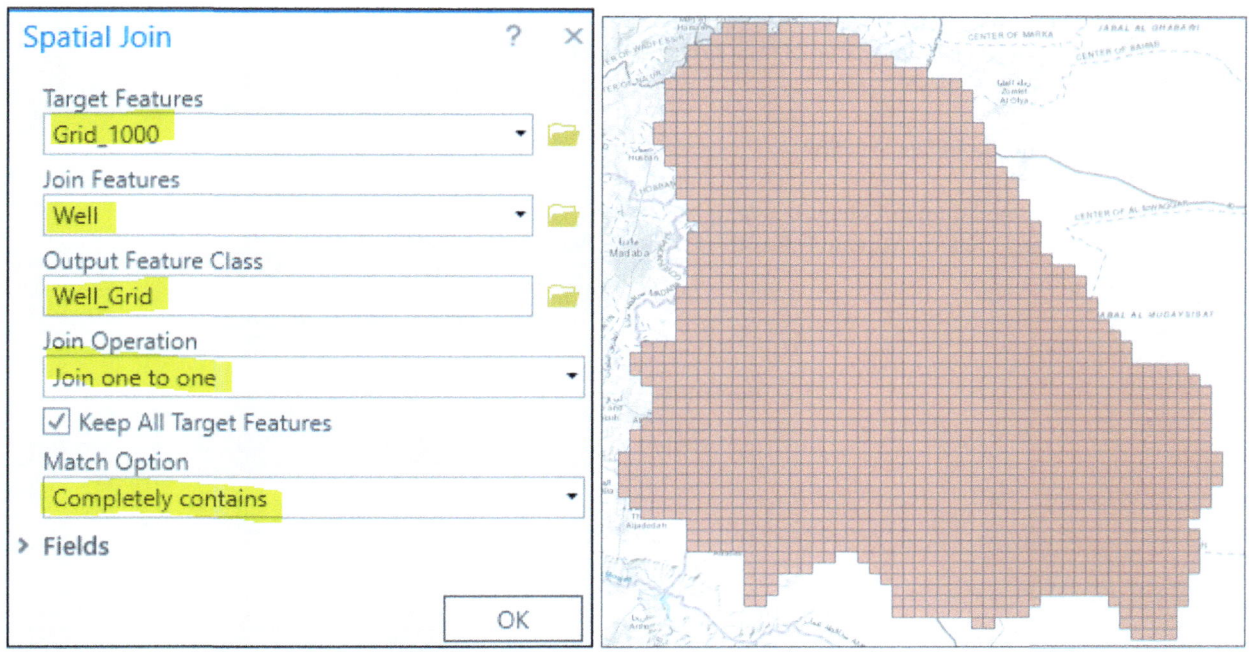

Note The cells that have a "0" value should be hidden before running the statistics.

Create a definition query for **Well_Grid**

12. The query should read as "Where **Join_Count** is greater than 0"
13. Apply the query

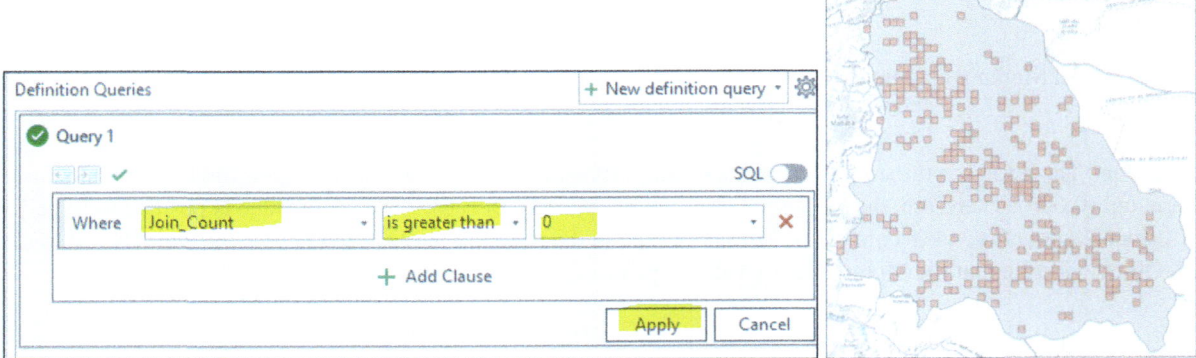

Result **Well_Grid** will only show the cells in the grid that have at least one well.

Spatial Autocorrelation (Global Moran's I)

14. Click the back arrow in the **Geoprocessing** pane open the **Spatial Statistics tools** and then the **Analyzing Patterns**
15. Click "**Spatial Autocorrelation (Global Morans I)**" and fill the dialog box as follows:
16. Input Feature Class: **Well_Grid**
17. Input Field:**Join_Count**

18. Check the **Generate Report**
19. Conceptualization of spatial relationship**ZONE_OF_INDIFFERENCE**
20. Distance Method**EUCLIDEAN**
21. Standardization:None
22. Distance Band or Threshold Distance**500**
23. Run
24. Click View Detail at the bottom of Geoprocessing pane
25. Open Ch13 folder and click on **MoransI_Result_11852_13792_.html** to open it

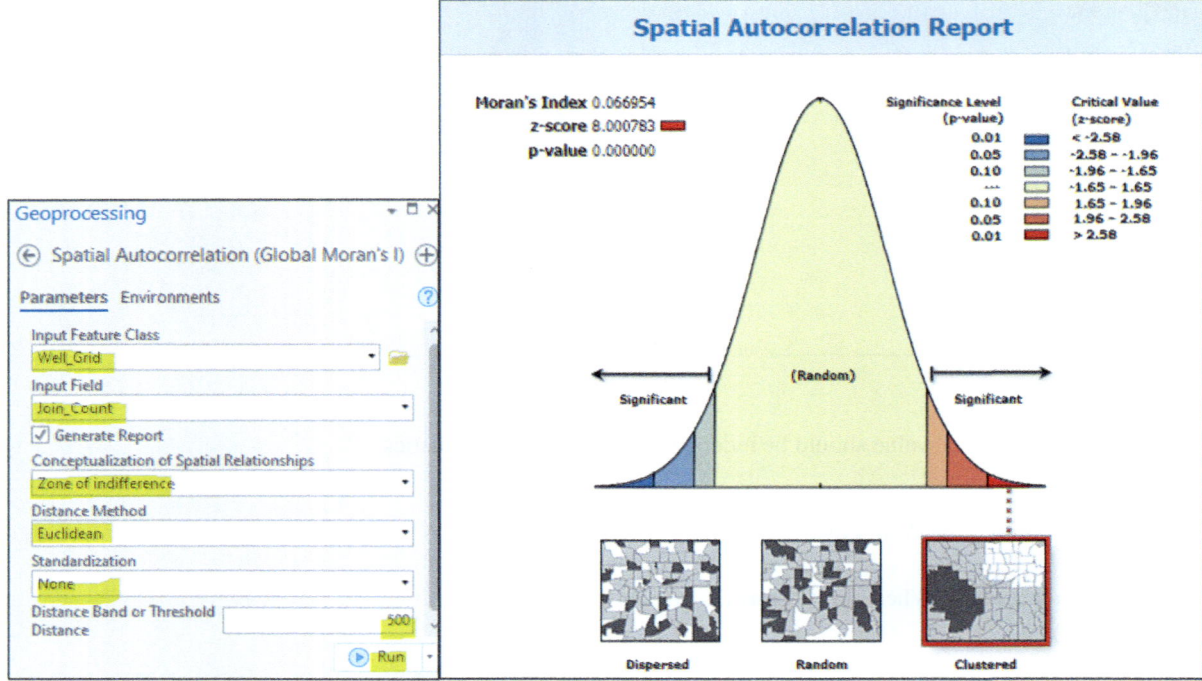

Result The spatial Moran's I summary displays 5 values and a normal curve.

Interpretation The p value returned by this tool is small (0.0), which is much less than the **p-value** of the significance level of the right tail (0.005). Therefore, the null hypothesis can be rejected. Because the z score value is positive (8.0) and higher than the critical value of the 99% confidence limit presented in the right tail in the graph (2.58), the Null Hypothesis can be rejected. Other evidence that the Moran's Index value is positive (0.06) indicates clustering.

26. Repeat the steps above by using 1000, 2000, 4000, 6000, and 8000 and record all the results in the table below.
27. Save the project

Distance	p value	Z score	Moran's index	Cluster
500	**0.00**	**8.00**	**0.06**	**Yes**
1000	0.00	3.42	0.28	Yes
2000	0.00	5.03	0.24	Yes
4000	0.00	5.87	0.14	Yes
6000	0.00	7.94	0.13	Yes
8000	0.00	7.62	0.09	Yes

Conclusion The most significant clustering occurs at the distance where the Z score is the highest and Moran's index (I) is the lowest. In this situation, significant clustering occurs at a distance of 500 meters.

Cluster and Outlier Analysis (Anselin Local Moran I)

The cluster analysis will examine a dataset of features (such as wells) with a value associated with the features (such as depth or salinity). The output result of the analysis will be displayed as a feature class, and the clustering will be highlighted. The generated output feature class will have the following fields in the attribute table: Local Moran's I index (LMiIndex), z score (LMiZScore), pseudo p value (LMiPValue), and cluster/outlier type (COType), in addition to other fields from the original input layer. The z scores and p values are measures of statistical significance that tell users whether to accept or reject the null hypothesis. The interpretation of the result will be based on the following fields in the attribute table:

A high positive z score in the attribute table indicates that the surrounding features have similar values (either deep wells or shallow wells).

The COType field will be HH for a statistically significant cluster of high values (deep wells) and LL for a statistically significant cluster of low values (shallow wells).

A low negative z score (less than -1.4) for a well indicates a statistically significant spatial data outlier. The COType field indicates whether the well has a deep well and is surrounded by a well with shallow depth (HL) or if the well has a shallow depth and is surrounded by wells with deep depth (LH).

No permutations are used to determine how likely it would be to find the actual spatial distribution of the wells you are analyzing. For each permutation, the neighborhood values around each feature are randomly rearranged and the **Local Moran's I** value is calculated. The result is a reference distribution of values that is then compared to the actual observed Moran's I to determine the probability that the observed value could be found in the random distribution. The default is 499 permutations; however, the random sample distribution is improved with increasing permutations, which improves the precision of the pseudo p value.

Scenario 6 In the Amman-Zarqa basin, there are many groundwater wells drilled for agricultural development, and they tap two aquifer systems: the carbonate and basalt aquifers. The wells that penetrate the basalt aquifer are in general deeper than the wells penetrating the carbonate aquifer. Your task is to identify if there is a clustering based on the depth of the wells in the study area.

1. Insert a New Map and rename the Map "**Clustering**"
2. From Catalog pane, integrate **Geology.shp**, **Well.shp**, and **WWTP.shp** from **Q6**
3. In the **CP**, right click the **Well** layer and in the "**Properties**" tab click **Definition Query**
4. Click "**+ New definition query**"
5. The query should read as "**Where Well_Depth is greater than 0**"
6. Click Apply and then OK to exit the **Definition Query** dialog box

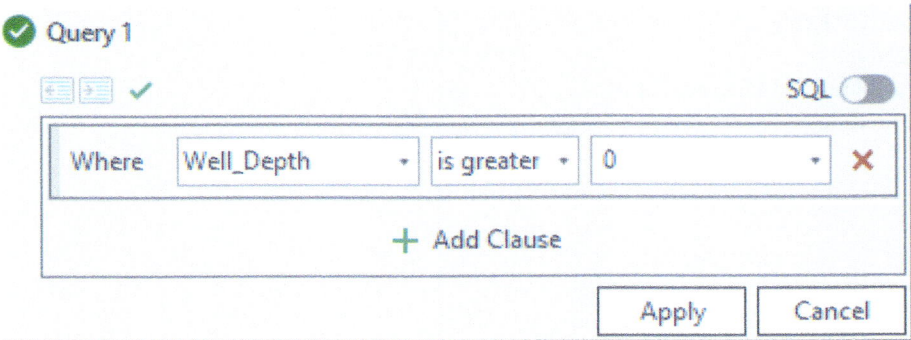

Result The number of records decreases from 2039 to 1787.

7. Set the symbology of the **Geology** layer to "Unique Values" based on the **Lithology** field and "Basic Random" color scheme
8. Hide "All Other Values" by unchecking the option from the "**More**" dropdown menu

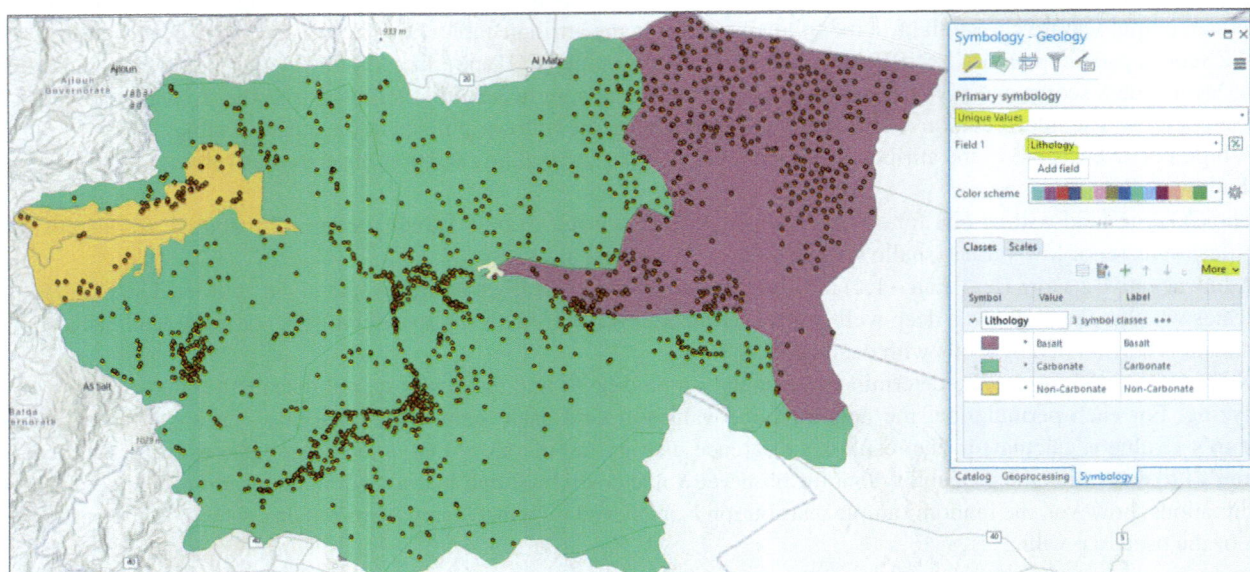

Run Cluster and Outlier Analysis (Anselin Local Moran I)

This method allows you to use a distance of your choice to find a significant number of neighbors. A 1,000 m Euclidean distance will be used, and wells outside the 1,000 m for a target well are ignored in the analysis for any given well.

9. Click the back arrow in the **Geoprocessing** pane open the **Spatial Statistics tools** and then the **Mapping Clusters**
10. Click "**Cluster and Outlier Analysis (Anselin Local Moran I)**" and fill the dialog box as below
11. Input Feature Class: **Well**
12. Input Field: **Well_Depth**
13. Output Feature Class: **MICluster1000**
14. Conceptualization of spatial Relationships **Fixed distance band**
15. Distance Method **EUCLIDEAN**
16. Standardization: None
17. Distance Band or Threshold Distance **1000**
18. Number of Permutations 999
19. Click Run

Cluster and Outlier Analysis (Anselin Local Moran I)

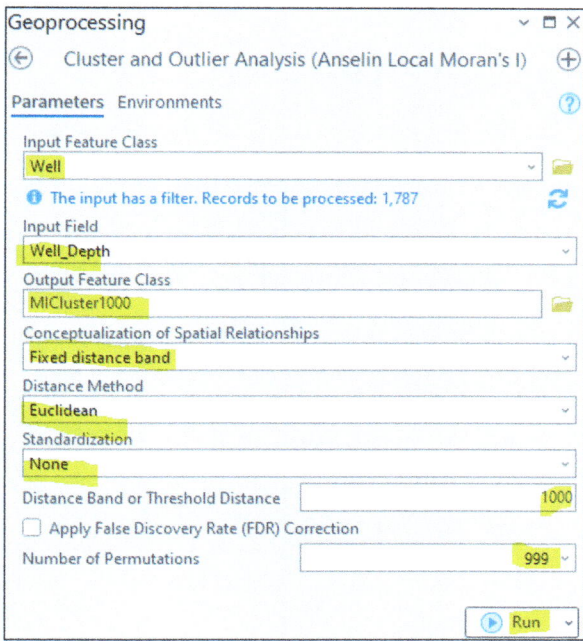

Result The new output feature class "**MICluster**" was created, and its attribute table includes new fields such as local Moran's I index, z score, p value, and COType. These fields are important for the interpretation of the results. The z scores and pseudo p values represent the statistical significance of the computed index values. The "**MICluster**" is automatically added to the **Content** pane with default classification applied to the COType field.

20. Change the symbols of the **MICluster** to more distinguishable colors and to be slightly larger at your own discretion
21. Save the project

Interpretation A total of 193 wells have HH records, which means a statistically significant cluster of deep wells and surrounding deep wells. The depths of the wells in this dataset range from 230 to 675 meters. The number of wells in the LL records is 385 wells, which means a statistically significant cluster of shallow wells surrounded by shallow wells. The depths of the wells in this dataset range from 5 to 217 m. The HL is an outlier in which a high value is surrounded primarily by low values (HL), and the LH is an outlier in which a low value is surrounded primarily by high values.

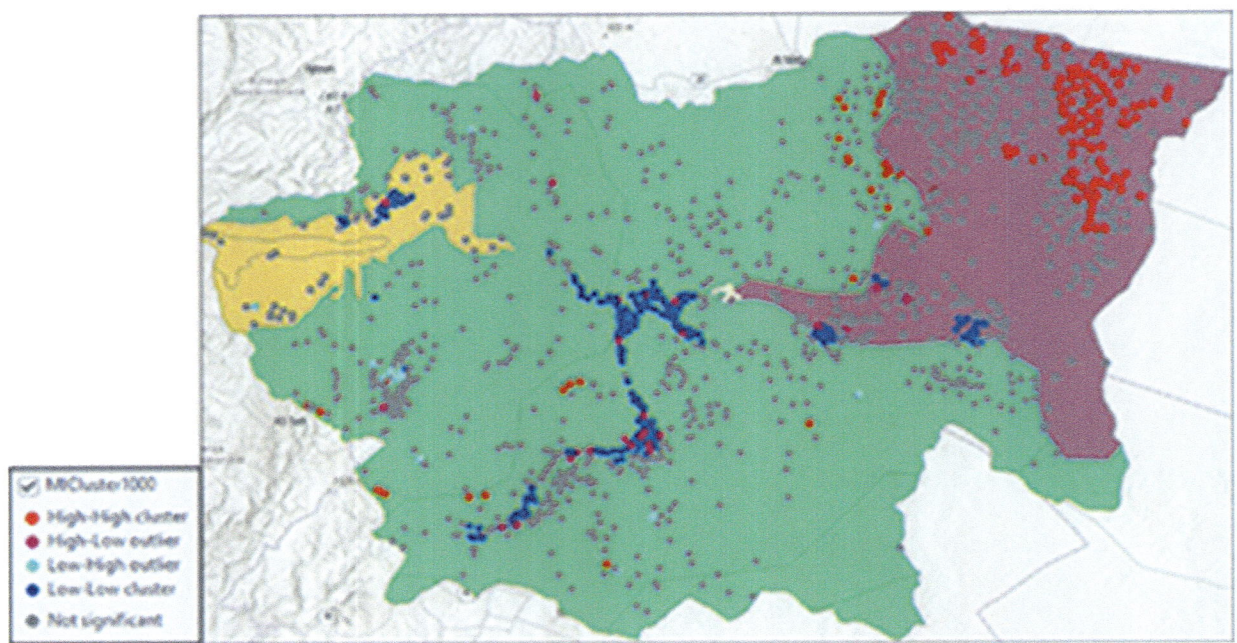

Hot Spot Analysis (Getis-Ord GI*)

This is another method to identify statistically significant spatial clusters of wells of high depth (hot spots) and shallow depth (cold spots) using the Getis-Ord Gi* statistic. The tool creates a new output layer with a z score, p value, and confidence level bin (Gi_Bin) for each well in the input layer. The z scores and p values are measures of statistical significance that tell the users whether to accept or reject the null hypothesis. The Gi_Bin field also identifies statistically significant hot and cold spots as follows:

- Wells (+3 bins) reflect "Hot Spot" statistical significance with a 99% confidence level.
- Wells (+2 bins) reflect "Hot Spot" statistical significance with a 95% confidence level.
- Wells (+1 bins) reflect "Hot Spot" statistical significance with a 90% confidence level.
- Wells (-3 bins) reflect "Cold Spot" statistical significance with a 99% confidence level.
- Wells (-2 bins) reflect "Cold Spot" statistical significance with a 95% confidence level.
- Wells (-1 bins) reflect "Cold Spot" statistical significance with a 90% confidence level.
- Well with 0 bin indicates no apparent spatial clustering

Scenario 7 You are going to use the wells from the previous scenario to identify the hot and cold spot based on the groundwater wells depth in the Amman-Zarqa basin

1. Insert a New Map and rename the Map "**Hot and Cold Spot**"
2. Above the Map View activate the Clustering tab, in the CP, r-click **Well.shp** and **Geology.shp** and click **Copy**
3. Above the Map View activate the **Hot and Cold Spot** tab
4. In the CP, r-click the **Hot and Cold Spot** map and click Paste
5. Click the back arrow in the **Geoprocessing** pane open the **Spatial Statistics tools** and then the **Mapping Clusters**
6. Click "**Hot Spot Analysis (Getis-Ord GI*)**" and fill the dialog box as follows:
7. Input Feature Class:**Well**
8. Input Field:**Well_Depth**
9. Output Feature Class:**HotSpot**
10. Conceptualization of spatial Relationships**Fixed distance band**
11. Distance Method**EUCLIDEAN**
12. Distance Band or Threshold Distance**1000**

13. Accept the rest of the default
14. Click Run
15. Change the symbols of the **HotSpot** to more distinguishable colors and to be slightly larger at your own discretion

Result The **HotSpot** feature class is automatically added to the Contents pane with default classification applied to the **Gi_Bin** field. It was classified into seven classes: Cold Spot, Hot Spot, and no clustering. The cold and hot spots each consist of three groups. A cold spot is depicted by a 99% – 90% confidence interval, which shows that they are shallow wells and have a negative Gi_Bin value with a low ZScore. A hot spot is depicted by a 99% – 90% confidence interval, which shows that they are deep wells with a higher ZScore and a positive Gi_Bin. Wells with ZScore close to 0 and 0 bin reflect no clustering. In addition, the tool also generates a histogram charting the value of the **Well_Depth** field.

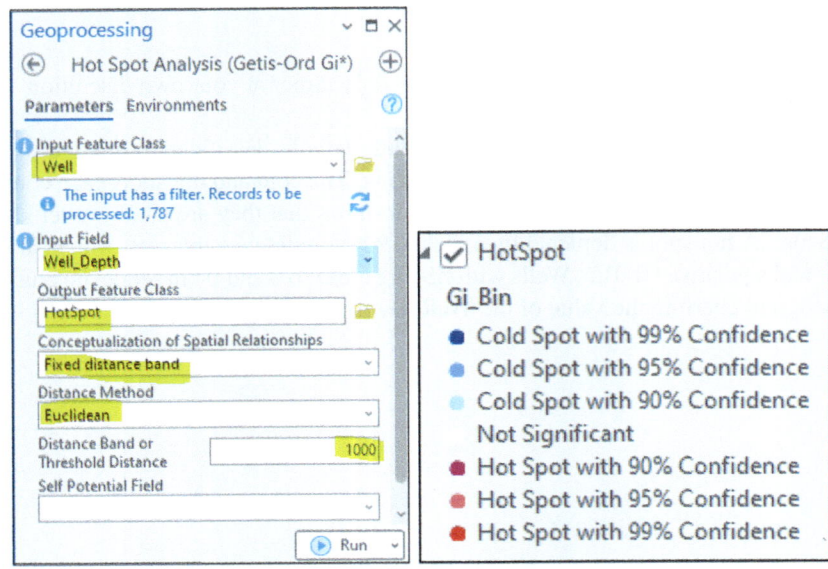

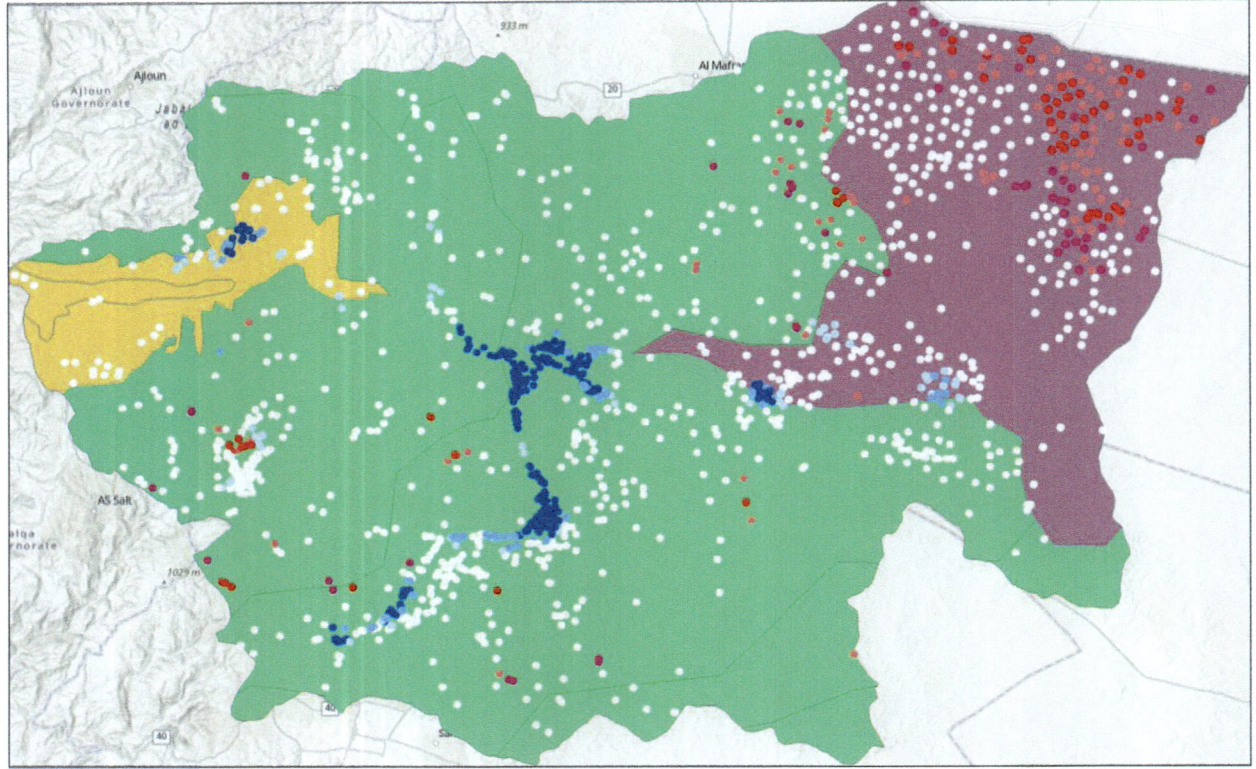

Summary Statistics

The tool will summarize the attribute table of the **HotSpot** feature class into seven classes, and the output will be a table showing the summary

16. Click the back arrow in the **Geoprocessing** pane open the **Analysis Tools** then the **Statistics**
17. Click the **Summary Statistics** tool to fill the dialog box as follows:
18. Input Table:**HotSpot**
19. Output Table:**HotSpot_Statistics**

20. Under **Statistics Fields**
21. Well_Depth: **Mean**
22. GiZScore Fixed 1000 **Mean**
23. Case Field: **Gi_Bin Fixed 1000**
24. Click Run
25. Save the project

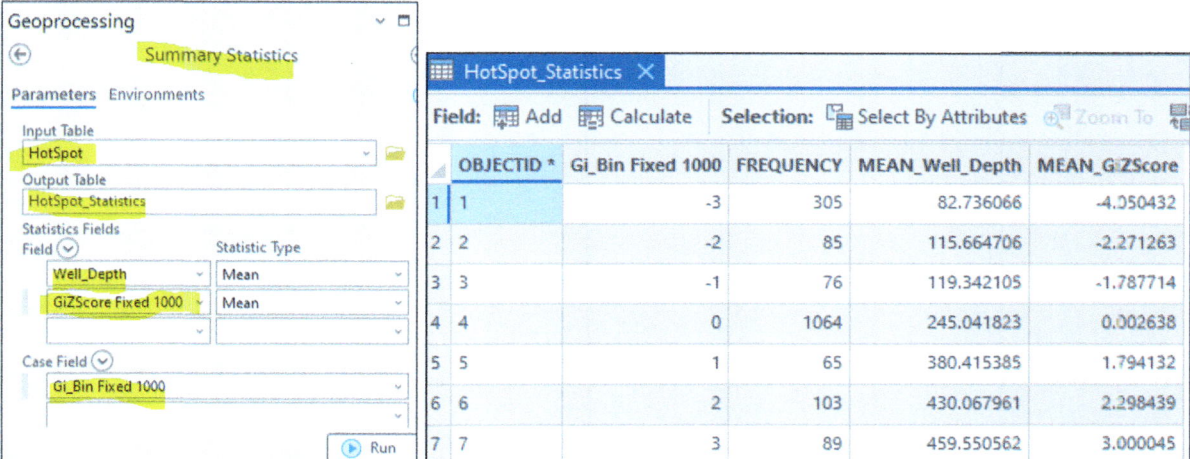

Table Interpretation A positive z score higher than zero for the wells indicates spatial clustering of deep wells, while a low negative z score indicates spatial clustering of shallow wells. The higher (or lower) the z score, the more intense the clustering. A z score near zero and with a 0 bin indicates no clustering.

Proximity and Network Analysis

Part I – Proximity Analysis

Proximity analysis is an important function in GIS because it covers a wide range of topics that help in answering many spatial questions, such as

1. How close is the observation well to a treatment plant?
2. Do any wells fall within 500 meters of a fault system?
3. What are the distances between the wells and the treatment plant?
4. What is the nearest or farthest well from the dam?
5. What is the shortest street network route from the water tower reservoir to the towns?

Proximity tools can be applied in vector and raster formats. The vector-based tools vary in the types of output they produce and can be explained briefly in this chapter.

Proximity Analysis in Vector Format

Buffer analysis is used for identifying areas surrounding any type of feature, whether it is point, line or polygon. The buffer polygon is created to a specified distance around an input feature. The output polygon features can be used as an input to overlay tools (union, intersect, erase, and spatial join). **Multi-Ring Buffer** creates a new feature class of buffer features using a set of buffer distances. Buffer function does not take into consideration any physical obstacle that might exist in the area of buffering.

Select by Location After generating the buffer, the user can use the **select by location** using different relationships between the buffer and the source feature under investigation. The selection by location does not draw a boundary but selects the features that are determined by the relationship between the source and target layers.

The near function selects one feature of a set and then calculates the distance to all other features in the same set. The Near tool adds a new field called "distance" in the attribute table of one of the input layers. The distance will be calculated based on the map unit of the coordinate system of the map document.

Generate Near Table calculates distances and other proximity information between features in one or more feature classes or layers. Unlike the Near tool, which modifies the input, Generate Near Table writes results to a new stand-alone table and supports finding more than one near feature.

The spider diagram will draw a line from each record to the one selected feature to identify the exact location. The Desire lines tool is part of the business analyst and shows which customers visit which stores. A line is drawn from each customer point to its associated store point, making it easy to see the actual area of influence of each store. This tool can be used in environment-related problems.

Supplementary Information The online version contains supplementary material available at https://doi.org/10.1007/978-3-031-42227-0_14.

Scenario 1 The region of **Dhuleil-Samra** in Jordan is considered an arid area, and groundwater, which is scarce and has low water quality, is the only source for domestic use. As a hydrogeologist working for the Water Authority, you have been asked to explore the possibilities of finding two wells with good water quality. One well in each region: Dhuleil and Samra. The two selected wells were used for water supply in the two regions. The two wells should have the following criteria:

1. The well in the Samra region should be 2.5 km away from the wastewater treatment plant (WWTP) and serve only the towns in the Samra region.
2. The well in the Dhuleil region should be 2.5 km away from the stream and serve only the towns in the Dhuleil region.
3. Both selected wells should have total dissolved solids (TDS) and nitrate (NO_3) less than 1000 mg/l and 20 mg/l, respectively.

GIS Approach to Solve Scenario 1

1. Launch ArcGIS Pro
2. Click **Open another project** (upper-right) browse to **Env_Water\Ch14** select **Ch14.aprx** and click OK

The **Ch14.aprx** open and the CP includes the Map, which is empty. The Map View displays both the World Topographic Map and World Hillshade.

3. In the **Catalog** pane, r-click Folder and click Add Folder Connection and browse to **Env_Water\Data_Ch14** (or **Database\ Data_Ch14**) open it and highlight **Data_Ch14** and click OK.
4. In the Content pane rename the Map to "**Proximity Analysis**"
5. In the Catalog pane, expand **Q1** under **Folder\Data** and select the **Region, Road, Stream, Town, Well,** and **WWTP** and drag them to the Map View.
6. Perform the symbology for the following layers as below:

Layer	Symbol	Color	Size	Line Width
Town	Square 1	Pink	6	
Stream	Water (line)			2
Road	Minor road			
WWTP	Water (area)			

7. Click the **Region** symbol, click the back arrow in the Symbology pane, under Primary symbology select "**Unique Values**", Field 1 = **Name**, click Add All Values, click the **More** drop-down arrow and uncheck all other values
8. In CP select the **Region** and click the **Labeling** tab, in the Label Class group make sure the Field = Name
9. In the Layer group, click the Label button
10. Save the project

Quiz Make the font of the label Time New Roman and size 14?

Buffer the WWTP in the SAMRA Region

11. Click **Analysis** tab on the ribbon, in the **Geoprocessing** group and click the **Tools** button
12. In the **Geoprocessing** pane, click the **Favorites** tab and click the **Pairwise Buffer** tool.
13. Fill the **Pairwise Buffer** dialog box as follows:
14. Input Features: **WWTP**
15. Output Feature Class: **WWTP_Buffer**
16. Distance: 2.5 Kilometers
17. Accept the other default
18. Click Run
19. In CP select **WWTP_Buffer** and click the **Feature Layer** tab in the **Effects** group, slide the layer Transparency to 50% (or you can type 50 in the Transparency window)
20. To see the result of the Transparency, in the CP, uncheck the **Region**, **Stream**, and **Road**

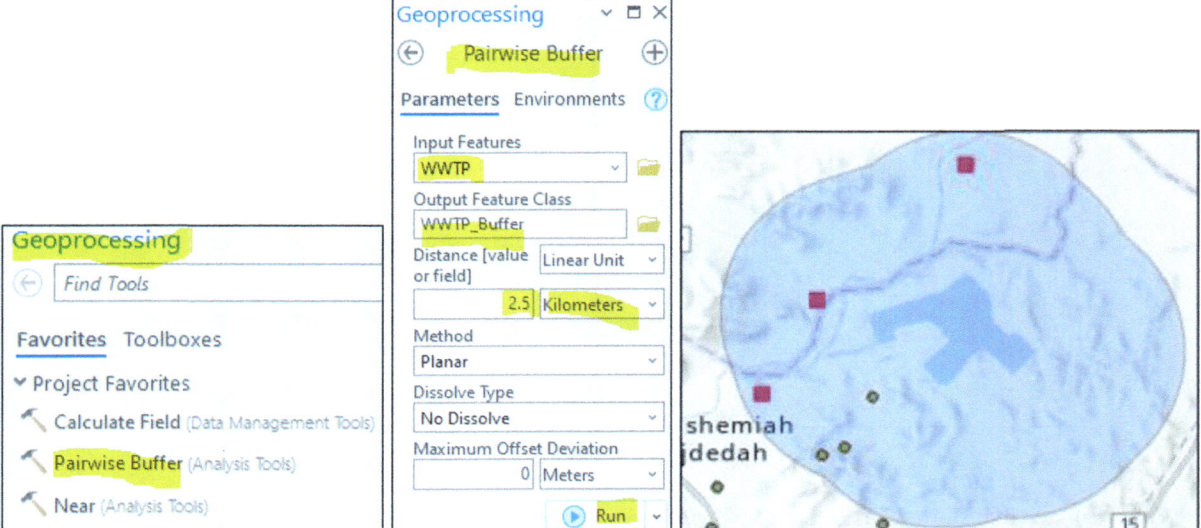

Select Wells Outside the WWTP_ Buffer in the Samra Region

In this step, we will select **wells** located outside the **WWTP_Buffer** zone in the **Samra** region that have **TDS** and **NO₃** concentrations less than **1000** and **20 mg/l,** respectively.

Select by Location

21. In the CP select **Well** layer and click the **Map** tab, in **Selection** group, and click **Select By Location** button
22. Fill the Select By Location as follows:
    (a) Input Features: Well
    (b) Relationship: Completely within
    (c) Selecting Features: WWTP_Buffer
    (d) Selection type: New selection
23. Check *Invert spatial Relationship*
24. Click **Apply** and then **OK** to close the Select By Location

Result All **wells** outside the **WWTP_Buffer** layer were selected.

18. Open the Attribute Table of the **Well**

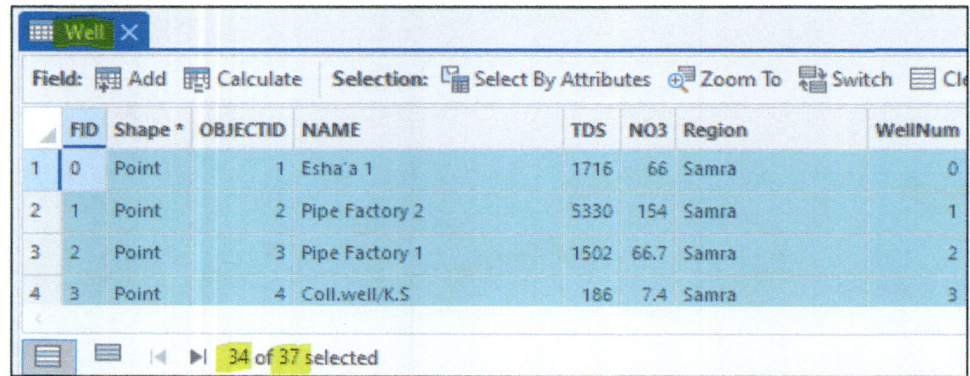

Result The attribute table shows that 34 out of 37 wells are selected.

The next step is to select only one well that is located outside the buffer zone in the **Samra region** and has **TDS** and **NO3** contents less than **1000 and 20 mg/l,** respectively.

Select by Attributes

18. In the CP select **Well** layer and click the **Map** tab, in **Selection** group, and click **Select By Attributes** button
19. Fill the Select By Attribute dialog box as follows:
20. Input Rows: Well
21. Selection type: **Select subset from the current selection**
22. Where *Region is not equal Dhuleil*
23. Click + Add Clause
24. **The** *TDS is less than* type *1000.*
25. Click + Add Clause
26. **In addition,** *NO3 is less than* type *20*

Add New Field to the Well Layer

Note Before clicking Apply, if you click the **SQL** icon you will get the following argument.

Region <> 'Dhuleil' And TDS < 1000 And NO3 < 20

27. Apply and then OK to exit the Select By Attributes dialog box
28. Save your project

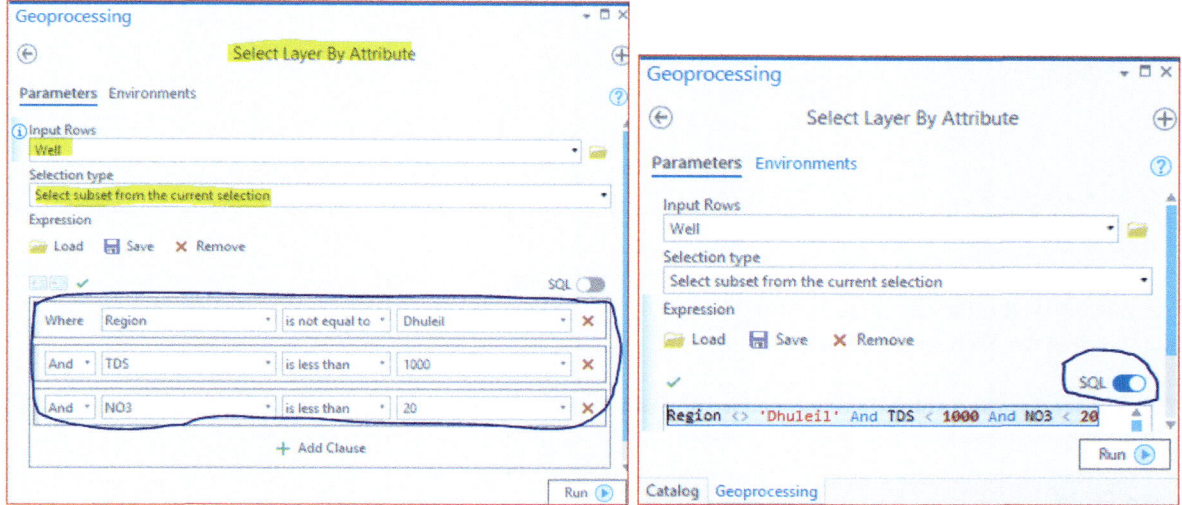

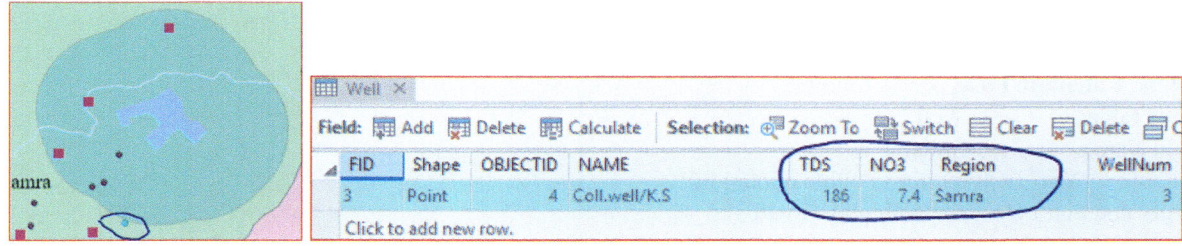

Result Only the **Coll.Well/K.S** well is selected in the **Samra** region outside the buffer zone. The TDS and NO3 in the attribute table of the selected well are **186** mg/l and **7.4** mg/l, respectively.

To proceed with the analysis, you must add a new field in the attribute table of the **Well**.

Add New Field to the Well Layer

29. In the CP r-click **Well** point to **Data Design** and click **Fields**
30. The Fields: **Well** table open
31. At the bottom of the table "*Click here to add a new field*"
32. Under **Field Name** type "**Suitable**" and click Tab in the keyboard and type again "**Suitable**" under Alias
33. Under **Data Type** select "**Text**", and under **Length** type 12
34. In the **Field** tab on the ribbon in the **Changes** group click **Save** button

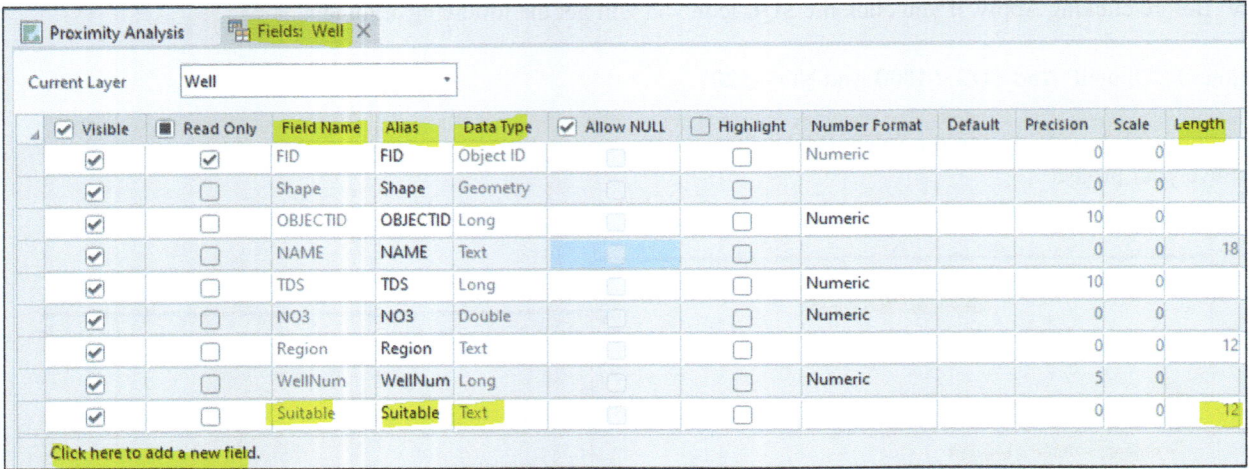

35. Close **Fields: Well table** tab above the Map View
36. Make sure that the **Well** Attribute Table opens

Note You will see the "**Suitable**" field is added to the **Well Attribute Table** and one well is selected from the previous step.

Calculate Field

Now you are going to use the "**Calculate Field**" tool to populate the "**Suitable**" field

37. R-click "**Suitable**" field and click **Calculate Field**
38. Fill the **Calculate Field** pan as follows:
39. Input Table: Well
40. Field Name: Suitable
41. Suitable = "Yes" (text requires double quotation)
42. **Apply** and don't close the **Calculate Field**

Result The word **Yes** is added under Suitable to the well "**Col.well/K.S**"

43. In the **Well** attribute table click Switch button
44. In the **Calculate Field** pane
45. Repeat the previous step but replace the "Yes" with "NO" (see below)
46. Apply and click Ok to close the **Calculate Field** pane
47. Close the Well attribute table
48. Save your project

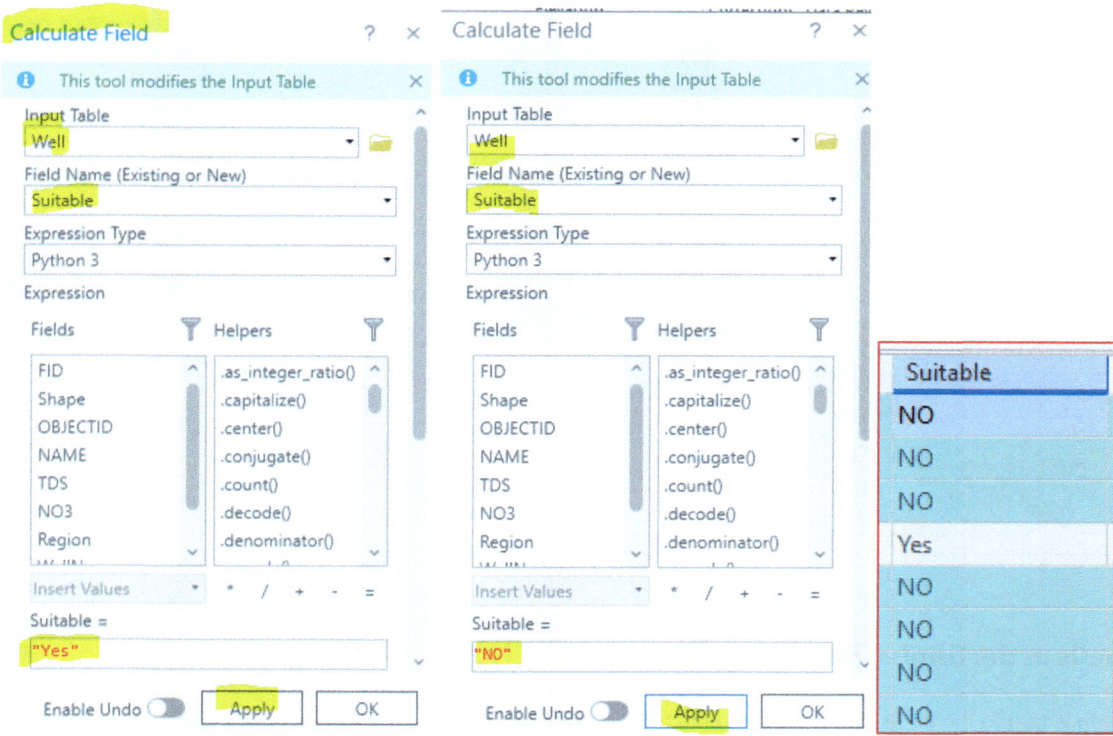

Result The attribute table of the well that has a TDS and NO3 less than 1000 and 20 mg/l populated with "**YES**" and the rest with "**NO**" in the "**Suitable**" field.

Buffer the Stream in the Region

49. In the CP, check the **Stream** and select it
50. If you close the Geoprocessing from the previous step
51. Click **Analysis** tab, **Geoprocessing** group and click the **Tools** button
52. In the **Favorites** tab click the **Pairwise Buffer** tool
53. The Geoprocessing pane fill the **Pairwise Buffer** as follows:
54. Input Features: **Stream**
55. Output Feature Class: **Stream_Buffer**
56. Distance: 2.5 Kilometers
57. Accept the other default
58. Run
59. In CP select **Stream_Buffer** and click the **Feature Layer** tab in the **Effects** group, slide the layer Transparency to 50%

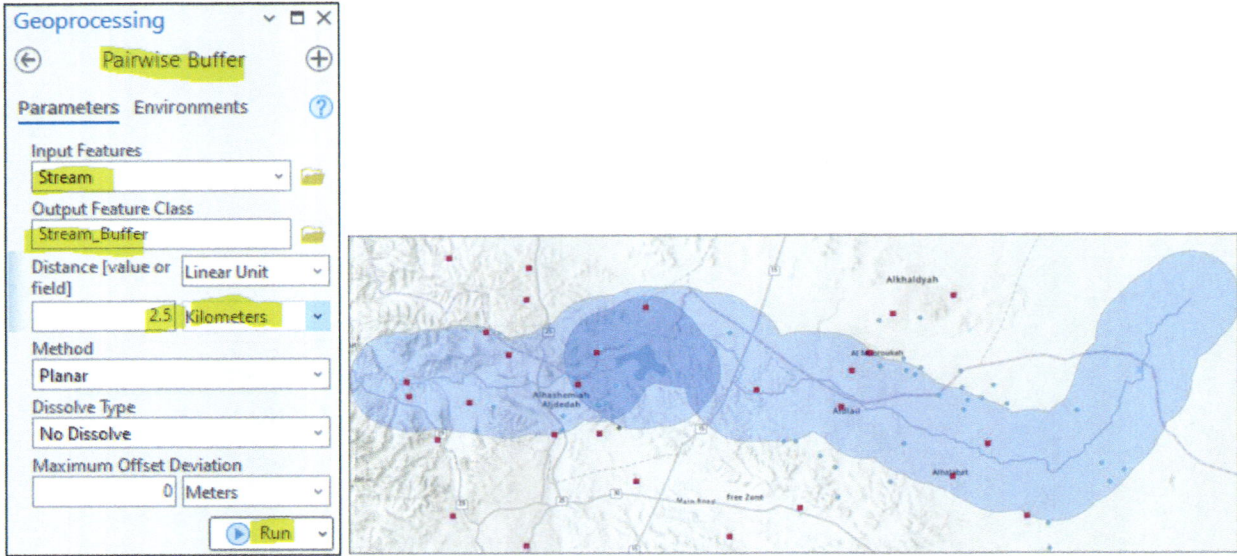

Select Wells in the Dhuleil Region with Low TDS and NO3

The next step is to find the **wells** that are located outside the **Stream Buffer** zone in the **Dhuleil area** that have **TDS** and **NO3** less than **1000** and **20 mg/l,** respectively, using the **Select By Location** and then using the **Select By Attributes**.

60. In the CP select **Well** layer and click **Map** tab and in the **Selection** group, click **Select By Location** button
61. Fill the Select By Location dialog box as follows:
    (a) Input Features: Well
    (b) Relationship: Completely within
    (c) Selecting Features: Stream_Buffer
    (d) Selection type: New selection
62. Check *Invert spatial relationship*
63. Click **Apply** and then click **OK** to close **Select By Location** dialog box

Result Seventeen wells from **both regions** were selected **outside** the **stream buffer**.

Now find **wells** in the **Dhuleil are**a that have **TDS** and **NO3** less than **1000** and **20 mg/l,** respectively.

64. Make sure the Well Attribute Table is open
65. In the CP, select **Well** layer and click **Map** tab and in the **Selection** group, click **Select By Attributes** button
66. Fill the Select By Attribute dialog box as follows:
67. Input Rows: Well
68. Selection type: Select subset from the current selection
69. Click the SQL icon and type the following statement:
    Region = 'Dhuleil' And TDS < 1000 And NO3 < 20.
70. Click to verify SQL
71. The SQL expression is valid display
72. Click Apply and then click OK to close the Select By Attributes dialog box

Select Wells in the Dhuleil Region with Low TDS and NO3

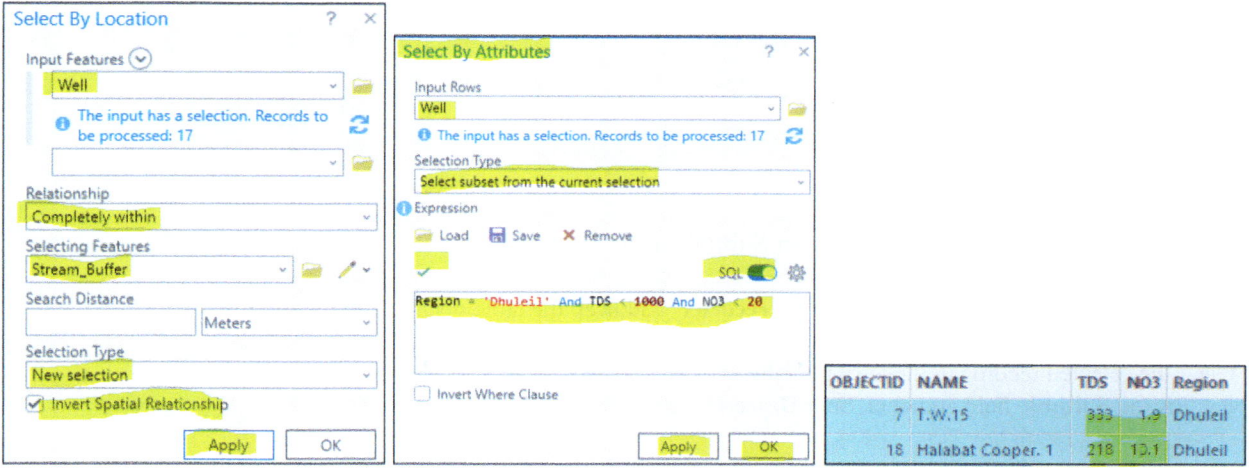

Result Two wells have been selected outside the buffer, therefore, you are going to choose the **Halabat Cooper 1** well because it has a lower TDS.

73. In the **Well** attribute table, click the first cell of the record that has a TDS = 333, and the color of the record will turn to yellow.

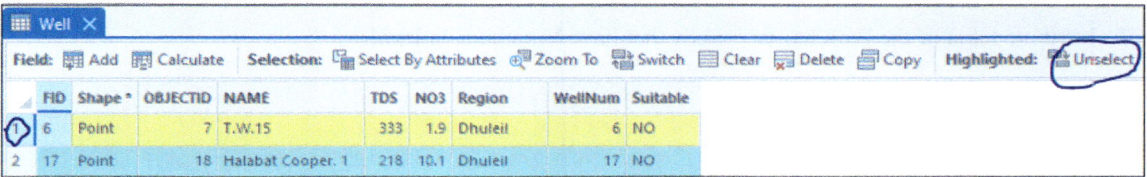

74. In the Attribute Table click the **Unselect** button

Result The **T.W.15** well that has TDS 333 will be deselected, and the **Halabat Cooper 1** well remains selected.

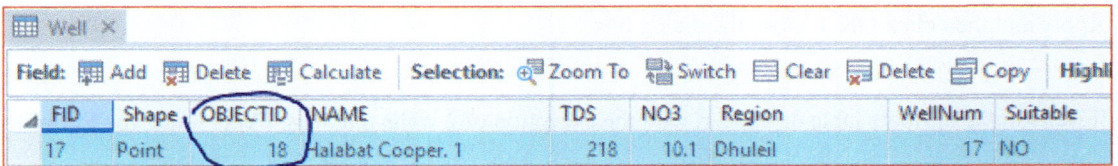

75. In the **Well** attribute table, r-click **Suitable** field and click **Calculate Field** and type "**YES**" under Suitable and click Apply and then OK to close the dialog

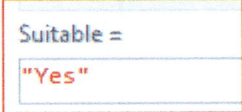

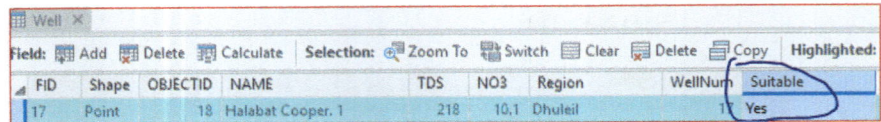

Result The "**Yes**" will replace the "**NO**" under **Suitable** field

76. Clear the selected and then click "**Show all records**" and keep the table open
77. At the bottom at the **Well** attribute table, click "**Show All Records**"
78. Clear the selected records in the attribute table
79. R-click the **Suitable** field and click **Sort Descending**

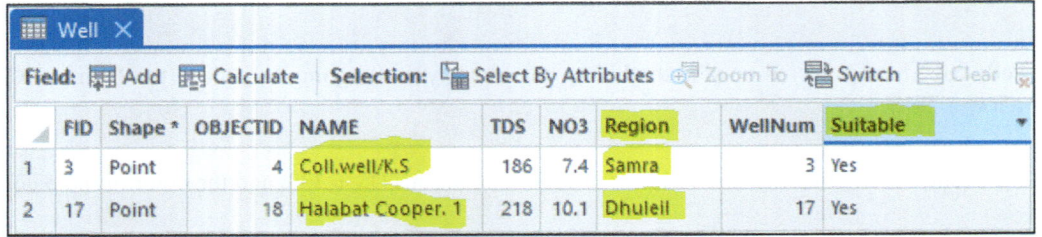

Result The attribute table of the Well has two wells that have **Yes** under **Suitable**.

Hide Wells Using Definition Query

To proceed with the analysis, we must hide all the wells and keep only the two selected wells in the Samra and Dhuleil regions that have low TDS and NO3.

80. In the CP, r-click **Well** layer and select **Properties**, click the **Definition Query**
81. Click "**+ New definition query**"
82. Type the following SQL statement
    Where Suitable is equal to Yes
83. Click Apply and then click OK to close the dialog box

Result All wells in the attribute table disappear, with the exception of **2 wells** that have "**Yes**" under the **Suitable** field.

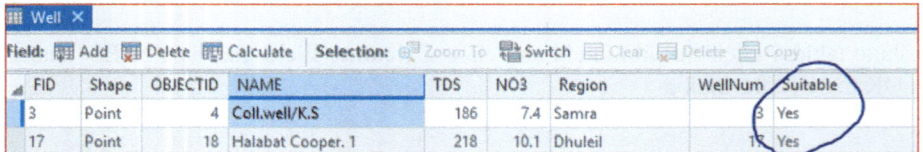

84. In the CP, click the symbol of the **Well** layer, use Circle 1, size 10, ultra-blue color and click Apply
85. Close the Symbology pane and save the project

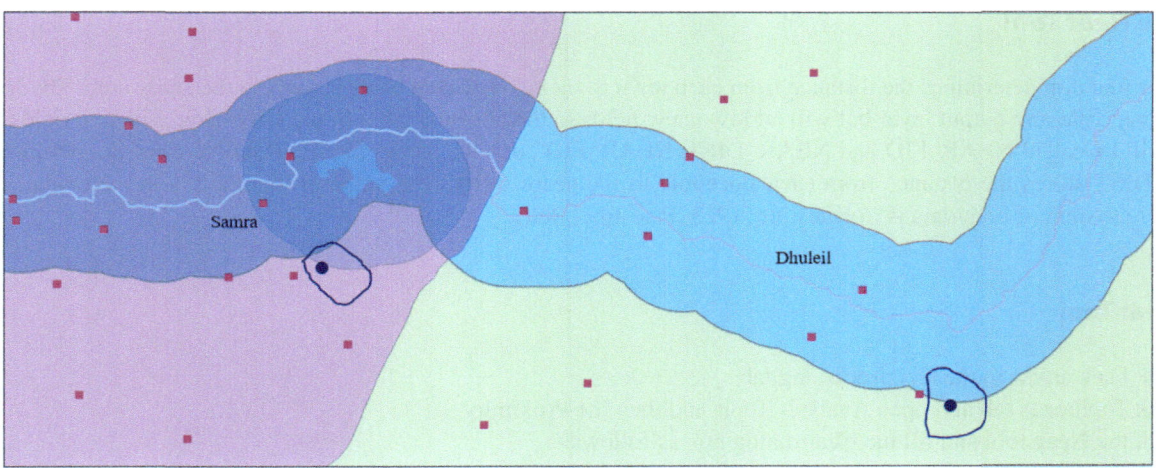

Convert Well Shapfile into Feature Classes in Geodatabase

86. In the CP, r-click **Well** click **Data** and select **Export Features**
87. Fill the Export Features dialog box as follows:
88. Input Features   Well
89. Output Feature Class: Borehole
90. Click OK

Repeat the previous step, import the **Town** and call the output feature class "**City**".

Comment Importing each feature class individually in order to provide different name for the output feature class from the input feature class.

Result The **Borehole** and **City** are added into the **CP** and **Ch14.gdb**. The symbol of booth feature classes is preserved.

91. In the CP, remove the **Well** and **Town**

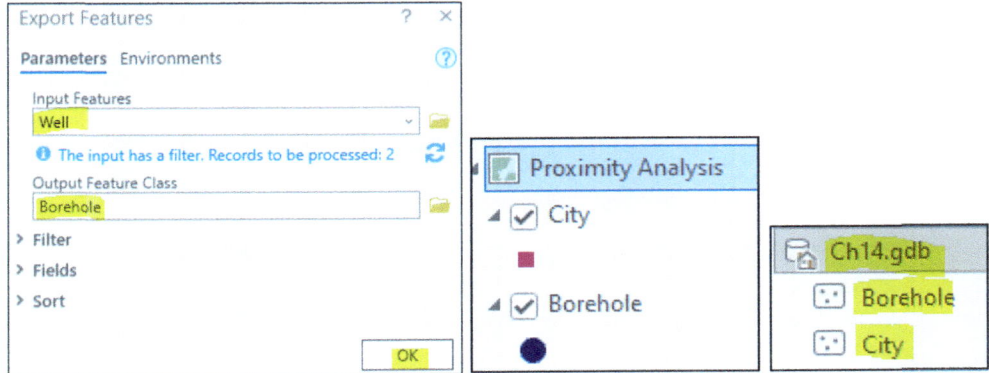

Use the Near Tool

The Near function determines the distance from each town to the two selected nearest wells in the study area. The Near tool will not generate any output layer but will add two new fields to the attribute table of the **City** layer. The two fields that are added will be called NEAR_FID and NEAR_DIST. NEAR_FID contains the feature ID (WellNum) of the borehole layer. NEAR_DIST stores the distance from each **borehole** to the nearest **city**. The value of this field is in meters because the coordinate system of both data is in Palestine_1923_Palestine_Belt.

Run Near Tool

92. Click back arrow in the Geoprocessing tab
93. Click Toolboxes tab and open Analysis Tools and then the Proximity
94. Click the **Near** tool and fill the Near dialog box as follows:
    (a) Input Features: City
    (b) Near Features: Borehole
    (c) Accept the rest as a default
95. Run
96. Save your Project

Result If you open the attribute table of the **City**, you will see two new fields are added **NEAR_FID** and **NEAR_DIST**.

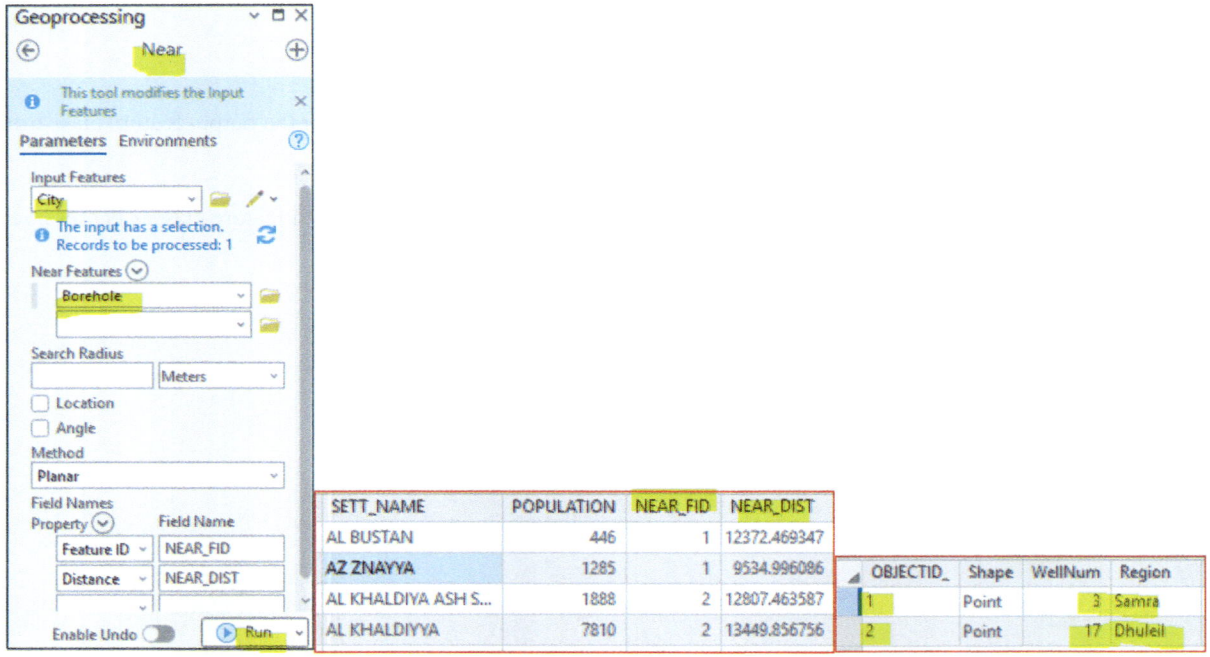

Result Explanation First note: In the attribute table of the **City** two fields are added: **Near_FID** and **Near_Dist**. **Near_FID** has two variables (1 and 2). **Near_FID # 1** corresponds to **WellNum** 3, and number 1 corresponds to **OBJECTID_1** in the **Borehole** layer located in the **Samra** area. **Near_FID # 2** corresponds to **WellNum** 17 under and number 2 under **OBJECTID_1** in the **Borehole** layer located in the **Dhulei**l area. The **NEAR_DIST** field shows the distance between each **borehole** and the cities in both **Samra** and **Dhuleil** regions.

Second note: The **Near** distance will associate each city with the borehole that has a shorter distance. For example, **Hay Arnous city** is in the **Dhuleil area** and is associated with well No. 3, which is located in the **Samra region**. The distance between **Arnous city** and **borehole No. 3** is shorter than the distance between **Arnous city** and borehole No. 17.

Quiz: Label the **Borehole** based on the **WellNum** field and the **City** layer based on the **SETT_Name** and measure the distance in meters between **well No 3** and well **No 17** and **Al Bustan city**.

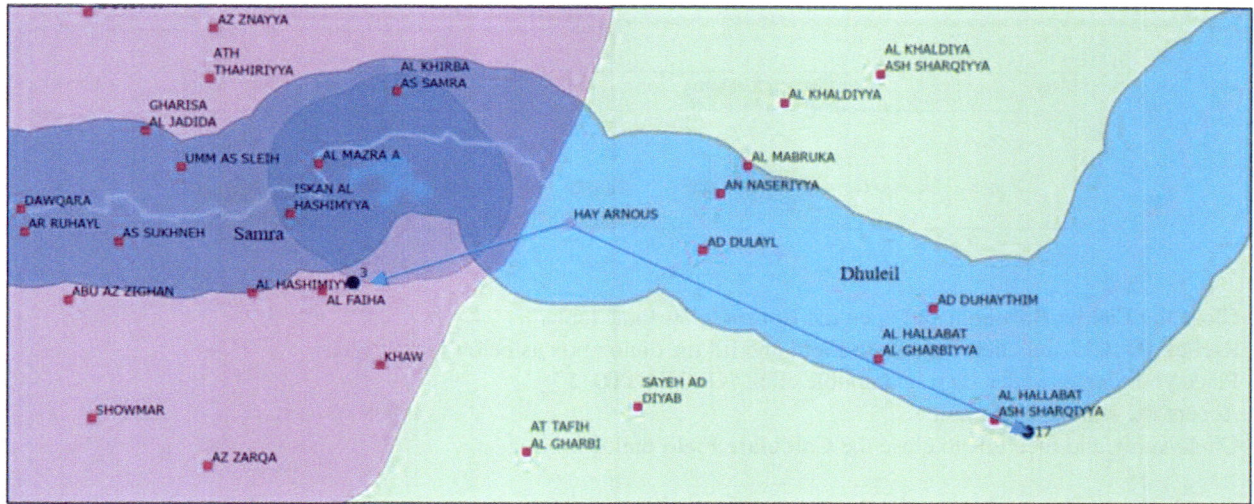

Save Region, City, and Borehole as a Layer

A layer can exist outside of your map or project as a **layer file** (.lyrx). This allows other users to access the layer you have built. You can share any layers over the network or by email.

When you add a layer file to a map, it draws exactly as it was saved, provided the data referenced by the layer are accessible. The layer file stores many properties of the input layer, such as symbology, labeling, and custom pop-ups.

Note Layer files saved from ArcGIS Pro cannot be used in ArcMap

1. In the CP, select the **Region** layer
2. Click the **Share** tab on the ribbon, in the **Save As** group, click **Layer File** (or right-click **Region**, point to **Sharing** and click **Save As Layer File**)
3. On the Save Layer File dialog box, click **New Item** and under Data_Ch14 in the **Folder** create a new folder call it **Layers**
4. D-click the **Layers** folder and accept the default name "**Region.lyrx**" and click **Save**
5. Repeat the previous step and create a City layer (**City.lyrx**) and Borehole layer (**Borehole.lyrx**).

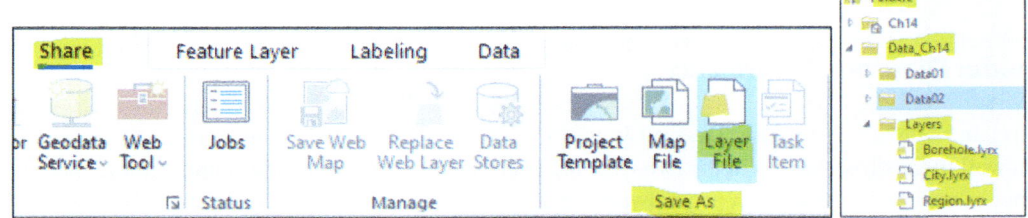

Create a New Field in the Borehole

You are going to create a new field called "**ID**" in the **Borehole** layer.

6. In the attribute table of the **Borehole**, add **Field** (like in the step above)
7. Call the field **ID**, Alias ID, Data Type "**Short**" and **Numeric** and **Save**

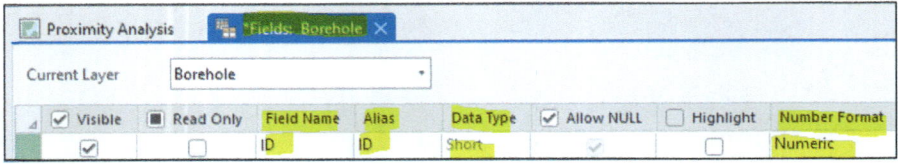

8. Close the **Fields: Borehole** and open the **Borehole** attribute table
9. R-click **ID** field and click **Calculate Field** and fill the dialog box as below
10. Place your cursor under **ID =** and double click "**OBJECTID_1**"
11. Accept the rest of the default
12. Click Apply and then OK to close the **Calculate Field** dialog box

Result The ID field in the borehole will be populated with two values, 1 and 2. ID 1 corresponds to well # 3 in the Samra Region, and ID 2 corresponds to well # 17 in the Dhuleil Region.

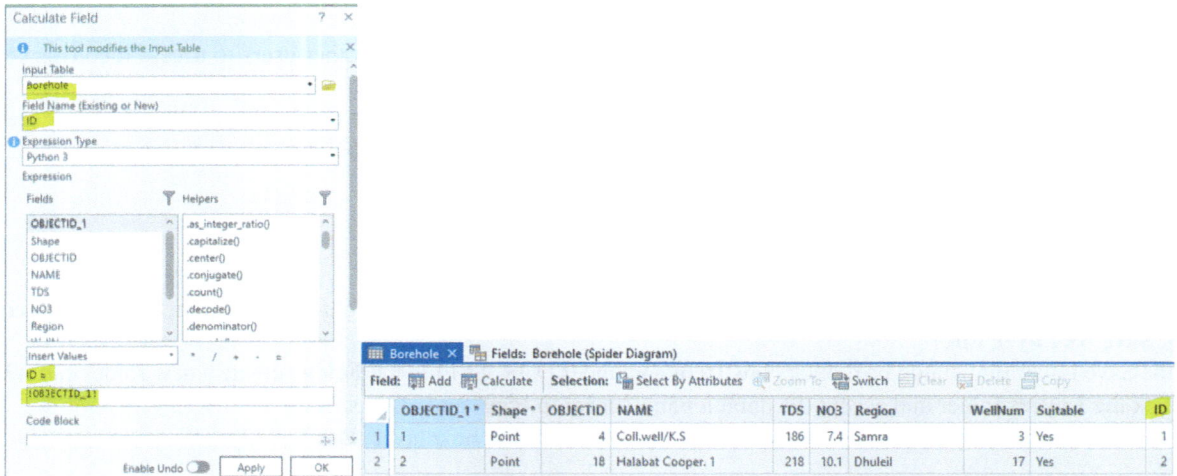

Generate Spider Diagram (Desire Lines)

Spider diagrams can be created using the "**Desire Lines**" tool, which is part of the Business Analyst. **Desire Lines** will draw lines from each well to the closest cities in the two regions. The tool can be used in environmentally related problems to see, for example, the influence of a landfill on groundwater observation wells. Therefore, a line is drawn from the landfill to its nearest wells, making it easy to see the actual area of influence of the landfill.

To run the **Spider Diagram**, you must design that both the **Borehole** and the **City** attribute tables have two fields related to each other. The **City** attribute table has a field "**NEAR_FID**", and the **Borehole** layer has a field "**ID**" and both fields are related.

Generate Spider Diagram (Desire Lines)

Note To run this function make sure you have the **Business Analyst license** and you have logged on to your account in **ArcGIS Online**.

13. Click **Insert** tab on the ribbon, in the **Project** group, click **New Map** button
14. In the CP rename the **Map** to "**Spider Diagram**"
15. In the **Catalog** pane, select the **Borehole.lyrx**, **City.lyrx** and **Region.lyrx** layers from **Layers** in the **Data_Ch14** folder and integrate them into the Map View.
16. In the Geoprocessing pane, click the back arrow and in the Toolboxes open the **Business Analyst Tools** and then open the **Analysis** and click the **Generate Desire Lines** tool
17. Fill the **Generate Desire Lines** dialog box as follows:
    - (a) Store Layer: City
    - (b) Customer Layer: Borehole
    - (c) Output Feature Class: SpiderDiagram
    - (d) Store ID Field: Near_FID
    - (e) Associate Store ID Field: ID
    - (f) Distance Type: Straight Line
    - (g) Measure Units: Kilometers
    - (h) Check Create Report
    - (i) Open the Report Options
    - (j) Report Title: Water Distribution
    - (k) Report Format: PDF
18. Open the Report Options
19. Under Report Title type "**Water Distribution**"
20. Under Output Report Folder type "**DesireLines**"
21. Report Format choose PDF
22. Click Run

Result A line will be drawn from each well in each region to the closest town. Five cities in the Dhuleil area will be connected to well # 3 in the Samra area. This is because the distance between well # 3 and these cities is shorter than the distance between them and well # 17 in the Dhuleil area.

The Output Report Folder "**DesireLines**" is created and saved in **Ch14** folder. It contains the **WindRose.pdf** file, which facilitates a better understanding of the city distribution. It summarizes the information in the output attribute table and highlights some key facts about the distribution of the city location points.

23. Save your Project

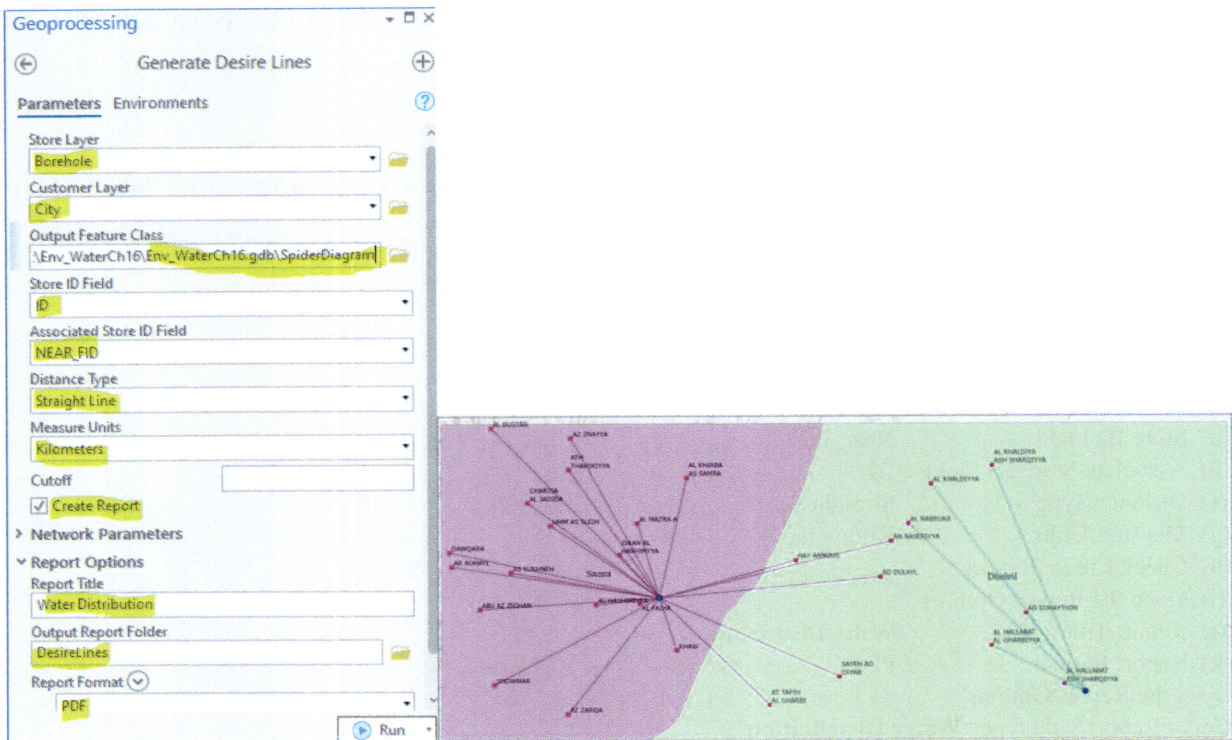

Scenario 2 Your boss asked you to choose the "**Hay Arnous**" city and build a big **water supply tower** (**WST**) in it. The **WST** will receive water from pumping the two selected wells in the Dhuleil and Samra regions. The water in the **WST** will be distributed into all cities in the two regions by gravity. To carry the work, you need to know the number of cities that will be served by water and their population. Therefore, you should do the following:

(a) Buffer **Hay Arnous** city into 3 rings with radii of 4, 11, and 18 km.
(b) Find the number of cities in each ring and their total population

Multi-Ring Buffer Around Hay Arnous City

Hay Arnous city will be buffered using 4, 11, and 18 km radii.

24. In the CP, uncheck the **Spider Diagram** layer
25. In CP select **City**, click **Map** tab, in **Selection** group, click **Select By Attributes** button
26. where **SETT_NAME is equal to HAY ARNOUS.**
27. Click Apply and then OK to exit the **Select By Attributes** dialog box

Result The 'HAY ARNOUS' town is selected.

Multi-Ring Buffer Around Hay Arnous City

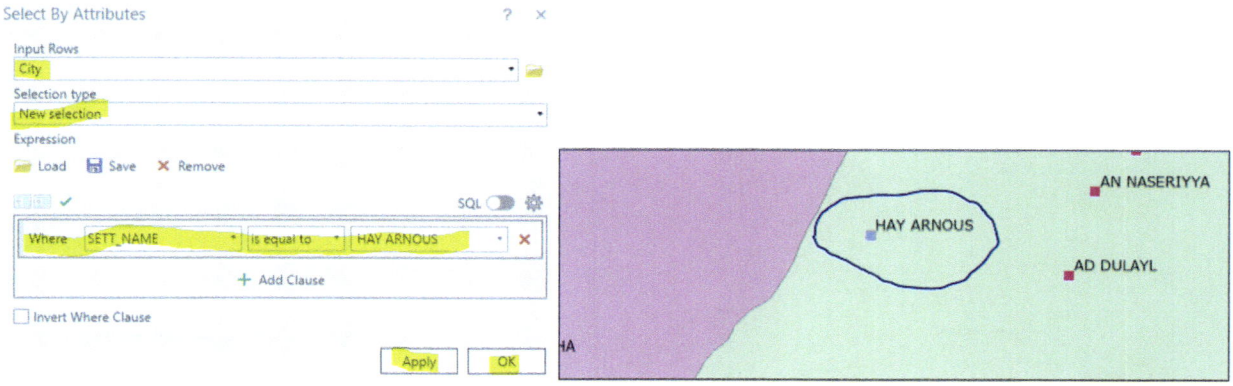

28. Click the back arrow of the **Geoprocessing** pan, in the **Toolboxes** tab, open the **Analysis Tools**, then the **Proximity** and click **Multiple Ring Buffer**
29. Fill the **Multiple Ring Buffer** dialog box as follows:
30. Input Features: **City**
31. Output Feature Class: **Arnous_Buffer**
32. Distance: type 4 click the + Add another, Type 11 click the + Add another, Type 18,
33. Distance Unit: **Kilometer**
34. Dissolve Option: Non-overlapping (rings)
35. Method: Planar
36. Accept the rest of the default
37. Click Run

Result A 3-ring buffer is created around the town of 'HAY ARNOUS'.

38. In CP/r-click **Arnous_Buffer** select Symbology, choose Unique Values and Field: Distance, Open more and uncheck **Show all other values**
39. Close the Symbology pane
40. Deselect the **Hay Arnous** in the City attribute table
41. Save the project

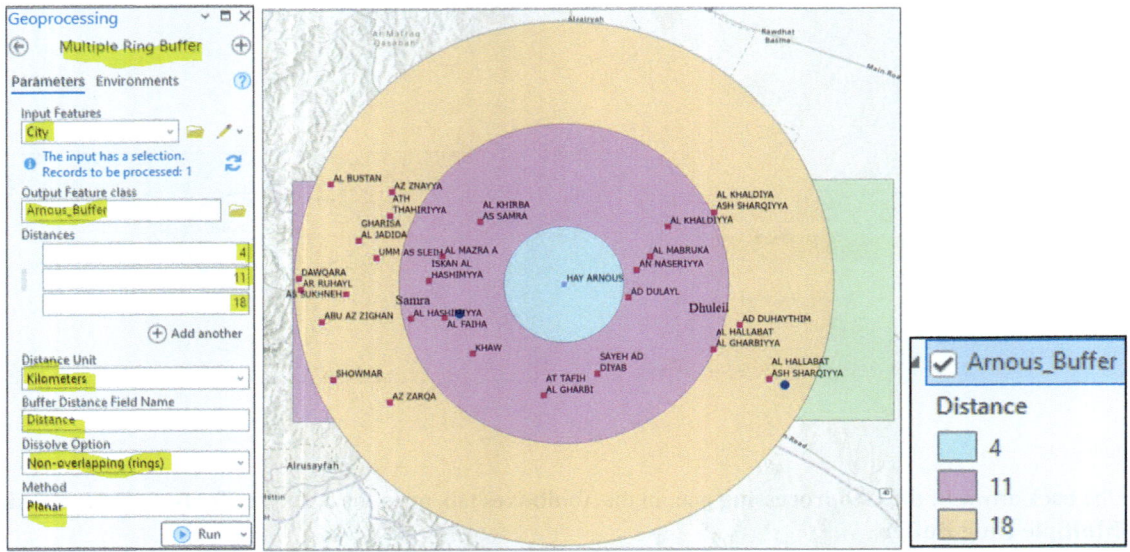

Select Cities Inside Each Buffer Zone

42. Open the attribute table of the **Arnous_Buffer** and select the third record (18 km distance)

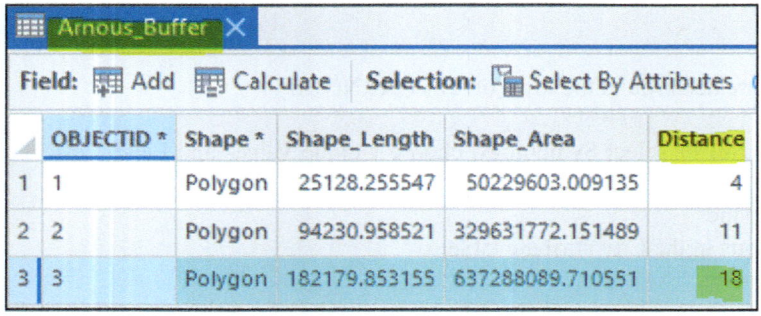

43. In Cp select City, Map tab in the Selection group, **Select By Location**
44. Fill the **Select By Location** dialog box as follows:
    (a) Input Features: City
    (b) Relationship: Completely within
    (c) Selecting Features: Arnous_Buffer
    (d) Selection type: New selection
    (e) Click Apply and then OK to close the dialog box
45. Open the attribute table of the **City**

Result There are **14 cities** selected out of 28.

46. R-click the field Population and click Statistics

Result The total population of the 14 cities is 365,341.

Proximity Analysis in Raster and Vector Formats   305

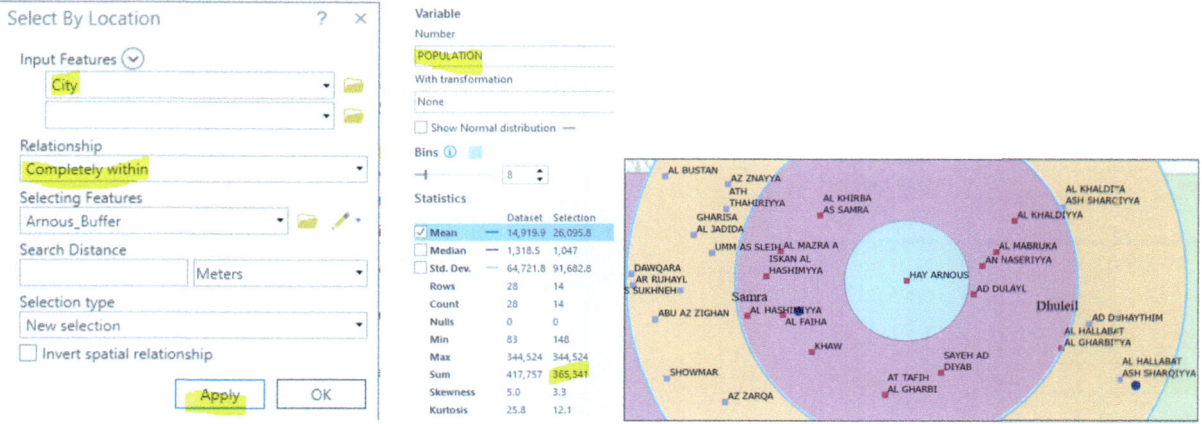

47. In the table below, type **14** under **Total Cities** and type **365,341** under the **Total Population**
48. Repeat the previous steps and write down the total cities and population of the 4- and 11-kilometer buffer zones.
49. Save your project

Ring Radius (km)	Total Cities	Total Populations
4		
11		
18	14	365,341

Scenario 3 You are working in Dhuleil region as a hydrogeologist and you have been given an assignment by your supervisor to verify the argument that the dam in the north–east of the region is playing a role as an artificial recharge. Therefore, the water stored behind the dam in the rainy season will infiltrate into the subsurface aquifer and improve its water quality. The water quality will be checked through two parameters, TDS and NO3, in the wells that are located within a 10 km radius from the dam. Low concentrations of TDS and NO3 in the wells close to the dam mean that the argument is true; otherwise, it is false.

Proximity Analysis in Raster and Vector Formats

In this section, two approaches will be applied in the raster and vector formats. In the **Raster** format, the **Euclidean Distance** approach will be used. In the **vector-based** application, the **Point Distance** will be used. Both approaches aim to evaluate the effect of the dam on the quality of groundwater.

Euclidean Distance

The raster-based Euclidean distance tool measures distances from the center of source cells to the center of destination cells. The Spatial Analyst extension can perform analysis where the output layer is in raster format. One of the analyses that can be applied in Earth sciences is the distance surface. This method creates a continuous layer from a vector input layer. The vector layer can be a point (groundwater well), line (stream), or polygon (treatment plant).

1. Click **Insert** tab on the ribbon, in the **Project** group, click **New Map** button
2. In the CP rename the **Map** to "**Recharge**"
3. In the **Catalog** pane, select the **Dam**, **Region**, **Stream**, and **Well** from \\Data\Q2 folder and integrate them into Map View
4. In the CP, r-click the **Region** layer and Zoom to layer
5. Symbolize the layers in Content pane as in the table below.

6. Close the Symbology pane after finishing

Layer	Symbol	Color	Size	Line Width
Well	Circle 1	Blue	6	
Dam	Dam	Black	14	
Stream	Water (line)	Apatite blue		2
Region	Black outline			1 pt

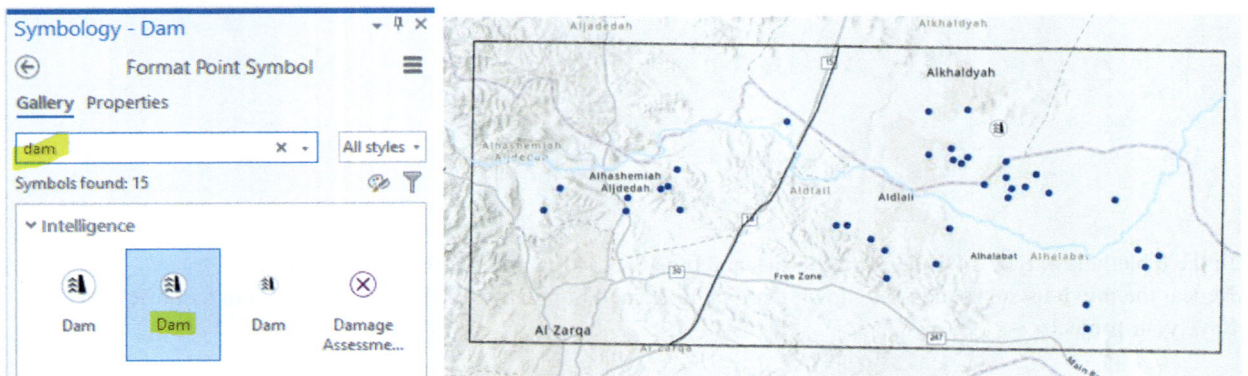

Euclidian Distance

You are going to use a 10 km distance and a cell size of the output raster to be 100 meters

7. Click the back arrow of the Geoprocessing pane and click the **Toolbox** tab
8. Open the **Spatial Analysis Tools**, then the **Distance** and then the **Legacy**
9. Click the **Euclidian Distance** and fill it as follows:
10. Input raster or feature source data: Dam
11. Output distance raster: Dam_Distance
12. Maximum distance: 10,000
13. Output cell size: 100
14. Accept the rest of the default
15. Click Environmental tab
16. Output Coordinate System: Region
17. Extent: Region
18. Click the Parameter tab
19. Click Run

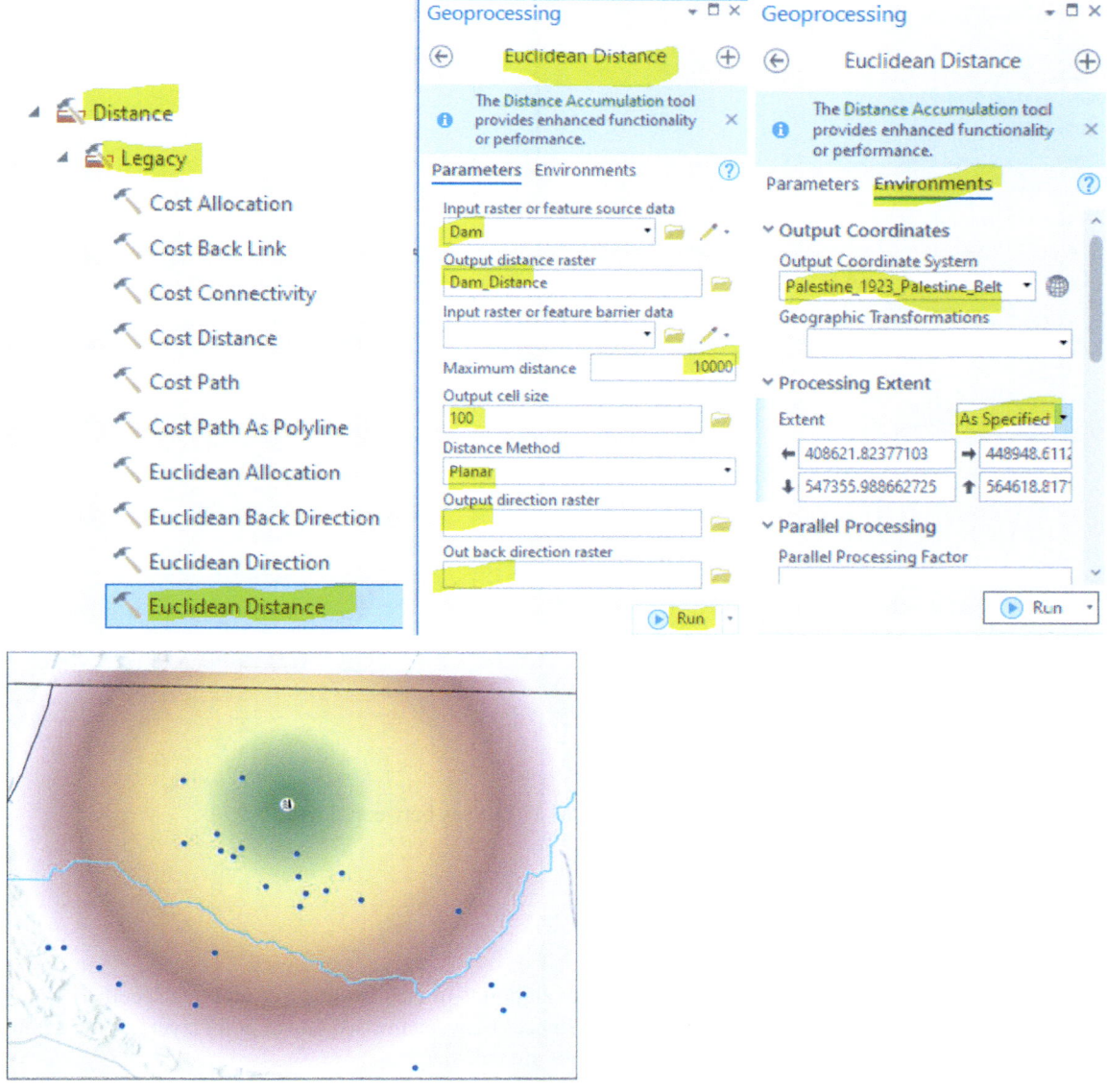

Classify the Dam Distance Raster

20. In the CP, r-click **Dam_Distance** and select **Symbology**
21. In the Symbology pane under Primary symbology select Classify choose 10 classes and the Manual Interval for the Method, and the Color scheme choose Prediction
22. In Classes tab change the numbers under Upper value into 1000, 2000, 3000, 4000, 5000, 6000, 7000, 8000, 9000, and 10,000
23. Under label change, the value in the first class from 0.001 to 0

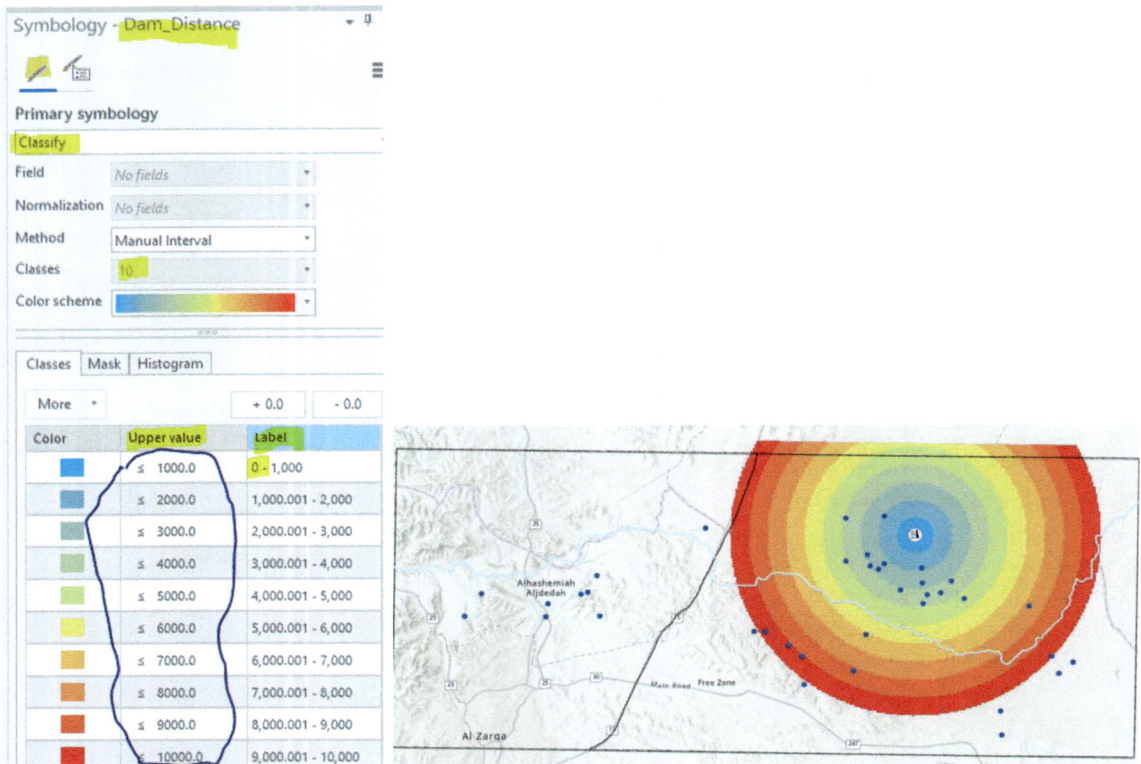

Result The output **Dam_Distance** raster is displayed and has 10 classes between 0 and 10,000.

Generate Near Table Tool

The tool determines the distances from input point features to all points in the near features within a specified search radius. The tool is similar to the **Near** tool but creates a table with distances between the two sets of point layers. If the default search radius is used, distances from all input points to all near points are calculated.

50. Click the back arrow in the Geoprocessing pane and in the Toolboxes tab
51. Open **Analysis** Tools, then the **Proximity** and click **Generate Near Table** tool
52. In the **Generate Near Table** dialog box under **Parameters** tab fill it as below
    - (d) Input Features: Dam
    - (e) Near Features: Well
    - (f) Output Table: Dam_Well (saved in geodatabase)
    - (g) Search Radius: 10,000 Meters
    - (h) Uncheck the find only closest feature
    - (i) Accept the rest of the default
53. Click **F** tab, in the Extent select Region
54. Run

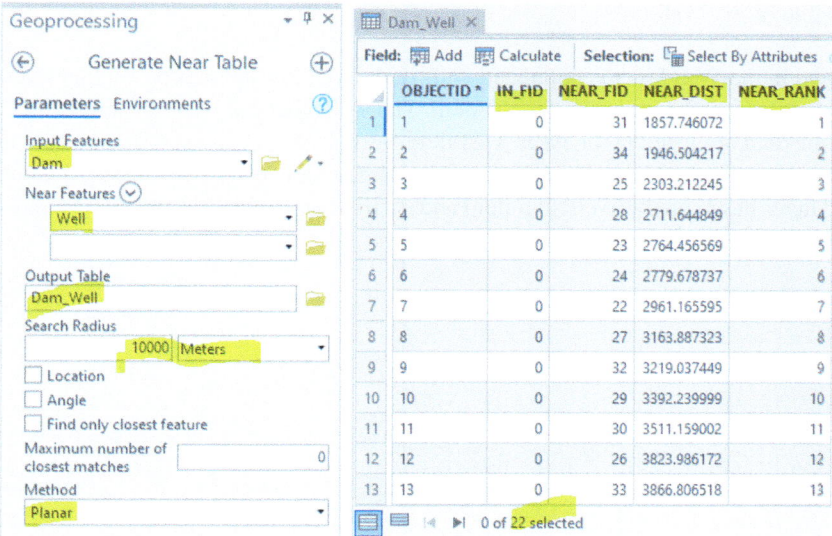

Result The File Geodatabase Table "**Dam_Well**" is added into the CP, and it has only 22 wells. The original number of wells in the region is 37. Only 22 wells are displayed because these wells are within 10,000 meters of the dam.

Note If you open the attribute table of **Dam_Well**, you will three identical fields that relate the dam to the wells located in the Region layer

- IN_FID: The Dam feature ID, which has a value of 0
- NEAR_FID: The feature ID of all Wells located in the Region layer
- NEAR_DIST: The distance between the wells and the dam
- NEAR_RANK: The wells are ordered based on their distance from the dam; the closest well has 1 rank, and the far distance well has a rank of 22.

Join Two Tables Based on Common Field

The next step is to join **Dam_Well** with the **Well** layer based on a common field. The **WellNum** field in the **Well** layer is identical to the **NEAR_FID** field in **Dam_Well**.

55. In the CP, r-click **Well** and select **Join and Relates** and click **Add Join**
56. Fill the **Add Join** dialog box as follows:
57. Input Table: **Well**
58. Input Join Field: **WellNum**
59. Join Table: **Dam_Well**
60. Join Table Field: **NEAR_FID**
61. Check   **Keep All Target Features**
62. Check   **Index Join Fields**
63. Click Validate Join, then click Close
64. Click OK

Result **Dam_Well** is joined to the **Well** layer.

65. Open the attribute table of the **Well** layer

The **NEAR_DIST** field in the **Well** layer is populated with zero <NULL> and numbers larger than zero. The zeros are the wells that are located more than 10,000 meters from the dam. The wells that are located within 10,000 meters from the dam and their distance from the dam are listed in the Near_Dist field. Well # 31 is the closest to the dam, and its distance from the dam is 1857.75 meters, while well # 19 is the farthest from the dam, and its distance is 9980.18 meters.

Quiz: Label the Well layer based on the WellNum field and verify that well # 31 is the closest and well # 19 is the farthest from the dam as its distance is 9980.18 meters.

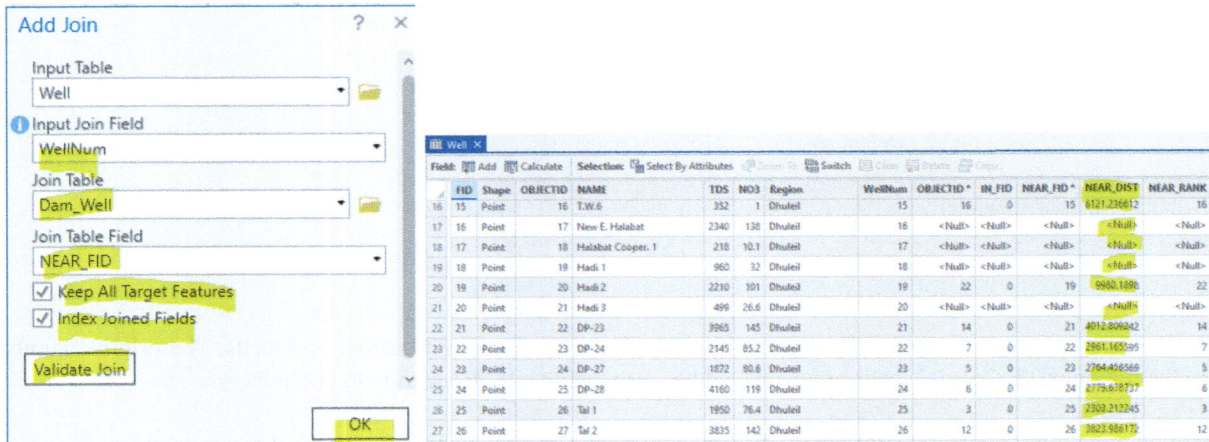

Definition Query

This step is necessary to hide wells that are located more than 10,000 meters from the dam.

66. In the CP, r-click Well layer and point to Properties and click Definition Query
67. Click **+ New definition query**
68. Type Where NEAR_DIST is greater than 0
69. Click Apply and then click OK to close the dialog box
70. Save the Project

Result All the wells located farther than 10,000 meters will disappear.

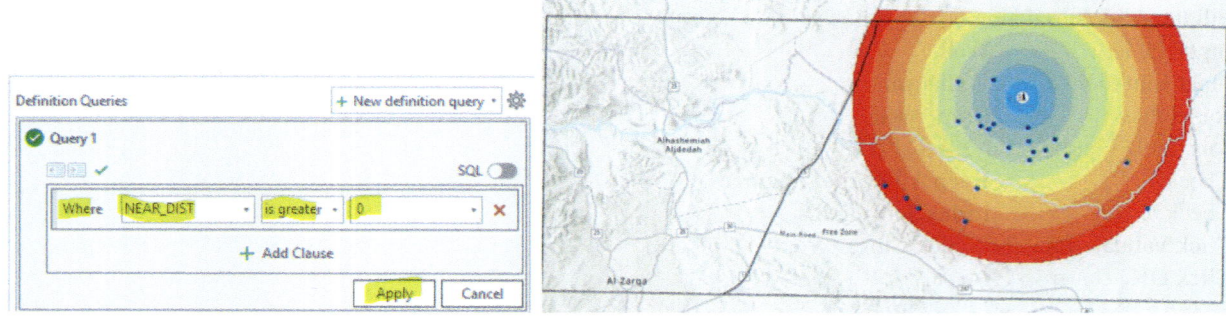

Well Classification

To verify the effect of the dam on the quality of the groundwater, the total dissolved solid (TDS) and nitrate (NO3) concentrations of the wells will be classified.

Salinity (TDS) Classification

71. In the CP, r-click Well Layer and select Symbology
72. In the Symbology pane under Primary symbology, select Graduated symbols
73. In the Field select TDS, classes 5, Method select Manual Interval/
74. In Classes tab, under Upper Value, type from top to bottom 500, 1000, 2000, 3000, and keep the last value 4160 and click Enter
75. Change the color of the TDS symbol from highest to lowest to red, pink, green, blue, then cyan
76. In the Symbology pane, click Advance symbology options (last icon), open Format Labels
77. Check Show thousands of separators

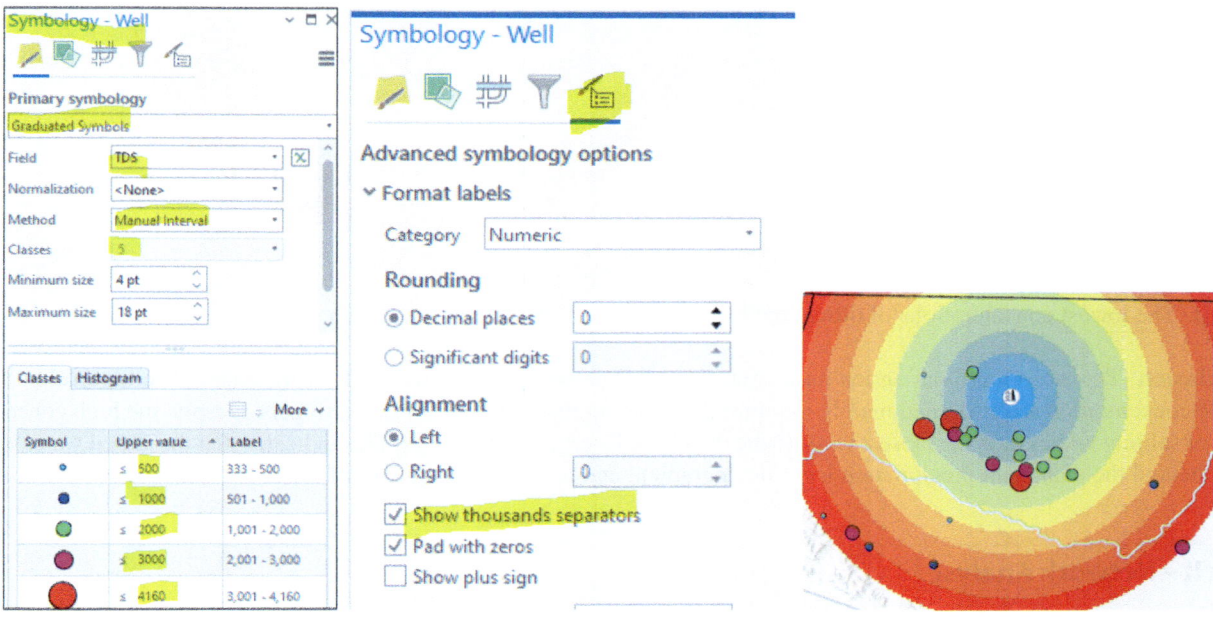

Result The closest wells to the dam have salinities ranging between 1000 and 2000 mg/l. Wells that are located further away from the dam have higher salinity.

Nitrate (NO$_3$) Classification

78. In the CP, r-click the Well layer and click Copy
79. R-click Recharge map and click Paste
80. Rename the copied **Well** Layer into **Well_Nitrate**
81. In the Symbology pane under Primary symbology
82. Change the Field to NO3
83. In Classes tab under Upper value, type from top to bottom, 20, 45, 70, 100, and keep the last value 145 and click Enter
84. In the Symbology pane/click Advance symbology options (last icon), open Format Labels
85. Under Rounding make the Decimal places = 0

86. Close the Symbology pane
87. Save your Project

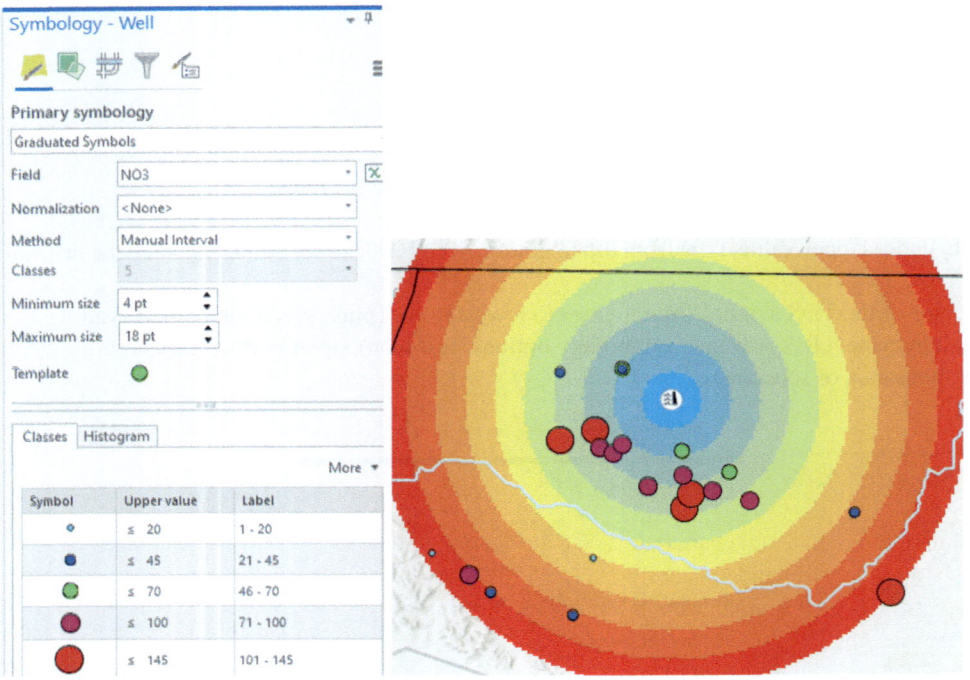

Result The lowest concentrations of nitrate are found in the wells that are near the dam.

Conclusion The wells are in an agricultural area, where groundwater is mainly used for irrigation. The high salinity and nitrate concentration are attributed mainly to return flow and application of chemical fertilizers. Despite the high concentrations of salinity and nitrate, the closest wells to the dam have lower concentrations. This indicates that the stored water in wet areas behind the dam from precipitation leaks downgradient and improves the quality of water.

Part II – Network Analyst

This section will perform different GIS functions in environment-related problems using the Network Analyst. Network Analyst is a powerful tool that deals with proximity analysis but is different than the proximity analysis performed in the previous section. The applications in the Network Analyst tools overcome the concept of straight distance related to some applications, such as buffering and Euclidian distance calculation. The network analyst will overcome any natural barrier such as hills, lakes, or where there is absolutely no network of a street system. The network analyst will use the actual distance that is associated with the street feature, which is an important feature in the application. This approach is more accurate than using the near function or spider diagram model.

To use the Network Analyst, you need to have a line feature that has connectivity such as a street, pipe, railroad, etc. If, for example, you have a street, you need to calculate the time from the length of the street and the speed of each segment of the street. The time will be used as the cost in the Network Analyst.

You will perform **two scenarios** in the **Dhuleil-Samra** region related to the water supply problem. The data include a street feature class that consists of many segments. The length and speed are documented for each street segment. A new field "MINUTES" will be added to the street feature class, which will allow network routing.

Part II – Network Analyst

Scenario 4A Samra - Dhuleil region has a shortage of water supply in summer time. The Water Authority decided to use two good quality water wells to supply the towns in the region with potable water using water trucks to distribute the water. Your duty is to find how long time it requires the truck to supply the towns in the study region with water.

Scenario 4B After finding the time required to cover the towns in the region, you want to find the actual path and time that the water truck will take from each well to each town.

To solve **Scenario 4A**, you should do the following:

(a) Create Network Dataset (**Driving_ND**)
(b) Build the **Driving_ND** network dataset

1. Click **Insert** tab on the ribbon, in the **Project** group, click **New Map** button
2. In the CP rename the **Map** to "**Service**"
3. In the **Catalog** pane, select the **Region, Town,** and **Well_Supply** feature classes from the **Data\Q3\Region.gdb** file geodatabase and integrate them into the Map View.
4. Classify the **Region** based on the Unique Value using the **Name** field
5. Symbolize the **Town** layer (Square 1, size 6, and ginger pink color) and click Apply
6. Symbolize the **Well_Supply** layer (Circle 1, size 8, and blue color) and click Apply
7. Close the Symbology pane

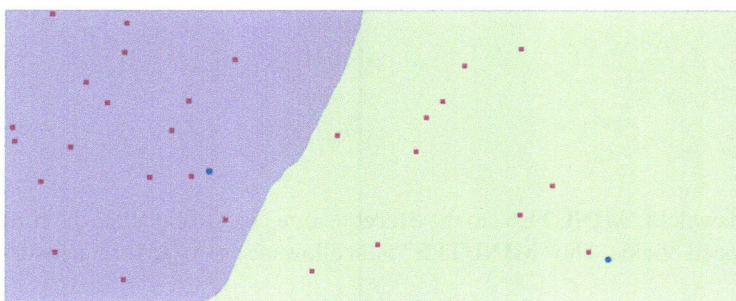

8. In Catalog pane open the **Database** folder, r-click **Ch14.gdb 6**,click **New**, and select **Feature Dataset**
9. In the **Create Feature Dataset** dialog box, fill it as follows:
10. Feature Dataset Name: **SamraDhuleil**
11. Click **Select Coordinate System**, click the drop-down arrow of Add Coordinate System and click Import Coordinate System, browse to \\Data\Q3\Region.gdb and select **Region**
12. Click OK twice and then click **Run**

Note The coordinate of the **Region** feature class is **Palestine_1923_Palestine_Belt,** and this coordinate system is now assigned to the **SamraDhuleil** Feature Dataset.

13. In the **Catalog** pane, open **Ch14.gdb** in the **Geodatabase**, r-click **SamraDhuleil** Feature dataset and click **Import** and select **Feature Class(es)**
14. Fill the **Feature Class To Geodatabase** dialog box as follows:
15. Input Features: **Street.shp** (Data\Q3)
16. Output Geodatabase: **SamraDhuleil**
17. Click Run

Result The **Street** feature class created in the **SamraDhuleil Feature Dataset** in the Geodatabase

18. Drag the **Street** feature class into the Map View
19. Open the attribute table of the Street feature class, and you can see that the table has two fields: "**Speed**" and "**Shape_Length**".

Note The **Speed** field is in **kilometers per hour**, and the **Shape_Length** field is in meters.

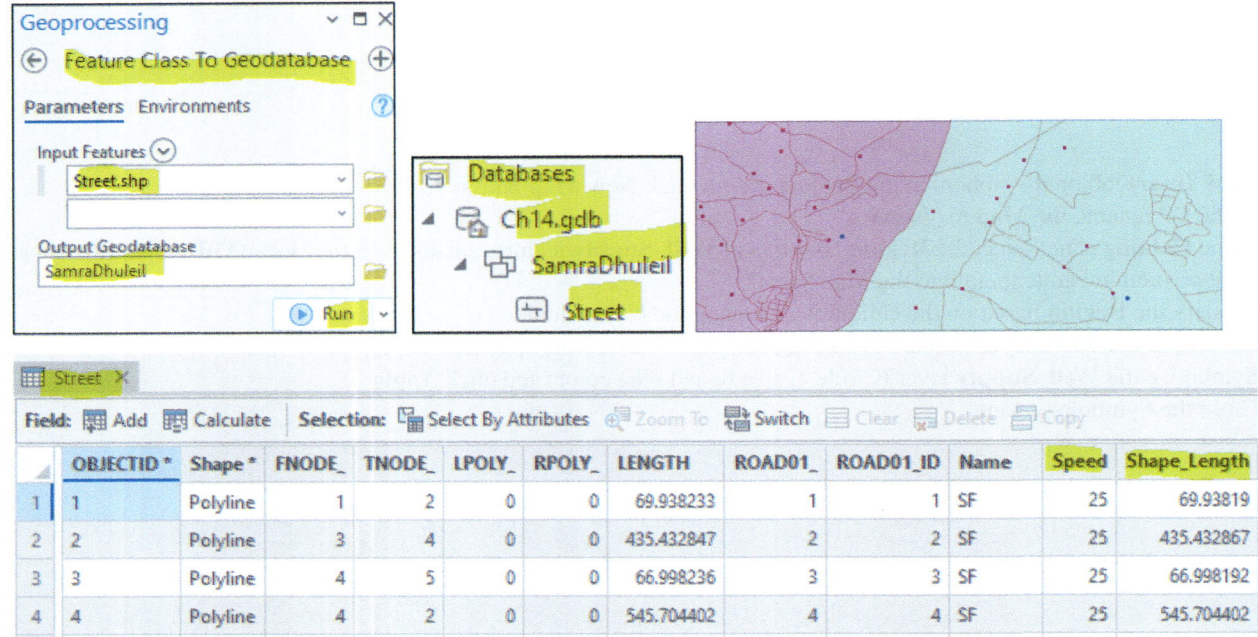

The next step is to add a new field "**MINUTES**" to the **Street** feature class to calculate the **time** from the "Shape_Length" (length of the street) and "Speed" fields. The "**MINUTES**" field allows users to perform network routing.

$$Time = \frac{Distance(Shap\_Length)}{Velocity(Speed)}$$

20. In CP, r-click **Street** feature classes point to **Data Design** and click **Fields**
21. At the bottom of the table "**Click here to add a new field**"
22. Under **Field Name** type "**MINUTES**" Data Type "**Double**"
23. Number Format select **Numeric** from **Category**
24. Accept the rest of the default and click OK
25. In the **Field** tab, in the **Changes** group, click **Save** button
26. Save the project
27. Close the **Field: Street** and make sure the attribute table of the **Street** is open
28. In Attribute table of **Street**, r-click **MINUTES** and click **Calculate Field**
29. In the **Calculate Field** dialog box, under MINUTES = type the following Statement:

Create Network Dataset

MINUTES = (!Shape_Length! / (!Speed! * 1000) * 60)

$$\text{Explanation}: Time = \frac{Length}{Velocity} = \frac{69.938\,m}{\frac{25\,km}{h} \times \frac{1{,}000\,m}{1\,km} \times \frac{1\,h}{60\,\min}} = 0.16\,\min$$

Click **Verify** (green arrow) at the bottom of the dialog box.
Click **Apply** and then **OK** to close the **Calculate Field** dialog box.

Result The Minutes field is now calculated.

Close the **Calculate Field** and attribute table of the **Street** remove the **Street** from the CP.

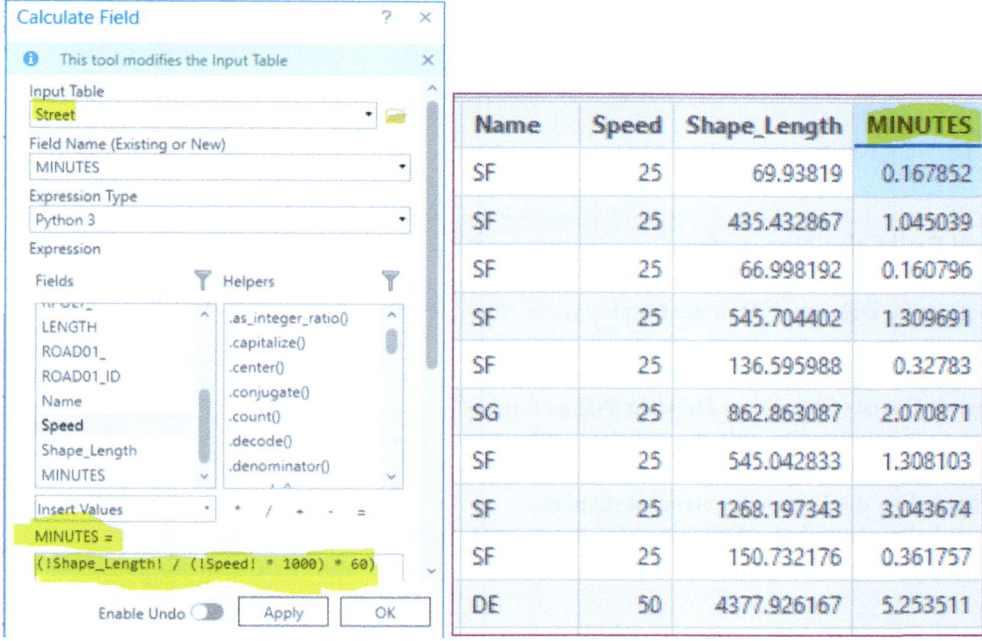

Create Network Dataset

30. In Catalog pane, r-click **SamraDhuleil** Feature Dataset, click **New** and choose **Network Dataset**
31. Fill the Create Network Dataset dialog box as follows:
32. Target Feature Dataset: **SamraDhuleil**
33. Network Dataset Name: **Driving_ND** (abbreviated for driving network dataset)
34. Check the **Street** (this will participate in the networking)
35. Elevation Model: **No Elevation** (this requires if there are two roads above each other, such as a highway above a bridge, so the driver cannot turn left or right into the road below it or above it)
36. Click Run

Result The network datasets **Driving_ND** and **Driving_ND_Junction** are created in the **SamraDhuleil** feature dataset. Only **Driving_ND** is added to the CP.

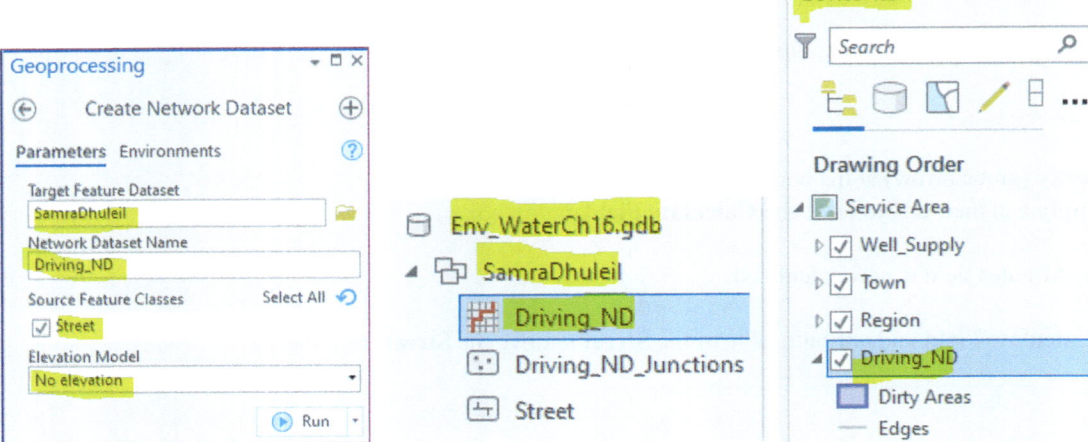

37. Remove the Driving_ND from the CP

Customize and Build the Network

Now you must set up the **Driving_ND** in the Catalog pane.

38. In Catalog pane, r-click the **Driving_ND** and click **Properties**
39. Fill the Network Dataset Properties: **Driving_ND** as follows:
40. Click the **General** tab open the **Indexes** and check the **Service-Area Index**
41. Next click the **Source Settings** tab
42. Make sure that below the **Edges** the **Street** is display
43. Make sure that below **Junctions: Driving_ND_Junctions** is display
44. Next, click the **Travel Attributes** tab, which has several tabs.

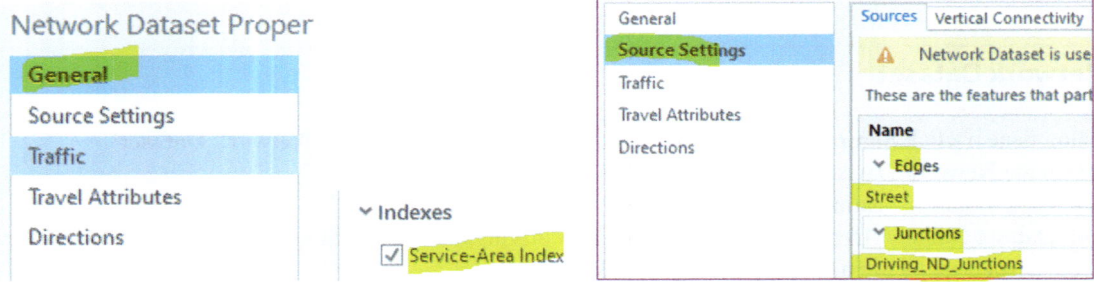

45. Click the **Travel Modes** tab
46. Click on the **Menu** button ▤ (3 lines to the right) and click **+ New**
47. **New Travel Mode** is added, rename it to "**Travel Time**"
48. **Description:** Driving from the wells to different towns using water trucks
49. Under **Type** make sure **Driving** is selected
50. Under **Cost**: **Impedance** TravelTime (minutes)
51. Under Distance Cost: Length (meters)

Customize and Build the Network

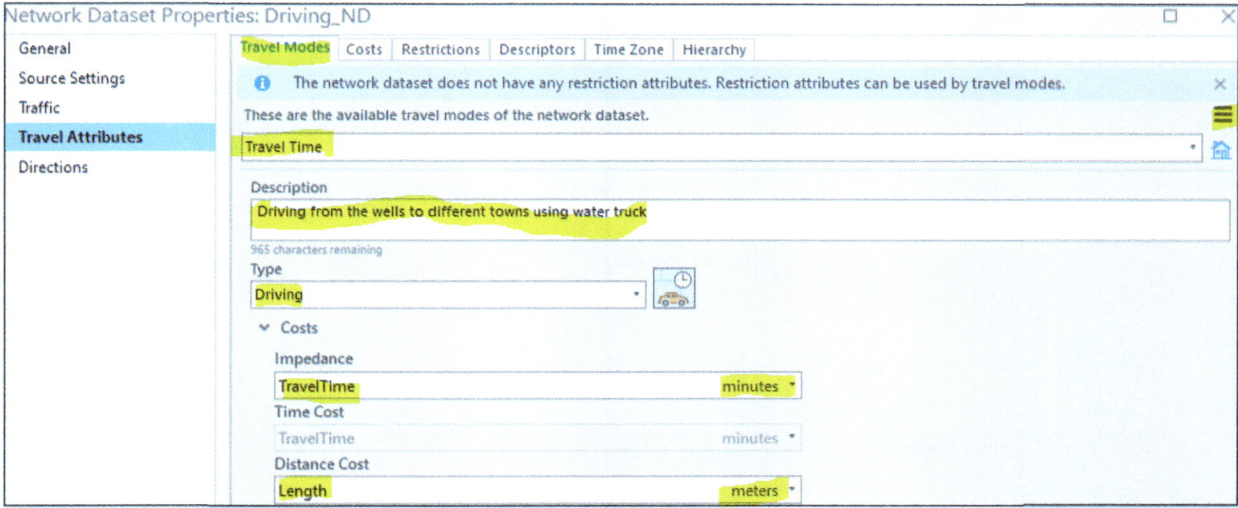

52. Click the Cost tab, click on the **Menu** button and click **+ New**
53. Under Properties change the Name to TravelTime
54. Units: Minutes
55. Under **Evaluator** in the **Street (Along)** and under **Type** Select **Field Script**

56. D-click under **Value**, click **Field Script setting** and change Language to **Python**
57. Under **Result** type**!MINUTES!** (exclamation mark should be before and after the MINUTES and click OK
58. Repeat for the next row **Street (Against)** by repeating exactly what you did with the **Street (Along)**.

Note MINUTES is a field in the Street attribute table.

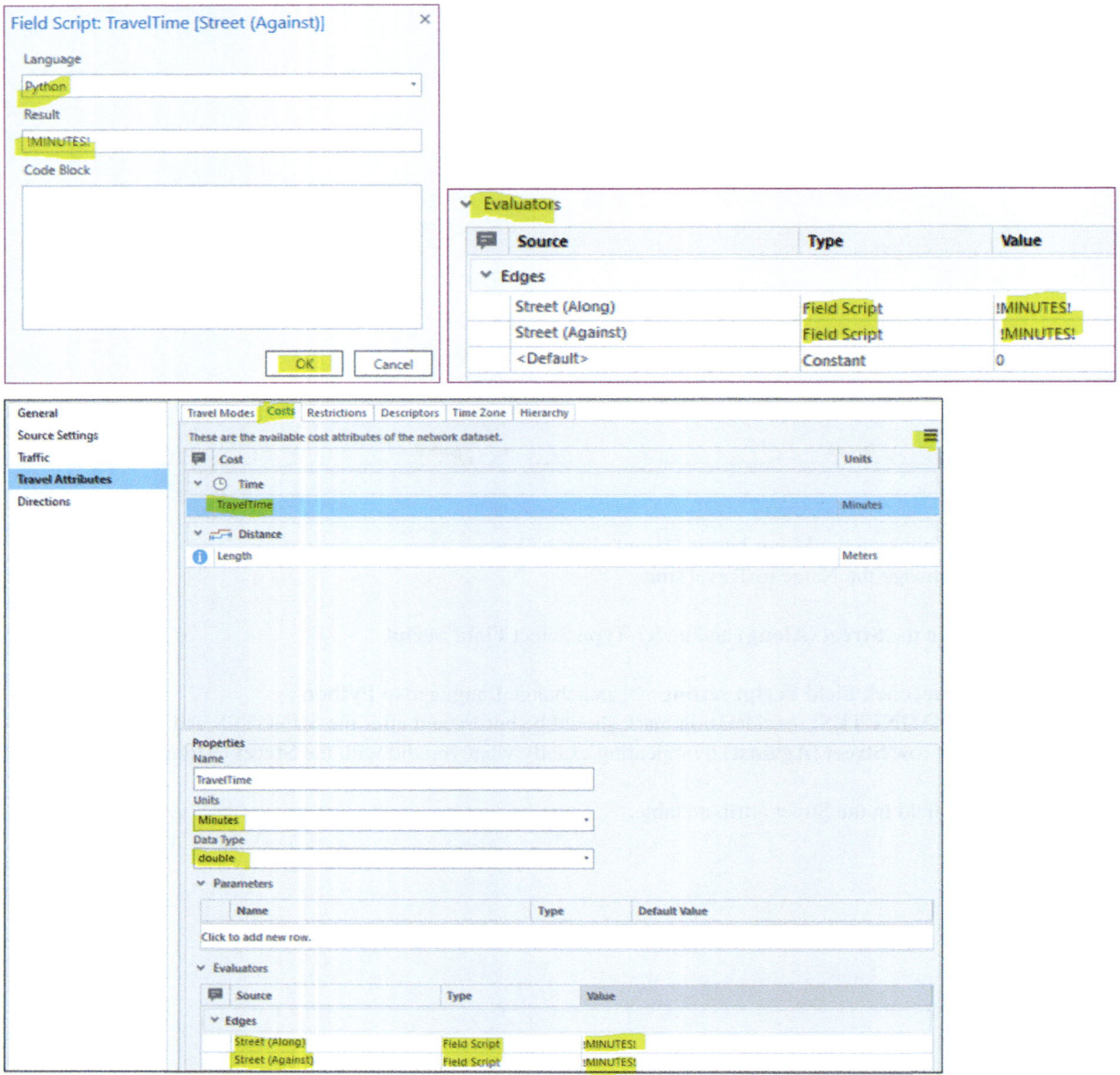

59. Under **Turns**, click under **Type** and select **Turn Category**
60. Under Value click the Turn Category Setting icon
61. The **Turn Category Evaluator** dialog box will appear.
62. Fill the Turn Category dialog box by changing the **Left Turn** to 3 seconds and **Reverse Turn** to 4 seconds
63. Click OK and OK to exit the Network Dataset Properties

Explore the Network

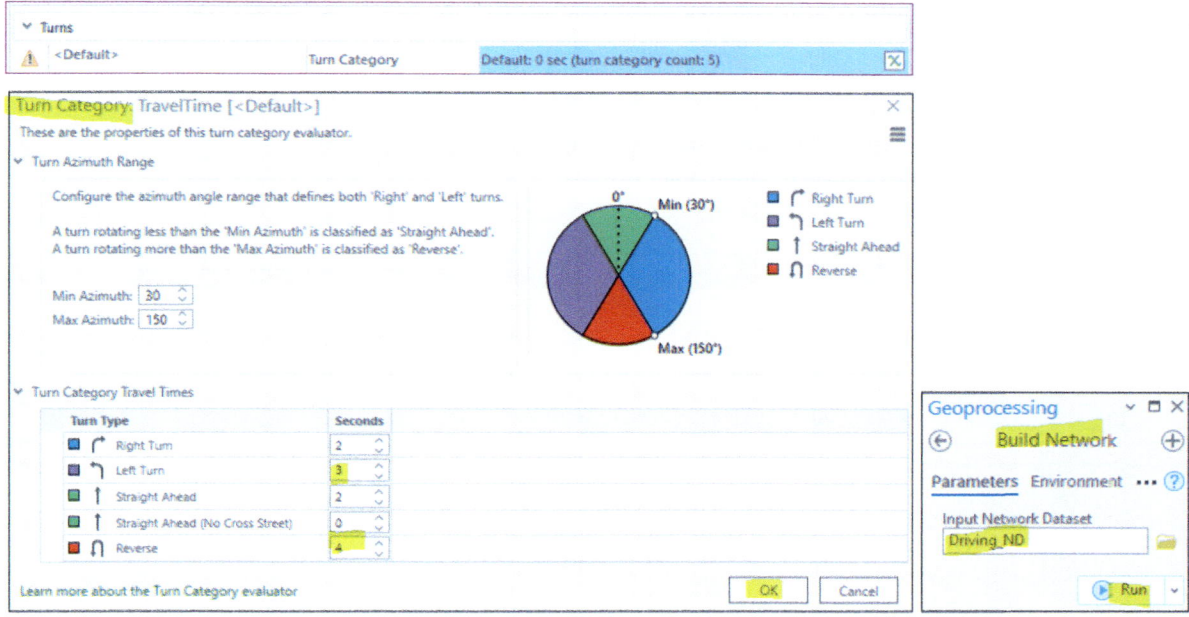

Result **Driving_ND** is filled with the proper direction.

64. In the Catalog pane, r-click **Driving_ND**, select **Build** and click **Run**

Result The **Driving_ND** and **Driving_ND _Junction** are built. **Driving_ND** will be added to CP.

To see the **Driving_ND** in the CP drag it above the **Region** and uncheck the **Street**.

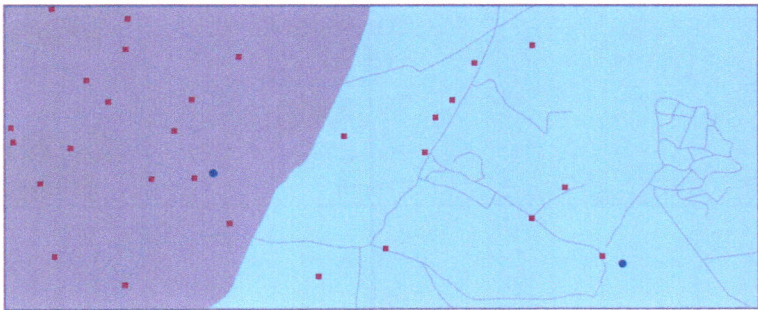

65. In the CP, remove the **Driving_ND** and save the project

Explore the Network

There are two approaches that allow you to explore how the **Driving_ND** network is working. You want to check if the street segments are connected and the travel time along the street segments.

First Approach

66. In Catalog pane drag the **SamraDhuleil** feature dataset into the Map View
67. In the CP Highlight, **Driving_ND**.
68. Click **Data** tab on the ribbon, in the **Explore** group, click the **Explore Network** button
69. Click on any road segment that connected more than street

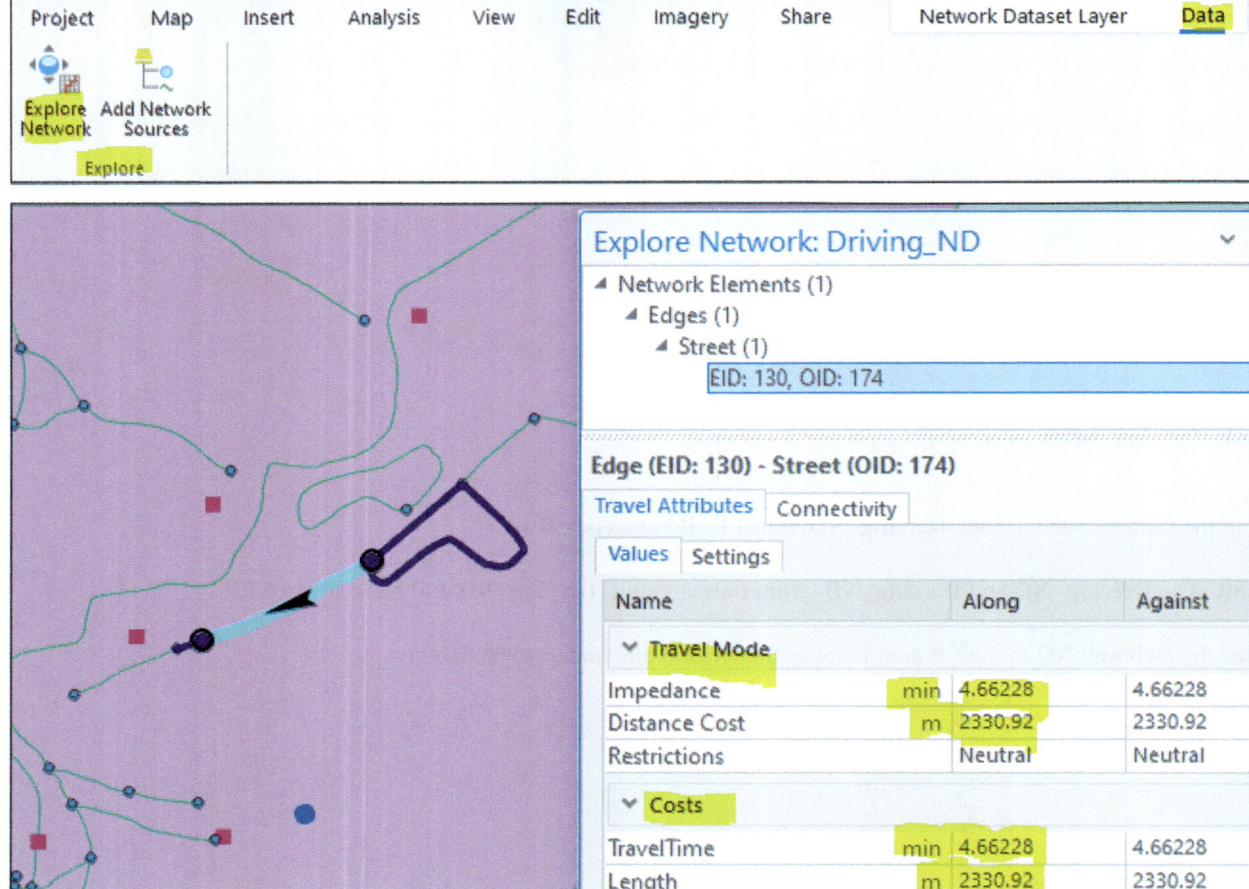

70. Close the Explore Network dialog box to deselect the segment of the street

Second Approach

The other way to check the connectivity and travel time is to do the following:

71. Click the **Analysis** tab, in the **Workflows** group, click the **Network Analyst**
72. You will see the network **Driving_ND** that you set
73. Choose **Route**
74. The **Route** analysis layers are added to the CP and will allow you to use a network analysis tool to assess whether you have connectivity or not.
75. In CP, select **Route** and click on the street segment as in the first approach. It shows the same information as the previous step

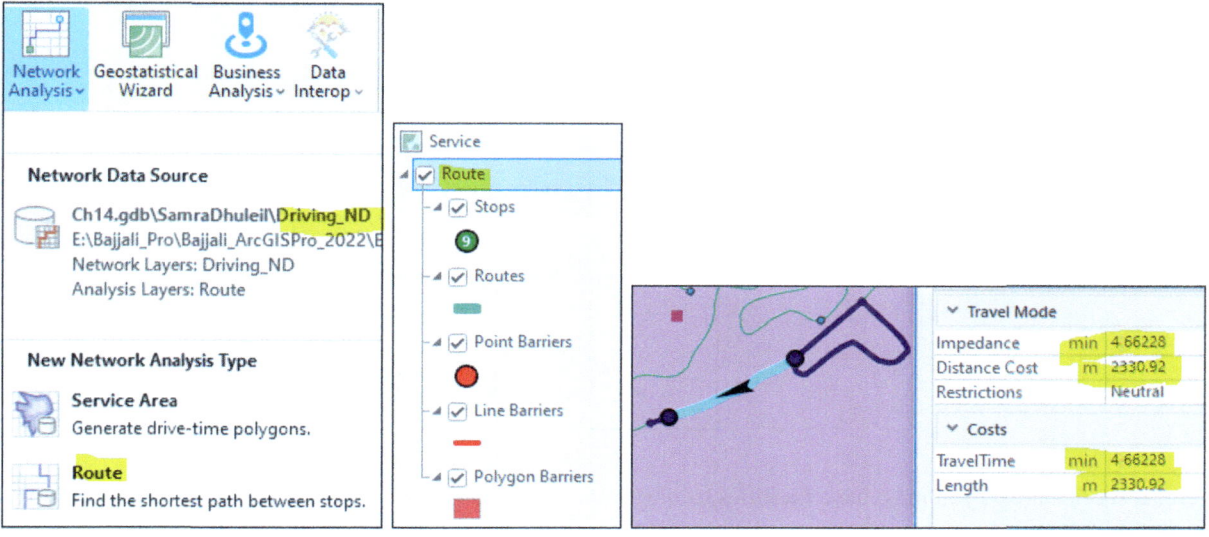

Result The two approaches show the same result about the connectivity between the streets, direction and the length of the street segment and the travel time in minutes.

Run New Service Area

The **Network Service Area** is a region that covers all accessible streets in the study region. The tool shows, for example, that within 10 minutes, the water truck will travel from the water supply well along the street network and will include all the streets that can be reached within **10 minutes**.

Service areas created by Network Analyst also help evaluate accessibility. Concentric service areas show how accessibility varies with impedance. Once service areas are created, you can use them to identify how much land, how many people, or how much of anything else is within the neighborhood or region.

76. Click the **Analysis** tab, in the **Workflows** group, click the **Network Analyst** drop Arrow and choose **Service Area**

Result The Service Area is added to Ch14.gdb and to the CP and has six new service area layers. The attribute table of all these feature class layers is empty.

1. Facilities
2. Polygons
3. Lines
4. Point Barriers
5. Line Barriers
6. Polygon Barriers

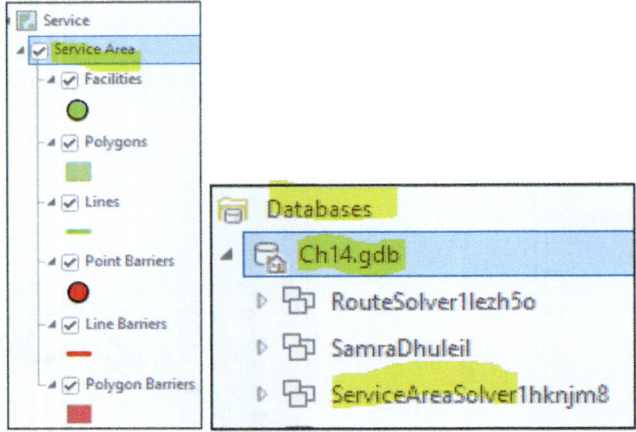

To create service areas, you must have **facilities** to start from. The service area solver simulates all possible paths that a vehicle, in this scenario, a water truck can travel when departing from the facility. Because water trucks are typically parked at **Well_Supply**, you will import **Well_Supply** into the facilities sublayer. The **facilities** represent the "**Town**" that will receive the water supply from the wells "**Well_Supply**" by the **water truck**.

Add Facilities

The facility is the starting location of a water truck. The service area solver simulates all possible paths the vehicle can travel within an elapsed time when departing from the facility. This adds **Well_Supply** to the **Network Dataset** so the **Well_Supply** feature class can be used in the Network Service Area analysis.

77. In CP, highlight the **Service Area**, click on **Service Area Layer** tab on the ribbon, in the **Input Data** group, click **Import Facilities** button
78. The Add Location pane

79. Fill the **Add Locations** as follows:
    (a) Input network Analysis Layer: **Service Area**
    (b) Sub Layer: **Facilities**
    (c) Input Locations: **Well_Supply**
    (d) Field Name: **Name**
    (e) Check **Append to Existing Location** and **Snap to Network**
    (f) Snap offset   5 meters
    (g) Click Advance
    (h) Search Tolerance   2000 meters
80. Accept the rest of the default
81. Click Apply and OK to exit the dialog box

Result The "facilities" that are under the service area in the content pane are now created with two features that represent the **2 wells** (Well_Supply). The two wells (facilities) are now snapped to the network of the Driving_ND_Junctions, which indicates that they have now become part of the network dataset and can participate in network analysis.

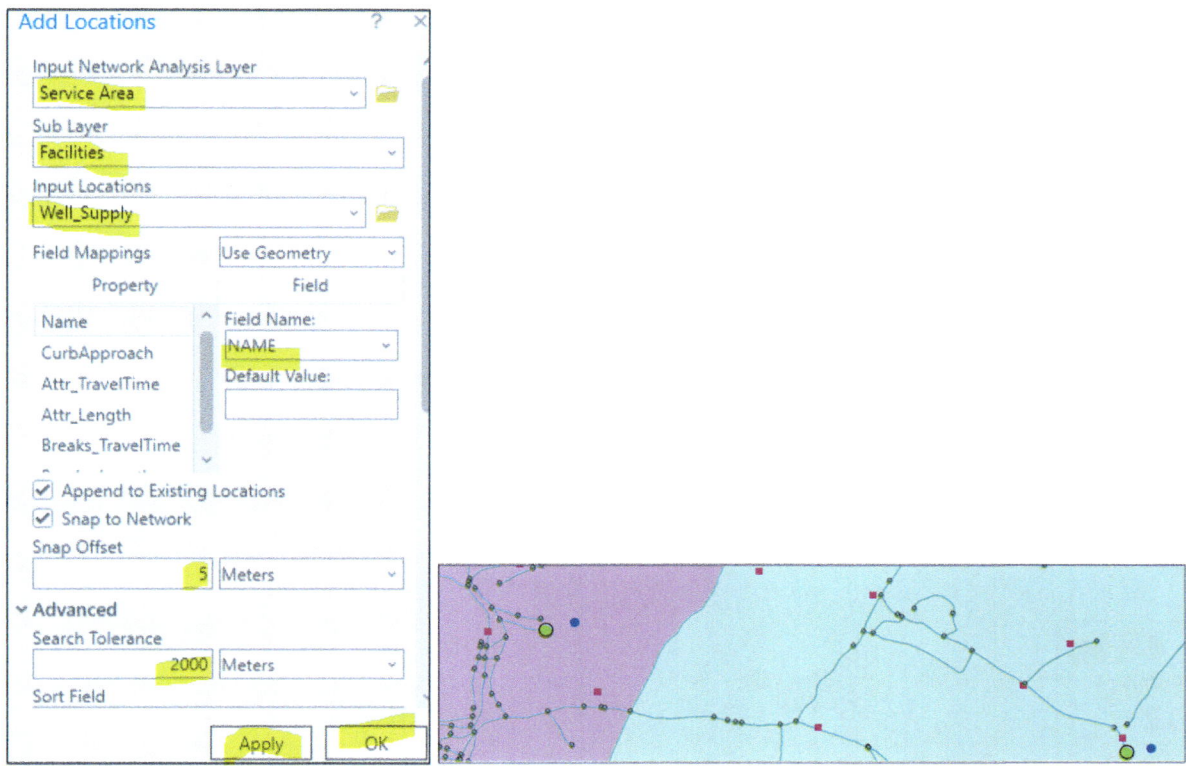

Set the Cutoff Time

After loading **Well_Supply** into the Facilities sublayer, the water truck travel distance should be set by setting a cutoff value. The Water Authority now wants to see the areas that will be served with water supply within different times by the water truck and to see which would be more effective.

82. Make sure the **Service Area** in the CP is selected
83. Click **Service Area Layer** in **Travel Setting** group, click the **Mode** drop-down list and select **Travel Time**
84. Direction: **Away from Facilities**
85. Cutoffs: type 5, 10, 15

The 5, 10, and 15 cutoffs are minutes, and they represent how much the water truck can cover within these minutes from the water supply wells to the towns in the region.

86. In the **Output Geometry** group, click the drop-down arrow and select **Polygons**
87. Click the **Polygon Trim Distance** and choose **500 m**
88. Select **High Precision**, **Sissolve**, and **Rings**
89. In the Analysis group, click **Run**

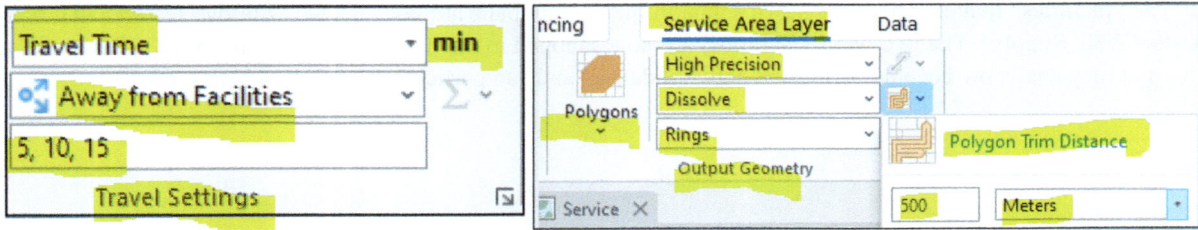

90. Change the color of the 3 polygons in the CP by assigning cyan, green, and pink for 5, 10, and 15, respectively.
91. Save the project

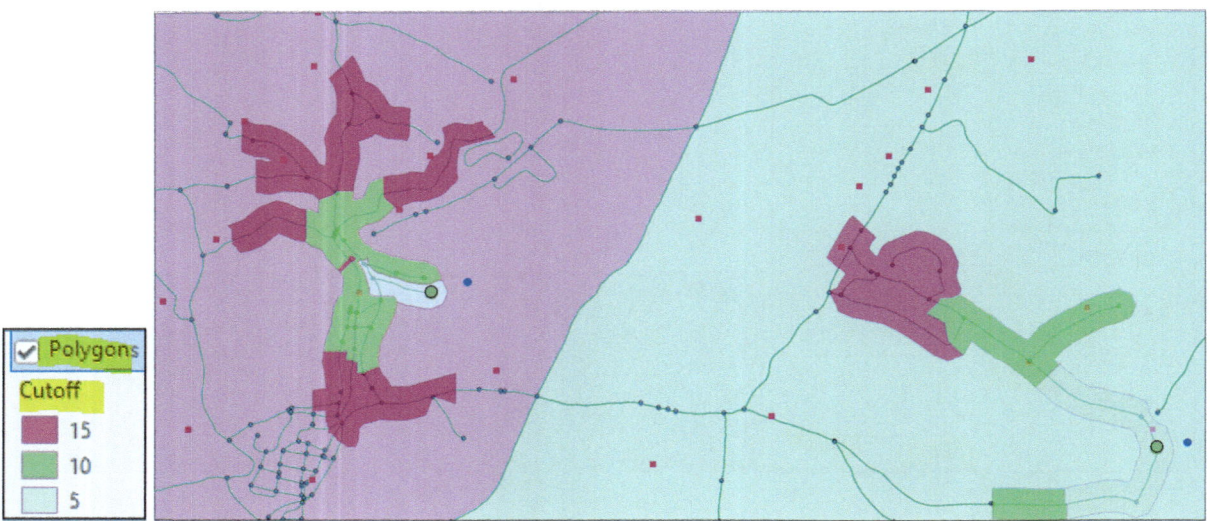

Result The three polygons that are created represent the area that the water truck will cover in 5, 10, and 5 minutes.

True Path and Total Time Between the Wells and each Town

Scenario 4B After finding the time required to cover the towns in the region, you want to find the actual path and time that the water truck will take from each well to each town. The actual path can be carried out using the "Closest Facility" tool on the Network Analyst.

The "**Closest Facility**" tool is similar to the "Near" tool that has been used earlier, as both measure the distance between two locations. However, they are different, as the "Near" tool measures the straight line distance, while the "Closest Facility" tool measures the distance along a network.

To solve **Scenario 4B,** you have to do the following:

(a) Use the **Driving_ND** network dataset
(b) Calculate the True Path and Total Time between Wells and Towns

1. Click **Insert** tab on the ribbon, in the **Project** group, click **New Map** button
2. In the CP rename the **Map** to "**True Path**"
3. In the **Catalog** pane, select the **Borehole, City,** and **Region** layers from the **Layer** folder under the **Data_Ch14** in the **Folder** and drag them into Map View

Add Facilities   325

4. Remove the labels from the **City** and **Region** layers (challenge yourself)
5. In Catalog pane drag the **SamraDhuleil** feature dataset into the Map View
6. Click the **Analysis** tab, in the **Workflows** group, open the **Network Analysis** drop-down arrow and select the **Closest Facility**

Result   A set of **closest facility** layers are added to the CP. There are 6 layers:

(a) Facilities
(b) Incidents
(c) Routes
(d) Point Barrier
(e) Line Barriers
(f) Polygon Barriers

7. In the CP, change the color of the **Facilities** layer into green and the **Incidents** layers into blue

Add Facilities

You will use the **Borehole** layers to load the **Facilities** sublayer network analysis class. The facilities should be located on the network to perform the analysis.

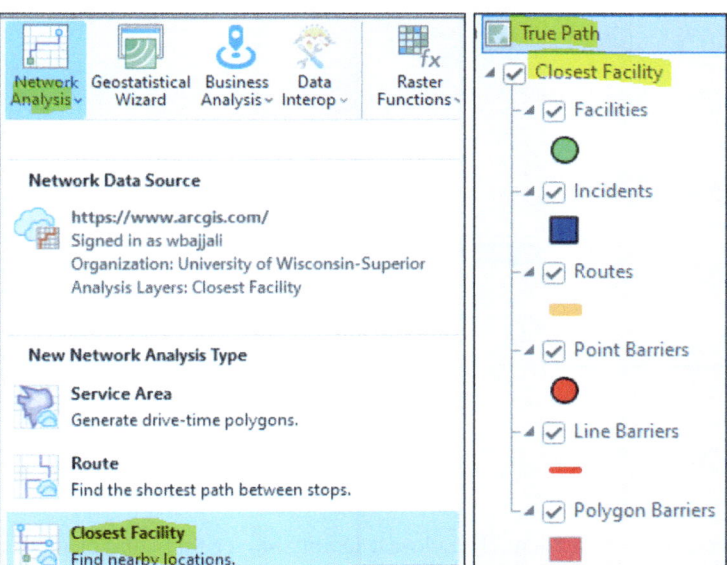

8. In CP, highlight the **Closest Facilities**, click the **Closest Facility Layer** tab on the ribbon, in the Input Data group, click **Import Facilities**
9. Fill the **Add Location** pane dialog box as below
    (a) Input Network Analysis Layer: **Closest Facility**
    (b) Sub Layer: **Facilities**
    (c) Input Locations: **Borehole**
    (d) Field Name: **WellNum**

10. Check **Append to Existing Locations**
11. Check **Snap to Network** (5 meters)
12. Open Advance and type **2000 meters** under **Search Tolerance**
13. Click **Apply**
14. Click **OK** to close the dialog box

Result A total of 2 boreholes were loaded as facilities. The attribute table of the Facilities includes the two boreholes: Number 3 and 17. The Facilities that represent the two boreholes are snapped to the Network of the street, which indicates that they are now part of the Network Dataset and can participate in network analysis.

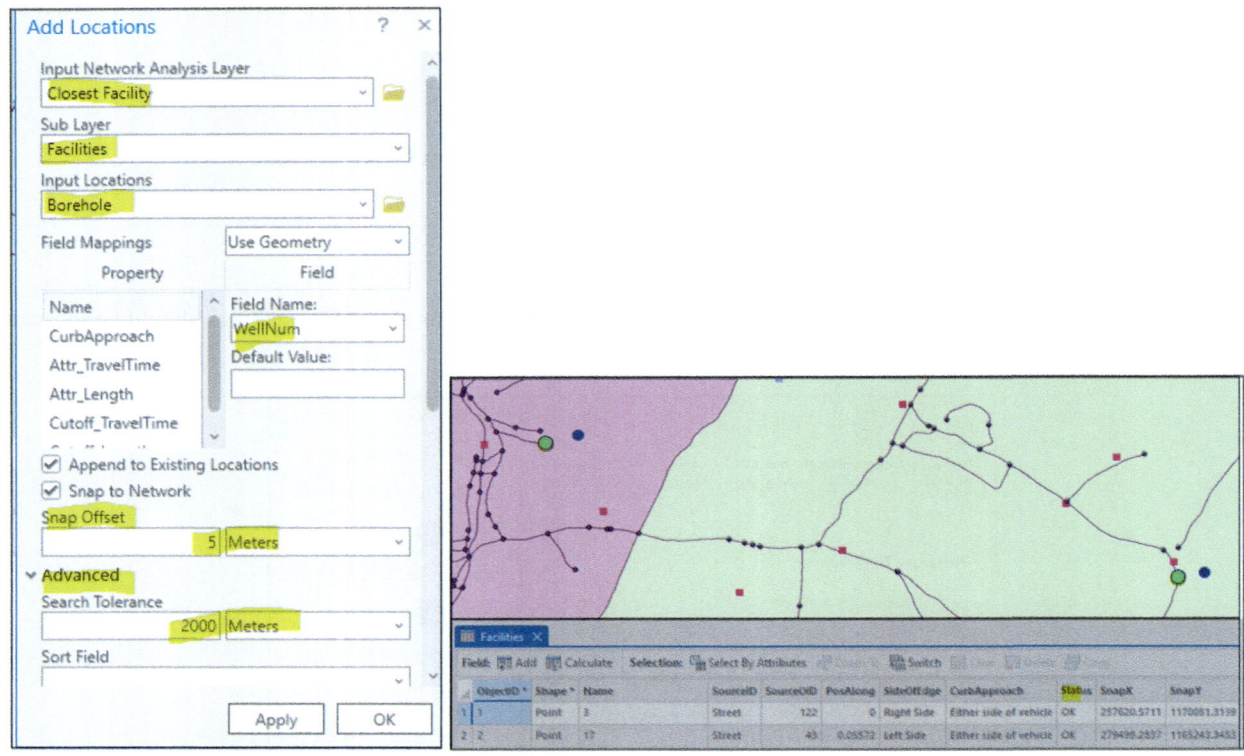

Add an Incident

The location of the boreholes site is an incident. The closest facility solver finds one or more cities that are closest to the borehole location.

15. In CP, select **Closest Facilities** and click the **Closest Facility Layer** tab, and in the Input Data, click **Import Incidents**
16. Fill the **Add Location** pane as follows:
    (a) Input Network Analysis Layer: Closest Facility
    (b) Sub Layer: Incidents
    (c) Input Locations: City
    (d) Field Name: SETT_NAME
17. Check Append to Existing Locations
18. Check Snap to Network, Snap Offset 5 meters
19. Advanced Search Tolerance   2000 meters
20. Accept the rest of the default
21. Click Apply

22. Click OK to close the dialog box

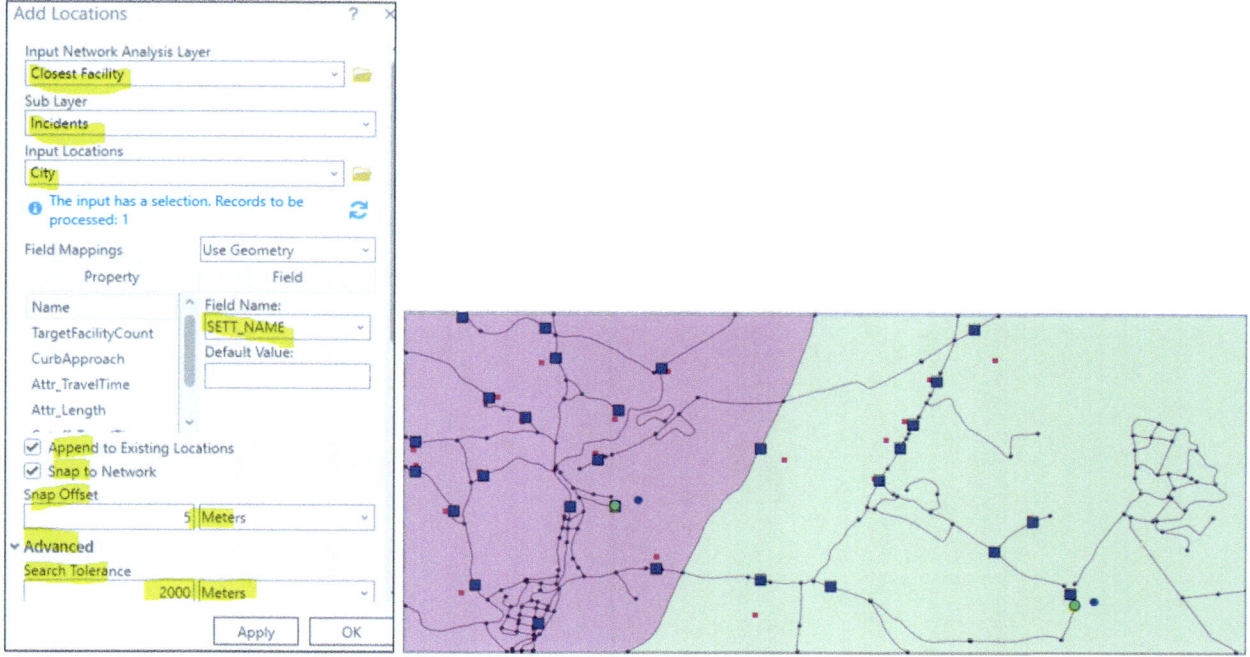

Result The incidents are connected in the network, and the number of incidents is 28, which is similar to the number of towns.

23. In CP highlight the **Closest Facilities** and click **Closest Facility Layer** tab, in the **Travel Setting** group AND SELECT THE FOLLOWING
24. Mode: Travel Time
25. Direction: **Away from Facilities**
26. Click drop-down **Cost Attributes to Accumulate along Output Lines** check the **TravelTime (minutes)**
27. **Facilities** = 1 (because we have only 2 wells)
28. In the **Output Geometry** group, choose **Along Network**
29. Click Run
30. Output Geometry group, choose Straight Lines
31. Click Run

Result When you run Along Network, it will generate Routes along the network street. When using straight lines, it generates a straight line, which is easy to see.

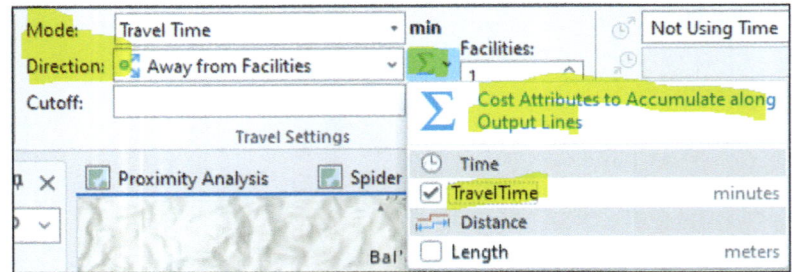

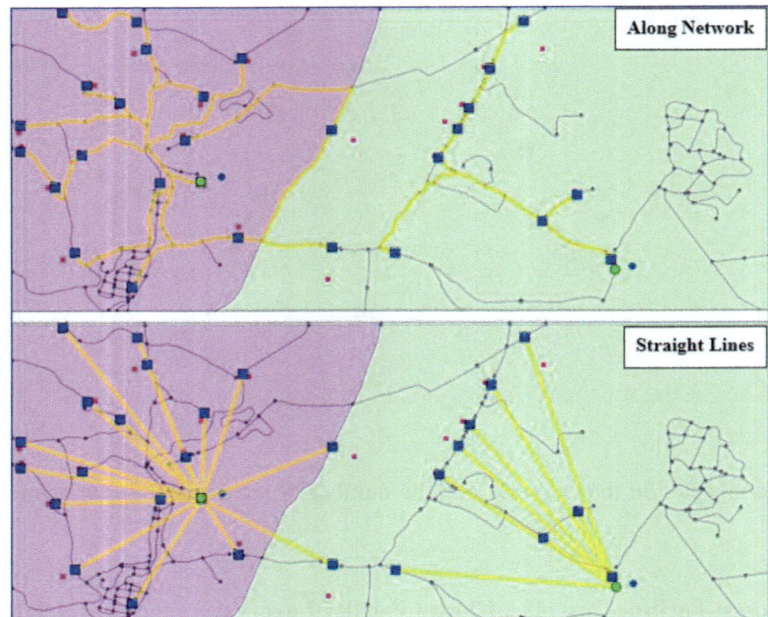

Classify the Straight Lines Routes

32. Open the attribute table of the Routes; you will see a field called **FacilityID.**
33. In CP r-click Routes, click Symbology, Unique Values, Field 1=Facility ID, click Add All Values
34. Change the first color to red, width 2, change the second line green, width 2, click Apply
35. Close the Symbology pane
36. Save the project

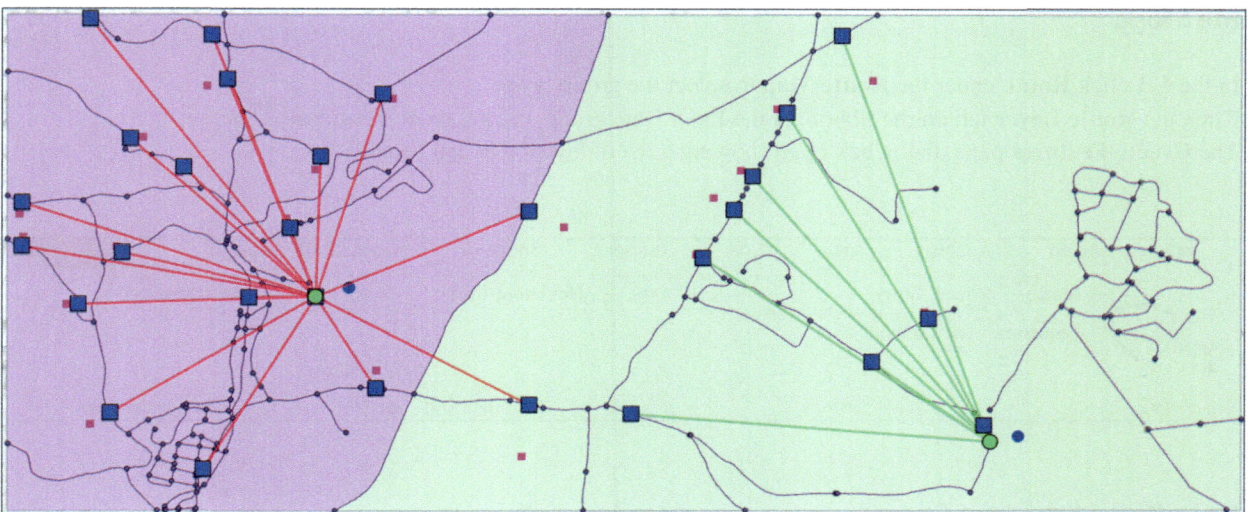

Result The image shows the direction from each borehole to the cities.

Create the Route Layer

A route layer provides the structure and properties needed to set up and solve route problems. It also contains the results after solving.

36. Click **Insert** tab on the ribbon, in the **Project** group, click **New Map** button
37. In the CP rename the **Map** to "**Route**"
38. In the **Catalog** pane, select the **Borehole**, **City**, and **Region** layers from the **Layer** folder under the **Data_Ch14** in the **Folder** and drag them into Map View.
39. Remove the labels from the **City** and **Region** layers (challenge yourself)
40. In Catalog pane drag the **SamraDhuleil** feature dataset into the Map View

41. Click the **Analysis** tab, in the **Workflows** group, open the **Network Analysis** drop-down arrow and select the **Route**

Result The **Route layer** is added to the **CP**, and it includes several sublayers that hold the following sublayers:

1. Stops
2. Routes
3. Point Barriers
4. Line Barriers
5. Polygon Barriers

The route references the **Driving _ND** network dataset because the network was in the CP when the route layer was created.

Create Stops

42. In the CP, click **Route** under the **Route Map** to select the group layer.
43. Click the **Route Layer** tab on the ribbon, in the Input Data group, click **Create Features** button.
44. The **Create Features** pane dialog box open showing a list of layers that can be edited

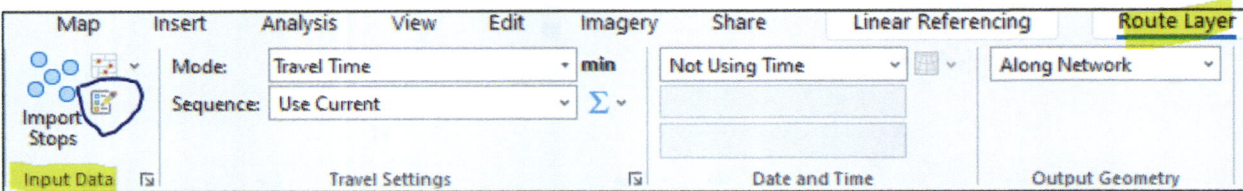

45. Under **Route: Stops**, click **Stops**.
46. Use the **Point** tool to create a 4-stops on the **Driving _ND** network dataset on the map. Create 2-stops in Samra and 2-stops in Dhuleil regions

47. On the **Edit** tab, in the Selection group, click **Attributes**.
48. In the **Attributes** pane appears.
49. Select the lower stop you created in **Samra** region and in the Attributes tab change the name to **Samra1**, then click the upper stop in **Samra** region and change the name to **Samra2**.
50. Select the lower stop you created in **Dhuleil** region and in the Attributes tab change the name to **Dhuleil1**, then click the upper stop in **Dhuleil** region and change the name to **Dhuleil2**
51. In the **Manage Edits** group, click **Save**, and **Yes** to save all Edits

Run Analysis

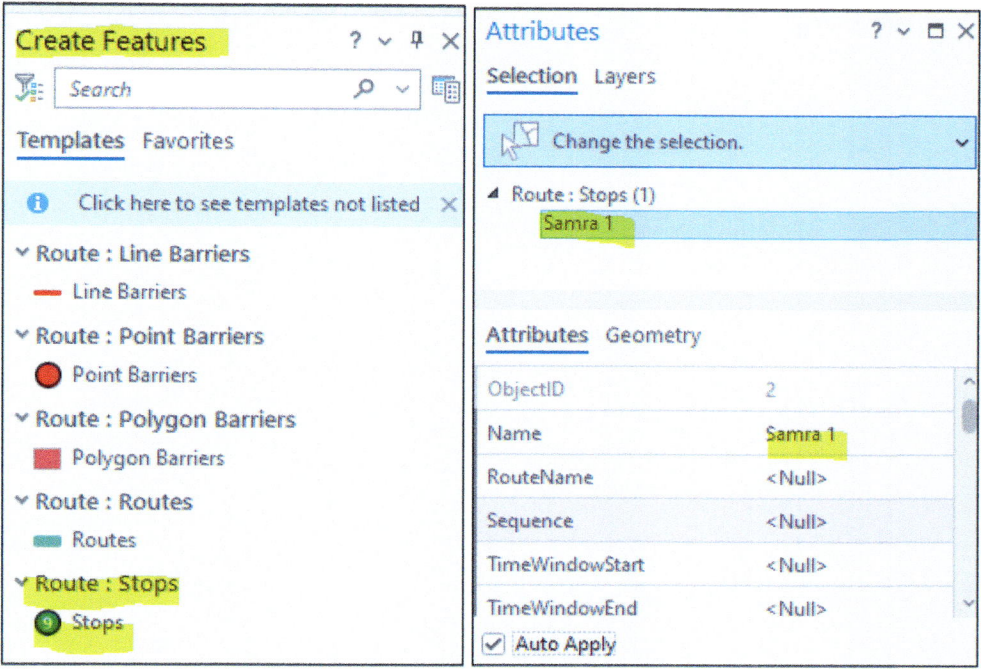

Run Analysis

52. Click the Route Layer tab, in the Analysis group, click **Run** .

Result The fastest path through the network, connecting all the stops you created, will be displayed. The stop symbol on the map shows the sequence number in the order in which the stops were entered and visited by the route solver.

3-D Visualization

The ArcGIS 3D Analyst extension in ArcGIS Pro provides tools for creating, visualizing, and analyzing GIS data in a three-dimensional (3D) context. The ArcGIS 3D Analyst extension toolbox provides a collection of geoprocessing tools that enable a wide variety of analytical, data management, and data conversion operations on surface models and three-dimensional vector data. The toolbox is conveniently organized into toolsets that define the scope of tasks accomplished by the tools they contain. You must enable ArcGIS 3D Analyst extension before you can work with it in ArcGIS Pro. ArcGIS Pro requires that an administrator of an organization assign software licenses and extensions. Once you have been assigned ArcGIS 3D Analyst extension, you gain access to the 3D Analyst tools. ArcGIS Pro allows you to work with data in 2D and 3D environments from within the same application. Having both 2D and 3D in the same application gives you more freedom to work in 3D than in working with ArcMap. The 2D and 3D of the same data can be analyzed together in ArcGISPro, which allows for new understandings into that data. A 2D map is viewed from above at a 90-degree angle to the map; a 3D scene can be viewed from any angle or perspective. The perspective used to view and navigate the scene is called the camera. 2-D maps are always useful in GIS applications, but 3-D maps are also useful in some applications. It allows users to visualize features and objects above the ground surface, such as buildings, fences, mountains, or trees, or visualize subsurface features, such as wells, pipelines, storage tanks or geological faults. In this condition, the 3-D maps and scenes can be helpful for visualizing the real world and being aware of your surroundings. ArcGIS Pro can examine and browse through your data in more than one way at once. Users can link 3D scenes to 2D maps or to other scenes to enable simultaneous views of their data. Although multiple linked scenes and maps reference the same source data, they are separate items within your project. If you change the visibility of a layer in the 2D map or remove it altogether, that change will not appear in a linked 3D scene. However, if you change the source data in a way that affects visualization, that change will be shown in both the 2D map and 3D scene.

Surfaces and Z Values

A 3D surface model is a digital representation of features, either real or hypothetical, in 3-D space. Some simple examples of 3D surfaces are a landscape, an urban corridor, well depths drilled under the ground to determine water quality or plume of contamination. These are all examples of real features, but surfaces could be derived or imaginary.

Functional surfaces are most commonly used to model terrestrial data representing the Earth's surface, although they can also be used to model many other types of surfaces, such as bathymetric data, individual geologic strata, or statistical surfaces describing geographic concentrations. Terrain datasets, Triangulated Irregular Network (TIN), Digital Elevation Model (DEM) Raster, and LAS datasets (LAS is a file format for the interchange of 3-dimensional point cloud data) are all examples of functional surfaces.

To work with the 3-D Analyst, you need a 3D dataset that consists of (x, y, z). x and y are the locations, and z represents a value in the attribute other than the location. z could be the elevation of points above sea level, chemical concentration of groundwater, precipitation, or any phenomenon that varies across a specific location. Table 15.1 below shows that x and y are

Supplementary Information The online version contains supplementary material available at https://doi.org/10.1007/978-3-031-42227-0_15.

Table 15.1 Groundwater well locations and other information

Well-ID	Northing	Easting	Elevation	Depth	Salinity
840	3872.70	1027.00	626.99	17	456
841	3872.50	1024.20	663.03	38	567
842	3871.50	1021.80	662.91	58	435
843	3928.50	1358.20	662.73	14	289
846	3572.20	941.30	662.38	35	987
847	3554.10	2416.40	664.35	14	888
848	3038.90	964.70	663.61	16	846
853	2501.00	808.80	668.19	55	484
856	2133.00	2387.70	673.90	35	503
857	2129.10	2390.20	673.95	61	359

the easting and northing of the projected coordinate, and z represents either the elevation, well depth, or salinity of the groundwater wells.

Create Triangular Irregular Network From Contour Line

A triangular irregular network (TIN) layer is commonly an elevation surface that represents height values across a study area. TIN is vector-based digital geographic data constructed by triangulating a set of vertices (points). The vertices are connected to a series of edges to form a network of triangles. The edges of TINs form contiguous, nonoverlapping triangular facets and can be used to capture the position of linear features that play an important role in a surface, such as a ridge line or stream course. Because nodes can be placed irregularly over a surface, TINs can have a higher resolution in areas where a surface is highly variable and a lower resolution in areas that are less variable. The TIN can be created from features such as points, lines and polygons that contain elevation information. Using either feature class to construct the TIN will generate many nonoverlapping triangles that cover the entire study area, and the land surface is designated with these triangles.

There are different methods of interpolation to form these triangles, such as Delaunay triangulation or distance ordering. ArcGIS supports the Delaunay triangulation method. The resulting triangulation satisfies the Delaunay triangle criterion, which ensures that no vertex lies within the interior of any of the circumcircles of the triangles in the network. If the Delaunay criterion is satisfied everywhere on the TIN, the minimum interior angle of all triangles is maximized. The result is that long, thin triangles are avoided as much as possible. The TIN units can be in feet or meters, not decimal degrees. Delaunay triangulations are not valid when constructed using angular coordinates from geographic coordinate systems. Therefore, when constructing the TIN, you should use a projected coordinate system and avoid using the latitude-longitude coordinate systems. TIN layers are available in both map and scene views in ArcGIS Pro.

Scenario 1 You have been asked by your advisor to generate TIN from contour lines to provide a presentation about the water resources in the Dhuleil area in a 3-D setting.

Data Connection and Integration

1. Launch ArcGIS Pro
2. Click **Open another project** (upper-right) browse to **\\Env_Water\Ch15** select **Ch15.aprx** and click OK

The **Ch15.aprx** open and the CP include the Map, which includes both the **World Topographic Map** and **World Hillshade**.

3. Make sure that the **Catalog** pane is open
4. Click the **Insert** tab on the ribbon, in the **Project** group, click **Add Folder**
5. Browse to **\\Env_Water\Ch15 (or \\Database\ Data_Ch15)** open it and highlight **Data_Ch15** and click **OK**.
6. In the CP, name your **Map "TIN"**
7. In the **Catalog Pane** expand the **Folders, Data_Ch15**, and then **Q1** folder to integrate the **Contour.shp** layer into the **TIN** map
8. The **Contour** layer display
9. In the CP, r-click the **Contour** layer and open the Attribute Table

Data Connection and Integration

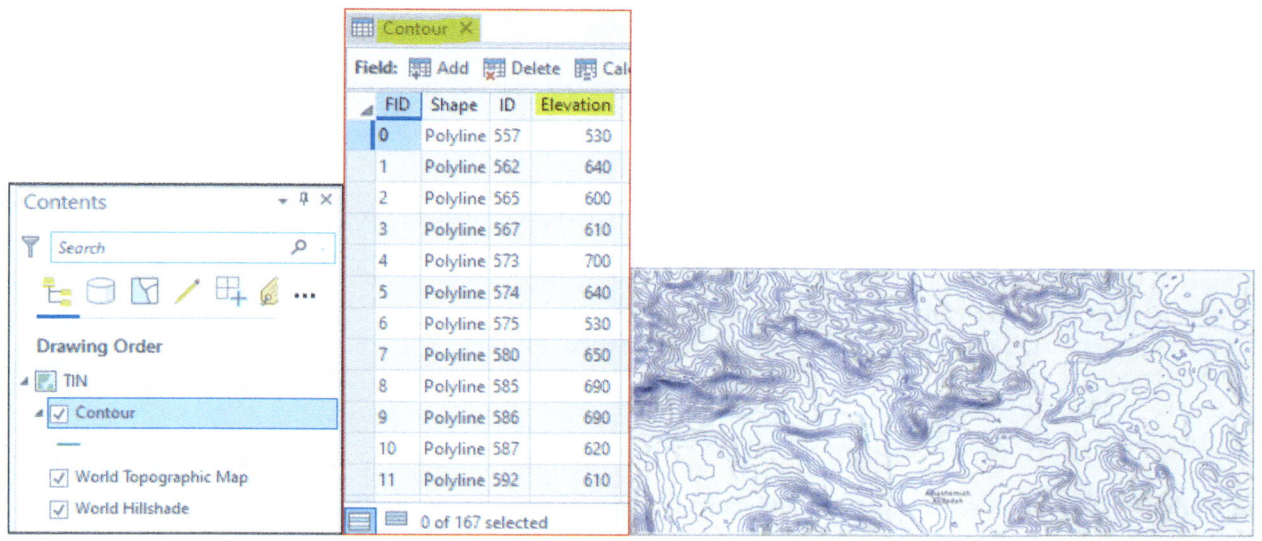

Note The table has an "Elevation" field that will be used to create the TIN

10. Close the **Contour** attribute table
11. Click the **Analysis** tab on the ribbon, in the **Geoprocessing** group, click the **Tools** button
12. In **Geoprocessing** pan, click the **Toolboxes** tab and open the **3D Analyst Tools**
13. Open **TIN Dataset** and click **Create TIN**
14. Fill the Create TIN pane as follows:
15. Output TIN: **Dhuleil_TIN** (save it in Data_Ch15 under Folder)
16. Click Save
17. Coordinate System Select **Contour** (from the drop-down arrow)
18. Input Features: **Contour**
19. Height Field: **Elevation**
20. Type: **Hard_Line**
21. Accept the reset of the default
22. Check Constrained Delaunay
23. Click Run

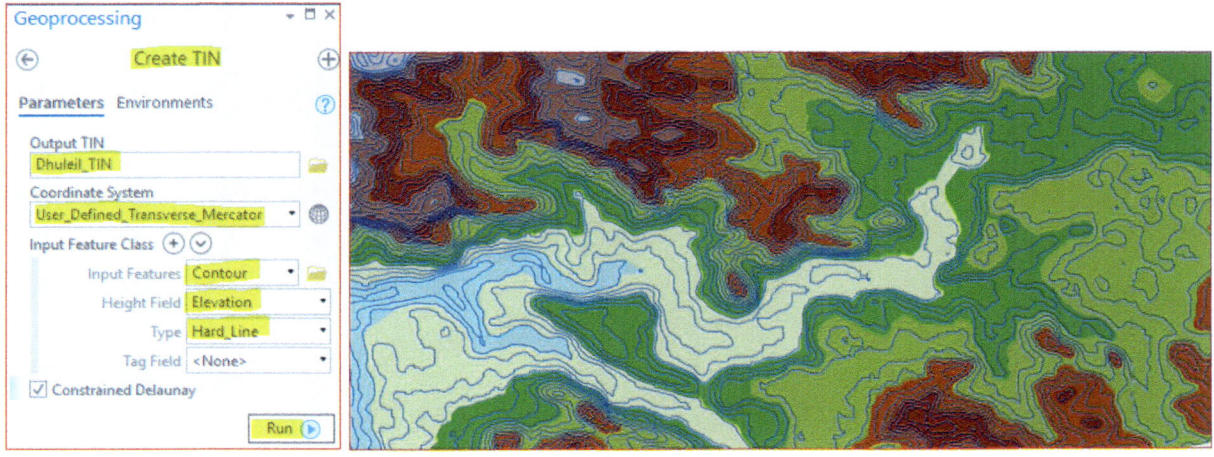

Result **Dhuleil_TIN** is created and added to CP.

Change the Symbols of the TIN

24. In the CP, highlight "**Dhuleil_TIN**" and click the **TIN Layer** tab on the ribbon, in the **Drawing** group, open the drop-down arrow of the **Symbology**, select "**Slope**"
25. The **Symbology – Dhuleil_TIN** pane open

26. Make sure the "**Symbolize your layer using a surface**" is selected (fourth icon)
27. Under **Surface** from **Draw using** select "**Simple**"
28. Under "**Current symbol**", click the **color** symbol
29. Under **Format Polygon Symbol**, select the **Properties** tab
30. Under Appearance click the color symbol and select the **Mango** (c4, r2)
31. Click Apply
32. Keep the Symbology pane open
33. In the CP, r-click the Contour layer and click Remove

Result Dhuleil_TIN now has a mango color and shows the topography of the study area.

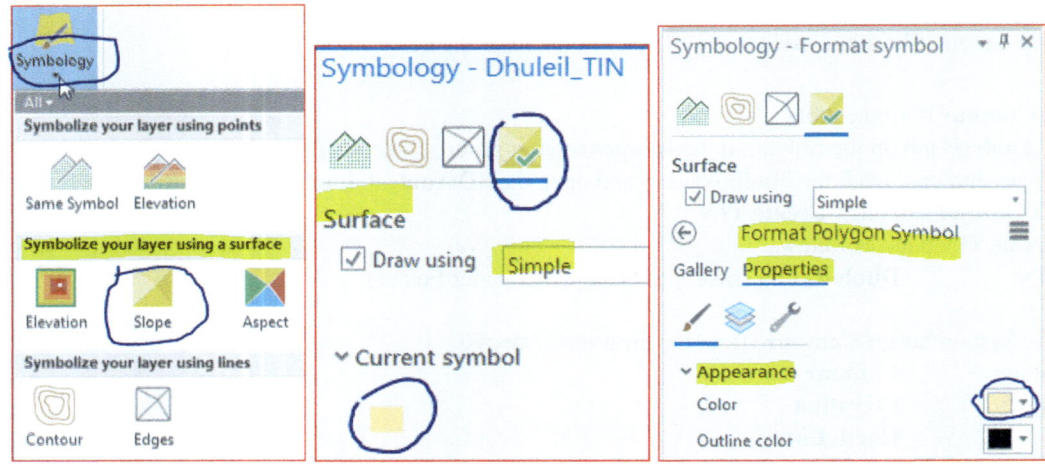

Drape Layers onto Dhuleil_TIN

Now, you will drape some layers over the TIN. The layers do not have an elevation field in their attribute tables, so the base height is taken from **Dhuleil_TIN** and applied to the layers.

34. In Catalog pane expand the **Data_Ch15**, and then **Q1** folder, select **Building, Farm, ObserbationWell, Street, Tree, Valley,** and **WWTP** layers and drag them into **Map View**
35. In the CP, drag the **Building** and place it above the **WWTP** layer
36. In **CP,** change the color of some layers based on the table below
37. Classify the **Building** based on the **Height** field

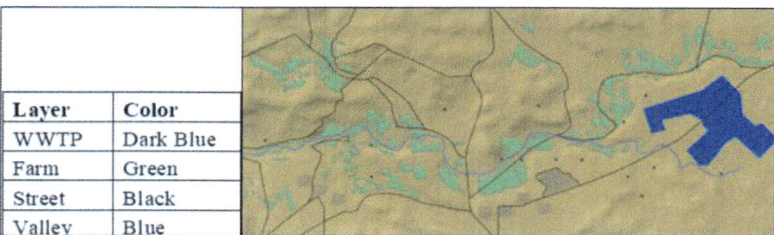

Layer	Color
WWTP	Dark Blue
Farm	Green
Street	Black
Valley	Blue

Add Data to the Scene

The **3D scene** helps to visualize geographic data in 3D, which is more intuitive than viewing the data in 2D. The difference between data displayed in 2D and 3D is that the 3D data must have an elevation associated with it. That elevation value may be

- Stored as a part of the data itself, or
- The data may be draped over an elevation surface.

Scenes in ArcGIS Pro are 3D maps that can be either **local** (for small areas) or **global** (for large areas). In ArcGIS Pro, the scene has a default surface called the **WorldElevation3D/Terrain3D (Ground)**, which comes from ESRI's World Elevation service. You can define other surfaces in addition to the **Ground,** for instance, **Dhuleil_TIN** or any other **DEM**. Data displayed in a 3D scene must have an **Elevation** value, that is, a **height** (i.e., building) or depth (i.e., well) at which to be displayed. Elevation information might be stored within the feature's geometry as a **z value** or as a value in the attribute table. Alternatively, the feature might be assumed to be on, or relative to, an elevation surface.

38. Click **View** tab on the ribbon, in the **View** group, click **Convert** drop-down arrow and choose **To Local Scene**

Result Two things happen.

A new map, called **TIN_3D,** was established in the CP, and a new tab called **TIN_3D was established** above the **Map View**. **TIN_3D** consists of the following:

(a) **3D** Layers: empty
(b) **2D** Layers: contain **Tree, ObserbationWell, Valley, Street, Building, WWTP,** and **Farm** layers

Note In the CP, the **Elevation Surfaces** contain the **Ground** (**WorldElevation3D/Terrain3D**).

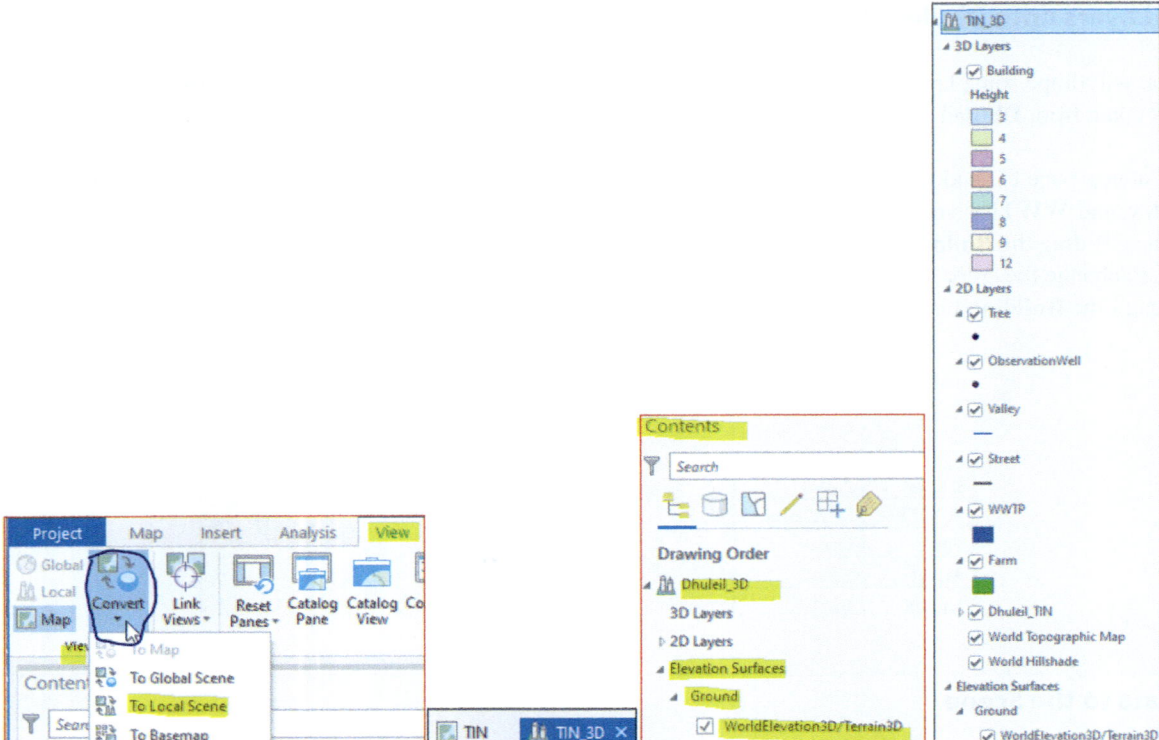

Extrude Layers

Extrusion is the process of stretching a flat, 2D shape vertically to create a 3D object in a scene. The points, lines, and polygons support extrusion; for example, building polygons can be extruded by a height value to create three-dimensional building shapes. When you extrude a layer in the 2D Layers category in the Contents pane, it is moved to the 3D Layers category, as the Extrusion is only available in scenes. In ArcGIS Pro there are various feature extrusion types and, in this section, you will use the "**Base Height**".

39. In CP, drag the **Building** from **2D Layers** to **3D Layers**
40. In CP, under the **3D layer** highlight the **Building**, click **Feature Layer** tab on the ribbon, in the **Extrusion** group, click the **Type** drop-down arrow and choose "**Base Height**",

41. In the **Field**: Choose **Height** from the drop-down arrow, and click the **Extrusion expression** button
42. In the **Expression** Builder box, below the **Expression** after **$feature. Height**, type ***10**
43. Click the Verify button
44. If expression is valid, click OK
45. Zoom in to the eastern **Building**

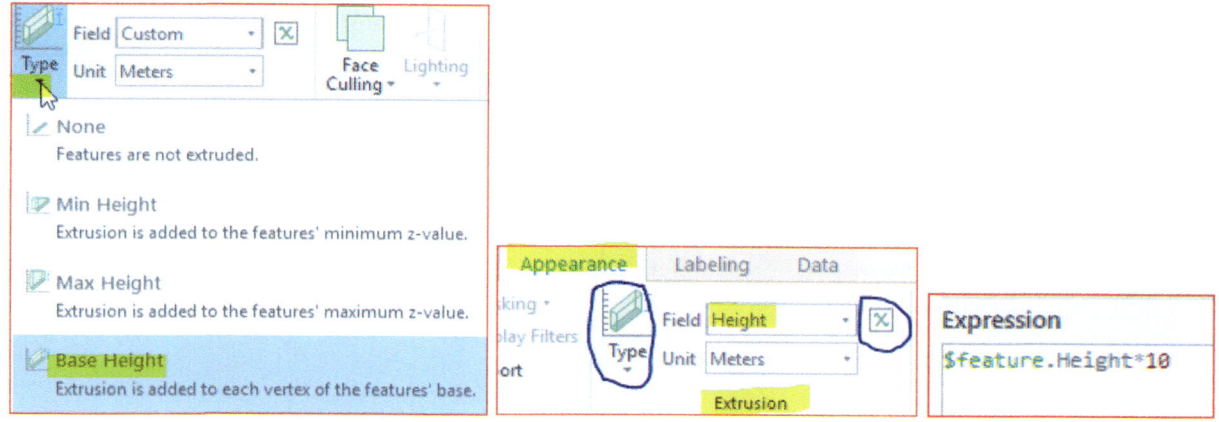

Result The **Building** was extruded based on the "**WorldElevation3D/Terrain3d**" raster.

On-Screen Navigator

The on-screen navigator exposes many camera navigation commands in a single control in the lower left of a view. Using the navigator is optional but provides quick access to controls that help refine camera movements when moving through maps and scenes. The navigator has two modes: a smaller display mode providing a north indicator and pan function and a full-control display state for raising and lowering the camera, looking around, zooming, and rotating. The control appears by default in 3D but not in 2D. However, you can set preferences for when maps and scenes are opened, if you want the control to appear at all, and in which display mode.

Heading Mode and Full Control Mode

The smaller display state of the on-screen navigator shows the heading as you pan using the ring. Once the panning ring changes color, panning can begin by dragging in any direction. The closer you are to the center of the navigator, the slower the pan speed. A semitransparent arrow follows your pointer to help indicate direction as you move away from the ring. Click the **North** button to reset the view back to facing north.

The full-control navigator has a larger set of functions for manipulating the camera. Hover the pointer over the controls to highlight controls you can interact with. Precise camera control gives you a more intuitive navigation experience if you are new to navigating in 3D. The following table lists the navigator control options and their descriptions:

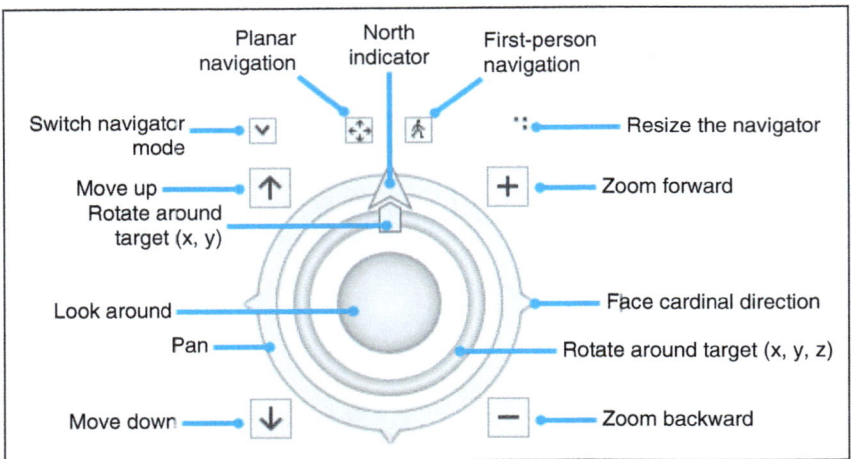

Navigator control option	Description
North indicator	Rotate to indicate north direction. Click the arrow to return the view to face north
Switch navigator mode	Change the display view of the navigator to heading mode or full control mode
Move up/Move down	Specific to 3D, move the camera vertically higher or lower. The camera's X,Y and heading are maintained
Resize the navigator	Drag to resize the navigator. Larger sizes may be more suitable for touch screen use
Zoom forward/zoom backward	Continuously zoom in closer or farther away from the view
Face cardinal direction	The arrows are clickable on the outer ring. Click to rotate the view to face the specific cardinal direction (N/S/E/W)
Pan (outer ring)	Continuously pan the camera horizontally in 2D and 3D. The ring is clickable for short pan movements, or you can click and hold to continuously pan across the view following the direction of the pointer
Rotate around target (inner ring)	Rotate the camera around the center target point. Click and hold to rotate in a motion similar to a dial
Look around	The camera remains stationary but can look in all directions. (same behavior as using the B keyboard with the **explore** tool)
Planar navigation	When you navigate with the mouse, the camera is constrained to maintain the current viewing angle and distance
First-person navigation	The camera navigates the scene using the perspective of a person walking. Roam in 3D using the keyboard to move the camera and the mouse pointer to look around

Apply a Realistic Layer

There are a variety of symbols for the **tree** layer, and some symbols can be displayed only in maps or only in scenes. In this section, you will use the **3D Vegetation - Realistic Trees**. Realistic trees display a point feature class as realistic-looking 3D trees in a scene.

46. In CP, drag the **Tree** from **2D Layers** to **3D Layers**
47. In the CP, click the symbol of the **Tree**, In the **Symbology – Tree** pane, in the Gallery tab, type "**tree**" in the search window and enter
48. Scroll down and select "**Date Palm**" under **3D Vegetation – Realistic**
49. Click Properties dam, change the size to 30 and click Apply

Result The **Tree** display in the **Map View** as a 3D layer.

Extrude the Well and the WWTP Layers

The well layer "**ObservationWell**" has a field in the attribute table called "**Height**". The height field represents the depth of the wells below the ground surface.

50. In CP, drag the **ObservationWell** from **2D Layers** to **3D Layers**
51. In the CP, r-click the **ObservationWell** and Zoom To Layer
52. In the CP, click the symbol of the **ObservationWell**
53. In the **Symbology** pane in the **Properties** tab, under Appearance
54. Change the **color** to **blue** and the **size** to **6**
55. Click Apply
56. In CP ensure that the **ObservationWell** is selected, click **Feature Layer** tab on the ribbon, in the **Extrusion** group
57. Click the **Type** drop-down arrow and choose "**Base Height**", in the Field: choose **Height**
58. Click the **Extrusion Expression** button
59. In the Expression Builder dialog box, after **$feature. Height**, type ***10**
60. Click the Verify button
61. OK
62. In the CP, under the "**Ground**" uncheck the **WorldElevation/Terrain3D** and the **World Hillshade**
63. Rotate the image to see the **ObservationWell**

Result The wells will be displayed as features drilled below the ground.

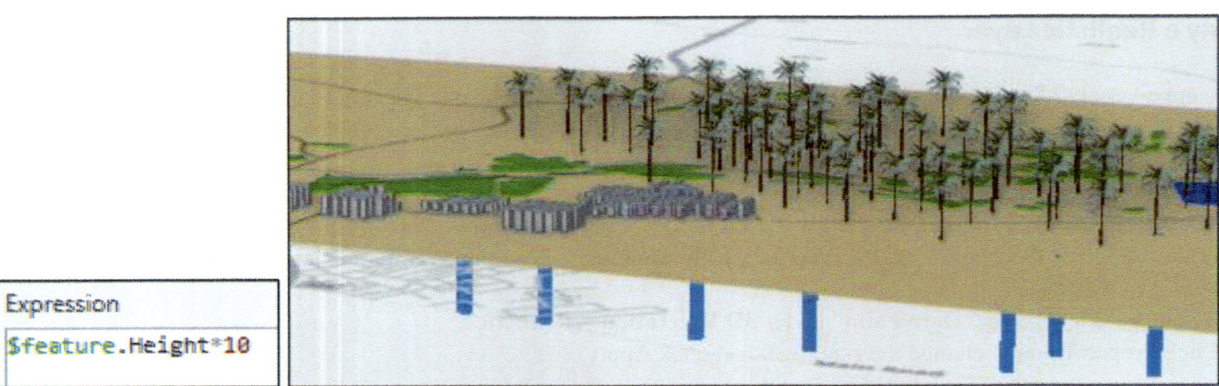

64. In CP, drag the **WWTP** from the **2D Layer** into the **3D Layers**
65. In the CP, click the symbol of the **WWTP** in **3D Layers**
66. In CP, ensure that the **WWTP** is selected, and click **Feature Layer** tab on the ribbon, in the **Extrusion** group
67. Click the **Type** drop-down arrow and choose "**Base Height**", in the Field: choose **Depth**
68. Click the **Extrusion Expression** button
69. In the Expression Builder dialog box, after **$feature. Depth**, type ***20**
70. Click the Verify button
71. OK

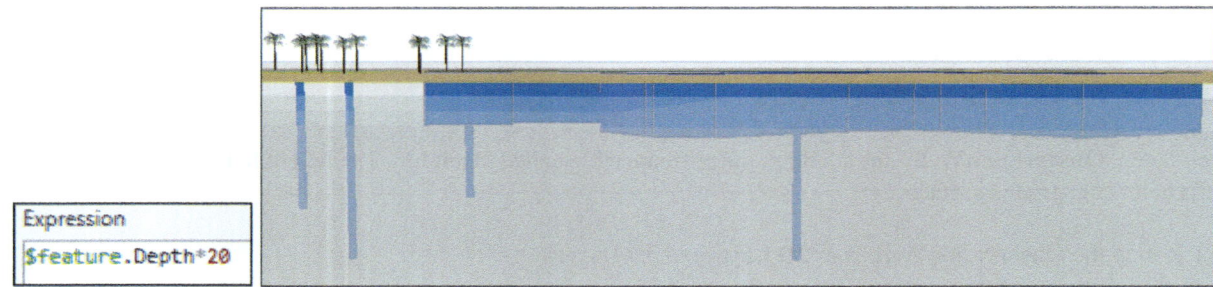

Display the Street, Valley and Farm Layers in the Scene

72. In CP, drag the **Street** from under 2D layer into to 3D Layers
73. In CP ensure that the **Street** is selected, click **Feature Layer** tab on the ribbon, in the **Extrusion** group, click the **Type** drop-down arrow and choose "**Base Height**",
74. Repeat the above steps for the **Valley** and **Farm** layers

Change Elevation Units

When working with a scene, the viewing height relative to the ground is listed in the height list in the lower corner of the display. To change the elevation unit displayed in the list do the following:

75. In the CP, right-click **TIN_3D** (scene) and click **Properties**.
76. On the **Map Properties: TIN_3D** dialog box, on the **General** tab, choose **Meters** from the **Elevation Units** drop-down menu
77. Click OK

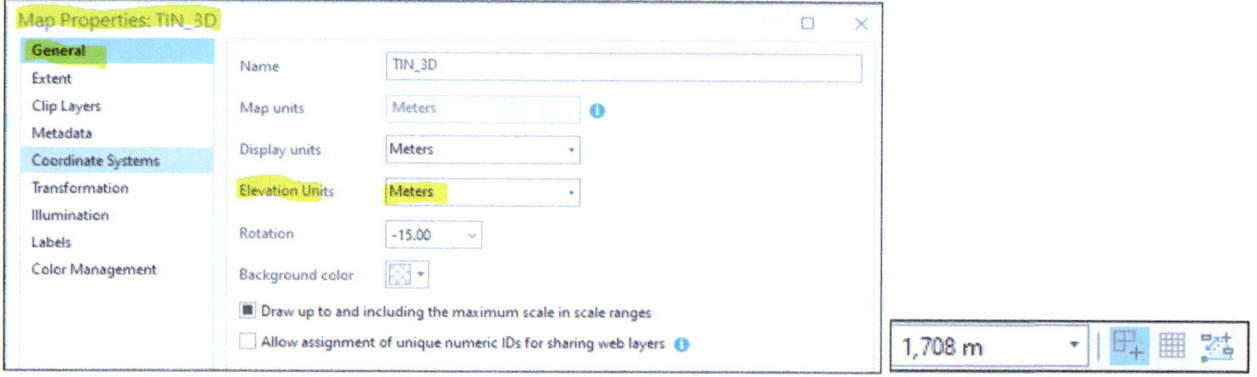

Result The unit changes from feet to meter.

Navigate Underground and Change Vertical Exaggeration

Belowground navigation is disabled by default for layers in a scene to avoid navigating underground accidentally. Since the scene (TIN_3D) contains underground data "wells", you should enable this capability.

78. In the CP, highlight the **Ground** under the **Elevation Surfaces**
79. On the **Elevation Surface Layer** tab, in the **Surface** group, check the **Navigate Underground** and **Shade to Relative To Light Position**.
80. In the **Drawing** group, for **Vertical Exaggeration**, type **4** and for **Surface Color** choose gray 10%

Note Make sure that the **World/Elevation/Terrain3D** is checked under the **Ground.**

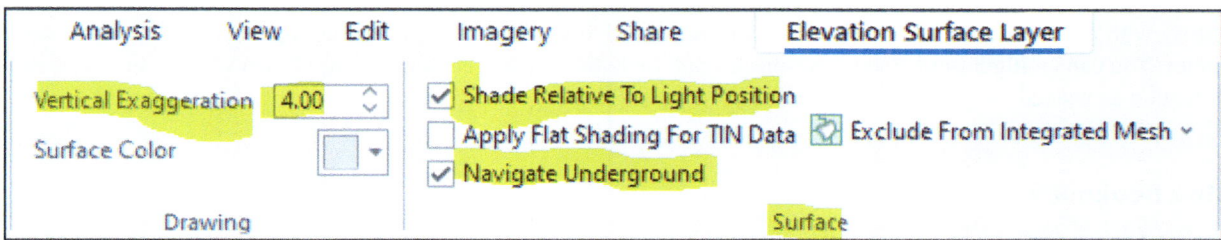

Result Vertical exaggeration is a visual effect that enhances the 3D appearance of the scene.

Use the Dhuleil TIN as an Elevation Source

An elevation source layer references the data that contribute height values to the elevation surface layer. An elevation surface layer can have more than one elevation source layer, and different types can be combined. Local elevation source layers can be a single-band raster that contains elevation information or a TIN dataset. The **Dhuleil_TIN** that created earlier is considered a source of an elevation and you are going to use it by replacing the "**WorldElevation3D/Terrain3D**".

81. In CP, drag the **Dhuleil_TIN** below the Ground under the Elevation Surfaces **OR**
82. Right-click the **Ground** under the **Elevation Surfaces** and choose **Add Elevation Source Layer**
83. Browse to **Data_Ch15** under **Folder** select **Dhuleil_TIN** and click OK **OR**
84. **Map** tab on the ribbon in the **Layer** group and open the **Add Data** drop down arrow and click "Elevation Source Layer" and choose **Dhuleil_TIN**
85. In the CP uncheck the "**WorldElevation3D/Terrain3D**"
86. In the CP uncheck the **World Topographic Map** and keep the **World Hillshade** checked

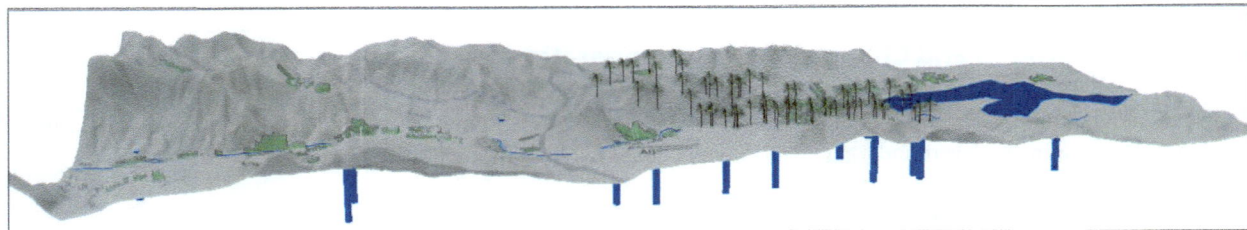

Create Animation From Bookmarks

The animation in ArcGIS Pro can be created either using bookmarks or fly through a scene.

A bookmark is a navigation shortcut to a position on a map or perspective in a scene to return to later or share with others. Bookmarks can be used to create keyframes in an animation or pages in a bookmark map series. Animations can help you tell a story with a map or a scene and then export it as a video to share. You can create animations by capturing a series of keyframes. You can configure how the transitions are interpolated between each keyframe. When you're done, you can edit the animation as needed.

Create a Bookmark

1. In the CP, under the **Ground**, r-click the **Dheleil_TIN** and click **Zoom To**
2. On the **Map** tab, in the **Navigate** group, click **Bookmarks** and click **New Bookmark**.
3. On the **Create Bookmark** dialog box, type **Basemap**, in the **Description**, type **Dhuleil Region**.
4. Click **OK**.
5. Zoom in to the eastern part of the region around the "**WWTP**" and create a bookmark and call it **WWTP** and in the Description type "**Wastewater Treatment Plant**" and click OK

Create Animation From Bookmarks

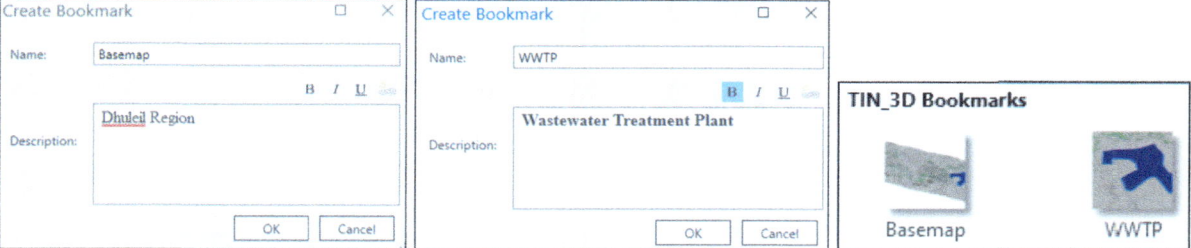

6. Zoom to other locations such as eastern Dhuleil, central Dhuleil, northern highland, and western Dhuleil and show the wells, trees, buildings and create more bookmarks.
7. After finishing creating the r-click the **Dheleil_TIN** and click **Zoom To** and create bookmark

Result Ten bookmarks created, but you can more if you desire.

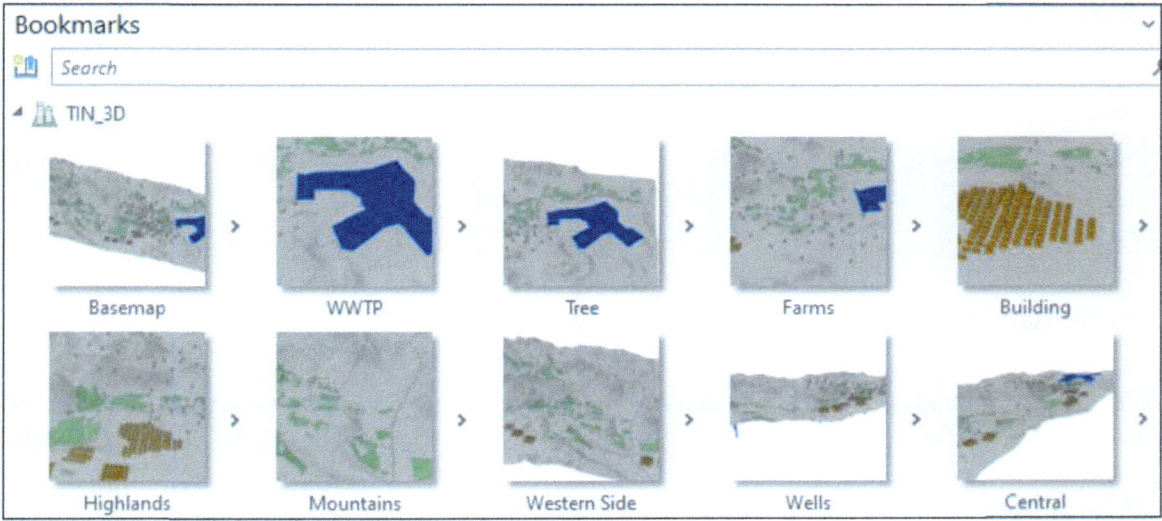

Use Bookmarks to Create an Animation

In this step, you will import the 10 bookmarks to create an animation by converting the bookmarks to keyframes. Each bookmark represents a location on the scene; it will start from the basemap that covers the whole region and then go to different locations and end at the beginning location.

8. Click the **Map** tab, in **Navigate** group, click the **Bookmark** and click **Manage Bookmark**
9. Make sure that the bookmarks are organized in order
10. Click the **View** tab, in the **Animation** group, and click **Add**.
11. The **Animation Timeline: Animation** dialog box display
12. In the **Animation** tab on the ribbon, in the **Create** group, click the **Import** drop-down arrow and select **Bookmarks to Fly Through**

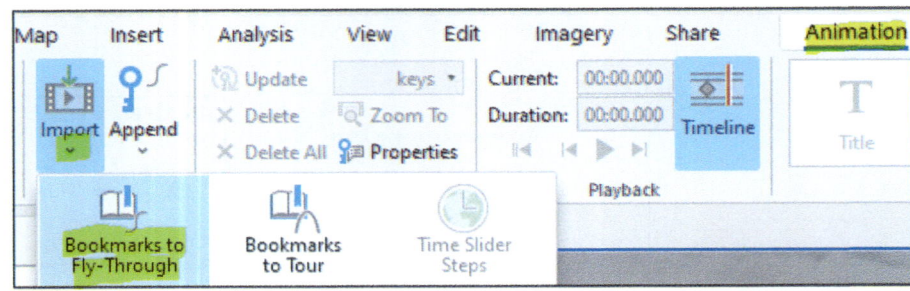

Result All the bookmarks the way they organized will create the keyframes and establish the camera fly path

13. In the **Animation Timeline: Animation** click the **first keyframe** and in the **Animation** tab on the ribbon, in the **Playback** group, click inside the **Duration** box and type **00:30:000**, then click the **Play** button

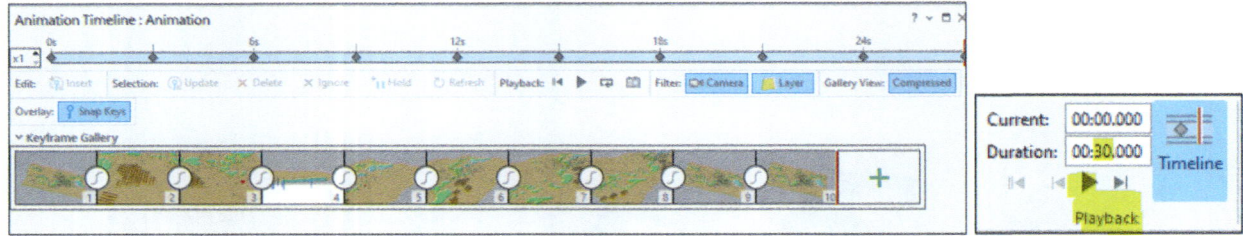

Result Now, the keyframes from the bookmarks generated the animation.

14. If you closed the **Animation Timeline: Animation**, you can open it by clicking on **Animation** tab, in the **Playback** group, and click **Timeline**

Create a Video

Once you are happy with your 3D map animation, you can make a video.

15. Click the **Animation** tab in the ribbon, in the **Export** group, click the **Movie** button
16. In the **Export Movie** dialog box, select **YouTube**
17. File Name: **Dhuleil.mp4** and save it in **Data_Ch15** under **Folder** and click **Save**
18. Click **Advanced Movie Export Setting** and accept the default (720p HD Letterbox (1280 x 720)
19. Click **Export**

Note To create the video, it takes a few minutes depending on the computer configuration.

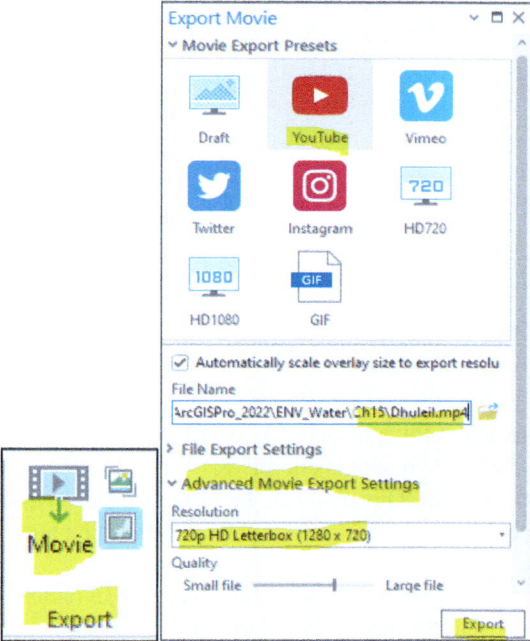

20. To view the **Dhuleil.mp4**, open the folder **Data_Ch15** in window explorer and double click it

Fly Through a Map or Scene

A fly through animation simulates the camera moving through a map or scene and mimics what it is like to be physically present in the view. Examples include a 3D fly although along a parade route or 2D panning along a proposed electrical transmission line path. A fly through animation is usually best served by smooth curves between keyframes using a fixed transition. In a few cases, the linear transition may work well, but one must be aware that this creates simulated bumps when the camera changes direction. In this section, you will delete the keyframe created by the bookmarks and start over with the fly through a map or scene. You can use different camera transition types, as 6 types are available: fixed, adjustable, linear, stepped, and hold. In this section, you will use the fixed transition.

21. Click the **Animation** tab on the ribbon, in the **Edit** group, click **Delete All** to delete all keyframes in the track to start over.
22. In the **Animation** tab, in the **Create** group, click the **Append** drop-down menu and verify that **Fixed** is the transition type.
23. In the CP, r-click the **Dhuleil_TIN** and Zoom To
24. In the **Create** group, click the **Append** to create the first keyframe.
25. Navigate the camera to the first keyframe location

Note The camera location is stored in the first keyframe at zero seconds. You can verify that the first keyframe was created by confirming the thumbnail appears in the Animation Timeline pane. You can also verify it using the Keyframe List drop-down menu in the Edit group.

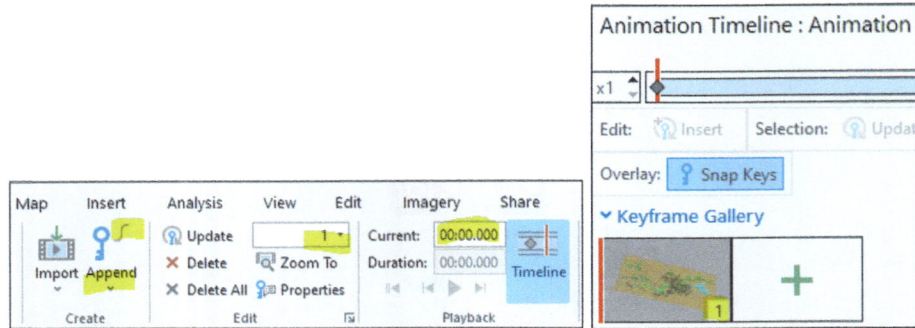

26. Zoom to the eastern part of the **Dhuleil_TIN** that show the **WWTP** and click **Append**.
27. Repeat the previous step until the fly through path is complete.
28. Make sure the **WorldElevation3D/Terrain3D** is uncheck the to see the subsurface wells
29. In the **Animation** tab, in the **Playback** group, click inside the **Duration** box and type 30, then click the **Play** button to play through the entire animation.
30. Create a YouTube movie as in the previous step, and save it in **Data_Ch15** under **Folder**
31. Close the **Export Movie** dialog box, and the **Animation Timeline**, the **Geoprocessing** pane and any other pane
32. Save your project

Time Tracking

Time tracking is a visual representation that uses the time field to show how the events are changing over time. The Plume attribute table layer contains two fields that represent a time, which allows the user to visualize the events at various locations over time.

1. Click **Insert** tab on the ribbon, in the **Project** group, click **New Map** and call it **Plume**
2. In Catalog pan, select **GasStation**, **Plume** and **SupplyWell** from **Q2** under **Data_Ch15** and drag them to the **Map View**
3. In the CP, r-click **Plume** select **Symbology**
4. In the **Symbology** pane, under **Primary symbology,** choose **Unique Values** from the drop-down arrow, and in **Field 1,** choose **PlumeLevel**
5. Click the **Symbol layer drawing tab**, and turn on **Enable symbol layer drawing**.
6. On the **Basic** tab, drag symbol classes to reorder their drawing order if necessary (Plume 1 at the top and Plume 5 at the bottom)
7. Change the color of the plumes from top to bottom into **light blue**, **yellow**, **mango**, **pink**, and **red**
8. Click **More** and uncheck "**Show all other values**"

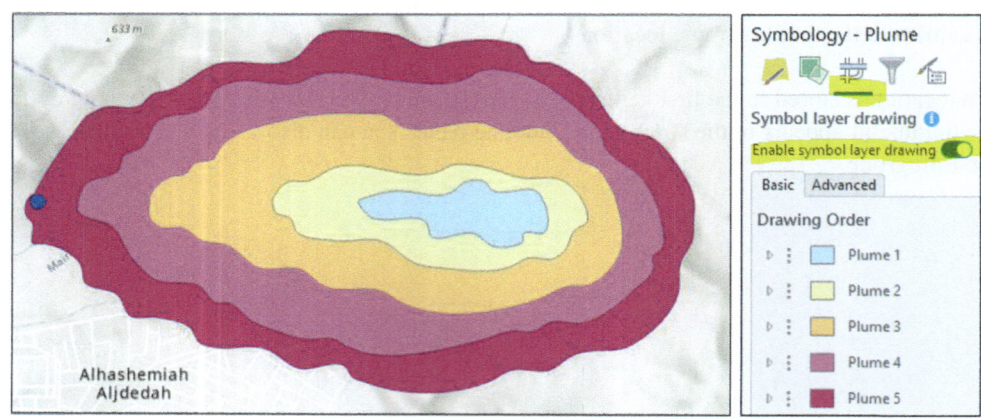

9. In the CP, click the symbol of **SupplyWell**, select **Circle1** symbol, **blue** color and size 8
10. In the CP, click the symbol of **GasStation** and change the color to **red**, outline color (no color, outline width 0, and make sure the **GasStation** is above the **Plume** in the CP
11. Click **View** tab on the ribbon, in the **View** group, click **Convert** drop-down arrow and choose **To Local Scene**
12. In CP, r-click the **Plume** and Zoom To Layer
13. In CP, drag the **SupplyWell** from **2D Layers** to **3D Layers**
14. Under the **3D layer** highlight the **SupplyWell**, click **Feature Layer** tab, in the **Extrusion** group, click the **Type** drop-down arrow and choose "**Base Height**", in the **Field**: Choose **Height** from the drop-down arrow, and click the **Extrusion** expression button ⊠, In the **Expression** Builder box, below the **Expression** after $feature. Height, type *10
15. Click the Verify button; if the expression is valid, click OK
16. In CP, drag the **GasStation** from the 2D layer into the 3D layers
17. In the CP, click the symbol of the **GasStation** in **3D Layers**, click **Feature Layer** tab on the ribbon, in the **Extrusion** group, click the **Type** drop-down arrow and choose "**Base Height**", in the Field: choose **Height**, click the **Extrusion Expression** button, in the Expression Builder dialog box, after **$feature. Height**, type **25*
18. Click the Verify button, then OK
19. Make sure the **WorldElevation3D/Terrain3D** and the **World Hillshade** are unchecked
20. Use the **Navigator** and the **Rotate around target** to rotate the image

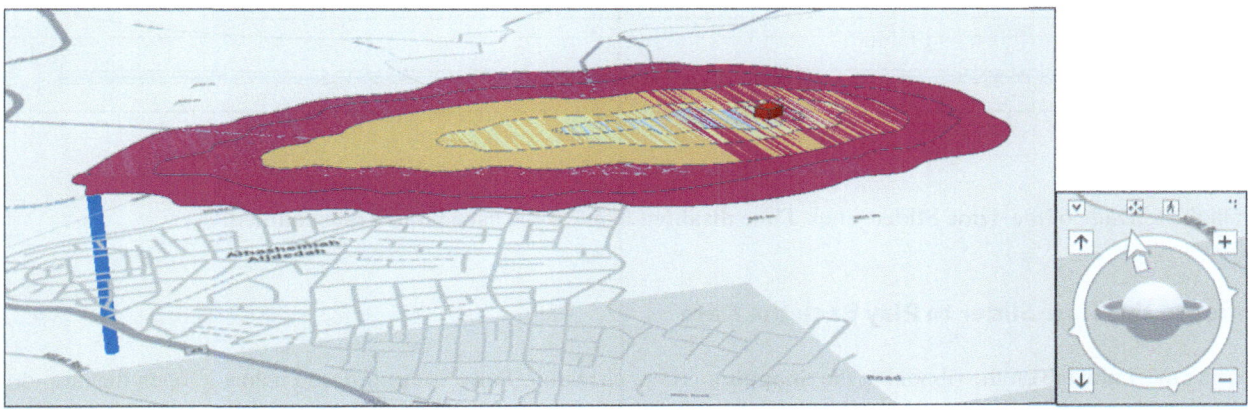

21. In the CP, r-click the **Plume** to open the Attribute Table, you see two date fields: **Coll_Date** and **End_Date**, familiarize yourself with the time and close the table.

Enable Time and Visualize the Data

22. In the CP, d-click the **Plume** to open the **Properties** dialog box.
23. Click the **Time** tab
24. Click the **Layer Time** drop-down arrow and click **Each feature has start and end time field**.
25. The **Time Field** value is automatically set to **Cool_Date** and **End_Date**
26. The **Time Extent** values are set to the range of the data.
27. Accept the rest of the default and click OK
28. The **Time Slider** appears at the top of the map view.
29. Hover over the **Time Slider** to see its controls.

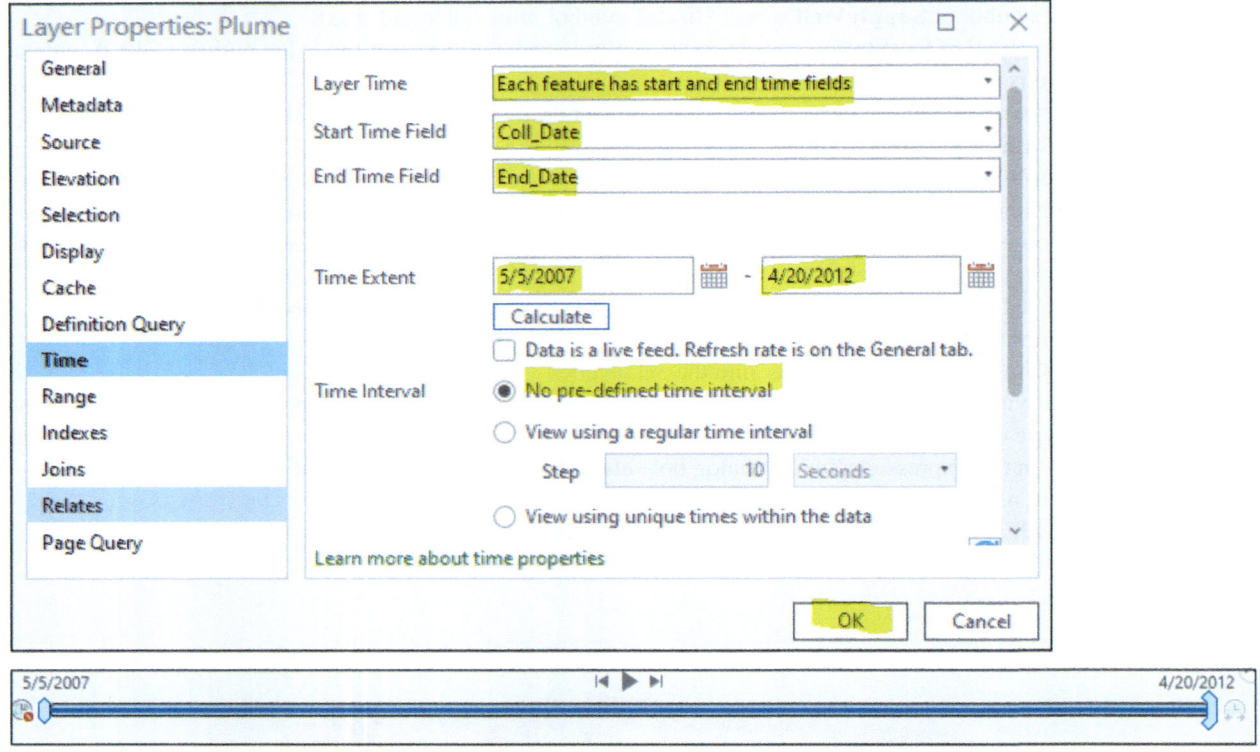

28. On the left side of the **Time Slider**, click **Time disabled** icon to change it to **Time enabled**.

Configure the Time Slider to Play Back the Data

29. Click the **Time** tab on the ribbon, in the **Snapping** group, check the **Time Snapping**, and below it, open the drop-down arrow and click **Months**
30. In the **Current Time** group, change the **Span** setting to 4 and press the Enter key.

This ensures that only four months' worth of data is displayed at a time.

31. In the **Playback** group, click **Repeat** and then click **Play All Steps**
32. In the **Playback** group, the Adjust Playback Speed slider was used to decrease the speed.
33. Play the data again.

34. To stop the plume, on the left side of the **Time Slider**, click **Time enabled** icon to change it to Time **disabled**.

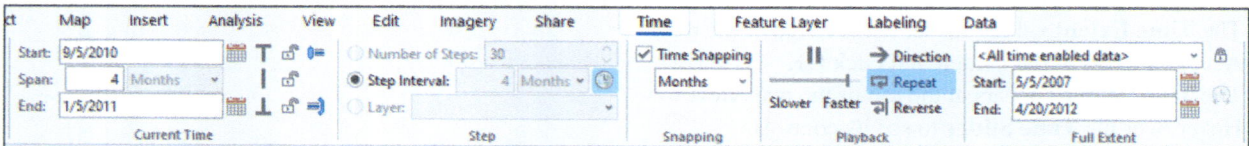

35. Close the Symbology pane
36. Save the project

Create an Animation and Export a Video

36. Click the **View** tab on the ribbon, in the **Animation** group, click **Add**.

Result An empty **Animation Timeline: Animation** pane appears under the map view. On the ribbon, an **Animation** tab appears. Before you create keyframes for the animation, you will zoom to the correct map extent.

37. In the **Contents** pane, right-click the **Plume** layer and click the **Zoom To Layer**.

38. Click the **Animation** tab on the ribbon, in the **Create** group, click **Import** drop-down arrow and click **Time Slider Steps**.

Result The **animation timeline** pane populates with few keyframes. (It takes a few moments for the thumbnail images to display in the keyframes.) The first keyframe represents the animation at zero seconds (current: 00:00.000) and duration: 00:48:00. Each subsequent keyframe corresponds to the data for 4 months.

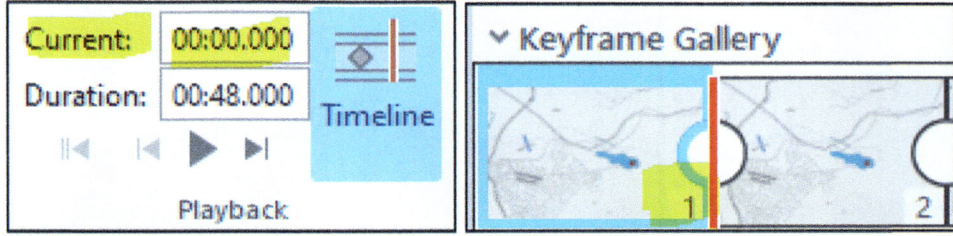

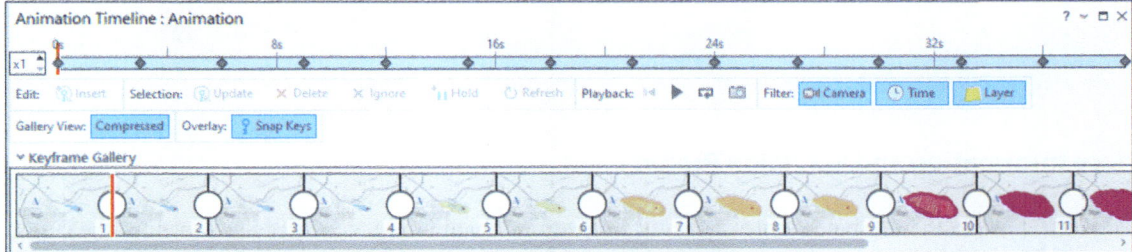

40. On the **Animation** tab, in the **Playback** group, the **Duration** setting is forty-eight seconds (00:48.000). There are 17 slide transitions, which means that each slide is visible for 2. 28 seconds. You want to make the transition time to 2 seconds.
41. On the **Animation** tab, in the **Playback** group, replace the **Duration** value with 00:34 and press the **Enter** key. The time stamps in the **Animation Timeline** pane are adjusted. Each transition now takes one second.
42. In the **Playback** group, click **Play**. (Do not click **Play** on the time slider changes you have made to the animation are not reflected in the time slider playback.)

43. On the **Animation** tab, in the **Export** group, click **Movie** .
44. The **Export Movie** pane appears. The default settings output a video configured for **YouTube** in **mp4** media format.
45. In the **Export Movie** pane, in the **File Name** box, change the default file name **Plume.mp4**, and save it in **Data_Ch15** under **Folder**
46. Click **Export**
47. To view **Plume.mp4,** open the folder **Data_Ch15** in the window explorer and double click it.

Result The video will play.

48. Close the **Animation Timeline** and the **Export Movie** dialog box
49. Save the project

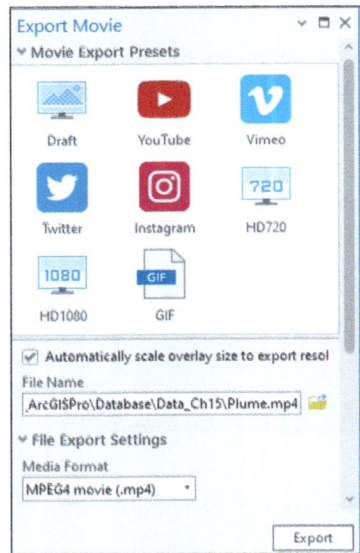
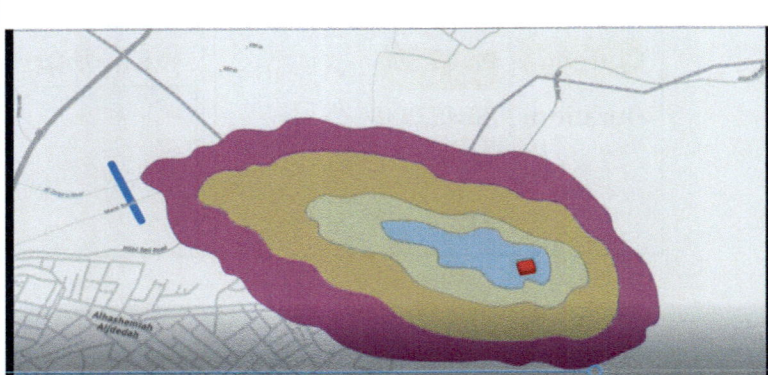

Add Surface Information and Elevation Profile

The tool adds a **Z** field to the attributes table of the "**RainStation**" feature with statistics derived from the **duluth** digital elevation model. The Z field in the attribute table is calculated based on the **duluth** surface, which is used to interpolate Z information for the input features.

Scenario 2 You have been asked to use the **duluth** DEM to identify the elevation of the nine "**RainStation**" feature class.

48. Click **Insert** tab, in the **Project** group, click **New Map** and call it **Rain Station**
49. In Catalog pan, select **City**, **duluth**, **RainStation,** and **Stream** from **Q3** under **Data_Ch15** and drag them to the **Map View**.
50. In the CP, r-click **duluth** raster and Zoom To Layer
51. Symbolize the layers in the CP as shown in the table below.
52. In the CP, r-click **RainStation** and open the attribute table. The attribute table has only 3 fields: FID, Shape, and Id. Then, close the attribute table

Create a Video

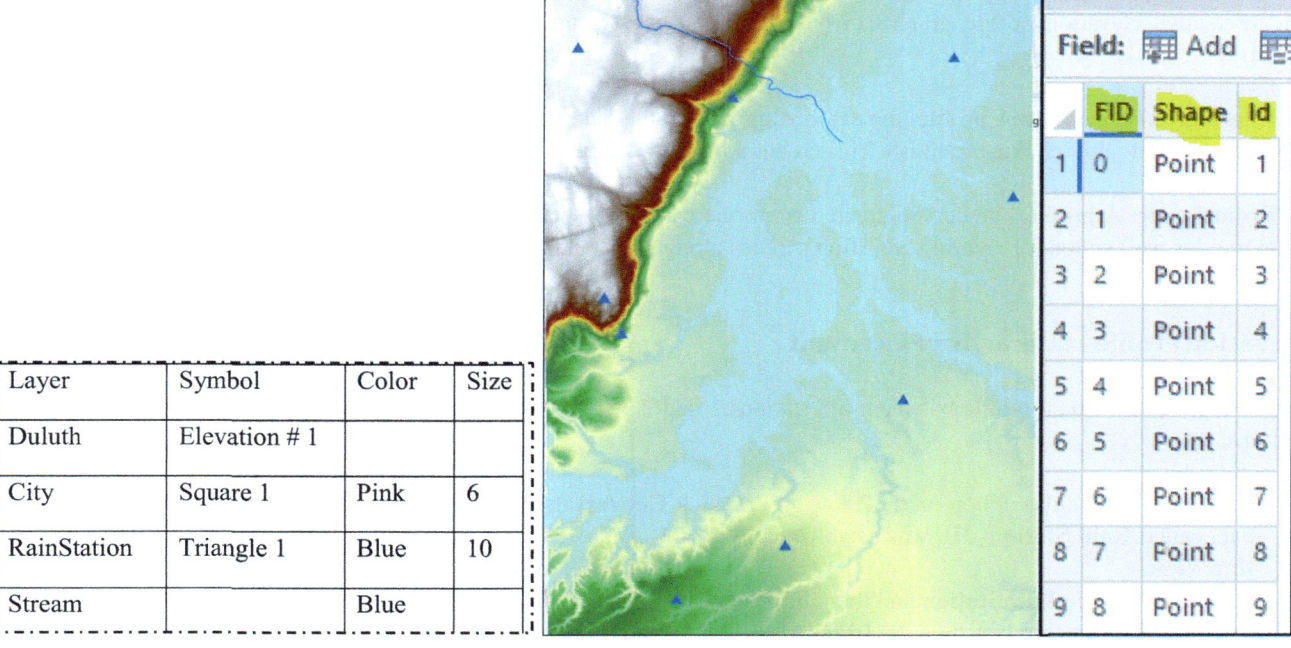

Layer	Symbol	Color	Size
Duluth	Elevation # 1		
City	Square 1	Pink	6
RainStation	Triangle 1	Blue	10
Stream		Blue	

52. Click the **Analysis** tab on the ribbon, in the **Geoprocessing** group, click the **Tools** button
53. In **Geoprocessing** pan, click the **Toolboxes** tab and open the **3D Analyst Tools**, then open the **Statistics** and click **Add Surface Information**
54. Fill the **Add Surface Information** dialog box as follows:
55. Input Features: **RainStation**
56. Input Surface: **duluth**
57. Check the **Z** under Output Property
58. Method: **Bilinear**
59. Accept the rest of the default
60. Click **Run**

Result Z is added to the attribute table of the **RainStation** feature. The elevation of Z is between 183.5 and 397.2 meters.

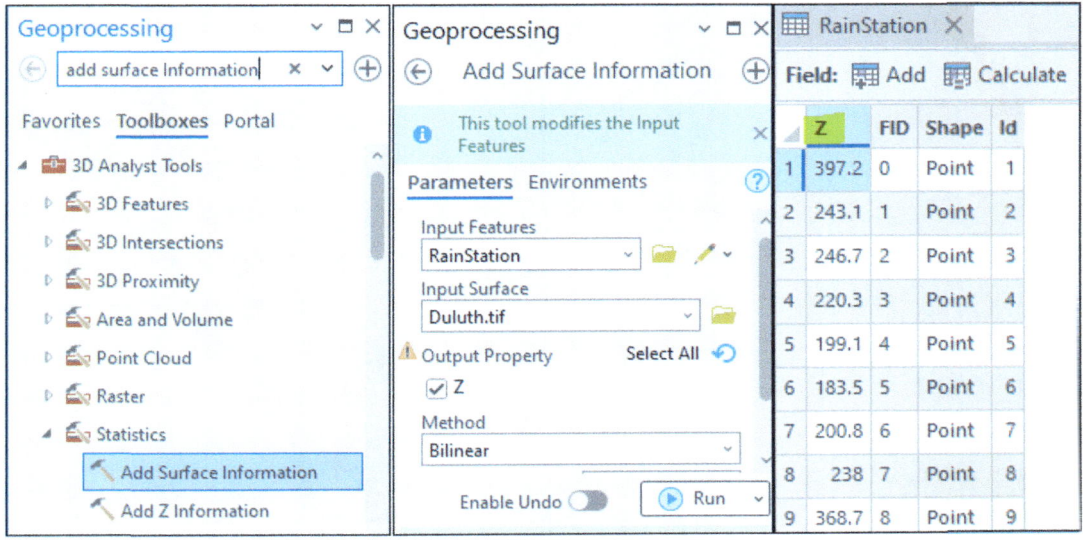

Elevation Profile

An interactive elevation profile graph can be created in either a map or a scene using one of the following creation methods:

- **Interactive Placement**: A profile line can be digitized in the map or scene by interactively clicking to define the path's vertices and double-clicking to finish. This creation method is well suited for maps and scenes that do not contain existing line features.
- **From Layer**: A profile line can be created from one or more selected line features in the map or scene. This creation method is well suited for creating elevation profiles along paths that already exist as data in the view, such as a stream.

Elevation Profile - Interactive Placement

You will create a vertical profile between rain stations 1 and 2 by digitizing by clicking to define a path between these two stations.

61. Click **View** tab on the ribbon, in the **View** group, click **Convert** drop-down arrow and choose **To Local Scene**. A new map, called **Rain Station_3D, was** established in the CP, and a new tab called **Rain Station_3D was added** above the **Map View**.
62. In the CP, drag the **RainStation** and then the **Stream** from the **2D Layers** into **3D Layers**
63. Label the **RainStation** based on the **Id** field (make the label top right)
64. Click the **Analysis** tab on the ribbon, in the **Workflows** group, expand the **Exploratory 3D Analysis** drop-down list and choose **Elevation Profile**.
65. The **Exploratory Analysis** pane appears and the **Create** tab is active
66. Click the **Interactive Placement** tool ; when selected, the option is highlighted blue. When active, hover over the map or scene, and the pointer should now be a crosshairs icon.
67. Under **Distance Units**, select **Meters** to be used by the elevation profile.
68. Click the "**RainStation**" labeled 1 in the scene to set the starting point of the profile line.
69. Continue to add vertices to digitize the path between 1 and 2 and double-click on the "**RainStation**" labeled 2.

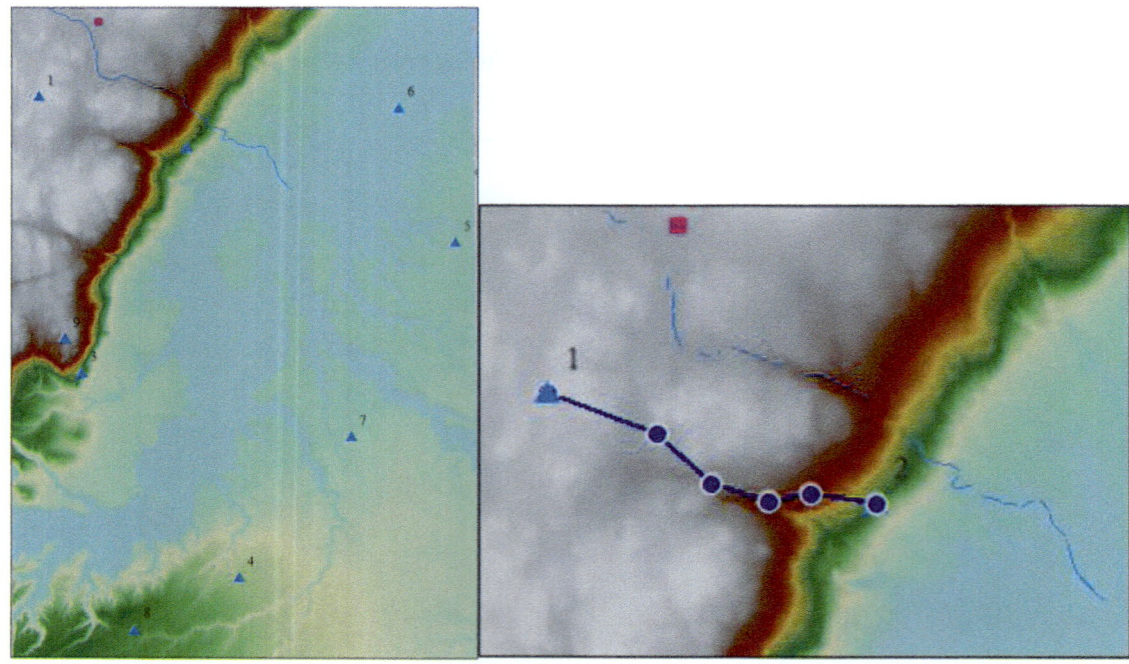

Create a Video

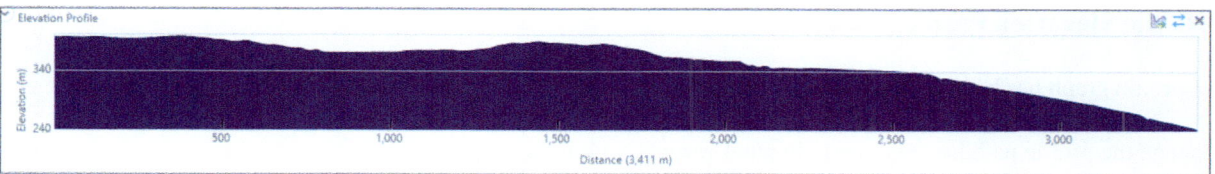

Result The elevation profile overlay window appears at the bottom of the active view.

Elevation Profile - From Layer

A profile line will be created from **Stream** feature in the scene.

70. In the **Exploratory Analysis** pane, under **Creation Method**, click **From Layer**. When selected, the option is highlighted blue and provides an option to choose an input line layer.
71. In the **Line Layer** text box, select the **Stream**,
72. Click the **Map** tab on the ribbon, **Selection** group and click **Select** button, click the **Stream** to make it selectable.
73. Under **Distance Units**, make sure **Meters** is selected.
74. Click **Apply**.

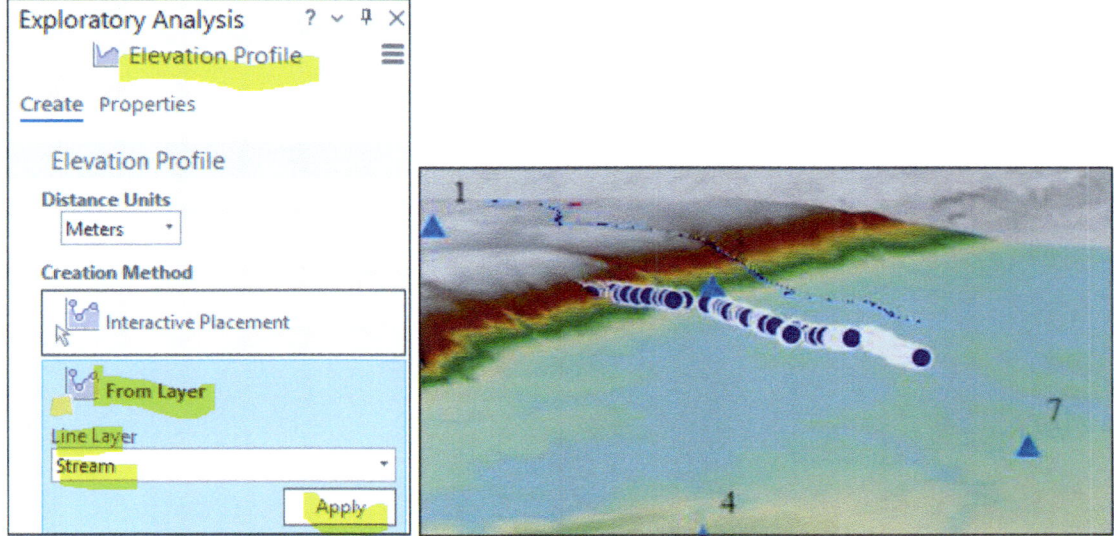

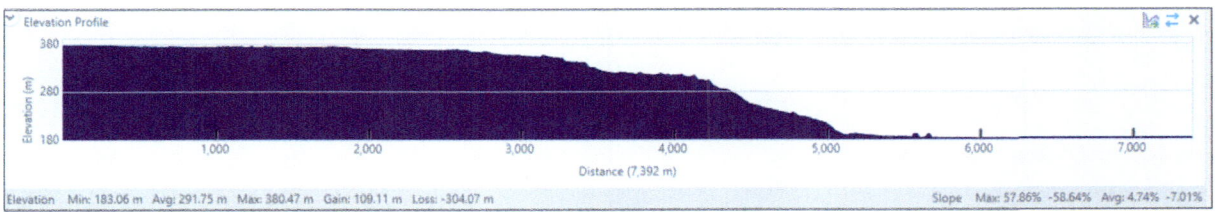

Result The elevation profile overlay window appears at the bottom of the active view.

Update the Elevation Profile Graph

Once a profile graph has been created, you can update it in the following ways (optional):

75. Change the profile path by dragging individual vertices within the map or scene.
76. Flip the direction of the graph in the **Elevation Profile** window by clicking the **Reverse Direction** button ⇌.
77. Change the units used to label the axis of an existing profile graph by changing the **Distance Units** property on the **Create** tab of the **Exploratory Analysis** pane.
78. Close the Geoprocessing pane, the Exploratory pane and Elevation Profile
79. Save the project
80. Exit ArcGIS Pro

Working with ArcGIS Online and StoryMap App

16

ArcGIS Online is a cloud-based mapping and analysis solution. It is used for creating and sharing GIS maps, apps, and content. The content can be accessed at any time on any device. It provides access to critical information throughout your organization (i.e., University of Wisconsin).

In this chapter, you will learn the following:

1. Manage your data
2. Publish it as web layers
3. Create web maps using services in ArcGIS Online
4. Create StoryMap (web apps)
5. Share this content with your organization and publicly

Section 1: Publish Data to ARCGIS Online

Scenario 1 You have studied the Newton Creek in Superior, WI and you would like the result of your study to be published as web app (Story Map) in ArcGIS Online to be used by researchers and students dealing with environmental problems.

You are going to integrate data as a zipped shapefile and publish the data as **web layers**.
To perform these tasks, you need the following:

- ArcGIS Online Organizational Account
- Publisher role or equivalent

Step 1: Sign in to ArcGIS Online

The first step to publish content to ArcGIS Online is to sign in to your **organizational account**.

1. In a web browser, go to www.arcgis.com
2. Sign in with your organizational account username and password.

After you sign in, the organization's home page appears.

Supplementary Information The online version contains supplementary material available at https://doi.org/10.1007/978-3-031-42227-0_16.

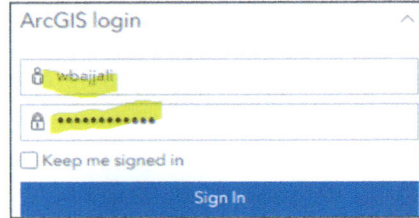

Step 2: Publish a Feature Service

After you have signed in, you will add a **zipped shapefile** to ArcGIS Online and publish it as a **feature service**.

Note To obtain the feature data into ArcGIS Online, you can upload a shapefile, or you can upload a file geodatabase to upload several feature classes. To upload these, you need to zip them and upload the zip file to ArcGIS online.

3. In ArcGIS Online at the top of the page, click Content (My Content will be active).

What Is My Content?

The **My Content** page is your storage space for items that you have added to the site or created. You can add files here and publish these files as services (if they are the appropriate file type).

4. On the **Content** page, next to **Folders**, click the **Create New Folder** button.
5. In the Create A Folder dialog box, type **Creek** and click OK.

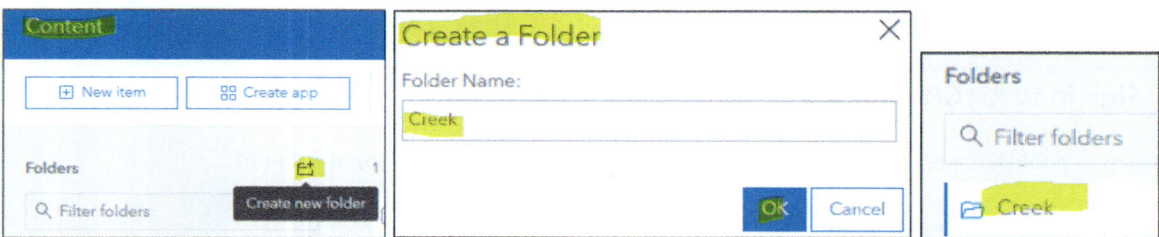

Result The **Creek** folder is created and will be used to store contents.

6. Click on **Creek** folder, under **Content**, click **New Item**.
7. The **New Item** dialog box open, click **Your Device** tab
8. Browse to the **Story** folder under \\Env_Water\Ch16 folders on your computer, select **Superior.zip**, and click Open.

Note You can also click **Google Drive** or **OneDrive** if you saved your data in one of them.

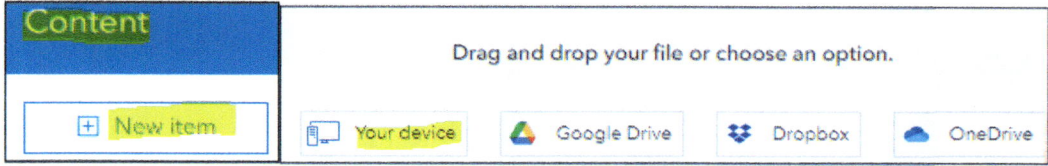

Note **Superior.zip** is a zipped shapefile that includes four shapefiles that will be used as feature layers.

9. In the New item dialog box, under *How would you like to add this file?* Check the "Add Superior.zip and create a hosted feature layer" box.

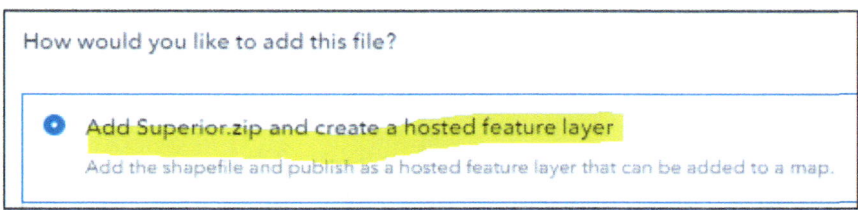

Note The zipped shapefiles already have the *geographic location* configured in the file, so ArcGIS Online does not need to geocode the file before publishing the layer.

10. Click Next

    In the **New Item** dialog box, fill it as follows:

Title: type **Newton Creek.**
Folder: **Creek.**
Tags: type **Newton and Faxon Creeks, Rivers, Water Quality, Murphy Oil, Hog Island, Impact Assessment, Superior, Wisconsin, USA** and press Enter after each tag.
Under Summary, type **Water quality evaluation in Newton Creek in Superior, WI.**

11. Click Next

Note1 It may take a few minutes for the zipped shapefile to publish.

Note2 If you encounter an error that the service name already exists, rename the title by adding your initials to the title name and try publishing it again.

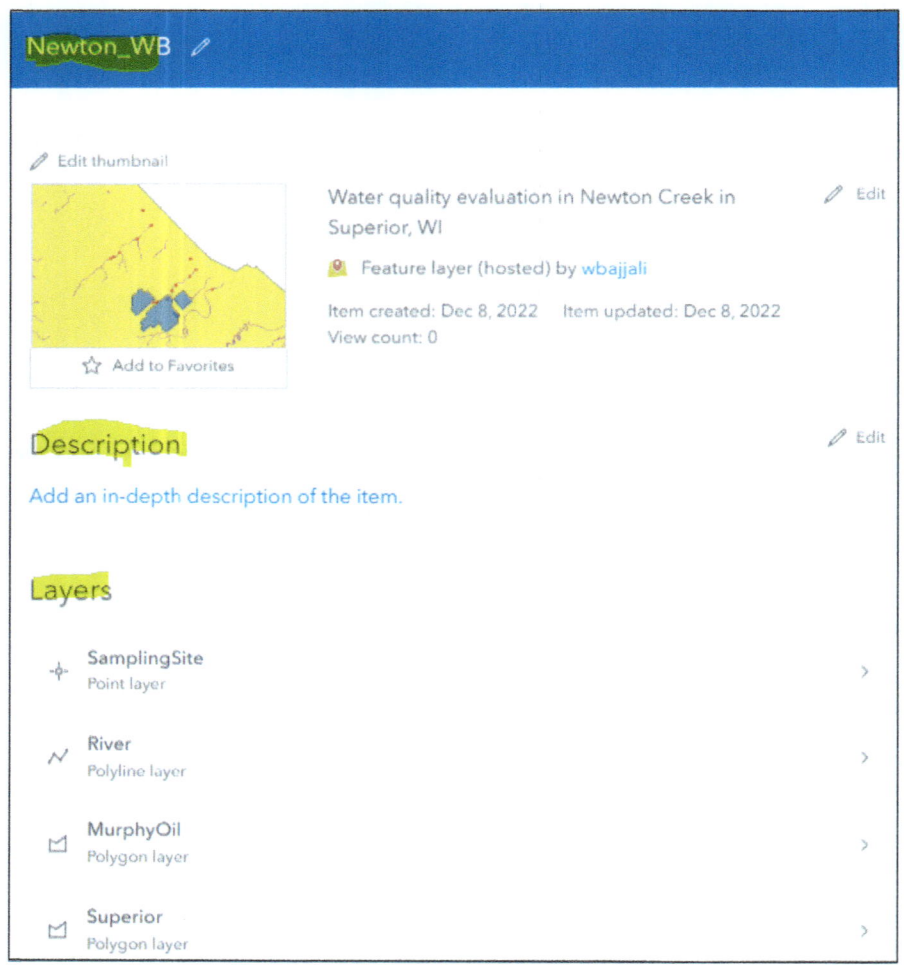

Step 3: Edit the Description and Term of Use

Result After the file has been successfully published, the item's details appear. The item page provides descriptive information about the item "**Newton_WB**", as well as options to use or to configure the item, such as providing a description and term of use. Below the **Layers** it shows that the item includes 4 layers" **SamplingSite**, **River**, **MurphyOil**, and **Superior**.

Step 3: Edit the Description and Term of Use

You will complete the **Newton_WB** details to help other site members understand what this information represents.

12. To the right of the **Description**, click **Edit**.

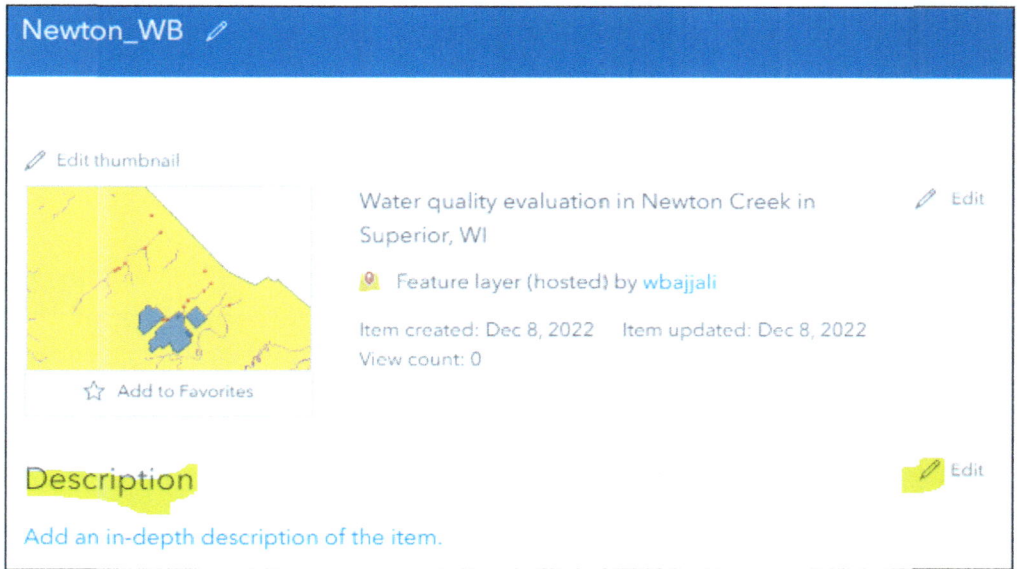

13. In the **Edit Description**, type **Water quality evaluation in Newton Creek, Superior, WI**
14. Highlight the text and choose "**medium**"
15. Click Save.

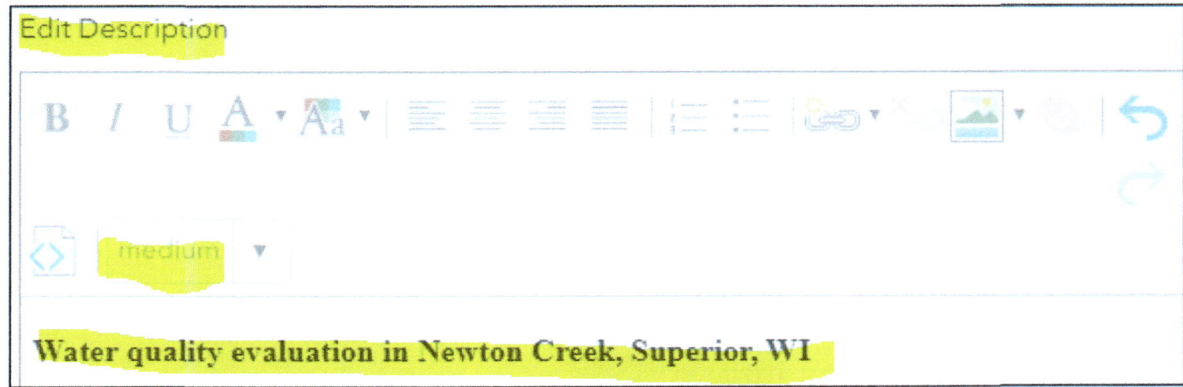

Because the content of the work is taken from a published paper, you should cite the paper.

16. Under **Terms of Use**, click **Edit** and type

Bajjali, W. Water Quality Assessment of Newton Creek and Its Effect on Hog Island Inlet of Lake Superior. *Water Qual Expo Health* **4**, 123–135 (2012). https://doi.org/10.1007/s12403-012-0071-1

17. Highlight the text and make it medium
18. Click Save

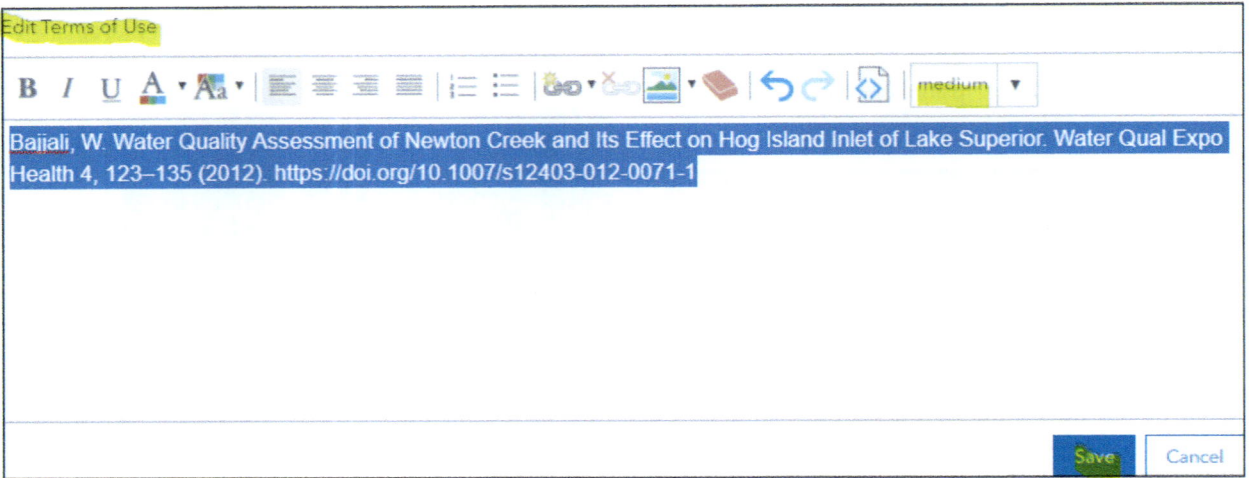

19. Click **Share** tab on the right.
20. In the **Share** dialog box, for the Set sharing level, select **Everyone (Public)** and click **Save**.

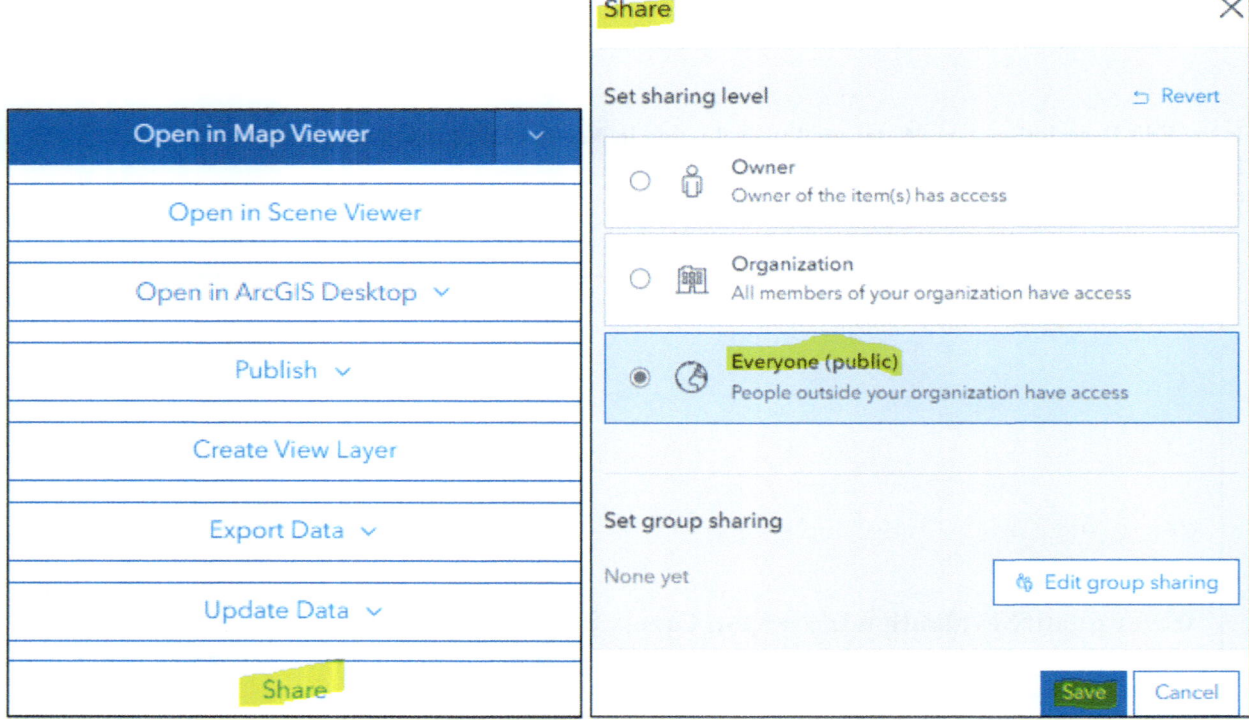

Step 4: Open the Feature Service   363

Note If you are unable to share with everyone, share with your organization or a group within your organization.

Step 4: Open the Feature Service

After the item's details are complete, you will add the **layer** to a new **web map**.

21. Click the back **arrow** on the top-left of the page or click the **Content** tab
22. In the **Content** tab (**My Content**), make sure that the **Creek** folder is selected.

    You will see two files:

- **Newton_WB   Feature Layers (hosted)**
- **Newton_WB   Shapefile**

23. To the right of the **Newton_WB feature layer (hosted)** item, click the **Options** button ...
24. Choose **Open in Map Viewer**.

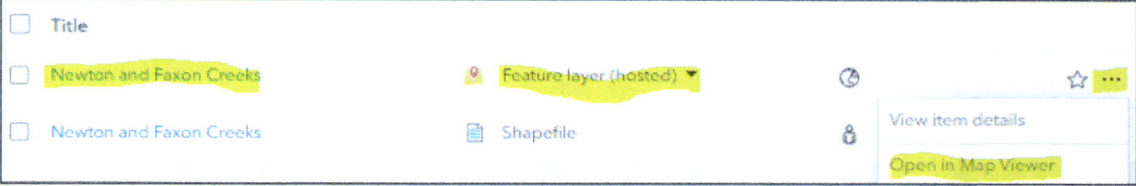

Comment The **Map Viewer** is the primary map-making tool for ArcGIS Online. In this section, you will use the **Map Viewer** to create your **web map**. You will add layer to the web map that have been shared in ArcGIS Online, the layers that have been shared with the community (layers that have been publicly shared on ArcGIS Online), organization or your group.

Map Viewer opens, and the **Newton_WB feature layer (hosted)** is added to the map.

25. To the left below the **Layers** expand the **Newton_WB**, you will notice it consists of 4-layers
    (a) *SamplingSite*
    (b) *River*
    (c) *MurphyOil*
    (d) *Superior*

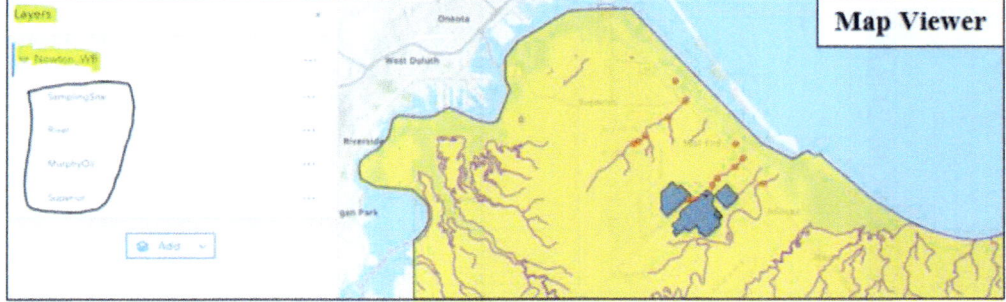

Content and Setting Toolbar

Each layer in the **Map Viewer** is associated with **two tollbars**

1. Setting toolbar
2. Content toolbar

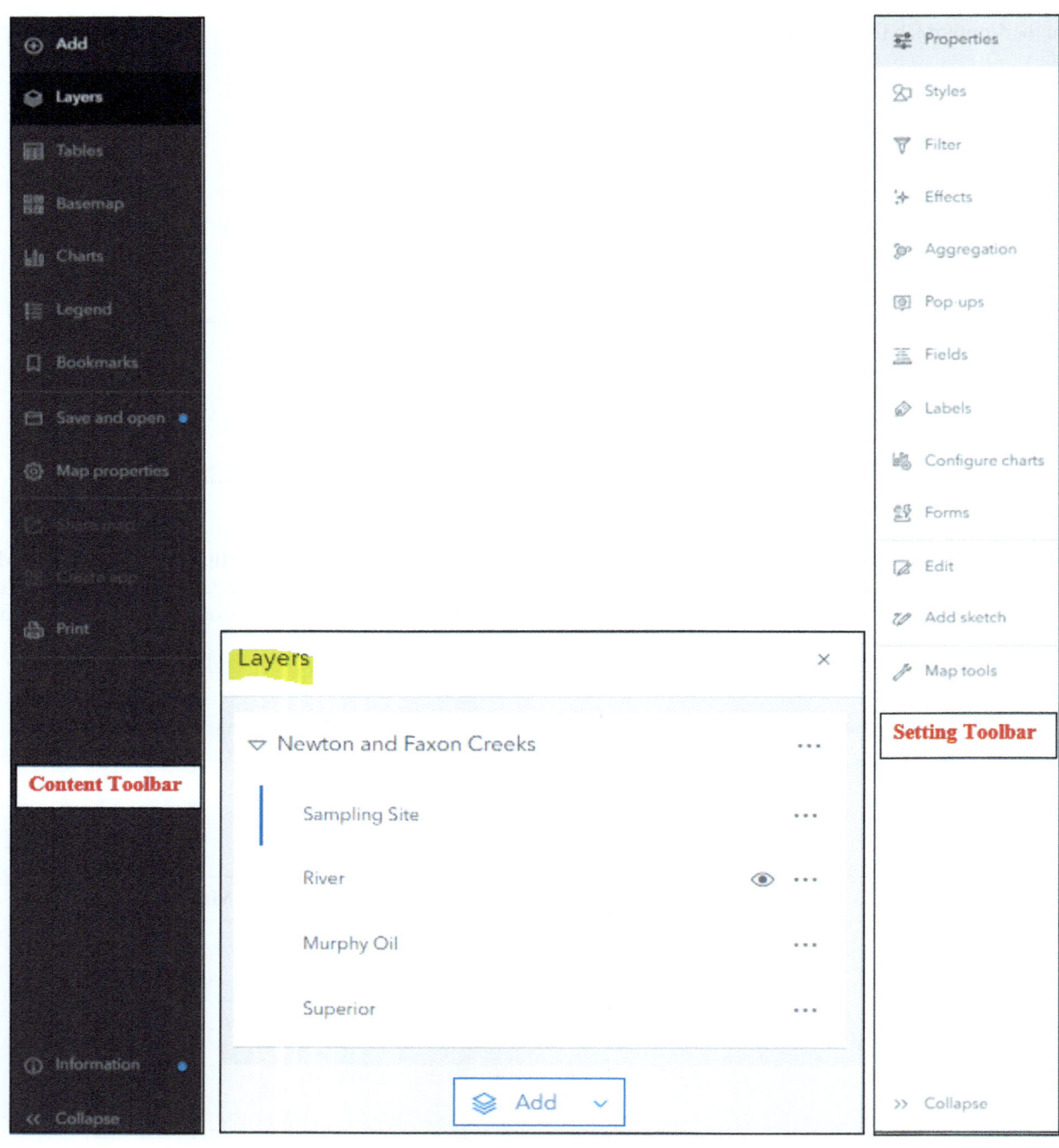

Rename the Layers

Settings toolbar (light color) on the *right side of Map Viewer* and includes tools and options available for the selected layer, such as properties, styles, filtering, effects, and aggregation. The toolbar is contextual, which means that these options only appear when there is a selected layer in the map and the menu depends on the layer that you select under **Layers**.

Contents toolbar (dark color) on the left side of **Map Viewer** and includes options such Add, Layers, Basemap, Charts, Legend, Bookmark, Save and Open, Map Properties, and others.

The **Content** and **Settings** toolbars can be expanded or collapsed by clicking the Expand arrows or the Collapse arrows at the bottom of each toolbar.

26. In the **Map Viewer**, click on one of the river layers.
27. A pop-up window appears with information from the attribute table of the **layer**.
28. Close the pop-up window.

Rename the Layers

29. In the **Layers** pane, expand the **Newton_WB layer**,
30. Select the sublayer **SamplingSite** and click the **Options** button (…) and choose **Rename**.
31. Change the layer name to **Sampling Site** and click OK.

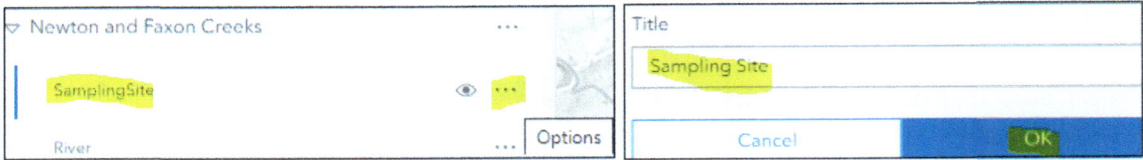

Change the name of the sublayer **MurphyOil** to **Murphy Oil.**

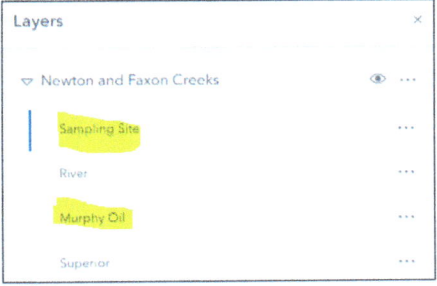

Change Layer Symbology

Next, you will change the symbology of the Sampling Site and Murphy Oil layers to make it easier to differentiate between the **Sampling Sites**, **Murphy Oil**, and **Superior.**

First, you will change the symbology for the **Sampling Site** to different colors

32. Select the sublayer **Sampling Site** and click the **Options** button and choose **Show table**
33. The attribute table opens, and the **Location** field is used to symbolize the layer
34. Close the table by click on the **X** on the top right

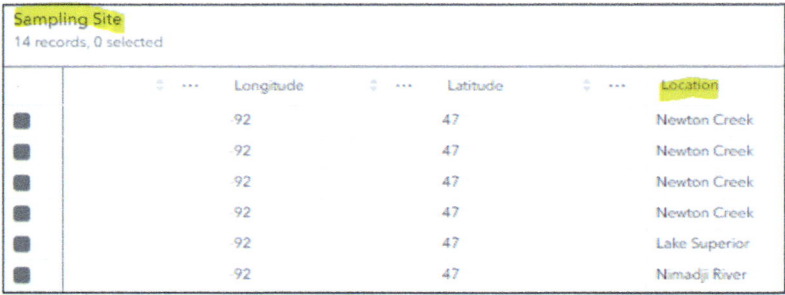

35. On the right, in the **Settings toolbar**, click the **Styles** button.
36. Under **Choose attributes**, click **Field**, select **Location** and click **Add**.

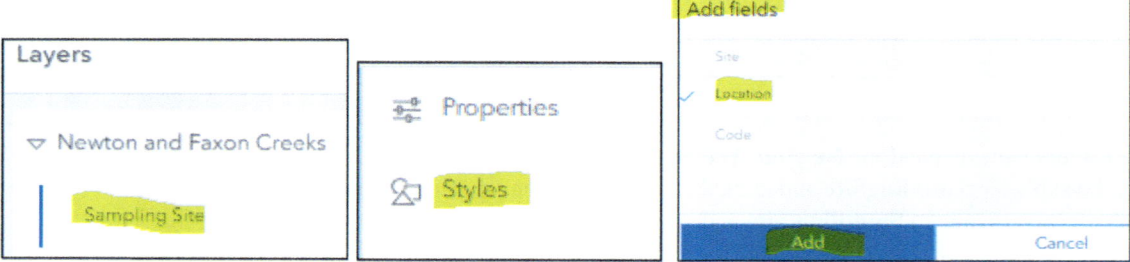

37. Under **Pick a style**, click **Style Options**.
38. To the right, under the **Symbol style**, click the **Edit** button.
39. Click the **Symbol** under **Current symbol**

Change Layer Symbology

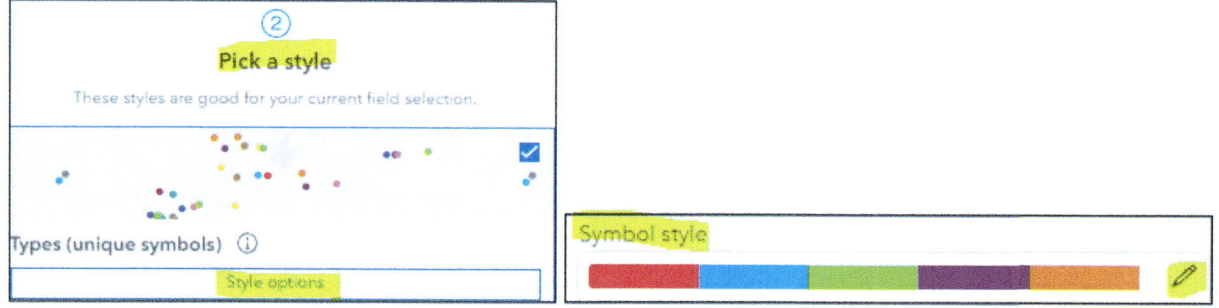

40. Choose **circle** (first symbol) and click **Done**
41. Under **Size**, change the size to **14**
42. Under **Fill transparency**, change it to **35**
43. Under **Outline color**, change it to **black**
44. Under **Outline transparency**, change it to **0**
45. Under Outline width make it **1**
46. Click Done then Done

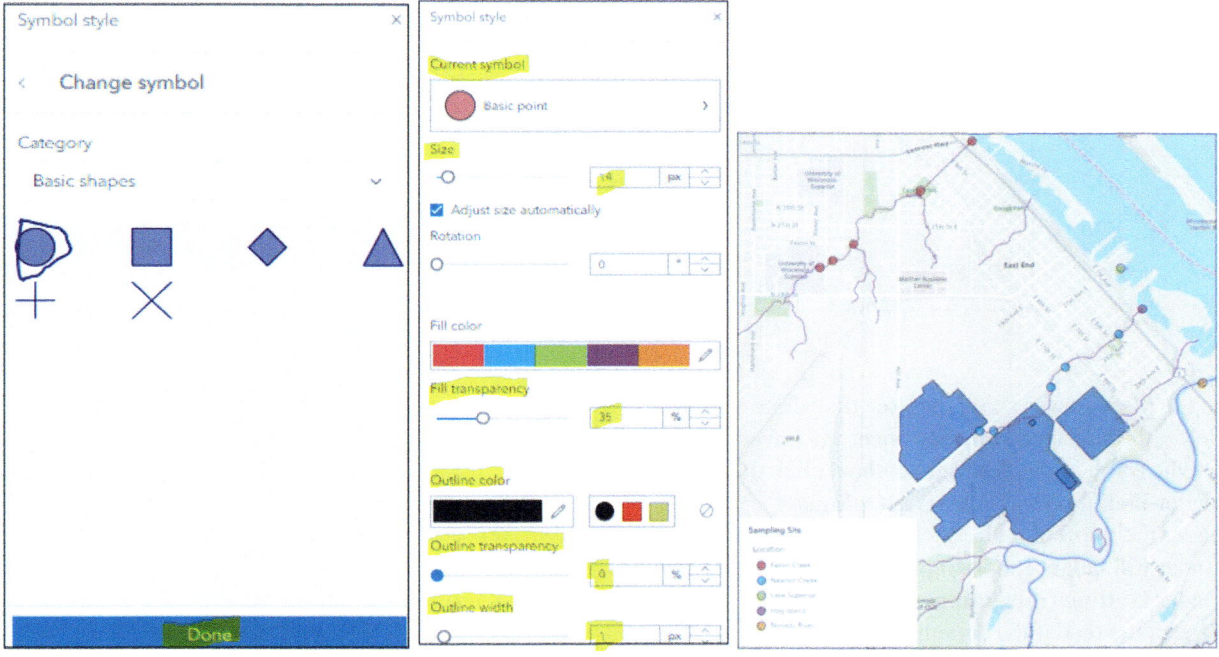

Result The Sampling Site sublayer is displayed as a circle with different colors in Map Viewer.

47. Click **Done** to Exit the Styles

Result The legend disappears in the Map Viewer.

48. Select the sublayer **Murphy Oil** under the Layers pane
49. On the right, in the **Settings toolbar**, click the **Styles** button
50. Under **Pick a style**, click **Style Options**
51. Under **Symbol style**, click the **Edit** button.
52. Under **Current symbol**, click the **Basic polygon**

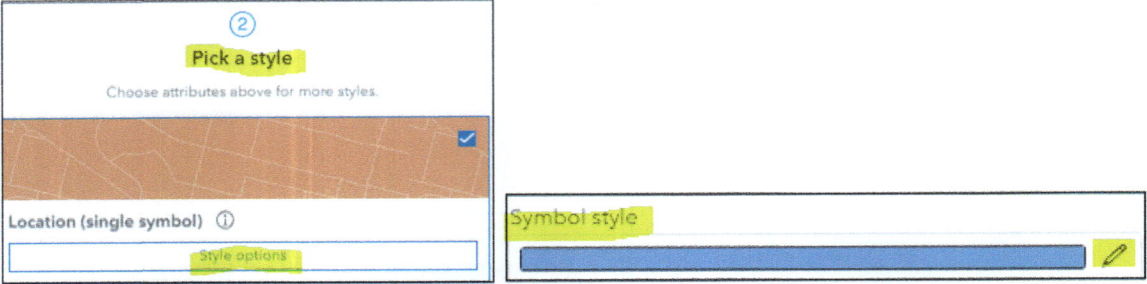

53. Under **Category**, select from the drop-down arrow **Hatch Fill** and choose a hatch of your taste Click **Done**
54. Click **Fill color** and choose **green color**
55. Accept the rest of the default
56. Click **Done** twice to exit the **Styles**

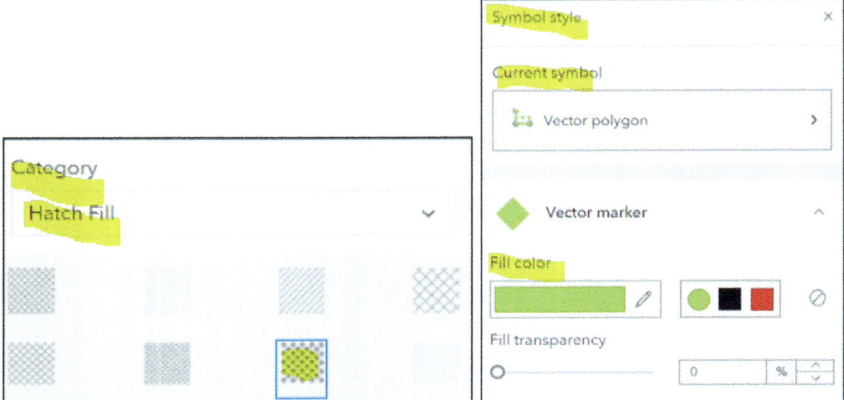

57. In the Map Viewer zoom in to see the whole **Superior** layer
58. Select the sublayer **Superior** under the Layers pane
59. On the right, in the **Settings toolbar**, click the **Styles** button .
60. Under **Pick a style**, click **Style Options**
61. under the **Symbol style**, click the **Edit** button.
62. Under **Fill color**, choose **light gray** and click **Done**
63. Under **Outline transparency** make it 75
64. Accept the rest of the default
65. Click **Done** twice to exit the **Styles**

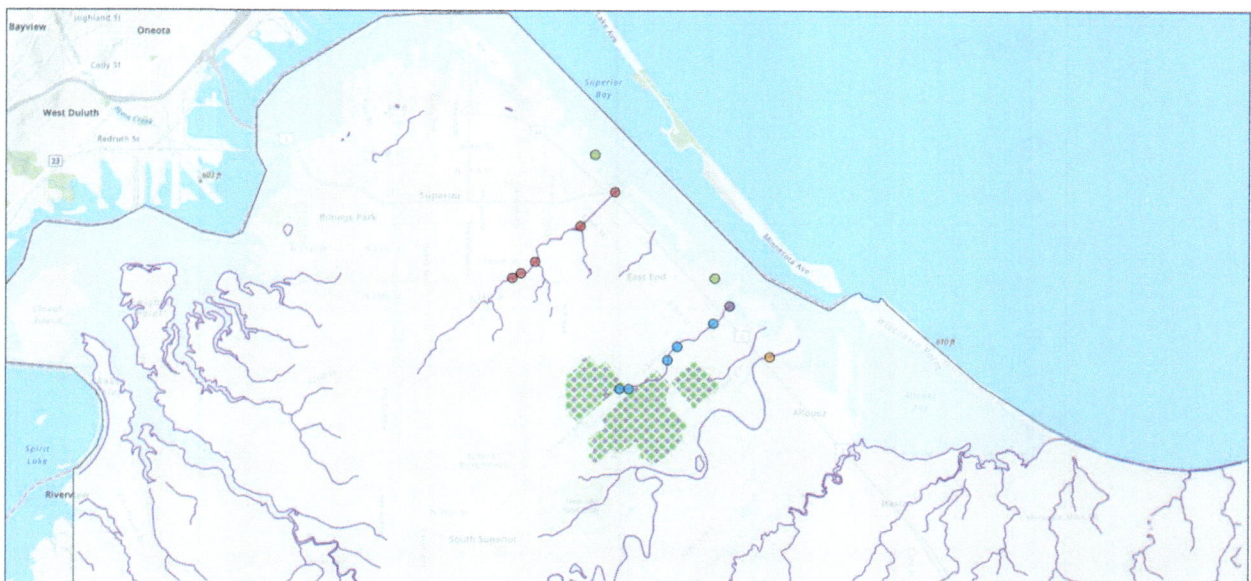

Save the Web Map

Now that you have set up the map, you are ready to save it as a **Web Map**.

66. On the left, in the **Contents toolbar**, click the **Save and open** button, and click **Save as**.

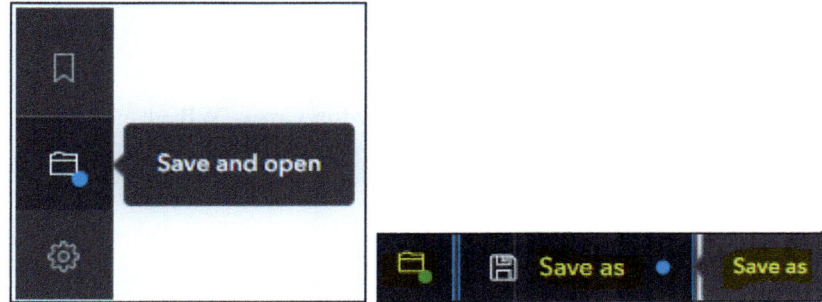

In the **Save Map** dialog box, specify the following information:

- For Title, type **Newton Creek_WB** (type your initials, i.e., WB).
- For Folder, choose Creek
- For Tags, type **Newton Creek, Faxon Creek, Murphy Oil, Hog Island, Site Location, Superior, Wisconsin**
- For Summary, type **Sampling Sites of Newton and Faxon Creeks and Lake Superior, WI, USA**

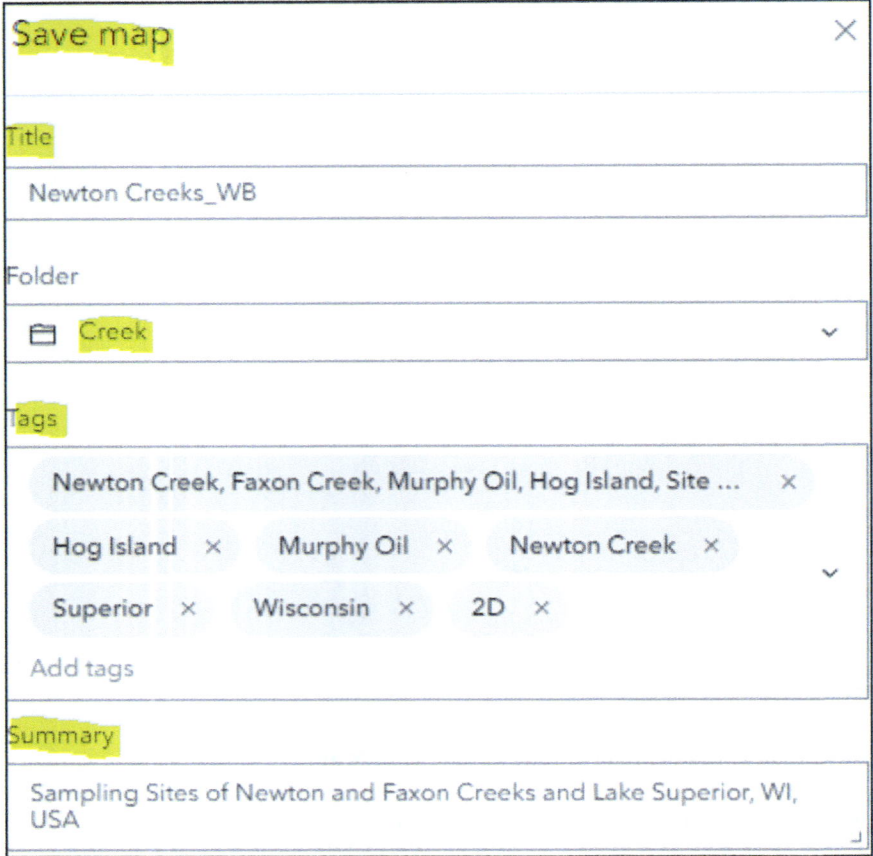

67. Click Save Map

Result If you go to **Content**, you will see the new **Web Map** "Newton Creek_WB" display under **Creek** folder.

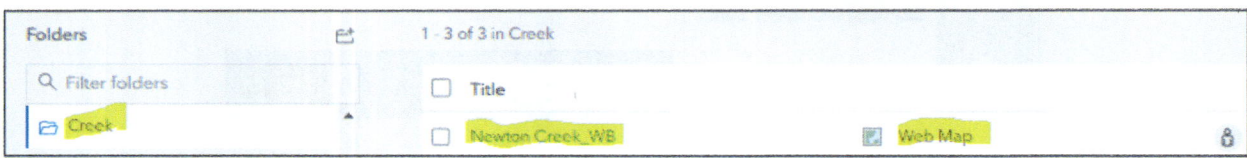

Share an App Using ArcGIS Online

You have created the **web map** and want to share it with community members. You want to share the map in a way that focuses on the **site locations** so the map will be used in any geo app such as **Story Map** or another app.

Share the Web Map

You want to share the **Newton Creek_WB** web map with the world community members. **Everyone** (Public) is the most appropriate level. Remember, the **web map** *contains layers*, which will not be visible to community members if they are not shared at the **same level** as the web map, so the **layers** will need to be shared with the same permission level as the map.

Share the Web Map

Note You can update sharing any item in the **Content** pane.

68. From the **Content** page, open your **Creek** folder if necessary.
69. To the right of the **Newton Creek_WB**, click the **More Options** button … and choose **View item details**.

67. The map's item page opens
68. On the item page, on the right, click **Share**.
69. In the Share dialog box, select **Everyone (Public)**
70. Click **Save**.

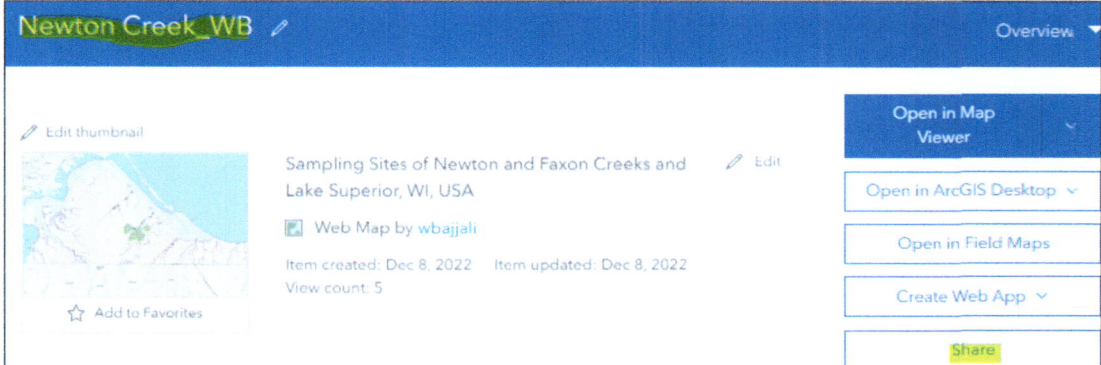

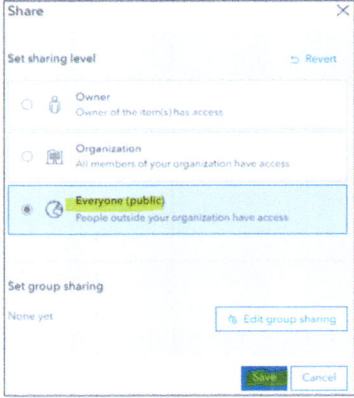

Note If you are unable to share with everyone, share with your organization instead.

Create Story Map

Making a story can be done in different ways and from different locations in ArcGIS Online. One way is from **Map Viewer**, the second way is using the **ArcGIS StoryMaps app** (http://**storymaps.arcgis.com**), and the third way is creating the **StoryMap app** from **App Launcher** in ArcGIS Online. In this section, we will use the **App Launcher** to create the **StoryMap app**. In Chap. 17, we will use the **Map Viewer** to create the **Instant app**.

ArcGIS **StoryMap** is a web map that has been created by integrating maps, text, photos, and others and provides functionality, such as swipes, pop-ups, and time sliders, that helps users explore this content of the story. Before starting the story, you should have the story text and media, you will start building the story using content blocks. Each title, paragraph, image, and media type are added as a separate block from the block palette. In this chapter you will create a story about Newton Creek in Superior, WI and publish with your organization or if you desire with the whole world. After creating the story, you can share your maps in the context of narrative text and other multimedia content. You can use ArcGIS StoryMaps to publish and share your stories. Published stories each have their own URL, and you can use these URLs to share your stories within your organization, to specific groups, or with everyone.

You will start from scratch to create your story using the **ArcGIS StoryMaps app.**

1. Launch ArcGIS Online (if you signed out)
2. Sign in to ArcGIS Online
3. Click the App Launcher in the top-right corner
4. Scroll down and click the **ArcGIS StoryMaps**

5. Click the **New story** button in the top right and then click Start from scratch.

Result A blank story template appears.

Note Ch16 includes "**Image**" and "**Picture**" folders that will be used to add to your story as well as a word document "**Newton Text**" that you can use to fill in content.

6. In this chapter folder, open the "**Newton Text**" document, you are going to use the document to build your story by copy and paste.
7. For **Title your story**, type or paste **Water Quality of Newton Creek** from the "**Newton Text.docx**"
8. For Start with a short introduction or subtitle, type or paste the following text: **in Superior, WI. USA**

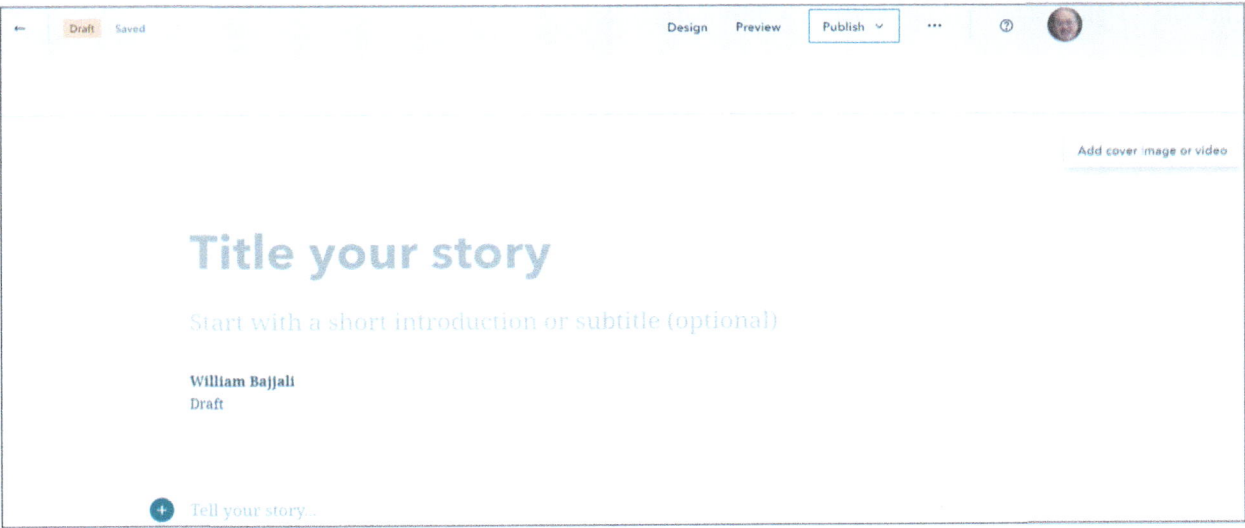

The ribbon updates with the new title.

Note You can change the title at any time.

You can see above the title the **Draft** badge and next to it, you will see evidence of the autosave feature in ArcGIS StoryMaps; any time you edit your story, that text will let you know your story is saving and then confirm that your changes have been saved.

9. Click the + sign next to Tell your story, click Text under Basic.

Comments The block palette contains options for adding content. You will tell your story using these blocks to add sections of content to the body of your story.

There are a few types of blocks you can choose from.

- **Basic** elements such as

  - **Text** blocks (paragraph, heading, quote, and so on)
  - **Button**
  - **Separator**

- **Media** blocks such as

  - Map
  - Image
  - Image Gallery
  - Audio
  - Embed
  - Swipe
  - Timeline

- **Immersive** blocks such as

  - Slideshow
  - Sidecar
  - map tour

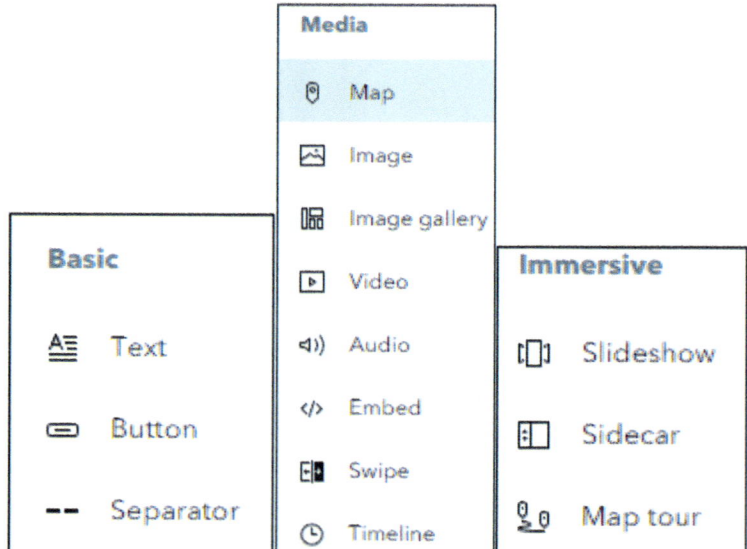

10. Type or paste the following: **Introduction**
11. Highlight the **Introduction**, click **Paragraph** and choose **Heading**

12. Click the **+ sign** below the **Introduction**, click **Text** under **Basic**.

Type or paste the following: *Newton Creek is a 1.6-mile water-way running through a mostly residential area of Superior, WI. It originates from an artificial pool that receives its main source of water from the treated wastewater of the Murphy Oil Refinery. It discharges to Hog Island Inlet, which is part of the greater Saint Louis River System's Superior Bay and connects to Lake Superior. (Figure 1).*

13. Highlight the first paragraph. In the text menu, click the Italic button.

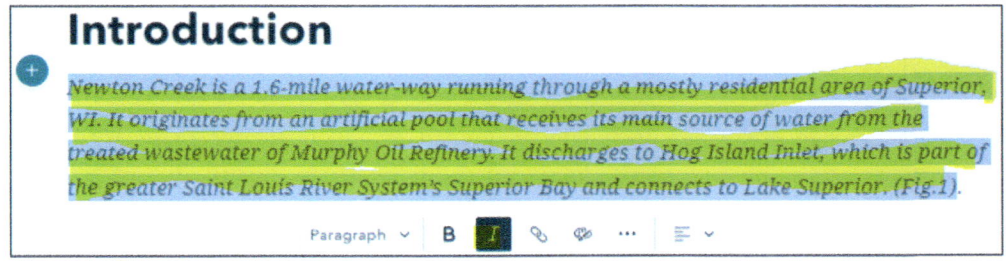

Note You can also use keyboard shortcuts such as **Ctrl + I** for italics and **Ctrl + B** for bold.

Work with Media

Once you have added your first text, you will add an "**Image**" under "**Media**" to represent **Fig. 1** under first paragraph. Images, video and other media are important because they break up a long narrative and provide context.

14. Below the text under the "**Introduction**", hover over the left margin below **Fig. 1** and click the **+ sign** (Add content block), the block palette open, choose **Image** under **Media**.
15. From the Add an image dialog box, click "Browse your files" and browse to the directory where you saved the images (Ch16\Story\Images) and select **Fig. 1.jpg**
16. Click open and then click Add

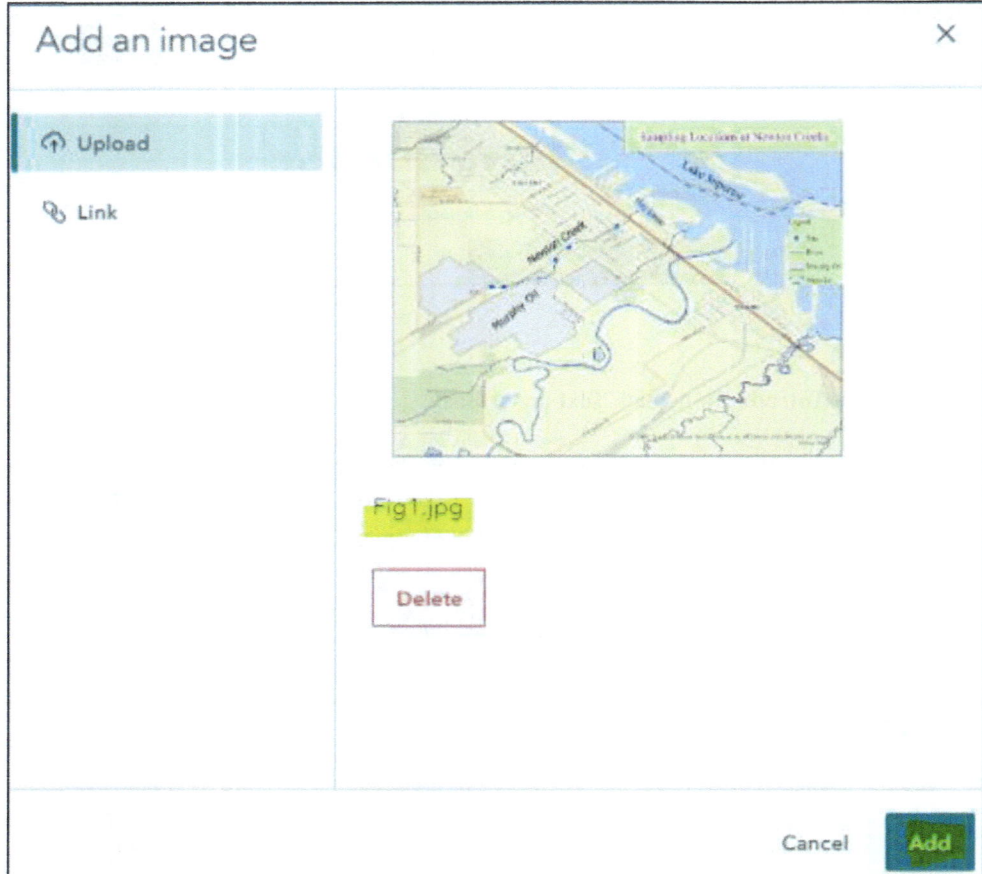

Result The image is added to the story.

17. Hover over **Fig. 1 image**. In the toolbar, click **Medium.**

Depending on the width of the image you upload, larger sizes may not be available, ensuring that your content is never stretched beyond its maximum width. Full-width media, for example, must be at least 2001 pixels wide.

18. Hover over the image again and click the **Options** button.

19. The **Image options** window appears

You will add alternative text, or alt text, that describes the image so that anyone reading your story can see the study area of your work. You can specify alt text in the properties menu for each media item using the **Options** button. Unlike captions, which appear below an image, the alt text associated with an image can be viewed by hovering over it.

20. For Alternative text, type Sampling Locations of Newton Creek
21. In the Attribution type "Your Name", i.e., "William Bajjali"
22. Click Save

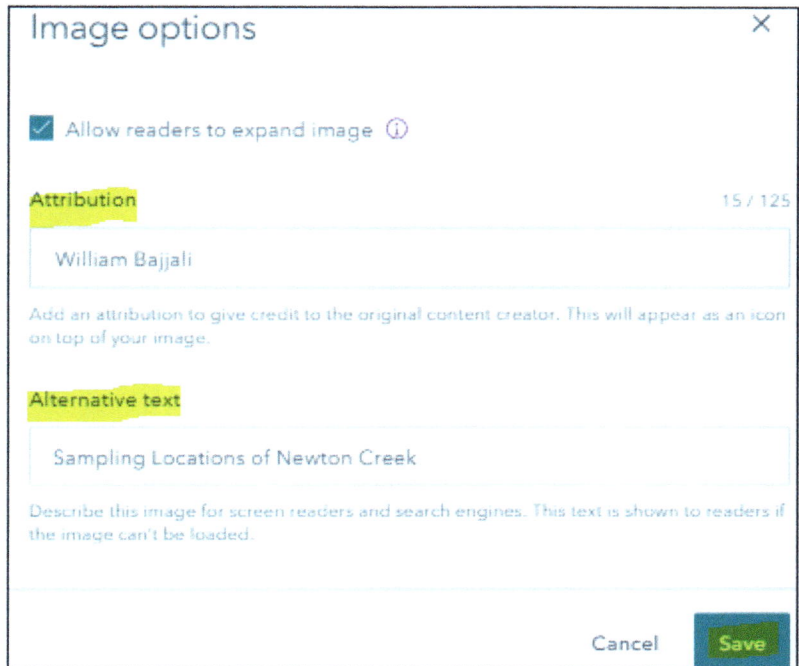

23. Under **Fig. 1** image, click Add a caption for this image (optional) and type or paste the following text: **Sampling Locations at Newton Creek**

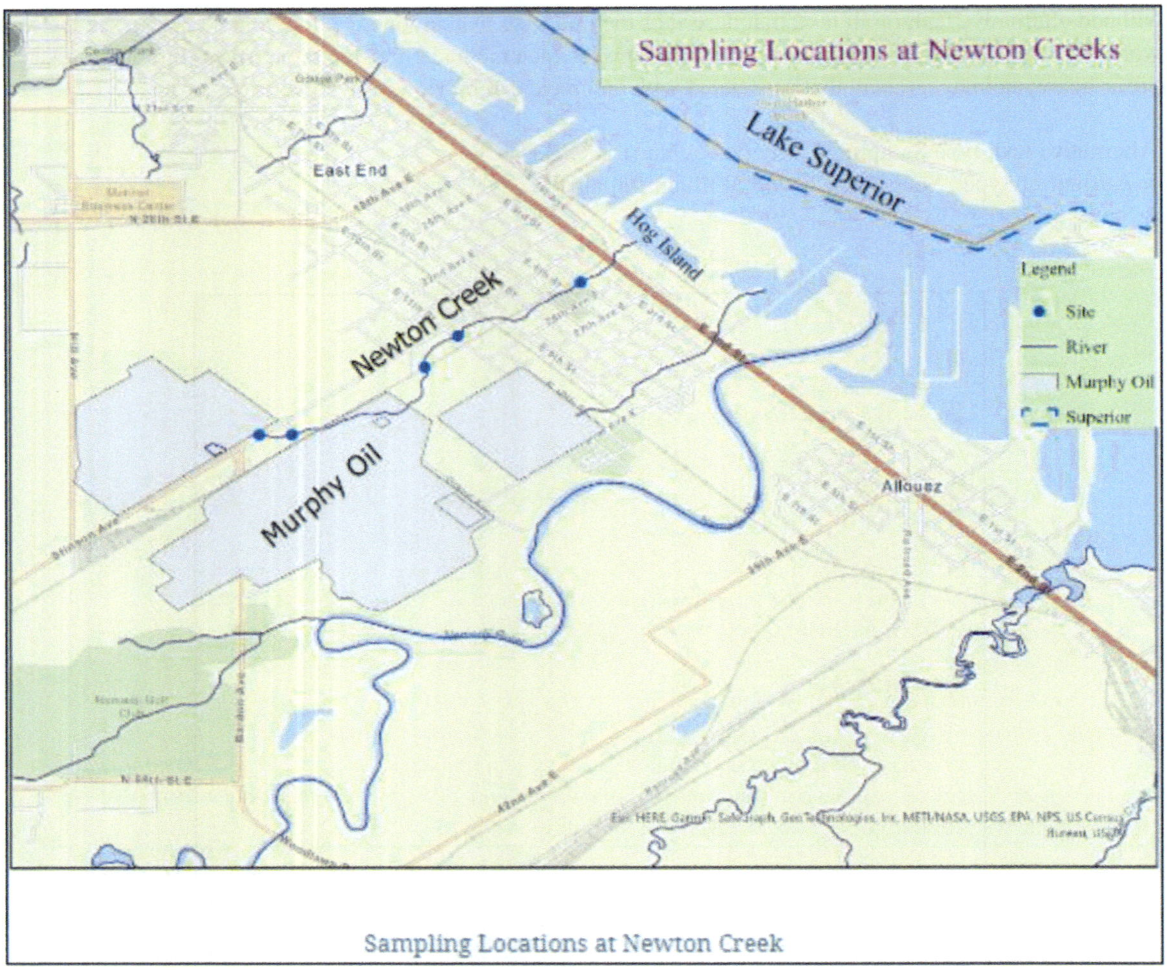

Sampling Locations at Newton Creek

24. Click the **+ sign** below **Fig. 1 image**, click **Text** under **Basic**.
25. Type or paste the following: "**The Problem**"
26. Highlight "**The Problem**", click the **Paragraph** and choose **Heading**.
27. Click the **+ sign** below the "**The Problem**", click **Text** under **Basic**.
28. Type or paste the following:
    The Newton Creek water, surrounding soil, and HII were found to be highly contaminated with hydrocarbon byproducts. The contaminated site was classified by the EPA as an **"area of concern"**. Remediation was conducted to remove contaminants, and restoration began. Between 1997 and 2005, a total of 60,175 tons of contaminated sediment was removed from the creek and HII. Despite this remediation, contaminates continued to exist within the creek environment.

29. Highlight the pasted paragraph. In the text menu, click the **Italic** button

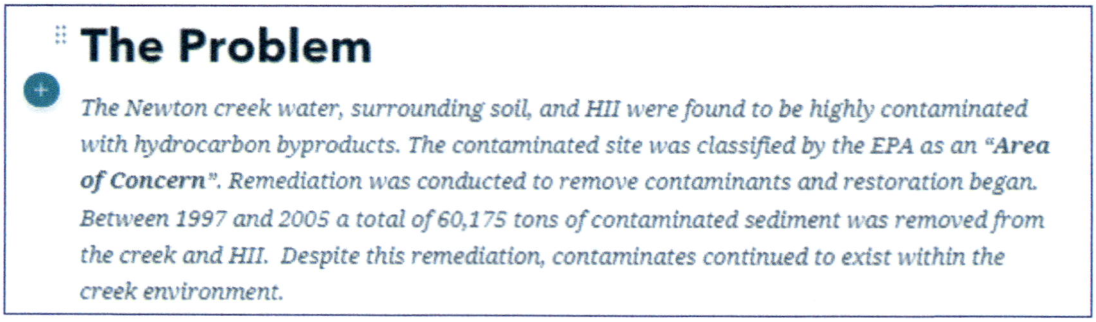

The Problem

*The Newton creek water, surrounding soil, and HII were found to be highly contaminated with hydrocarbon byproducts. The contaminated site was classified by the EPA as an "**Area of Concern**". Remediation was conducted to remove contaminants and restoration began. Between 1997 and 2005 a total of 60,175 tons of contaminated sediment was removed from the creek and HII. Despite this remediation, contaminates continued to exist within the creek environment.*

30. Click the **+ sign** below the pasted text below "**The Problem**", click **Text** under **Basic**.
31. Type or paste the following: "**Objective**"
32. Highlight the "**Objective**", click the **Paragraph** and choose **Heading**.
33. Click the **+ sign** below the "**Objective**", click **Text** under **Basic**.
34. Type or paste the following:
    Determine if the water quality of Newton Creek improved after remediation.
35. Highlight the pasted paragraph. In the text menu, click the **Italic** button

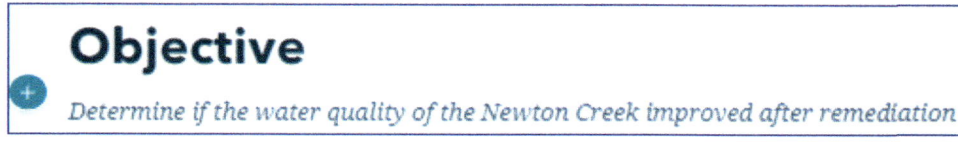

36. Click the **+ sign** below the pasted text below "**Objective**", click **Text** under **Basic**.
37. Type or paste the following: "**Methodology**"
38. Highlight the "**Methodology**", click the **Paragraph** and choose **Heading**.
39. Click the **+ sign** below the "**Methodology**", click **Text** under **Basic**.
40. Type or paste the following:
    Between 2005 and 2011, Dr. Bajjali from the University of Wisconsin - Superior and his students conducted a study of water quality for Newton Creek. The creek water was sampled for DO, EC, pH, and temperature. The creek water was measured on a weekly basis using Omega devices, dual pH/conductivity handheld instruments and portable digital dissolved oxygen temperature meters. The water at the five sites was measured from the bank of the stream (Table 1).
41. Highlight the pasted paragraph. In the text menu, click the **Italic** button.

Methodology

Between 2005 and 2011, Dr. Bajjali from University of Wisconsin - Superior and his students conducted a study of water quality for Newton Creek. The creek water sampled for the DO, EC, pH, and temperature. The creek water was measured on a weekly basis using Omega devices; Dual pH/conductivity Handheld Instruments and Portable Digital Dissolved Oxygen Temperature Meters. The water at the five sites was measured from the bank of the stream (Table 1).

42. Below the text under the "**Methodology**", hover over the left margin below **Table 1** and click the **+ sign** (Add content block), the block palette open, choose **Image** under **Media**.
43. From the **Add an image** dialog box, click "Browse your files" and browse to the directory where you saved the images (Ch16\Story\Images) and select **Table 1**
44. Click Open and the click Add
45. Hover over the image again and click the **Options** button.

46. The **Image options** window appears.
47. For **Alternative text**, type **Average EC, pH, D.O., and Temperature in Newton Creek**
48. In the **Attribution** type "Your Name", i.e., "**William Bajjali**"
49. Click Save
50. Under the **Table 1** image, click Add a caption for this image and type or paste the following text: **Table 1 Average EC, pH, D.O., and Temperature of the Newton Creek Water**

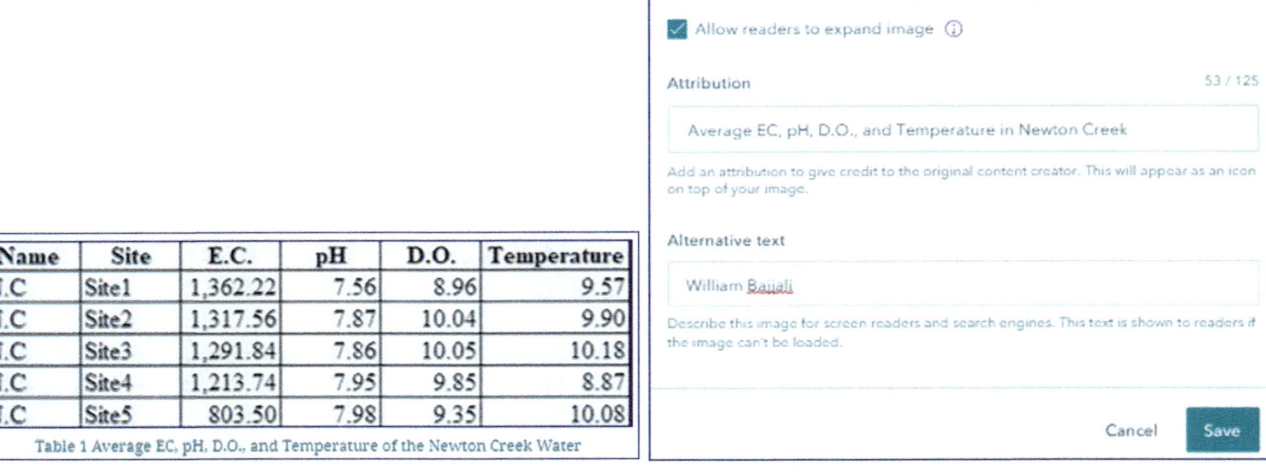

51. Click the **+ sign** below the "**Table 1**", click **Text** under **Basic**.
52. Type or paste the following: "**Sampling Sites**"
53. Highlight the "**Sampling Sites**", click the **Paragraph** and choose **Heading**.
54. Click the **+ sign** below the "**Sampling Sites**", click **Text** under **Basic**.
55. Type or paste the following:

The data set was conducted at five sites along the flow of the creek starting from the discharge point of the treated water from the Murphy Oil treatment plant and ending on Hog Island. The following images show the locations of Murphy Oil that form the head water of the creek, the sampling site along Newton Creek, Hog Island, which is considered the mouth of the creek, and a picture of one student sampling the creek.

56. Highlight the pasted paragraph. In the text menu, click the **Italic** button.

Work with Immersive – Sidecar Docked

Immersive blocks are unique in that they become full-screen takeovers of your story, providing different, interactive reading experiences. You will add a "**Sidecar**" and add five pictures as the media in your sidecar. You will customize the pictures to best fit the story. The first picture is the head of the creek that starts from the discharge point of the treated water from the Murphy Oil treatment plant. The second, third, and fourth are the sampling locations. The fifth picture is the mouth of the creek that ends on Hog Island.

1. Click the **+ sign** below the "**Sampling Sites**" text
2. Click **Sidecar** below the **Immersive**
3. The **Choose a layout** dialog box display
4. Choose **Docked** panel, and then click **Done**

Result The sidecar display, but empty and you have to fill the sidecar by adding the 5-pictures that represent the Newton Creek sampling sites

5. At the top, click the **Add** drop-down arrow and choose **Image or Video** (you can also select "**Map**" and this will allow you to select a web map that you created previously and saved on ArcGIS Online)
6. The **Add an image or video** dialog box display

7. Make sure the **Upload** tab in the left is selected and click **Browse your files**

8. Browse to picture folder in Chapter16/Story and select **1_NewtonSite1. JPG**
9. Click Open, and then click Add

Result Picture 1, which represents sampling site 1 at the headwater of the creek in the area of Murphy oil, has been added to the story.

10. At the top of the picture click Options ⚙
11. In the Image options
12. Check Fit (do not crop)
13. In the Attribution, type your name (i.e. William Bajjali)
14. Click Save

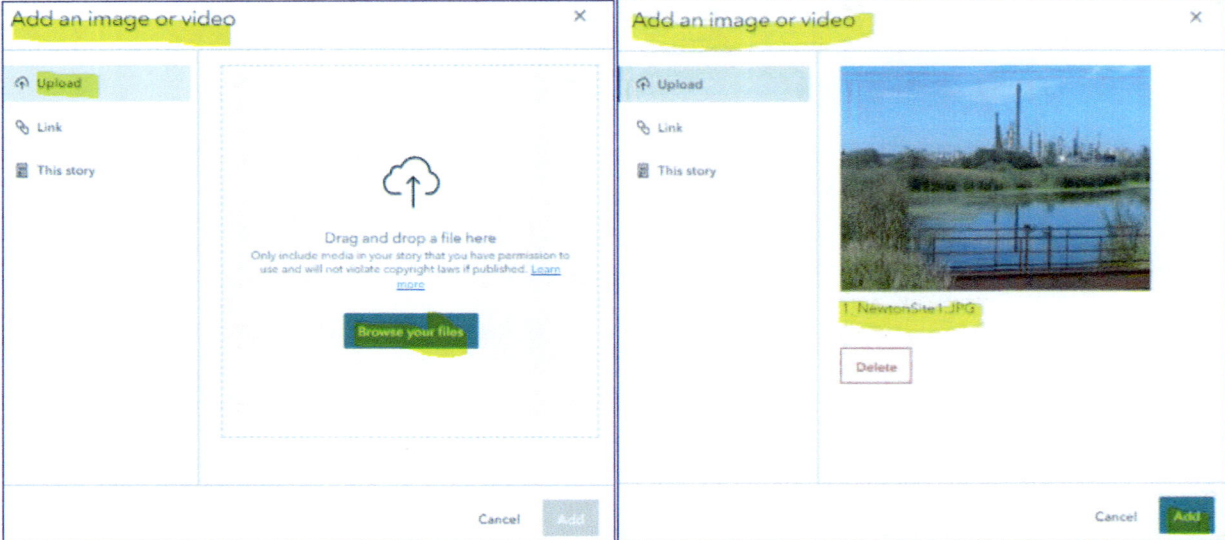

15. Next to the **Picture**, click **Continue your story** and type or paste the following: **Site 1 of sampling Newton Creek that originates from an artificial pool that receives its main source of water from the treated wastewater of Murphy Oil Refinery.**

16. Highlight the text, click **Paragraph**, and click **Quote**.
17. Add another text block and type **Photo**: type your name (i.e., **William Bajjali**).
18. Highlight the **Photo** text and click the **Bold** and then **Italic** button.

Result The first sidecar slide is complete.
Add Second Picture

19. At the bottom of the sidecar block is the slide panel.

Comment In the slide panel, you can add or remove slides, adjust the transitions between them, or remove the sidecar block from your story entirely. You can change which side the narrative panel is on by using the double-arrow button where the media and narrative panel meet.

20. In the slide panel, click the **+ sign** (**New slide**) button.

Result A blank slide is added to the pane.
21. Click Open, and then click Add

22. Click **Add** and click **Image or video**.
23. Browse to Chapter16\Story\Pictures and choose **2_NewtonSite2.jpg**.
24. Click Open, and then click Add

25. At the top of the picture click Options
26. In the Image options, check Fit (do not crop)
27. In the **Attribution**, type your name (i.e. William Bajjali)
28. Click Save
29. Next to the **Picture**, click **Continue your story** and type or paste the following sidecar text:

Newton Creek Sampling Site 2.

30. Highlight the text, click **Paragraph**, and click **Quote**
31. Add another text block and type **Photo**: type your name (i.e., **William Bajjali**).
32. Highlight the **Photo** text and click the **Bold** and then **Italic** button.

33. Repeat the previous steps by adding the rest of the pictures, as seen in the table below.

Picture	Text	Photo
3_NewtonSite3.jpg	Newton Creek Sampling Site 3	Your Name
4_NewtonSite4.jpg	Luai (student) Sampling Site 4	Your Name
5_HogIsland.jpg	Newton Creak discharge in Hog Island	Your Name

34. **Click the + sign below the Sidecar, click Text under Basic.**
35. Type or paste the following: "**Fluctuation of EC, pH, DO, and Temperature**"
36. Highlight "Fluctuation of EC, pH, DO, and Temperature", click the Paragraph and choose Heading.
37. Click the **+ sign** below the "**Fluctuation of EC, pH, DO, and Temperature**", click **Text** under **Basic**.
38. Type or paste the following:
    The DO, EC, pH, and temperature parameters of the sampling water along the flow path of Newton Creek are important to aquatic life in the stream.
39. Highlight the text and in the text menu, click the **Italic** button

Work with Immersive – Sidecar Floating

You will add a "**Sidecar**" with floating panel and add 4 graphs that represent the fluctuation of the **EC, pH, DO, and Temperature** from 2005 to 2011.

40. Click the **+ sign** below the text of the "**Fluctuation of EC, pH, DO, and Temperature**"
41. Click **Sidecar** below the **Immersive**, the **Choose a layout** dialog box display
42. Choose the **Floating** panel, and then click **Done**
43. At the top, click the **Add** drop-down arrow and choose **Image or Video**
44. The **Add an image or video** dialog box display
45. Make sure the **Upload** tab in the left is selected and click **Browse your files**
46. Browse to **Images** folder in Chapter16/Story and select **EC. JPG**
47. Click Open, and then click Add

Result EC images that represent the fluctuation of the EC are added to the story.

48. At the top of the **EC image** click Options

49. In the Image options, check **Fit (do not crop)**
50. In the **Attribution**, type your name (i.e. William Bajjali)
51. Click Save
52. Next to the EC, click Continue your story and type or paste the following text:

Electrical Conductivity

High EC indicates pollution from different sources, such as urban runoff, road salt, wastewater, and others. The graph represents the average EC values on a monthly basis. The creek water fluctuates between the winter, summer, fall, or spring throughout the sampling years at all sites, except for site 5. The similar trends of EC seasonal fluctuation emphasize that the input sources of EC to the creek water system behave in a similar way throughout the year.

53. Highlight the title **Electrical Conductivity**, click **Paragraph**, and click **Quote**.
54. Highlight the text below the **Electrical Conductivity**, click **Paragraph** twice
55. Add another text block and type **Graph**: type your name (i.e., **William Bajjali**).
56. Highlight the **Graph** text and click the **Bold** and then **Italic** button.

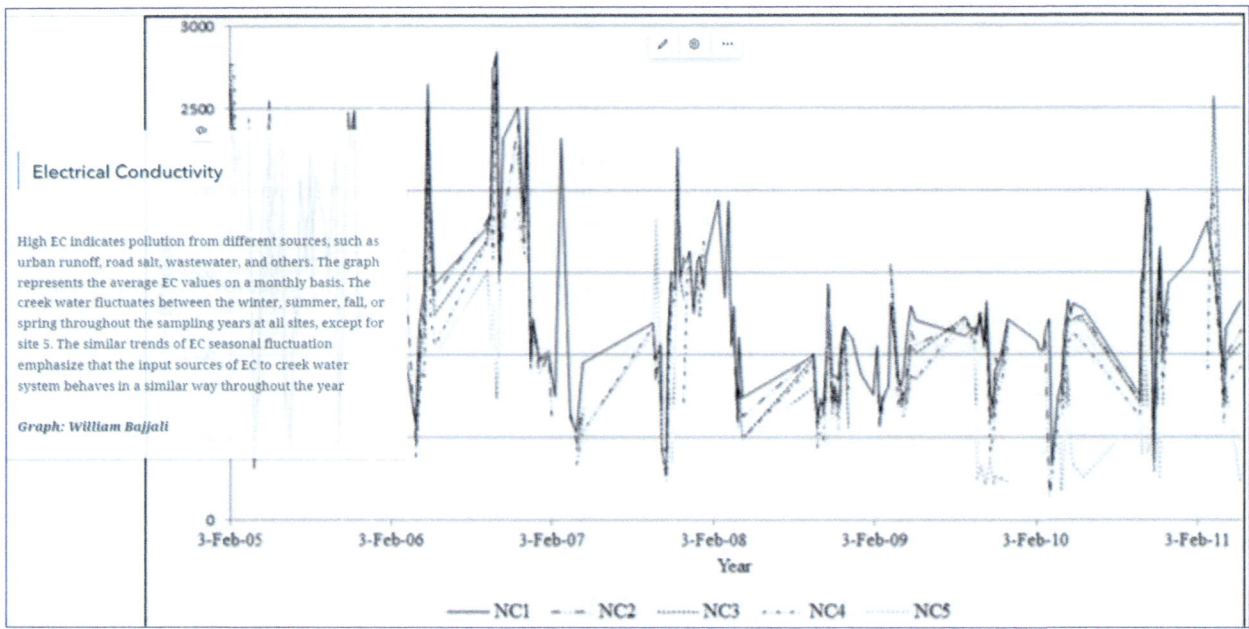

Add Second Graph

57. At the bottom of the sidecar block is the slide panel.
58. In the slide panel, click the + sign (New slide) button.
59. Repeat the previous step and add the **DO graph** and type the following text:

Dissolve Oxygen (DO)

The presence of **DO** in water is essential to aquatic life, such as fish vertebrates. DO is a function of atmospheric P, water T, and other dissolved substances in the water. To sustain life, DO should be above the chronic criterion for growth: 4.8 mg/l. Clear variation in the DO concentration was recorded at the five sites. Several sample sites recorded levels below 4.8 mg/l.

There is clear variation in the DO concentration at the five sites. This could be due to warmer summer temperatures and increased biological activity in terms of organic matter decomposition. Several sample sites recorded levels below the chronic criterion for the growth of aquatic life. The graph shows that extreme values below 4.8 mg/l were measured at all sites on different dates.

60. Customize it as in the previous step

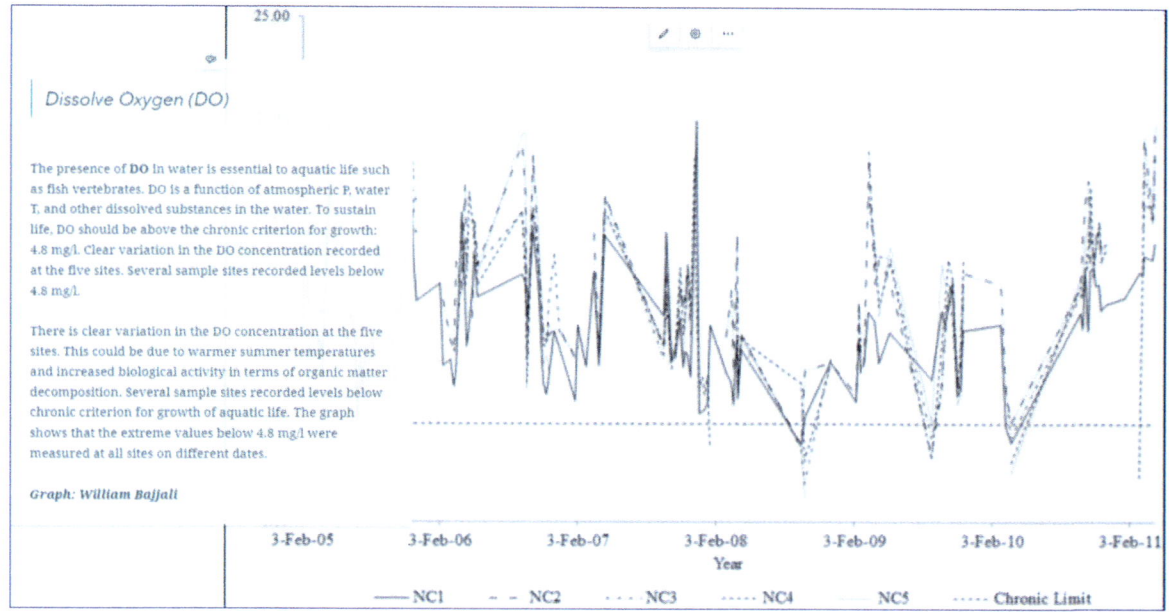

Add the Third and Fourth Graphs
61. Add the third graph "**pH.jpg**" and add the following text:

Hydrogen ION Concentration (pH)

pH affects many chemical and biological processes in the water, which is important to the survival and reproduction of aquatic life such as fish. A pH higher than 8 causes stress and limits growth, while a pH lower than 5 can speed up the mobility of certain trace elements that could harm plants and animals. The pH affects many chemical and biological processes in the water. Different organisms thrive within different ranges of pH (6.5–8.0). Newton Creek water shows pH fluctuation. pH fluctuation reflects contamination by a strong base (high pH). Excess aquatic plant growth (high pH). Low pH) pH in the stream shows that at times, the water is harmful to aquatic animals. Low pH could cause some toxic trace elements to be released, which may end up in the fat tissue of the organism on Hog Island.

62. Add the fourth graph "**Temp.jpg**" and add the following text:

Temperature

Water temperature is a critical parameter for aquatic life and has an impact on DO concentration and bacterial activity in water. Temperature can determine which fish and macroinvertebrate species can survive in a given stream. Macroinvertebrates are sensitive to T and will change location to select the optimal T. Site 1 never recorded a T value below zero compared to the rest of the sites. The graph shows that the temperature in the graph increases in summer and decreases in winter.

63. Customize the "**pH.jpg**" and the "**Temp.jpg**" as you did with the **EC graph**

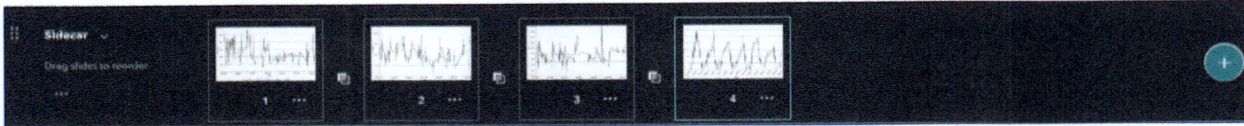

Conclusion

64. Click the + sign below the **Sidecar**, click Text under Basic.
65. Type or paste "**Conclusion**"
66. Highlight the "**Conclusion**", in the text menu, click the Paragraph and select Heading.
67. Click the + sign below the "**Conclusion**", click Text under Basic.
68. Type or paste the following text:

Some of the recorded parameters of EC, DO, and pH were observed to be much greater than the acute criteria. The relatively high EC value demonstrates that the source of water in the creek originates from polluted water. This source could originate mainly from the effluent of the treated wastewater of the oil refinery. There is clear variation in the DO concentration at the five sites. This could be due to warmer summer temperatures and increased biological activity in terms of organic matter decomposition. The creek water showed fluctuation of the pH between acidic and alkaline along the stream at the five sampling sites. Water with pH values above 10 was exceptional and may reflect contamination by a strong base such as NaOH and $Ca(OH)_2$. The creek demonstrates fluctuation of the temperature during the study period, and the temperature of the water increases in summer and decreases in winter.

69. Highlight the text. In the text menu, click the Italic button.
70. Click the + sign below the text of the "**Conclusion**", click Text under Basic.
71. Type or paste **Reference**
72. Highlight the **Reference**, in the text menu, click the Paragraph and select Heading
73. Click the + sign below the text of the "**Reference**", click Text under Basic.
74. Type or paste the following text

    Bajjali, W. Water Quality Assessment of Newton Creek and Its Effect on Hog Island Inlet of Lake Superior. ***Water Qual Expo Health* 4**, 123–135 (2012). https://doi.org/10.1007/s12403-012-0071-1

75. Highlight the text. In the text menu, click the Italic button.

Conclusion

Some of the recorded parameters of the EC, DO, and pH were observed to be much greater than the acute criteria. The relatively high EC value demonstrates that the source of water in the creek originates from polluted water. This source could originate mainly from the effluent of the treated wastewater of oil refinery. There is clear variation in the DO concentration at the five sites. This could be due to warmer summer temperatures and increased biological activity in terms of organic matter decomposition. The creek water showed fluctuation of the pH between acidic and alkaline along the stream at the five sampling sites. Water with pH value above 10 were exceptional and may reflect contamination by a strong base such as NaOH and $Ca(OH)_2$. The creek demonstrates fluctuation of the temperature during the study period, the temperature of the water increases in summer and decreases in winter.

Reference

Bajjali, W. Water Quality Assessment of Newton Creek and Its Effect on Hog Island Inlet of Lake Superior. Water Qual Expo Health 4, 123–135 (2012). https://doi.org/10.1007/s12403-012-0071-1

Review the Story

The next step is to review the story, as this step allows you to view your final product before sharing it with the organization or the community. In addition, if you see something that needs change, it is time to change it or proofread and edit the story.

1. On the ribbon, click the **Design** tab
2. The **Design** pane appear
3. Under Cover, accept the default "**Minimal**"
4. Under **Optional story sections**, check **Navigation**

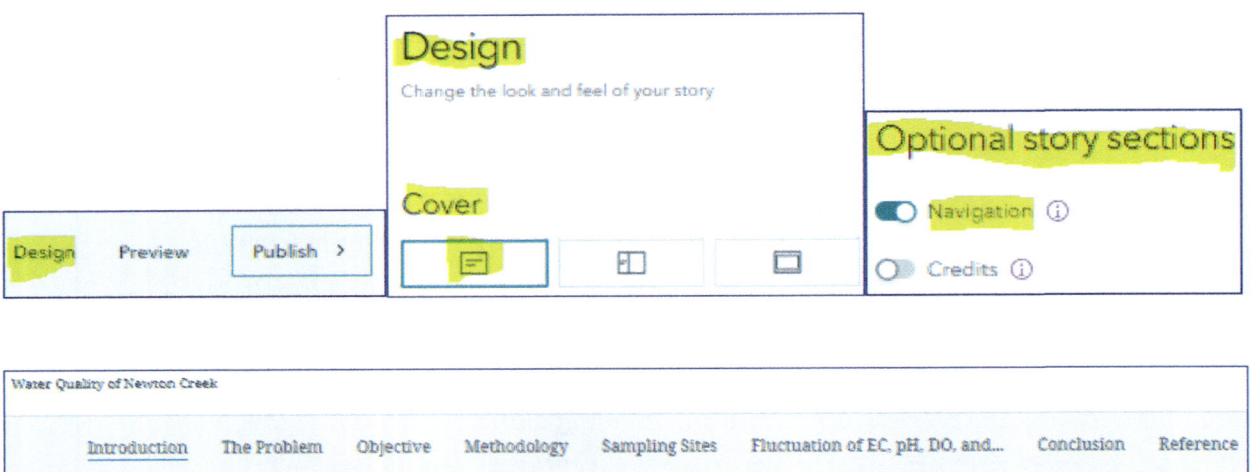

Result A Navigation ribbon display at the top of your Water Quality of Newton Creek story. The ribbon shows all the headings in the story. The ribbon lets the reader jump to a specific heading in the story.

5. For **Theme**, switch between the themes to see which one best matches the tone of your story
6. Select the Tidal theme

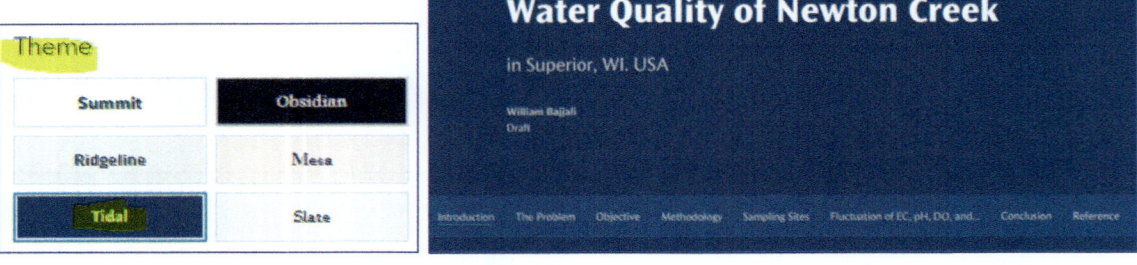

7. On the ribbon, click **Preview**.

8. The Preview mode allows you to see the story as
   (a) Preview on phone
   (b) Preview on tablet
   (c) Preview on desktop
   (d) Preview full screen
9. Click the Preview on phone

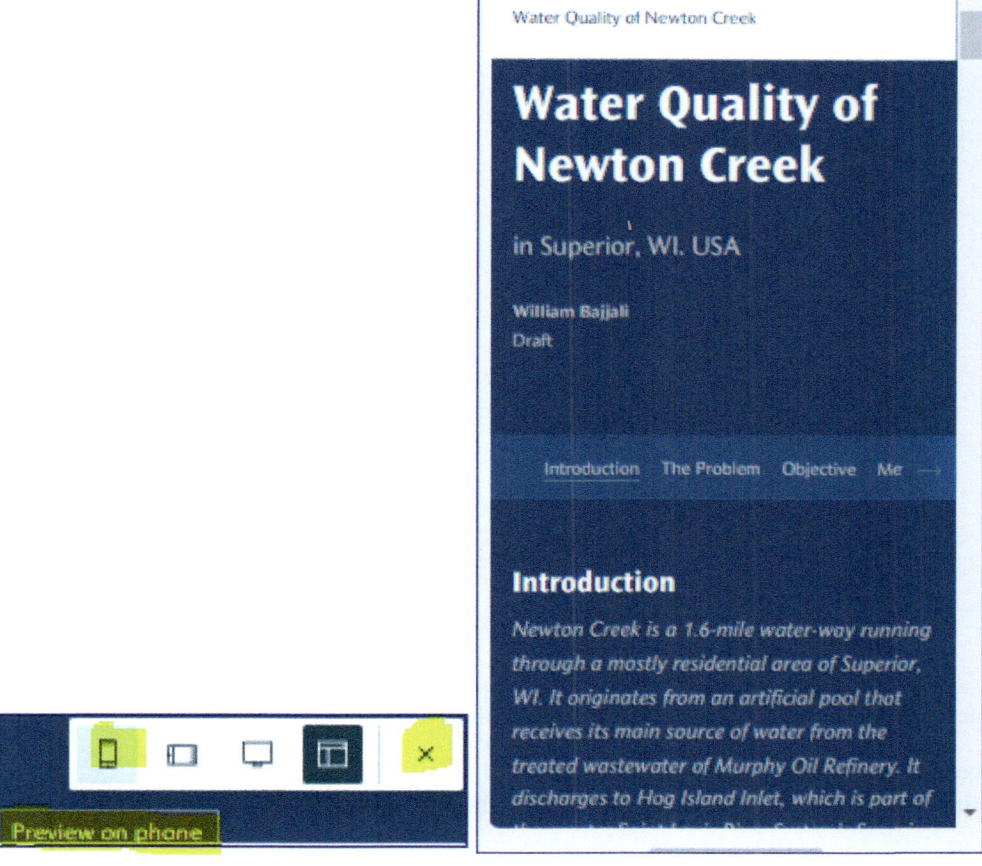

10. Review the story for typos, missing text, or other changes before publishing the story
11. When you are done reviewing, click Close preview "x" to return to the editing mode

Publish and Share

Now, you will publish your final story and share it with others. Once it is published, you can choose to share the story with only people in your organization, or you can share it publicly with the world community.

12. On the ribbon, click **Publish**
13. The Publish options dialog box display, which allows you to edit the story and share it.
14. In the Share panel, open the Private drop-down arrow and choose Organization
15. Click Publish

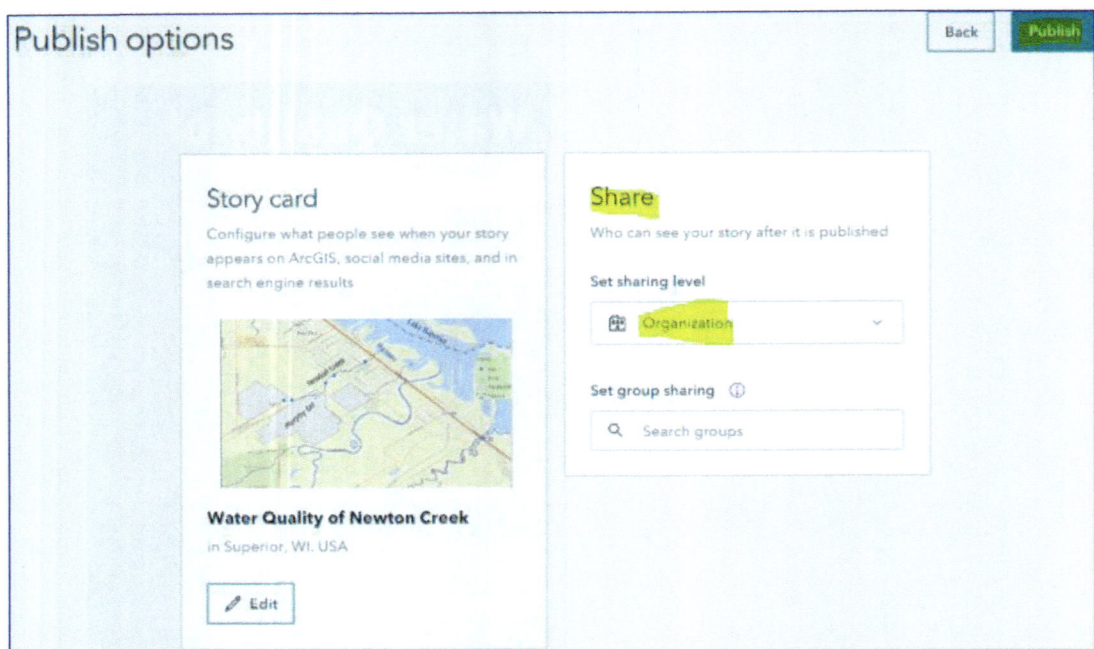

Result The Water Quality of Newton Creek Story is now published and displayed.

16. In the Ribbon, click the Share button
17. Click copy link (you can send the link to anyone who is interested in your research)

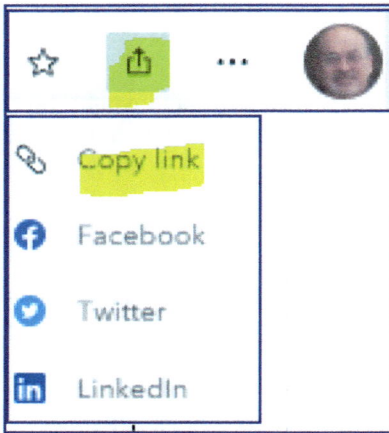

https://arcg.is/qSTvm

Note You can share your story also on Facebook, Twitter, or LinkedIn.

17 Instant App – Emergency Situations in California

Background California faces threats from different natural disasters, and citizens need a variety of solutions. Specifically, earthquakes due to the San Andras fault. California is also a coastal state and therefore is at risk of tsunamis. Throughout California, there are citizens in danger of these natural events. In the event of an earthquake or tsunami, hospitals can treat the injured and be a storm shelter, schools can also be a storm shelter, and groundwater wells can be used as drinking water.

ArcGIS Pro Approach

1. Launch ArcGIS Pro
2. Create a New Project from scratch
3. Name Project **Emergency_WB** (Emergency with your initials, i.e., WB)
4. Save it in Ch17

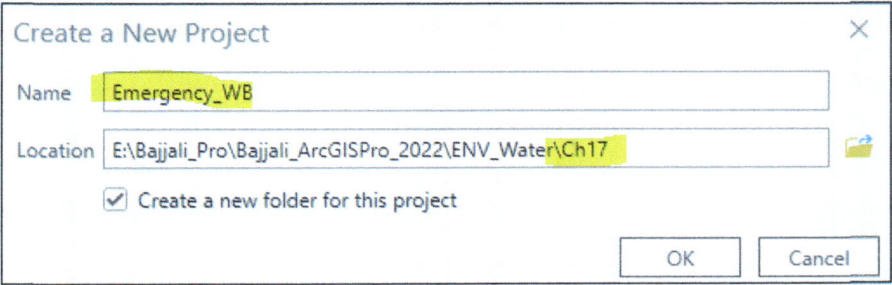

Change the Name and the Basemap

5. Rename the Map and **Emergency_CA**
6. Click the **Map** tab in the ribbon, in the **Layer** group, click the **Basemap** drop-down arrow.
7. Choose "**Light Gray Canvas**"

Supplementary Information The online version contains supplementary material available at https://doi.org/10.1007/978-3-031-42227-0_17.

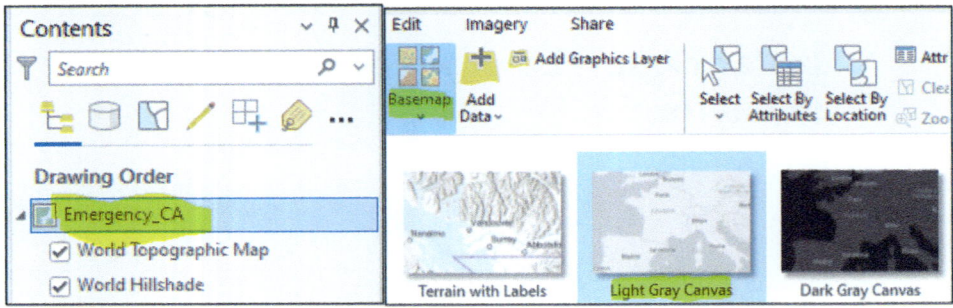

Connect to Data and Add Layers to the Map

8. From the **Catalog** pane, right-click **Folders** and choose **Add Folder Connection**.
9. Browse to **\\Env_Water\Ch17** highlight **Data** and click OK

Note The **Data** folder consists of **Layer** folder and file geodatabase (**Disaster.gdb**).

10. Open the Layer folder under Folder\Data and drag the Earthquake.lyrx, CA.lyrx, Fault_Andreas.lyrx, Hospitals.lyrx, Schools.lyrx, Tsunami.lyrx, and Wells.lyrx to the Map View

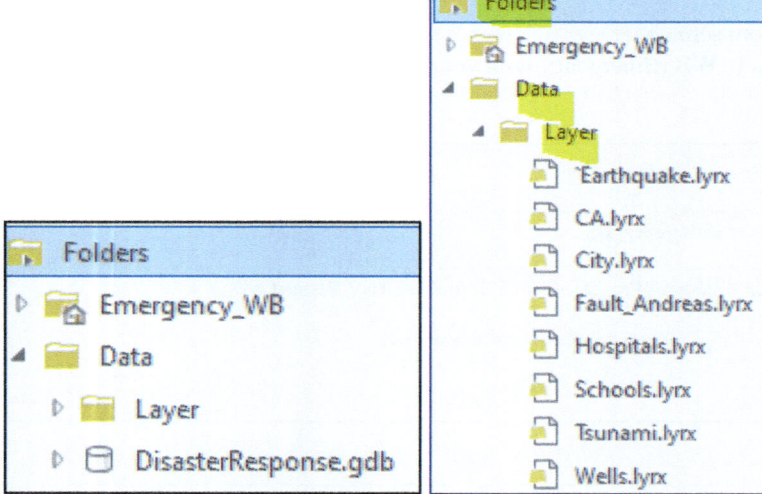

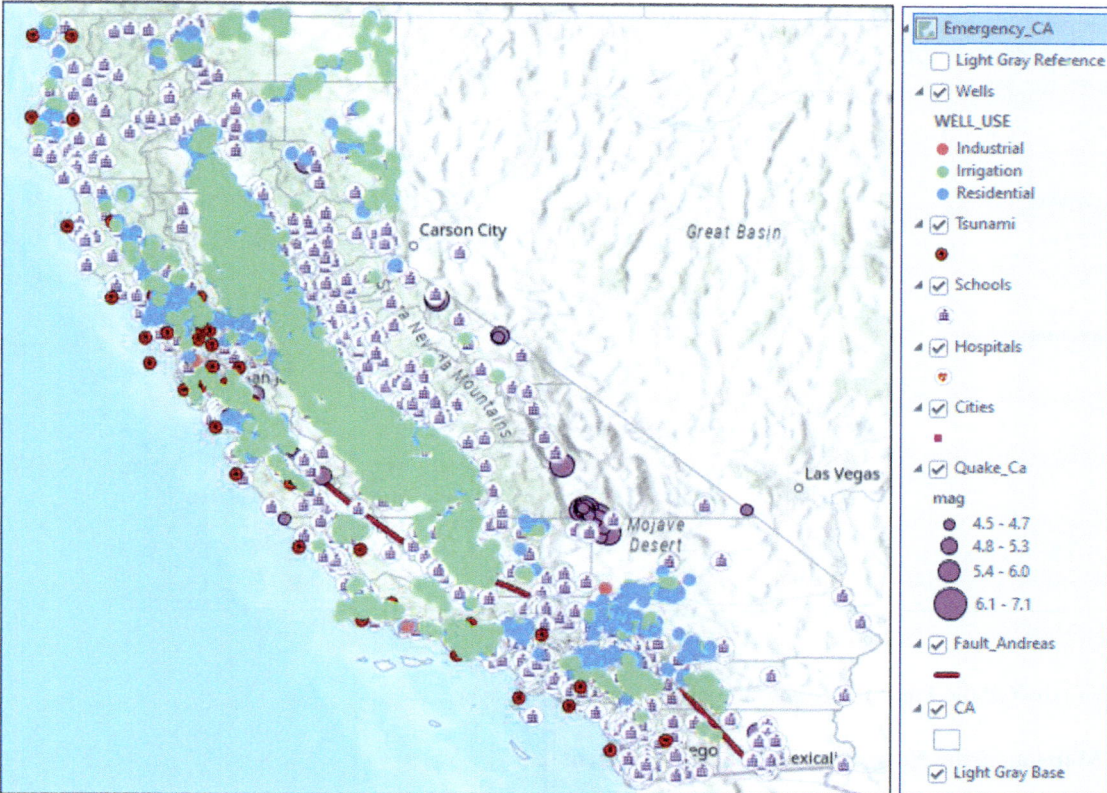

Important When you add a layer file to a map, it draws exactly as it was saved, provided the data referenced by the layer is accessible. The **Earthquake.lyrx** is displayed as **Quake_Ca** because the original feature class in the **DisasterResponse**.gdb.is called "**Quake_Ca**".

Set the Extent

You will set the map extent to **CA** layer using custom extent
11. R-click **Emergency_CA** map, point to **Properties**, select the **Extent** tab
12. Click **Use a custom extent**, under **Extent of a layer**, click CA
13. Click OK

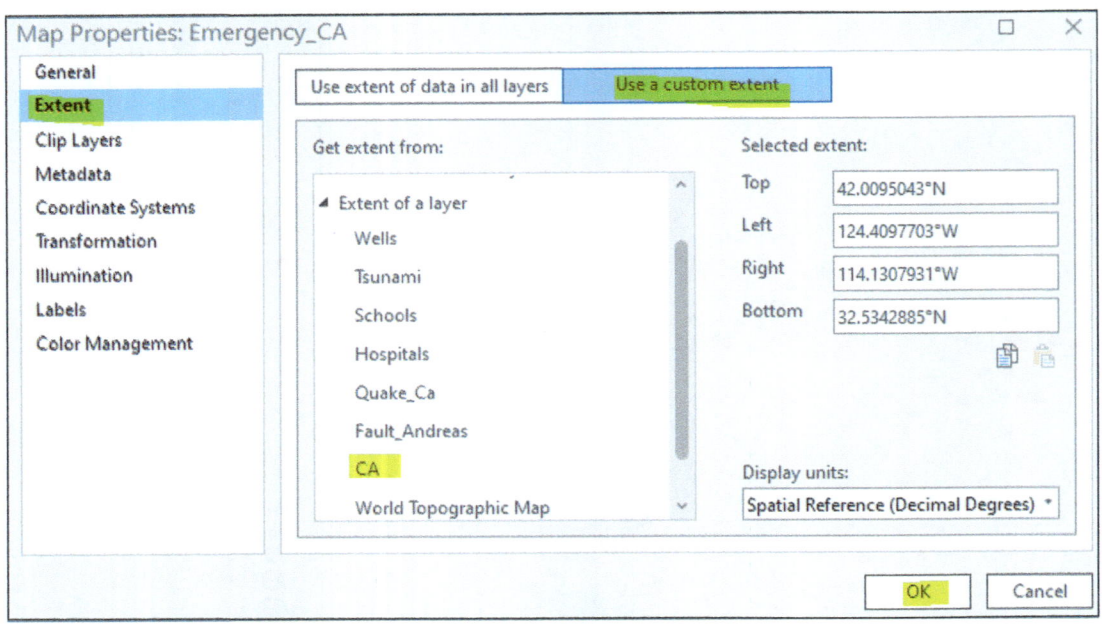

14. R-click in the **CP** the **Fault_Andreas**, and point to Zoom To Layer

15. In the **Map** tab, in the **Navigate** group, click **Full Extent**

Update Label Properties and Configure Pop-up Windows

The symbology of all the added layers is based on the layer file that was added to the current map. These symbols do not require updating because they already meet the web map design requirements.

Before sharing the web map, you will update the labels and configure the Pop-up information.

16. Click **Shift** on the keyboard and in the CP, and uncheck the **Quake_Ca**, **Hospitals**, **School**, and the **Wells** layers
17. In the CP, highlight the **Tsunami** layer and click the **Labeling** tab on the ribbon. In the **Label Class** group, choose **County** from the **Field**, then click the **Expression** to the right of the **County**
18. In the **Label Class – Tsunami** pane, click the **Symbol** tab, and then expand **Appearance**.
19. Change **Size** to 11 pt., click the **Color** down arrow (under Text fill symbol) and choose **Mars Red** (R3, C2).
20. Click Apply.
21. Click the **Position** tab.
22. Click the **Conflict Resolution** button (third icon), expand **Remove Duplicate Labels**.
23. From the drop-down list, choose **Remove All**.
24. Close the Label Class pane.
25. In the Labeling tab, in the Layer group click the Label button

Quiz How to remove the label?

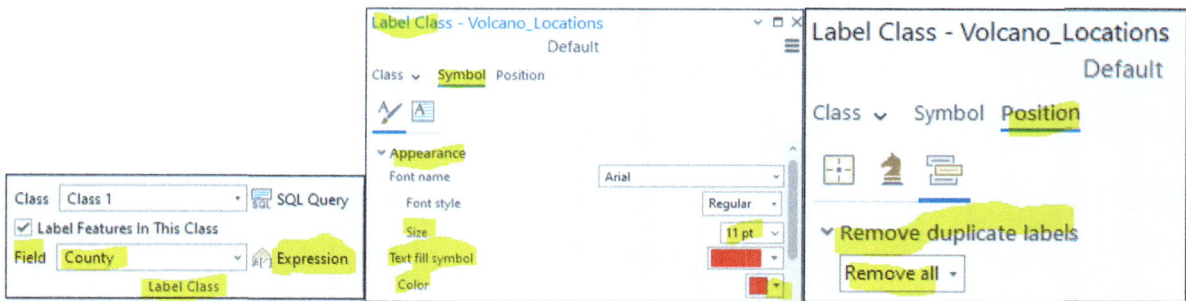

Configure Pop-ups

You will now customize the **pop-up** window for the "**Schools**" layer by removing unnecessary fields from the display.

26. In the CP, r-click **Schools** layer and open Attribute Table (69 fields and 10,043 rows)
27. In the CP, check the **Schools** layer and then right-click it and choose **Configure Pop-ups**.
28. In the **Configure Pop-ups – Schools** pane, in the list of locations, point to Fields (69) and click the **Edit Pop-Up Element** Button.
29. Uncheck the "**Only Use Visible Fields and Arcade Expressions**" box.
30. The **Display** column tells you the columns that will be displayed in your pop-up window.
31. In the **Display** column header, uncheck the **Display** box to deselect all fields.
32. Check the display option for the following fields only:
    (a) SchoolName
    (b) SchoolType
    (c) City
33. At the top of the pane, click the Back button
34. In the map, zoom around the **San Francisco** area and click any school feature to confirm that the pop-up window has been properly configured.
35. Close the pop-up window, and then close the **Configure Pop-ups** pane.
36. In the Map tab, in Navigate group, click Full Extent
37. Close the **Schools** attribute table.
38. Save the project.

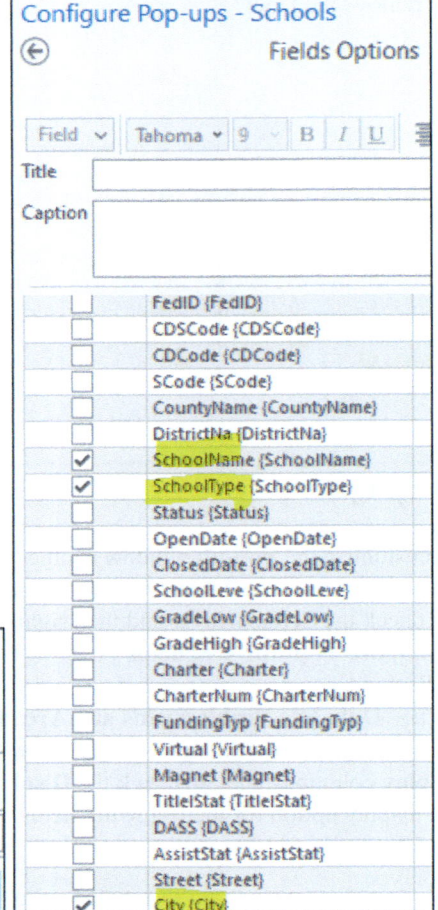
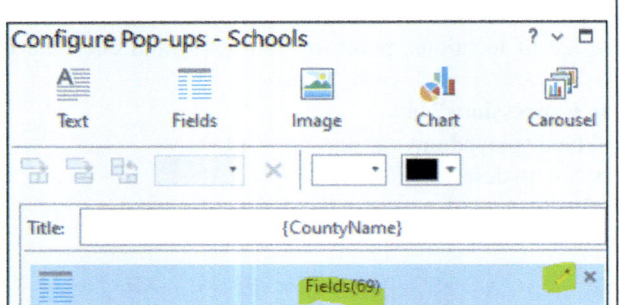
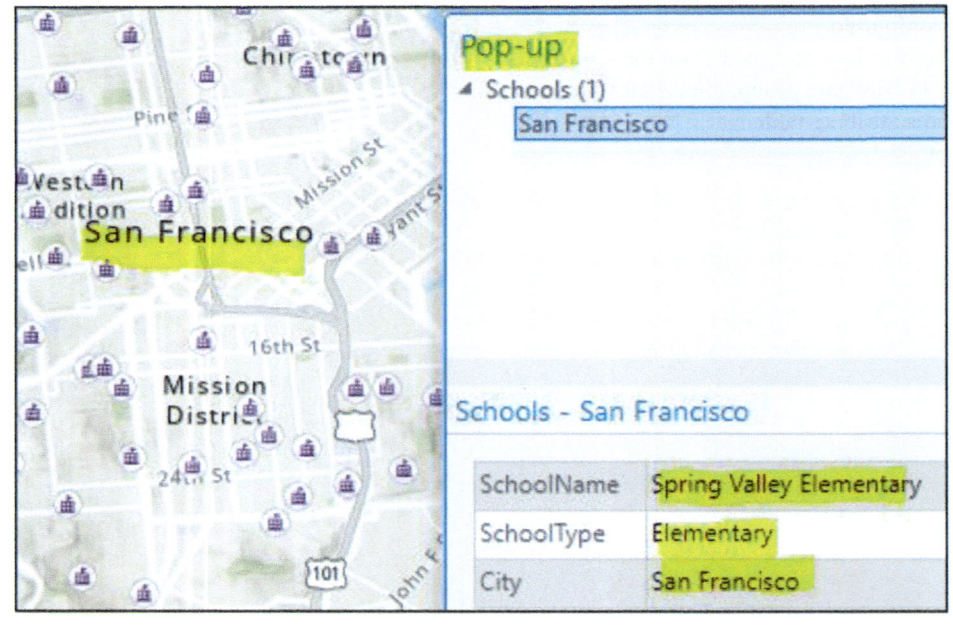

Tip As an alternative to pop-up windows, you can use the Attributes pane for editing workflows where you want to visualize and update attributes of multiple features at the same time.

Challenge Task

Configure the Pop-Up for the following layers **Quake_Ca**, **Hospitals**, **School**, and **Wells** by removing the unnecessary fields and keeping the field as shown in the table below.

Layer	Field	Field	Field
Quake_Ca	Depth	Mag	Place
Hospitals	Name		
Wells	WELL_NAME	WELL_USE	

Share a Web Map and a Web Feature Layer

You will share a web map with the public so that anyone can use it to generate a web or native app using an app builder. You will configure a hosted web feature layer to support data export by other users.

39. In the CP, right-click **Emergency_CA** and choose **Properties**.
40. In the **Map Properties** dialog box, on the **General** tab, check "**Allow assignment of unique numeric IDs for sharing web Layers**"
41. Click **OK**.

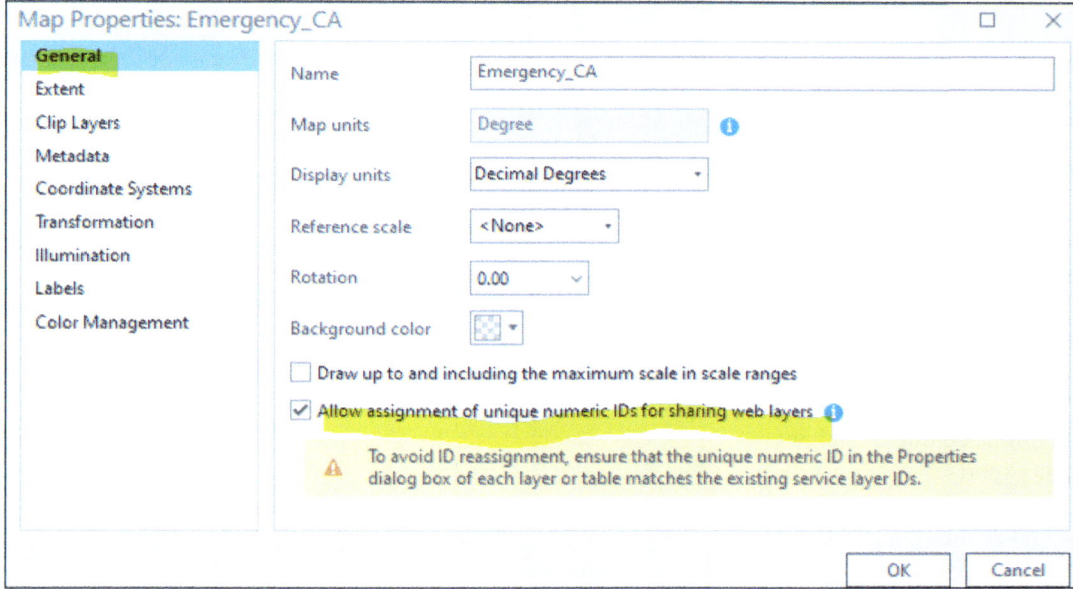

Question When a map contains both vector and raster data layers, should you share it as a web map or a web layer?

Answer You should share the map as a web map. You can only share a web map as a web layer if all the layers in the map are of the same type.

The **Emergency_CA** map has

- A basemap (raster) that consists of tiles (Light Gray Canvas)
- Many layers contain point and line features (Wells, Quake_Ca, School, Tsunami, Hospitals, Fault_Andreas).

Therefore, the **Emergency_CA** map should be shared as a **web map**.

Important To **share** the **web map** on **ArcGIS Online**, you must be signed in ArcGIS Online.

42. In the CP, highlight the **Emergency_CA** map, click **Share** tab on the ribbon, in the **Share As** group click **Web Map**
43. In the **Share As Web Map** pane, for Name, verify that the field is set to **Emergency_CA**
44. For Summary, type **Earthquake, Tsunami, Hospital, School, Cities, and Saint Andreas fault in California**.
45. For Tags, type **Earthquake, Tsunami, Hospital, School, Cities, Wells, Fault, California, USA** and then press Enter.

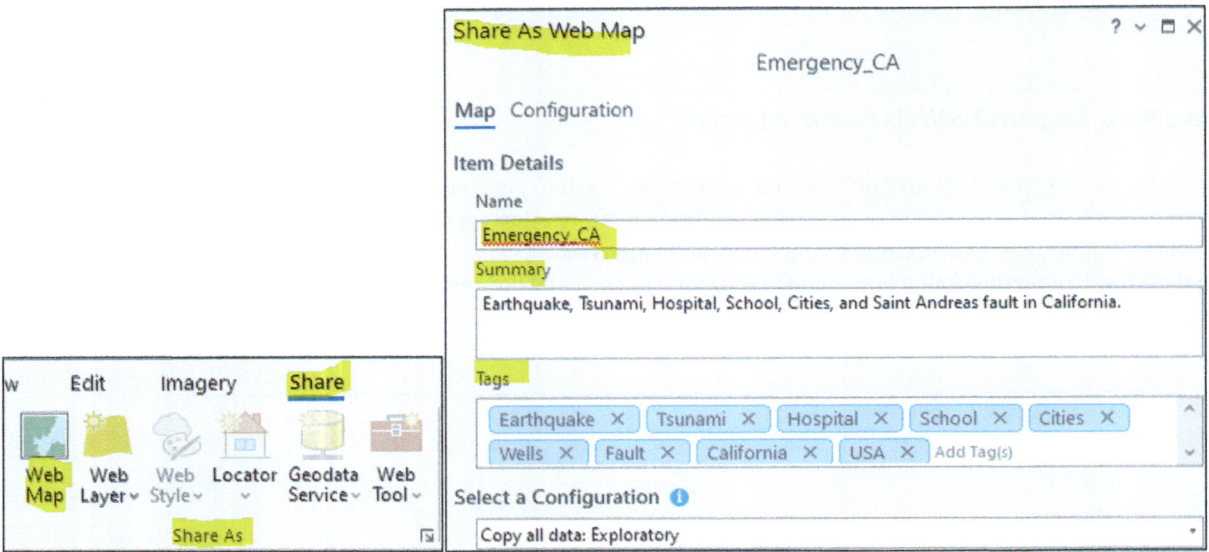

Configure the Web Feature Layer

When you share a **web map**, individual layers are also shared as **web layers**.
The different layers in the **ShareWeb** map can support

- Online and offline editing workflows
- Support data export.

You will now update the configuration of some layers to support **data export**.
46. In the "**Share As Web Map**" pane, click the **Configuration** tab.
The **web feature layer** (WFL) will be shared as **Emergency_CA_WFL1** by default. These web feature layers contain several layers in the map (**Wells, Tsunami, School, Hospital, Quake_Ca, Fault_Andreas,** and **CA**). A web feature layer can contain multiple data layers in the map if they are of the same type (features or tiles).
47. Click **Emergency_CA_WFL1** to select the layer and click **Properties**

- The **Layer Type** property is set to **Feature** by default because all layers in the **Emergency_CA** map are points, line, and polygon features (vector structure).
- **Web feature layers** can be queried without further configuration.
- **Editing and data export** workflows can be supported by enabling more operations. You will enable the **Export Data operation** on the web feature layer.

48. Click the **Configuration** tab and check the **Export Dat**a box.

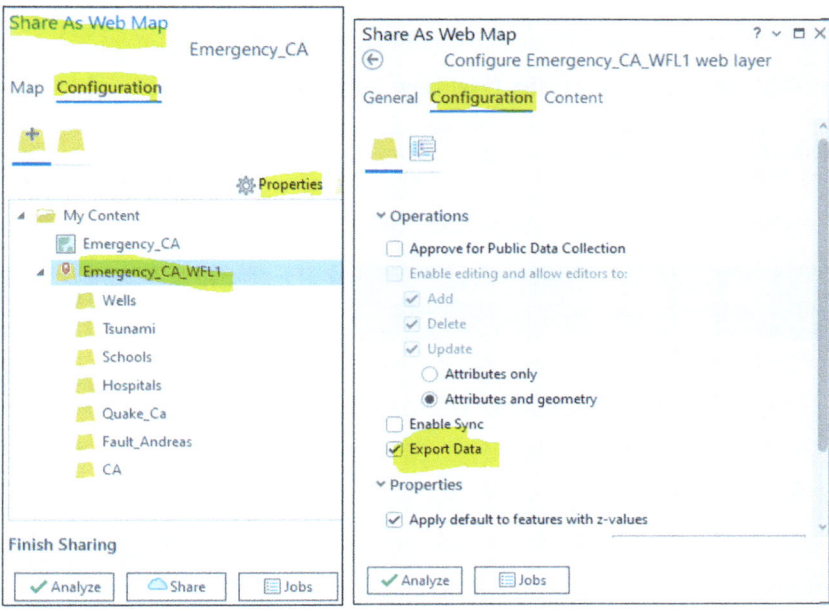

Share a Web Map

You are now ready to analyze the map and share it as a web map in ArcGIS Online and make sure that there are no errors before you share it.

49. At the upper left of the **Share As Web Map** pane, click back button.
50. From the **Share As Web Map** pane, click **Analyze**.

If you see a warning message, you can ignore it. There are no errors, so you can proceed to share the web map. The error is that the service layer has a different projection. To correct the error, you can expand the error and double click on each one, and the error will disappear.

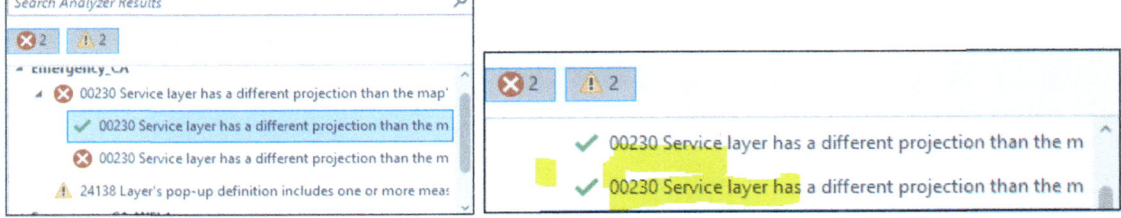

51. Click **Share** tab to the right of the **Analyze** table
    A message indicates that the web map was created successfully.

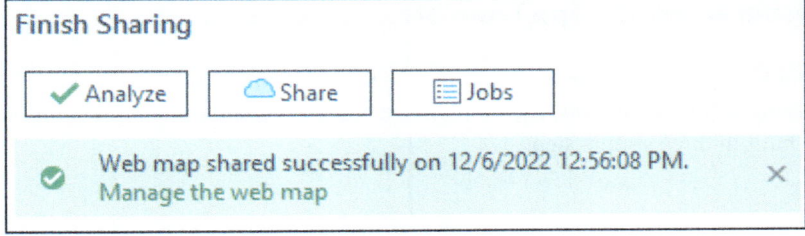

52. Click the **Manage the web map** link (green color text).

Note If necessary, sign in to your organization's portal.

Result The **Emergency_CA** web map display under **Overview** Tab.

53. Review the item details of the **web map** in your organization.

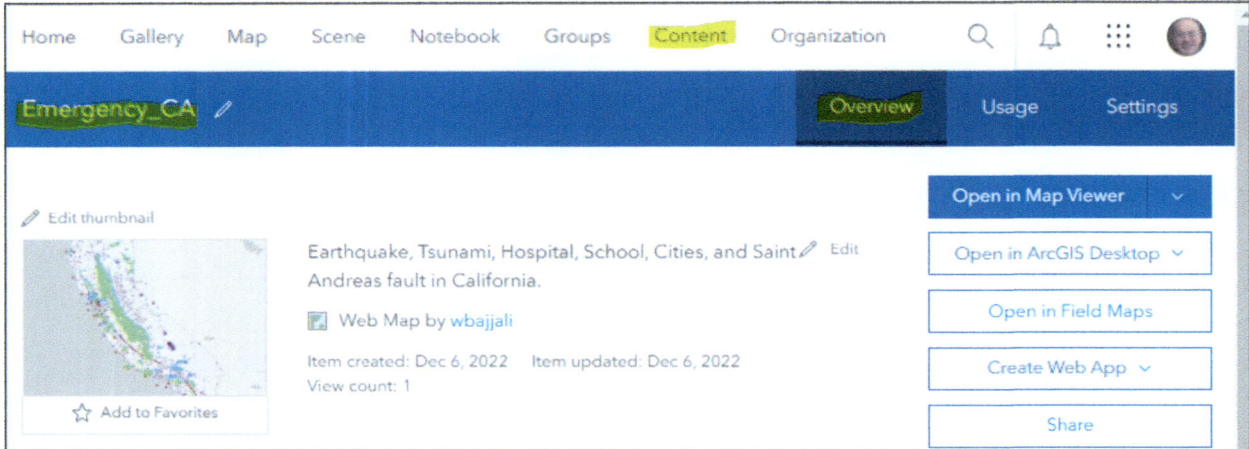

54. Click the **Content** tab

Result You will see the **Emergency_CA** web map, **Emergency_CA_WFL1** feature layer (hosted) and **Emergency_CA_WFL1** service definition is listed under your folder (i.e., wbajjali).

You notice also that the web map and the layers are shared with yourself only (Owner) .

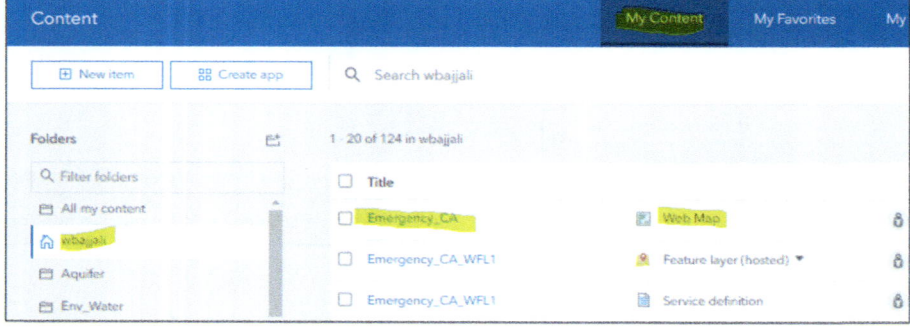

Publish a Geo APP Using an Instant App Template

In this section, you will publish a geo app from the **Emergency_CA** web map that you created in ArcGIS Pro. The geo app will highlight features such as hospitals, schools, and wells within a certain distance from an earthquake or tsunami location. Web app template can be built in three simple steps.

- Add data from online in the form of feature layers such as hosted feature layer that has been published to ArcGIS Online. These features (point, line, polygon) support querying, editing, and visualization. They can be added to more than web map
- Web maps can be created by adding hosted feature layers.
- From the web map you create a web app, which is the best way to share information

The Geo app can be created in two different approaches.

First Approach to Build Geo App

Continue using the **Emergency_CA** web map by performing the following steps:
55. Under **My Content** tab click on the **Emergency_CA** web map
56. To the right of the **Emergency_CA** web map, click "**Open in Map Viewer**"

Comment Map Viewer is the primary map-making tool for **ArcGIS Online**. In this chapter, you will use the **Map Viewer** to create the geo app.

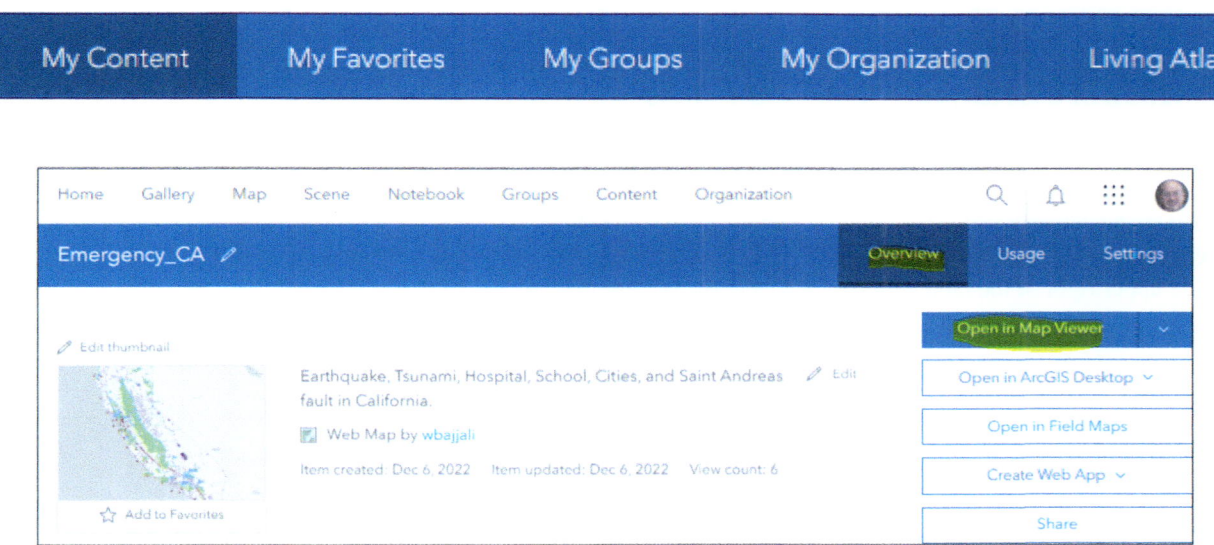

Result The web map opens and has different layers, and the Wells layer is selected. To the right if each feature there is a symbol, when you click it show and hide the layer.

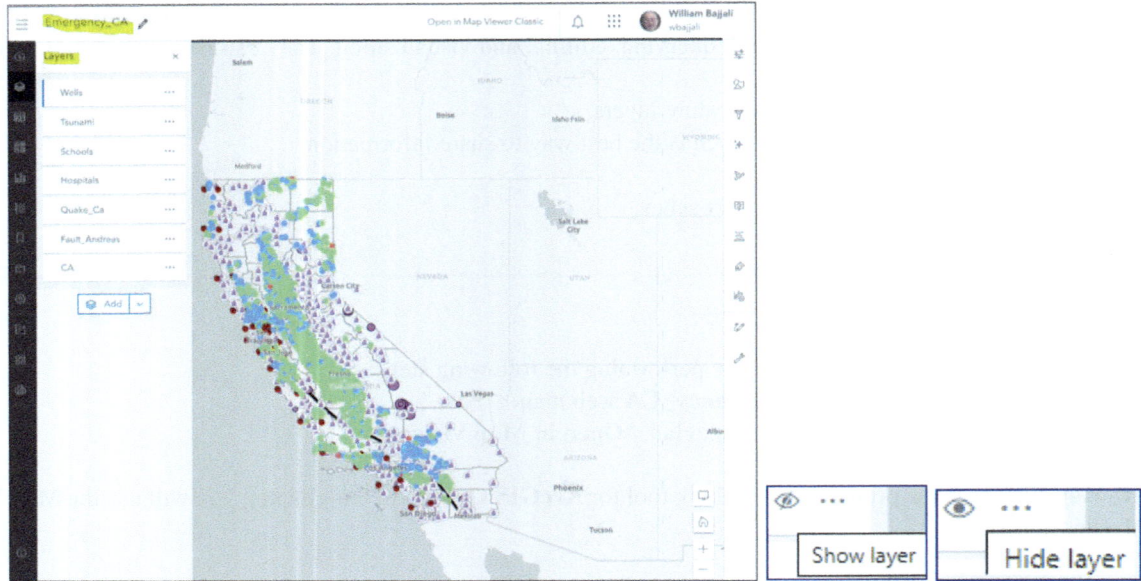

Examination of the Attribute Table of the Wells

57. In the **Layer** pane, for the **Wells** layer, click the Option button (…) to the right
58. Click Show table
59. The table has many capabilities, such as sorting, changing column width, applying filters, and editing records, if editing is enabled.
60. In the upper right of the table, click the "**X**" button to close the table

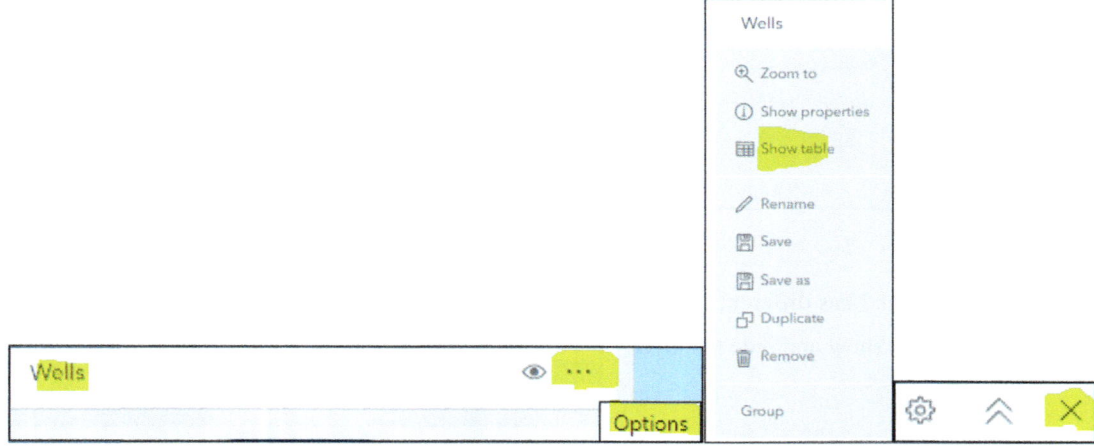

Configure the Pop-up Window of the Earthquake

You will configure the pop-up window of the **earthquake** (**QUAKE_CA**) in California between 2011 and 2022 by using a subset of the available attributes. The pop-up windows are useful because they are how users receive information from the map. The pop-up configuration of the web map's layers influences the behavior of the geo app that you publish

61. In the **Map Viewer**, zoom in to an area where you see the **Quake_Ca** feature (pink circle) and click on one circle to open its pop-up window.
62. The pop-up window opening showing different information

63. Close the pop-up window

Configure the Pop-up of the Earthquake

You will use the **Quake_CA** layer attribute data to control what information is shown when a feature is selected on the web map and how certain geo app components behave.

64. In the **Layer** pane, ensure that the **Quake_Ca** layer is selected
65. On the right, on the **Setting Toolbar**, click the **Pop-ups** button
66. In the Pop-ups pane of the **Quake_CA** layer, next to **Title**, click **Expand** button
67. From the **Title Field**, delete the word **time**
68. To the right of the **Title field**, click the **Add Field** button { }
69. From the list that opens, scroll down and select (place)
70. In the **Fields list**, click on the **Fields list** to Expand it

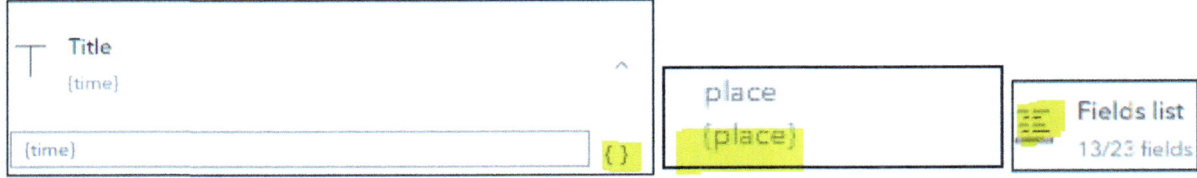

All the fields in the attribute table will be displayed, and all the display fields except **time**, **depth**, **mag**, and **place** will be removed.

71. Click the X to the right of each field except **time**, **depth**, **mag**, and **place**

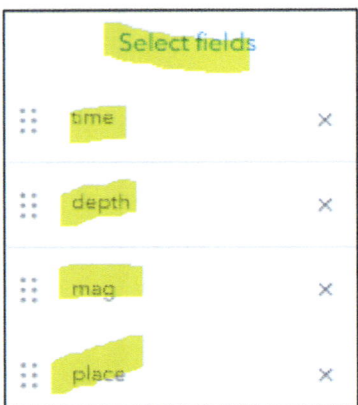

72. On the **Setting Toolbar**, click the configure **Fields** button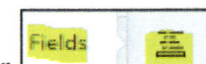
73. In the **Quake_Ca** panel, click the **place** field
74. The **Formatting** dialog box display
75. Under **Display name**, delete the **place** and type **Earthquake Location**, click **Done**

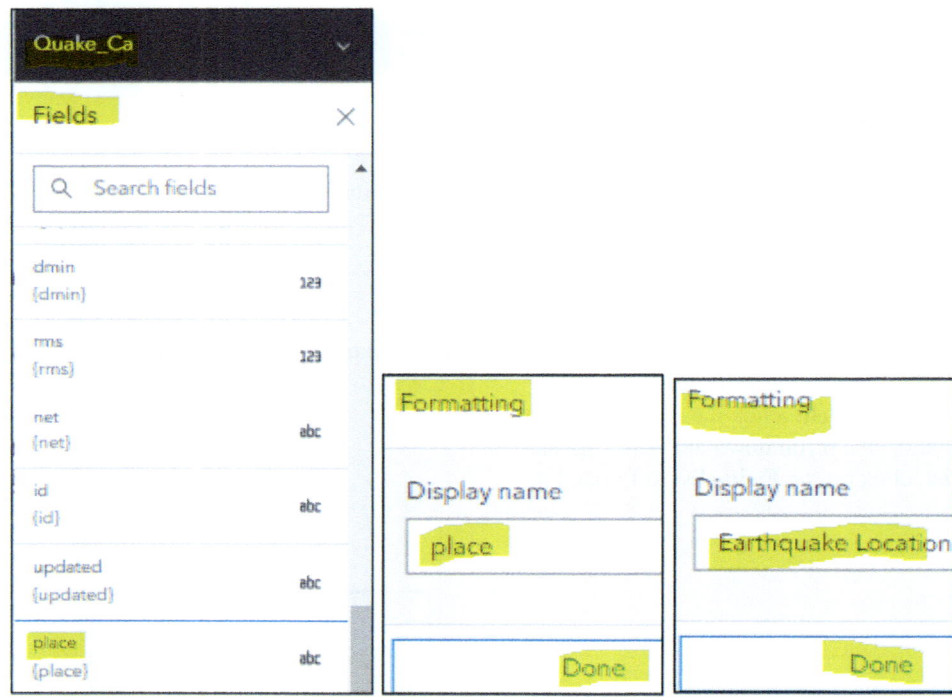

76. Click the **mag** field
77. Under Display name, delete the **mag** and type **Magnitude**
78. Under the Significant digits, open the drop-down arrow and make it 2 decimal places.
79. Click Done
80. Repeat and make the **depth** to be **Depth** with two decimals
81. Make the **time** as **Time**
82. On the map, click a **Quake_CA** feature, you will see that names are updated

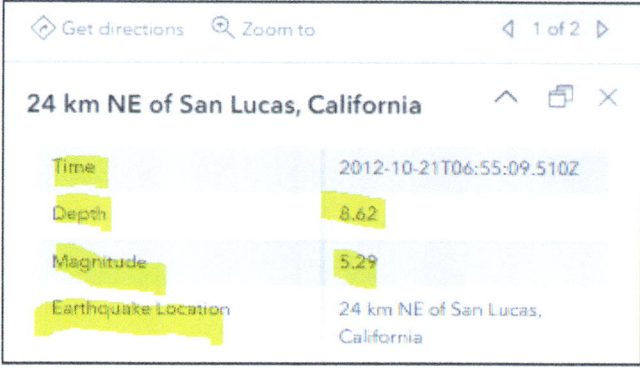

Save the Web Map

Before proceeding and creating the web app, make sure that that all layers are active, and the web map is saved.

83. On the **Content** toolbar, located on the left, click **Save and open** and choose **Save as**
84. In the **Save map** dialog box, fill it as follows:
85. **Title**: **California Emergency Map_WB** (use your initials instead of WB)
86. **Folder**: Save the web map in a folder in ArcGIS Online (i.e., wbajjali)
87. **Tag**: California, Earthquake, Tsunami, Groundwater, Wells, Hospitals, Schools, Fault
88. **Summary**: California Map showing the locations of the earthquake, tsunami, fault, groundwater wells, hospitals, and schools that will be used during a hazard event.
89. Click **Save**

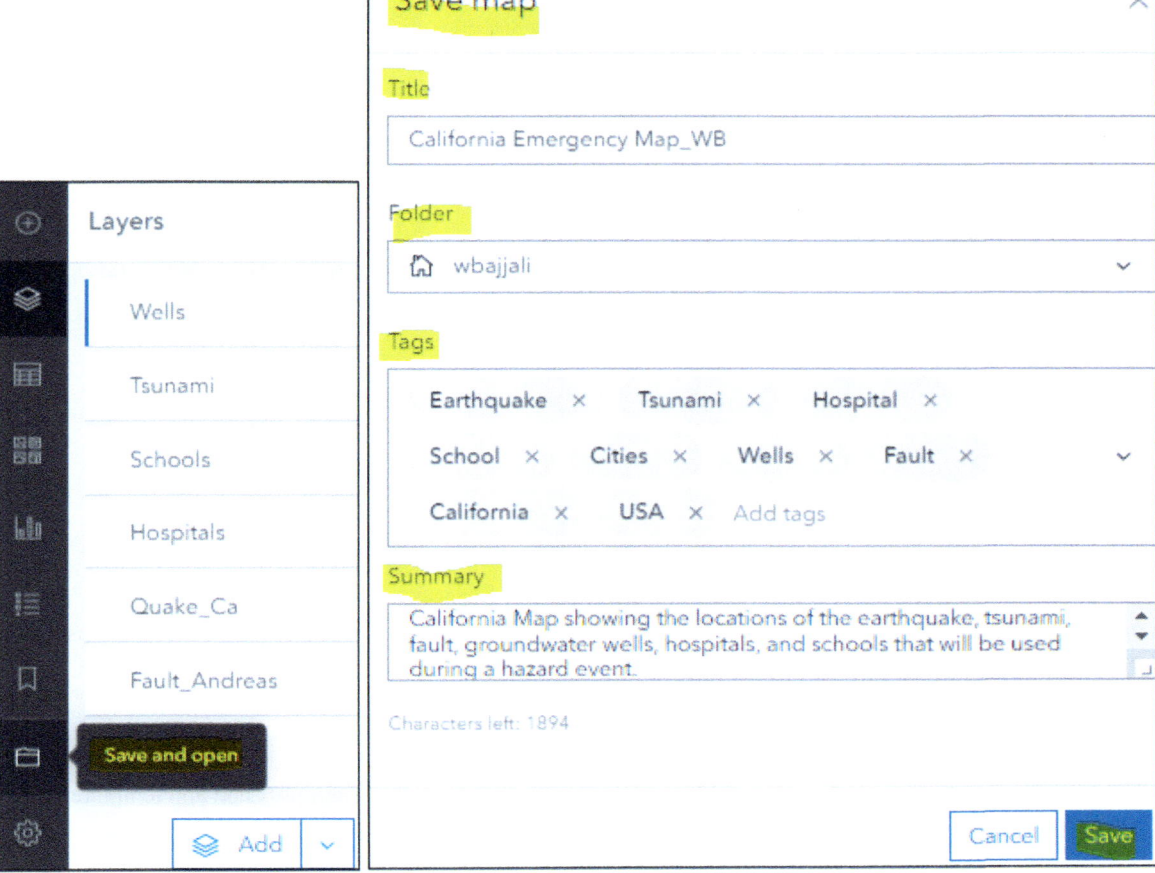

Result the new web will be called "**California Emergency Map_WB**" and appear at the top of the Layer pane.

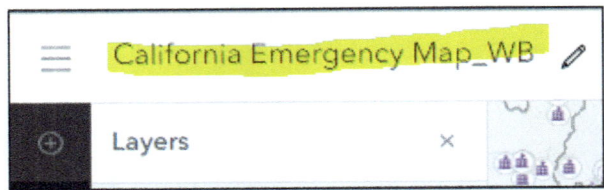

Create a Web App

Geo app is a good way for sharing spatial data with your organization or with public, especially non-GIS users. The app allows users to combine maps, scenes, multimedia, and more to educate and inspire others to many topics related to different subjects. A geo app will be created from the "**California Emergency Map_WB**". The created geo app will pinpoint the locations of earthquakes, tsunamis, wells, schools, and hospitals that can be used as emergency facilities in terms of shelter, medical treatment and potable drinking water or identify what exists within a distance from a specific address in the state of California.

1. In the **Content** toolbar on the left, click **Create app** button
The menu shows some options for creating geo app from the web map that you created. The menu shows four options
   (a) Instant Apps
   (b) Experience Builder
   (c) ArcGIS Story Maps
   (d) Dashboard

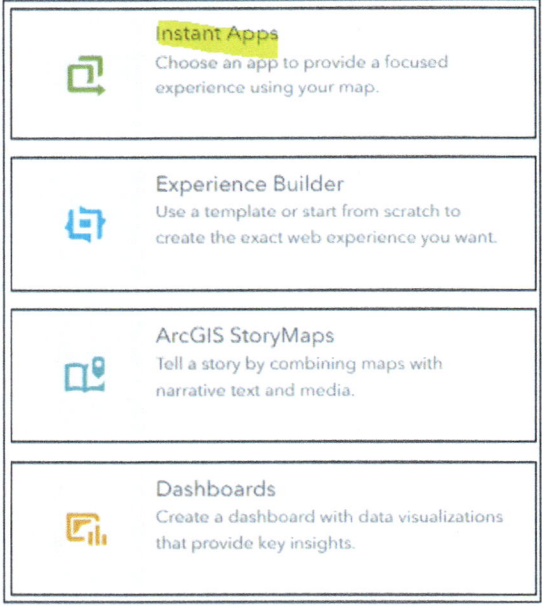

Each option offers different functionalities. In this chapter, you will create an **Instant App** that will allow users to interact with the map and data.
2. Click **Instant Apps**
3. A gallery of app templates opens

Result The instant apps include a gallery no-code app template and configuration options. Each app template has a specific purpose, such as finding some features nearby, which we will use in this section.

Create a Web App

Note You want to create an app that allow viewer to explore the location of earthquake, tsunami, and the other features that are closest to the hazard event or specific location or address.

The Instant Apps allow you to preview and test the app before creating it.
4. In the gallery try several templates and when you finish click **Back** button
5. In the gallery, find the **Nearby** template and click the **Preview** button
6. Zoom in or zoom out to find a tsunami location
7. Click a **tsunami** location on the map

Result The "**California Emergency Map_WB**" displays a circle and features that exist within a certain radius inside the circle. The left side of the map shows the nearby features. In this example, you see within 2 km some features (i.e. one tsunami and 5 schools).

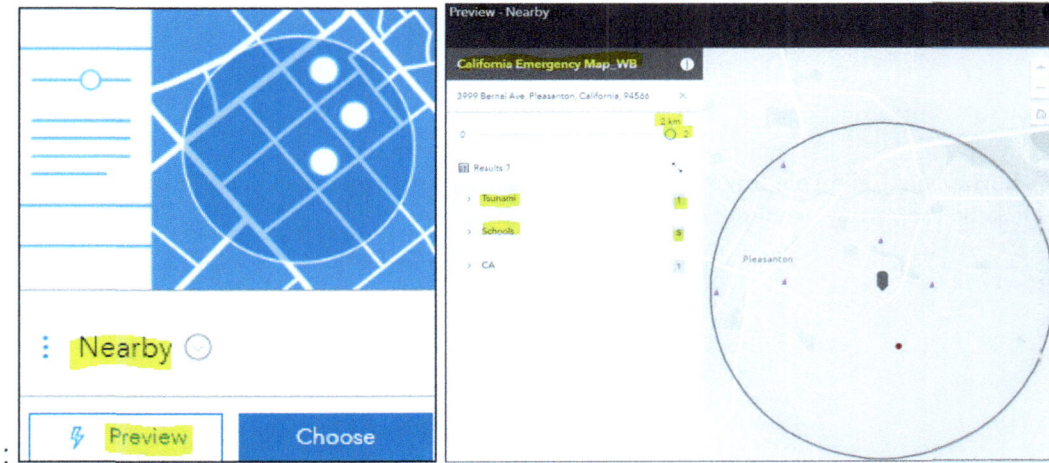

8. From the map click Clear Search Location
9. Click the Default Map View button
10. Click another earthquake location to explore the nearby feature
11. Click then Clear Search Location, and click **Choose**

12. Give your app a title: **California Emergency App_WB** (use your initials instead of WB)
13. The tag field automatically populated with the tag from the web map
14. Click **Create App**

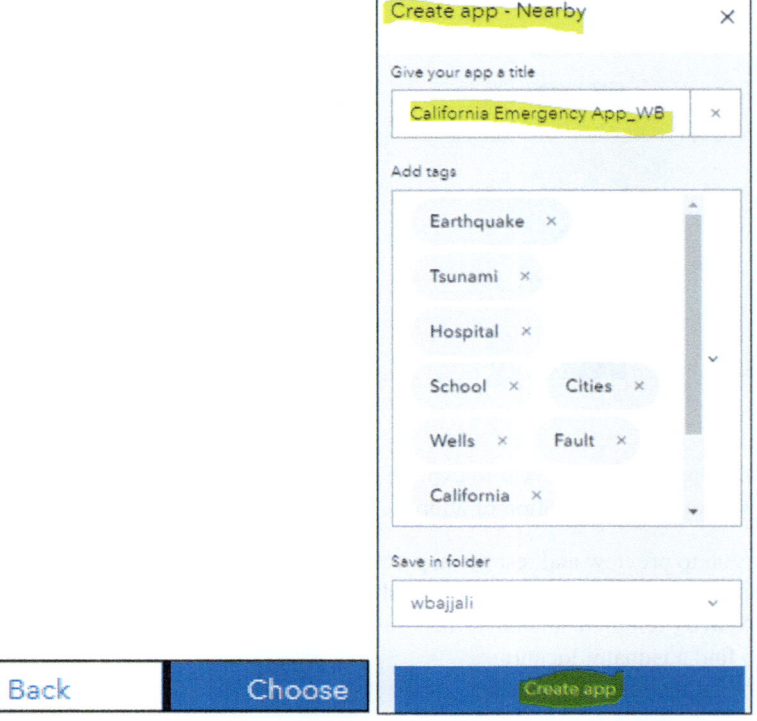

Configure Your Instant App

You will use the **Express** setup to configure your **Instant App**. The Express setup consists of five steps:
(a) Map
(b) About
(c) Nearby
(d) Interactivity
(e) Theme & Layout

15. Click **Step 1. Map**, the **California Emergency Map_WB** is the current web map in use. Because you are not going to change it click the back arrow to return to the **Express**.

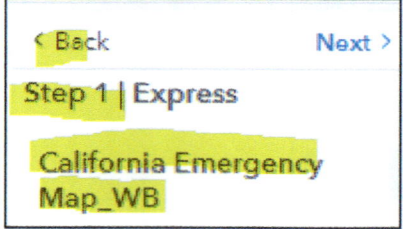

16. Under Express, click Step 2. About
17. App title: Type of Natural Disaster Emergency Resources in California

Configure Your Instant App

18. Click Next

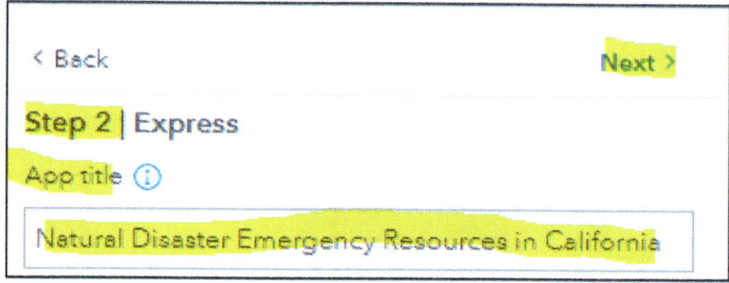

18. Click **Step 3. Nearby** This step allows you to configure the setting to search for what is nearby, therefore, the minimum and maximum distance unit should be configured.
19. Under **Step 3 | Express**: Under **Layers to include in results**, check all boxes, with exception CA
19. Under the **Maximum search distance**, replace the value with 20
20. Under **Search units**, choose **miles** from the drop-down arrow

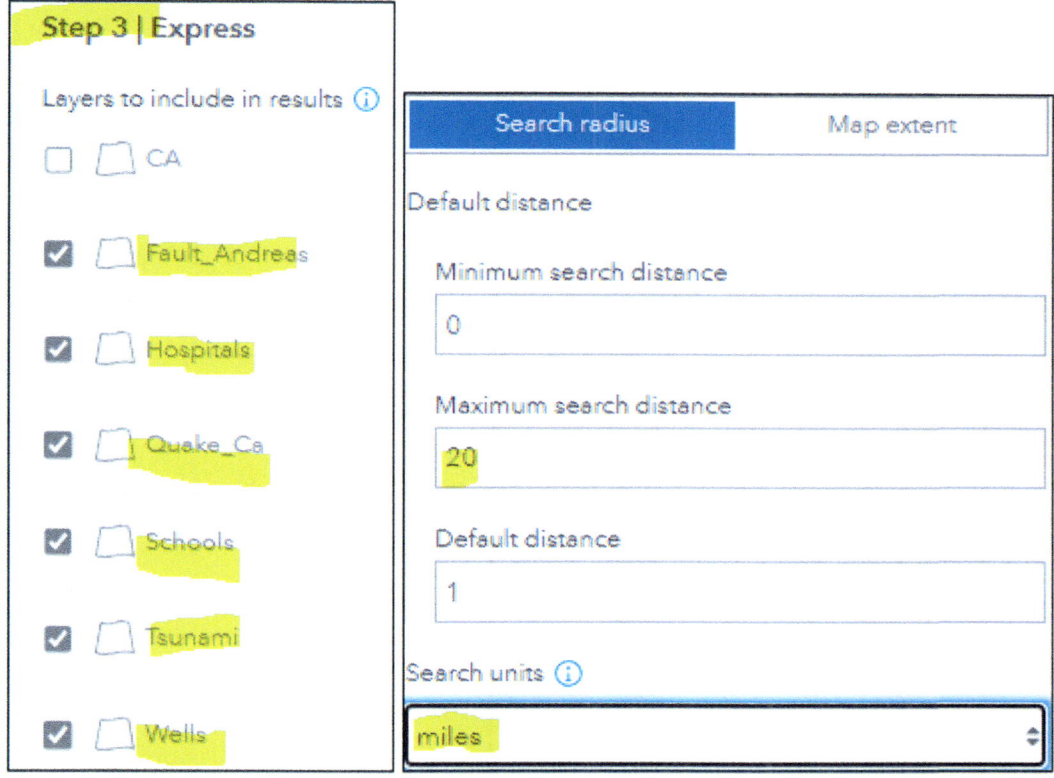

21. Click Next to go to Step 4 (Interactivity)

22. Under the Search Configuration in the "All sources window", click the ArcGIS World Geocoding Service to select it

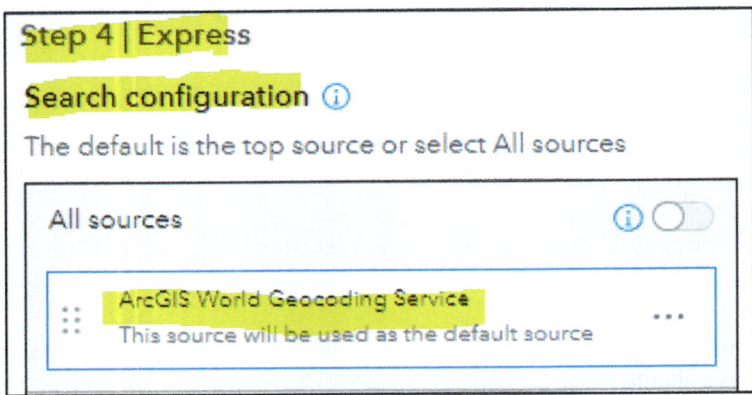

23. Click next o go to **Step 5 (Theme & Layout)**
24. Select a mode: **Light**
25. Select a preset theme: open drop-down arrow and select **Ocean** (blue color)
26. Manage widget positions: accept the default

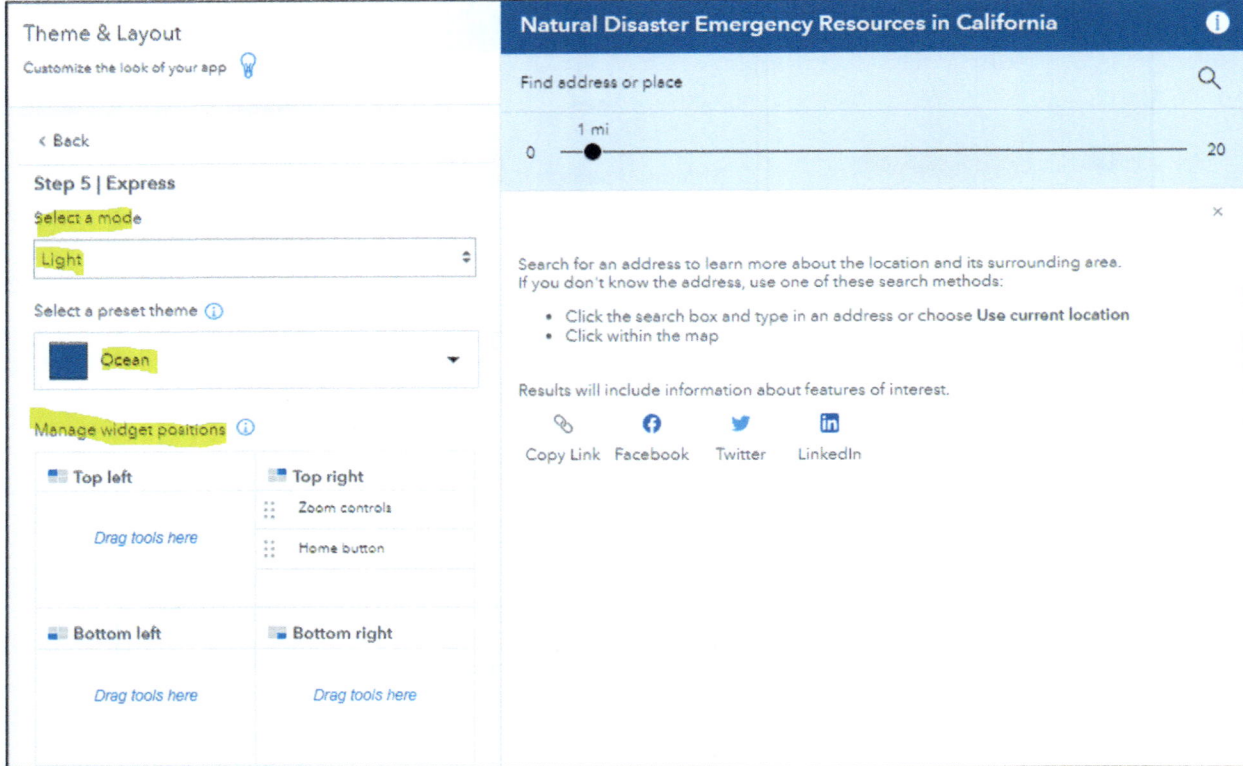

Configure Your Instant App

The final step is to view the Instant app and explore how the app works and see the effect of the configuration and the pop-ups on the map.

27. At the bottom, click **Publish**, and then click **Confirm**

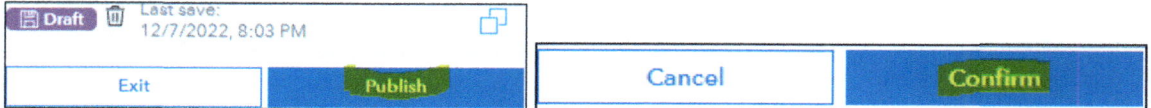

28. In the **Share** dialog box, click **Copy Link**
29. https://arcg.is/0iSqH4
30. Click **Launch**
31. To the left of the map under **Natural Disaster Emergency Resources in California**
32. Drag the distance scale to 20 miles
33. Type **San Francisco, CA, USA** and press Enter

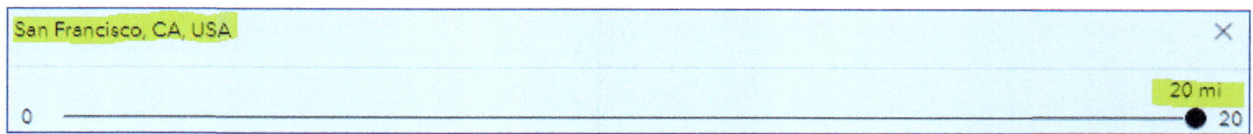

Result The following were found: 28 wells, 4 tsunamis, 606 schools, 43 hospitals, and a section of Fault Andreas.

34. From the top of the map click "**Clear search location**"

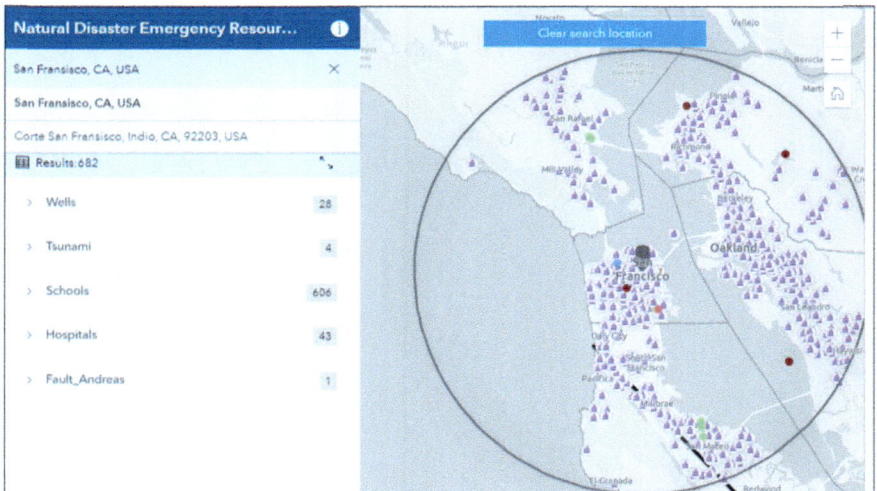

Type a Specific Address

35. Under **Natural Disaster Emergency Resources in California**
36. Type: **15581 Hortense Dr, Westminster, CA, 92683, USA** and press Enter

Result The search result updates and shows information about the emergency features near the address. It finds 8 schools within 1 mile of the address.

37. Change the distance to 20 miles, and press Enter

Result The search found the following features within 20 miles from the address:

Features	No
Earthquake	1
Tsunami	2
Hospitals	39
School	978
Wells	91

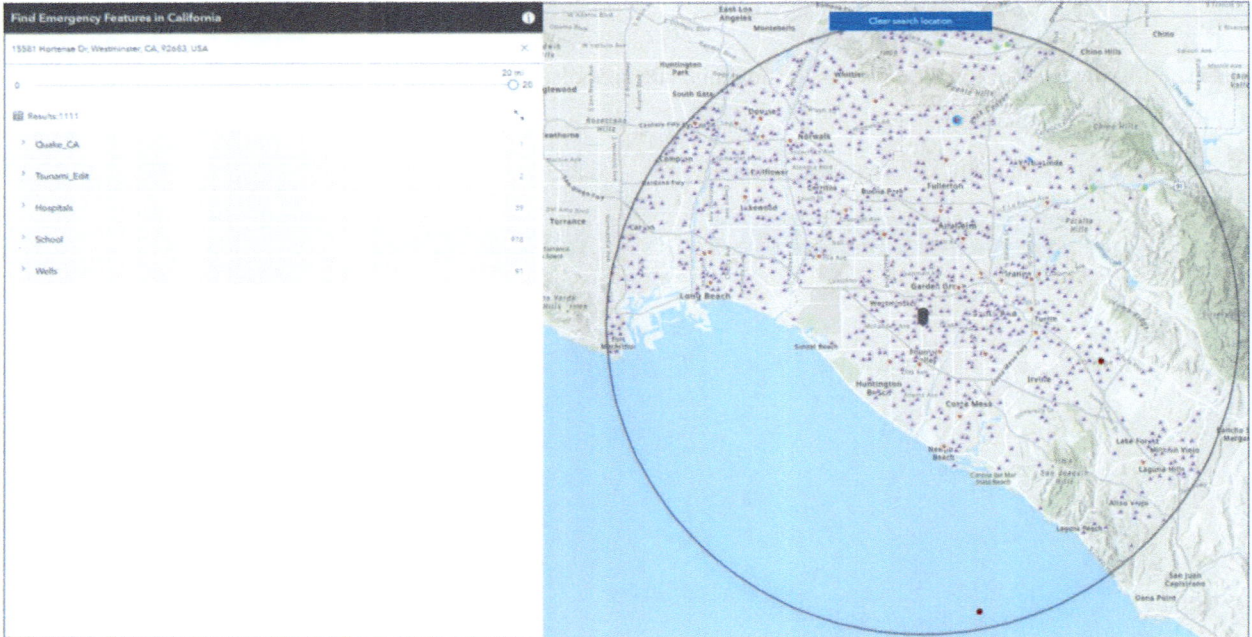

38. From the top of the map click "**Clear search location**"
39. Zoom in the map in the west coast of **California** around **San Jose** and click on a "**Tsunami**" north of **San Jose**

Result Five features were found: 41 wells, 4 tsunamis, 640 schools, 14 hospitals, and a section of Fault Andreas.

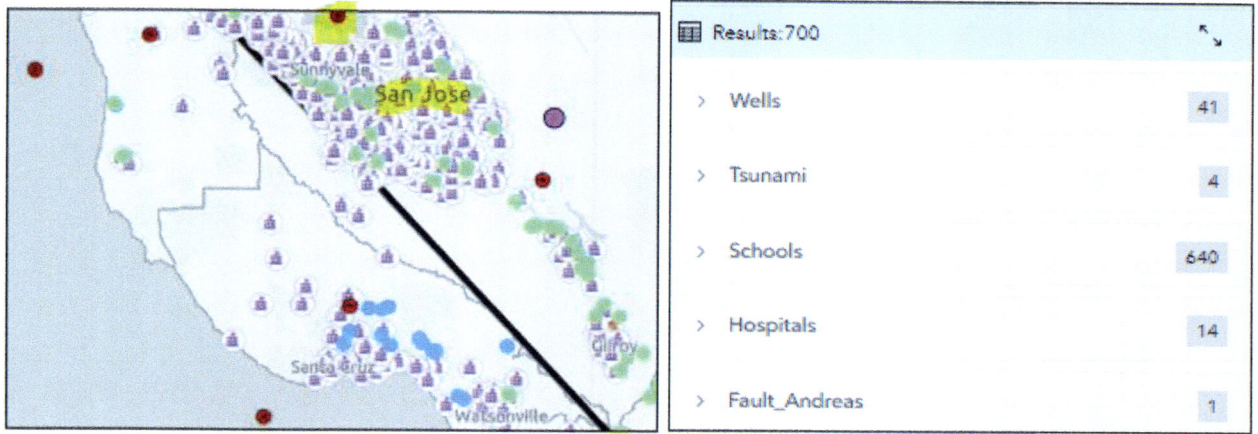

40. From the top of the map click "**Clear search location**"
41. Exit ArcGIS Online and ArcGISPro

References - Data Source Credits

The textbook was created using ArcGIS Pro and ArcGIS Online. "**Content is the intellectual property of Esri and is used herein with permission. Copyright © 2022 Esri and its licensors. All rights reserved.**"

Chapter 1 (INTRODUCTION TO ArcGIS Pro) **Data sources include the following: the following:**

Data (SamplingSite, NewtonCreek, MurphyOil, and HogIsland) are derived from the author article: Bajjali, W. Water Quality Assessment of Newton Creek and Its Effect on Hog Island Inlet of Lake Superior. Water Qual Expo Health 4, 123–135 (2012). https://doi.org/10.1007/s12403-012-0071-1. All graphs are created by the author.

The DEM of Superior, WI downloaded from USGS Web Page: The National Map – Data Delivery. GIS Data Download | U.S. Geological Survey (usgs.gov)

https://www.esri.com/en-us/what-is-gis/overview

Ahmad, Karam , & Laituri, Melinda (2017). Use of GIS in Environmental Science. obo in Environmental Science. doi: 10.1093/obo/9780199363445-0081 (Oxford Bibliographies)

Applications were performed using ArcGIS Pro. Copyright © 2022 Esri and its licensors. All rights reserved.

Chapter 2 (WORKING WITH ArcGIS Pro) **Data sources include the following:**

Data (Rivers, Streets, Parks, Hospitals, Airports, Cities, and County layers) are downloaded from the Minnesota Geospatial Commons web page Resources - Minnesota Geospatial Commons (mn.gov)

Applications were performed using ArcGIS Pro. Copyright © 2022 Esri and its licensors. All rights reserved.

Chapter 3 (MAP CLASSIFICATION AND LAYOUT) **Data sources include the following:**

Data (Fault, Geology, and Well layers) are derived from the author article: Recharge origin, overexploitation, and sustainability of water resources in an arid area from Azraq basin, Jordan: case study, 2006. http://hr.iwaponline.com/content/37/3/277

Applications were performed using ArcGIS Pro. Copyright © 2022 Esri and its licensors. All rights reserved.

Chapter 4 (COORDINATE SYSTEMS AND PROJECTIONS) **Data sources include the following:**

Data (streets of Superior, street of Duluth, and WI layers) are derived from TIGER/Line Shapefiles from United States Census Bureau web page: Partnership Shapefiles (census.gov)

Data (Lake.tif clipped from ortho_1-1_1n_s_wi078_2015_1.sid) downloaded from the Wisconsin View web page: https://bin.ssec.wisc.edu/pub/wisconsinview/NAIP_2015/Menominee/IP_2015/Menominee/

Data (GPS layer) created a new feature by the author using Garmin GPS

The image (Aquifer.jpg) representing the eastern dolomite aquifer of Wisconsin is downloaded from the web page: http://wgnhs.uwex.edu/water-environment/wisconsin-aquifers/

Applications were performed using ArcGIS Pro. Copyright © 2022 Esri and its licensors. All rights reserved.

Chapter 5 (GEODATABASE) **Data sources include the following:**

Data (image_rectify, Dam, Stream, StudyArea, Well and Dhuleil layers) are derived from the author article: Water quality and geochemistry evaluation of groundwater upstream and downstream of the Khirbet Al-Samra wastewater treatment plant/Jordan, 2015. http://link.springer.com/article/10.1007/s13201-014-0263-x,

Data (Well, Fault, and Plant layers) are created as a new feature using geodatabase forms designed by the author.

Data (Well, Catchment, Table 1, and Table 2 layers) are created as new features by the author.

Applications were performed using ArcGIS Pro. Copyright © 2022 Esri and its licensors. All rights reserved.

Chapter 6 (DATA EDITING AND TOPOLOGY) **Data sources include the following:**

Data (Farm and LandB layers) created new by the author for editing purposes

Data (Geology and Field_Geology layers) are derived from the author article: Water quality and geochemistry evaluation of groundwater upstream and downstream of the Khirbet Al-Samra wastewater treatment plant, Jordan 2015 http://link.springer.com/article/10.1007/s13201-014-0263-x

Data (Street_Mn, River_MN, Lake_MN, and Fault layers) are downloaded from the Minnesota Geospatial Commons web page Resources - Minnesota Geospatial Commons (mn.gov)

The image (North_Duluth.jpg) is an aerial photograph of St. Louis County derived and downloaded from the MN-DNR web page: http://www.dnr.state.mn.us/maps/landview.html

Data (Watershed_1, and Watershed_2 layers) are derived from the author article: Bajjali, W. Water Quality Assessment of Newton Creek and Its Effect on Hog Island Inlet of Lake Superior. Water Qual Expo Health 4, 123–135 (2012). https://doi.org/10.1007/s12403-012-0071-1. All graphs are created by the author.

Applications were performed using ArcGIS Pro. Copyright © 2022 Esri and its licensors. All rights reserved.

Chapter 7 (Geoprocessing) **Data sources include the following:**

Data (watershed, stream, soil, and Aflaj layers) are derived from the author article: Evaluation the Groundwater Salinity throughout Sultanate of Oman Using GIS, Ministry of Water Resources, Oman 1999.

Data (watershed, stream, soil, FarmA, FarmB, FarmC, and FarmD layers) are created as new layers by the author.

Applications were performed using ArcGIS Pro. Copyright © 2022 Esri and its licensors. All rights reserved.

Chapter 8 (SITE SUITABILITY AND MODELING) **Data sources include the following:**

Data (Landuse, Pipeline, StudyArea, and Vegetation layers) are digitized on screen from an image that had been downloaded from Google Earth (https://www.google.com/earth). The image was then clipped and georeferenced using local projection.

Data (Well, Fault, Stream, KSWTP, and GEOL_KS layers) are derived from the author article: Water quality and geochemistry evaluation of groundwater upstream and downstream of the Khirbet Al-Samra wastewater treatment plant/Jordan, 2015. http://link.springer.com/article/10.1007/s13201-014-0263-x,

Applications were performed using ArcGIS Pro. Copyright © 2022 Esri and its licensors. All rights reserved.

Chapter 9 (Geocoding) **Data sources include the following:**

Data (ZipCode_WI and Street of city of superior layer) are derived from ESRI ArcGIS Data and Maps (2020) ArcGIS Pro | My Esri. Copyright © 2017 Esri (include any data providers). All rights reserved

Data (Well_Owner layer and Well table) are created new by the author for the purpose of geocoding.

Applications were performed using ArcGIS Pro. Copyright © 2022 Esri and its licensors. All rights reserved.

Chapter 10 (WORKING WITH RASTER) **Data sources include the following:**

The Burnsville DEM image downloaded from the web page GIS Data Download | U.S. Geological Survey (usgs.gov)

The images (AZ_DEM, Dhuleil.tif and KTDam grids) are downloaded and clipped from SRTM 90 m Digital Elevation Data downloaded from CGIAR-CSI HOME http://srtm.csi.cgiar.org

References - Data Source Credits

Data (Geology, stream, and Luhfi_Dam layers) are derived from the author article: Water quality and geochemistry evaluation of groundwater upstream and downstream of the Khirbet Al-Samra wastewater treatment plant/Jordan, 2015. http://link.springer.com/article/10.1007/s13201-014-0263-x,

Applications were performed using ArcGIS Pro. Copyright © 2022 Esri and its licensors. All rights reserved.

Chapter 11 (SPATIAL INTERPOLATION) **Data sources include the following:**

Data (dam, stream, watershed and well layers) are derived from the author article: Model the effect of four artificial recharge dams on the quality of groundwater using geostatistical methods in GIS environment, Oman, 2005 http://www.spatialhydrology.net/index.php/JOSH/article/view/39

Applications were performed using ArcGIS Pro. Copyright © 2022 Esri and its licensors. All rights reserved.

Chapter 12 (WATERSHED DELINEATION) **Data sources include the following:**

Data (Burnsville.tif and CreditRiver.tif DEM) downloaded from the web page GIS Data Download | U.S. Geological Survey (usgs.gov)

Data (CreditRiver layer) are downloaded from the Minnesota Geospatial Commons web page Resources - Minnesota Geospatial Commons (mn.gov)

Data (Bridge layer) are created as new layers by the author

Applications were performed using ArcGIS Pro. Copyright © 2022 Esri and its licensors. All rights reserved.

Chapter 13 (GEOSTATISTICAL ANALYSIS) **Data sources include the following:**

Data (Governorate, Town, Well, WalaWatershed, Grid_1000, Geology, and WWTP layers) are derived from GIS Workshop in Hydrogeology at Water Authority of Jordan, Ministry of Water and Irrigation, September 2 - 6, 2012 created and instructed by the author.

Data (Dam, Geology, Stream, and Well layers) are derived from the author article: Water quality and geochemistry evaluation of groundwater upstream and downstream of the Khirbet Al-Samra wastewater treatment plant/Jordan, 2015. http://link.springer.com/article/10.1007/s13201-014-0263-x,

Applications were performed using ArcGIS Pro. Copyright © 2022 Esri and its licensors. All rights reserved.

Chapter 14 (PROXIMITY AND NETWORK ANALYSIS) **Data sources include the following:**

Data (Dam, Region, Road, Stream, Street, Town, Well, Well_Supply, and WWTP layers) are derived from GIS Workshop in Hydrogeology at Water Authority of Jordan, Ministry of Water and Irrigation, September 2 - 6, 2012 created and instructed by the author.

Applications were performed using ArcGIS Pro. Copyright © 2022 Esri and its licensors. All rights reserved.

Chapter 15 (3-D ANALYSIS) **Data sources include the following:**

Data (Building, Farm, GasStation, ObserbationWell, Plume, Street, SupplyWell, Tree, Valley, and WWTP layers) are derived from GIS Workshop in Hydrogeology at Water Authority of Jordan, Ministry of Water and Irrigation, September 2 - 6, 2012 created and instructed by the author.

The DEM of the city of Duluth, MN (Duluth.tif) is downloaded from the USGS web page: http://viewer.nationalmap.gov/. The data (Contour and Duluth layers) are obtained from the DEM (Duluth.tif)

Data (city and stream layers) are derived from ESRI ArcGIS Data and Maps (2020) ArcGIS Pro | My Esri. Copyright © 2017 Esri (include any data providers). All rights reserved

Data (RainStation layer) is created as a new layer by the author

Applications were performed using ArcGIS Pro. Copyright © 2022 Esri and its licensors. All rights reserved.

Chapter 16 (WORKING WITH ARCGIS ONLINE AND StoryMap APP) **Data sources include the following:**

Data (SamplingSite, River, and MurphyOil) are derived from the author article: Bajjali, W. Water Quality Assessment of Newton Creek and Its Effect on Hog Island Inlet of Lake Superior. Water Qual Expo Health 4, 123–135 (2012). https://doi.org/10.1007/s12403-012-0071-1.

Data (StoryMap maps) created by the author

Data (StoryMap- Photo) taken by the author

Applications were performed using ArcGIS Pro and ArcGIS Online. Copyright © 2022 Esri and its licensors. All rights reserved.

Chapter 17 (ARCGIS ONLINE - INSTANT APP) **Data sources include the following:**

Data (Wells, School, Hospitals, Quake_Ca, Fault) are downloaded from www.ca.gov | California State Portal

Data (Tsunami) are downloaded from the California Tsunami Maps and Data California Tsunami Maps and Data

Applications were performed using ArcGIS Pro and ArcGIS Online. Copyright © 2022 Esri and its licensors. All rights reserved.

Index

A
Addresses, vii–ix, 22, 169–173, 175–182, 188, 408, 409, 414–415
Analyze data, vii, 2
Animation, ix, 31, 344–348, 351–352
Anselin Local Moran I, 279–284
Append, viii, 117, 129, 130, 135, 322, 326, 347, 348
App templates, 402–403, 408
ArcGIS Online, vii, ix, 2, 3, 8, 9, 40, 48, 130, 169, 172, 199, 200, 301, 357–392, 400, 401, 403, 407, 415
ArcGIS Pro, vii–ix, 1–23, 26, 35, 36, 39, 40, 42, 49–51, 58, 68–70, 81–83, 98, 110, 113, 129, 130, 144–146, 151, 154, 169, 170, 177, 185, 186, 188, 196, 221, 226, 234, 243, 245, 246, 258, 261, 288, 299, 333, 334, 337, 338, 344, 356, 393, 402
Aspects, viii, 39, 42, 146, 151, 184, 209–211, 214–215, 272
Average Nearest Neighbor, 268–270, 273
Azraq basin, 1, 36, 37, 40, 46, 48, 50

B
Base height, 337, 338, 341, 342, 349
Basemaps, 2, 8, 10, 16, 17, 180–181, 344, 345, 365, 393–394, 400
Bookmarks, ix, 20, 31–33, 344–347, 365
Buffer, viii, 129, 130, 139–141, 147, 156, 157, 159, 166, 264–268, 287, 289, 291, 293–295, 303, 305, 312

C
California, ix, 393–415
Capture data, 1, 2, 83
Catalog pane, 9–11, 16–19, 28, 30, 36, 37, 39, 40, 42, 45–47, 69, 70, 73, 78, 83–85, 90–93, 99, 102, 104, 109, 110, 114–116, 121, 122, 124, 131, 133, 137–139, 142, 147, 152, 154, 163–165, 168, 170, 171, 175, 185, 186, 188, 191, 192, 194, 197, 202, 206, 210, 216, 227–229, 233, 234, 236, 237, 239, 240, 242, 245, 246, 255, 261, 264, 270, 272, 276, 279, 288, 301, 305, 313, 315, 316, 319, 320, 324, 325, 329, 334, 337, 394
Cells, 5–7, 12, 183, 186–188, 192–195, 209, 214, 215, 217, 220, 227, 228, 234, 235, 237, 238, 241, 243–247, 249, 250, 252, 253, 276, 277, 295, 305, 306
Central feature, 259, 260, 264
Classifications, vii, 35–61, 193, 281, 283, 311–312
Classify, vii, 30, 38, 43–48, 133, 183, 193–194, 210–211, 228, 236–237, 239, 242, 307–308, 313, 328–329, 337
Clip, viii, 129, 130, 137, 183, 189–190, 235, 238
Content toolbar, 364, 407, 408
Contents pane (CP), 9–11, 16, 17, 19, 23–31, 33, 37, 38, 40–43, 45–47, 51, 52, 54–61, 69–72, 74–78, 83, 84, 86–90, 93, 94, 99, 101–105, 107, 111–113, 115–117, 131, 133–141, 147, 150, 152, 154, 161, 170, 173–176, 179–181, 185, 186, 188, 189, 191–195, 197–199, 201–203, 206, 209, 210, 213, 220, 227–231, 233–242, 245, 246, 254, 255, 261, 262, 264, 266, 270, 272, 279, 282, 283, 288–291, 293, 294, 296, 297, 299, 301–303, 305, 307, 309–311, 313–316, 319–330, 334–338, 341–344, 347–349, 352, 354, 396, 397, 399, 400
Contours, viii, 184, 196–199, 334–336
Convert, 3, 72, 169–171, 176, 183, 187–188, 194–196, 235–236, 241–243, 245, 251, 252, 256–257, 297–298, 337, 349, 354
Coordinates, vii, 2, 4, 5, 13, 23, 32, 33, 42, 63, 64, 66–76, 79, 80, 82, 84, 85, 87, 97, 102, 108, 114, 121, 124, 142, 154, 169, 170, 186, 188, 189, 191, 198, 208, 223, 224, 228–231, 235, 238, 241, 247, 259, 262, 287, 298, 306, 313, 334, 335
Copy link, 392, 413
Create video, 346
Criteria, 2, 40, 145–160, 162, 264, 270, 272, 288, 334, 386, 388

D
Definition query, 270, 272, 277, 279, 296–297, 310–311
Delete, 23, 32, 37, 47, 77, 98, 100, 103, 347, 405, 406
Digital elevation model (DEM), viii, 5, 7, 183–186, 188–189, 193, 196–198, 200, 205, 243, 244, 246, 247, 333, 337, 352
Disasters, 393, 394, 410, 413, 414
Dissolve, viii, 129, 130, 132, 133, 140, 173, 303, 386–387
Distortions, 64, 65, 72

E
Earthquakes, ix, 393–395, 400, 402, 404–409, 414
Editing, viii, 8, 60, 81, 87, 97–127, 151, 161, 391, 399, 400, 403, 404
Emergency, ix, 175, 393–415
Equal interval, 42, 46, 50–52, 54, 56, 61
Erase, viii, 129, 130, 143–144, 159–161, 164, 287
Euclidian distance, 276, 306–307, 312
Excel table, 175, 176
Explore tool, 21
Extrude, 338–339, 341–342

F
Facilities, ix, 321–328, 408
Fault, 83–87, 89, 114, 115, 117–121, 152, 156, 159, 287, 333, 393, 400, 407, 413, 414
Feature classes, vii, viii, 18, 19, 21, 23, 25, 26, 41, 71–73, 75, 81–92, 114–116, 121, 122, 124, 125, 129, 132, 133, 136–139, 141–145, 148–150, 156–159, 169, 170, 172, 173, 175, 179–182, 189, 194, 198, 201, 203, 206–209, 229, 237, 256, 262–267, 270, 272–274, 276, 277, 279–284, 287, 289, 293, 297–298, 301, 303, 312–314, 321, 322, 334, 341, 352, 358, 395
Feature datasets, vii, 84–87, 114–116, 121, 122, 124, 125, 209, 313, 315, 320, 325, 329

Feature service, 3, 358, 363–364
File geodatabase, 16, 18, 19, 81, 83–87, 114, 133, 165, 309, 313, 358, 394
Fill, viii, 16, 43, 45–47, 59, 61, 70–72, 74, 75, 78, 79, 83–85, 90–93, 107–109, 114, 115, 120, 121, 124, 132–134, 137–142, 144, 150, 154–159, 165, 172, 176, 177, 179, 187, 189–191, 193, 195, 197–208, 210, 211, 214, 216, 217, 220, 227, 229, 231, 234, 235, 238, 241, 243, 246–251, 254–258, 262–267, 270, 273, 274, 277, 280, 282, 284, 289, 290, 292–294, 297, 298, 300, 301, 303, 304, 306, 308, 309, 313, 315, 316, 318, 322, 325, 326, 335, 353, 359, 367, 368, 372, 381, 396, 407
Float raster, 252
Flow accumulations, viii, 244, 245, 250–253
Flow directions, viii, 243–244, 246–247, 249–251, 254, 255, 258
Fly, 344–348
Folder, 8, 15–19, 28, 30, 36, 37, 39, 41, 42, 45–49, 69–71, 73, 74, 78, 81, 83, 85, 90, 91, 98, 99, 102, 104, 109, 110, 114, 121, 124, 130, 131, 133, 137–139, 146, 147, 152, 154, 163, 169, 170, 175, 177, 184–186, 188, 190, 191, 193–195, 197–199, 202, 205, 206, 216, 226, 227, 229, 233, 234, 237, 240, 245, 246, 255, 261, 264, 270, 272, 274, 276, 278, 288, 299, 301, 305, 313, 324, 329, 334, 335, 337, 344, 346–348, 352, 358, 359, 363, 369–372, 381, 385, 394, 402, 407

G
Geo app, 370, 402–405, 408
Geocoding, viii, 22, 169–182, 412
Geodatabase, vii, viii, 2, 3, 8, 19, 75, 77, 81–95, 97, 98, 107, 113–115, 117, 118, 121, 122, 124–127, 133, 169, 177, 191, 192, 194, 297–298, 308, 313
Geographic coordinate system (GCS), vii, 63, 67–69, 72, 78, 85
Geographic distributions, viii, 259–262, 264, 265
Geographic information system (GIS), vii–ix, 1–5, 8, 15, 23, 30, 32, 35, 63, 64, 68, 73, 81, 82, 97, 104, 110, 114, 129, 130, 145, 146, 150, 151, 154, 169, 183, 184, 199, 201, 223, 224, 227, 243, 259, 287–289, 312, 333, 357
Geoprocessing, viii, 3, 9, 48, 71, 72, 75, 79, 83, 129–137, 140, 144–154, 156–160, 163–166, 171–173, 176, 177, 187, 189–191, 193, 195, 197, 198, 200, 202, 205, 206, 210, 211, 214, 216, 217, 219, 227, 246, 251, 258, 262–264, 266, 267, 270, 273, 274, 276–278, 280, 282, 284, 289, 293, 298, 301, 303, 306, 308, 333, 335, 348, 353, 356
Georeferencing, 67, 73–77
Geostatistical analysis, viii, 259–285
Getis-Ord general G, 272–275
GIS description, 2
Global Moran's I, 275–278
Global polynomial (GP), viii, 225, 234–235
Greenhouse, viii, 146–150
Grid, 5–7, 51, 63, 75, 81, 146, 183, 188, 189, 192, 215, 223, 236, 243, 244, 246, 247, 276–277
Groundwater, viii, 1, 4, 36, 43–48, 51–53, 55, 61, 83, 91, 92, 129, 130, 151, 152, 157, 170, 196, 216, 227–229, 234, 243, 245, 256, 259, 260, 264, 269, 276, 279, 282, 288, 300, 305, 311, 312, 333, 334, 393, 407
Guides, vii, 53–55, 89, 110

H
High/low clustering, 274–275
Hillshade, viii, 10, 16, 77, 83, 115, 184, 196–197, 199, 226, 245, 261, 288, 334, 341, 344, 349
Hospitals, ix, 19, 23, 24, 393, 394, 396, 399, 400, 402, 407, 408, 413, 414
Hot spots, 282–284

I
Immersive, 374, 380–385
Import, 8, 39–40, 71, 72, 83–85, 90–91, 114, 115, 121, 124, 142, 297, 313, 322, 325, 326, 345, 351
Instant app, ix, 372, 393–415
Integer raster, 187–188, 211, 249, 252
Intermediate data, 164
Intersects, viii, 63–65, 111, 130, 138, 231, 254, 287
Inverse distance weighting (IDW), viii, 223–225, 234, 237–240, 242, 243

J
Join, 159, 166, 170, 216, 276, 309

K
Kriging, viii, 223–226, 234, 240–243

L
Labels, 15, 26–30, 33, 35, 39, 44, 45, 59, 74, 87, 92, 99, 102–104, 106, 117, 134–137, 141, 143, 163–164, 229, 236, 239, 242, 288, 289, 299, 307, 310, 311, 325, 329, 354, 356, 396–397
Latitude, 4, 23, 63, 64, 67, 68, 72, 73, 170, 188
Layer, viii, ix, 3, 8, 10, 11, 13, 15, 16, 19, 21, 23, 26–33, 35, 36, 38, 41–43, 48, 52, 56–58, 87, 97, 101, 102, 104, 105, 107–110, 112, 113, 129, 132, 138, 141, 143, 146, 148, 149, 152, 154, 159–161, 164, 168, 170, 174, 175, 180, 181, 183, 186–188, 193, 194, 197, 199, 201, 203, 209, 220, 225, 227, 230–231, 233, 235, 237, 238, 241, 246, 261–265, 268, 276, 279, 280, 282, 287–290, 293, 294, 296, 298–301, 305, 306, 309–311, 321–323, 325, 326, 329–331, 333, 334, 336–338, 341, 342, 344, 348, 349, 352, 355, 357, 359, 361, 363–365, 368, 370, 393–397, 399–405, 407, 408, 411
Layouts, vii, 8, 15, 35–62, 156–161, 163, 166, 380, 385, 410, 412
Legends, 51, 55–58, 61, 365, 367
Line features, viii, 4, 5, 83, 86, 199–201, 206, 207, 312, 354, 400
Line of sight, 184, 205–209
Locators, viii, 17, 169, 171–173, 177–179
Longitude, 4, 23, 63, 64, 67, 68, 72, 73, 170, 188
Lysimeter, 184, 215, 217–221

M
Manual classification, 51, 56
Map extent, 32–33, 106, 351, 395
Maplex, 15, 26
Map scales, 28–30, 58
Map template, 16
Map viewer, ix, 184, 363–365, 367, 368, 372, 403, 404
Match, 40, 87, 92, 169, 173, 174, 177, 179–182, 262, 276, 390
Mean center, 259, 260, 262–265
Media, 228, 352, 361, 362, 372, 374–380, 384
Merge, viii, 98, 109–110, 120, 121, 129, 130, 139, 183, 190–192
Meridians, 63–65, 67, 188
ModelBuilder, viii, 8, 130, 145, 146, 150–152, 154, 161–165
Model parameters, 164–168
Model tools, 151, 164–166, 168
Model window, 153, 155, 161
Modifies, 65, 98, 100, 104, 109, 111–113, 120, 129, 145, 164, 169, 233, 287
Modify, 35, 57, 60
Mosaic, viii, 183, 190–192
Move, 20, 21, 30, 61, 71, 76, 98, 100, 120, 160, 184, 339, 340
Multi-ring buffer, 287, 302–305
My Content, 358–361, 363, 403

N

Natural Breaks, 42–44, 50, 52, 54–61
Navigator, 339, 340, 349
Near, viii, 130, 152, 264, 266–268, 270, 272, 276, 285, 287, 298–301, 308–310, 312, 324, 414
Neighborhood, 223, 225, 227, 228, 240, 244, 279, 321
Network dataset, 84, 313, 315–316, 318, 322–324, 326, 329, 330
Nitrate, 152, 170, 173–175, 227, 242, 288, 311–312
North arrow, 51, 55, 60, 61
Nuclear power plant (NPP), viii, 151–161, 163–165, 168
Null hypothesis, 269–274, 276, 278, 279, 282

O

Oman, viii, 114, 132, 226–228, 232–234
Outliers, 237, 279–284
Overshoots, viii, 98, 110–113

P

Package Project, 48–51
Pane, 9, 15, 23–25, 28, 30, 32, 69–72, 79, 106, 108, 109, 111–113, 115, 147, 154, 157, 159, 187, 189–195, 197, 198, 201, 202, 204, 206, 209–211, 213, 214, 216, 226–228, 231, 233, 235, 241, 246–249, 251–254, 258, 292, 293, 296, 301, 303, 306, 307, 311, 322, 328, 330, 336, 341, 347, 348, 351, 354–356, 365, 367, 368, 371, 384, 389, 396, 397, 399–401, 405, 408
Parallels, 1, 63–67
PDF file, 61–62, 301
Places, 2, 23–26, 43, 52, 54–56, 58, 60, 66, 70, 71, 76, 89, 100, 103, 134, 138, 156, 157, 160, 169, 171–172, 179, 181, 182, 198, 203, 220, 229, 230, 233, 236, 239, 242, 272, 300, 311, 337, 399, 405, 406
Point features, 4, 83, 87, 169–171, 175, 179, 227, 228, 262, 308, 341
Polygon feature, viii, 6, 83, 86, 97, 109, 121, 139, 142, 148, 159, 169, 189, 194, 256, 287, 400
Pop up, 20–22, 160, 178, 299, 365, 372, 396–399, 404–407, 413
Pour points, 245, 254, 255
Project, vii, ix, 2, 3, 8–10, 13, 15, 16, 18–20, 29, 31, 32, 36, 37, 39–42, 44–52, 61, 68, 70, 72, 73, 78–80, 83, 85, 90, 98, 101, 103, 104, 106, 107, 109, 110, 112, 114, 119, 124, 127, 146, 150, 152, 153, 168, 170, 174, 175, 182–185, 188, 189, 191, 192, 194, 195, 197, 201, 202, 205, 207, 210–212, 214, 216, 219, 221, 226, 227, 229, 233, 234, 236, 237, 239, 240, 242, 245, 254, 255, 258, 261, 263, 264, 267, 270, 275, 278, 281, 285, 288, 291, 292, 296, 298, 299, 301, 303, 305, 310, 312–314, 319, 324, 329, 333, 334, 348, 350, 352, 356, 393, 397
Projection, vii, 7, 13, 63–80, 146, 183, 188, 189, 191, 228, 233, 401
Projection on the fly, vii, 67, 69–71
Projection parameters, 67
Proximity analysis, 130, 147, 287–288, 305–306, 312
Publish and share, 9, 372, 391–392
P values, 269, 272, 274–276, 278, 279, 281, 282

Q

Quantile, 42, 44–46, 50–52, 54, 56, 61, 225

R

Raster, vii, viii, 6, 7, 10–14, 69, 72–76, 78–80, 82, 145, 151, 183, 184, 186–195, 197, 198, 202, 205, 209, 210, 214–220, 227, 228, 234–236, 238, 239, 241–258, 287, 305, 306, 308, 333, 339, 344, 352, 399, 400
Raster dataset, viii, 6, 8, 75, 76, 78, 79, 82–84, 87, 90, 183, 188–195, 256
Raster models, 3, 6, 7
Raster projection, viii, 67, 78

Realistic layer, 341
Relationship classes, vii, 91–95
Resample, 183, 192–193
Reservoir, 141–144, 151, 266, 267, 287
Reshape, 98, 102, 111–113
Ribbon, 9, 13, 15, 18, 20, 21, 23, 42, 45–49, 51, 52, 54–56, 58, 60, 61, 71–75, 78, 79, 83, 87–90, 100, 102–105, 109–112, 114, 119–122, 124–127, 132, 134, 136–140, 143, 144, 147, 152–154, 161–165, 171, 175, 178, 180, 187, 188, 191, 192, 194, 197, 199, 201, 202, 208, 210, 216, 227, 229–231, 234, 237, 240, 245, 246, 251, 255, 261, 262, 264, 289, 291, 299, 301, 305, 313, 320, 322, 324, 325, 329, 330, 334–338, 341, 342, 344–351, 353–355, 373, 389–393, 396, 400
Route, viii, 97, 287, 320, 325, 327–331, 347
Run the model, 151, 166, 168

S

Saint Andreas fault, 400
Salt intrusion, viii, 227
Scale bar, 51, 55, 58, 59, 61
Scenes, ix, 8, 15, 16, 20, 21, 26, 28, 31, 32, 39, 333, 334, 337–345, 347–349, 354–356, 408
Schools, ix, 4, 393, 394, 396, 397, 399, 400, 402, 407–409, 413, 414
Search, 411, 412, 414
Search location, 409, 413–415
Select, 15, 19, 21–25, 29–32, 35–38, 41, 43, 49, 51, 54, 55, 57–61, 68, 70–72, 74, 79, 83–85, 88, 89, 92, 94, 99–101, 103, 104, 107–109, 111–114, 116, 117, 121, 124, 125, 131, 133, 134, 136, 139, 140, 142–145, 147, 149, 150, 152, 158, 160, 161, 163, 164, 170, 175, 180–182, 185, 189, 190, 194, 197–199, 201, 203, 210, 219, 226–228, 230–232, 234, 236, 238–242, 244, 245, 250, 253, 254, 261, 264–266, 287–290, 292–294, 296, 299, 302, 304, 307, 309, 311, 313, 314, 317–320, 323, 325, 329, 330, 334–337, 344, 348, 352, 354, 355, 358, 362, 365, 367, 368, 375, 379, 381, 385, 388, 390, 395, 403, 405, 412
Service areas, 316, 321–323
Setting toolbar, 364–365, 405, 406
Shapefile, 2, 3, 81, 85, 90, 97, 98, 107, 110, 111, 190, 194, 226, 245, 255, 358, 359, 363
Share, viii, ix, 2, 8, 15, 31, 39, 41, 48, 49, 61, 84, 97, 145, 299, 344, 357, 362, 363, 370–372, 391, 392, 399–403, 413
Sidecar, 374, 380–386, 388
Sinks, viii, 243, 244, 246–249
Site suitability, viii, 145–168
Slopes, viii, 184, 209–217, 219, 336
Smooth feature, 98
Soil, 121–123, 138, 139, 378
Source raster, 77, 252–254
Spatial interpolation, viii, 223–242
Spatial join, viii, 130, 276–277, 287
Spider diagrams, viii, 287, 300–302, 312
Split, viii, 98, 101, 102, 129, 130, 169
Standard distance, 264–265
StoryMaps, ix, 357–392
Stream Link, 244, 254
Stream networks, 244, 250–253, 257, 258
Street addresses, viii, 170, 175–182
Structured query language (SQL), 27, 81, 82, 93, 109, 149, 155, 158, 204, 218, 219, 266, 291, 294, 296
Styles, 17, 25, 27, 33, 35, 39–40, 114, 177, 365–368
Symbolizing, 15, 35, 174–175, 261–262
Symbols, 4, 11, 23–28, 30, 35, 36, 39, 41, 43–47, 55–59, 70, 74, 87–89, 93, 102, 106, 111, 115, 119, 135, 137–140, 143, 147, 150, 152, 174, 175, 179, 190, 198, 199, 201, 202, 206, 218, 227, 229, 231, 232, 255, 261, 262, 264, 281, 283, 288, 296, 297, 306, 311, 331, 336–337, 341, 342, 348, 349, 366–368, 396, 403

T

Table, viii, 15, 16, 25–27, 30, 35, 38, 40, 41, 43, 47, 54, 56, 77, 81, 82, 87–92, 94, 97, 101, 107–109, 117–119, 121, 125, 129, 131, 133, 134, 136, 139, 140, 143, 145, 148, 150, 152, 154, 169, 171–173, 175, 179–182, 186, 187, 195, 203, 204, 211, 213–217, 220, 225, 228, 230, 231, 234, 237, 243, 249–252, 254, 259, 260, 263, 267, 269, 270, 272–274, 276, 278, 279, 281, 284, 285, 287, 290–292, 294, 296, 298, 300, 301, 304, 305, 308, 309, 314, 317, 328, 335, 337, 340, 341, 348, 349, 352, 353, 365, 366, 385, 397, 399, 404, 405
Tag, 48, 49, 359, 369, 400, 407, 410
3-D, viii, ix, 8, 63, 196, 333–356
Thumbnail, 31–32, 347, 351
Time, viii, 1–3, 7, 9, 26–28, 55, 57, 59, 65, 97, 134, 135, 169, 183, 211, 232, 249, 260, 268, 272, 289, 312–314, 316, 319–325, 327, 348–351, 357, 372, 373, 387, 389, 399, 405, 406
Time tracking, ix, 348–349
Title, 51, 55, 56, 201, 232, 301, 359, 360, 369, 372, 373, 386, 405, 407, 410
Topologies, viii, 4, 82, 84, 97–127
Topology rules, viii, 114, 117, 121, 125
Travel Modes, 316
Trend surface analysis, 223–225, 229
Triangulated Irregular Network (TIN), 214, 333–337, 342–345, 347, 348
Tsunamis, ix, 393, 394, 396, 400, 402, 407–409, 413, 414

U

Undershoots, viii, 98, 110–113
Union, viii, 130, 148, 149, 159, 164, 166, 287

Universal Transverse Mercator (UTM), vii, 64, 67, 69, 72, 73, 84, 85, 115, 121, 124, 198, 227, 228, 231
Unmatched addresses, 180–182

V

Validates, 117–119, 125–127, 161, 162, 166, 309
Vector model, 3, 7
Vertical exaggeration, 343
Vertical profile, viii, 184, 199–201, 354
View window, 10, 16, 28
Visibility, 26–31, 184, 202–203, 205, 206, 208, 209, 333

W

Watershed, viii, 1, 97, 124–127, 130–137, 184, 226–229, 233–241, 243–258, 272, 274, 276
Web feature, 3, 399–401
Web Map, ix, 184, 357, 363, 369–372, 381, 396, 399–405, 407–408, 410
Wells, vii–ix, 1, 4, 7, 9, 35–37, 42–48, 58, 82–85, 87, 90–94, 114, 129, 130, 151, 152, 154, 157, 159, 162–164, 166, 169, 170, 172–176, 179–182, 224–229, 259–274, 276–277, 279–283, 285, 287–302, 305, 306, 308–313, 316, 321–325, 327, 333, 334, 337, 341–343, 345, 347, 348, 354, 361, 372, 393, 394, 396, 399, 400, 402–404, 407, 408, 413, 414

Z

Zip codes, viii, 170–175
Zipped shapefiles, 357–360
Zonal statistics, 248–249
Z scores, 268, 269, 271, 272, 274–276, 278, 279, 281, 282, 285

SPRINGER NATURE

GPSR Compliance

The European Union's (EU) General Product Safety Regulation (GPSR) is a set of rules that requires consumer products to be safe and our obligations to ensure this.

If you have any concerns about our products, you can contact us on ProductSafety@springernature.com

In case Publisher is established outside the EU, the EU authorized representative is:

Springer Nature Customer Service Center GmbH
Europaplatz 3
69115 Heidelberg, Germany

The manufacturer's authorised representative in the EU is Springer Nature Customer Service Centre GmbH, Europaplatz 3, 69115 Heidelberg, Germany. If you have any concerns regarding our products, please contact ProductSafety@springernature.com

Printed and bound by CPI Group (UK) Ltd, Croydon, CR0 4YY

26/03/2026

02079000-0001

Medieval Garments
Reconstructed

Medieval Garments Reconstructed

Norse Clothing Patterns

By Lilli Fransen, Anna Nørgaard and Else Østergård

Aarhus University Press

Medieval Garments Reconstructed
Norse Clothing Patterns
© Aarhus University Press and the Authors 2011
Cover: Grafisk SIGNS
Cover photo: Peter Danstrøm
Cover illustration: Lilli Fransen
Photos in chapter 3 by Roberto Fortuna
Layout: Grafisk SIGNS
Typeface: LinotypeSyntaxOsF
Paper: Arctic Silk
Printed in Denmark by Narayana Press
ISBN 978 87 7219 871 2

Paperback edition, 1st printing 2022

Aarhus University Press
Langelandsgade 177
DK-8200 Aarhus N
www.unipress.dk

International distributors
Oxbow Books Ltd., oxbowbooks.com
ISD, isdistribution.com

Published with the financial support of

Dronning Margrethe II's Arkæologiske Fond
KULTURFONDEN DANMARK-GRØNLAND
Manufakturhandler-Foreningen i Kjøbenhavns Almene Fond
VELUX FONDEN

Preface

The 'cut' and 'fit' of a garment are terms that we use today in connection with the cutting and sewing of clothes. We know what size we use and we expect that a garment is cut and formed so that it fits our body.

In the Early Middle Ages the cutting and production of a piece of clothing was associated with a great deal of mystery, and how the Norse, who lived on the edge of the world's society, so to speak, could carry out this profession under such primitive conditions is just as mysterious.

As the photographs and measurements in this book illustrate, several of the Norse garments are sewn to fit closely to the body, but with a large fullness at the bottom of the garment and sleeves with 'set-in' sleeve seams that are formed to give ease of movement. The practical liripipe hoods with shoulder cape, and stockings (either with or without feet) resembled the prevailing fashion further south in Europe. In the Patterns Section of the book, the 800 year old garments are spread out side by side with the more recently sewn reproductions.

MEDIEVAL GARMENTS RECONSTRUCTED – NORSE CLOTHING PATTERNS is the result of a cooperation between three textile experts: Pattern Constructor, Lilli Fransen, MSc Clothing Product Development; Weaver, Anna Nørgaard; and Conservator, Else Østergård. Because of our different backgrounds, each of us has of course taken a different approach to the Herjolfsnes garments, but common to us all is the joy of working with these garments.

Our gratitude goes to the National Museum's Department of Conservation in Brede, which, among other things, has contributed economically to the photography in the book. Our thanks must also go to photographer Robert Fortuna from the Department of Conservation for an inspiring cooperation and for taking splendid photographs of the new garments. Also, museum conservator Irene Skals deserves much thanks for her illustrative material. We are indebted to TEKO Design and Business School in Herning for their generosity in sponsoring the fabric to be used for the sewing of the many new garments, hoods and stockings; and to specialist-teacher Ingrid Andersen, who has sewn the named garment parts. We wish also to thank photographer Werner Karrasch from the Viking Ship Museum in Roskilde. And, last but not least, we are extremely thankful to Chief Curator and the Clinical Faculty, Shelly Nordtorp-Madson, from the University of St. Thomas in St. Paul, Minnesota, USA, who has had the rather awesome task of translating the text from Danish to English.

Lilli Fransen, Anna Nørgaard, and Else Østergård
September, 2010

Contents

Preface	5
Chapter 1	
Introduction · *Else Østergård*	9
The historic textile discovery	9
The Herjolfsnes garments are sent to Denmark	10
The study	11
The Norse Greenlanders' patterns	12
Technical information	13
Garment types	15
Garments	15
Hoods	16
Caps	16
Stockings	16
Notes	16
Chapter 2	
Producing a hand-made reconstruction · *Anna Nørgaard*	17
Treatment of the wool prior to spinning	19
Spinning/yarn	20
The fabric's quality	22
Dyeing/colors	22
Weaving	26
Sewing	28
Footweaving and tablet-woven piping	33
Braided cords	34
Buttons and buttonholes	35
Using the tables	35
Table: Color and thread	35
Table: Seams and stitching	36
Notes	38

Chapter 3
Reconstruction of Patterns · *Lilli Fransen*   39

    Table of Reconstructed Patterns   40
    Garments: 41
    Hoods: 42
    Caps: 42
    Stockings: 42

Garments:

    Museum No. D5674   44
    Museum No. D10580   50
    Museum No. D10581   58
    Museum No. D10584   66
    Museum No. D10585.1   74
    Museum No. D10586   82
    Museum No. D10587   88
    Museum No. D10593   96
    Museum No. D10594   100

Hoods:

    Museum No. D10596   106
    Museum No. D10597   110
    Museum No. D10600   114
    Museum No. D10602   118
    Museum No. D10606   122
    Museum No. D10608   126

Caps:

    Museum No. D10608   126
    Museum No. D10610   130

Stockings:

    Museum No. D10613   134
    Museum No. D10616   138

Literature   141
List of Abbreviations   143

Kalaallit nunaat *is the Greenlandic name for Greenland. It means the land that belongs to the people who call themselves kalaallit.*

Chapter 1

Introduction

By Else Østergård

The many garments, hoods, and stockings described in *Woven into the Earth: Textiles from Norse Greenland*, (Aarhus University Press, 2004), were discovered during an archaeological excavation at the site of Herjolfsnes in Greenland nearly 100 years ago. At that time the find was described as the single-most greatest historical textile event in Europe. Here in the far north European fashion was followed, just as it was in the far south of Europe. With the finds from Herjolfsnes it became possible to see well-preserved examples of medieval clothing and gain an insight into how children and adults had dressed 800-900 years ago.

Readers of *Woven into the Earth* have, since its publication in 2004, made it clear that they desired additional pattern drawings, with instructions on how to produce a garment either as an exact reconstruction or as an adapted reconstruction. Therefore, in this latest work, *Medieval Garments Reconstructed: Norse Clothing Patterns*, which contains significantly more measurements and illustrations, we have endeavoured to meet these requests.

To produce a garment as an 'exact reconstruction' means that the garment must be constructed of hand-spun and hand-woven wool, and sewn with the kind of stitches used in the original garment. However, should one wish to sew a garment as an 'adapted reconstruction', one is free to choose both cloth and production methods.

Instructions are included for reconstructing one of the Herjolfsnes garments: the pattern pieces must to be laid out and cut from the hand-woven cloth to be sewn by hand. The result is a very durable garment – just as the originals were. There are also instructions for machine-sewn garments in other types of fabric: linen, for example, which when constructed in the "Norse Greenland Style", can become an accurate-looking copy.

The pattern book can be seen as a supplement to *Woven into the Earth*, but can also be read and utilized without previous exposure to it.

The historic textile discovery

It was archaeologist Poul Nørlund from the National Museum of Copenhagen who made the momentous discovery in the summer of 1921. He had been chosen to lead an excavation at the ruined church at Herjolfsnes, which lies in the southwestern part of Greenland in Nanortalik Municipality. The ruin was about to be lost to the encroaching sea, and a large portion of the cemetery had already vanished, leaving behind human bones and textiles that from time to time were gathered up from the beach below the ruins.

Nørlund's excavation was not, however, the first at that site; digs were conducted as early as the 1830s after a garment was found on the beach, which was believed to be the jacket of a sailor lost at sea. It was not until Nørlund's 1921 excavation however, that it was discovered that the so-called jacket did not belong to a modern, drowned sailor.[1]

The background of the above excavations is found in the *Icelandic Sagas* as well as other medieval manuscripts, which tell how the Vikings braved the dangerous journey of exploring Greenland's coasts. We know of Erik the Red and Herjolf Bårdson, who in 981 sailed southwest from Iceland to Greenland, to settle permanently with their households and livestock. Their descendants, later known as Norse Greenlanders, lived there for nearly 500 years. And it was not just a small group of expatriates who survived;[2] at the beginning of the 14th century, when the population was at its largest, there were at least 3,000 people residing in Norse Greenland.

Fig. 1
A find from the "Farm Beneath the Sand" in the Western Settlement shows that in addition to sewing clothes the Norse had many other skills: here is an example of a coiled basket probably made of willow root. The original height was c. 30 mm and the diameter c. 60 mm. Greenland National Museum and Archive. Photo: Erik Holm.

Written chronicles and the oral tradition, as well as more or less trustworthy sources, have kept alive the romantic history of the Norse people living in Greenland. It has continued to fascinate people around the world, particularly the mystery of their disappearance in the 15th century, which incited subsequent explorers to seek the answers for themselves.

The Herjolfsnes garments are sent to Denmark

Poul Nørlund concluded his excavation of the Herjolfsnes cemetery in August of 1921, and all the textiles were sent from Greenland to Denmark on the last ship to leave that year. In Copenhagen, the garments were prepared for exhibition, while Nørlund began work on the manuscript *Buried Norsemen at Herjolfsnes*,[3] which was published three years later. The book is an exciting account of how the deceased Norse Greenlanders had been interred in their everyday clothing because there was not enough timber for coffins. He describes a difficult excavation that was only possible because sun and rain thawed the permafrost, turning the dig site into one big mud mire; and of how one piece of textile after the other was lifted carefully from the mire. In all, seventy articles of clothing were recovered, including complete outer-garments, hoods, pill-box caps, and stockings. Other pieces were too fragmentary to be preserved and had to be abandoned.

Poul Nørlund's book about the Herjolfsnes garments was never translated into Danish. It was, in part, because of this that the original book, *Som Syet til Jorden: Tekstilfund fra det Norrøne Grønland*, was written. To reach a wider audience, the book was translated into English and released simultaneously under the title *Woven into the Earth: Textiles from Norse Greenland*. Additionally new technical methods of analyzing textiles, now considered commonplace, have made it possible to 'tease' surprising amounts of new data from the Herjolfsnes find. As an example, the examination of sewing and weaving techniques resulted in tangible evidence of medieval construction methods that had been hitherto unknown, and which gave indication of a sophisticated textile tradition.

The study

In preparation for *Woven into the Earth*, each and every textile, however fragmentary, was measured and examined. Small thread samples were taken for analysis of possible dye traces and determination of fiber type. Additionally, samples were taken for radiocarbon dating, which in the case of archaeological textiles first became practicable in the 1980s with the advent of the AMS-technique[4] (Accelerator Mass Spectrometry). With this process it is possible to date textiles with a very small amount of fiber – only about 1 mg. pure carbon. The Herjolfsnes garments have existed for at least 800 years and have been subjected to significant amounts of wear, use as burial shrouds directly in the earth, alternating freezing and thawing, excavation, cleaning, and lining (attached by gluing and sewing). At the museum they were on display both inside cases as well as outside without protection against dust and "investigating fingers". The finest of them were exhibited almost constantly from their exhumation in 1921 until just a few years ago. It was therefore crucial that the fabrication techniques used were documented as accurately as possible, since it was possible that it was the last time such a thorough investigation would take place.

All of the garments and fragments were photographed, and many details of these unique textiles are now documented photographically as well as with technical drawings. The textiles, being fragile, are equally damaged by daylight or artificial light, and lessons were learned from many years unfortunate experience.

Fig. 2
In a niche in the northern wall of the "Farm Beneath the Sand" a circlet made of hair was found. The hair is from a fair-haired Norse Greenlander. Two by two twisted strands of hair follow parallel, crossing each other in an "over-under" pattern. An exception to this can be seen in some few places as in "ply-splitting". Greenland National Museum and Archive. Photo: Peter Danstrøm.

Fig. 3
Buckle made of walrus tooth. Note the attractive carvings on the 27 mm wide buckle. Qaqortoq Museum. Photo: Geert Brovad.

The Norse Greenlanders' patterns
– the medieval garments' silhouettes

The garments' cuts have been measured as precisely as possible under the circumstances. The woven cloth had been stretched through usage; it had alternately been frozen and thawed throughout the centuries it had lain in the ground. The roots of plants had grown through the upper layers of clothing, as well as microbial breakdown of the wool. The removal from the excavation in Greenland was difficult and damaging to the textiles. On their arrival in Copenhagen, the garments were cleaned and lined with sateen, which has remained since the 1920s and has maintained the shape that the garments were given, after they were cleaned and their style interpreted in the first quarter of the 20th century. The condition of the textile material must be respected when handled. A few garments are exceptionally well preserved, while others are so fragile, that the fibers, and therefore the individual threads nearly disintegrate with the lightest touch. As it was a foregone conclusion that it would not be possible to stretch or straighten the weave so that the warp and weft threads could once more run perpendicular to each other, *this was not attempted*.

Patterns have been drafted for only the most well-preserved garments, but there is also information regarding the cutting and sewing techniques of others that one might use as inspiration for garments not included, but that one might wish to reproduce, even though precise measurements are not in this volume. There are other items which are in such poor condition that it was impossible to draft reliable patterns. Therefore those garments have been excluded.

The authors hope that the large audience of medieval enthusiasts, who will sew apparel based on that period, will find gratification from the measurements and descriptions that are assembled here and are presented as: *Medieval Garments Reconstructed: Norse Clothing Patterns.*

Technical information

The Norse Greenlanders wove with single-ply wool yarn, hand-spun of wool from sheep they brought to Greenland. This type of sheep belongs to the "Northern Short-tail", a breed that also includes Norwegian Spaelsau, Swedish Landrace, and a number of others.[5] However, goat hair and hair from various non-domesticated animals – the arctic hare, among others – were used for weaving cloth. There is also weaving done in linen, a material otherwise not expected to have been cultivated in Greenland; but it could, of course, be imported.

Why is all the Herjolfsnes clothing brown? That is a question that a countless number of interested guests at the National Museum of Copenhagen have asked, as they stood alongside the exhibition cases containing the apparel. That question was answered in connection with the new technical investigations that were recently completed. Most of the colors come directly from the fleece: natural white, gray, brown, and black colors, which

Fig. 4
Decorated four-hole weave-tablets made of bone (50x50 mm) for tablet weaving. In Greenland tablet-woven piped edging on garments has been found, but an independent piece of tablet weaving has not yet been registered. Greenland National Museum and Archive.
Photo: John Lee.

Fig. 5
Buttons made from the same Greenlandic vadmel as garment D10583, with which they were found. The buttons measure 10-12 mm in diameter. They are formed so that the upper surface is smooth, while the cloth is gathered on the bottom surface. One can see that the buttons have had small stitches in concentric circles. The thread has disappeared, but the holes from the stitching are still there. National Museum, Copenhagen.
Photo: John Lee.

the Norse women used to provide many color variations. It was not a surprising result, although the analysis showed that some of the garments had also been colored with a dye either before or after weaving and sewing. The Norse Greenlanders had, for the most part, a limited number of natural dye-stuffs to choose from. Of the raw materials that were easy to find locally were the many lichen varieties found on Greenland, and which produce red-violet colors. One can mention Evernia and Ocholechia among other possibilities. Red dye madder, which derives from roots of the plant Rubia tinctoria, was very common in medieval Europe, but in Greenland it is represented by only two examples, one is a fine diamond twill, which is clearly an import and the other is an edging on a gown (D10594). But also the blue colors – from woad or indigo – are registered, but only on small fragments. The woad plant (Isatis tinctoria) is most likely the plant they were able to obtain, even though it probably did not grow in Greenland. Finally, a non-organic coloring agent has been identified – derived from iron, apparently found in high concentration, which in this case has been given a red-brown nuance. The brown color that the garments have today is caused by tannins that are found naturally in the soil and that have produced the color changes that have occurred throughout the many centuries the clothing has been interred.[6]

Analysis of the weaving shows that the warp threads – with few exceptions – are all Z-spun, very tight, and thin; however, since it was spun from the long hair of the so-called double-coated sheep, the yarn is consequently quite strong. The weft threads are spun in the opposite direction – S-spun – of the wool, that had been separated from the rougher long hair during combing. This produces a softer yarn, which, in this case, is slightly thicker, as it is also spun more loosely than the warp yarn.[7] The loom that was used was an upright, warp-weighted loom, sometimes known by its Norwegian name, *oppstadvev*: a loom leaning against a wall, where the weaver performs the work while standing.

The finished cloth, called Greenlandic *vadmel*, or frieze, usually has a thread count of 8-10 threads per cm in the warp and 10-12 threads per cm in the weft, which is the opposite to most other medieval weaving. Frequently the weft completely covers the warp and, when newly woven, must have been thick and warm.

Only one garment (D5674) is not completely lined, therefore making it possible to get some idea of its original weight, even though the piece has a sleeve missing and one of its gores was replaced in the 1920s by a sateen lining. Its current weight is ca. 500 grams. An unlined hood weighs approximately 125 grams. Both the outer garment and the hood have presumably weighed more, when they were new and the wool was fresh (not dehydrated).

Nearly all the Greenlandic weavings are done in 2/2 twill, but both 2/1 and tabby weave are found. Many of the pieces have significant width along their bottom hems, but it has not been possible to find even one single length of cloth with two intact selvedges, so the actual width of the weaving is unknown. During the excavation of "The Farm Beneath the Sand" (*Gården under sandet*) in Greenland, an intact loom beam was found, which measured 188 cm in its complete length.[8] From that measurement, it is possible to calculate that a cloth width of about 120 cm was probably the largest measurement that could be woven, since the warp must necessarily end at appropriate length from the side beams. None of the panels that make up the gored skirts of the garments measure more than 95 cm at their bottom edges, so it is possible to hypothesize that cloth at Herjolfsnes was woven in approximately 1 meter breadths.

The clothing is sewn using several different stitches and sewing techniques. Hidden seams that are only visible on the inside, are sewn differently than those that are discernable on the outside. The visible seams have a double function: they are most often both decorative as well as reinforcing, and are found along edges that have been subject to significant wear, such as a neck opening, a pocket slit, or along the hemline. Finally, there are also different "decorative edges" that are a combination of sewing and weaving, or in one example, perhaps a combination of sewing and twisting (see Fig. 26). Both finishing approaches are ideal as a solution to the technical difficulty of turning under and sewing a curved edge in such a thick, coarse frieze, but the Norse woman solved this problem in her own elegant manner.

Garment types

The selected items – those that are most well-preserved and therefore most adaptable for reconstruction/copying are garments: D5674, D10580, D10581, D10584, D10585.1, D10586, D10587, D10593, and D10594.

Hoods: D10596, D10597, D10600, D10602, D10605, D10606 and D10608.
Caps: D10608 and D10610.
Stockings: D10613 and D10616.

The above named items are listed by type, a classification that follows Poul Nørlund's organization in his book: *Buried Norsmen at Herjolfsnes* and also used in *Woven into the Earth: Textiles from Norse Greenland*.

Garments

Type 1a describes an outer garment that is pulled on over the head with identical front and back widths, as well as side pieces; all of them are slightly outwardly curving, which gives them a moderate width at the hem. There are set-in sleeves eased into the armhole, as well as gussets at the underarms. The neckline is oval, with its greatest depth at the front.

Type 1b is a closely fitted outer garment that is pulled on over the head. These garments feature slits in both front and back where two-piece gores are sewn in, as well as four to eight triangular side gores that give an outward curving fullness from approximately the waist downward. There are set-in short or long sleeves which are eased into the armholes, as well as gussets at the underarms. The neckline is either oval or round, and

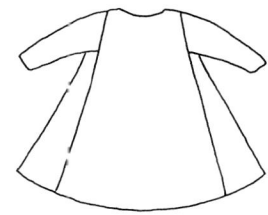
Type Ia

Type Ib

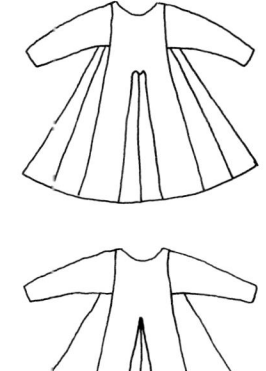
Type Ic

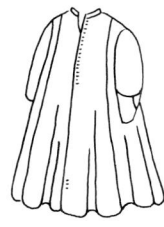
Type II

is cut most deeply in the front. On a few of the garments, of which one is only partially preserved, there is a slit in the front of the neck line with a button closure.

Type 1c has almost the same shape as Type 1b, but it is not as closely fitted, as the side gores slope outward from chest height. There are set-in short or long sleeves which are eased into the armholes and gussets at the underarms. The necklines are either round or oval and cut most deeply in front. A garment for a child between the age of 8 – 10 is also found under this type. Children's clothing is cut and sewn in the same manner as that of adults.

Type II is the designation for two garments with button closures, of which only one is relatively well preserved. It has eight gores, each having one straight grain and one bias side, which together give the garment a particularly pleasing drape, as can be seen in reconstructions. The width curves out gently from the shoulder seam. Along the tightly-fitted neckline is a small, stand-up collar, a so-called Mandarin Collar. Both garments are open in front with button holes in the left side. The buttons were sewn on the exterior of the right side, but they are not preserved.

Hoods

Type I describes liripipe hoods with shoulder capes. The capes are so large that they cover the shoulders completely. There are four of this type, of which one is cut on the bias. The hoods are meant to be pulled over the head.

Type II. Under this type are the short hoods with liripipes. The hoods are short and terminate at the shoulder. They are meant to be pulled over the head.

Caps

These are small pill-box hats with rounded crowns that are used by both children and adults.

Stockings

There is a pair of over-the-knee stockings and a footless stocking.

Notes

1. Østergård 2004, pp. 21-27.
2. Arneborg 2004, pp. 221ff.
3. Nørlund 1924.
4. Østergård 2004, pp. 253-255.
5. Walton Rogers 2004, pp. 79-89.
6. Walton Rogers 2004, pp. 89-92.
7. Østergård 2004, pp. 53-57.
8. Østergård 2004, p. 30 & pp. 58-60.

Chapter 2

Producing a hand-made reconstruction

By Anna Nørgaard

There can be many reasons to create a reconstruction of an historical garment, so before one begins this kind of project, it is important to be completely sure what the garment will be used for. This is crucial in order to know how accurate the garment needs to be in relation to the original.

In the case of the textile finds from Norse Greenland, i.e. the Norse Greenlanders' clothing, there are three reconstruction possibilities. One can choose to create a garment as it looked:

1. when it was excavated
2. when it was interred in the cemetery
3. when it was completely new, and worn for the first time by its owner

Thirty to forty years ago, the possibility of doing a color and fiber analysis did not exist as it does today. Therefore, the garment reconstructions done before that time, including those, for example, by Margrete Hald (1897-1982) for the exhibition at the National Museum of Copenhagen, were often sewn of cloth of the same brown color that the originals had on excavation.

Today, with the help of fiber analysis, it is possible to see whether the clothing is made of wool, linen, or something completely different, making it feasible to come much closer to the original appearance of the individual garments. It is also viable to determine what the initial colors were.

As the original garments have lain in the ground for several hundred years, much of the material has degraded, resulting in finds of many incomplete pieces: there might be a missing sleeve, a front piece, etc. In such cases, when producing a reconstruction, one must assume that both sleeves, for example, were identical. There is nothing, however, that we know with certainty; we do not know the people for whom the clothing was sewn; we do not know if they were fat or thin, or if both their arms were the same length.

When producing a reconstruction, one's point of departure must be the person(s) who will be wearing the garment, with as few deviations from the original cut as possible. If the garment is to be used in a museum exhibition, it is crucial that the reconstruction is as close to the original as possible.

When the pieces of clothing were interred along with the dead, many of them were extremely worn, having been patched and mended. Are these repairs something that should be included in a reconstruction? These signs of wear and tear are, as it were, 'traces' of the individuals who had worn the original garments, whereas those who will be wearing these reconstructions lead a completely different life and therefore the signs of wear on *their* clothing will also be different.

The making of a reconstruction that resembles the original as closely as possible when it was new must be the optimal goal for anyone who desires a historically correct garment, and there are indeed great possibilities to achieve this today. However, there are a number of other questions that must be taken into consideration:

- Should the yarn be spun on a spindle, or on a spinning wheel, and woven on an upright warp-weighted loom, or a horizontal treadle loom?

Fig. 1
Raw wool staples from "The Farm Beneath the Sand". The longest staple is 150 mm. At the top of the three staples to the left, soft underwool that is mixed with hair can be seen. The long hair on its own becomes evident further down.
Photo: John Lee.

- Should one purchase machine-spun yarn and weave the cloth on an upright warp-weighted loom or a horizontal treadle loom?
- Should the fabric be hand-woven on a horizontal loom and the garment hand-sewn?
- Should one purchase a piece of fabric and hand sew the garment?
- Should one purchase a piece of fabric and sew the garment on a sewing machine?

The answers to these questions are often dependent on how much the individual can afford to spend on the project. It is a time-consuming process, and requires experienced artisans to create a reconstruction that is as close as possible to the original, with hand-spun yarn and fabric woven, for example, on an upright warp-weighted loom – which is the type the Norse Greenlanders used and one which would provide the greatest satisfaction. If the garment is to be used in an educational context – especially if the audience is children and the budget is limited – it can be advantageous to purchase machine-woven fabric and sew the garment

Fig. 2
Start combing with hand-held combs, where the wool is placed in the supporting comb. The free comb should be pulled down through the wool in front of the teeth of the supporting comb, and drawn through the wool as many times as may be necessary to ensure that all the impurities and knots are removed. Under this process all the wool fibers will be divided evenly in both combs.
Photo: Roberto Fortuna.

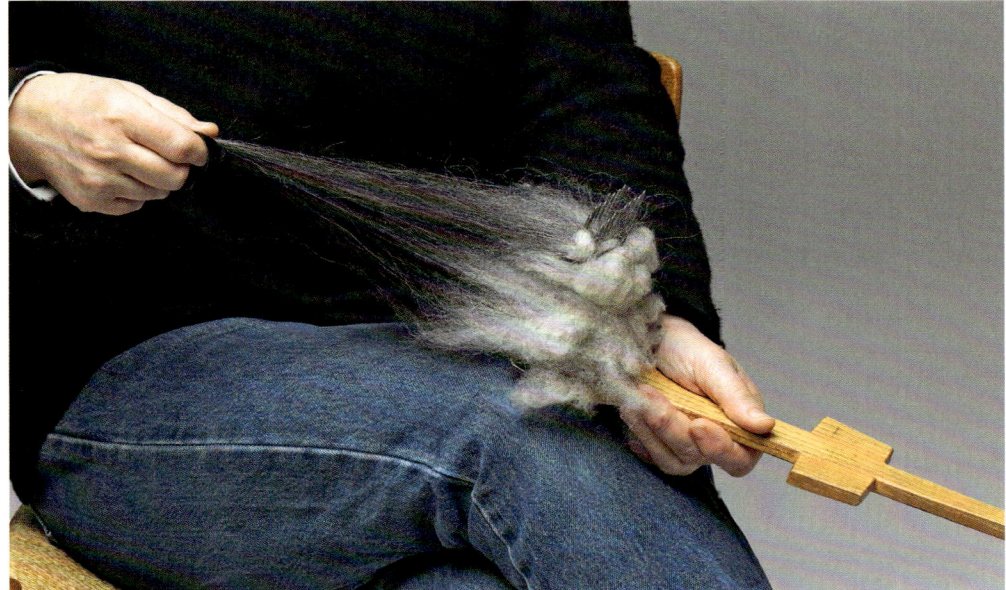

Fig.3
When the wool fibers are combed into a uniform mass, they are first carefully pulled out of the one comb, and then out of the other, in two long, thin, identical bands, where the hair comes out first.
Photo: Roberto Fortuna.

by hand. It is not particularly difficult to find a commercially woven fabric that resembles hand-woven, but there is no way of duplicating hand-sewing with a sewing machine.

Treatment of the wool prior to spinning

The fabric for the original articles of clothing that are shown in this book is all woven from one-ply yarn, spun from wool from 'Northern Short-tailed' sheep. This breed of sheep is still found in a number of places such as the Faroe Islands and Iceland and is also related to the Norwegian Spelsau and Swedish Landrace breeds. This type of sheep has a fleece consisting of two types of wool: a very soft and fine underwool that keeps the sheep warm, and an outer coat of long-stapled hair, smooth, and considerably longer than the underwool. This hair helps to keep the sheep clean and dry (Fig. 1).

Before the wool is spun, the wool and the hair are separated, either by hand, which is a lengthy process, or with the help of woolcombs. Woolcombs with both one row and two rows of teeth have been recovered from archaeological finds.[1] Hand cards are irrelevant to the discussion, as they were first introduced to Scandinavia at the end of the medieval period.

Woolcombs are used in pairs (Fig. 2). An abundant amount of fiber is placed between the teeth of one comb, which can be fixed to a solid surface, such as a table. The teeth of the other comb are pulled downwards in front of the teeth of the stationary comb containing the fiber, drawing out and straightening out the fiber. By repeating this process the fiber becomes divided evenly onto the teeth of both combs and is combed smooth, so that dirt and knots are eliminated. If the combs are small enough to be hand-held (rather than having one comb fixed to a surface), it can be advantageous to switch hands several times during the combing process. When all the dirt and knots have been removed, and the wool in the teeth is in a uniform mass, the fiber is pulled out with the fingers in a long, thin, continuous band, where all the fibers lie parallel. One will now see that the hair comes out first (Fig. 3), and this long, thin band of hair fibers can be spun directly with no further treatment.

After the hair has been removed, the underwool is removed from the combs and is gently 'teased' apart. It is now ready for spinning.

Along with the wool from sheep, there are other yarn types found in the Norse Greenlandic materials. These include yarn spun from the fur of the arctic hare, and the hair of

Fig. 4
Soapstone spindle whorls and wooden spindles from farms in the Western settlement. The large weight, of which only a half is preserved, is from Abel's farm in the Eastern Settlement and it has been used as a weight for flywheels in drills and not for spinning.
Photo: John Lee.

dogs/wolves. These yarn types appear as stripes in the weavings. These alternative yarns are found so infrequently in the extant textiles that it is difficult to determine how common their use was.

Spinning/yarn

The yarn for the Norse Greenland garments was spun on spindles, a time-intensive method; it might therefore be a good idea to use a spinning wheel for a modern recon-

Fig. 5
Yarn can be spun either to the right = Z-spun or to the left = S-spun. The twist-angle in the yarn tells us how hard the yarn is spun and it can be measured with the help of a diagram like this. You place the yarn on top of the diagram, so that the direction of the fibers in the thread is the same as the line which marks the angle of the protractor. The warp-yarn in the original costume from Norse Greenland is almost all spun with an angle of about 45 degrees and Z-spun. The weft is S-spun with an angle of about 35 degrees.
Drawing: Irene Skals.

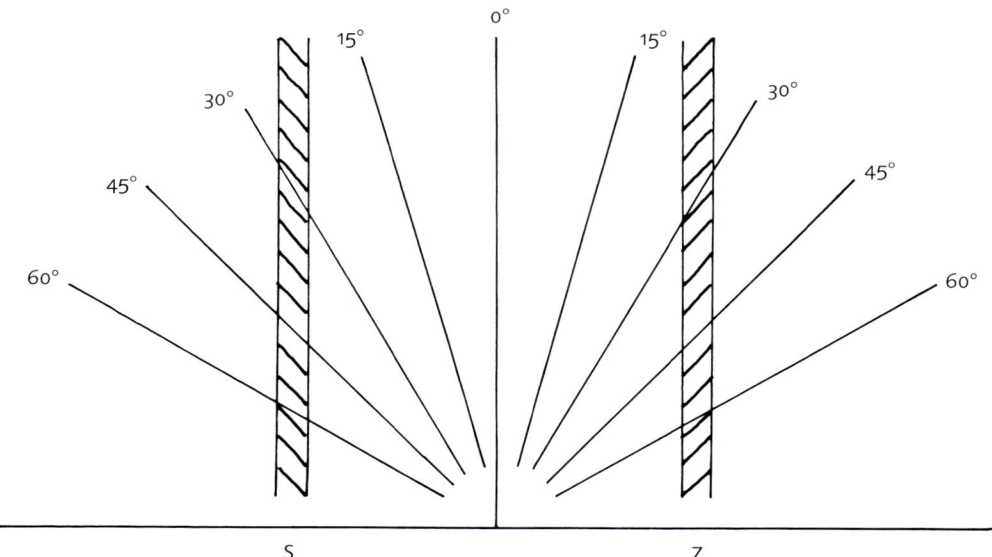

Fig. 6
In a part of this textile a harder spun yarn has been used for the weft than in the rest of the textile. This makes the textile wrinkle when it is washed or merely put in water.
Photo: Anna Nørgaard.

struction. The spindle has the advantage that it is inexpensive to make and can be taken anywhere like other handwork (Fig. 4).

All of the warp threads for the weavings are spun from the long band of hair, which produces a smooth, firm, and strong yarn with a spinning twist of 40-50°. This is almost always right-twisted = Z-spun.

The weft yarn is spun from the teased underwool, with a spinning twist of 30-40°, i.e., a little looser and with fewer twists per centimeter than the warp threads. These are always left-twisted = S-spun (Fig. 5).

If one chooses to spin the yarn oneself, it is important that the yarn has a consistent thickness, but it is even more crucial that there are the same number of twists per centimeter throughout. If the number of twists per centimeter is irregular, there can be some nasty surprises when the cloth is taken off the loom and placed in a water bath to take out the stiffness, or when the fabric needs to be washed. I have personally seen cloth with a width of 80cm that has shrunk up to 10cm in width. When this shrinkage is dispersed over the entire woven cloth, it is not only the width that is reduced in size, but the textile is wrinkled over its entire surface (Fig. 6).

This problem can be alleviated by weaving with two shuttles, where each shuttle holds yarn from each spinning. With this weaving process the weft threads cross each other in the selvedges. A number of textiles from the middle ages have this type of selvedge. Of the ten selvedges that are registered among the garments from Herjolfsnes, seven have crossed selvedges[2] (Fig. 7a + b).

It is not the intention to elaborate on spinning techniques, but instead, the reader is referred to some of the many books written on this topic.[3]

Clothing woven from yarn spun from wool, which contains both hair and wool, is very long-wearing and warm, but since yarn containing hair can also be quite stiff, most

Fig. 7a and b
Selvedges on the textiles in 2/2 twill. The weft threads turn one after the other, (ordinary turning), and a turning with crossing weft threads at the edge. Drawing: Irene Skals.

Fig. 8a
Textile in 2/2 twill, the most commonly used weave in Norse Greenland. Drawing: Irene Skals.

Fig. 8b
The characteristic diagonal structure is emphasised by the dark warp threads and the light weft. Fragment from "The Farm Beneath the Sand". Photo: John Lee.

modern people would consider the apparel made from such wool to be uncomfortable and scratchy. If one therefore chooses to use yarn spun from soft wool, the garment will be more comfortable, but it will never have the same drape as the original.

If one decides to purchase machine-spun yarn, there are several spinning mills in Norway that specialize in yarn spun from spelsau wool, including Rauma and Norsk Kunstvevgarn A/S. However, one can also check on the net and search under *weaving yarn* (*vævegarner*, *vevgarner*) as there are undoubtedly many more yarn companies with appropriate types of wool. Today, when weaving goes in and out of fashion with considerable frequency, new spinning mills appear while others close.

The fabric's quality

The quality of woven fabric is broadly determined by two things: the thickness of the yarn and the thread count per centimeter. If, for example, a yarn that is 1mm thick is used in both the warp and weft, and both warp and weft have a thread count of ten threads per centimeter, the finished cloth will be very tightly woven. If, on the other hand, the yarn is only ½ mm thick and the cloth is still woven with ten threads per centimeter, the finished fabric will be very loose and open, almost like gauze.

With regard to the yarns used in the garments that are discussed here, most of them are very close to an average of 1mm thick, with the maximum of 2mm (the weft threads for stockings D10613) and minimum of .7mm (the weft threads for hood D10602). The thread count per centimeter, on the other hand, has a greater variation, although there are always more weft than warp threads in the same piece of textile. With six warp and eight weft threads per centimeter, D10613 stockings, is the garment with the thickest woven cloth. The stockings are also woven with the thickest yarn, while the short-sleeved, garment D10581, has the finest, with 11 warp and 15 weft threads per centimeter (see Table: *Color and thread*, p. 36).

It is characteristic of all the Norse Greenland textiles that they are very tightly woven, and the clothing made from them, when new, must have been stiff and inflexible. They are also almost all woven in the weave called 2/2 twill. In this weave, each weft thread goes over and under two warp threads, with a shift by one warp thread with each new weft row. This gives a diagonal striped effect, which can be enhanced if one color is used for the warp and another used for the weft (Fig. 8a + b).

It can almost be impossible to attain these exact measurements, both in terms of yarn thickness and the density of the weave. If one chooses to purchase machine-spun yarn, Norsk Kunstvevgarn A/S carries a single-ply yarn that runs 7,000 meters per kilogram = 7/1, which is tightly spun and therefore quite heavy in relation to its thickness. It is available in both S- and Z- spun versions. It is recommended for textiles with a thread count of over 10 threads per centimeter. An alternative could be a weaving yarn of either number 12/1 or 9/1, or possibly a worsted yarn 16/2 or 20/2, even though these are two-ply.

Cloth with fewer than 10 threads per centimeter can be woven with 6/1 or 5/1, which is a quality that most yarn companies that produce weaving yarn carry.

Dyeing/colors

The reason the original Norse Greenland garments appear today in a range of brown colors is due to the chemical processes that have taken place in the ground in which the garments were lying for many centuries. Contemporary color analyses show that the majority of the garment pieces were originally woven in two colors, where the darkest had a brown warp and a light – in many cases, white – weft. The colors derive from the fleece's natural white, gray, brown, and black colors.

This combination could be due to the use of wool from the same sheep. Wool, for example, from a gray sheep of the Nordic Short-tailed breed is usually composed of black

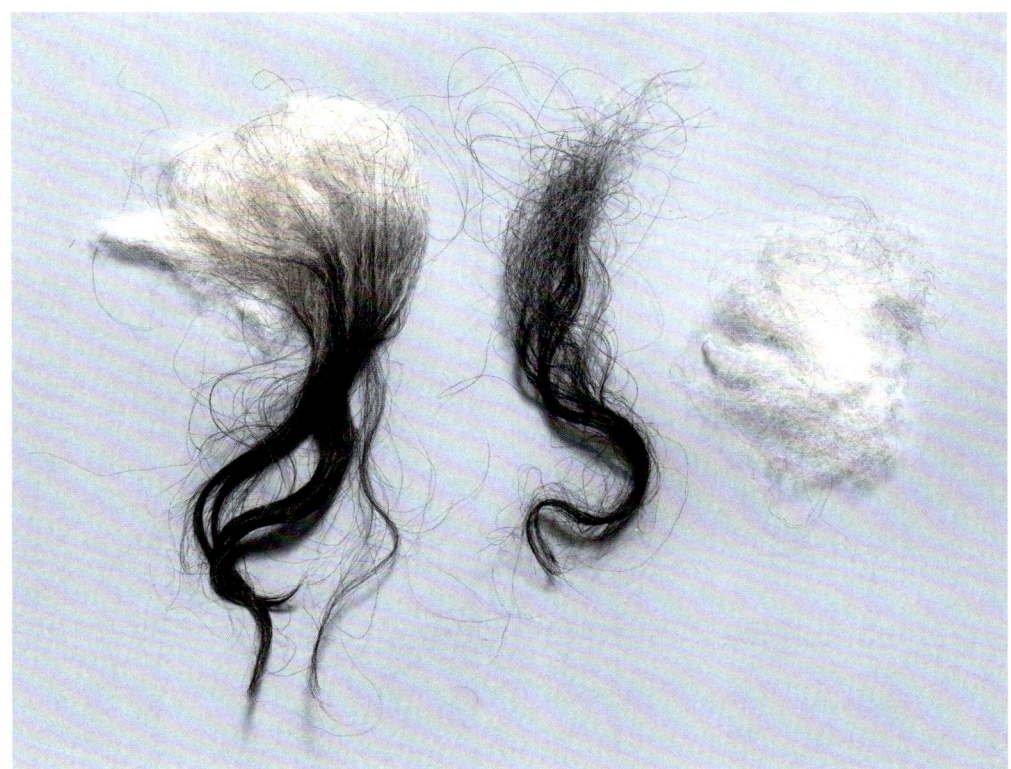

Fig. 9a
Wool from Norwegian Spelsau. Note the different colors after the gray fleece is separated into 'almost'-white underwool and black hair.
Photo: Roberto Fortuna.

Fig. 9b
This small garment for a child is a reconstruction of D10593 woven on a horizontal loom from hand-spund yarn. The different colors of the yarn in the warp and weft accentuate the twill-weaving.
Photo: Roberto Fortuna.

Fig. 10
Diagram of the vertically-slanted, upright warp-weighted loom: A=uprights, B=cloth-beam, C=heddle-rod, D= heddle-rod bracket, E=heddles, F=fixed shed-rod, G=spacing cords, H=loom weights. This loom is from the Faroe Islands and belongs to the National Museum of Copenhagen. Photo: Niels Erik Jehrbo.

hair and white underwool. When the hair has been combed out, what remains is a white underwool, possibly with a few short, black hairs, which, when mixed, give the appearance of light gray. When the hair is used for the warp and the underwool is used for the weft, an automatic color contrast appears with a dark warp and lighter-colored weft (Fig. 9a + b).

There are also traces of an unidentified brownish coloring agent with a significant tannin/tannic acid content. This dye has often been used over already natural-brown wool, which might, at first seem odd, but since natural black and brown wool fades appreciably when exposed to light, it might have been done to avoid this loss of color. It could also have been that the "raw" wool was very uneven in color, and over-dyeing mitigated the difference. Garments D10581 and D10584 are woven with a natural-brown warp yarn over-dyed with a tannin-rich color, and a completely white weft yarn. In hood D10597 it is the weft yarn that is over-dyed with tannin.

This combination has made the 2/2 twill effect in the weaving very obvious (See Fig. 8). Garment D10594 is woven with brown hair and underwool in both the warp and weft. All of the yarn has been over-dyed with tannin, although it is unknown if the fabric was over-dyed before or after it was woven.

Even though most of the garments were made in brown, gray, white, or black wool, there are a few examples where the Norse Greenlanders used other colors. A small fragment from Narsaq shows traces of indigo (Indigofera), but whether it was dyed with woad (Isatis Tinctoria) or imported indigo is not known.[4]

A decorative border that is sewn along the front edge of the buttoned garment D10594 is dyed red with madder (Rubia tinctorum), a well-known dye plant. This piece is woven in a 2/1 twill, and the yarn is Z-spun in both warp and weft, which is extremely atypical for the Norse Greenland textiles, so it is highly probable that it was imported.[5]

Fig. 11
"Dressing" the warp-weighted loom: The warp is either sewn to the cloth-beam or to a "helping" stick. Loom weights are mounted with the required number of warp-threads to each weight: first in the back layer, and then in the other layers. This warp-weighted loom is a reconstruction made from wooden implements from one or more looms found at "The Farm Beneath the Sand". The loom belongs to the Viking Ship Museum in Roskilde.
Photo: Werner Karrasch.

Aside from the above-mentioned dyed pieces, there are seven textile fragments from Narsaq that show a red-violet color extracted from lichens. The same color is also seen on a pile weave from Narsarsuaq. The child's cap D10608 from Herjolfsnes has traces of the same color, but in a small concentration. Since many of the lichen types that grow in Greenland contain this red-violet color, it is likely that these textiles are locally dyed. It should be mentioned that this color is highly sensitive to light, and becomes quickly "fugitive"; i.e. it fades, and in the old textiles, appears as an odd muddy-gray color.

For a Norse Greenlandic reconstruction one needs to consider whether it should be in the original color or if one's point of departure should be in the colors from textile finds in the area where the garment will be used. In the same vein, one must also decide if the reconstruction should be dyed with plants from Greenland, or from suitable plants that grow in the environs where the garments will be worn. In looking at the various church frescoes from the Middle Ages, it appears that colored garments were far more common in the rest of Scandinavia and Europe than in Greenland, even though the cut of the garments was very similar.

Should one wish to dye one's own reconstruction, a number of good books on plant dyeing are available, although most of them are not of current date. In these books, therefore, it is not unusual to find chemicals mentioned for the pre-treatment and mordanting of wool (before the actual dyeing), that are so poisonous that they are no longer available through retail outlets. It is therefore recommended that, if a mordant is needed, that alum be used.[6]

Fig. 12a
When all the warp-threads are mounted on the weights, they are secured to the heddle-rods by heddles (loops). Different sheds in the weave are obtained by lifting the heddle-rods. For a 2/2 twill, for example, three heddle-rods are needed. The opening of the fourth shed is achieved by hanging the warp-thread for this shed in front of the fixed shed-rod, at the bottom of the weave = the natural shed. When the loom is in use it has to be placed vertically slanted, for example against a wall.

Fig. 12b
The heddles should be made from a firm, smooth thread, as they become worn during weaving. A three-ply worsted-thread, spun from long hairs, is used here. An extra knot around the heddle-rod between each heddle is also useful to prevent the heddle from turning around during weaving.

Weaving

The fabric for the Norse Greenland garments are woven on an upright, warp-weighted loom (in Scandinavia, the Norwegian term, *oppstadvev*, is generally used) (Fig. 10).

During the excavation of "The Farm Beneath the-Sand" (see the Introduction), remnants of an upright loom were found in Room number 1. This room was in use from the beginning of the 13th century to the end of the 15th century. Apart from this incomplete loom, many loom weights made of soapstone – that have had a hole bored out for hanging the loom weights up – have been found in the Norse Greenland excavations.

Weaving on an upright warp-weighted loom is extremely time-consuming, but the advantage is that the loom does not take up much space, which may be one of the reasons that it remained the preferred loom type in Iceland right into the 19th century.

The upright warp-weighted loom consists of two uprights, between and above which the cloth-beam is placed. There are holes drilled into the uprights along their entire lengths. Mounted with wooden pegs, these holes can be used for measuring the length of the warp threads. During the weaving process they are used to hold the heddle-rod bracket. Between the uprights at the bottom of the loom there is a fixed shed rod.

The warp is sewn to the cloth-beam at the top, and the threads are divided in two layers: one layer hangs vertically from the beam and the other is brought in front of the fixed shed rod. Loom weights are hung from the bottom of the warp threads in each of the two layers (Fig. 11). These weights hold the warp threads in tension. The number of threads tied to each weight is dependent both on how heavy the individual loom weight is and the thickness of the threads. The majority of loom weights found in Denmark are made of clay, which has the advantage that the clay can be formed into a consistent size and weight. If each loom weight has the same mass and weight, it is possible to gather

Photo: Werner Karrasch

Fig. 13a
A good swordbeater is indispensable when the wefts have to be beaten together, which is done in small parts all the way along the edge of the woven fabric and after each weft. A swordbeater with a long handle, where the weaver can place the hand between the blade and the handle, is a great help. This makes the sword work like a lever and spares the wrist. The swordbeater in the photo is a reconstruction of one found in an excavation near Löddeköpinge in Sweden, and is dated to around the year 1000.

Fig. 13b
Under the excavation of Norse Greenland, pieces of swordbeaters were found which were made of whalebone (also known from Norway). This swordbeater was made in Norway from the jaw of a whale and has a weight of 384 grams – only 6 grams less than my swordbeater from Löddeköpinge.

Fig. 14a
A useful tool when weaving is a little comb that is employed to press the wefts tighter together, and also to keep an even-distance between the warp-threads. It should be no bigger than the palm of your hand. The comb here is round, like the bone from which it has been made, but fits the dimensions of the weaver's hand. Most of the combs that have been excavated are flat and made of bone or antlers.

Fig. 14b
The small bone-comb in use when the weft is placed up against the woven fabric before a new shed. The teeth on the comb do not have to be very long and it is an advantage if the comb has a slight curve upwards but actually an ordinary fork for eating can just as easily be used as long as it has a smooth surface.

the correct number of threads for each weight from the beginning of the warping process. It is crucial that all the warp threads have the same number of grams per thread or the woven fabric will have raised indentations. When using weights made of soapstone or other form of natural stone, each loom weight must be separately weighed to determine how many threads should be tied to it.

When the weights are fitted, heddles/loops had to be tied around the warp threads that, in sequence, are lifted to produce the weaving pattern: here a 2/2 twill. The heddles are secured to the heddle-rods, which, when they are lifted together with the heddles and laid into the heddle-rod brackets, help form the shed through which the weft is

Fig. 15a
A 'singling' seam is used for reinforcing and is found on many finishing edges. The stitches are pulled 'flat' into the textile and are invisible from the right side.
Drawing: Irene Skals.

Fig. 15b
Magnified section of the bottom edge of the garment D5674, reverse side. The singling, for reinforcement, is seen as bows of thread 10-12 mm in height, drawn through the textile from the edge inwards. The edge has been finished with decorative stitching.
Photo: Peter Danstrøm.

Fig. 16
Long seams were probably sewn from the right side and usually with small invisible stitches, and the seam allowance was sewn down to the cloth with tight overcast stitches.
Drawing: Irene Skals.

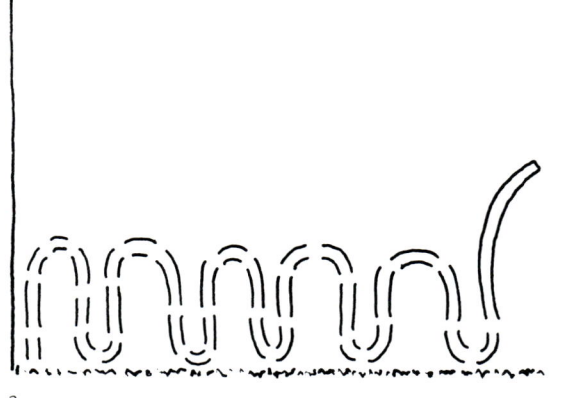

a   b

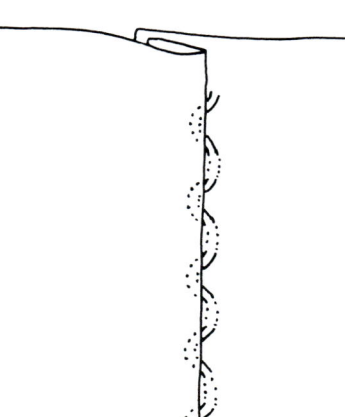

passed. The weaving can begin after the heddles are tied. (Figs 12a and b)

When weaving, it is important to use a good sword beater for beating the weft threads together. For fine weaves, a wooden sword is sufficient, but for thicker and tighter weaves, such as those used in the Norse Greenland garments or for sailcloth, a heavier sword is needed. There are several archaeological finds of sword beaters made of whalebone from Norway and Greenland. When I weave, my preferred sword beater is one that is reconstructed from an iron sword beater found in Löddeköpinge, Sweden (dated to c. 900-1050).[7] It weighs 390 gr. (13.82 oz.), while my whalebone sword weighs 384 gr. (13.54 oz.), and my wooden sword is nearly 200 gr. (7.05 oz.) lighter. (Figs. 13a and b) The iron sword beater's long handle gives a good balance when it is held at the juncture of blade and handle, and only a small wrist movement is necessary to beat the weft upwards.

Another essential weaving tool is a small comb made of bone [Figures 14a and b]. I carry this on a cord around my neck; it is used to move the weft threads up to the level of the weaving – before the shed is changed – and pounded into place with the sword beater. The comb is also useful for extra force along the weft threads and to keep an even distance between the warp threads. Instead of the comb, a pinbeater – a small, pointed stick made of wood or bone – can be used; it is particularly known from Iceland.[8]

There is a significant difference between weaving on an upright warp-weighted loom and a horizontal treadle loom, and it isn't only that for the first one stands up and for the latter one sits down. Since there isn't a reed to hold the warp threads in place on a warp-weighted loom, the weaver has to be aware of where the warp threads are at all times. The width of the weaving must not change; it must not become wider or narrower. It requires a great deal of practice to weave on an upright loom.[9] It is therefore probably easier to use a treadle loom if a hand-woven fabric for a reconstruction is desired.[10]

Sewing

In looking at the entirety of the excavated Norse Greenland textiles, the level of professionalism is impressive. The spinning is perfect, the weaving is perfect, but nothing is as beautiful or well-executed as the stitching; in the needlecraft, nothing is left to chance. The sewing alone shows what value the textiles and garments had for their owners. As an example, there is not one stitch longer than five mm.

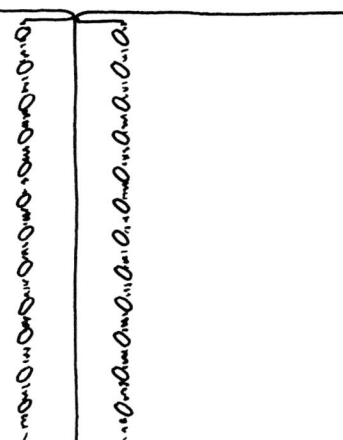
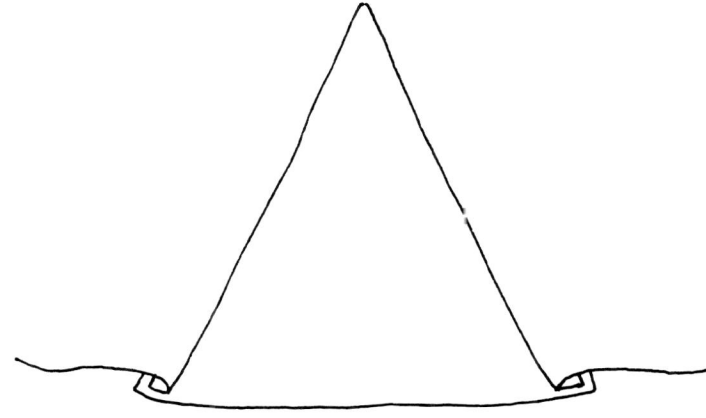

The sewing threads in the original garments were specifically made for stitching. They were S-twisted of two Z-spun threads and are rarely over one mm in diameter. They are spun from fine-combed hair that was undoubtedly singled out expressly to be used for sewing thread. The thread might have been waxed before using, but this has not been proven.

It is virtually impossible to purchase this kind of thread for sewing a reconstruction. Alternatively, excess warp threads can be used, if one is weaving the fabric oneself. If the fabric is purchased, one could perhaps unravel some threads from the textile to be used for stitching. But wool embroidery yarn whose threads have been separated into the appropriate thickness will also work. All three types of yarn will require extra finger twisting in order for it to be smoother and stronger. Rubbing the thread with wax will also smooth and strengthen it.

Today, after the garment pieces have been cut out of the woven fabric, one would either machine zig-

Fig. 17
Shoulder seams are often finished in this way.
Drawing: Irene Skals.

Fig. 18a
Gussets on sleeves and hoods are inserted to lie under the cloth in such a way that the seam allowance is folded away from the gusset. The seam is the same as the long seams on the garment. b: The panels on the garment (D10584) are sewn to each other so that the seam allowances face towards/away from each other. On the middle gusset seams, as well as the side seams, the seam allowances face 'inwards'. In order to clarify the principle, the drawing shows unevenness in the width of the seam allowances, but in reality they have the same width.
Drawing: Irene Skals.

Fig. 18b

Fig. 19
A false seam has no function but is solely for the purpose of giving the garment symmetry, so that the number of panels is even. Right side.
Drawing: Irene Skals.

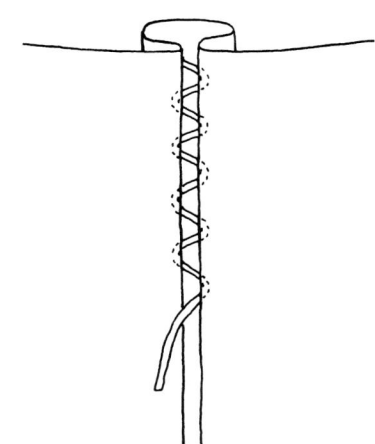
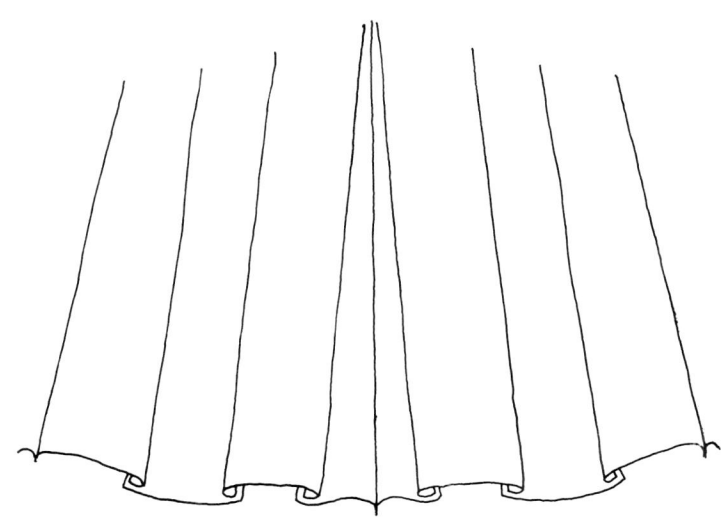

Fig. 20a og b
Stab stitches seen in cross section and from the right side. This stitching is found mostly along seams at the top of garments – probably to mark the cut, and as reinforcement for the finishing edges.
Drawing: Irene Skals.

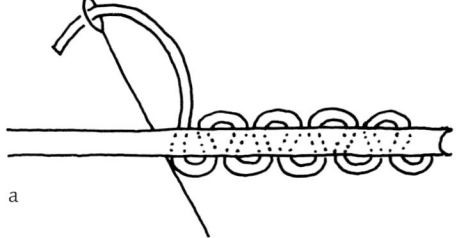
a

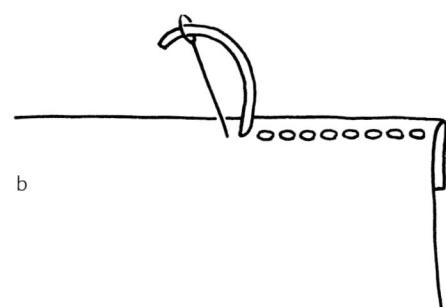
b

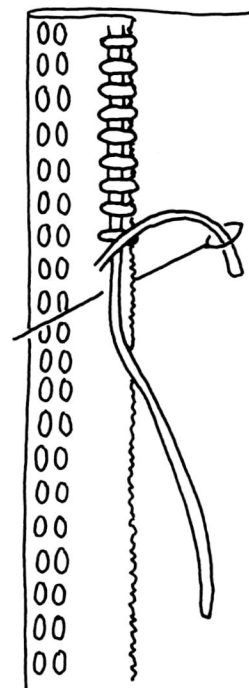
a
b

Fig. 21a og b
A turned back hem, with overcast stitches sewn on top of one or several (filler) threads that cover the raw edge, was prevalent in Norse Greenland. This type of needlework can be found around face-openings on hoods, and in neck-openings; almost always seen together with one or two rows of stab stitches placed some few millimetres from the outermost edge.
Drawing: Irene Skals.

zag or cast the raw edges over by hand to prevent fraying. This is a problem the Norse Greenlanders solved by using an almost invisible stitching on the reverse side of the fabric – so-called *singling* (Fig. 15a + b) – that is sewn with a tack-stitch between the threads in the outermost layer of the fabric, and which is not visible on the right side. The stitches are laid in curves one to two centimeters from the outside edges and towards the body of the fabric. The stitches are never longer than 3 mm.

Many of the seams show evidence of being sewn from the right side (Fig. 16). The stitches are sewn into the fold and are therefore invisible, which might be difficult to see on the illustration, since it has been enlarged to make it more comprehensible.

On the reverse side it is most common that both seam allowances are turned the same way except for the shoulder seams (Fig. 17). To prevent unraveling there are tight overcast stitches over the outer edges of the fabric, which are sewn down into the garment but are invisible on the right side of the garment. On the panels of the garment the seam allowances are sewn together so that they are turned alternately towards and away from the next set of seams (Figs 18a + b).

Aside from the seams that are laid into the fabric, there are seams where the two raw edges are overcast with very close overcast stitches, but instead of lying flat, the edges stand up.

A great deal of attention has been paid to symmetry in the garments, and to this end a number of the pieces of fabric have been "divided" with a false seam since the facing panel on the opposite side is sewn of two pieces of cloth and therefore have natural center seams. A false seam is created by making a pleat down the middle of a gusset or side piece, which is sewn together with small, invisible stitches (Fig. 19).

All of the seams described until now are invisible on the right side of the garment parts. It is something entirely different with stab-stitches (Fig. 20a + b). Stab-stitches are used on all edges that are turned over, such as at the wrists, the neckline, at the bottom of the garment, and at the edges of the hoods. The stitches are very tight and close together and are never more than three mm. In a number of cases there are two rows closely adjacent. The choice of one or two rows does not adhere to a specific system, but it is clear that these rows of stab-stitching help to strengthen the edges. The amount of fabric turned under is never greater than seven mm.

In connection with the stab-stitching, particularly at the neckline of the garments, a thread or two is laid on top of the fabric on the reverse side – but under the overcast-stitches (Fig. 21a + b). This gives a decorative 'fullness', and, as these threads are not sewn down by the overcast-stitches, they can be pulled tightly. This has the effect that a border that is curved, such as at the neckline, is stabilized and does not stretch, but lies flat especially where the seam lies on the bias (Fig. 22).

Stab-stitching is common on the Norse Greenland garments, since it is also used as a decorative stitch several places along seam lines. Button garment D10594 appears to have had all seams decorated with stab-stitching on the sides where the seam-allowance is

folded (Fig. 23). The stand-up collar also has stab-stitching where it attaches to the body of the garment, and the front edge, where the buttons must have sat, is also decorated with stab-stitching.

It might seem strange that there are many blank squares in the table: "Seams and stitching", page 36, particularly those squares that show the *flat and stand-up seams*, because there must have been seams! Unfortunately, records of the sewing techniques are not preserved in many cases. As examples: the sleeves of the button garment D10594 were removed from the body and wrapped around the feet of the deceased. Another garment was cut up the back for convenience in use as a burial shroud. As many of the sewing threads used on the seams had perished, many of the garments were, in connection with the exhibit in 1923, first cleaned with water; holes were patched; and the garments re-sewn, and completely lined.

Fig. 22
The hood is a reconstruction of D 10606, woven from hand-spun yarn on a horizontal loom. On the bottom edge by the shoulder, and on the outer-side of the hood's face-opening, stab stitches are visible. Inside the hood the turned-in hem with overcast stitching is visible.
Photo: Roberto Fortuna.

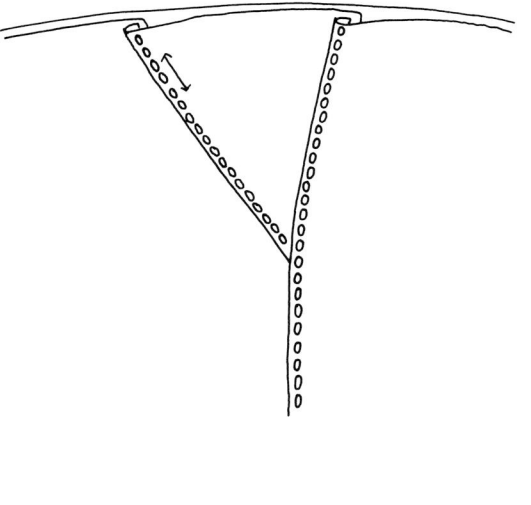
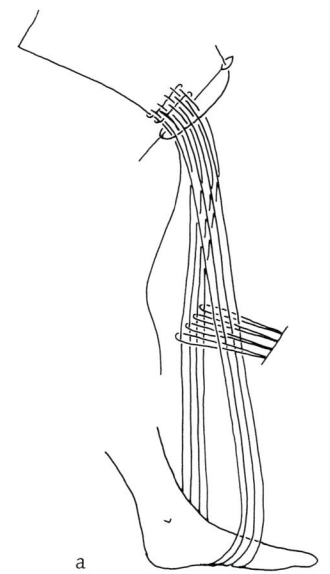
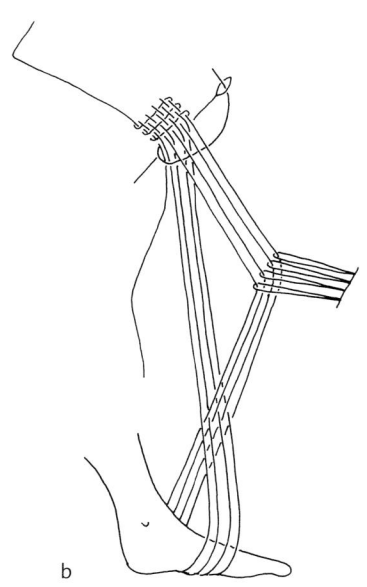

Fig. 23
The gussets of the sleeves, when sewn in, were laid under the front part of the sleeve where stab-stitches can be seen. The hindmost gusset seam (on the grain, marked with an arrow) lies on top of the sleeve. On the opposite sleeve the insertion of the gusset and the sewing are laterally reversed.
Drawing: Irene Skals.

Fig. 24a and b
Footweave or "slynging" is a combination of weaving and sewing, where the weft thread in the weave is also the sewing thread that secures the woven edge down to the cloth. On the right side, the footweaving can be seen as a tabby weave. On the reverse side, only the cross-threads can be seen.
Drawing: Irene Skals.

In the 1980s some of the articles of clothing were "re-conserved". In several instances, the condition of some of the pieces was so poor that the conservator did not dare remove the cotton linings that had been sewn into the garments and hoods in the 1920s. The old linings are still inside many of the garments, and because of this their seams are hidden by the linings. Therefore we can observe today that there are seams, but not how they were sewn.

Those places where there are notes in the table's squares are those where investiga-

Fig. 25a and b
Schematic presentation of tablet-woven piped edging. This is a combination of tablet weaving and stitching where the weft thread in the tablet weaving is also the sewing thread that secures the edging to the cloth. On the reverse side (a) only the cross-threads – possibly with filler threads underneath – can be seen. On the right side (b) the edging can be seen as parallel-lying cords, while the weft thread is invisible.
Drawing: Irene Skals.

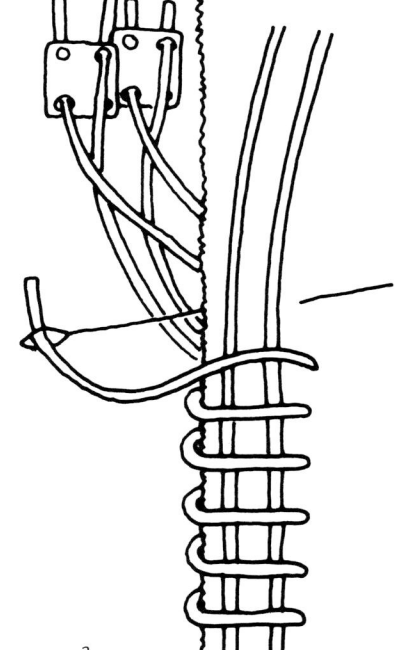
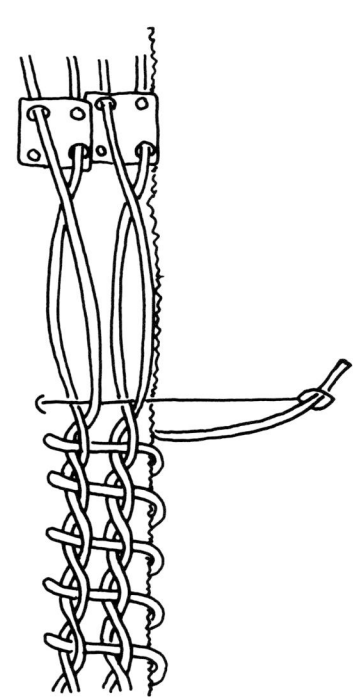

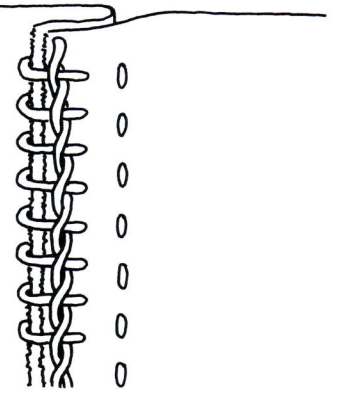

Fig. 26
A seam allowance secured with a tablet-woven piped edging. Two threads – probably turned around each other with the help of a two-hole tablet – produce a cord that is sewn down with overcast stitches. Alternatively, this can be achieved by twisting two threads together with the fingers at the same time as the overcast stitches are sewn. On some of the longitudinal seams the cord is seen innermost, at other times, outermost.
Drawing: Irene Skals.

Fig. 27
From garment D 10585.1. Section of the sewing on the middle gusset seen from the reverse side, back. The sewing is either tablet-woven piped edging with two threads in a tablet, or sewing where two threads are twisted together using the fingers, simultaneously with the overcasting of the raw edge. The twisted thread is seen outermost on the seam.
Photo: John Lee.

tions of the seams by The National Museum of Copenhagen were possible. The table only deals with those garments described in this book, but it should be mentioned that there are many other pieces that are so damaged that it is impossible to reconstruct the full garments. Of course, these fragments also have seams, tablet-woven piping, and footwoven borders, all of which tell us that these sewing techniques, braids, and weaving methods were widespread.

Footweaving and tablet-woven piping

Two types of sewing/weaving that the Norse Greenlanders used as decoration and to stabilize edges are called footweaving (or "slynging") and tablet-woven piping.

"Slynging" is still used today in Iceland.[11] It can be made with the help of a foot – with tabby weaves – where the foot is placed in one of the sheds. The other shed is created with the help of heddles, tied one at a time to the warp threads that are lifted to form the opposite shed. The weft threads are inserted through the shed in the weaving with a needle, which is then stuck through the fabric and around the back of it. The "woven" band lies flat on the right side of the fabric, where only the warp threads are visible. On the reverse side of the fabric, only the weft threads can be seen (Fig. 24a + b).

The other way of creating an edge resembles footweaving to a certain degree, in that also here the weft threads are inserted through the shed with a needle, which is then stuck through the fabric and around the back of it. In this case, however, two warp threads are wound two-by-two around each other, for each weft (Fig. 25a + b). It is not known if the threads are twisted solely by using the fingers or hands, but the easiest method would be to insert the threads through two holes in a weaving tablet and to turn the cards in the same direction after each weft, hence the name, "tablet-woven piping". With this method it is also the case that the warp threads are so compact that the weft threads are only visible on the reverse side of the fabric. On garments D10585.1 and D10587, tablet woven piping – here, though, with only two twisted threads – has also been found in several places in connection with the overcast stiching of the turned-under areas of seams (Fig. 26 and Fig. 27).

The tabby-woven "slynging" can also be achieved with the aid of weaving tablets with two threads in each tablet, but in this case the tablets must be turned in opposite directions after each weft, so that the warp threads come to resemble tabby weaving. A number of weaving tablets have been found in Greenland (see Introduction, p. 13, Fig. 4), so this is a tool that the Norse Greenlanders knew.

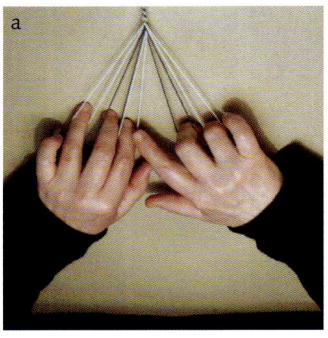 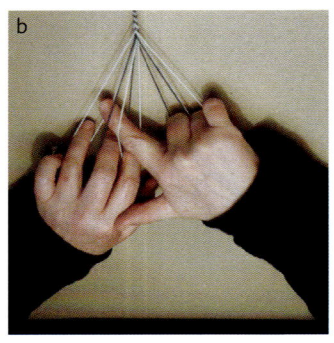 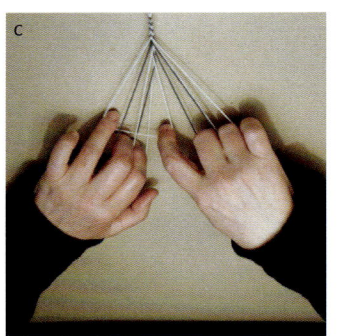 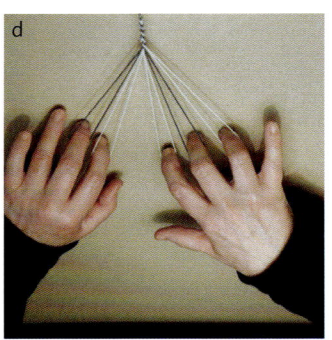

Fig. 28
a) There are three loops on the fingers of the left hand and two on the fingers of the right hand. The right hand's index finger is without loops and is moved downwards from above through the loop on the left hand's index finger. b) The right hand's index finger is then conveyed from above through the loop on the middle finger. c) The right hand's index finger is inserted into the loop on the third finger and pulls this loop back through the two loops on the long finger and index finger. d) There are now three loops on the right hand and two on the left hand. These two are moved over to the third finger and middle finger, so the left hand's index finger is without loops. Now the left hand's index finger is conveyed through the loops on the right hand fingers, as described under Figs a and b: i.e. the whole process is repeated, but this time from the opposite side. Each time a loop has changed hand, both arms must be stretched as far out to each side as possible, preferably 90 degrees in relation to the braided band, so that the braids become tighter. Photo: Andrea Otterstrøm Nørgaard.

Braided cords

This technique is also called Faroese cords. These types of cords are used in several places on the garments, both as decoration and reinforcement. The cord is braided with strands held by separate fingers; you need at least three strands to make a braid, five is the most common, but there are examples from other parts of the world where the plaits have so many strands that two or three people must have sat in a row and passed the strands back and forth to each other. The braided cords on the Norse Greenland garments are square and generally made from four, five, or seven strands.

A five-strand square cord is plaited as shown in Figure 28. The strands must be of equal length; they are gathered into one bundle and are attached to a fixed point at the opposite end of the strands.

(a) The strands are brought around three fingers on one hand: the third finger, the middle finger, and the index finger. On the other hand, the strands are brought around the middle and ring fingers; the index finger carries no strands at this time.

(b) This bare index finger is now brought from above, down through the strands on the opposite hand's index finger, and again from above, through the strand on the middle finger, and subsequently through the strand on the third finger.

(c) The bare index finger is now crooked like a hook around the strand on the third finger and pulls this strand back through the strands on the middle and index fingers. When the strand has moved onto the opposite hand's index finger, it is removed from the third finger.

Fig. 29
Method by which the buttons could have been made and also how they could have been attached to the edge of the garment. The buttons are very flat and under 10mm in diameter. Drawing: © Museum of London/ Christina Unwin.

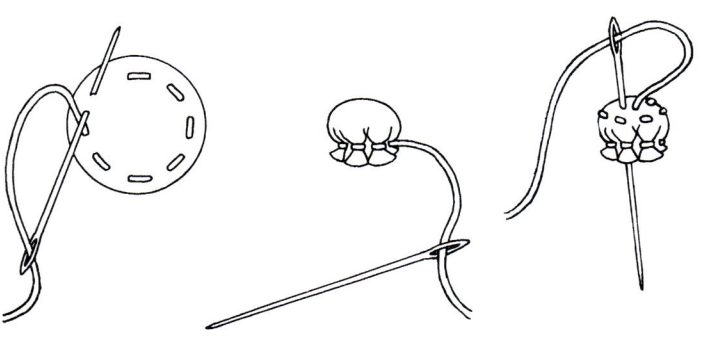 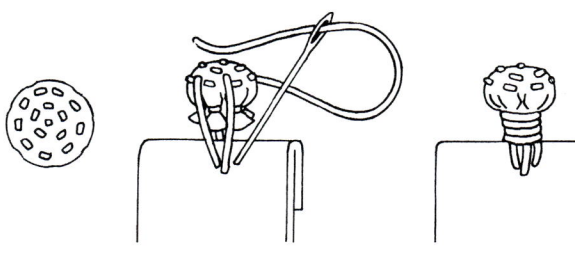

34 • MEDIEVAL GARMENTS RECONSTRUCTED – NORSE CLOTHING PATTERNS

(d) Now there are two strands on the hand that previously had three; they are moved to the ring and middle fingers so that the index finger is bare. Then, the entire process is repeated, but from the opposite side.

Each time a strand crosses from one hand to the other, the arms are spread out to their full extent, so that the braiding is pulled tight.

If one is working alone, it is a good idea that the strands are no longer than the span of one's own arm; otherwise it is difficult to tighten the plaits. If the cord needs to be longer, a helper is essential to tighten the braiding at the fixed point so it is firm and even.

Buttons and buttonholes

The only garment with buttonholes that is mentioned in this book is number D10594. It is open down the front and has buttonholes along the left edge. There is nothing remaining of the thread that was used to reinforce the buttonholes against unraveling so they could tolerate the stress of the buttons as closures. Therefore, we cannot know what kind of stitch was used around the buttonholes, or from what material the thread had been spun.

Two other garments with buttonholes are excluded from this book since they are rather fragmentary and therefore not suited for the production of a satisfactory pattern: D10595, a buttonhole garment like D10594, but less well-preserved, along with D10583, which has buttonholes on the sleeves from the wrist upwards to ca. 27cm. There are no stitches or thread preserved from these buttonholes either.

D10583 is the only garment where buttons are retained, with seven remaining (see p. 35, fig. 5). The buttons are sewn from the same fabric as the garment. The sewing thread that held the buttons has disappeared. It is impossible to determine if there was something put inside the buttons to make them more stable, but they are sewn through the fabric layers with small, tightly spaced stitches in concentric circles, and the sewing thread here is 2-ply, but very thin[12] (Fig. 29). The buttons are rather flat and under 10mm in diameter. Buttons made of bone or walrus tusk have also been found in the remains of the Norse Greenlandic settlements.

Using the tables

The two tables, "Color and thread" and "Seams and stitching" are conceived as a summary in which the reader can find information about the individual garments for such things as making weaving patterns or techniques for the hand-sewing of garments. The pattern for the garment can be found in the next part of this book.

Table: Color and thread

The table shows the garments for which there are patterns drafted in this book, with museum inventory numbers and garment type in the first two columns. In the subsequent four columns, the numbers or words on the left side of the slash refer to the fabrics' warp threads and those that are to the right of the slash refer to the fabrics' weft threads.

As an example, the textile with inventory number D5674 is a garment/gown, whose warp is light gray and weft is white. The fabric is woven with 10 warp threads and 11 weft threads per centimeter. The warp threads are a little more tightly spun with a grade of 40° in comparison to the weft threads that are spun with a grade of 35°. The warp thread has a diameter of 1.00 mm, while the weft is a little thicker, with a diameter of 1.5mm.

Since all the warp threads in the garment are spun to the right = Z-twist, and all the weft threads spun to the left = S-twist, there is no column designating spinning direction.

Table: Color and thread

Inventory Number	Garment Type	Color warp/weft	Number of threads per centimeter warp/weft	Gradation of twist in the treads warp/weft	Diameter of threads in mm warp/weft
D5674	Garment	Light gray/White	10/11	40°/35°	1.0/1.5
D15080	Garment	Dark gray/Light gray	10/14-16	45°-40°/35°	0.9/0.8
D10581	Garment	Medium Brown-tannin/ White	11/15	45°/35°	1.0/0.9
D10584	Garment	Dark Brown-tannin/Dark brown	8/9	40°/30°	1.2/1.0
D10585.1	Garment	Medium Gray/Light Gray	10/13	45°/35°	1.0/0.9
D10586	Child's Garment	White/White	8/9	45°/35°	1.0/1.2
D10587	Garment	Dark Brown/Dark Brown	9/8	40°/35°	1.0/1.2
D10593	Child's Garment	Light Gray/White	7/9	40°/30°	1.2/1.0
D10594	Button Garment / Decorative Border	Black-tannin/Black-tannin White – Madder red/White – Madder red	8/10 11/11	40°/35° 40°/40°	1.0/1.0 0.9/0.9
D10596	Hood	Brownish Black/Brownish Black	8/9-10	40°/30°	1.0/1.2
D10597	Hood	White/White-tannin	8/11	45°/30°	1.0/0.9
D10600	Hood	Dark Brown/ Dark Brown	9/10	45°/35°	1.0/1.0
D10602	Hood	Light Gray/White	8/14	45°/35°	1.2/0.7
D10606	Hood	Light Gray/White	9/12	40°/35°	1.2/0.9
D10608	Child's hood	White/White	10/12	45°/35°	1.0/1.0
D10608	Cap Crown 2 side-pieces 2 side-pieces	Medium Gray/Light Gray Medium Gray/Light Gray Medium Gray/Light Gray Whole cap over dyed with *korkje* (lichen purple)	9/21 8/9 7/10	40°/25° 45°/35° 45°/35°	1.0/0.5 1.1/1.0 1.0/1.0
D10610	Cap	White/White	10/14	55°/40°	1.0/0.8
D10613	Stocking	Black-brown/Black-brown	6/8	45°/50°	1.0/2.0
D10616	Footless stocking	White/White	8/9-10	40°/30°	1.0/1.0

Table: Seams and stitching

Museum no.	Singling	Abolished long seam	Upright long seam	Stab-stitching	False seams	Braided cords	Tablet weaving	Foot weaving
D5674 Garment	At the bottom			One row at the neck	Right side seam			
D10580 Garment	At the bottom	yes	yes	Two rows at the neck Two rows at the sleeves One row at the side widths	Both side-widths on the back	Neck Sleeves Pocket slits	At the bottom with three braids [cords]	
D10581 Garment		yes		One row at the neck One row on the sleeves One row at the gussets on the back One row at the bottom	Left side seam		Pocket slits with one braid	
D10584 Garment	At the bottom	yes		Two rows on the sleeves One row on the slits A row at the bottom	Right side seam	Neck Pocket slits	At the bottom	

Museum no.	Singling	Abolished long seam	Upright long seam	Stab-stitching	False seams	Braided cords	Tablet weaving	Foot weaving
D10585.1 Garment		yes		Two rows at the neck One row on the sleeves One at the side-widths One at the top, on the back middle gusset One row at the bottom			Middle gusset on the back, with one braid	
D10586 Child								
D10587 Garment		yes	yes	One row at the neck Two rows at the pocket slits One row at the bottom	Three on left side Two on the right and middle gussets on the back	Sleeve	Long seam with one braids [cords]	
D10593 Child					Back			
D10594 Buttoned garment		yes		One row at all seams One row at the bottom of the collar One row on the right front edge One row at the bottom				
D10596 Hood				Two rows at the face-edge				
D10597 Hood								
D10600 Hood								
D10602 Hood								
D10606 Hood				Two rows at the face-edge Two rows at the shoulder-edge				
D10608 Hood	Shoulder-edge	yes		Two rows at the face-edge				At the shoulder 1cm. wide
D10608 Hood								
D10610 Cap								
D10613 Stockings				One row at the top, on the hem				
D10 616 Stockings								

PRODUCING A HAND-MADE RECONSTRUCTION • 37

Notes

1. Walton, 2007, p. 16.
2. Østergård, 2004, p. 65.
3. Warburg, 1974; Crockett, 1977; Brown 1979; Ross 1988; Walton, 2001.
4. Østergård 2004, p. 90.
5. Østergård 2004, p. 69.
6. Brown 1979; Henningsen 1983.
7. Andersson 1999, p. 28.
8. Guðjónsson 1990.
9. Hoffmann 1964; Nørgård 2009.
10. Brown 1979; Henningsen 1983.
11. Magnússon 2006.
12. Crowfoot 1992, p. 171; Østergård 2004, p. 102.

Chapter 3

Reconstruction of Patterns

by Lilli Fransen

The patterns in this book are reconstructions based on measurements of the Norse garments excavated in 1921 by the Danish archaeologist Poul Nørlund on Herjolfsnes in Greenland. After initial repairs the garments have been restored several times and are today sewn onto a base of brown shirting or light wool fabric.

Looking at the condition of the garments it must be assumed that the Norse were not buried in their best garments but that those most worn and repaired were used for the purpose (Nørlund, 1934). When the garments were used as burial clothing the seams had often been partially or totally ripped open before the body of the deceased was wrapped in them. However, after arrival in Denmark in 1921, the garment pieces were sewn together again. On later examination it has been proved that, in a few cases, there is uncertainty about the original shape of the garment.

Several of the garments are asymmetrical, meaning that the left side is different from the right. It is unknown if, during the making of the garment, there have been anatomical or functional considerations made for the person who should wear the garment, or if this asymmetry has developed during wear, patching and repairs.

In the reconstruction process of the garment patterns great importance has been attached to trying to come as close as possible to the original pattern cut. The patterns are based on measurements of the original garments, hoods, caps and stockings (Østergård, 2004 and Nørlund, 1924) and are prepared in such a way that the garments appear "new". It has therefore not been taken into consideration that the garments might have been asymmetrical, neither are patches and repairs included.

The measuring survey of the museum, illustrated as gray shadows on the pattern, are digitalized in a CAD program (AccuMark), where they are subsequently modified so that the patterns are symmetrical and the seams that are to be sewn together have the same length. The patterns are reconstructed following the best-preserved side of the garment.

It is remarkable that most of the longer seams are sewn together in a piece where the edge is cut on the grain, and a piece where the edge is cut on the bias. This technique causes the seam to be stable, preventing the part that is cut on the bias from getting longer.

During the reconstruction of the sleeve patterns many have been made wider at the top to fit better into the armhole. The width of the neck on the hoods is made bigger as otherwise it would not be possible to pull the hood over the head. The same applies to the ankle width on the long stockings which has been made wider to make it possible to pull the stocking over the heel. Depending on the elasticity of the fabric, these measurements may be reduced.

The adult garments are all graded in sizes small, medium and large, of which the smallest size corresponds to the measurements of the museum. This procedure causes the size 'small' to vary in the different models of garments and hoods. The children's garments are graded using the centimeter system, where the height of the child indicates the size.

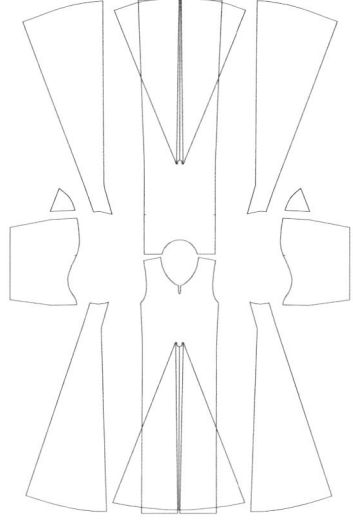

An overview of the specific garment measurements can be found in connection with the patterns for each model showing the following measurements:

- Garment – Length, Chest width and Sleeve length.
- Hood – Length, Face opening and Neck width.
- Cap – Head circumference.
- Stocking – Length, Thigh width, Calf width, Angle width and Foot length.

The patterns are in the scale 1:5, 2 mm = 1 cm.
The patterns are without seam allowance.
The pieces are sewn together according to the letter indications.

The false seams are marked on the patterns with dashed lines, providing multiple choices for cutting the garment.

1. No seam – ignore the dashed line.
2. False seam – add 1-2 cm fullness at the dashed line for seam allowance.
3. Real seam – divide the pattern at the dashed line.

In the pattern-cutting layout, a seam allowance of 1 cm on all parts of the patterns and a buffer of 1 cm between each pattern is estimated. The patterns are placed from the largest size, and in this way it will be possible to reduce the amount of fabric used for the smaller children's sizes by placing the patterns more expediently.

Table of Reconstructed Patterns

Nørlund No.	Museum No.	Garment type	Page	Template page
33	D5674	Ia	44	
38	D10580	Ib	50	
39	D10581	Ib	58	65
42	D10584	Ib	66	
43	D10585.1	Ic	74	80
44	D10586	Ic	82	
45	D10587	Ic	88	
62	D10593	If	96	
63	D10594	II	100	105

Nørlund No.	Museum No.	Hood type	Page
65	D10596	I	106
66	D10597	I	110
70	D10600	I	114
72	D10602	II	118
78	D10606	II	122
80	D10608	II	126

Nørlund No.	Museum No.	Cap	Page
86	D10608		126
83	D10610		130

Nørlund No.	Museum No.	Stockings	Page
88	D10613		134
91	D10616		138

Garments:

D5674 – The garment of type **Ia** is made to be pulled over the head. The front and the back are without any centre gusset, and the garment has only one side panel with a false seam, which indicates the side seam. In the reconstruction the side panel is divided into two at the side seam. The sleeve is divided into a front and a back piece and has one gusset. The sleeve has to be stretched approximately 5 mm when being sewn into the armhole.

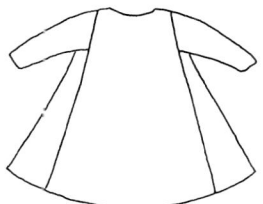

D10580, **D10581** and **D10584** – The silhouette of garment type **Ib** is relatively close-fitted in the upper part, flaring into fullness from about the waist line. The garments have centre gussets in the front and back piece and have four to six side panels. The garments are made to be pulled over the head and have necklines with an optional slit in the front. The one-piece sleeve has one gusset.

D10580 – In the original garment the side panel at the back is divided into two by a false seam. In the foremost side panel there is a 17 cm pocket slit. The sleeve has to be stretched approximately 15 mm when being sewn into the armhole.

D10581 – The curved seams at the upper part of the centre gussets are sewn according to the template page 65. The neckline has a 4 cm slit at the centre front. The sleeve has to be stretched approximately 15 mm when being sewn into the armhole.

D10584 – In the front side panel there is a 17 cm pocket slit. The sleeve has to be stretched approximately 15 mm when being sewn into the armhole.

D10585.1, **D10586** and **D10587** – In the garment type **Ic** the fullness starts around the chest line creating a wide silhouette. The garments have centre gussets in the front and back piece and have four side panels. The garments are made to be pulled over the head and have necklines with an optional slit in the front. The one-piece sleeve has one gusset.

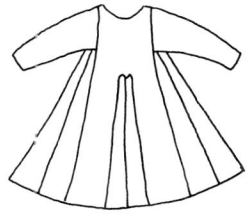

D10585.1 – The curved seams at the upper part of the centre gussets are sewn according to the template page 80. In the side seam there is a 15 cm pocket slit. The neckline has a 18.5 cm slit at the centre front. On the original garment the upper part of the sleeves is missing, so uncertainty exists regarding the shape of this part. The sleeves are therefore constructed following the same principle as the sleeves of the other garments of type Ib and Ic

D10586 – In the garment for children the sleeve has to be stretched approximately 12 mm when being sewn into the armhole.

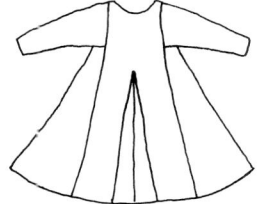

D10587 – In the original garment the centre gusset at the back, and both side panels, are divided into two by false seams. In the seam between the front piece and the front side

panel is a 18 cm pocket slit. The sleeve has to be stretched approximately 22 mm when being sewn into the armhole.

D10593 – The garment for small children of type **If** has centre gussets at the front and back but no side panels and no sleeve gussets. The centre gusset at the back is divided into two by a false seam in the original garment. The garment is made to be pulled over the head.

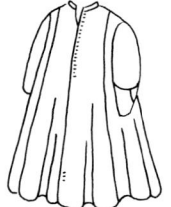

D10594 – The buttoned garment of type **II** has an open front with a narrow upright collar. The garment consists of eight panels, each with an edge cut on the grain and an edge cut on the bias. The sleeves have a seam at the elbow line and have two gussets. The buttonholes on the left side of the front opening are marked according to the templates on page 105.

Hoods:

D10596, **D10597** and **D10600** – The liripipe hoods of type **I** are made to be pulled over the head and all have shoulder capes with a gusset at the front. D10596 also has a gusset at the back.

D10602, **D10606** and **D10608** – The liripipe hoods of type **II** are short hoods ending at the shoulder. The hoods have a gusset at each side and are made to be pulled over the head.

Caps:

Both the caps have a round top with a rectangular sideband.

D10608 – The circumference of the crown is 65 mm bigger than the measurement of the edge. The crown has a false seam.

D10610 – The circumference of the crown is 30 mm smaller than the measurement of the edge.

Stockings:

D10613 – Each of the long stockings consists of a leg, a gusset and two foot pieces.

D10616 – The short footless stockings consist of a leg piece.

All the new garments, hoods, caps and stockings are made of heavy 2/2 twill fabric with 10/9 thread counts per cm. All seams are sewn together on a lockstitch machine. Depending on the placement of the seam, some of the seams have been over-locked or zigzagged before the pieces have been sewn together, while some of the seams have been over-locked or zigzagged after the pieces have been sewn together. The necklines and some of the sleeve hems are finished with a matching cotton bias binding, sewn on first by machine and afterwards blind-stitched by hand. The bottom hems of the garments have been blind-stitched by hand. There is therefore no visible stitch on the right side of the garment.

To trial initial sizes, it is recommended that a toile (in a less expensive fabric) be made of the upper part of the garment.

The garments and the original sewing techniques are described in detail by Else Østergård in her book *Woven into the Earth* (Østergård 2004).

The garment was originally sewn in Greenlandic vaðmál in 2/2 twill with a light gray warp and a white weft. (Østergård, 2004)

Museum No. D5674

Garment measurements in cm:

Size	Small	Medium	Large
Length from shoulder	111	119	127
Chest width	112	122	132
Sleeve length	56	57	58

Fabric consumption in cm:

Fabric width	80 cm	140 cm
Size Small	530	267
Size Medium	560	281
Size Large	589	296

Pieces to cut:

Front	1 piece
Back	1 piece
Side Panel Front	1 left piece and 1 right piece
Side Panel Back	1 left piece and 1 right piece
Sleeve 1	1 left piece and 1 right piece
Sleeve 2	1 left piece and 1 right piece
Sleeve Gusset	1 left piece and 1 right piece

Photo: Roberto Fortuna

New garment

Pattern-cutting layout:

Fabric width 80 cm

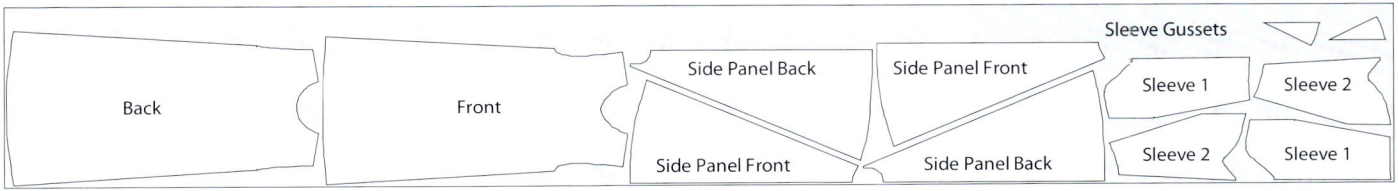

Fabric width 140 cm

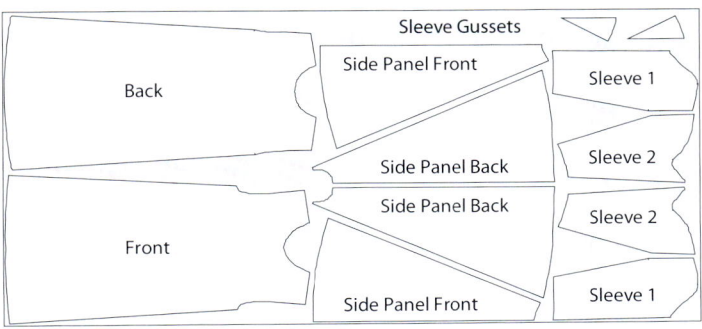

MUSEUM NO. D5674 • 45

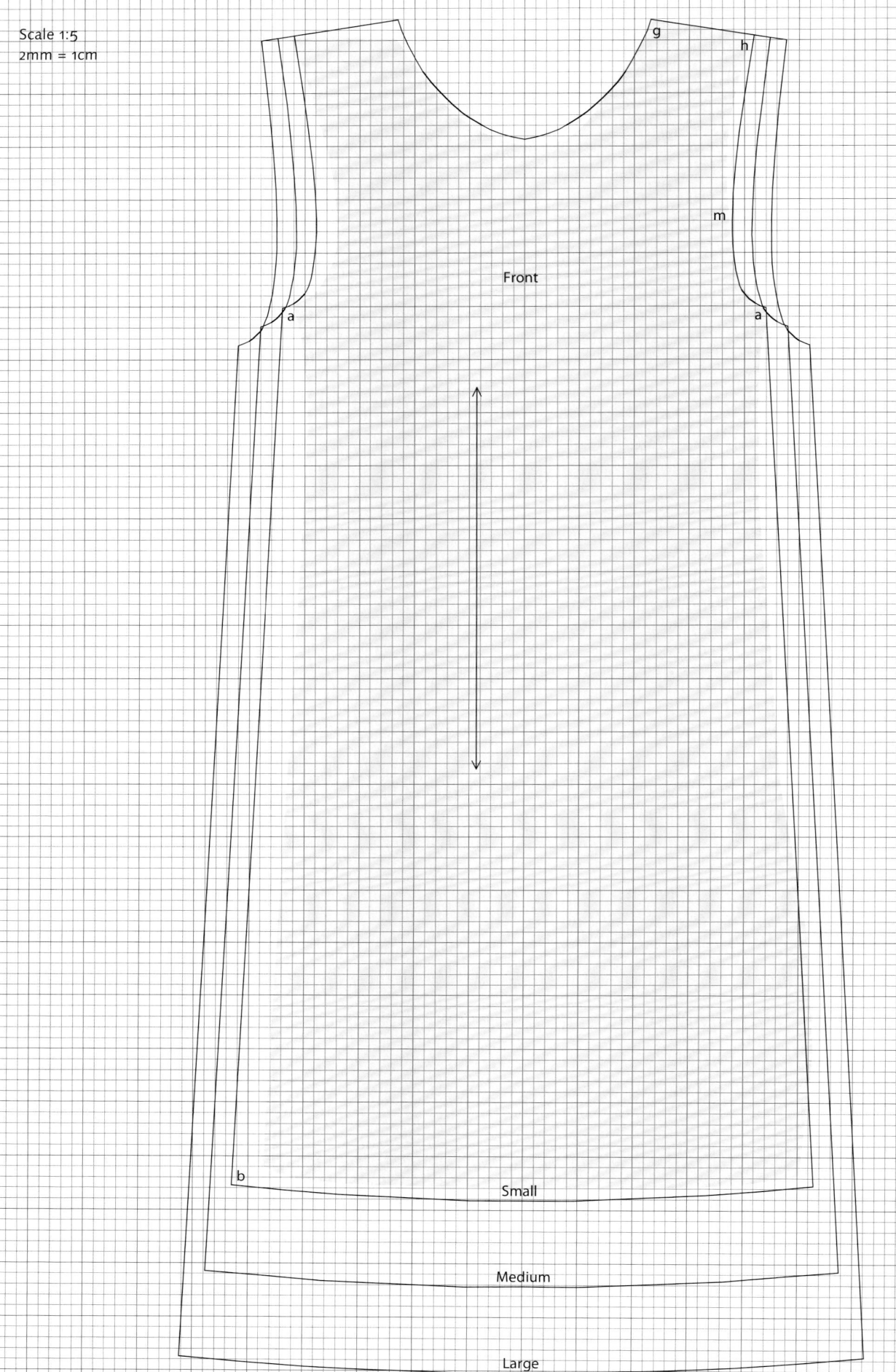

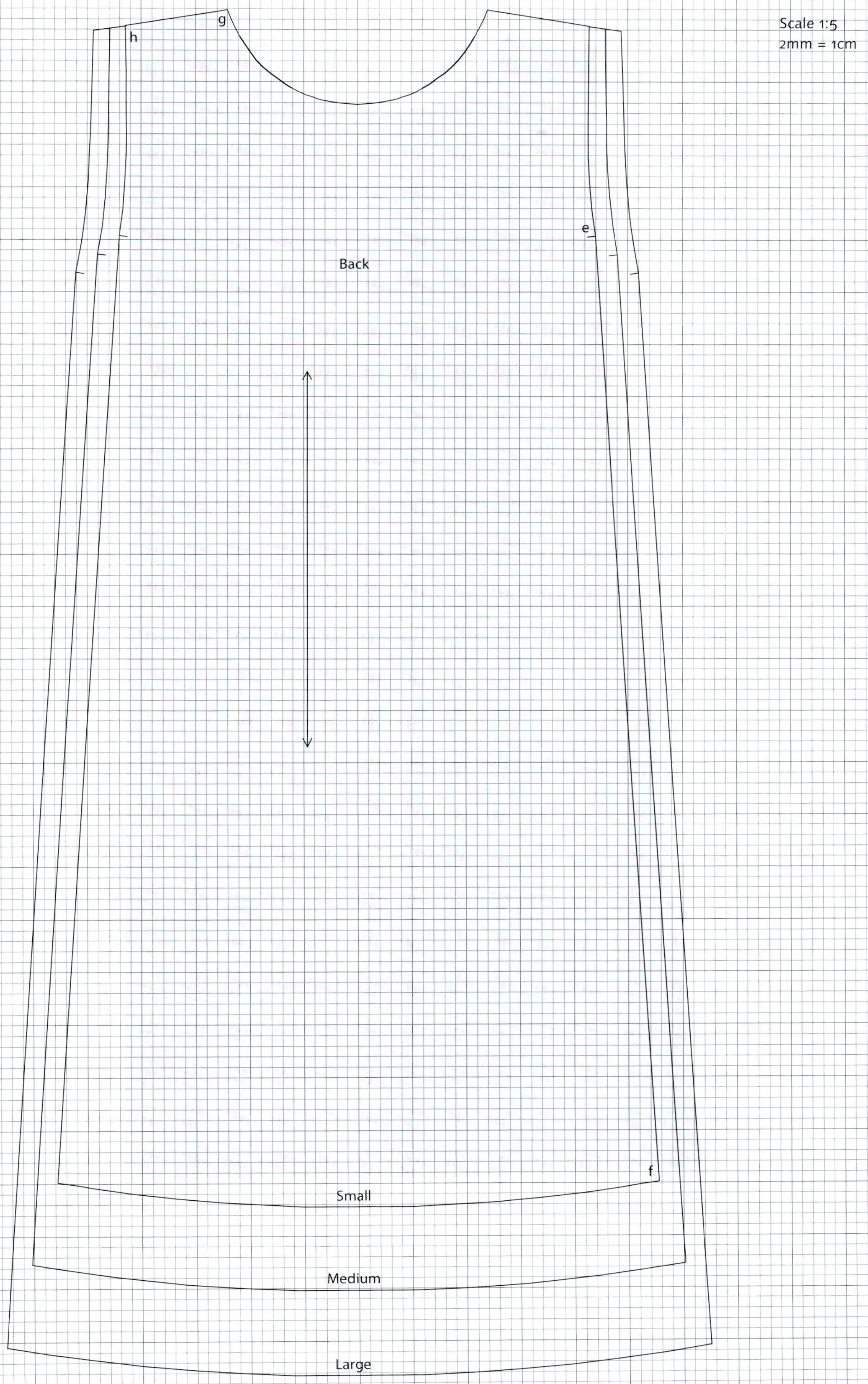

Scale 1:5
2mm = 1cm

D5674

h g

Back

e

f

Small

Medium

Large

Scale 1:5
2mm = 1cm

D5674

e
c a
c

Side Panel Back   Side Panel Front

f   b
Small   d   d   Small
Medium   Medium
Large   Large

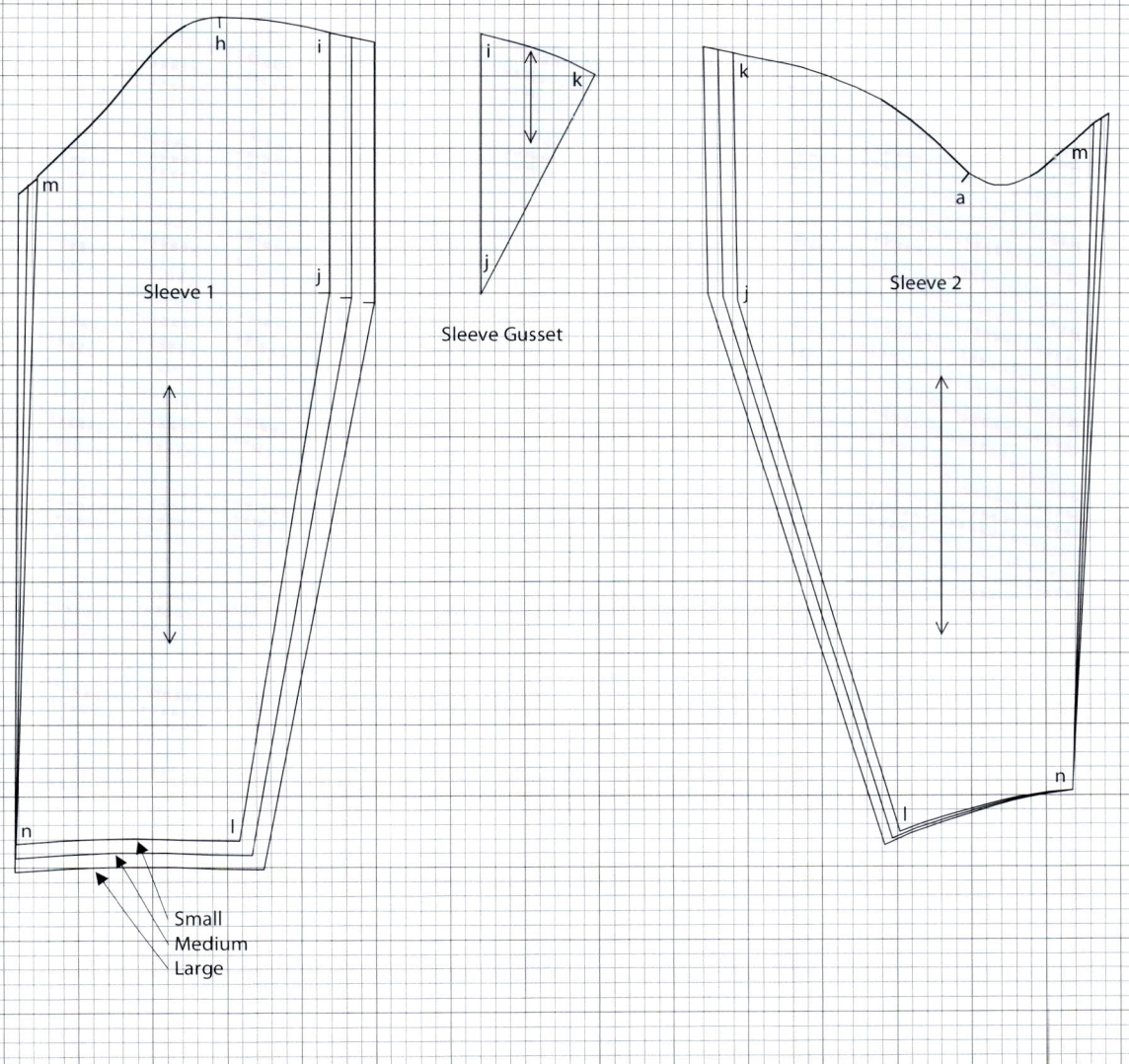

The garment was originally sewn in Greenlandic vaðmál in 2/2 twill with a dark gray warp and a light gray weft. (Østergård, 2004)

Museum No. D10580

Garment measurements in cm:

Size	Small	Medium	Large
Length from shoulder	123	133	143
Chest width	93	101	109
Sleeve length	50	51	52

Fabric consumption in cm:

Fabric width	80 cm	140 cm
Size Small	579	329
Size Medium	619	358
Size Large	661	387

Pieces to cut:

Front	1 piece
Back	1 piece
Centre Gusset	2 left pieces and 2 right pieces
Side Panel Front 1	1 left piece and 1 right piece
Side Panel Front 2	1 left piece and 1 right piece
Side Panel Back	1 left piece and 1 right piece
Sleeve	1 left piece and 1 right piece
Sleeve Gusset	1 left piece and 1 right piece

50 • MEDIEVAL GARMENTS RECONSTRUCTED – NORSE CLOTHING PATTERNS

Photo: Roberto Fortuna

New garment

Pattern-cutting layout:

Fabric width 80 cm

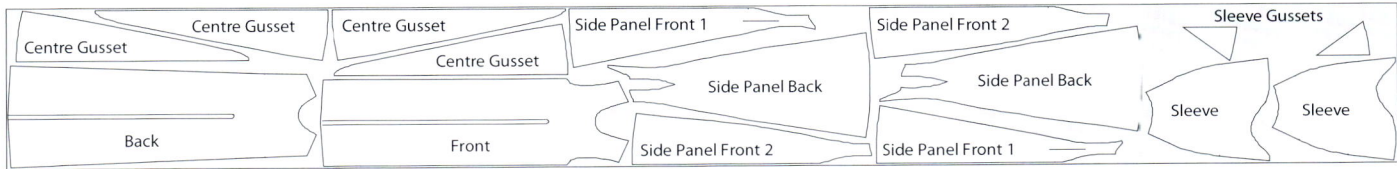

Fabric width 140 cm

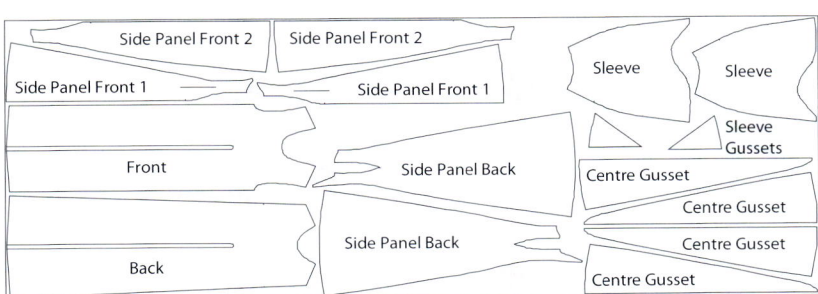

MUSEUM NO. D10580 • 51

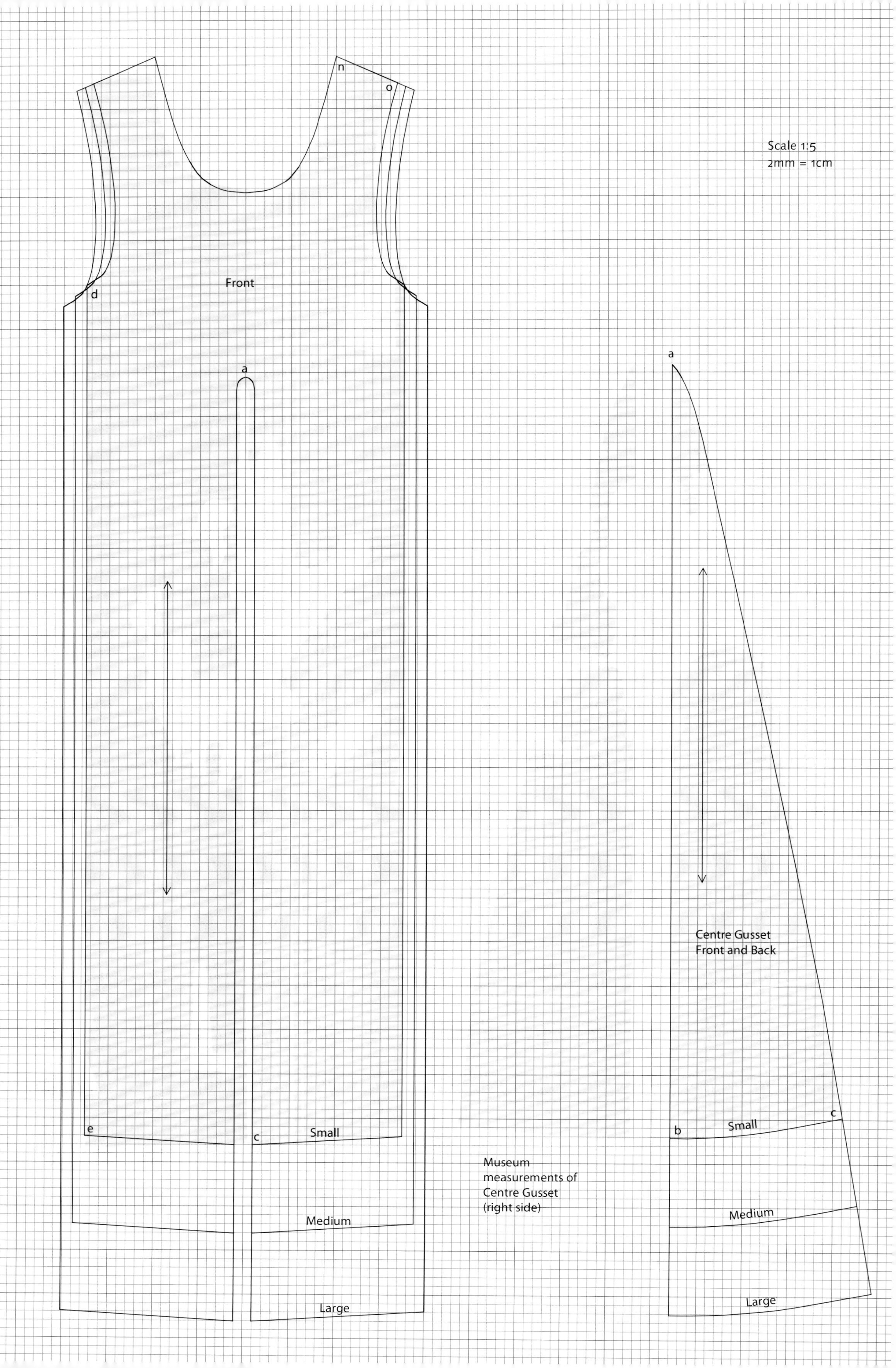

D10580

Front

Scale 1:5
2mm = 1cm

Centre Gusset
Front and Back

Museum
measurements of
Centre Gusset
(right side)

Small

Medium

Large

Scale 1:5
2mm = 1cm

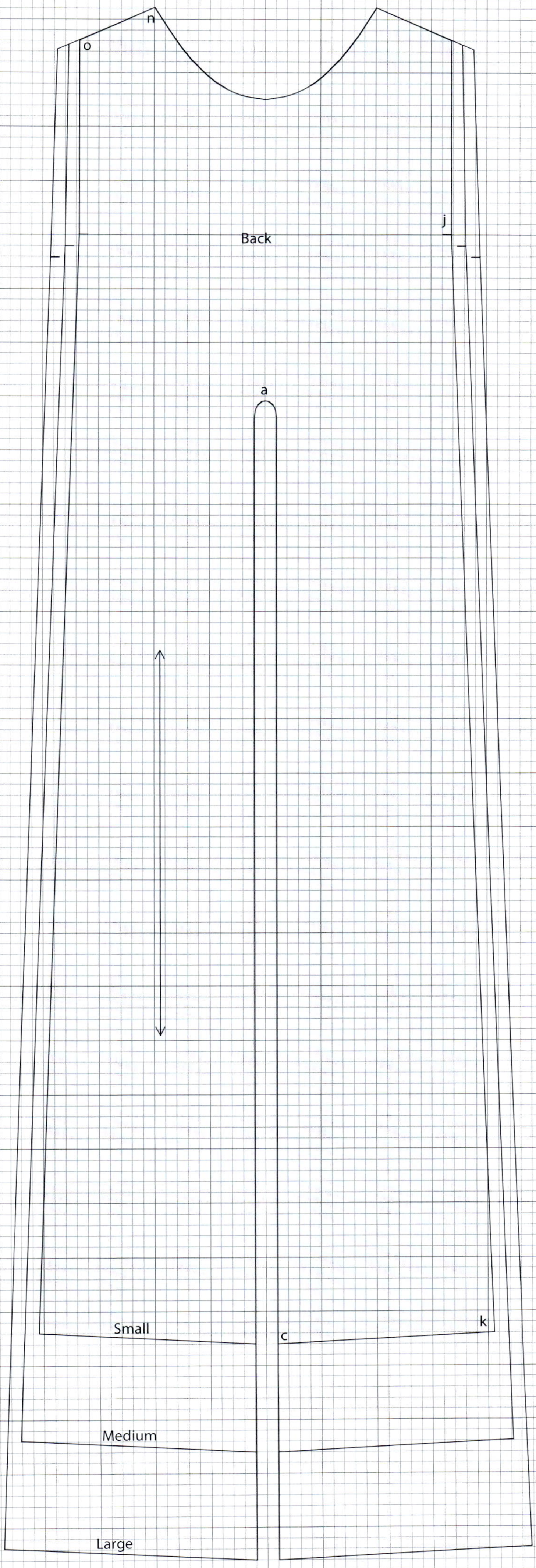

Back

n
o
j
a
Small
c
k
Medium
Large

D10580

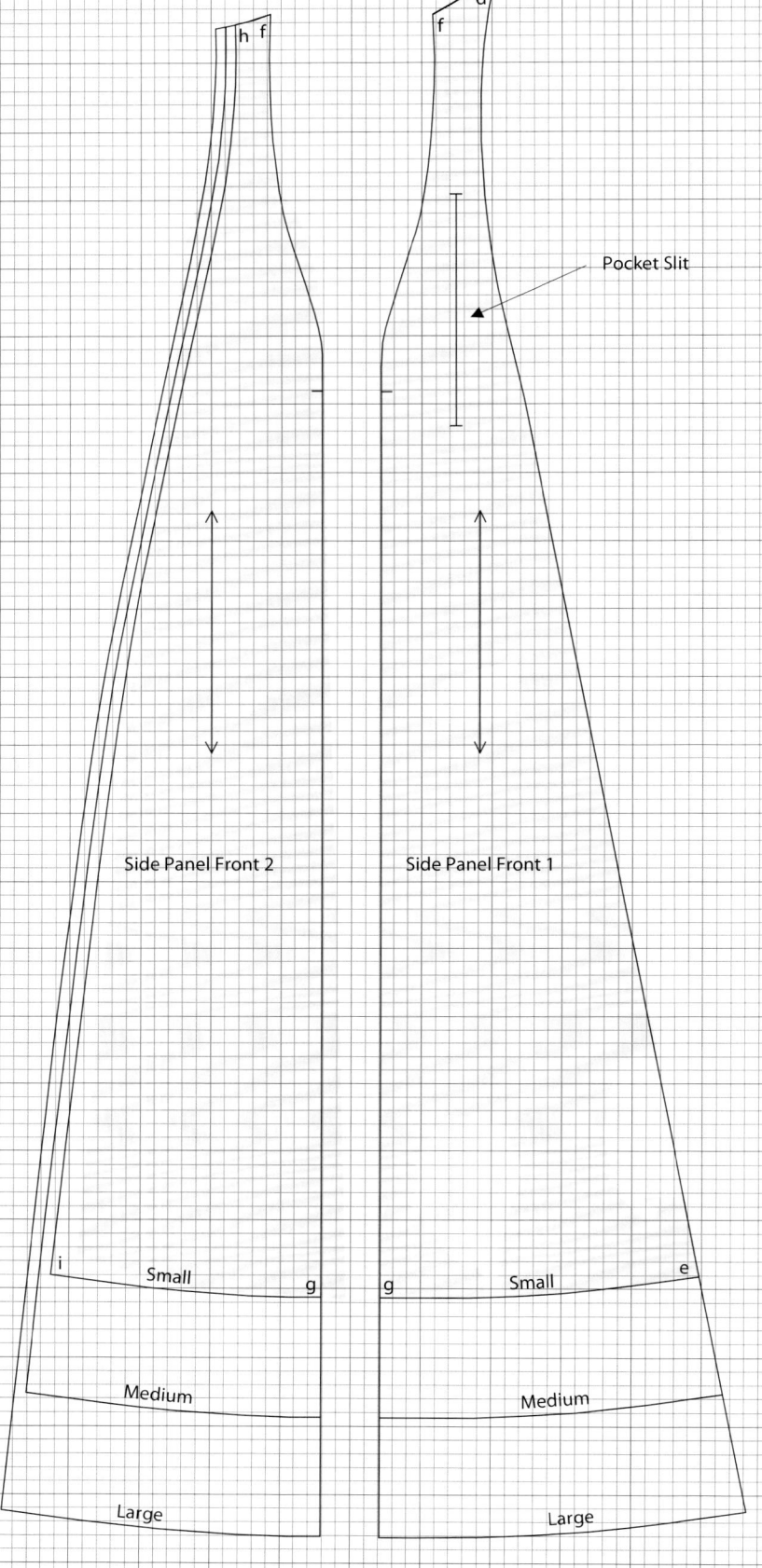

Scale 1:5
2mm = 1cm

D10580

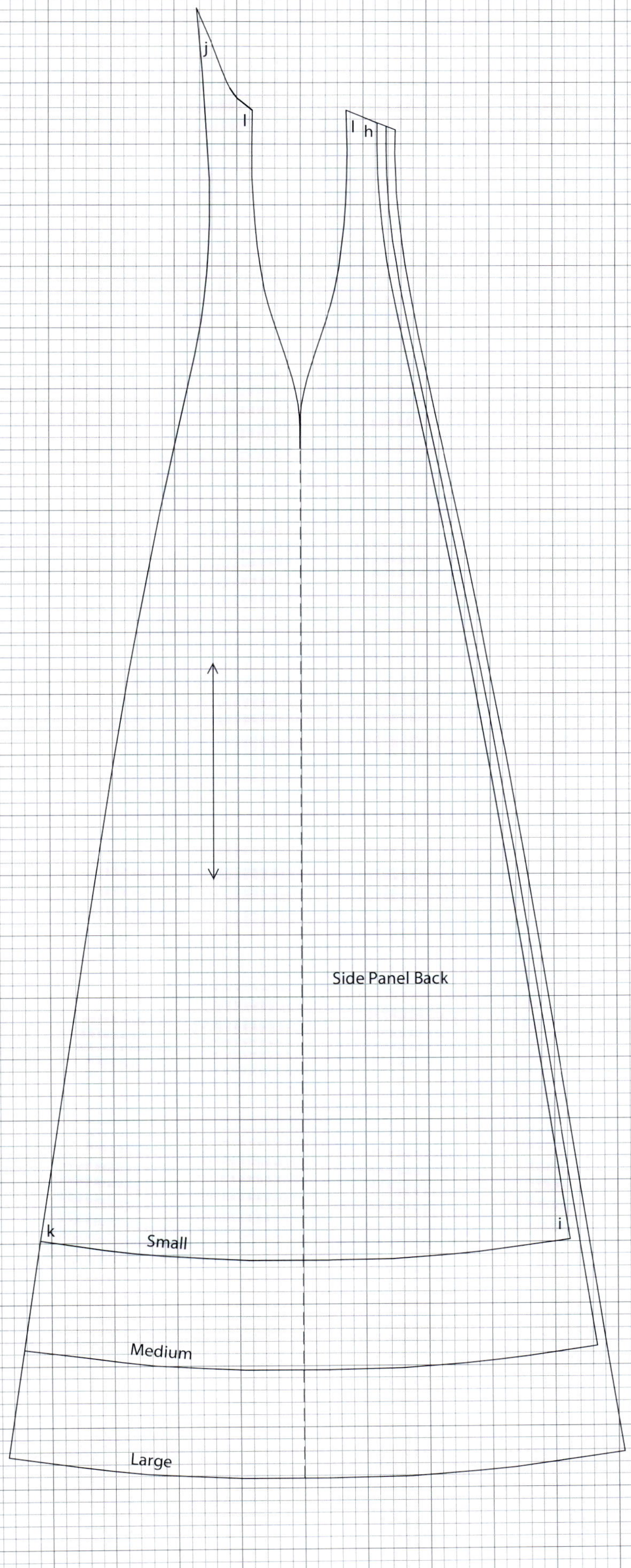

Side Panel Back

Small

Medium

Large

Museum measurements of
Side Panel Back (left side)

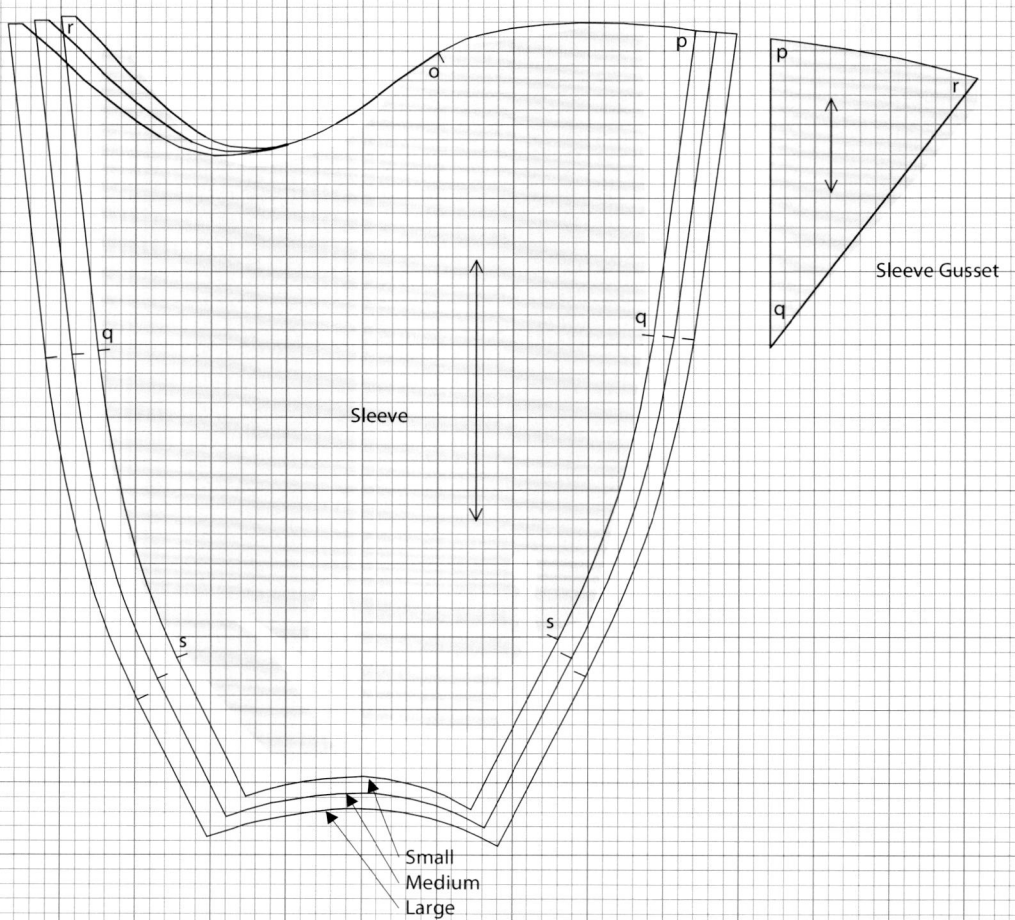

Museum measurements of
Sleeve and Sleeve Gusset (right side)

Scale 1:5
2mm = 1cm

The garment was originally sewn in Greenlandic vaðmál in 2/2 twill with a brownish, tannin-dyed warp and an undyed weft. (Østergård, 2004)

Museum No. D10581

Garment measurements in cm:

Size	Small	Medium	Large
Length from shoulder	123	133	143
Chest width	102	110	118
Sleeve length	30	31	32

Fabric consumption in cm:

Fabric width	80 cm	140 cm
Size Small	581	328
Size Medium	644	350
Size Large	719	375

Pieces to cut:

Front	1 piece
Back	1 piece
Centre Gusset Front	1 left piece and 1 right piece
Centre Gusset Back	1 left piece and 1 right piece
Side Panel Front	1 left piece and 1 right piece
Side Panel Back	1 left piece and 1 right piece
Sleeve	1 left piece and 1 right piece
Sleeve Gusset	1 left piece and 1 right piece

Photo: Roberto Fortuna

New garment

Pattern-cutting layout:

Fabric width 80 cm

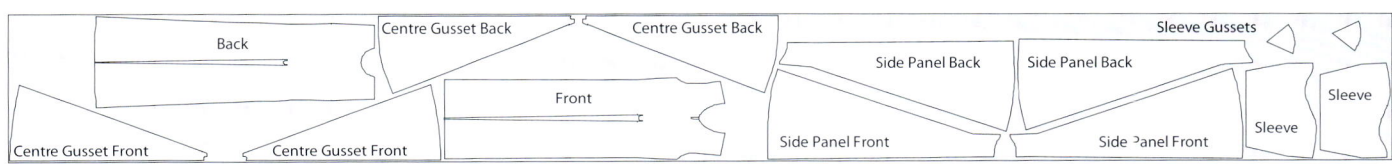

Fabric width 140 cm

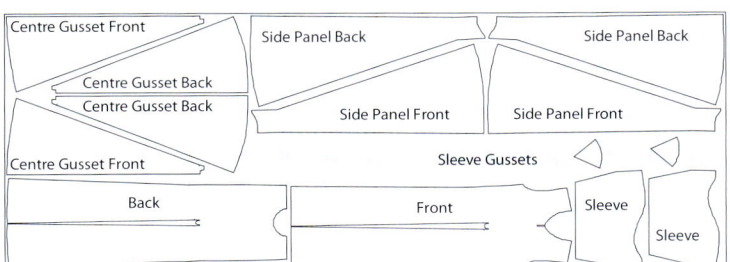

MUSEUM NO. D10581

D10581

Scale 1:5
2mm = 1cm

Museum measurements of
Centre Gusset Front (right side)

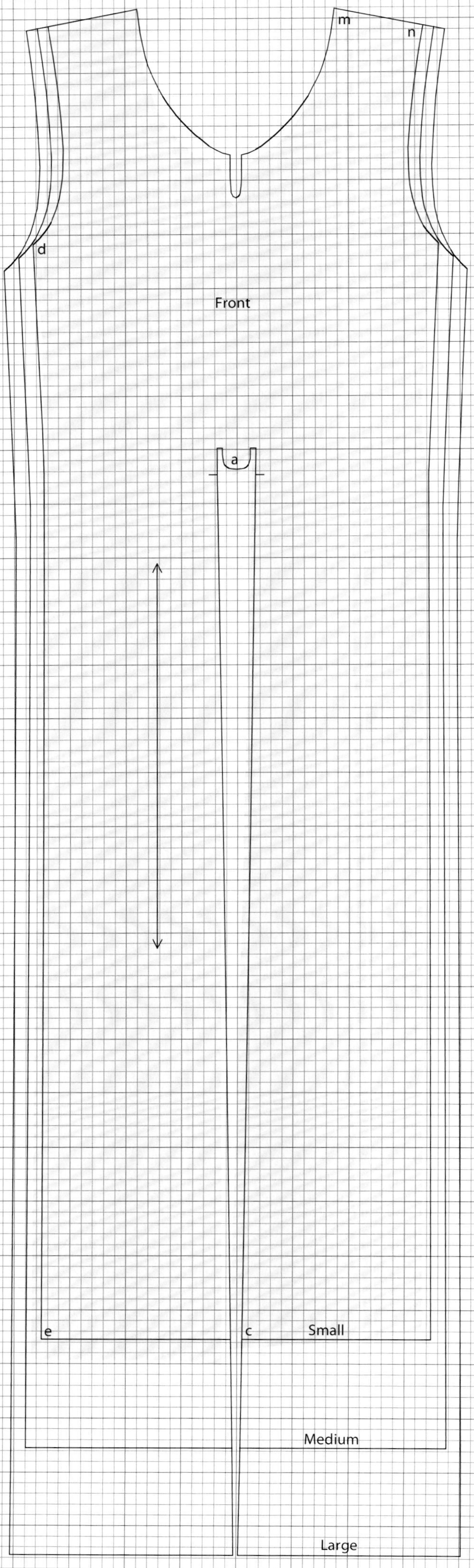

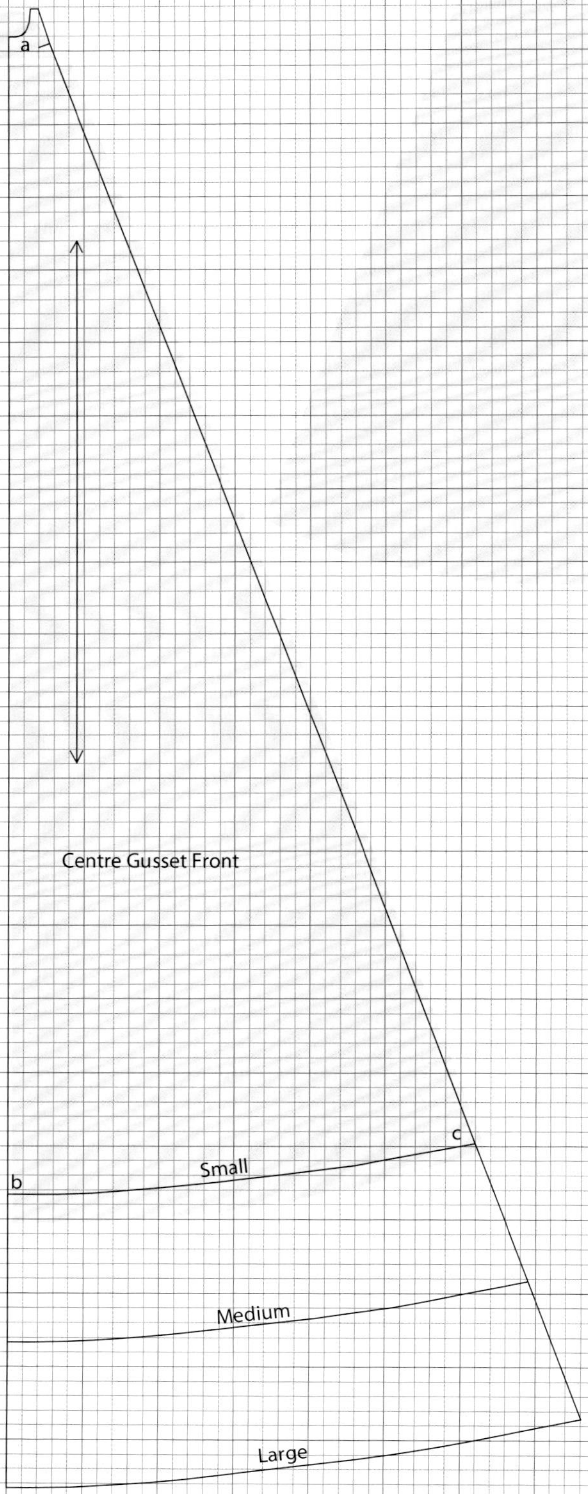

Front

Centre Gusset Front

Small
Medium
Large

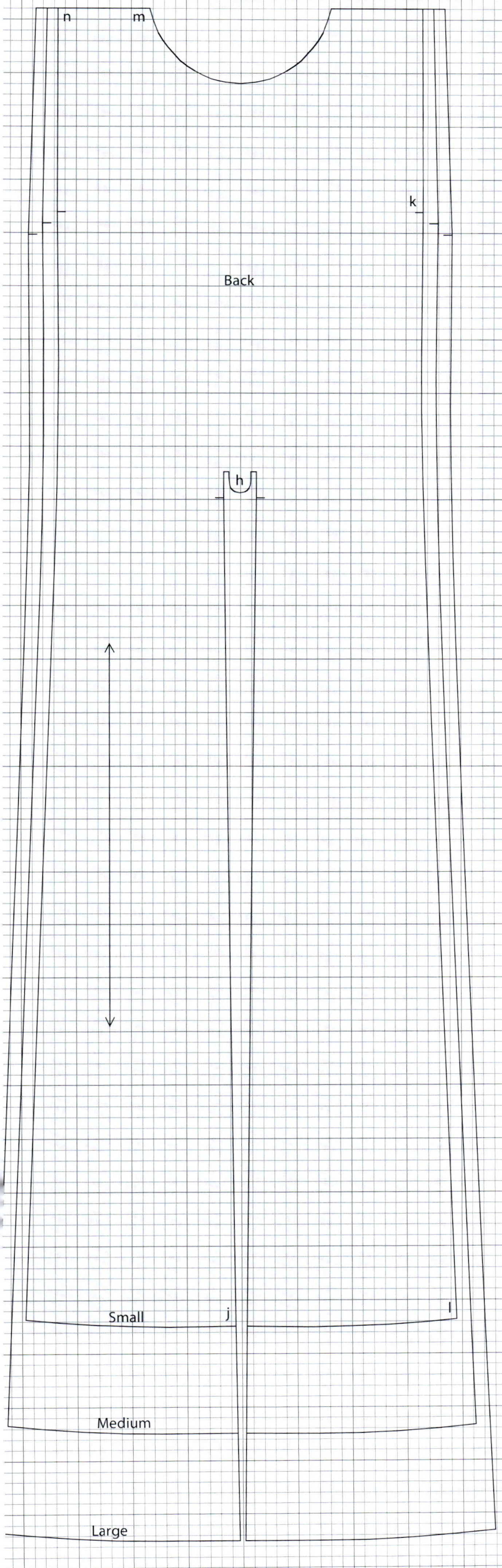

Scale 1:5
2mm = 1cm

Museum measurements of
Centre Gusset Back (right side)

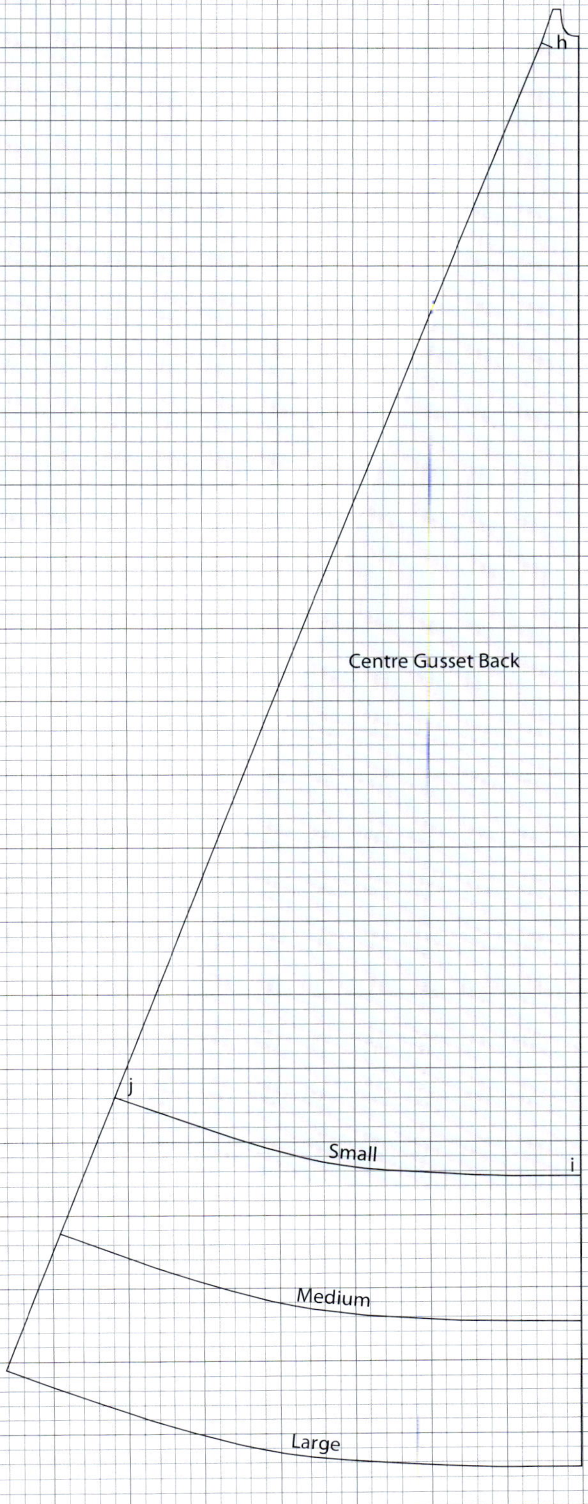

Back

Centre Gusset Back

Small

Medium

Large

Small

Medium

Large

D10581

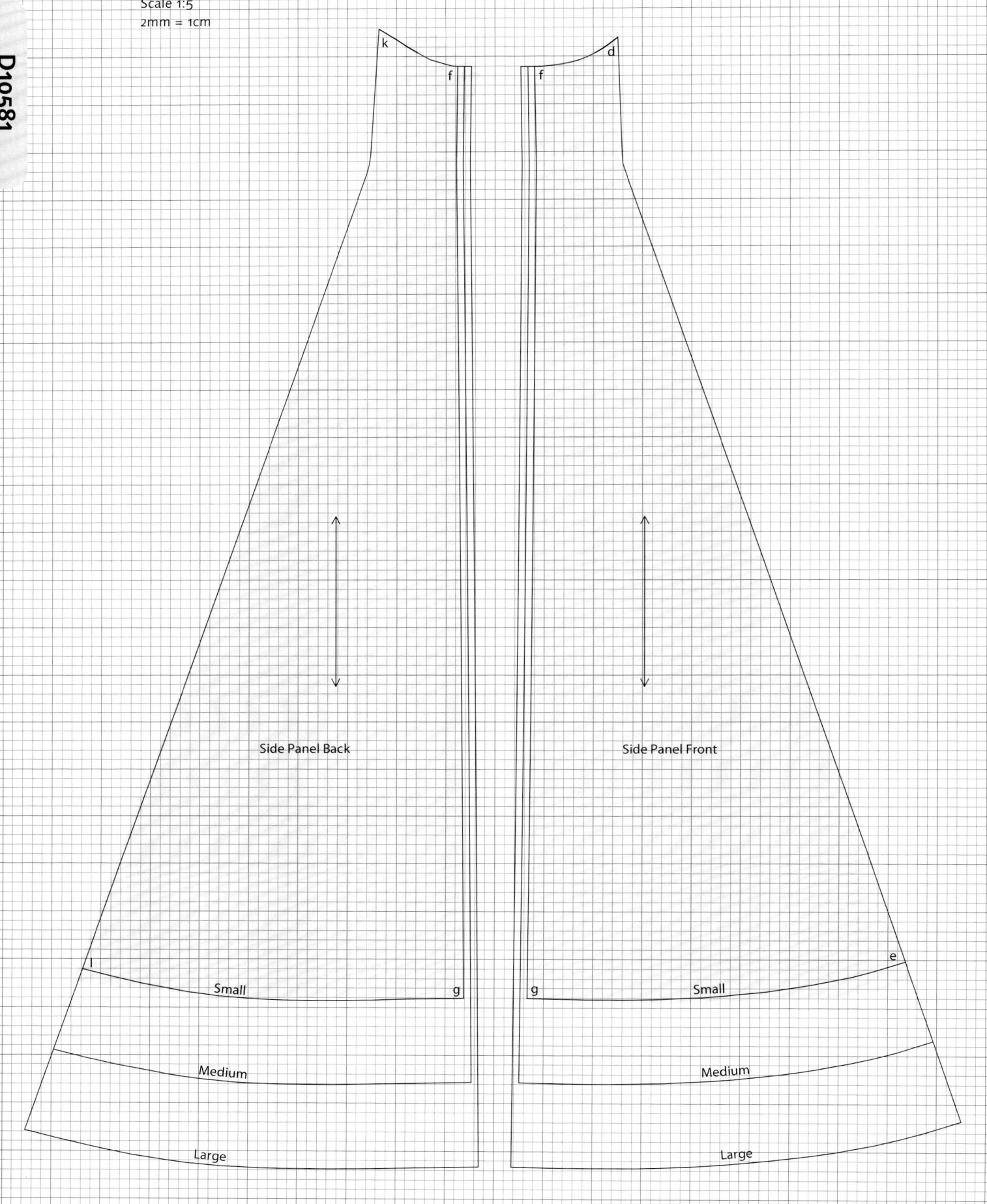

Scale 1:5
2mm = 1cm

D10581

Museum measurements of
Side Panel Front and Back (left side)

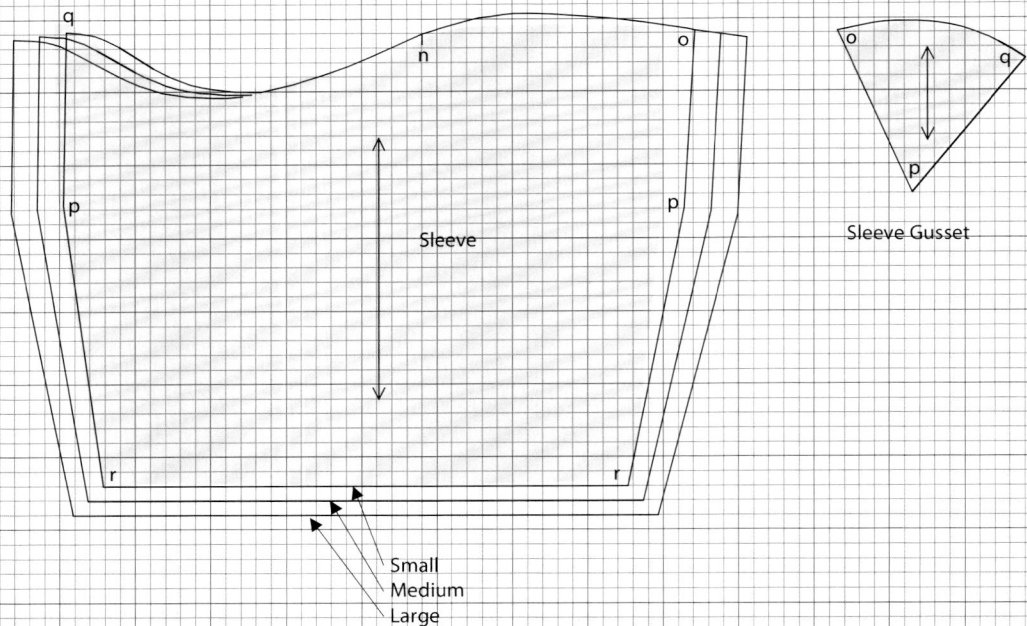

Museum measurements of Sleeve (right side)

Templates for D10581

Templates showing the seam allowance at the rounded points on the front, back and gussets

——————— Cut line
- - - - - - - Sew line

Front and Back

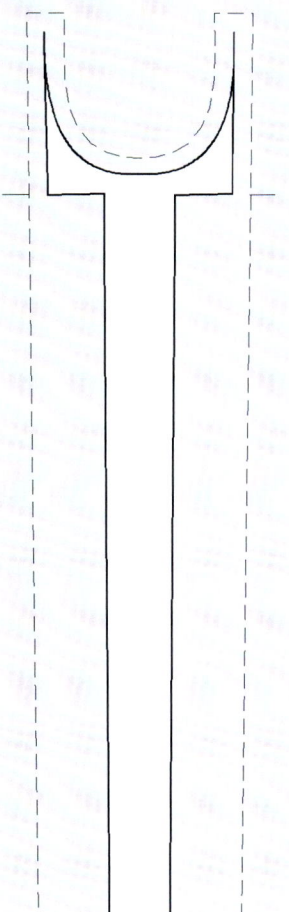

Gusset Centre Front and Back

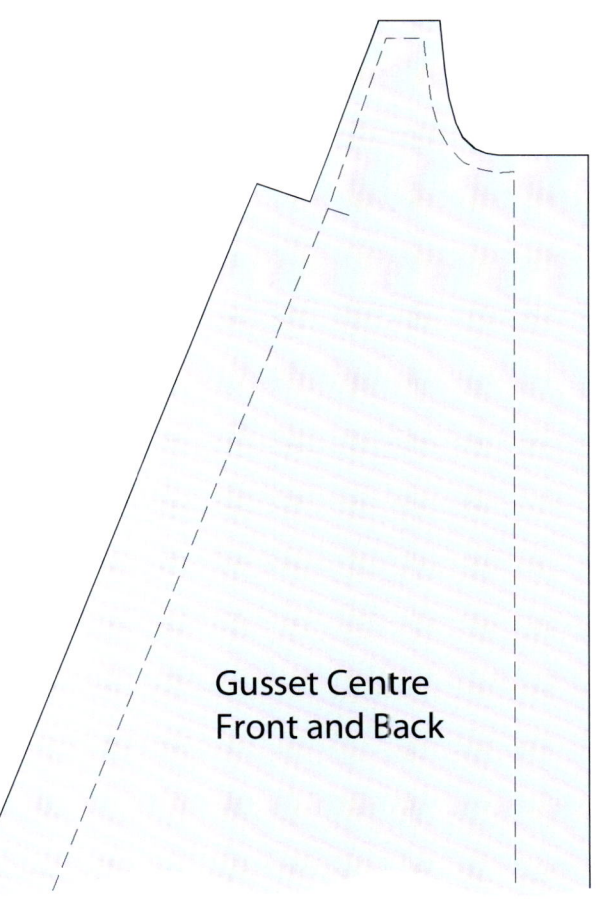

The garment was originally sewn in dark brown Greenlandic vaðmál in 2/2 twill. (Østergård, 2004)

Museum No. D10584

Garment measurements in cm:

Size	Small	Medium	Large
Length from shoulder	123	131	139
Chest width	79	87	95
Sleeve length	54	55	56

Fabric consumption in cm:

Fabric width	80 cm	140 cm
Size Small	535	297
Size Medium	584	326
Size Large	637	346

Pieces to cut:

Front	1 piece
Back	1 piece
Centre Gusset Front	1 left piece and 1 right piece
Centre Gusset Back	1 left piece and 1 right piece
Side Panel Front	1 left piece and 1 right piece
Side Panel Back	1 left piece and 1 right piece
Sleeve	1 left piece and 1 right piece
Sleeve Gusset	1 left piece and 1 right piece

Photo: Roberto Fortuna

New garment

Pattern-cutting layout:

Fabric width 80 cm

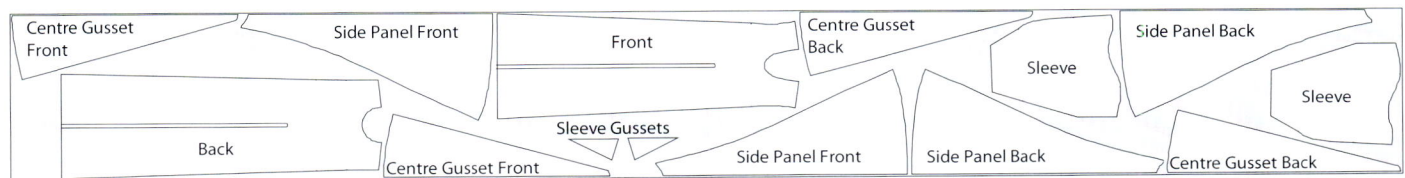

Fabric width 140 cm

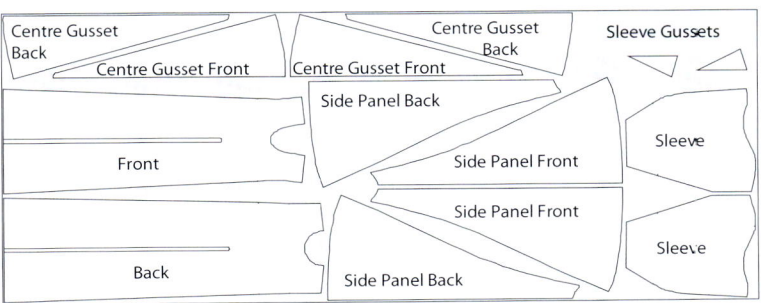

MUSEUM NO. D10584 • 67

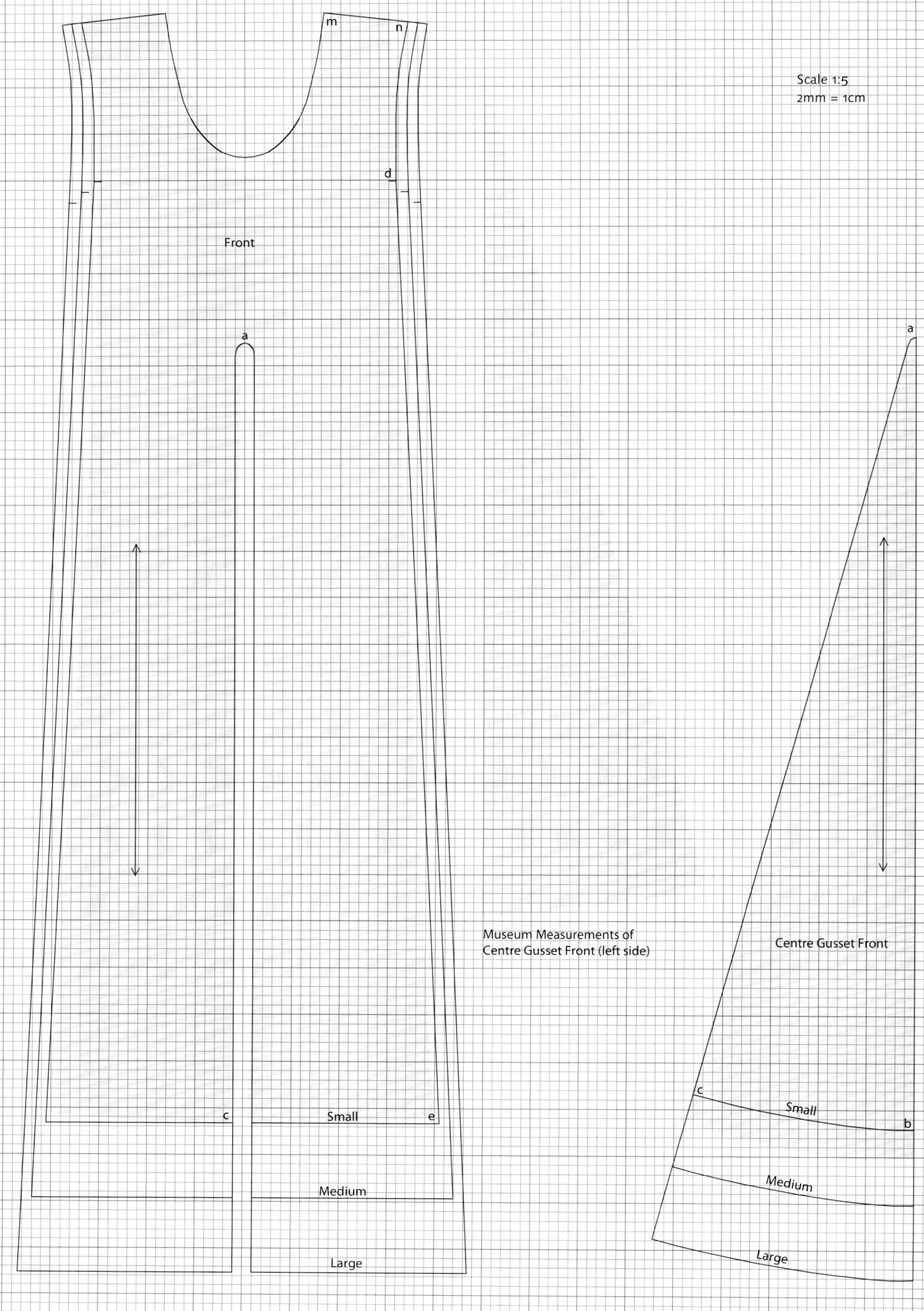

D10584

Scale 1:5
2mm = 1cm

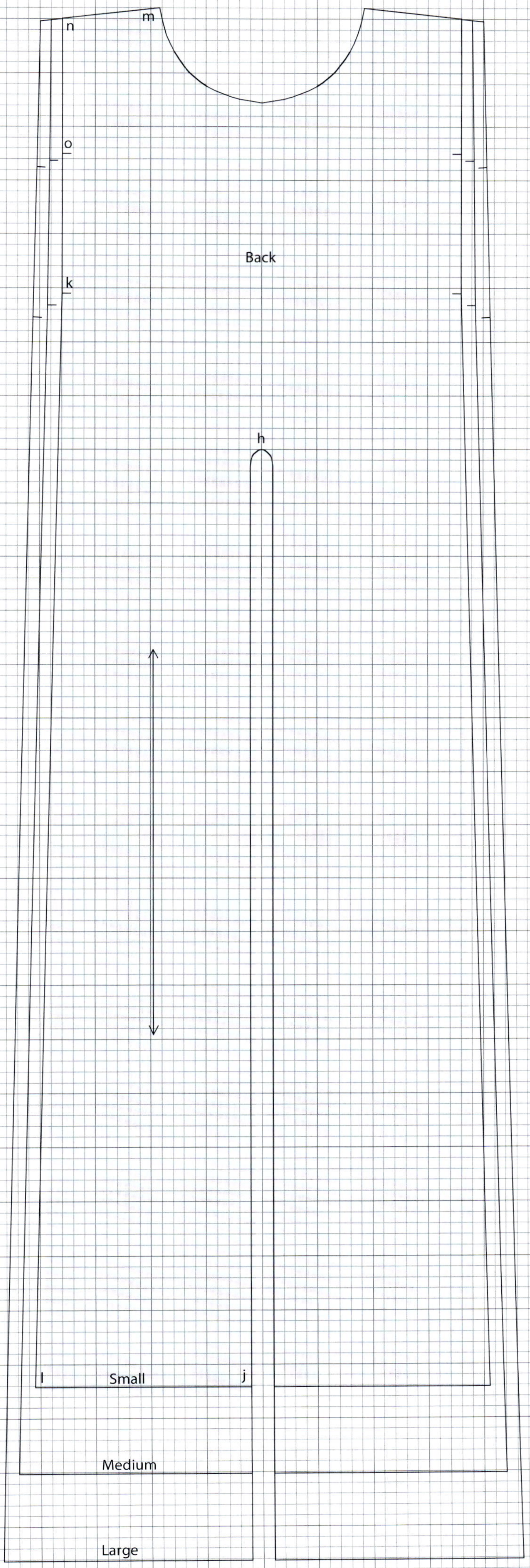
m
n
o
k
Back
h
l
Small
j
Medium
Large

Museum measurements of
Centre Gusset Back (right side)

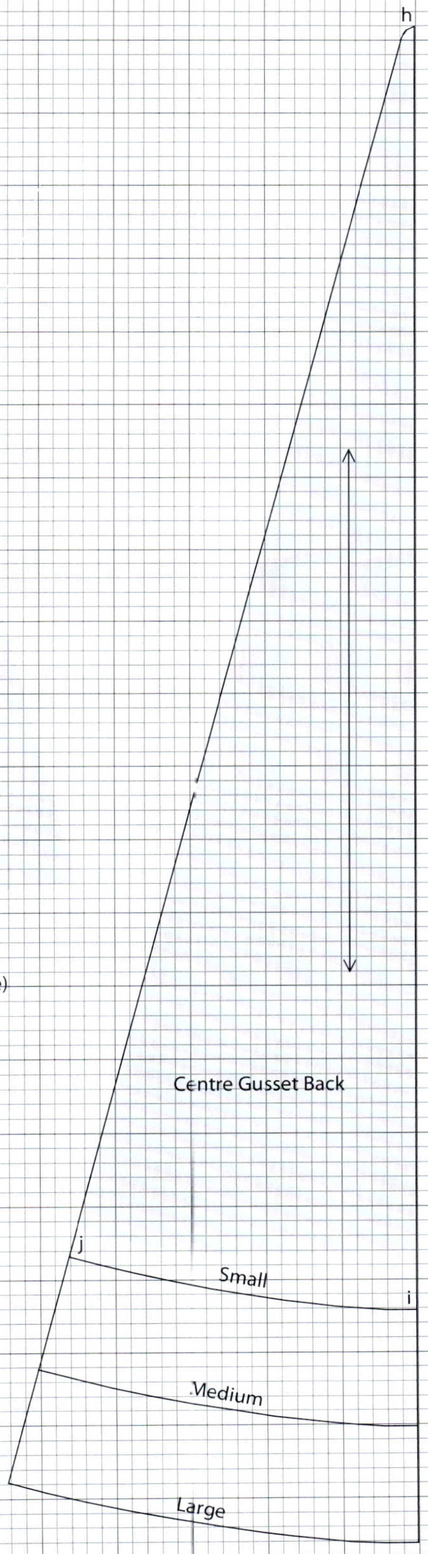
h
Centre Gusset Back
j
Small
i
Medium
Large

Scale 1:5
2mm = 1cm

D10584

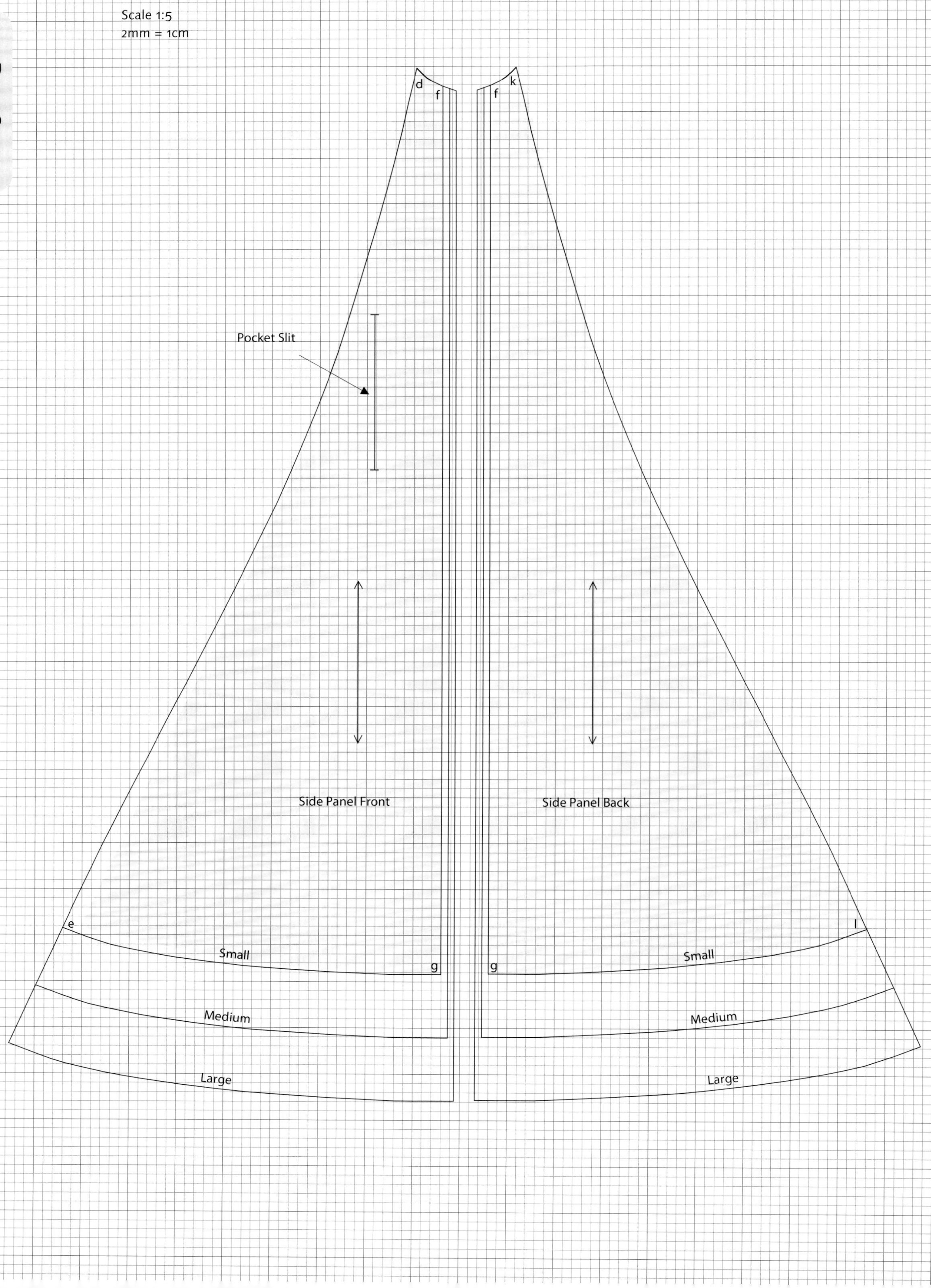

Pocket Slit

Side Panel Front

Side Panel Back

Small
Medium
Large

Small
Medium
Large

Scale 1:5
2mm = 1cm

D10584

Museum measurements of
Side Panel (right side)

Scale 1:5
2mm = 1cm

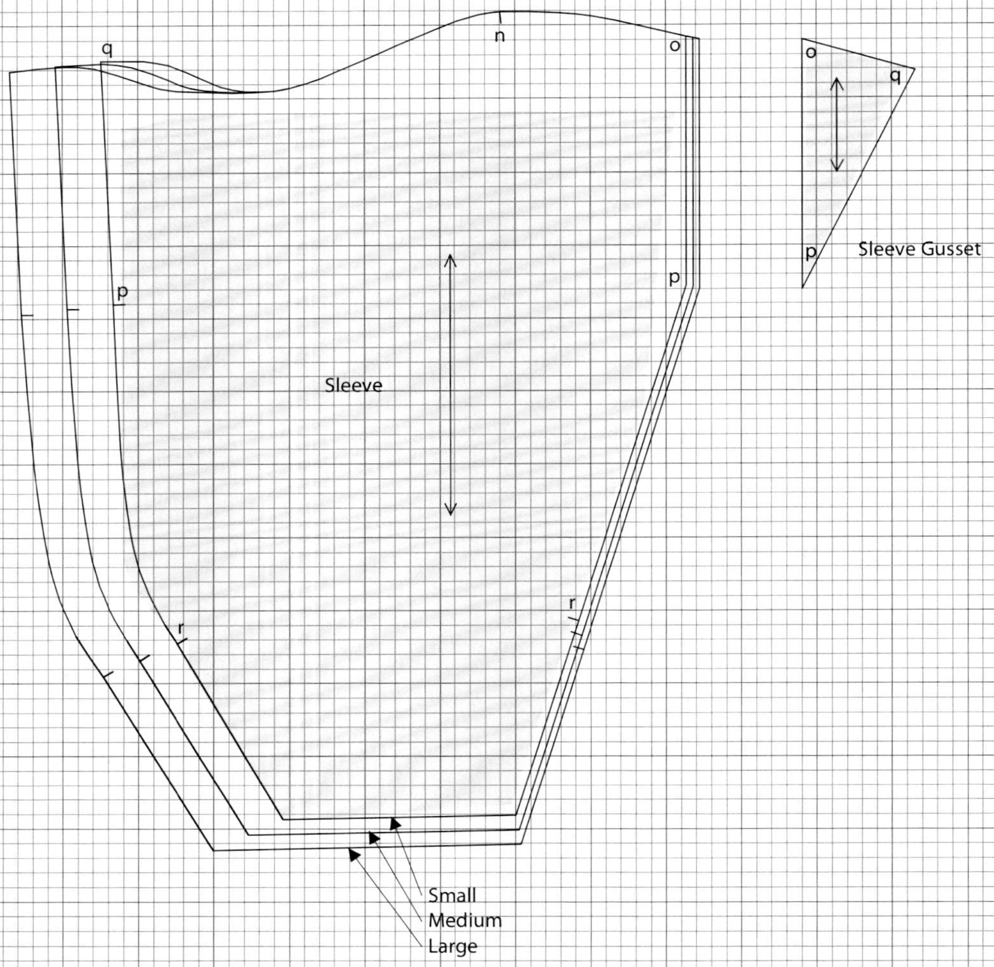

q   n   o

o   q

Sleeve Gusset

p

p   p

Sleeve

r   r

Small
Medium
Large

Museum measurements of
Sleeve and Sleeve Gusset (right side)

The garment was originally sewn in Greenlandic vaðmál in 2/2 twill with a gray warp and a light gray weft.
(Østergård, 2004)

Museum No. D10585.1

Garment measurements in cm:

Size	Small	Medium	Large
Length from shoulder	114	124	134
Chest width	98	106	114
Sleeve length	58	59	60

Fabric consumption in cm:

Fabric width	80 cm	140 cm
Size Small	443	256
Size Medium	483	276
Size Large	520	298

Pieces to cut:

Front	1 piece
Back	1 piece
Centre Gusset	2 left pieces and 2 right pieces
Side Panel Front	1 left piece and 1 right piece
Side Panel Back	1 left piece and 1 right piece
Sleeve	1 left piece and 1 right piece
Sleeve Gusset	1 left piece and 1 right piece

74 • MEDIEVAL GARMENTS RECONSTRUCTED – NORSE CLOTHING PATTERNS

New garment

Pattern-cutting layout:

Fabric width 80 cm

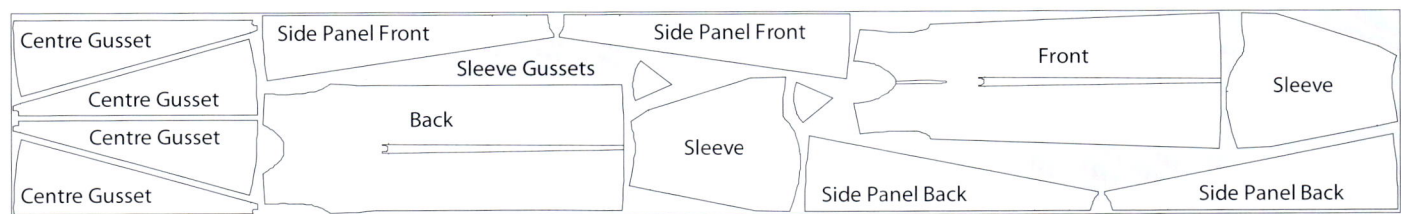

Fabric width 140 cm

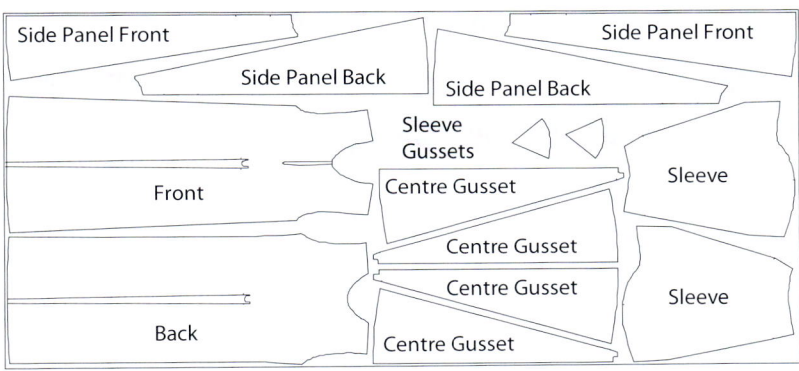

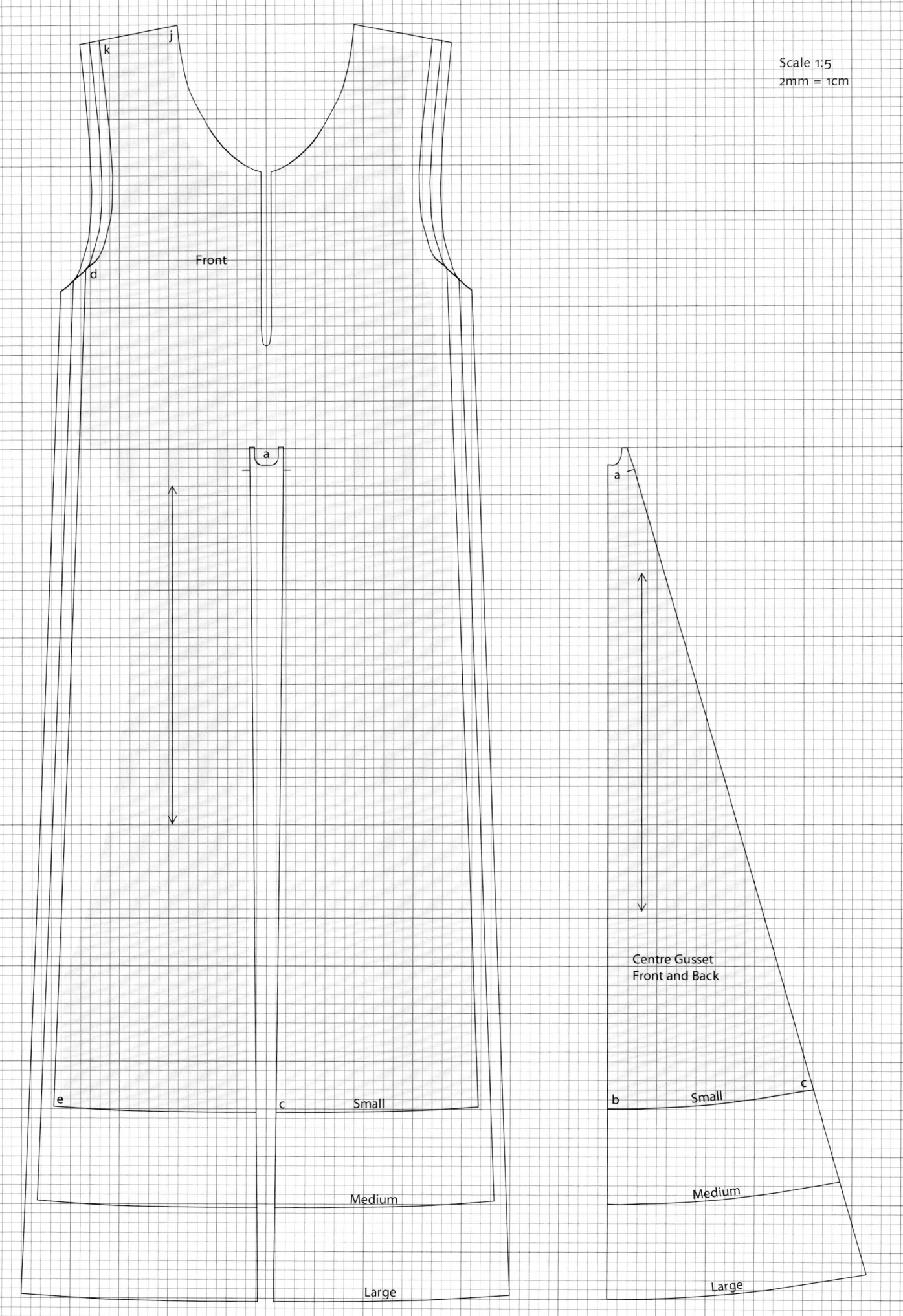

Scale 1:5
2mm = 1cm

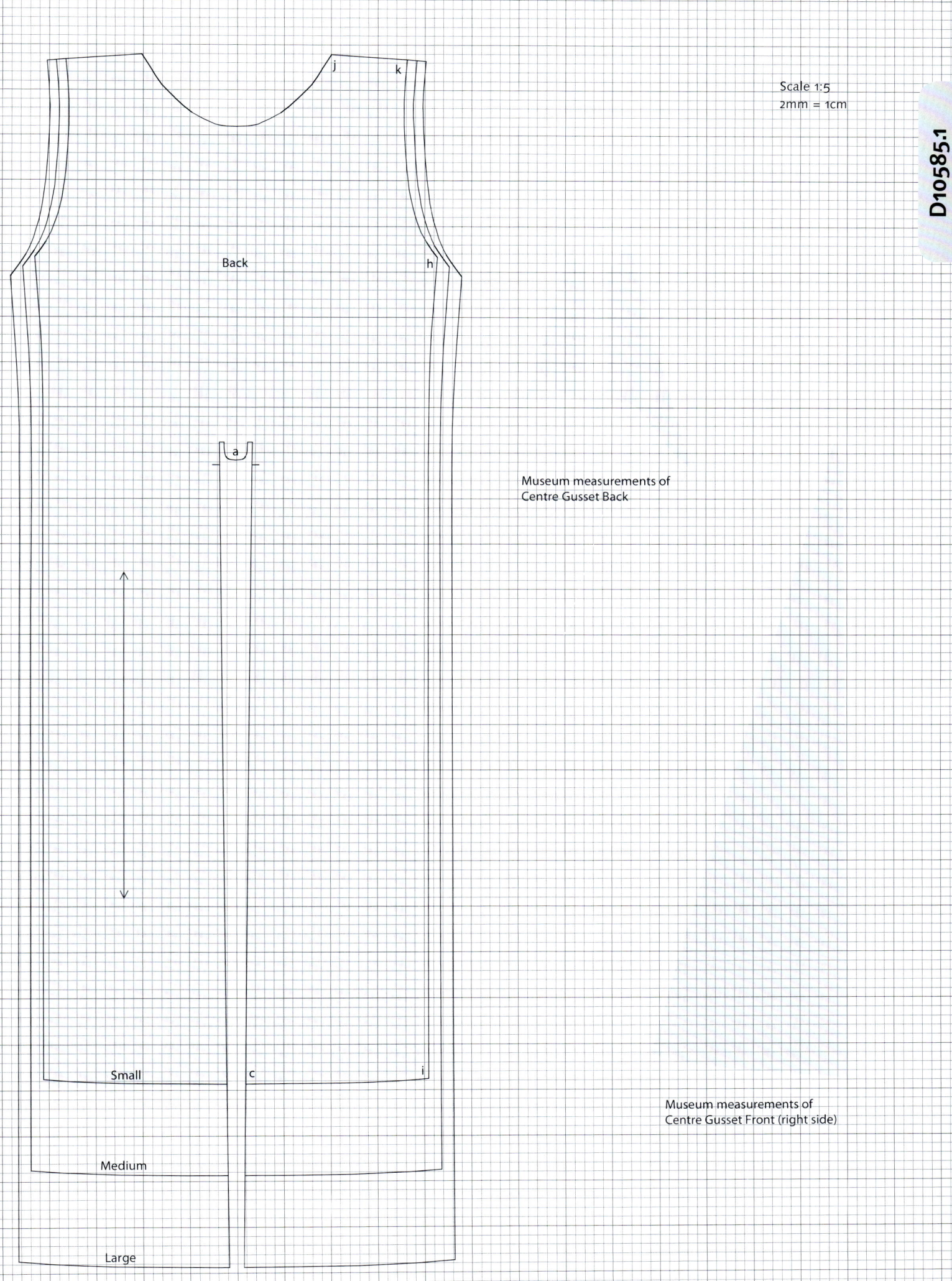

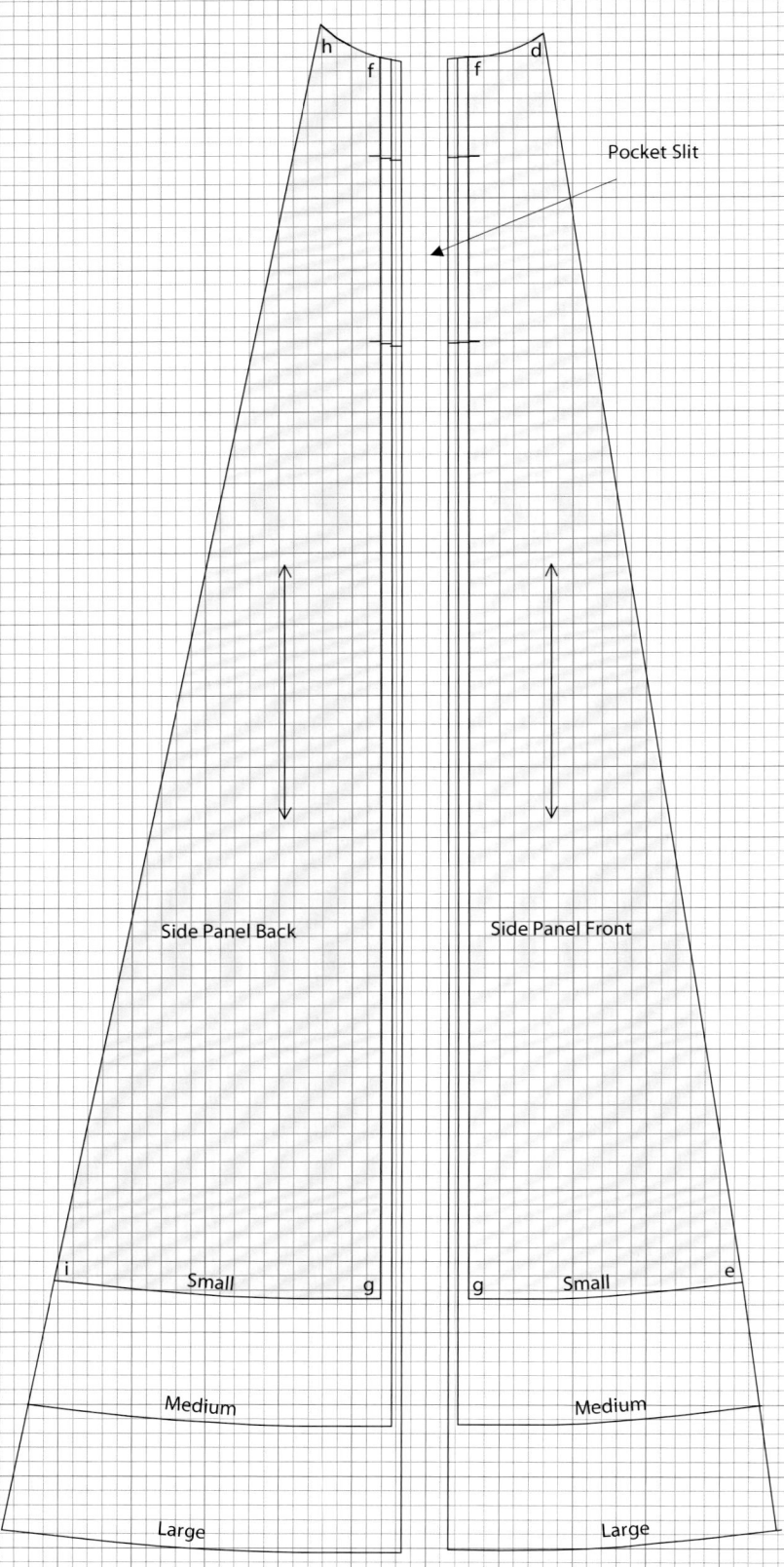

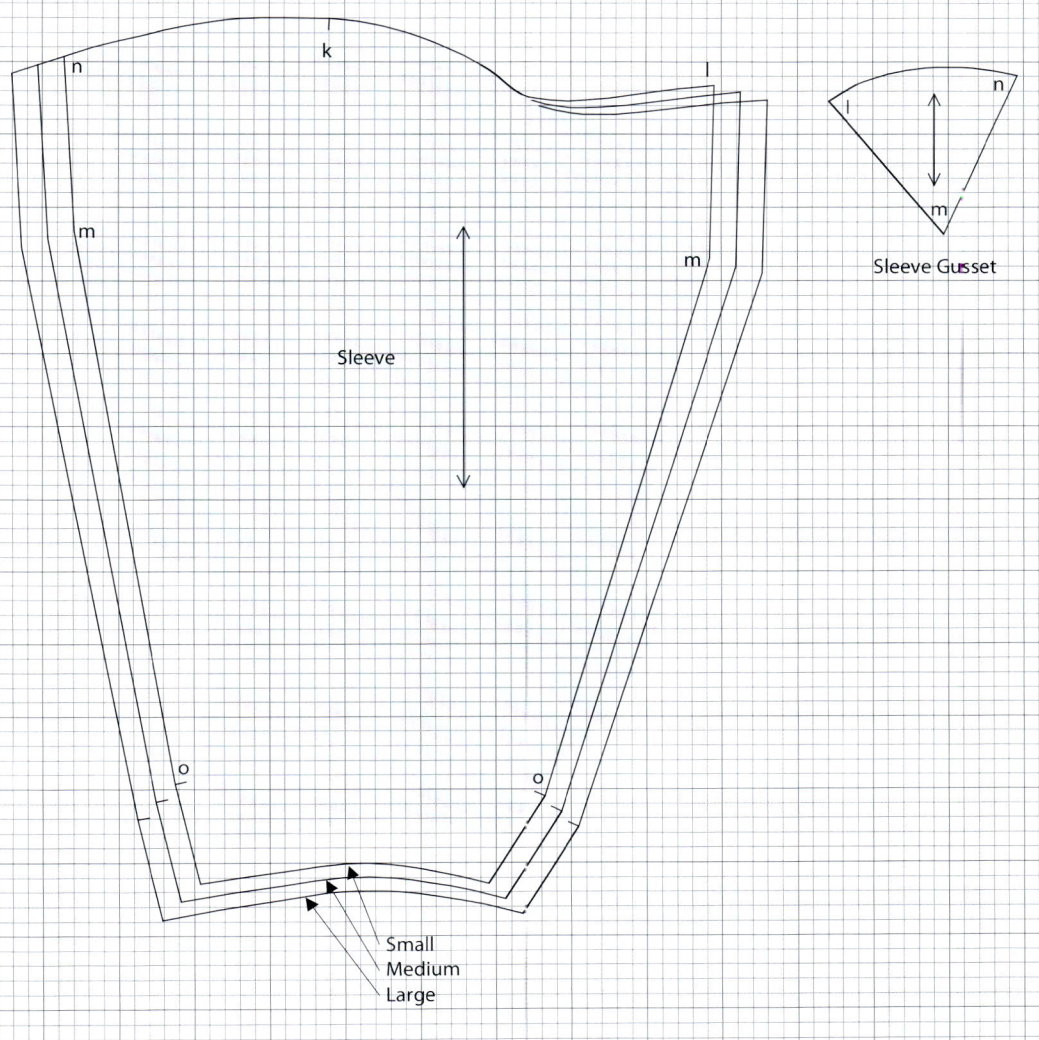

Templates for D10585.1

Templates showing the seam allowance at the rounded points on the front, back and gussets

——————— Cut line
- - - - - - - Sew line

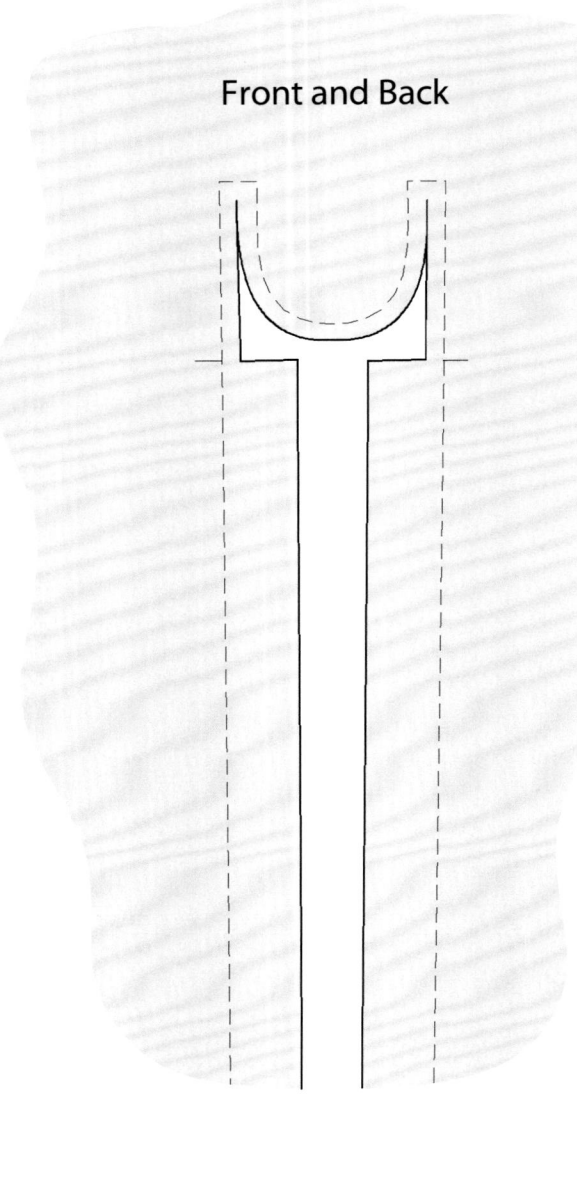

Front and Back

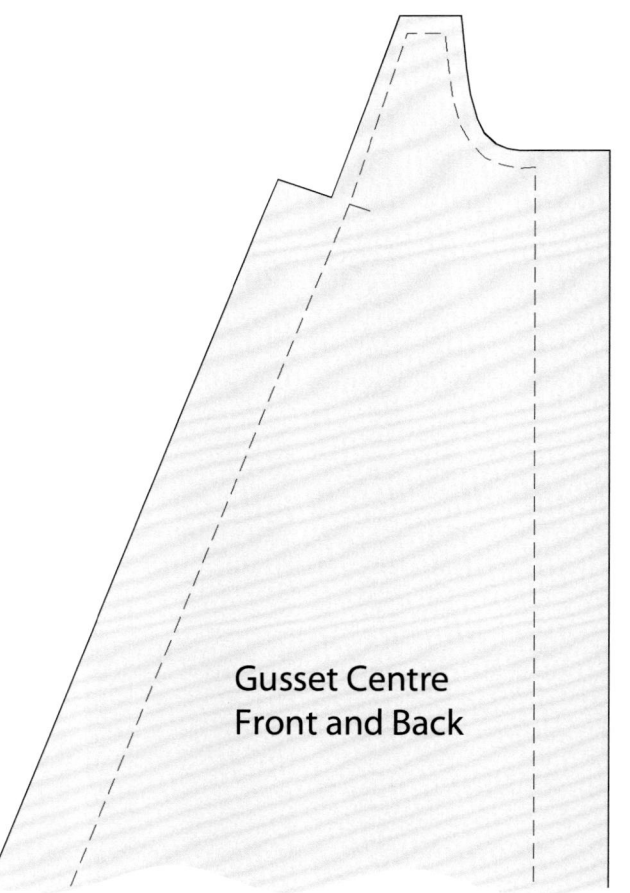

Gusset Centre Front and Back

The garment was originally sewn in white Greenlandic vaðmál in 2/2 twill. (Østergård, 2004)

Museum No. D10586

Garment measurements in cm:

Size	110	122	134	146
Length from shoulder	88	99	110	121
Chest width	81	87	93	99
Sleeve length	42	46	50	54

Fabric consumption in cm:

Fabric width	80 cm	140 cm
Size 110	344	197
Size 122	395	222
Size 134	452	251
Size 146	523	274

Pieces to cut:

Front	1 piece
Back	1 piece
Centre Gusset	2 left pieces and 2 right pieces
Side Panel Front	1 left piece and 1 right piece
Side Panel Back	1 left piece and 1 right piece
Sleeve	1 left piece and 1 right piece
Sleeve Gusset	1 left piece and 1 right piece

82 • MEDIEVAL GARMENTS RECONSTRUCTED — NORSE CLOTHING PATTERNS

Photo: Roberto Fortuna

New garment

Pattern-cutting layout:

Fabric width 80 cm

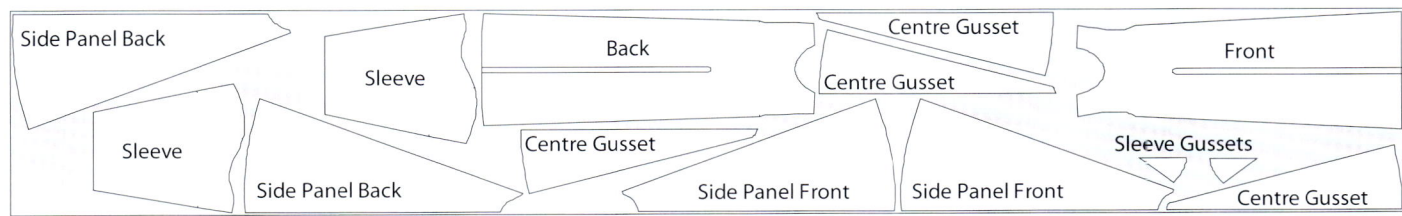

Fabric width 140 cm

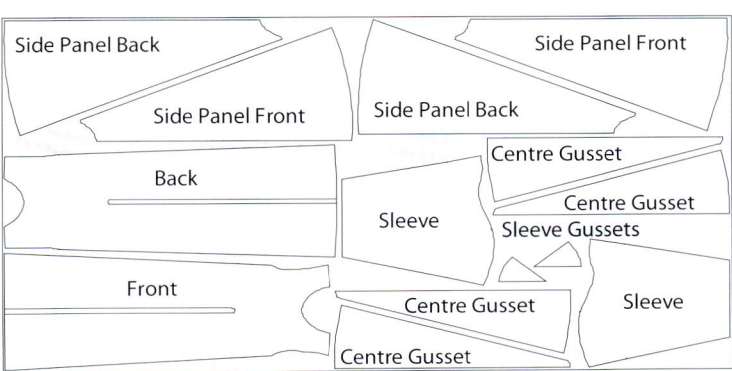

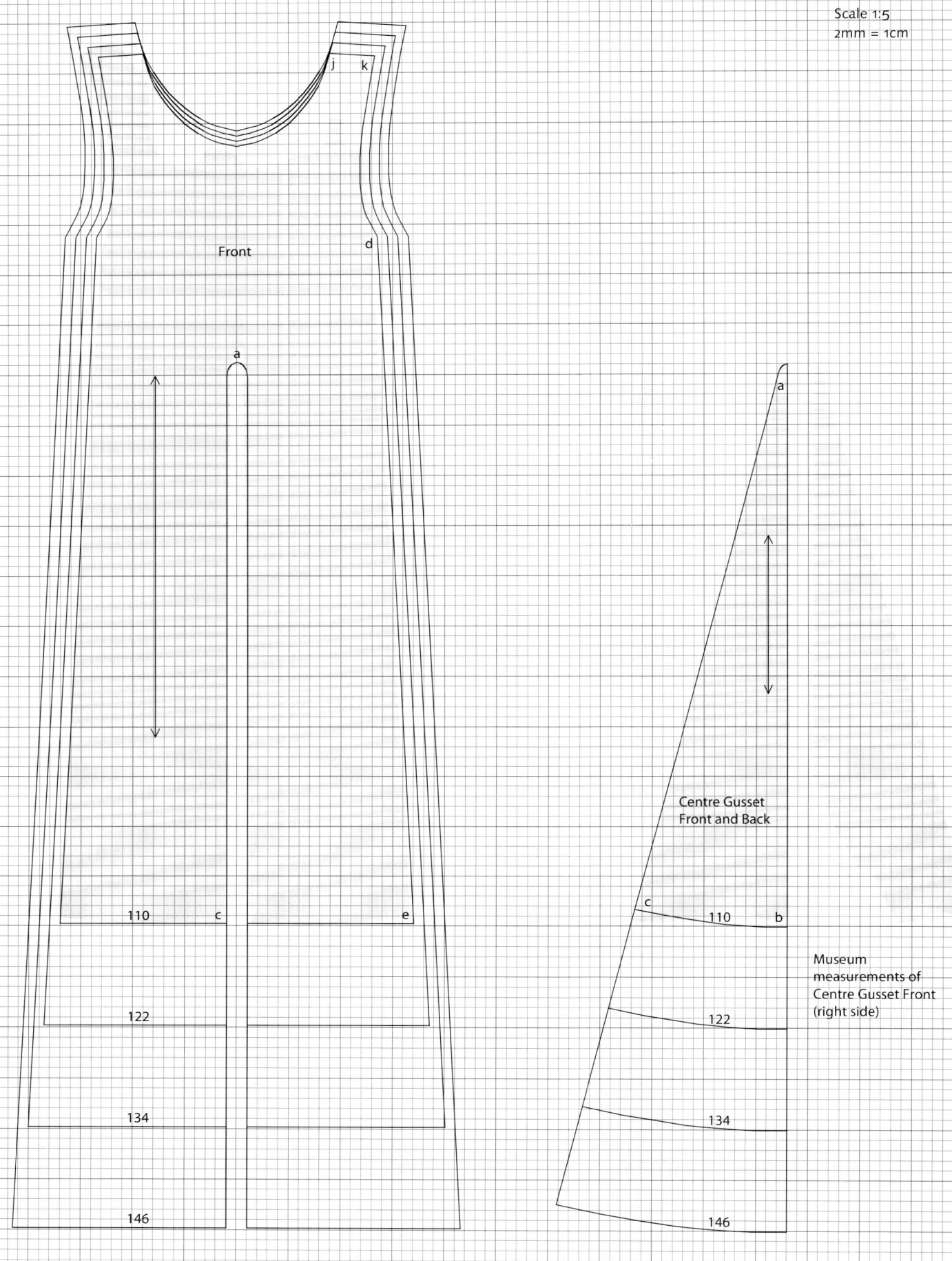

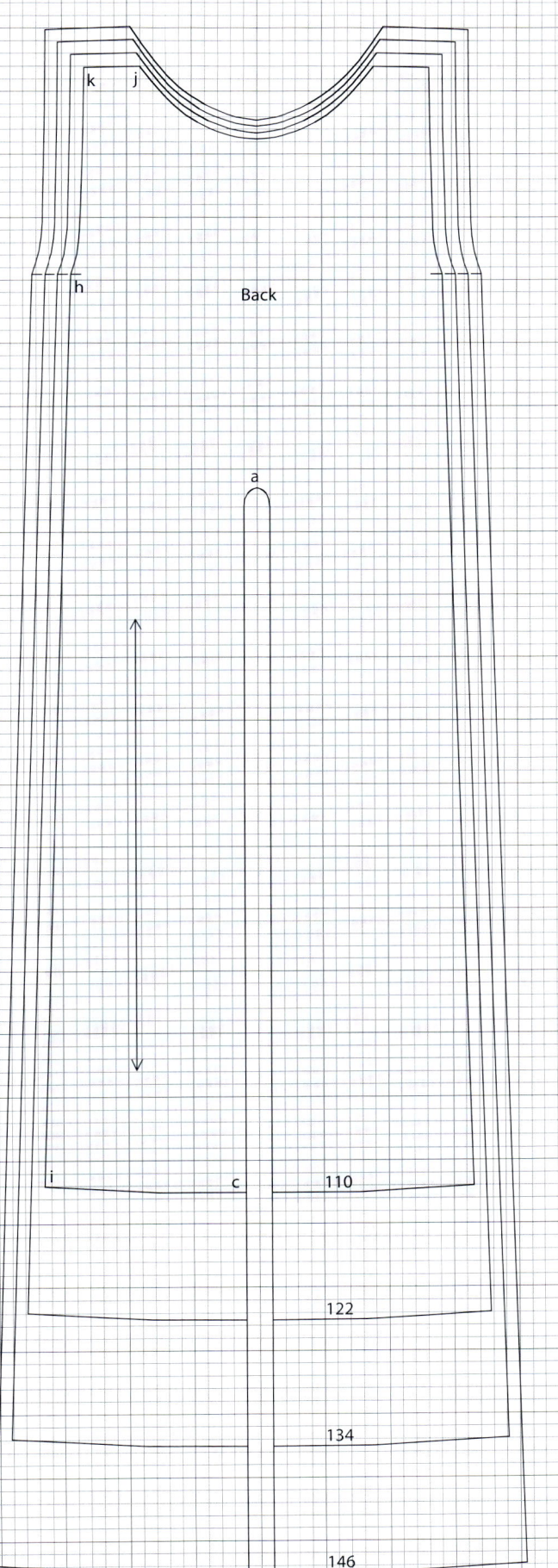

Scale 1:5
2mm = 1cm

D10586

Back

Museum measurements of Centre Gusset Back

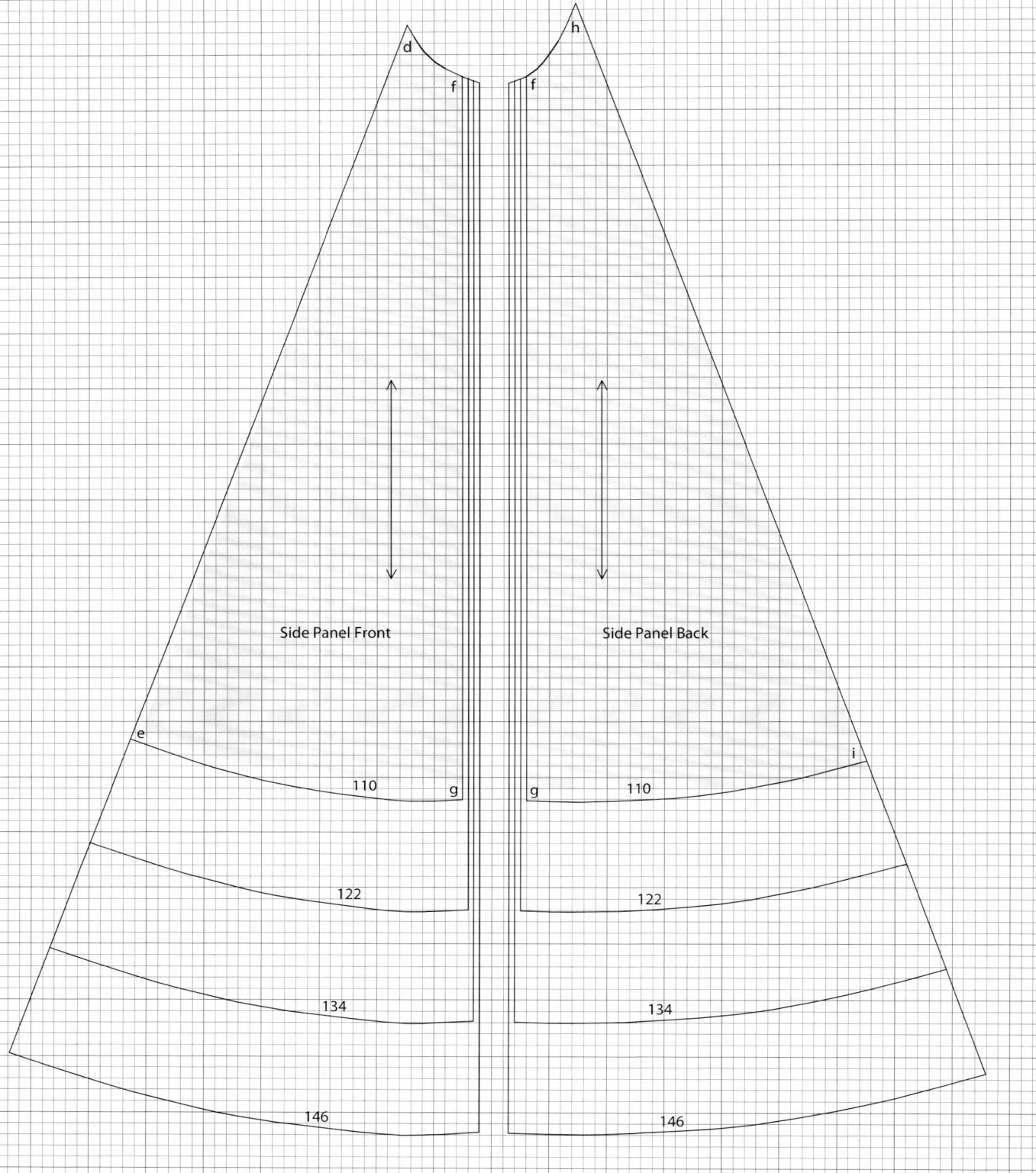

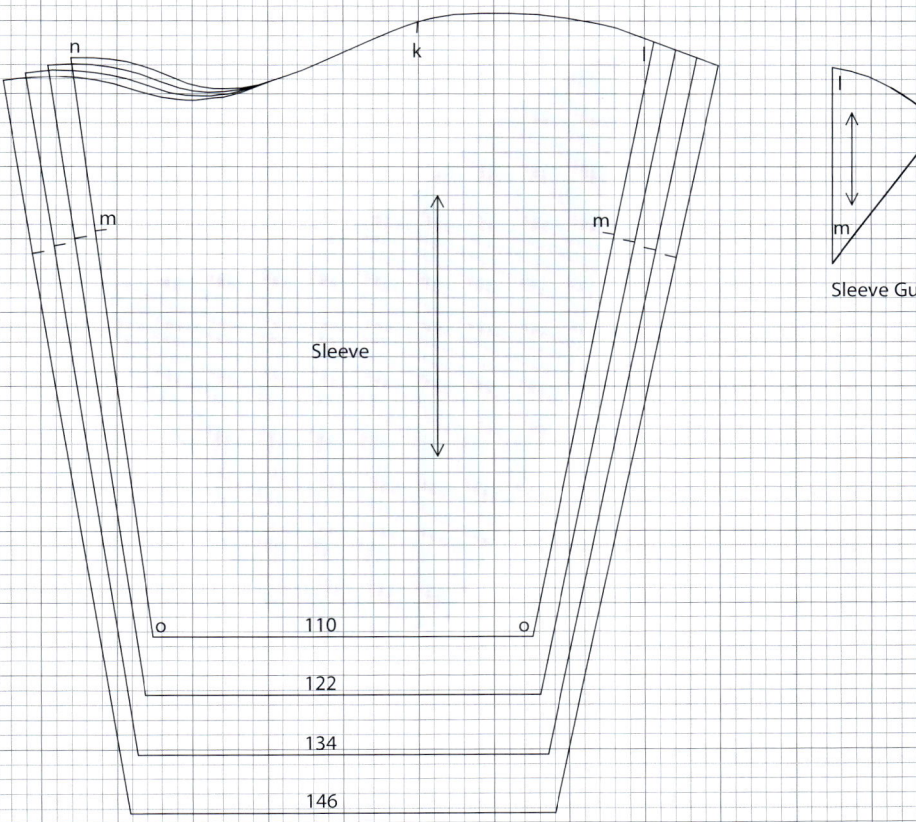

The garment was originally sewn in dark brown Greenlandic vaðmál in 2/2 twill. (Østergård, 2004)

Museum No. D10587

Garment measurements in cm:

Size	Small	Medium	Large
Length from shoulder	119	129	139
Chest width	115	123	131
Sleeve length	31	32	33

Fabric consumption in cm:

Fabric width	80 cm	140 cm
Size Small	534	280
Size Medium	575	308
Size Large	616	337

Pieces to cut:

Front	1 piece
Back	1 piece
Centre Gusset Front	1 left piece and 1 right piece
Centre Gusset Back	1 left piece and 1 right piece
Side Panel Front	1 left piece and 1 right piece
Side Panel Back	1 left piece and 1 right piece
Sleeve	1 left piece and 1 right piece
Sleeve Gusset 1	1 left piece and 1 right piece
Sleeve Gusset 2	1 left piece and 1 right piece

New garment

Pattern-cutting layout:

Fabric width   80 cm

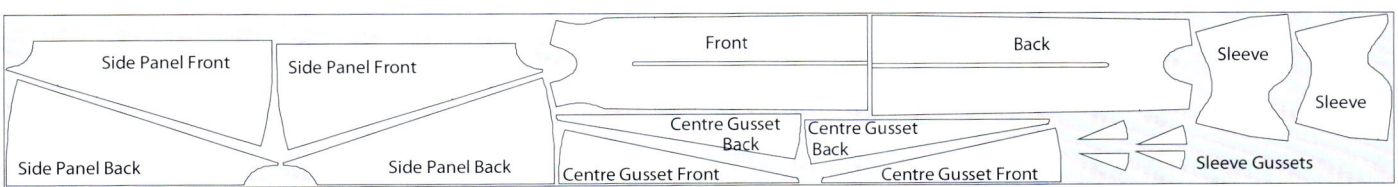

Fabric width   140 cm

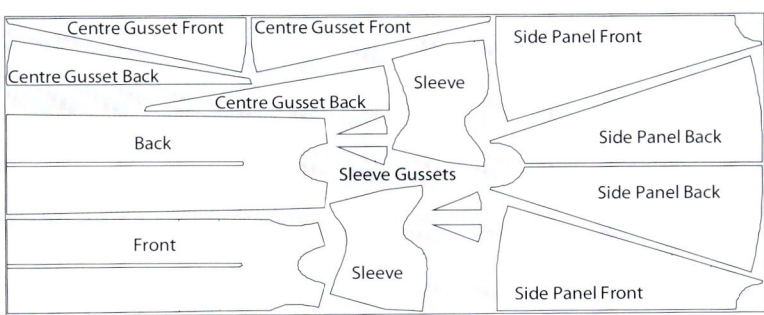

D10587

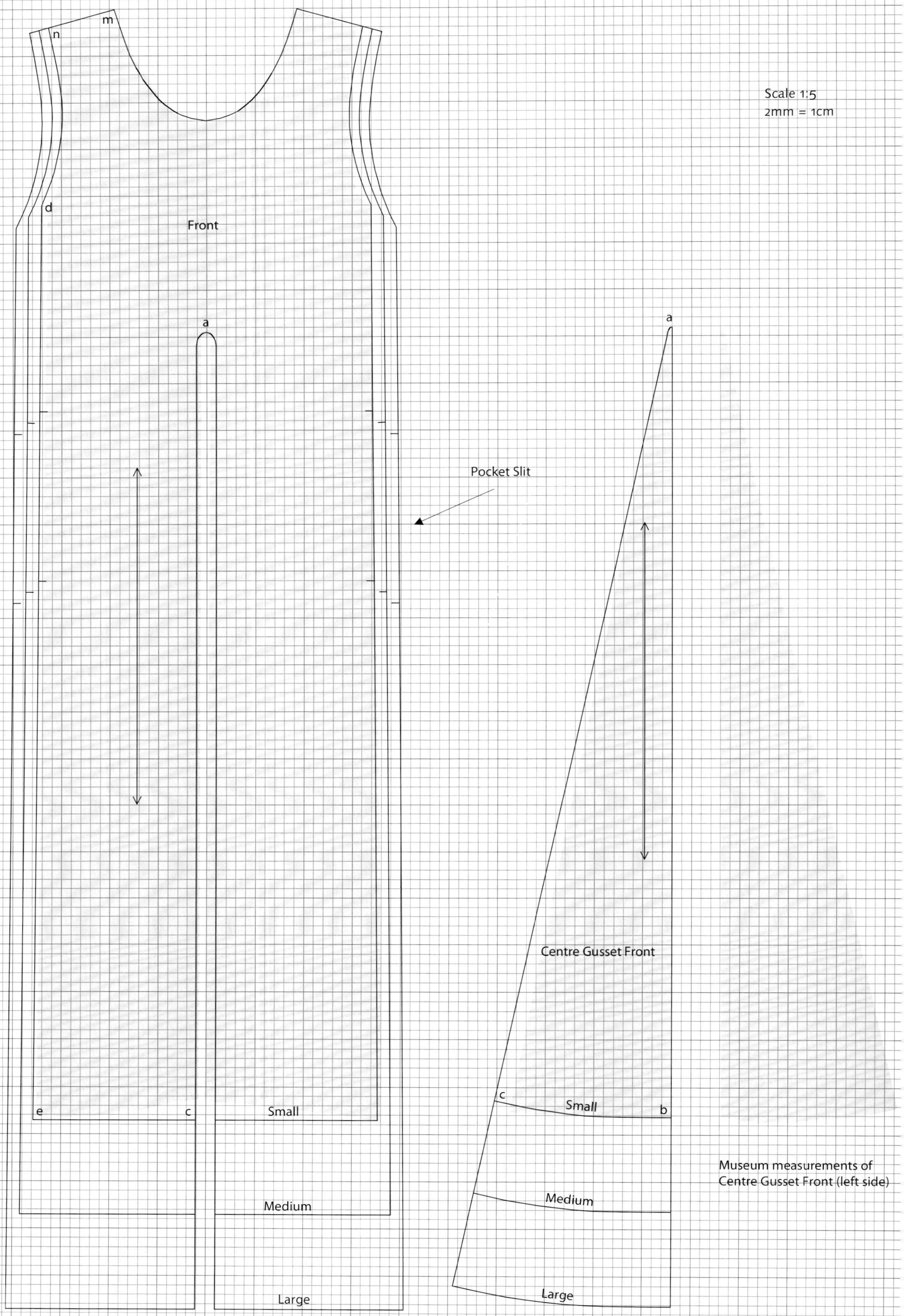

Scale 1:5
2mm = 1cm
m
n
d
Front
a
Pocket Slit
e
c
Small
Medium
Large
a
Centre Gusset Front
c
Small
b
Medium
Large
Museum measurements of
Centre Gusset Front (left side)

Scale 1:5
2mm = 1cm

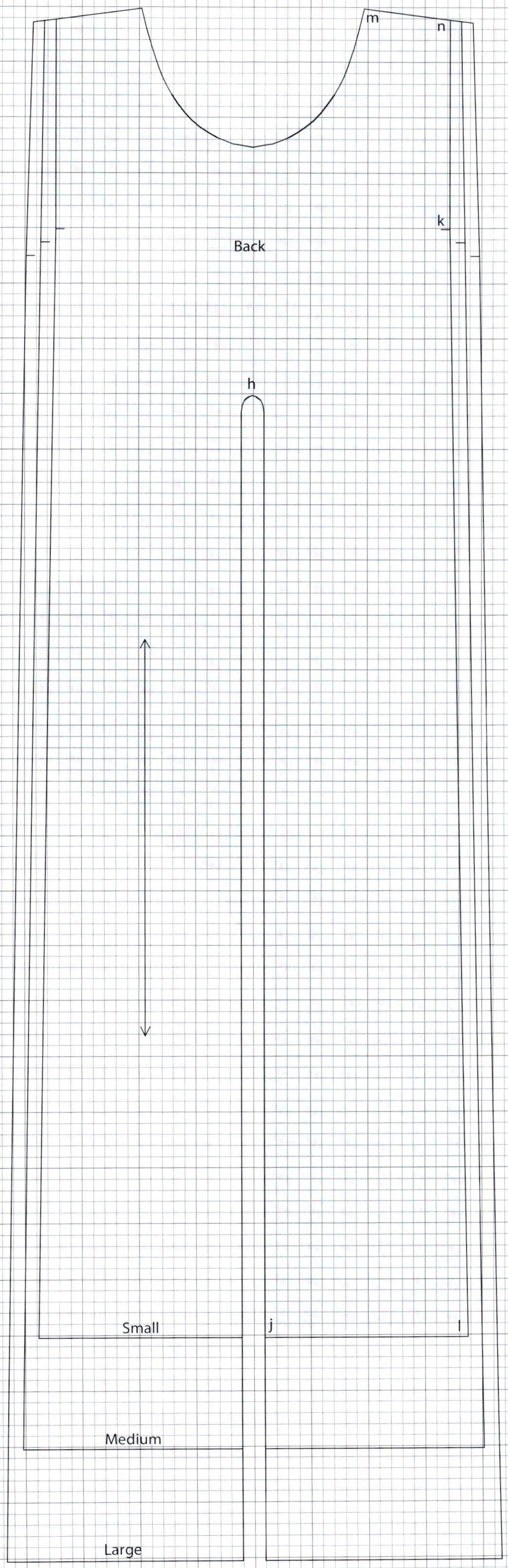

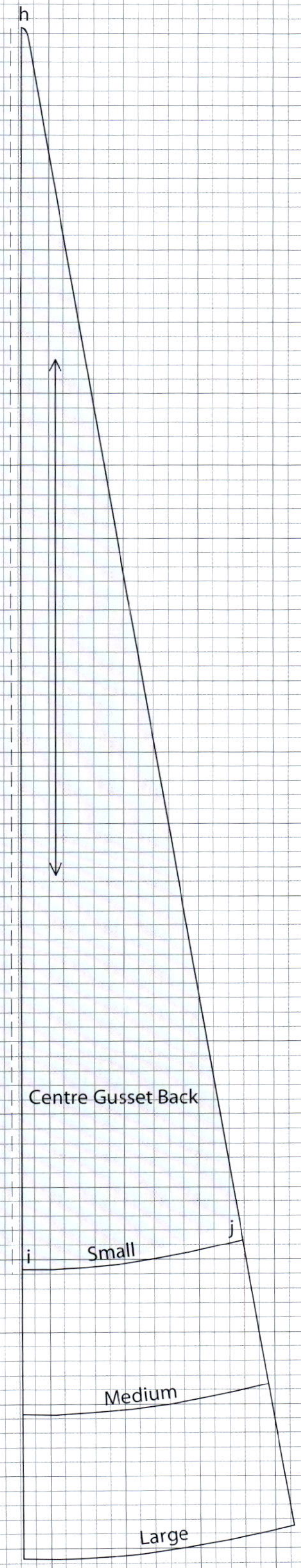

Back

m n
k
h

Centre Gusset Back

h

Small j l
i Small j

Medium

Medium

Large

Large

D10587

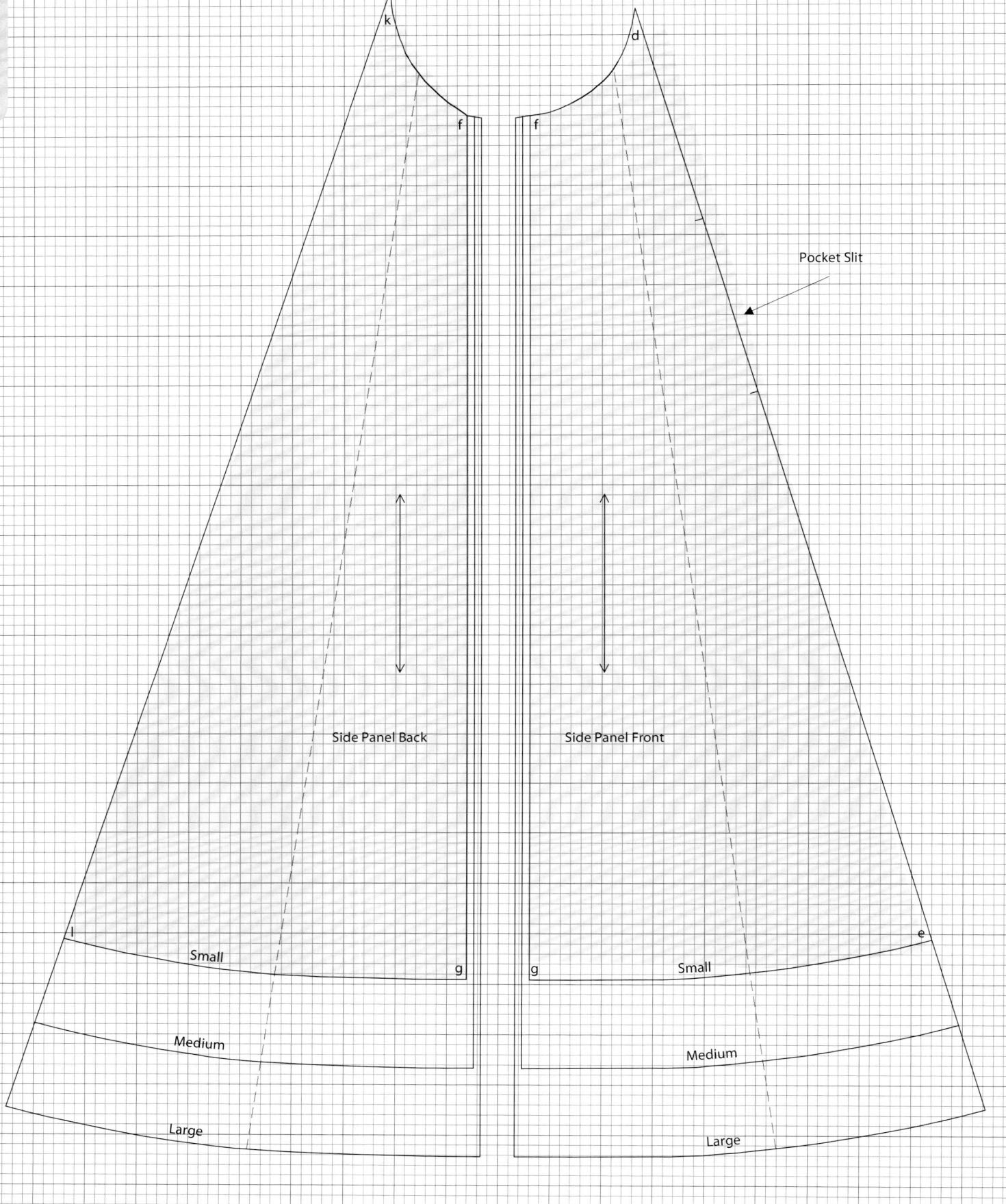

Scale 1:5
2mm = 1cm

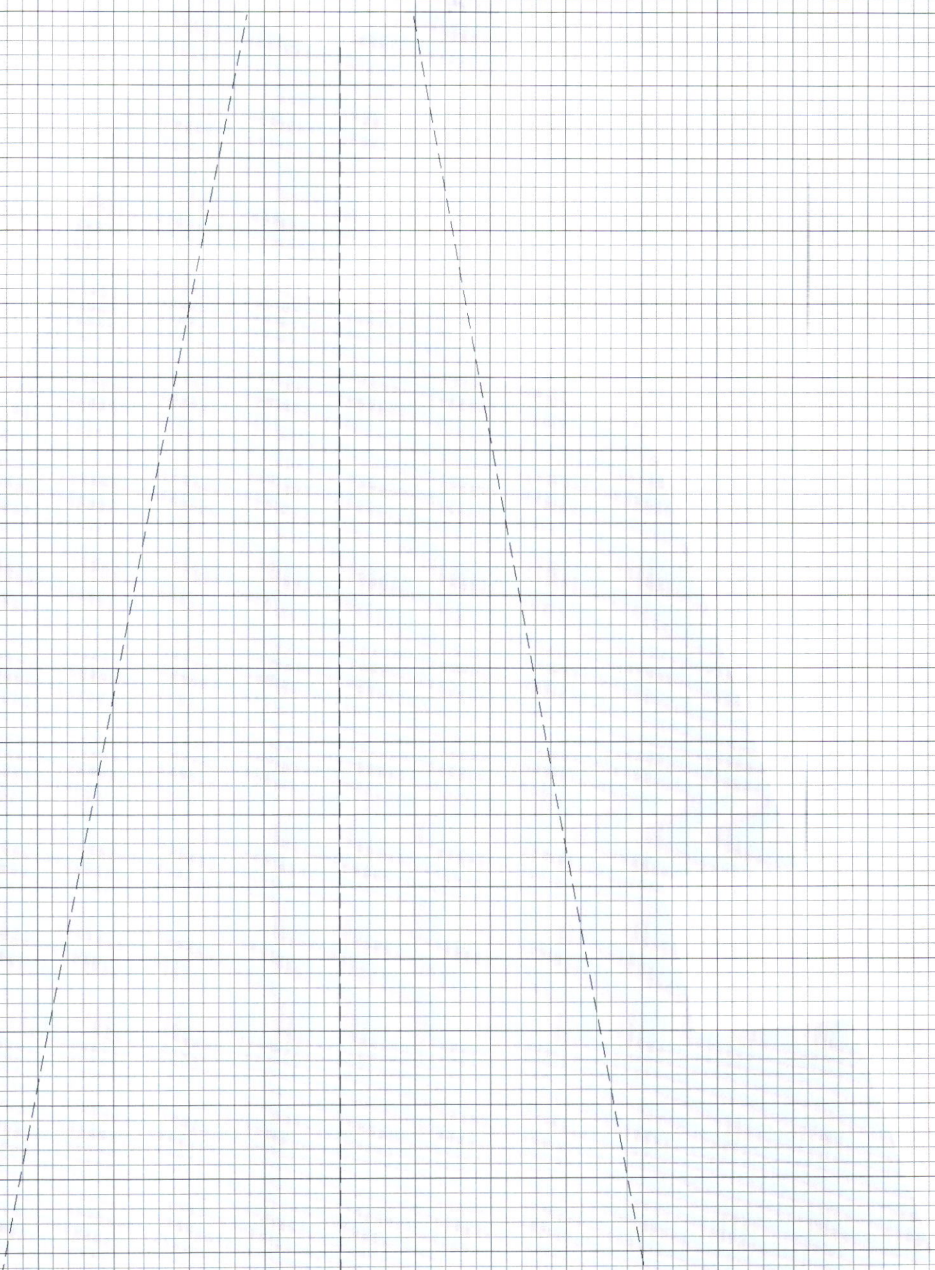

Museum measurements of
Side Panel Front and Back (left side)

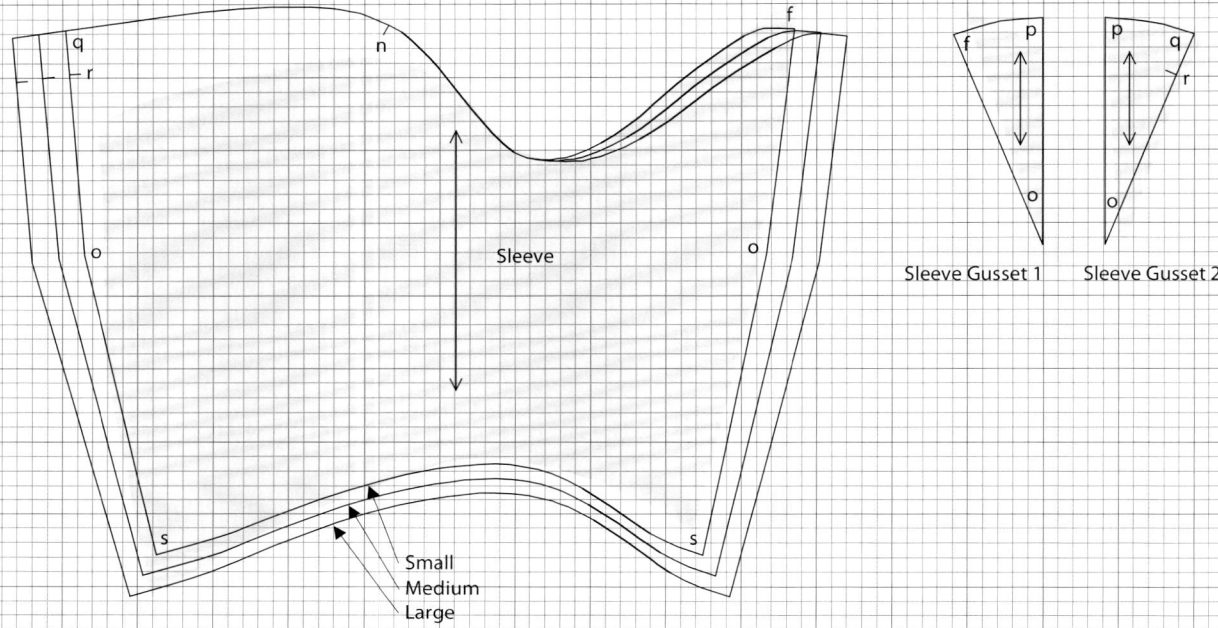

The garment was originally sewn in Greenlandic vaðmál in 2/2 twill with a light gray warp and a white weft. (Østergård, 2004)

Museum No. D10593

Garment measurements in cm:

Size	50	62	74	86	98
Length from shoulder	52	54	56	67	78
Chest width	42	46	50	54	60
Sleeve length	20	24	28	32	36

Fabric consumption in cm:

Fabric width	80 cm	140 cm
Size 50	121	58
Size 62	127	60
Size 74	133	64
Size 86	157	91
Size 98	184	113

Pieces to cut:

Front	1 piece
Back	1 piece
Centre Gusset Front	1 piece
Centre Gusset Back	1 piece
Sleeve	1 left piece and 1 right piece

New garment

Pattern-cutting layout:

Fabric width   80 cm

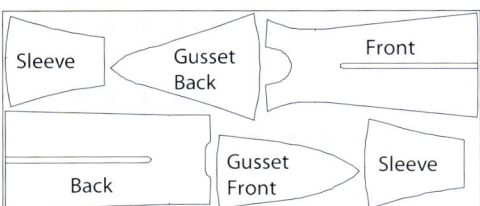

Fabric width   140 cm

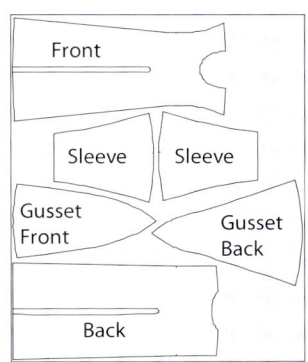

Scale 1:5
2mm = 1cm

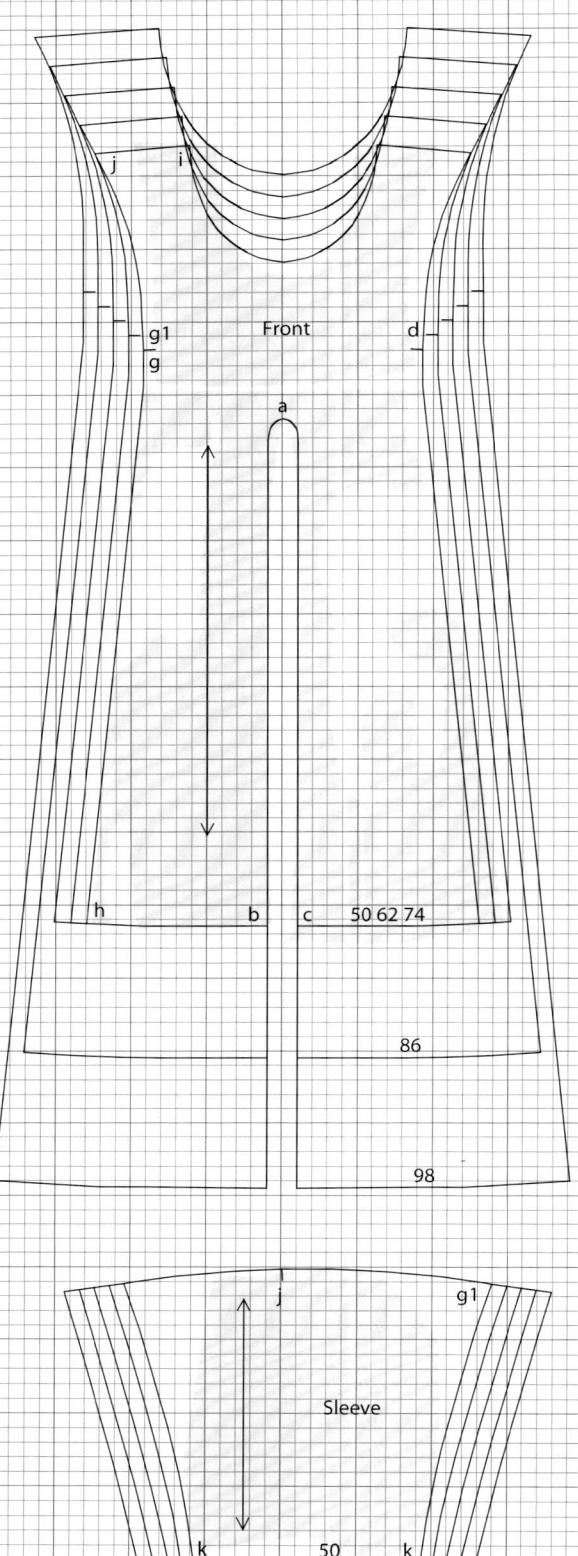

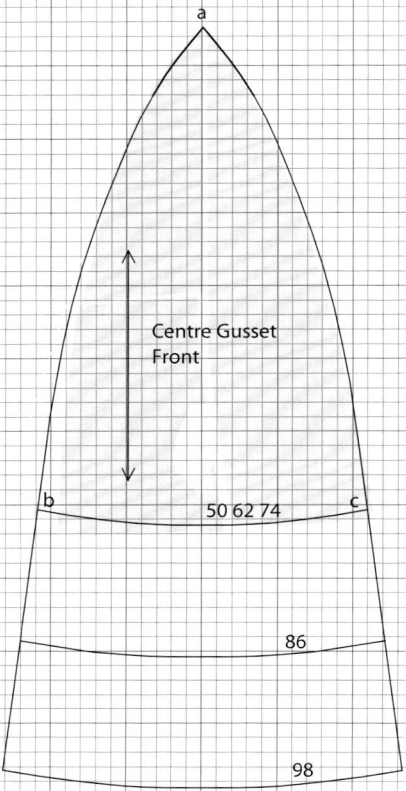

Centre Gusset Front

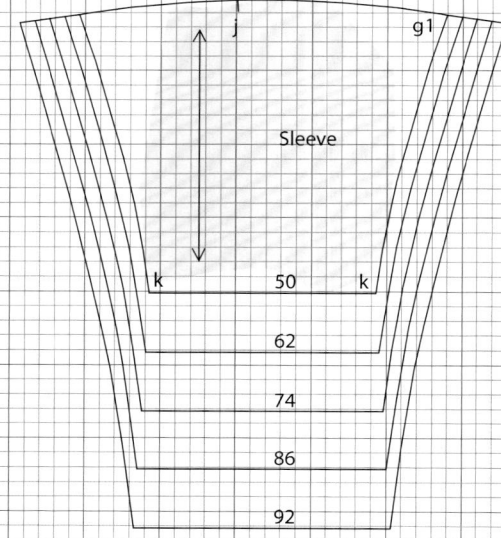

Museum measurements of Sleeve (right side)

Scale 1:5
2mm = 1cm

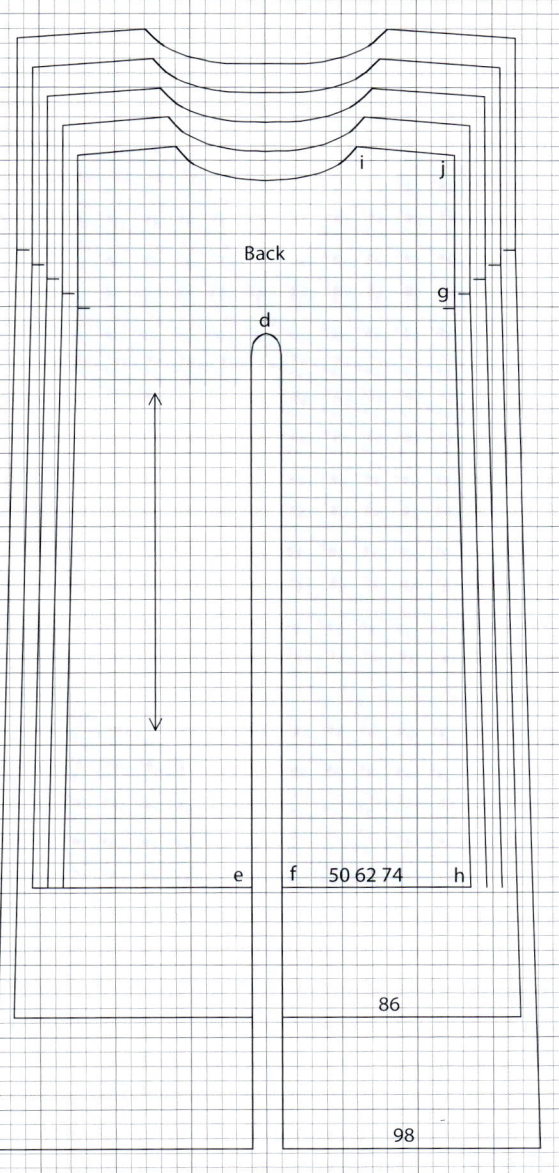
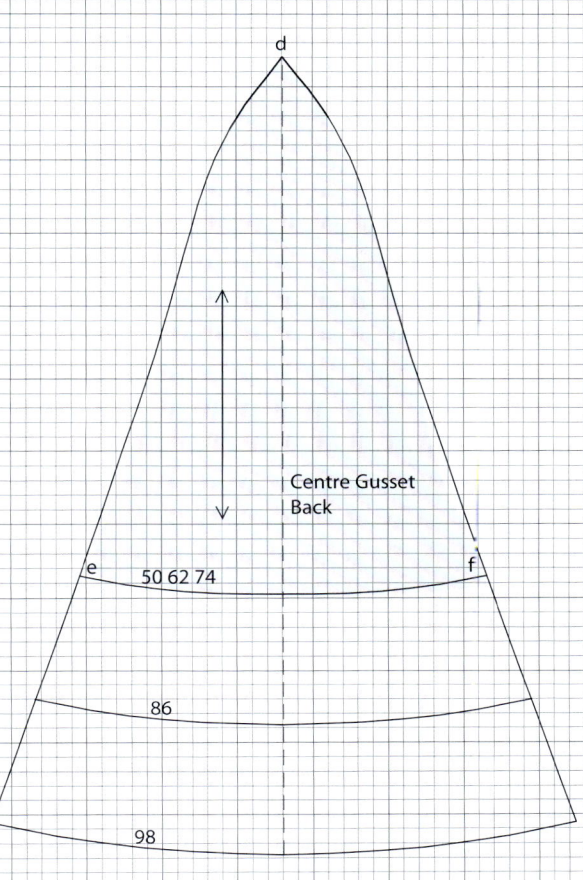

The garment was originally sewn in almost black Greenlandic vaðmál in 2/2 twill. (Østergård, 2004)

Museum No. D10594

Garment measurements in cm:

Size	Small	Medium	Large
Length from shoulder	108	116	124
Chest width	139	149	159
Sleeve length	59	60	61

Fabric consumption in cm:

Fabric width	80 cm	140 cm
Size Small	568	295
Size Medium	605	312
Size Large	641	330

Pieces to cut:

Centre Front Panel	1 left piece and 1 right piece
Side Front Panel	1 left piece and 1 right piece
Centre Back Panel	1 left piece and 1 right piece
Side Back Panel	1 left piece and 1 right piece
Top Sleeve	1 left piece and 1 right piece
Bottom Sleeve	1 left piece and 1 right piece
Sleeve Gusset 1	1 left piece and 1 right piece
Sleeve Gusset 2	1 left piece and 1 right piece
Collar	2 pieces

New garment

Pattern-cutting layout:

Fabric width 80 cm

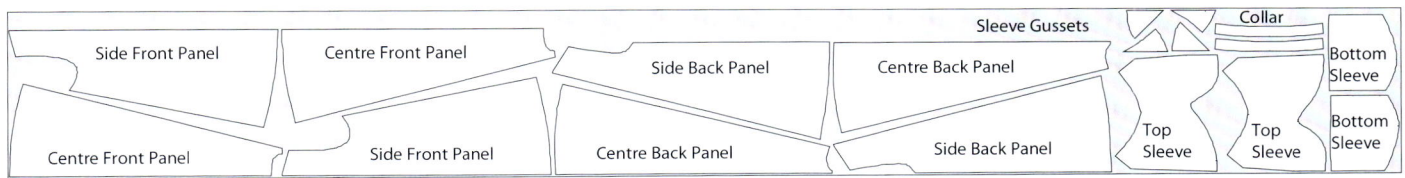

Fabric width 140 cm

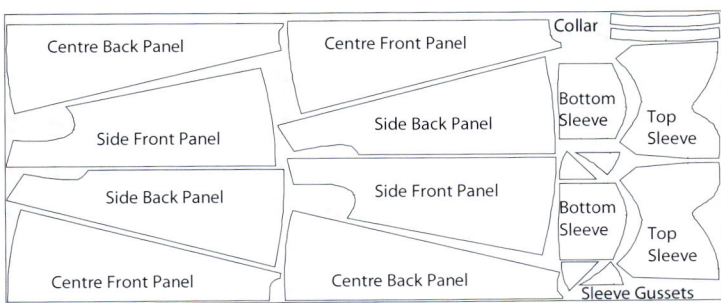

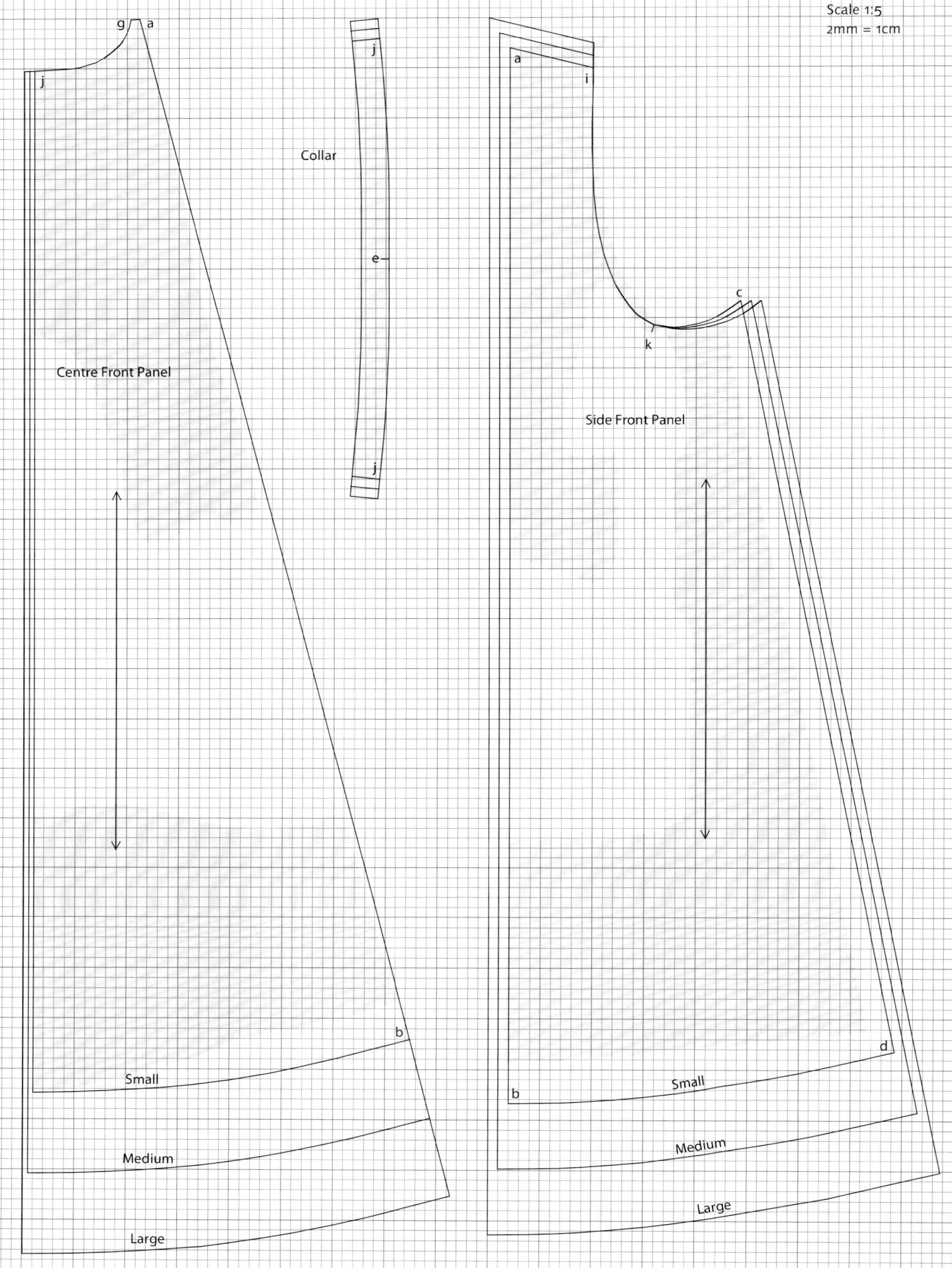

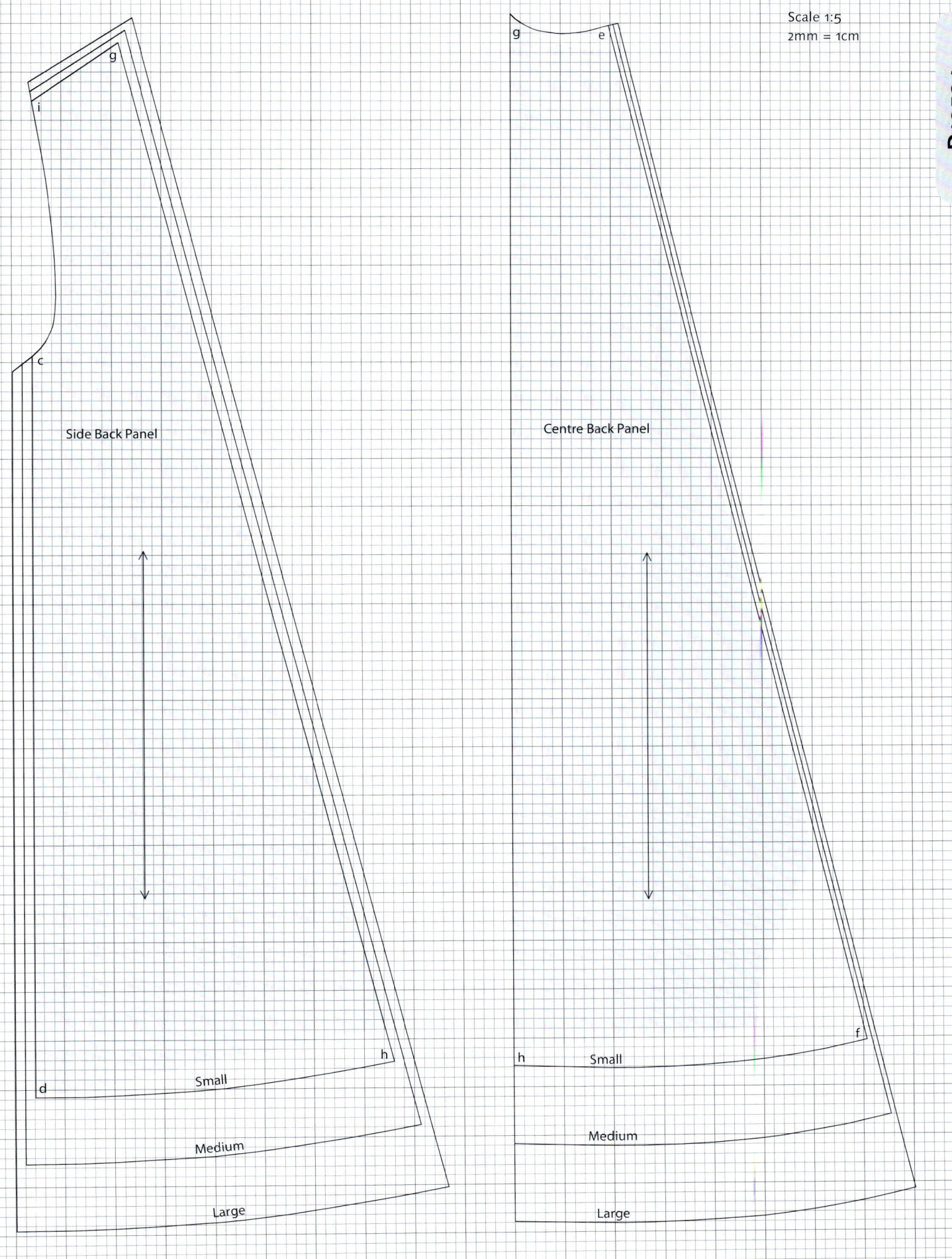

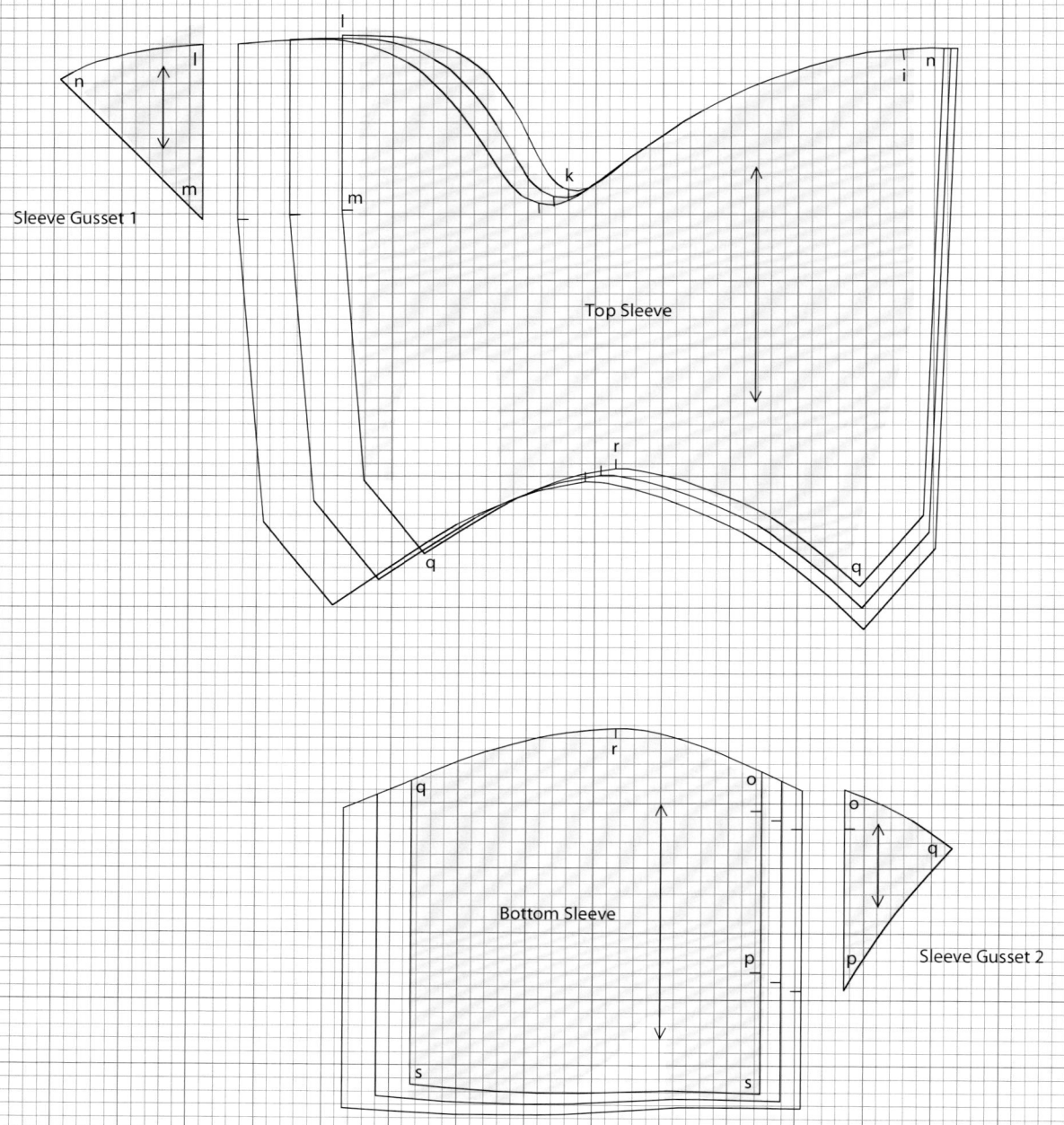

D10594 template for 1 buttonhole in the collar and the top 15 buttonholes at the centre front

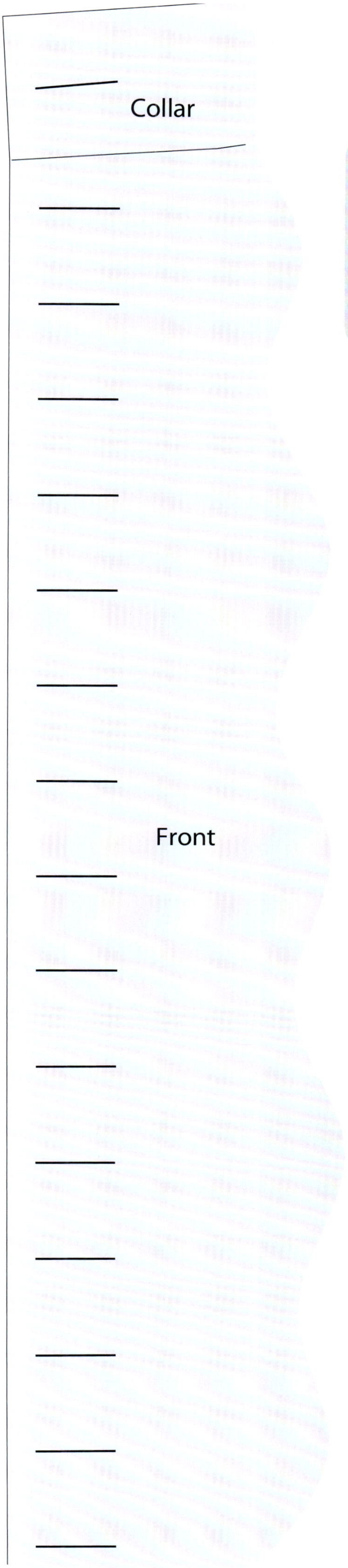

D10594 template for the lower 3 buttonholes at the centre front

Front

Hem line

Collar

D10594

Front

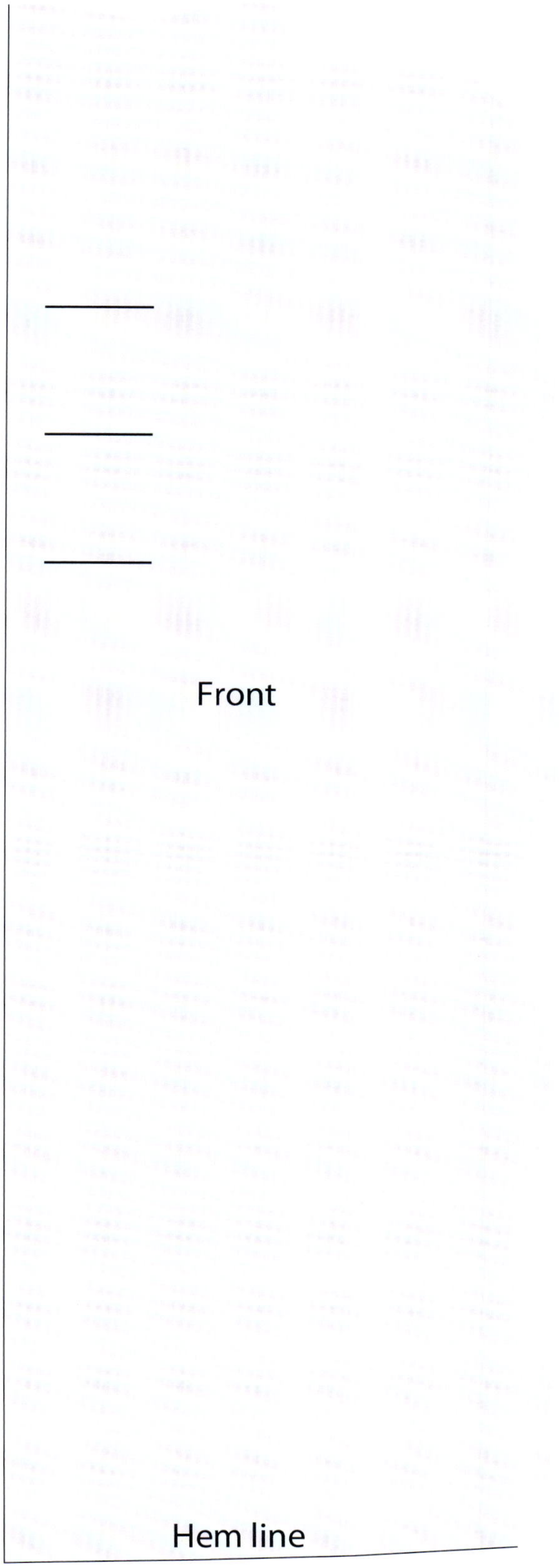

The original hood was originally sewn in brownish black Greenlandic vaðmál in 2/2 twill. (Østergård, 2004)

Photo: John Lee

Museum No. D10596

Hood measurements in cm:

Size	Small	Medium	Large
Face opening	56	59	62
Neck width	58	60	62
Length	58	59.5	61

Fabric consumption in cm:

Fabric width	80 cm	140 cm
Size Small	125	63
Size Medium	128	64
Size Large	131	66

Pieces to cut:

Hood	1 left piece and 1 right piece
Lirepipe	1 piece
Gusset Front	1 piece
Gusset Back	1 piece

Photo: Roberto Fortuna

New hcod

Pattern-cutting layout:

Fabric width 80 cm

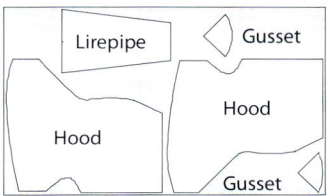

Fabric width 140 cm

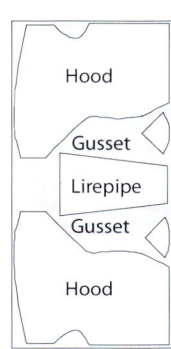

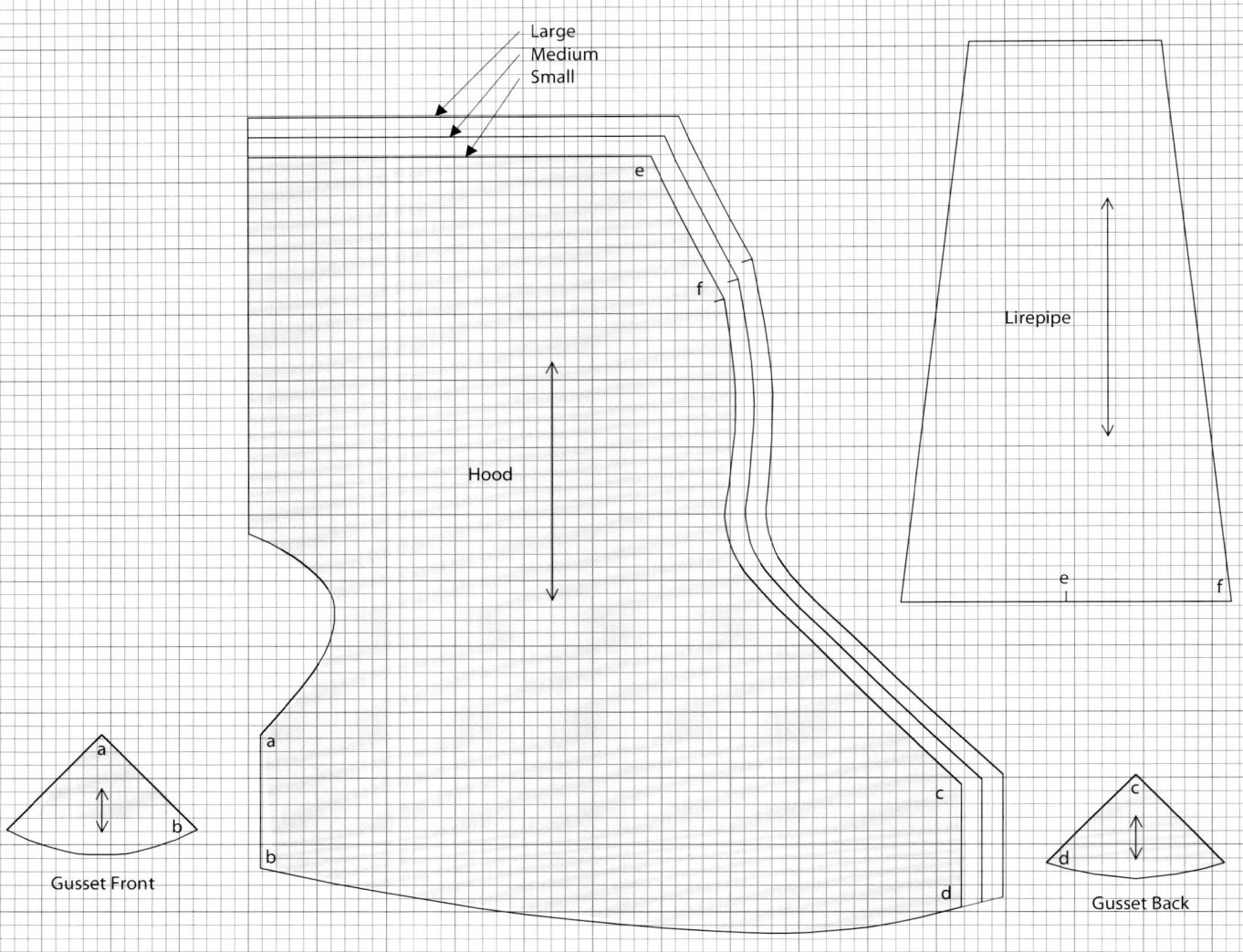

The hood was originally sewn in white Greenlandic vaðmál in 2/2 twill. The warp threads were possibly tannin-dyed. (Østergård, 2004)

Museum No. D10597

Hood measurements in cm:

Size	Small	Medium	Large
Face opening	66	68	70
Neck width	56	58	60
Length	68	69	70

Fabric consumption in cm:

Fabric width	80 cm	140 cm
Size Small	156	77
Size Medium	158	78
Size Large	160	79

Pieces to cut:

Hood	1 left piece and 1 right piece
Lirepipe	1 left piece and 1 right piece
Gusset Front	1 piece

New hood

Pattern-cutting layout:

Fabric width 80 cm

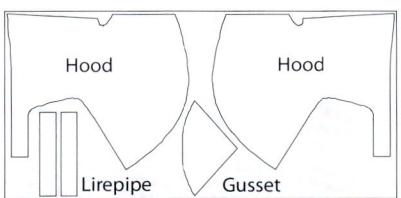

Fabric width 140 cm

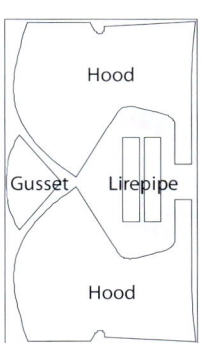

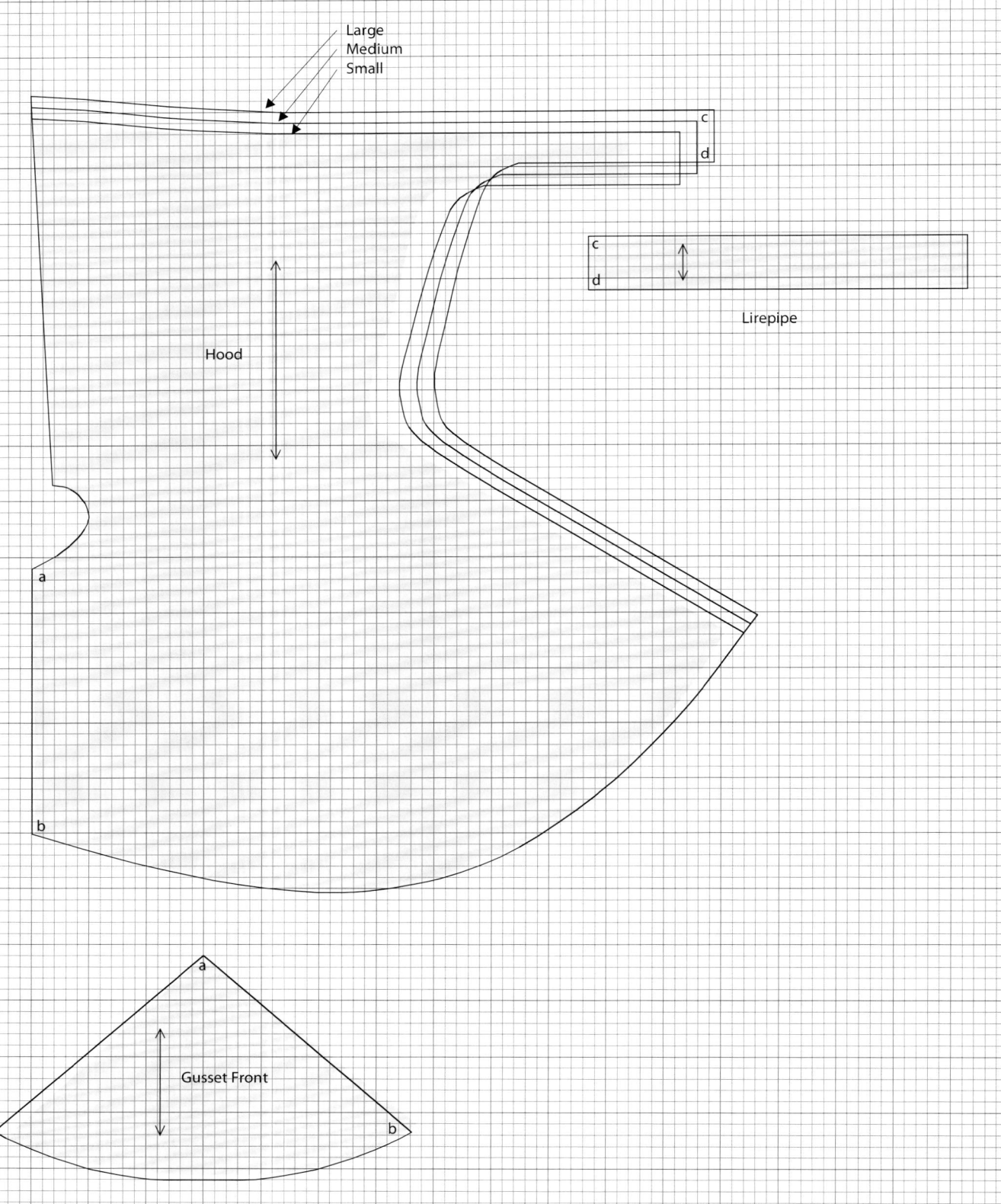

The hood was originally sewn in dark brown Greenlandic vaðmál in 2/2 twill. (Østergård, 2004)

Museum No. D10600

Hood measurements in cm:

Size	Small	Medium	Large
Face opening	55	58	61
Neck width	53	56	59
Length	51.5	53	54.5

Fabric consumption in cm:

Fabric width	80 cm	140 cm
Size Small	125	70
Size Medium	130	72
Size Large	136	78

Pieces to cut:

Hood	1 left piece and 1 right piece
Lirepipe	1 piece
Gusset Front	1 piece

New hood

Pattern-cutting layout:

Fabric width 80 cm

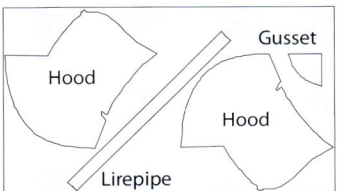

Fabric width 140 cm

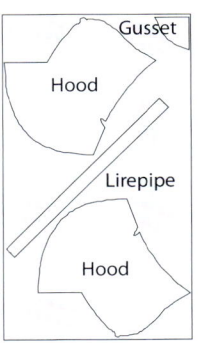

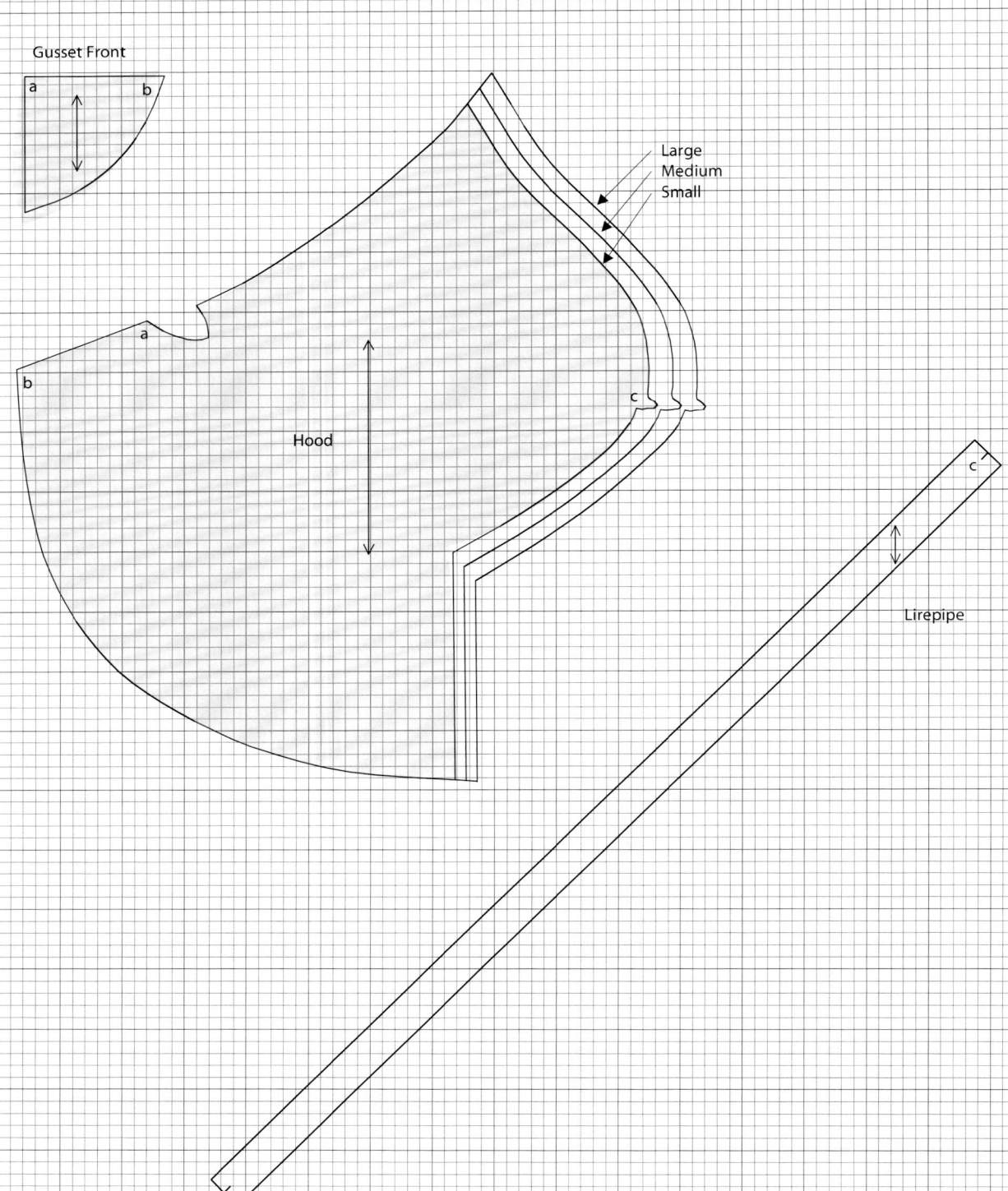

The hood was originally sewn in Greenlandic vaðmál in 2/2 twill with a light gray warp and a white weft. (Østergård, 2004)

Photo: John Lee

Museum No. D10602

Hood measurements in cm:

Size	Small	Medium	Large
Face opening	61	64	67
Neck width	56	58	60
Length	43.5	45	46.5

Fabric consumption in cm:

Fabric width	80 cm	140 cm
Size Small	109	61
Size Medium	112	62
Size Large	115	64

Pieces to cut:

Hood	1 left piece and 1 right piece
Lirepipe	1 piece
Gusset	1 left piece and 1 right piece

New hood

Pattern-cutting layout:

Fabric width 80 cm

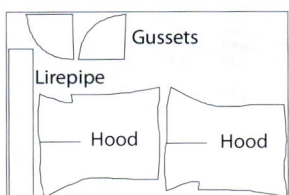

Fabric width 140 cm

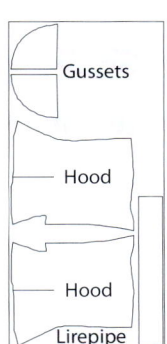

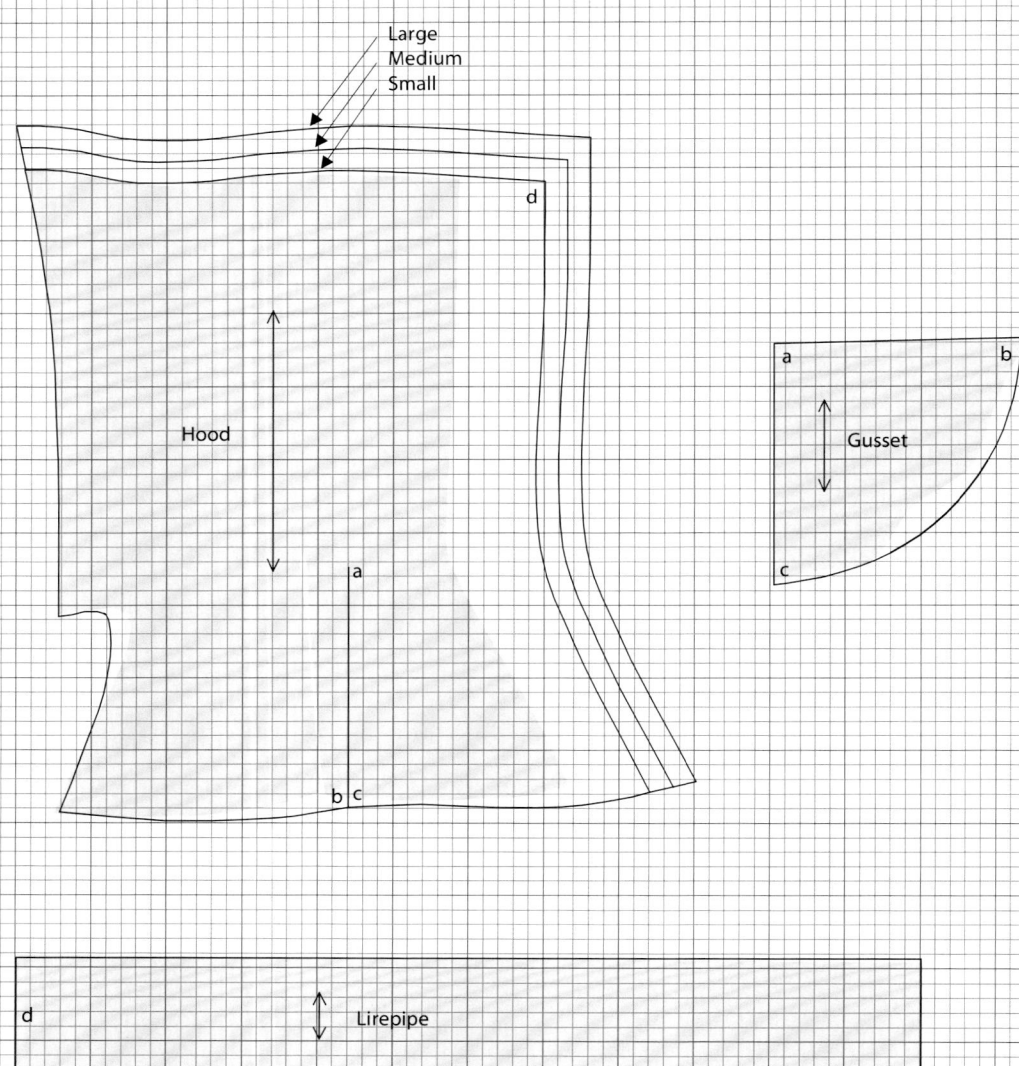

The hood was originally sewn in Greenlandic vaðmál in 2/2 twill with a light gray warp and a white weft. (Østergård, 2004)

Museum No. D10606

Hood measurements in cm:

Size	Small	Medium	Large
Face opening	66	68	70
Neck width	56	58	60
Length	38	39	40

Fabric consumption in cm:

Fabric width	80 cm	140 cm
Size Small	108	54
Size Medium	110	55
Size Large	112	56

Pieces to cut:

Hood	1 left piece and 1 right piece
Lirepipe	1 left piece and 1 right piece
Gusset	1 left piece and 1 right piece

122 • MEDIEVAL GARMENTS RECONSTRUCTED – NORSE CLOTHING PATTERNS

Photo: Roberto Fortuna

New hood

Pattern-cutting layout:

Fabric width 80 cm

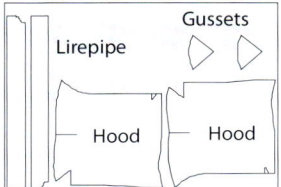

Fabric width 140 cm

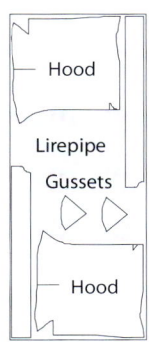

Scale 1:5
2mm = 1cm

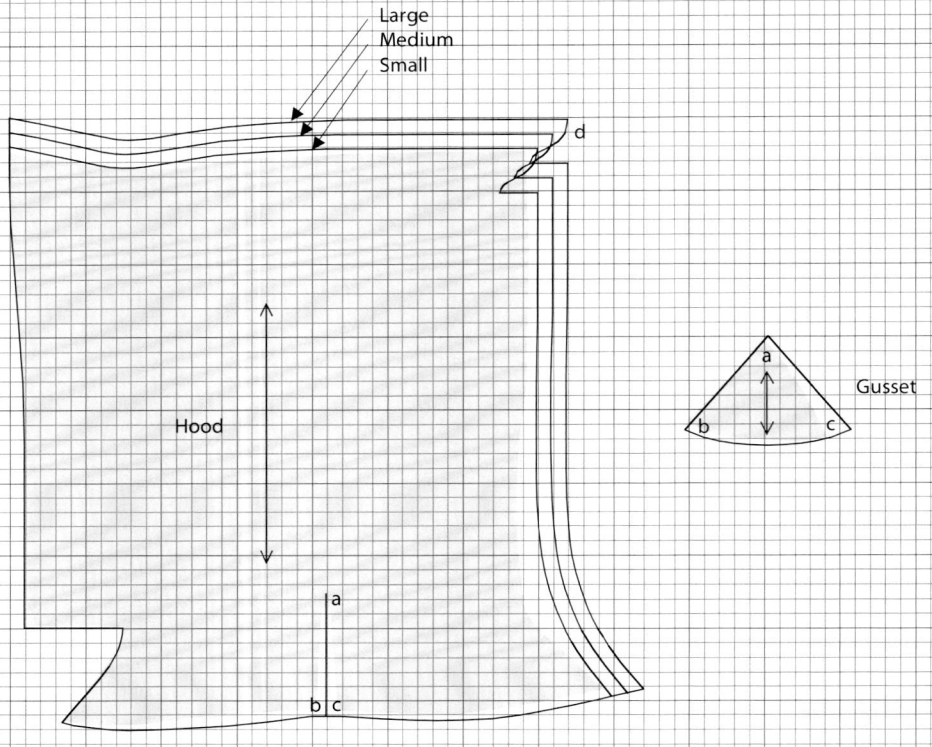

The cap was originally sewn in Greenlandic vaðmál in 2/2 twill with a gray warp and a light gray weft. (Østergård, 2004)

The hood was originally sewn in almost white Greenlandic vaðmál in 2/2 twill. (Østergård, 2004)

Museum No. D10608

Hood measurements in cm:

Size	110	122	134	146
Face opening	50	53	56	59
Neck width	42	43.5	45	46.5
Length	32.5	34	35.5	37

Cap measurements in cm:

Size	50	51	52	53	54	55	56
Head circumference	50	51	52	53	54	55	56

Fabric consumption in cm:

Fabric width	80 cm	140 cm
Hood		
Size 110 and 122	62	43
Size 134 and 146	65	46
Cap		
All sizes	37	24

Pieces to cut – Hood:

Hood	1 left piece and 1 right piece
Lirepipe	1 piece
Gusset	1 left piece and 1 right piece

Pieces to cut – Cap:

Top	1 piece
Sideband	1 piece

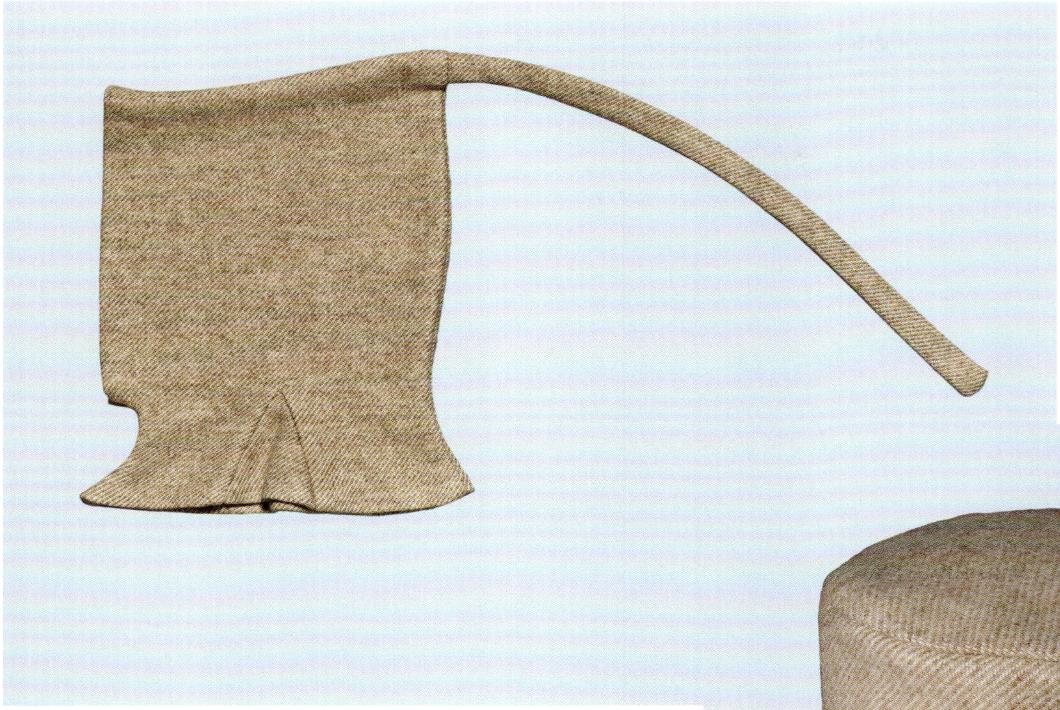

New hood and cap

Photo: Roberto Fortuna

Pattern-cutting layout:

Fabric width 80 cm

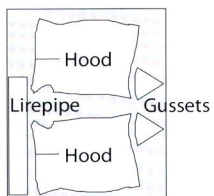 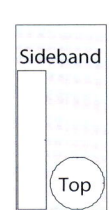

Fabric width 140 cm

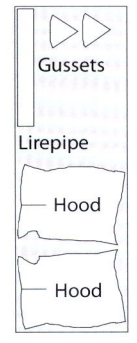 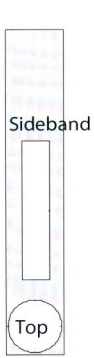

Scale 1:5
2mm = 1cm

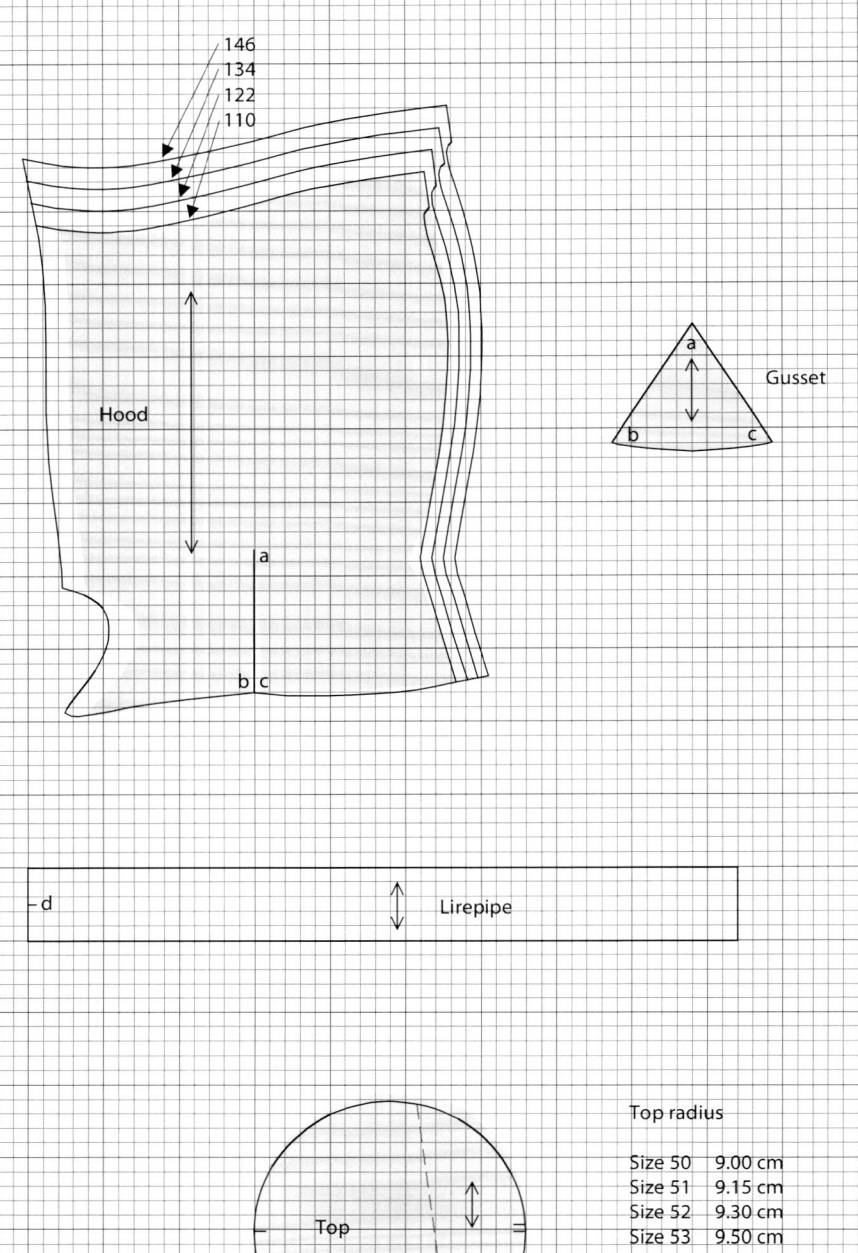

Top radius

Size 50	9.00 cm
Size 51	9.15 cm
Size 52	9.30 cm
Size 53	9.50 cm
Sice 54	9.65 cm
Size 55	9.80 cm
Size 56	9.85 cm

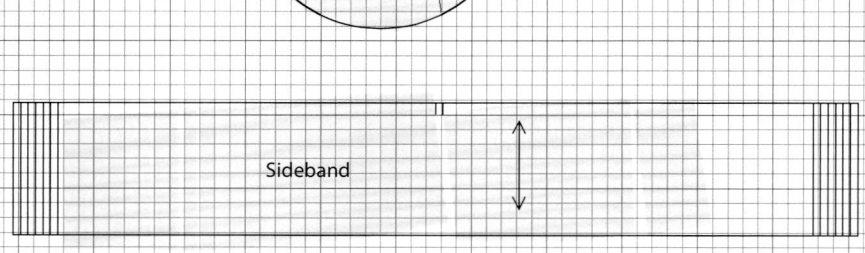

The cap was originally sewn in almost white Greenlandic vaðmál in 2/2 twill. (Østergård, 2004)

Museum No. D10610

Cap measurements in cm:

Size	58	59	60	61	62
Head circumference	58	59	60	61	62

Pieces to cut:

Top	1 piece
Sideband	1 piece

Fabric consumption in cm:

Fabric width	80 and 140 cm
Size 58 and 59	63
Size 60 and 61	65
Size 62	66

130 • MEDIEVAL GARMENTS RECONSTRUCTED — NORSE CLOTHING PATTERNS

New cap

Pattern-cutting layout:

Fabric width 80 and 140 cm

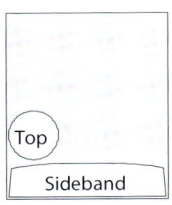

MUSEUM NO. D10610 • 131

Scale 1:5
2mm = 1cm

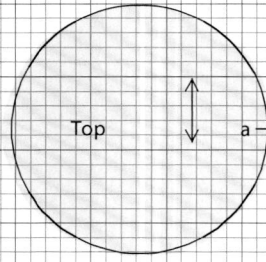

Top

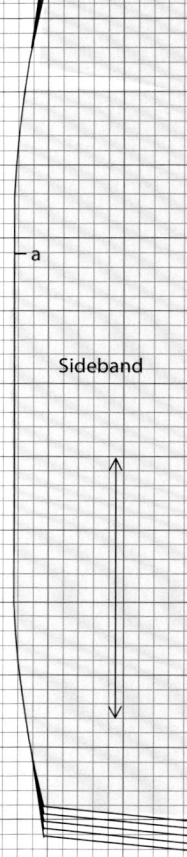

Sideband

Top radius

Size 58   8.50 cm
Size 59   8.65 cm
Sice 60   8.80 cm
Size 61   9.00 cm
Size 62   9.15 cm

The stockings were originally sewn in almost black Greenlandic vaðmál in 2/2 twill. (Østergård, 2004)

Museum No. D10613

Stocking measurements in cm:

Size	Small	Medium	Large	Extra Large
Length	90	91	92	93
Tight width	53.5	55	56.5	58
Ankle width	29	31	33	35
Foot length	23.5	26	28.5	31

Fabric consumption in cm:

Fabric width	80 cm	140 cm
Size Small	154	99
Size Medium	157	102
Size Large	163	106
Size Extra Large	182	111

Pieces to cut:

Stocking	1 left piece and 1 right piece
Outer Foot	1 left piece and 1 right piece
Inner Foot	1 left piece and 1 right piece
Gusset	1 left piece and 1 right piece

New stockings

Pattern-cutting layout:

Fabric width 80 cm

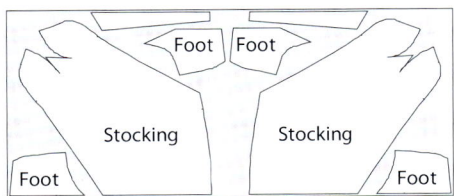

Fabric width 140 cm

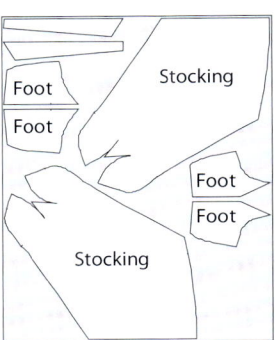

Scale 1:5
2mm = 1cm

Stocking

Gusset

Small
Medium
Large
Extra large

Inner Foot

Outer Foot

Small
Medium
Large
Extra large

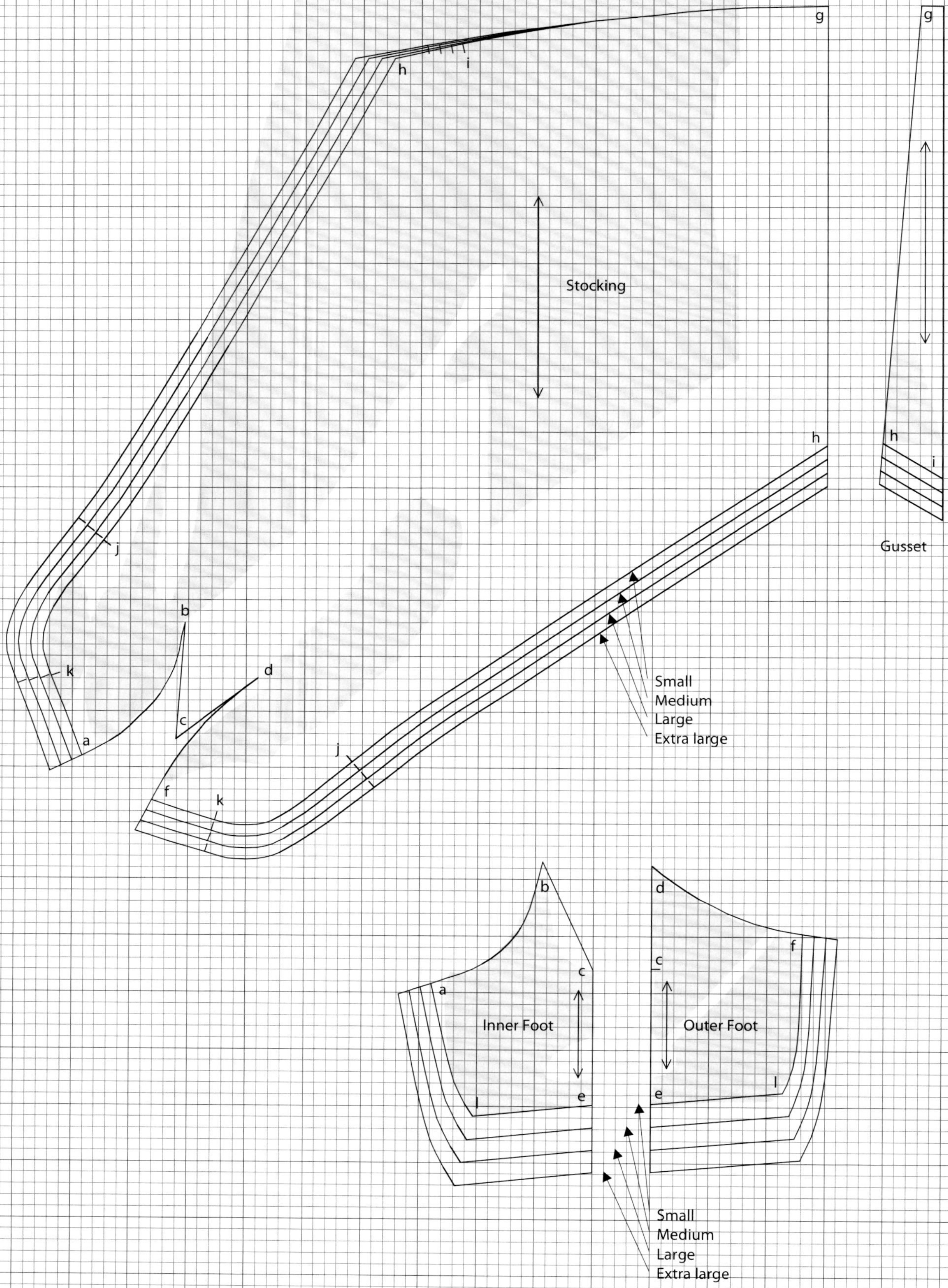

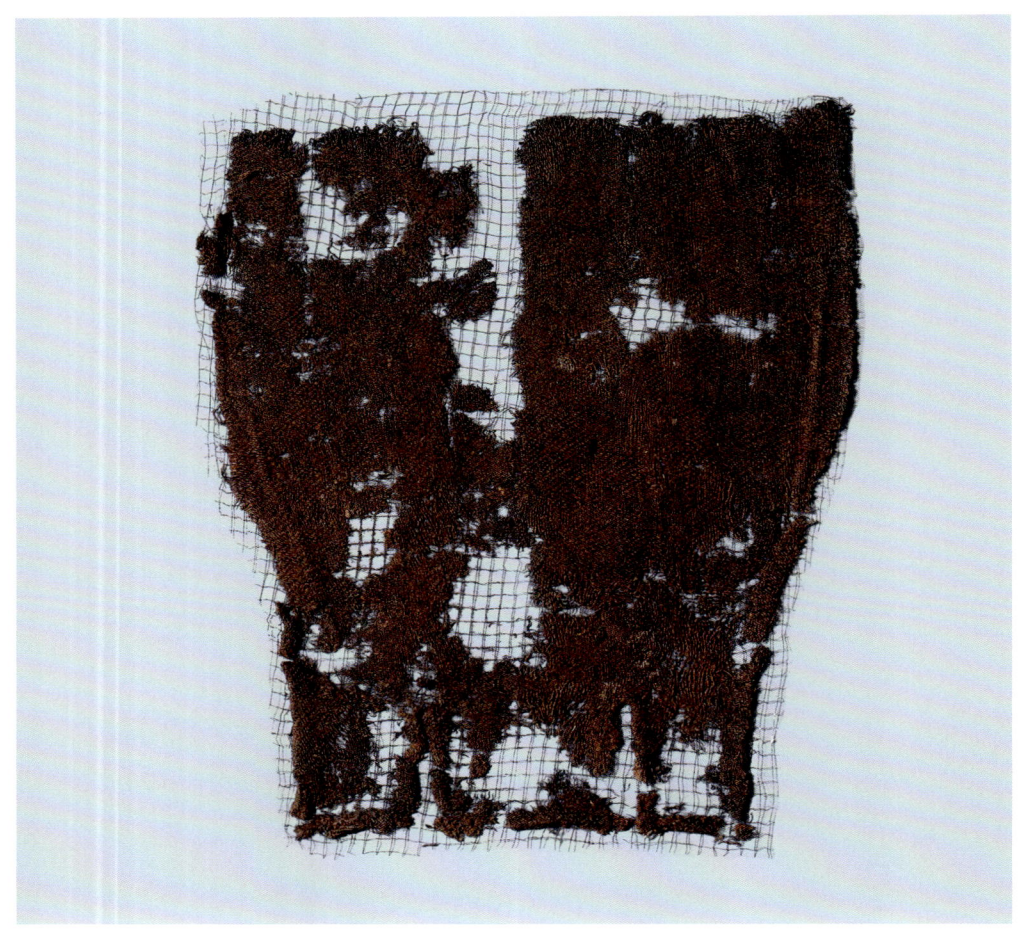

The stockings were originally sewn in white Greenlandic vaðmál in 2/2 twill. (Østergård, 2004)

Museum No. D10616

Stocking measurements in cm:

Size	Small	Medium	Large
Length	43.5	45	46.5
Calf width	39	42	45

Fabric consumption in cm:

Fabric width	80 cm	140 cm
Size Small	95	48
Size Medium	98	49
Size Large	101	51

Pieces to cut:

Stocking	1 left piece and 1 right piece

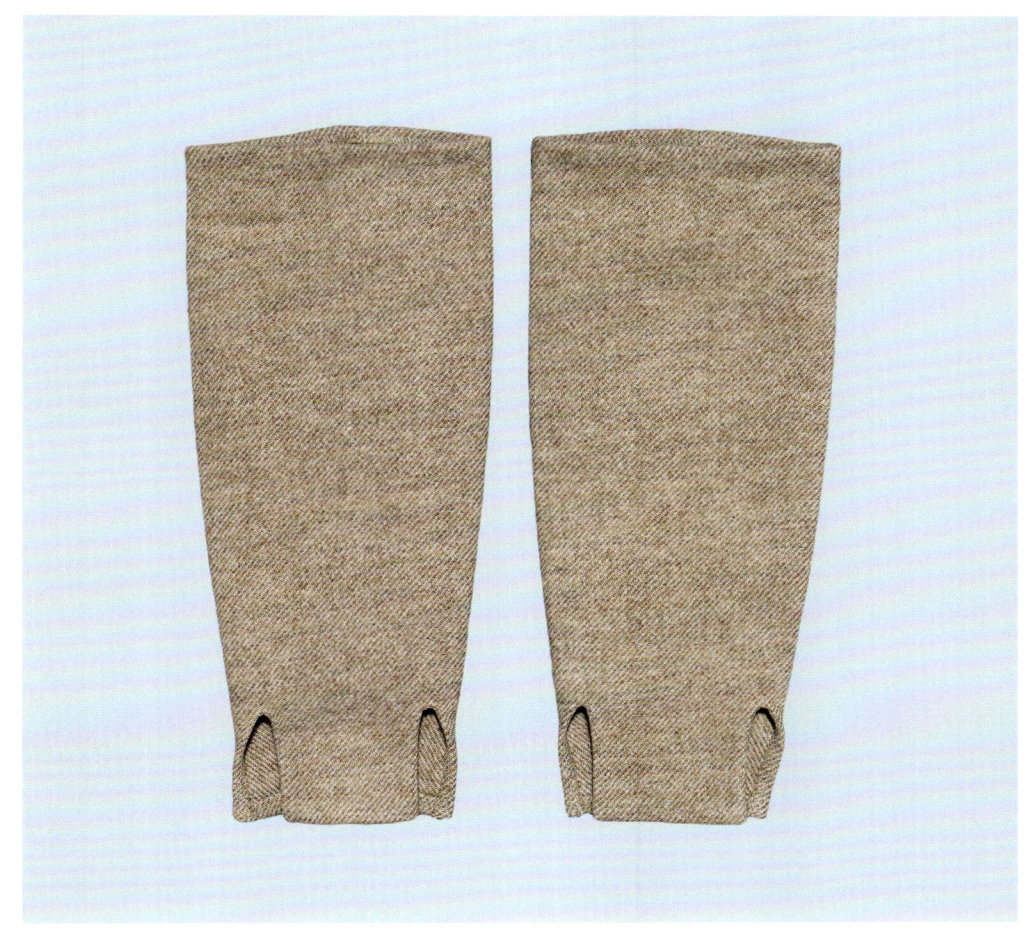

New stockings

Pattern-cutting layout:

Fabric width 80 cm

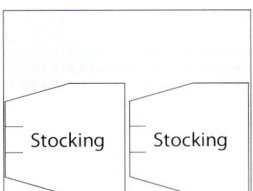

Fabric width 140 cm

Scale 1:5
2mm = 1cm

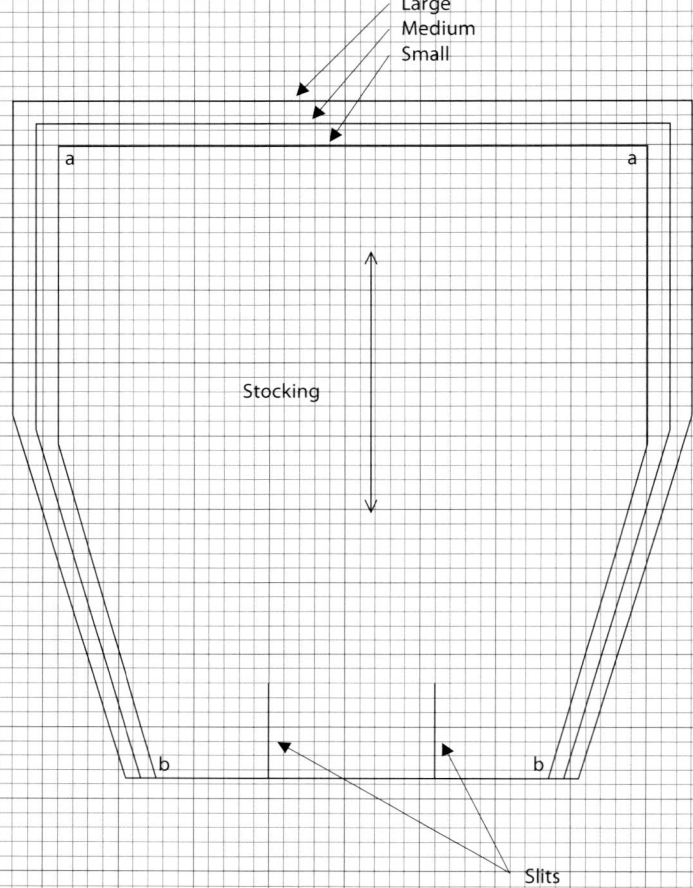

Literature

Andersson, E.: Invisible Handicrafts. The General Picture of Textile and Skin Crafts in Scandinavian Surveys. *Lund Archaeological Review* I. Lund 1995 pp. 7-20.

Andersson, E.: *The common Thread, Textile Production during Late Iron Age – Viking Age*. University of Lund, Institute of Archaeology, Report series No. 67, 1999.

Andersson, E.I.: *Kläderna och människan i medeltidens Sverige och Norge*. Avhandlingar från Historiska institutionen i Göteborg 47. Göteborg 2006.

Andersson-Wiking, C. & Wiking-Faria, P.: Bockstensmannen. Utställningskatalog. Varberg 2007.

Arneborg, J.: Burgunderhuer, baskere og døde nordboer i Herjolfsnes, Grønland. *Nationalmuseets Arbejdsmark*. København 1996 pp. 75-83.

Arneborg, J.: Nordboliv i Grønland. Else Roesdahl (ed.), *Dagligliv I Danmarks middelalder*. En arkæologisk kulturhistorie. København 1999 pp. 353-373.

Arneborg, J.: Det europæiske landnam – Nordboerne i Grønland, 985-1450 e.v.t.. H.C. Gulløv (ed.), *Grønlands forhistorie*. København 2004 pp. 221ff.

Bockstensmannen och hans tid. Pablo Wiking-Faria (red). Länsmuseet Varberg 2008.

Brown, R.: *The Weaving, Spinning and Dyeing Book*. Watson-Guptill. Publications/New York 1977.

Crockett, C.: *The Complete Spinning Book*, Watson-Guptill Publications/ New York. 1977.

Crowfoot, E., Pritchard, F. & Staniland, K.: *Textiles and Clothing c.1150-c.1450* (Medieval Finds From Excavations in London, 4). London 1992.

Danske Kalkmalerier. Tidlig Gotik 1275-1375, Ulla Haastrup (ed.). København 1989.

Danske Kalkmalerier. Gotik 1375-1475, Ulla Haastrup (ed.). København 1985.

Dansk Kvindebiografisk leksikon. Rosinante, København 2001.

De Alcega, J.: *Tailor's Pattern Book 1589*. Facsimile. Introduction and notes by J.L. Nevinson. Bedford 1979.

Fentz, M.: En hørskjorte fra 1000-årene. *Viborg Søndersø 1000-1300*. Byarkæologiske undersøgelser 1981 og 1984-85. Aarhus Universitetsforlag 1998 pp. 249-266.

Fentz, M.: Dragter. Else Roesdahl (ed.), *Dagligliv i Danmarks middelalder*. En arkæologisk kulturhistorie. København 1999 pp. 150-171.

Geijer, A., Franzén, A.M. & Nockert, M.: *Drottning Margaretas gyllene kjortel i Uppsala Domkyrka*. Kungl. Vitterhets Historie och Antikvitets Akademien. Stockholm 1985.

Gjessing, G.: Skjoldehamndrakten. En senmiddelaldersk nordnorsk mannsdrakt. *Viking. Tidsskrift for norrøn arkeologi*, 2. 1938 pp. 27-81.

Gudjónsson, E.E.: Icelandic Embroidery. Domestic Embroideries in the National Museum of Iceland. Reykjavik, 1983. 9 bls.

Guðjónsson, E.E.: *Nogle bemærkninger om den islandske vægtvæv, vefstaður*. By og Bygd, Norsk Folkemuseums årsbog 1983-84, Vol. XXX, Aurskog 1985.

Guðjónsson, E.E.: *Some Aspects of the Icelandic Warp-Weighted Loom, vefstaður*. Textile History, 21 (2), London 1990.

Gutarp, E-M.: *Hurusomman sig klädde*. Visby 2000.

Hald, M.: *Flettede Baand og Snore*. København 1975.

Henningsen, I. M.: *Plantefarvernes abc*. Høst og Søn. København 1983.

Henningsen, I. M.: *Vævernes ABC*, Høst og Søn. København 1983.

Hoffmann, M.: *The Warp-Weighted Loom*. Studia Norvegica nr. 14, Universitetsforlaget. Oslo 1964.

Magnússon, H.: *Icelandic Knitting*, Search Press. Reykjavik 2006.

Holck, P.: Myrfunnet fra Skjoldehamn – mannlig same eller norrøn kvinde? *Viking 1988. Tidsskrift for norrøn arkeologi*. Oslo 1991.

Holm-Olsen, I.M.: Noen gravfunn fra Vestlandet som kaster lys over vikingetidens kvinnedrakt. *Viking. Tidsskrift for norrøn arkeologi* Bd. XXXIX. Oslo 1976 pp. 197-205.

Hägg, I.: *Kvinnodräkten i Birka. Livplaggens rekonstruktion på grundval av det arkeologiska materialet*. Uppsala 1974.

Hägg, I.: *Die Textilfunde aus dem Hafen. Berichte über die Ausgrabungen in Haithabu. Bericht 20*. Neumünster 1984.

Ingstad, A.S.: Tekstilene i Osebergskipet. *Oseberg Dronningens Grav. Vår arkeologiske nasjonalskatt*. Oslo 1992 pp. 176-223.

Knudsen, L. Ræder: Høvding og præst. *Tidsskriftet SKALK*. No. 6. Århus 2007 pp. 3-9.

Kulturhistorisk leksikon for nordisk middelalder m.m. 1956-78 (reprinted 1980-82) Lin, Skrædder, Sömnad og Vadmål.

Mannering, U.: Sidste skrig. *Tidsskriftet SKALK* No. 4. Århus 1999 pp. 20-27.

Nockert, M.: Medeltida dräkt i bild och verklighet. *Den Ljusa Medeltiden. Studier tillägnade Aron Andersson.* (The Museum of National Antiquities. Stockholm Studies 4) 1984 pp. 5-11.

Nockert, M.: *Bockstensmannen och hans dräkt*, Halmstad och Varberg, Stiftelsen Hallands länsmuseer. 1985.

Nockert, M.: *Bockstensmannen och hans dräkt*. Borås 1997.

Nockert, M. & Possnert, G.: *Att datera textilier*. Södertälje 2002.

Nørgård, A.: Rekonstrueret barnekjole og hat. C. Hinsch (ed.), *Lille Margrete og andre børn i middelalderen*. Århus 1997 pp. 14-15.

Nørgård, A.: *Vævning af sejldugsprøver på opstadvæv*, Vikingeskibsmuseet, Roskilde. *Weaving samples of sailcloth on a warp-weighted loom*. 1999.

Nørgård, A.: *Et uldsejl til Oselver*, arbejdsrapport kan hentes på Vikingeskibsmuseets Hjemmeside, www.vikingeskibsmuseet.dk. 2009.

Nørlund, P.: Buried Norsemen at Herjolfsnes. *Meddelelser om Grønland*, Vol. 67. København 1924.

Nørlund, P.: *De gamle Nordbobygder ved Verdens ende*. Nationalmuseet 1934. Latest reprint 1967.

Nørlund, P.: Klædedragt i Oldtid og Middelalder. Dragt. *Nordisk Kultur* XV:B. København 1941 pp. 1-88.

Østergård, E.: Tøj til nordbobørn. C. Hinsch (ed.), *Lille Margrete og andre børn i middelalderen*. Århus 1997 pp. 12-13.

Østergård, E.: *Som syet til jorden. Tekstilfund fra det norrøne Grønland*. Aarhus Universitetsforlag 2003.

Ross, M.: *The Encyclopedia of Hand Spinning*. Interweave Press. Loveland Colorado 1988.

Ryder, M.L.: *Sheep and Man*. London 1983.

Staniland, K.: *Medieval craftsmen. Embroiderers*. London 1993.

Vedeler Nilsen, M.: Gravdrakt i østnorsk middelalder. Et eksempel fra Uvdal. *Collegium Medievale*, vol. 11, 1988 pp. 69-85.

Vedeler Nilsen, M.: Middelalderdrakt i lys av kirkekunst og arkeologisk materiale. *Spor -fortidsnytt fra midt-norge*, Nr. 2, 12. Årgang, 24, Hefte, 1997 pp. 20-22.

Walton Rogers, P.: Dyes and wools in Norse textiles from Ø 17a. Narsaq – a Norse landnáma farm (C.L. Vebæk). *Man & Society*, 18. København 1993 pp. 56-58.

Walton Rogers, P.: *Cloth and Clothing in Early Anglo-Saxon England*, AD 450-700, York 2007.

Walton Rogers, P.: The raw materials of the textiles from GUS, with a note on fragments of fleece and animal pelts (identification of animal pelts by H.M. Appleyard). J. Arneborg & H.C. Gulløv (eds.), *Man, Culture and Environment in Ancient Greenland, Report on a Research Programme*. The National Museum & Danish Polar Center. Viborg 1998 pp. 66-73.

Warburg, L.: *Spindebog*, Borgen. København 1974.

Østergård, E.: *Woven into the Earth. Textiles from Norse Greenland*. Aarhus University Press 2004.

Østergård, E.: *The Remarkable Clothing of the Medieval Norse Greenlanders*, Arctic Clothing, J.C.H. King, B. Pauksztat and R. Storrie (edt.) British Museum Press. London 2005 pp. 95-98.

Østergård, E.: Glimt fra hørrens 1000-årige historie – set ud fra arkæologiske og recente fund, Camilla Luise Dahl (ed.) www.dragt.dk *Dragtjournalen*. Årgang 3, NR. 4. Et internettidsskrift. Tidsskrift udgivet af Den danske Dragt- og Tekstilpulje, Januar 2009 pp. 3-18.

Østergård, E.: Glimt fra hørrens 10.000-årige historie – set ud fra arkæologiske og recente fund, *TENEN. Dansk Tekstilhistorisk Forening*. 20. årgang nr. 4. 2010 pp. 3-19.

List of Abbreviations

AccuMark

Computer program for pattern construction, grading and marker making, produced by Gerber Technology (www.gerbertechnology.com)

CAD

Computer Aided Design